P9-DVG-757

Introduction to
Differential Equations
with Boundary Value Problems

Introduction to Differential Equations

with Boundary Value Problems

Stephen L. Campbell
North Carolina State University

Richard Haberman
Southern Methodist University

HOUGHTON MIFFLIN COMPANY Boston Toronto
Geneva, Illinois Palo Alto Princeton, New Jersey

Senior Sponsoring Editor: Maureen O'Connor
Associate Editor: Dawn Nuttall
Production Team: Tamela Ambush, Kate Burden, Adriana Don, Christina Lillios
Senior Production/Design Coordinator: Jill Haber
Senior Manufacturing Coordinator: Priscilla J. Bailey
Marketing Manager: Charles Cavaliere

Cover design by Harold Burch Design, NYC.

Printed in the U.S.A.

Library of Congress Catalog Card Number: 95-76950

ISBN: 0-395-70828-1

23456789-QM-99 98 97 96

Contents

Preface

Philosophy This book was written with the student in mind. We have tried to follow certain basic principles including:

> "Do one thing at a time" and "Spend extra time where the student has difficulty."

We have tried to give as many examples as space allows. When given the choice between alternative approaches, we have chosen the one that best explains to the student how the concepts are used in practice. However, this book is not a cookbook.

> "Everything should be made as simple as possible, but not simpler."[1]

The careful development of not only techniques and theory but also applications and the geometry of differential equations will provide the student with the balanced background needed to go on in his or her chosen field, be it mathematics, engineering, the sciences, or something else. In an introductory calculus course, a student must learn not only how to manipulate the notation, but also how to interpret the derivative geometrically (tangent lines) and physically (velocity, acceleration). The same is true with differential equations.

Applications Many fundamental problems in science, engineering, and other areas such as economics are described by differential equations. We believe that many problems of future technologies will also be described by differential equations. Physical problems have motivated the development of much of mathematics and this is especially true of differential equations. In this book, we study the interaction between mathematics and physical problems. An understanding of the required mathematics aids in the solution of physical problems, and an understanding of the physical model often makes the mathematics easier. Thus, in our presentation, we devote some effort to deriving the governing differential equations from physical principles. We do not treat applications in a cursory manner. We take four major applications

[1] *Albert Einstein*

and in most cases carry them throughout. They are population growth, mixing problems, mechanics and electrical circuits. In this way, the student has the chance to understand the physical problem. Additional applications discussed include heat, radioactive decay, orthogonal trajectories, chemical reactions, economics, chemical diffusion, and vibrating strings.

Exercises Differential equations cannot be learned by reading the text alone. We included a large number of problems of various types and degrees of difficulty. Most are straightforward illustrations of the ideas in each section. Some exercises are more difficult, either because they are more theoretical or because they involve the application of physical principles. The answers are provided in the back for all the odd problems. Frequently, even problems are similar to neighboring odd problems so that the answers to an odd problem can often be used as guidance. An instructor's manual exists with the answers (and some discussion) to all the exercises.

Technology The increasing availability of technology (including graphing and programmable calculators, computer algebra systems, and powerful personal computers) has caused many to question the existing syllabi in university courses in differential equations. However, we believe that the importance of applications will continue to motivate the study of differential equations. This book has been written with that in mind. Instructors can use our book in several ways. Courses with a strong emphasis on applications can use our book. In addition, those that wish a greater presence of technology can use the increasingly available computational supplements with our book.

Organization, Applications, and Flexible Course Syllabi We have tried to make the text as flexible as possible. This book can be used for a one-quarter or one-semester introduction to differential equations, as well as for longer courses. At one time, the syllabi of differential equations courses were essentially the same everywhere. In the future, we believe that more variation will take place. Only Chapters 1 and 2 will be part of a universally-accepted core in introductory differential equations. Some schools now cover part of this core earlier in the Calculus sequence. First-order equations (Sections 1.1–1.6), especially first-order linear equations and their general solution by an integrating factor, are essential. Constant coefficient first-order equations, their applications to growth problems, and the inclusion of a constant input are also important (1.7–1.8). The more applications [mixture problems (1.9), electronic circuits (1.10), air resistance (1.11), and orthogonal trajectories (1.12) time permits the better. We consider the discussion (1.13–1.14) of special nonlinear first-order equations optional. Some applications to electrical circuits are introduced in Section 1.10, but these can be omitted and introduced in the discussion on applications of second-order linear differential equations to RLC circuits (Section 2.12). The material on second-order linear equations in Chapter 2 is also essential. Higher-order linear theory

(2.4) and constant coefficient examples (2.7) are interspersed so that they are optional to the instructor. If nth-order equations are covered, second-order equations serve to motivate and explain the corresponding nth-order section. The method of undetermined coefficients (2.9), which is important in applications, is presented using the roots of the characteristic equation since that is closest to how students will see it used. Many examples are given, often with the emphasis on the form of the particular solution. This method may be proved using annihilation operators in the optional Section 2.10. Applications of the mechanical vibrations of spring-mass systems are most important in this text. Spring-mass systems are the easiest second-order application for students to visualize. Applications are discussed immediately after the mathematical topic. For example, free and damped systems (2.8) follow homogeneous constant coefficient equations (2.6) and forced mechanical systems and resonance (2.11) follows undetermined coefficients (2.9). If an instructor wishes to do applications after Chapter 2, these applications have been written to allow for it. Constant coefficient linear equations are emphasized in Chapter 2. However, near the end of the chapter Euler equations (2.13) are presented because of their significance in the discussion of regular singular points in Chapter 4.

Chapter 3 (Laplace transforms) and Chapter 4 (Variable coefficients/series methods) can be taught in either order. Some prefer Laplace transforms first, since it is mostly applicable to the constant coefficient differential equations frequently solved in Chapters 1 and 2. However, Chapter 4 can be considered the natural continuation of the discussion of linear differential equations to the cases of variable coefficients. Engineering students often require Laplace transforms. If Laplace transforms are covered, Sections 3.1–3.3 are a basic introduction to the application of the Laplace transform to ordinary differential equations. More in–depth discussions of Laplace transforms are possible. For many, the inclusion of discontinuous forcing (3.4), periodic forcing (3.5), the convolution theorem (3.6), and/or impulses (3.7) are highly recommended. The book has been designed so that Chapter 4 (or Chapter 3) can be omitted if the instructor wishes. A short version of Chapter 4 emphasizing the approximate behavior of solutions near ordinary (4.3) and regular singular points (4.6) can be presented without series methods. Two somewhat alternative versions of series methods are presented (4.4) and/or (4.5). A more thorough discussion would include the method of Frobenius for series solutions near regular singular points (4.7) with or without the case of the roots of the indicial equation differing by an integer.

Further topics in differential equations can be selected from the remaining chapters in many different ways. We mention a few natural combinations:

I. Linear Systems (Chapter 5). A short version might emphasize elimination (Sections 5.2). A linear algebraic version (5.9) emphasizing eigenvectors (5.10) can be presented preceded by the necessary background material (5.7–5.8). If time permits, a more complete discussion of nonhomogeneous systems (5.11) and the matrix exponential (5.12) are given. With any version of linear systems, some (but probably not all) of the extensive applications [mixing problems (5.4), mechanical systems (5.5), and/or multiloop circuits (5.6)] can be discussed.

II. Numerical Methods Numerical methods (Chapter 6) can be covered at any point after, or even during, Chapter 1. A short treatment could begin and end with Euler's method (6.2). A more in depth development could include second-order methods (6.4), Runge-Kutta methods (6.5), or other topics.

III. Nonlinear Equations, Equilibrium, and Stability An introduction to nonlinear differential equations is presented in Chapter 7. For first-order equations, equilibrium solutions (7.2) and their linearized stability (7.3) are presented along with one-dimensional phase portraits (7.4). Applications to population dynamics (7.5), circuit theory (7.6), and chemical reactions (7.7) will be of interest to some. After Chapter 7, some may wish to discuss similar ideas for nonlinear difference equations in Section 8.3 and/or nonlinear systems of differential equations in Chapter 9. For nonlinear difference equations, a population dynamics example is thoroughly discussed (8.4) leading to the period doubling route to chaos.

IV. Boundary Value Problems, Fourier Series, and Partial Differential Equations An introduction to these topics is presented in Chapters 10 and 11. The most important examples of orthogonal functions (10.1) are Fourier series (10.2) and their important special cases, the Fourier sine and cosine series (10.3). A brief introduction to just Fourier series would consist of (10.1)–(10.3) followed possibly by their application to ordinary differential equations in (10.4). In Section 10.5 it is shown that the simplest example of a boundary value problem gives rise to a Fourier sine series and that there are significant generalizations. Chapter 11 presents a brief introduction to partial differential equations. The diffusion of a pollutant (heat equation) is mathematically modelled (11.2). The method of separation of variables is described for the one-dimensional time-dependent diffusion equation (11.2) and the steady-state two-dimensional diffusion (Laplace's) equation (11.3). The wave equation for a vibrating string is derived from physical principles in (11.4) and solved by separation of variables.

Acknowledgments Many people have had input into this book, including an excellent group of reviewers, users of an earlier text by one of us, and an experienced editorial staff. However, two groups have especially influenced us: the many students to whom we have taught differential equations and the colleagues with whom we have worked in differential equations, numerical analysis, and electrical and mechanical engineering.

For different reasons, we are appreciative of e-mail and William E. Hoffman.

Finally, the book could not have been written without the support and understanding of our wives, Gail and Liz, and our children (two each) Matthew and Eric and Ken and Vicki.

Stephen L. Campbell
Richard Haberman

Reviewers:
Stephen Agard, University of Minnesota
James E. Daly, University of Colorado
James Hassed, University of Colorado
Kenneth A. Heimes, Iowa State University
Arthur C. Heinricher Jr., Worcester Polytechnic Institute, MA
Seth Edward-Austin Hollar, Massachusetts Institute of Technology
Michael Longfritz, Rensselear Polytechnic Institute, NY
Dr. Frederick Norwood, East Tennessee State University
Kevin D. Oden, Harvard University, MA
Michel Smith, Auburn University, AL
Dalin Tang, Worcester Polytechnic Institute, MA
Dr. Junping Wang, University of Wyoming
Arwen Warlock Dixon, Rensselear Polytechnic Institute, NY
Ming Xue, Massachusetts Institute of Technology

CHAPTER

1

First-Order Differential Equations and Their Application

1.1 Introduction to Ordinary Differential Equations

Differential equations are found in many areas of mathematics, science, and engineering. Students taking a first course in differential equations have often already seen simple examples of differential equations in their mathematics, physics, chemistry, or engineering courses. If you have not already seen differential equations, go to the library and glance at some books or journals from your major. You may be surprised to see the way in which differential equations dominate the study of many aspects of science and engineering.

Applied mathematics involves the relationships between mathematics and its applications. Often the type of mathematics that arises in applications is differential equations. Thus, the study of differential equations is an integral part of applied mathematics. Applied mathematics is said to have three fundamental aspects, and this course will involve a balance of the three:

1. The **modeling process** by which physical objects and processes are described by physical laws and given mathematical formulations. Since so many physical laws involve rates of change (or the derivative), differential equations are often the natural language of science and engineering.

2. The **analysis** of the mathematical problems that are posed. This involves the complete investigation of the differential equation and its solutions, including detailed numerical studies. We will say more about this shortly.
3. However, the mathematical solution of the differential equation does not complete the overall process. The **interpretation** of the solution of the differential equation in the context of the original physical problem must be given, and the implications further analyzed.

When we have a differential equation, we attempt to obtain certain types of information. There are four general approaches to analyzing differential equations. In practice, solving a problem usually involves aspects of several of the four approaches. The approaches are:

1. Qualitative
2. Numerical
3. Analytic
4. Asymptotic or perturbation

Using the **qualitative** approach, we determine the behavior of the solutions without actually getting a formula for them. This approach is somewhat similar to the curve-sketching process in introductory calculus where we sketch curves by drawing maxima, minima, concavity changes, etc. Qualitative ideas will be discussed in Section 1.4 and Chapters 7 and 9.

Using the **numerical** approach, we compute estimates for the values of the unknown function at certain values of the independent variable. Numerical methods are extremely important and for many difficult problems are the only practical approach. We will discuss numerical methods in Chapter 6. The safe and effective use of numerical methods requires an understanding of the basic properties of differential equations and their solutions.

Most of this book is devoted to developing **analytical** procedures—that is, obtaining explicit and implicit formulas for the solutions of various ordinary differential equations. We shall present a sufficient number of applications to enable the reader to understand how differential equations are used and to develop some feeling for the physical information they convey.

Asymptotic and **perturbation** methods are introduced in more advanced studies of differential equations. These asymptotic approximations are introduced directly from the exact analytic solutions in order to get a better understanding of the meaning of the exact analytic solutions. Unfortunately, many problems of physical interest do not have exact solutions. In this case, in addition to the previously mentioned numerical methods, there are approximation methods known as perturbation methods that often are useful for understanding the behavior of differential equations.

In the first ten chapters, we will use the phrase "differential equation" to mean an ordinary differential equation (or a system of ordinary differential equations). An **ordinary differential equation** is an equation relating an unknown function of one variable to one or more functions of its derivatives. If the unknown y is a function of x, $y = y(x)$, then examples of ordinary differential equations are

$$\begin{aligned} \frac{dy}{dx} &= x^7 \cos y \\ \frac{d^2y}{dx^2} &= y\frac{dy}{dx} \\ \frac{d^4y}{dx^4} &= -5y^5. \end{aligned} \tag{1}$$

The **order** of a differential equation is the order of the highest derivative of the unknown function (dependent variable) that appears in the equation. The differential equations in (1) are of the first, second, and fourth order, respectively. Most of the equations we shall deal with will be of the first or second order.

In applications, the dependent variables are frequently functions of time, which we will denote by t. Some applications such as

1. Population dynamics
2. Mixture and flow problems
3. Electronic circuits
4. Mechanical vibrations and systems

will be discussed repeatedly throughout this text. Other applications, such as radioactive decay, thermal cooling, chemical reactions, and orthogonal trajectories, will appear only as illustrations of more specific mathematical results. In all cases, modeling, analysis, and interpretation are important.

One of the simplest examples of a differential equation, and one with significant application, is the following first-order differential equation:

$$\frac{du}{dt} = au. \tag{2}$$

We will show later that (2) describes the growth of a population $u(t)$ under the assumption that the growth rate a is a constant. If the population growth is retarded by crowding effects, then the logistic model is often introduced:

$$\frac{du}{dt} = au - bu^2. \tag{3}$$

In a typical application, physical laws often lead to a differential equation. As a simple example, we will consider later the vertical motion $y(t)$ of a constant mass. Newton's law says that the force F equals the mass m times the acceleration d^2y/dt^2:

$$m\frac{d^2y}{dt^2} = F. \tag{4}$$

If the forces are gravity $-mg$ and a force due to air resistance proportional to the velocity dy/dt, then the position satisfies the following second-order differential equation:

$$m\frac{d^2y}{dt^2} = -mg - c\frac{dy}{dt}, \tag{5}$$

where c is a proportionality constant determined by experiments. However, if in addition the mass is tied to a spring that exerts an additional force $-ky$ satisfying Hooke's law, then we will show that the mass satisfies the following second-order differential equation:

$$m\frac{d^2y}{dt^2} = -mg - ky - c\frac{dy}{dt}. \tag{6}$$

In the main body of the book, we will devote some effort to the modeling process by which these differential equations arise, as well as learning how to solve these types of differential equations easily.

We will also see that a typical electronic circuit with a resistor, capacitor, and inductor can often be modeled by the following second-order differential equation:

$$L\frac{d^2i}{dt^2} + R\frac{di}{dt} + \frac{1}{C}i = f(t). \tag{7}$$

Here the unknown variable is $i(t)$, the time-dependent current running through the circuit. We will present a derivation of this differential equation, and you will learn the meaning of the three positive constants R, L, and C (this is called an RLC circuit). For example, R represents the resistance of a resistor, and we will want to study how the current in the circuit depends on the resistance. The right-hand side $f(t)$ represents something that causes current in the circuit, such as a battery.

Physical problems frequently involve systems of differential equations. For example, we will consider the salt content in two interconnected well-mixed lakes, allowing for some inflow, outflow, and evaporation. If $x(t)$ represents the amount of salt in one of the lakes and $y(t)$ the amount of salt in the other, then under a series of assumptions described in the book, the following coupled system of differential equations is an appropriate mathematical model:

$$\begin{aligned} \frac{dx}{dt} &= \frac{1}{2} + 2\frac{y}{100} - 3\frac{x}{100} \\ \frac{dy}{dt} &= 3\frac{y}{100} - \frac{5}{2}\frac{y}{100}. \end{aligned} \tag{8}$$

1.1.1 Motivations from Partial Differential Equations

The first ten chapters concern ordinary differential equations. In Chapter 11 we will study partial differential equations. Physical problems such as electromagnetic radiation, chemical diffusion, wave diffraction, plate tectonics, seismic wave propagation, laser optics, semiconductor devices, acoustics, weather, global warming, aerodynamics, and biological fluid dynamics are formulated in terms of partial differential equations. A **partial differential equation** is an equation that involves partial derivatives. An example of a partial differential equation is the **heat equation**, well known to physicists and thermal engineers:

$$\frac{\partial u}{\partial t} = k\frac{\partial^2 u}{\partial x^2}. \tag{9}$$

In this example the temperature u depends on one spatial dimension x and on time t. Here, $\partial u/\partial t$ is the rate of change of the temperature with respect to time at a fixed spatial position, and $\partial^2 u/\partial x^2$ is the usual second derivative with respect to x keeping t fixed. This equation also describes the diffusion of a pollutant, where u is the concentration of the pollutant. Hence (9) is often called the **diffusion equation** instead of the heat equation.

Ordinary differential equations frequently arise in the study of partial differential equations. Some of the most important ways of reducing a partial differential equation to an ordinary differential equation are as follows:

1. The desired solution is independent of one of the variables, usually space or time. If it is independent of time, the solution is called a **steady state**.
2. Solutions are assumed to be in a product form in the method of **separation of variables**. This important method will be applied to simple partial differential equations in Chapter 11.
3. The solution is assumed to have **self-similar** structure under scalings as a result of physical symmetry in the problem.
4. The solution is assumed to be in the form of a **traveling wave**.

A brief example of the first way will be given.

Steady-State Distribution of a Pollutant

Let x measure the distance down a long, narrow lake. Let $u(x, t)$ be the density of a bacterial pollutant at position x at time t. The diffusion of a bacterial pollutant satisfies the diffusion equation (9). However, suppose that in addition to diffusing, the bacteria naturally dies off with decay rate a. Then the diffusion equation would be modified to become the following partial differential equation:

$$\frac{\partial u}{\partial t} = k\frac{\partial^2 u}{\partial x^2} - au. \tag{10}$$

Suppose that we are interested in the steady-state distribution of the bacteria. To find this distribution, we would assume that the concentration of bacteria only depends on x. That is, we assume that $u(x, t) = u(x)$. But then $\partial u/\partial t = 0$ and the partial differential equation (10) becomes the much simpler second-order ordinary differential equation

$$k\frac{d^2 u}{dx^2} - au = 0. \tag{11}$$

1.2 Definite Integral and the Initial Value Problem

This chapter is concerned with first-order differential equations. In first-order differential equations, the first derivative of a function $y(x)$ depends on the independent variable x and the unknown solution y. If dy/dx is given directly in terms of x and y, the differential equation has the form

$$\frac{dy}{dx} = f(x, y). \tag{1}$$

We will discuss a number of applications of first-order equations in later sections, such as growth and decay problems for populations, radioactive decay, thermal cooling, mixture problems, evaporation and flow, electronic circuit theory, and several others. In these applications, some care is given to the development of mathematical models.

A **solution** of the differential equation (1) is a function that satisfies the differential equation for all values of x of interest.

EXAMPLE 1.2.1 Showing a Function Is a Solution

Verify that $y = 3e^{x^2}$ is a solution of the first-order differential equation

$$\frac{dy}{dx} = 2xy. \tag{2}$$

SOLUTION Substituting $y = 3e^{x^2}$ in (2), we get

$$\frac{d}{dx}(3e^{x^2}) = 2x(3e^{x^2}).$$

Performing the differentiation and simplifying the right-hand side, we find that the equation becomes $6xe^{x^2} = 6xe^{x^2}$, which holds for all x. Thus $y = 3e^{x^2}$ is a solution of (2). ◄

Before beginning our general development of first-order equations in Section 1.4, we will discuss some differential equations that can be solved with direct integration. These special cases will be used to motivate and illustrate some of the later development.

1.2.1 Initial Value Problem and the Indefinite Integral

The simplest possible first-order differential equation arises if the function $f(x, y)$ in (1) does not depend on the unknown solution, so that the differential equation is

$$\frac{dy}{dx} = f(x). \tag{3}$$

Solving (3) for y is just the question of antidifferentiation in calculus. Although (3) can be solved by an integration, we can learn some important things concerning more general differential equations from it that will be useful later.

EXAMPLE 1.2.2 **Indefinite Integration Example**

Consider the following simple differential equation

$$\frac{dy}{dx} = x^2. \tag{4}$$

By an integration, we obtain

$$y = \frac{1}{3}x^3 + c, \tag{5}$$

where c is an arbitrary constant. From this example, we see that differential equations usually have many solutions. We call (5) the **general solution** of (4), since it is a formula that gives all solutions. Often, especially in applications, we are interested in a specific solution of the differential equation that is to satisfy some additional condition. For example, suppose we are given that $y = 7$ at $x = 2$. This is written mathematically as $y(2) = 7$ and called an **initial condition**. By letting $x = 2$ and $y = 7$ in (5), we get

$$7 = \frac{8}{3} + c,$$

so that the constant c is determined to be $c = \frac{13}{3}$. Thus, the unique solution of the differential equation that satisfies the given initial condition is

$$y = \frac{1}{3}x^3 + \frac{13}{3}.$$

◄

More generally, we might wish to solve the differential equation

$$\frac{dy}{dx} = f(x), \tag{6}$$

subject to the initial condition

$$y(x_0) = y_0. \tag{7}$$

We introduce the symbol x_0 for the initial value of x at which the solution is given. In the previous example, we had $x_0 = 2$ and $y_0 = 7$. We refer to (6) with the initial condition (7) as the **initial value problem** for the differential equation. We can always write a formula for the solution of (6) using an indefinite integral

$$y = \int f(x)\,dx + c. \tag{8}$$

Equation (5) is a specific example of (8). If we can obtain an explicit integral of $f(x)$, then this initial value problem can be solved like Example 1.2.2. Explicit integrals of various functions $f(x)$ may be obtained by using any of the various techniques of integration from calculus. Tables of integrals or symbolic integration algorithms such as MAPLE or Mathematica that are available on more sophisticated calculators, personal computers, or larger computers may be used. However, if one cannot obtain an explicit integral, then it may be difficult to use (8) directly to satisfy the initial conditions.

1.2.2 Initial Value Problem and the Definite Integral

A definite integral should usually be used to solve the differential equation $dy/dx = f(x)$ if an explicit integral is not used. The result can automatically incorporate the given initial condition $y(x_0) = y_0$. If both sides of the differential equation (6) are integrated with respect to x from x_0 to x, we get

$$\int_{x_0}^{x} \frac{dy}{d\bar{x}}\,d\bar{x} = \int_{x_0}^{x} f(\bar{x})\,d\bar{x},$$

where we have introduced the dummy variable $\bar{x}$. The left-hand side equals $y(\bar{x})|_{\bar{x}=x_0}^{\bar{x}=x} = y(x) - y(x_0)$, since the antiderivative of dy/dx is y. Thus we have

$$y(x) - y(x_0) = \int_{x_0}^{x} f(\bar{x})\,d\bar{x}, \quad \text{or equivalently,}$$

$$y(x) = y(x_0) + \int_{x_0}^{x} f(\bar{x})\,d\bar{x}. \tag{9}$$

Note that $y(x_0) = y_0$.

Any dummy variable of integration may be used besides $\bar{x}$. Note also that (9) satisfies the initial condition that $y = y_0$ at $x = x_0$.

EXAMPLE 1.2.3 Example with a Definite Integral

Solve the differential equation

$$\frac{dy}{dx} = e^{-x^2} \tag{10}$$

subject to the initial condition $y(3) = 7$.

Solution The function e^{-x^2} does not have any explicit antiderivative. Thus, if we want to solve (10), we use definite integration from $x_0 = 3$, where $y_0 = 7$. Then (9) is

$$y - 7 = \int_3^x e^{-t^2}\, dt,$$

and the solution of the initial value problem is

$$y = 7 + \int_3^x e^{-t^2}\, dt. \tag{11}$$

Any dummy variable of integration may be used. We have chosen t instead of $\bar{x}$. The function e^{-x^2} is important in probability, since $(1/\sqrt{2\pi})e^{-x^2/2}$ is the famous normal curve.

There are many situations in which it is desirable to use a definite integral rather than an explicit antiderivative.

1. It is sometimes difficult to obtain an explicit antiderivative, and a formula like (11) suffices.
2. For some $f(x)$ (as in the previous example), it is impossible to obtain an explicit antiderivative in terms of elementary functions.
3. The function $f(x)$ might be expressed only by some data, in which case the definite integral (9) represents the area under the curve $f(x)$ and can be evaluated by an appropriate numerical integration method such as Simpson's or the trapezoid rules.

It is difficult to give general advice valid for all problems. If an integral is an elementary integral, then the explicit integral should be used. However, what is elementary to one person is not necessarily elementary to another. With the wide availability of computers, a definite integral can usually be evaluated by a numerical integration.

General Solution Using a Definite Integral

Alternatively, in solving for the general solution of

$$\frac{dy}{dx} = f(x), \tag{12}$$

we may use the definite integral starting at any point a. In this case, we obtain the general solution of (12) to be

$$y = \int_a^x f(\bar{x})\, d\bar{x} + c. \tag{13}$$

There appear to be two arbitrary constants, c and the lower limit of the integral a in (13). However, it can be shown that this is equivalent to one arbitrary constant. In practice, the lower limit a is often chosen to be the initial value of x. To see that (13) actually solves the differential equation (12), it is helpful to recall the fundamental theorem of calculus:

$$\frac{d}{dx}\left(\int_a^x f(\bar{x})\, d\bar{x}\right) = f(x). \tag{14}$$

1.2.3 Mechanics I: Elementary Motion of a Particle with Gravity Only

Elementary motions of a particle are frequently described by differential equations. Simple integration can sometimes be used to analyze the elementary motion of a particle. For the one-dimensional vertical motion of a particle, we recall from calculus that

$$\begin{aligned} \text{Position} &= y(t) \\ \text{Velocity} &= v(t) = \frac{dy}{dt} \\ \text{Acceleration} &= a(t) = \frac{dv}{dt} = \frac{d^2y}{dt^2}. \end{aligned} \tag{15}$$

Newton's law of motion ($ma = F$) will yield a differential equation

$$m\frac{d^2y}{dt^2} = F\left(y, \frac{dy}{dt}, t\right), \tag{16}$$

where F is the sum of the applied forces, and we have allowed the forces to depend on position, velocity, and time. Equation (16) is a second-order differential equation, which we will study later in the course. Here y is a function of the independent variable t instead of the independent variable x of the previous examples.

There are no techniques for solving (16) in all cases. However, (16) can be solved by simple integration if the force F does not depend on y and dy/dt. As an example, suppose that the only force on the mass is due to gravity. Then it is known that $F = -mg$, where g is the acceleration due to gravity. The minus sign is introduced because gravity acts downward, toward the surface of the earth. Here we are taking the coordinate system so that y increases towards the sky. The magnitude of the force due to gravity, mg, is called the weight of the body. Near the surface of the planet earth, g is approximately $g = 9.8 \text{ m/s}^2$ in the mks system used by most of the world ($g = 32 \text{ ft/s}^2$ when feet are used as the unit of length instead of meters). If we assume that we are interested in a mass that is located sufficiently near the surface of the earth, then g can be approximated by this constant. With the only force being gravity, (16) becomes

$$m\frac{d^2y}{dt^2} = -mg,$$

or equivalently,

$$\frac{d^2y}{dt^2} = -g, \tag{17}$$

since the mass cancels. Integrating (17) yields

$$\frac{dy}{dt} = -gt + c_1, \tag{18}$$

where c_1 is an arbitrary constant of integration. We assume the velocity at $t = 0$ is given and use the notation v_0 for this initial velocity. Since

$$v(t) = -gt + c_1$$

from (18), evaluating (18) at $t = 0$ gives $v_0 = c_1$. Thus, the velocity satisfies

$$\frac{dy}{dt} = -gt + v_0. \tag{19}$$

The position can be determined by integrating the velocity (19) to give

$$y = -\frac{1}{2}gt^2 + v_0t + c_2, \tag{20}$$

where c_2 is a second integration constant. We also assume that the position y_0 at $t = 0$ is given initially. Then, evaluating (20) at $t = 0$ gives $y_0 = c_2$, so that

$$y = -\frac{1}{2}gt^2 + v_0t + y_0. \tag{21}$$

Equations (19) and (21) are well-known formulas in physics and are used to solve for quantities such as the maximum height of a thrown object. We do not recommend memorizing (19) or (21). Instead, they should be derived in each case from the differential equation (17). If the applied force depends only on time and is not constant, then formulas for velocity and position may be obtained by integration. If the applied force depends on other quantities, solving the differential equation is not as simple. We describe some more difficult problems in Section 1.11 and Chapter 2.

EXAMPLE 1.2.4 Motion with Gravity

Suppose a ball is thrown upward from ground level with velocity v_0 and the only force is gravity. How high does the ball go before falling back toward the ground?

SOLUTION The differential equation (as before) is (17):

$$\frac{d^2y}{dt^2} = -g. \tag{22}$$

The initial conditions are that $y = 0$ and $\frac{dy}{dt} = v_0$ at $t = 0$. By successive integrations of (22) and by applying the initial conditions, we obtain

$$\frac{dy}{dt} = -gt + v_0 \tag{23}$$

and

$$y = -\frac{1}{2}gt^2 + v_0 t. \tag{24}$$

From (24), the height is known as a function of time. To determine the maximum height, we must first determine the time at which the ball reaches its maximum height. From calculus, the maximum of a function $y = y(t)$ occurs at a critical point where $dy/dt = 0$. Scientists recognize that the velocity is zero at the maximum height. Thus, the time of the maximum height is determined from (23):

$$0 = -gt + v_0, \qquad \text{or equivalently,} \qquad t = \frac{v_0}{g}. \tag{25}$$

When this time (25) is substituted into (24), a formula for the maximum height is obtained:

$$y = -\frac{1}{2}g\left(\frac{v_0}{g}\right)^2 + v_0\frac{v_0}{g} = \frac{v_0^2}{g}\left(-\frac{1}{2} + 1\right) = \frac{v_0^2}{2g}. \tag{26}$$

◄

Exercises

In Exercises 1 through 8, verify that the given function y is a solution of the differential equation.

1. $y = 2e^{3x} + 1$, $\dfrac{dy}{dx} = 3y - 3$.

2. $y = x^2$, $\dfrac{dy}{dx} = 2\dfrac{y}{x}$.

3. $y = x - 1$, $\dfrac{dy}{dx} = \dfrac{y}{x-1}$.

4. $y = x^4$, $\dfrac{dy}{dx} = 4\dfrac{y^2}{x^5}$.

5. $y = e^{x^2}$, $\dfrac{dy}{dx} = 2xy$

6. $y = 3e^{4x} + 2$, $\dfrac{dy}{dx} = 4y - 8$

7. $y = e^{-2x}$, $\dfrac{dy}{dx} = -2e^{2x}y^2$

8. $y = x + 3$, $\dfrac{dy}{dx} = \dfrac{y-3}{x}$

In Exercises 9 through 16, find the general solution. If you cannot find an explicit integral solution, use a definite integral. Exercises with an asterisk (*) require definite integrals.

9. $\dfrac{dy}{dx} = 3e^x$

10. $\dfrac{dy}{dx} = 8e^{-x}$

11. $\dfrac{dy}{dx} = -5\cos 6x$

12. $\dfrac{dy}{dx} = 2\sin 7x$

13.* $\dfrac{dy}{dx} = 8\cos(x^{-1/2})$

14.* $\dfrac{dy}{dx} = 4\sin(x^{-1/3})$

15.* $\dfrac{dy}{dx} = \ln(4 + \cos^2 x)$

16.* $\dfrac{dy}{dx} = e^{x^2}$

In Exercises 17 through 22, find the solution of the initial value problem as an explicit or definite integral* as appropriate.

17. $\dfrac{dy}{dx} = x^4$ with $y(2) = 3$

18. $\dfrac{dy}{dx} = x^{3/2}$ with $y(3) = 7$

19.* $\dfrac{dy}{dx} = \dfrac{\ln x}{4 + \cos^2 x}$ with $y(2) = 5$

20.* $\dfrac{dy}{dx} = \dfrac{\sin x}{5 + \ln x}$ with $y(8) = 2$

21.* $\dfrac{dy}{dx} = \dfrac{e^x}{1+x}$ with $y(1) = 3$.

22.* $\dfrac{dy}{dx} = \cos(x^{-3})$ with $y(2) = 4$.

In Exercises 23 through 27, assume that when an automobile brakes, there is a constant deceleration.

23. At an accident scene, the police investigator is attempting to determine from the rubber marks on the road how fast the driver was going. Suppose it is known that this particular car brakes with a deceleration of 15 m/s^2. At what velocity was the car going at the moment it applied its brakes, if the car traveled 75 m before it stopped?

24. Repeat Exercise 23 with the observation that the car traveled 75 m before stopping, but assume that the car decelerates at only 10 m/s^2.

25. An overly cautious traveling distance between you and the car in front would be the distance it would take you to stop. At a speed of 60 km/h, how far does a car travel if it decelerates at 2500 km/h^2?

26. Referring to Exercise 25, with the same deceleration, how far does a car travel at 120 km/h?

27. Referring to Exercise 25, with the same deceleration, determine how far a car travels as a function of its speed before braking.

28. Suppose a car is going 50 km/h when the brakes are applied at $t = 0$. Determine the distance the car travels. Suppose the nonconstant deceleration is known to be $a = -6t$.

29. Suppose a car is going 50 km/h when the brakes are applied at $t = 2$. Determine the distance the car travels. Suppose the nonconstant deceleration is known to be $a = -6t$.

30. How fast must you slide a book in order for it to fall off a 15-m table if the book decelerates such that $a = -5\ \text{m/s}^2$?

31. Assume that a snowplow of width w meters moves forward along a road so that it clears a constant volume of snow per hour, Q m^3/h. Assume that snow started falling at 8 A.M. ($t = 0$) and is snowing at a constant rate of c m/h.

a) Show that $dx/dt = 1/(kt)$, where $x(t)$ is the position of the snowplow and $k = wc/Q$.

b) Where will the snowplow be at noon, if the snowplow did not start moving until 11 A.M.?

32. A toy rocket is fired upward from ground level at $t = 2$ s. Determine the maximum height of the rocket if the velocity at $t = 2$ is 76 m/s. Assume that the nonconstant acceleration of the rocket is known to be $a = -12t^2$.

33. At what velocity should a ball be thrown upward if it is to reach a maximum height of 100 m?

34. At what velocity should a ball be thrown upward if it is to reach its maximum height in 10 s?

35. Suppose a brick is dropped from a construction tower of height 200 m. How long will it take the brick to fall?

36. The general solution of (10) can be written $y = \int_0^x e^{-t^2}\, dt + c$. Determine c that satisfies the initial condition $y(3) = 7$, and show that (11) is valid.

37. The general solution of (12) can be written $y = \int_0^x f(\bar{x})\, d\bar{x} + c$. Determine c that satisfies the initial condition $y(x_0) = y_0$, and show that (13) is valid.

1.3 First-Order Separable Differential Equations

Another class of first-order differential equations that can be solved by integration is those that are separable. A first-order differential equation $dy/dx = f(x, y)$ is **separable** if the function $f(x, y)$ can be written as a product of a function of x and a function of y:

$$\frac{dy}{dx} = h(x)g(y). \tag{1}$$

There are two ways to describe solving (1). They both end up performing the same calculations. One is to divide (1) by $g(y)$ to get

$$\frac{1}{g(y)}\frac{dy}{dx} = h(x). \tag{2}$$

Integrating both sides with respect to x then gives

$$\int \frac{1}{g(y)}\frac{dy}{dx}\, dx = \int h(x)\, dx.$$

By the chain rule for integration, this equation is the same as

$$\int \frac{1}{g(y)}\, dy = \int h(x)\, dx. \tag{3}$$

Alternatively, the variables x and y in (1) can be separated using differentials:

$$\frac{dy}{g(y)} = h(x)\, dx. \tag{4}$$

Integration of (4) then gives (3).

The earlier examples $dy/dx = f(x)$ of Section 1.2 as well as differential equations in the form $dy/dx = g(y)$ are separable.

Note that both (2) and (4) assume that $g(y) \neq 0$. If $y = a$ is a constant such that $g(a) = 0$, then an easy calculation shows that $y = a$ is a constant solution of (1).

> If $g(a) = 0$, then $y = a$ is a constant solution of $\dfrac{dy}{dx} = f(x)g(y)$.

These constant solutions are sometimes lost in integrating (4), so it is necessary to check if they are all accounted for in the solutions.

EXAMPLE 1.3.1 Separable Equation

The first-order differential equation

$$\frac{dy}{dx} = y^2 \cos x \tag{5}$$

is separable. Separating variables gives

$$\int \frac{dy}{y^2} = \int \cos x \, dx.$$

Hence, by integration,

$$-\frac{1}{y} = \sin x + c. \tag{6}$$

Solving for y yields the solutions

$$y = \frac{-1}{\sin x + c}. \tag{7}$$

Since we divided by y^2, the calculations assumed that $y \neq 0$. Note that $y = 0$ satisfies the differential equation (5) but is not included in the solutions (7) for any finite value of c. The solutions of (5) thus consist of (7) and $y = 0$.

Suppose that we also had the initial condition $y(1) = 4$. Letting $x = 1$ and $y = 4$ in (6), we would determine that $c = -\frac{1}{4} - \sin 1$, so that the solution of the initial value problem would be

$$y = \frac{-1}{\sin x - (1/4) - \sin 1}.$$

Note that it is easier to determine c by applying the initial condition to (6) than to (7).

Some separable differential equations (see the next example) are more difficult because obtaining integrals may not be as easy as in the previous example.

EXAMPLE 1.3.2 Separable Example Using Partial Fractions

The differential equation

$$\frac{dy}{dx} = y(1 - y) \tag{8}$$

is called the logistic equation and is very important in the field of population dynamics. We will solve it by separation. We will analyze it again in a different manner in Chapter 7.

SOLUTION Equation (8) is separable:

$$\frac{dy}{y(1 - y)} = dx, \tag{9}$$

so that

$$\int \frac{dy}{y(1 - y)} = \int dx. \tag{10}$$

The integral on the left can be evaluated in several ways. Using partial fractions, for example, we get

$$\frac{1}{y(1 - y)} = \frac{A}{y} + \frac{B}{y - 1}.$$

Multiplying by $y(y - 1)$ yields

$$-1 = A(y - 1) + By. \tag{11}$$

Since $y(y - 1)$ has simple linear factors, y and $y - 1$, we may evaluate (11) at the roots, $y = 0$ and $y = 1$, to obtain $A = 1$ and $B = -1$. Thus, (10) becomes

$$\int \frac{1}{y}\, dy - \int \frac{1}{y - 1}\, dy = \int dx.$$

Carrying out the integrations gives $\ln|y| - \ln|y - 1| = x + c$, or equivalently,

$$\ln \frac{|y|}{|y - 1|} = x + c. \tag{12}$$

Keeping the absolute value is proper. Exponentiation of both sides of (12) yields

$$\frac{|y|}{|y - 1|} = e^{x+c} = e^c e^x,$$

using properties of the exponential. Here e^c is an arbitrary positive constant. The absolute value signs can be eliminated since

$$\left|\frac{y}{y - 1}\right| = e^c e^x \qquad \text{implies} \qquad \frac{y}{y - 1} = \pm e^c e^x.$$

Thus we have

$$\frac{y}{y-1} = ke^x,$$

where $k = \pm e^c$ is an arbitrary constant (positive or negative). We then multiply by $y - 1$ and solve for y to get the solution

$$y = \frac{ke^x}{-1 + ke^x}. \tag{13}$$

Since $y(y - 1) = 0$ if $y = 0$ or $y = 1$, it follows that $y = 0$ and $y = 1$ are solutions of (8). The solution $y = 0$ corresponds to $k = 0$. However, $y = 1$ is not included in (13) unless we allow $k = \infty$. Thus, the solutions of (8) are (13) and $y = 1$. The general solution involves one arbitrary constant k, which can be determined from any initial conditions. ◄

These algebraic steps involved in obtaining the general solution may be difficult for some students. For the integration step, tables of integrals will be helpful for some. It is not easy to understand the behavior of the solution of the differential equation (8) from this general solution (13). In particular, the cases $k < 0$, $0 < k < 1$, and $k > 1$ turn out to have different behavior.

Philosophically, we think of solving separable equations as being very easy, although this example shows that in practice there can be substantial complications. A different method for analyzing and understanding qualitatively the solution of differential equations such as (8) is introduced in Chapter 7.

Equation (12) is an **implicit solution** of (8), since it does not directly give y in terms of x. Equation (13) is an **explicit solution**, since y is explicitly in terms of x. For many problems, it is impossible to solve the implicit solution to get a formula for y. However, given values of c and x, we can usually solve the implicit equation numerically, say by Newton's method or bisection, to determine the value of y. Thus the implicit solution is still useful.

1.3.1 Using Definite Integrals for Separable Differential Equations

Definite integrals can also be used on the separable differential equation $dy/dx = h(x)g(y)$. Suppose that the initial condition is $y(x_0) = y_0$. Then using a definite integral we get

$$\int_{x_0}^{x} \frac{1}{g(\bar{y})} \frac{d\bar{y}}{d\bar{x}}\, d\bar{x} = \int_{x_0}^{x} h(\bar{x})\, d\bar{x}$$

or

$$\int_{y_0}^{y} \frac{d\bar{y}}{g(\bar{y})} = \int_{x_0}^{x} h(\bar{x})\, d\bar{x}. \tag{14}$$

We consider (14) to represent the solution to the separable differential equation. Often, a more explicit solution is desired. Whether a simpler representation of the solution of a specific separable differential equation may be obtained depends on the question from calculus of evaluating the integrals.

EXAMPLE 1.3.3 Separable with Definite Integral

The general solution of the initial value problem for (8) can be represented by a definite integral:

$$\int_{y_0}^{y} \frac{d\bar{y}}{\bar{y}(1-\bar{y})} = x - x_0.$$

This can be shown to be equivalent to (13). (Note to students: You will have to ask your instructor whether this will be acceptable.) ◄

Exercises

In Exercises 1 through 14, solve the differential equation. Definite integrals are necessary in problems with an asterisk (*).

1. $\dfrac{dy}{dx} = \dfrac{y+1}{x}$

2. $\dfrac{dr}{d\theta} = \dfrac{r^2 + r}{\theta}$

3. $\dfrac{dy}{dx} = e^x$

4. $\dfrac{dy}{dx} = e^{x+y}$

5. $\dfrac{dy}{dx} = xy + 4y + 3x + 12$

6. $\dfrac{du}{dt} = \dfrac{u^2 + 4}{t^2 + 4}$

7. $\dfrac{dy}{dx} = 3$

8. $\dfrac{dy}{dx} = x^2 y^2$

9. $\dfrac{dy}{dx} = y^5, \quad y(2) = 1$

10. $\dfrac{dy}{dx} = xy - 2y + x - 2$

11.* $\dfrac{dy}{dx} = y^2 \cos(x^2), \quad y(0) = 1.$

12.* $\dfrac{dy}{dx} = y^4 \cos(x^{-1/2})$

13.* $\dfrac{dy}{dx} = x \cos(y^{-1/2}), \quad y(1) = 2$

14.* $\dfrac{dy}{dx} = e^{x^2+y^2}, \quad y(2) = 4$

In Exercises 15 through 18, solve the differential equation.

15. $\dfrac{du}{dt} = \dfrac{t^2 + 1}{u^2 + 4}, \quad u(0) = 1$

16. $\dfrac{dr}{d\theta} = \sin r$

17. $\dfrac{dy}{dx} = x^2 y^2 + y^2 + x^2 + 1, \quad y(0) = 2$

18. $\dfrac{du}{dt} = u^3 - u$

The differential equations in Exercises 19 through 21 require partial fractions.

19. $y' = y(y-1)$

20. $y' = y^2(y-1)^2$

21. $y' = (y-1)(y-2)^2$

In Exercises 22 through 28, a first-order differential equation is written in differential form. Solve the differential equation.

22. $y^2\,dx + x^3\,dy = 0$

23. $(xy + y)\,dx + (xy + x)\,dy = 0$

24. $r \sin\theta\,dr + \cos\theta\,d\theta = 0$

25. $(t^2 - 4)\,dz + (z^2 - 9)\,dt = 0$

26. $(y^2 + y)\,dx + (x^3 + 4x^2)\,dy = 0$

27. $e^{x+y}\,dx + e^{2x-3y}\,dy = 0$

28. $xe^y\,dx + ye^{-x}\,dy = 0$

29. Show that if we make the substitution $z = ax + by + c$ and $b \neq 0$, then the differential equation

$$\frac{dy}{dx} = f(ax + by + c)$$

is changed to a differential equation in z and x that can be solved by separation of variables.

In Exercises 30 through 34, solve the differential equation using Exercise 29.

30. $\dfrac{dy}{dx} = (x + y)^2$

31. $\dfrac{dy}{dx} = (x + 4y - 1)^2$

32. $\dfrac{dy}{dx} = \tan(-x + y + 1) + 1$

33. $\dfrac{dy}{dx} = e^{x+y}(x + y)^{-1} - 1$

34. $\dfrac{dy}{dx} = \dfrac{x + y + 2}{x + y + 1}$

1.4 Direction Fields

We can often get a good pictorial or graphical idea of what the solutions $y(x)$ of a first-order differential equation,

$$\frac{dy}{dx} = f(x, y), \tag{1}$$

look like without actually solving the differential equation. The method we describe is based on tangent lines and is easily implemented on a computer. Software that does this is readily available (the software PHASER by H. Kocak, published by Springer-Verlag, has been used in preparing the direction fields in this text). It is highly recommended that students take advantage of the computer facilities that are available to them.

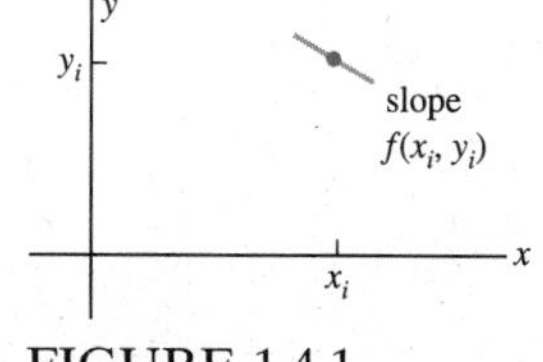

FIGURE 1.4.1

Slope $dy/dx = f(x, y)$ at a point.

We are interested in graphing y as a function of x, where $y(x)$ satisfies the differential equation (1). We first set up an x-y coordinate system, and note that at any point (x, y) the slope dy/dx of the tangent line to the solution y at that point is given by the differential equation $dy/dx = f(x, y)$. Through each point (x, y), we plot a short dash (line segment) with slope given by the differential equation, as shown for one point in Figure 1.4.1. Continuing in this manner at other points, we can build up a picture. The resulting plot is sometimes referred to as a sketch of the **direction field** or **vector field** of the solutions.

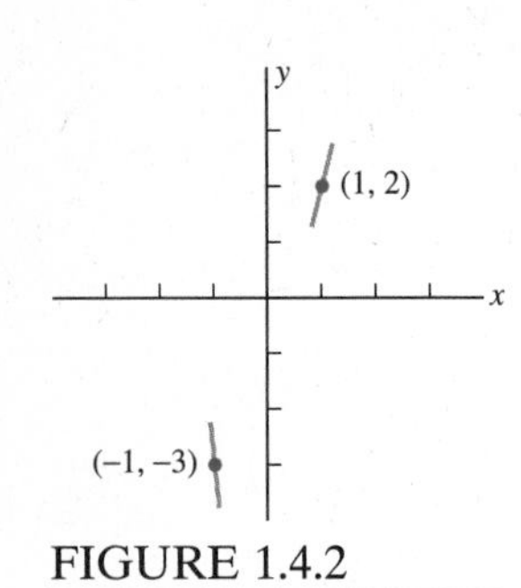

FIGURE 1.4.2

Slopes for Example 1.4.1

EXAMPLE 1.4.1 Slope at a Point

Suppose the differential equation is

$$\frac{dy}{dx} = xy^2.$$

Find the slope of the tangent lines to the solutions through $x = 1$, $y = 2$ and $x = -1$, $y = -3$, and plot short segments of both tangent lines.

Solution The differential equation says that the slope at (x, y) is xy^2. Thus the slope at $(1, 2)$ is $1(2)^2 = 4$ and the slope at $(-1, -3)$ is $(-1)(-3)^2 = -9$. These points and slopes are graphed in Figure 1.4.2. ◄

This example suggests a procedure that is fairly easy to implement on a computer with plot or graphic capabilities. The points would normally be chosen systematically on some grid in the xy plane. At each grid point (x_i, y_i), a short line segment is drawn with slope $f(x_i, y_i)$ given by the differential equation. The line segment is the tangent line at that point to the solution going through that point. We can connect the line segments in a smooth way to approximate the solution of the differential equation that goes through the point (x_i, y_i). This technique is most practical when the direction field is obtained on a computer. The computer can often sketch the general solution. The general solution is an infinite family of solutions, one corresponding to each initial condition, so that graphing all solutions is impossible. However, the computer can also be used to sketch solutions corresponding to many initial conditions. We discuss three examples.

Direction Field for Integrable Example

For the first-order differential equation, $dy/dx = f(x, y)$, in which the $f(x, y)$ only depends on x,

$$\frac{dy}{dx} = f(x), \tag{2}$$

the direction field and its solution are simpler than the general case. The sketch of the direction field is easier to obtain because the slope of the solution dy/dx at a given x is the same for all values of y. That is, along all vertical lines in the xy plane, the slope of the solution will be the same, $f(x)$. All solutions will differ only by a constant, since

$$y = \int f(x)\, dx + c \tag{3}$$

is the general solution. Thus, the graph of the solutions (obtained either exactly or from the direction field) will be an infinite family of curves, all parallel to one another, displaced vertically from one another. This displacement corresponds to the arbitrary additive constant in the general solution.

EXAMPLE 1.4.2 Direction Field

Consider

$$\frac{dy}{dx} = x(1 - x). \tag{4}$$

The direction field is graphed in Figure 1.4.3, along with some of the solutions. The exact general solution is

$$y = \frac{1}{2}x^2 - \frac{1}{3}x^3 + c. \tag{5}$$

In Figure 1.4.3, all solution curves are displaced vertically from one another. Note that the local extrema of the solutions are at $x = 0$ and $x = 1$, since from (4) that is where $dy/dx = 0$. ◄

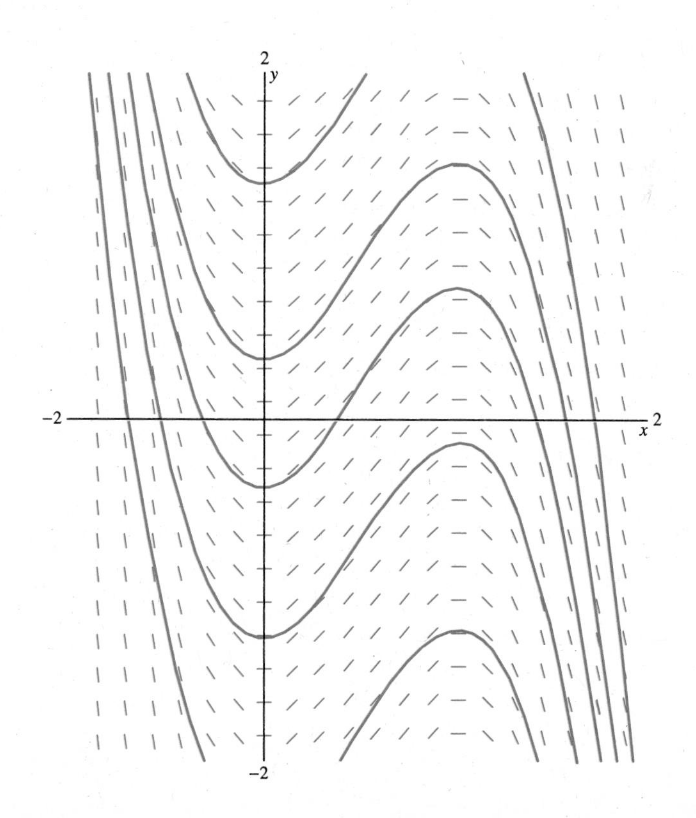

FIGURE 1.4.3

Direction field and solutions of (4), $dy/dx = x(1 - x)$.

Direction Field for Autonomous Example

We next will determine the direction field for the important class of first-order equations, $dy/dx = f(x, y)$, in which $f(x, y)$ does not depend on x:

$$\frac{dy}{dx} = g(y). \tag{6}$$

Such an equation is called **autonomous**. These equations are discussed extensively (including a brief explanation of the name) in Chapter 7. The direction field is simpler to obtain, since the slope of the solution dy/dx is the same value $g(y)$ along horizontal lines. By separation, the solution of (6) is

$$\int \frac{dy}{g(y)} = x + c. \tag{7}$$

Hence all solutions are parallel to one another. They are shifted horizontally from one another by a constant.

EXAMPLE 1.4.3 **Direction Field**

Consider

$$\frac{dy}{dx} = y(1 - y). \tag{8}$$

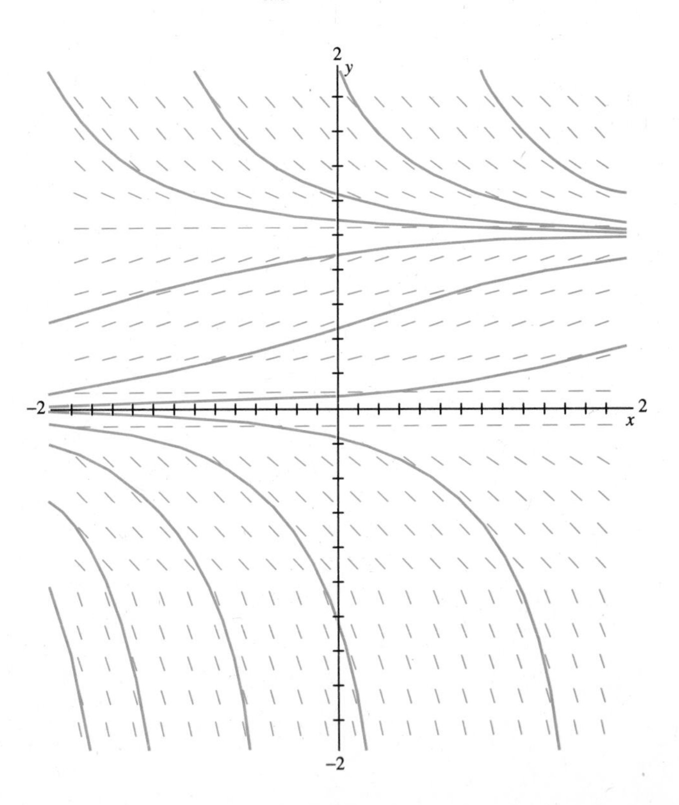

FIGURE 1.4.4

Direction field and solutions for (8), $dy/dx = y(1 - y)$.

The direction field and solutions are sketched in Figure 1.4.4 on page 22. Note that all solutions are shifted horizontally from one another. The general solution, which expresses the horizontal shifts, is given in (13) of Section 1.3. This equation is an important model in population dynamics and is discussed in Chapter 7. ◄

As a rule, the general solution of a first-order differential equation represents an infinite family of curves that are not just horizontal or vertical shifts of one another.

EXAMPLE 1.4.4 **Direction Field**

Consider

$$\frac{dy}{dx} = y^2 - x. \tag{9}$$

A sketch of the direction field obtained by a computer is shown in Figure 1.4.5. From Figure 1.4.5, we see that the differential equation has many solutions. However, there is only one solution (one curve) going through each

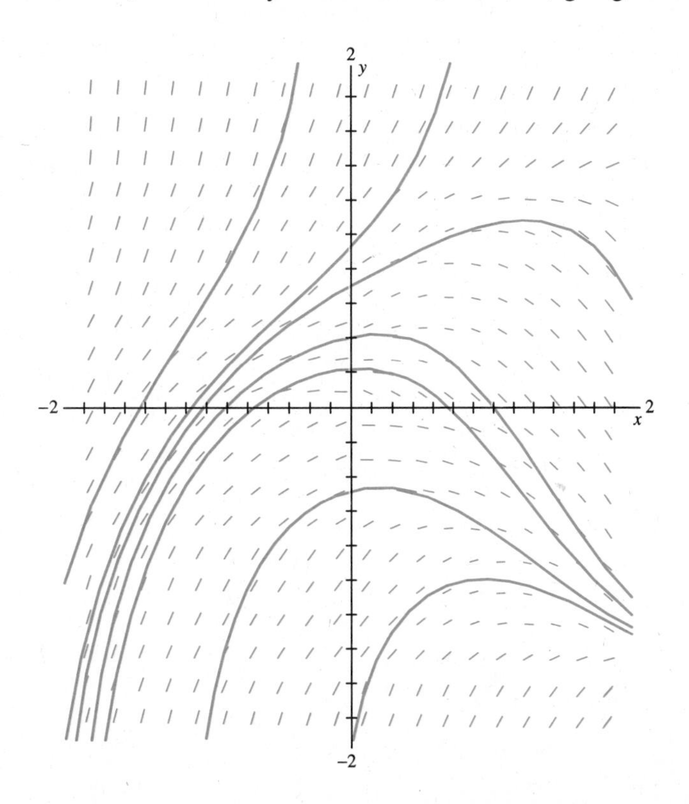

FIGURE 1.4.5

Direction field and solutions of (9), $dy/dx = y^2 - x$.

point. To find the solution of the initial value problem corresponding to the initial condition, $y(\frac{1}{2}) = 1$, we locate the specific curve that goes through the point $(\frac{1}{2}, 1)$. Unlike in the previous examples, the solution curves are not shifts of one another. ◄

From a plot such as that in Figure 1.4.5, we can often get a good idea of what the graph of the solutions of a first-order differential equation looks like. This procedure, however, is not much different from that of determining the graph of a function by plotting a large number of points. How do we know how fine a grid to choose? How do we know that the solutions do not change their behavior at a distance from the grid? What values of x or y are important? What is the behavior of the differential equation near these important points? Why do the solutions of the differential equations behave the way they are observed numerically to act? The direction field method is basically a numerical method, and, while useful, it is far from the best numerical method. For better numerical methods, see Chapter 6. We often need to get a better understanding of the solutions of differential equations than can be obtained this way. That is one reason we wish to study differential equations further.

Nullclines and Solution Sketching

In calculus we sketched the graphs of functions by finding where the tangent was horizontal (zero slope) or undefined. Then in between these points the function would be either increasing (positive slope) or decreasing (negative slope). Since the differential equation

$$\frac{dy}{dx} = f(x, y) \tag{10}$$

gives the slope of the tangent line to a solution at the point (x, y), we can use $f(x, y)$ to sketch the graphs of the solutions of the differential equation. This information can also be helpful in determining how big a grid to use with a computer-generated direction field.

The **nullcline** of $dy/dx = f(x, y)$ is a curve along which the slope of the tangent line to a solution is zero or undefined. Thus, nullclines correspond to $f(x, y) = 0$ or $f(x, y)$ undefined. At the points where $f(x, y) = 0$, there will be horizontal tangents. At points where $f(x, y)$ is undefined, there are many different kinds of things that can happen, including (sometimes) vertical tangents. These are discussed in Section 1.5. Here we just treat the points where $f(x, y)$ is undefined as a place where anything can happen, and so we do not graph there.

These curves (nullclines) split the xy plane into regions where $dy/dx = f(x, y)$ does not change sign. If $dy/dx > 0$ in a region, all solutions are increasing while in that region. We often plot an arrow (or arrows) with a positive slope in that region showing the direction in which the solution will change as x increases. If $dy/dx < 0$ in a region, then all solutions are decreasing while in that region, and we introduce an arrow (or arrows) with a negative slope. In summary, the procedure is as follows.

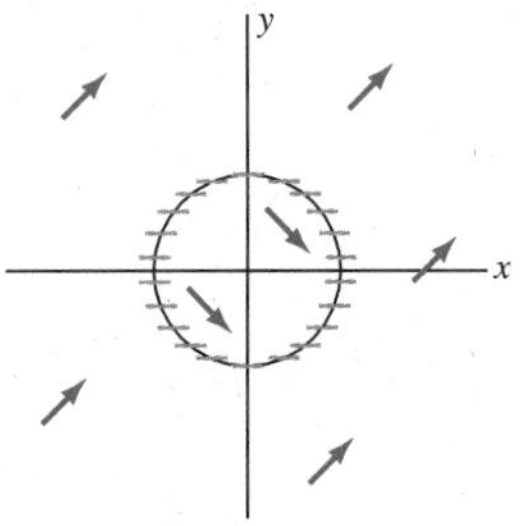

FIGURE 1.4.6

Nullclines for (11), $dy/dx = x^2 + y^2 - 1$.

Nullcline Procedure for Sketching Solutions of $\dfrac{dy}{dx} = f(x, y)$

1. Plot the curves where $f(x, y) = 0$ and $f(x, y)$ is undefined.
2. Determine whether $f(x, y)$ is positive or negative in each region.
3. Draw horizontal tangents on the $f(x, y) = 0$ curves and positive slope (negative slope) arrows in the region where $f(x, y)$ is positive (or negative).
4. Sketch solutions.

EXAMPLE 1.4.5 **Nullclines**

Using nullclines, sketch the solutions of

$$\frac{dy}{dx} = x^2 + y^2 - 1. \tag{11}$$

Solution We have that $f(x, y) = x^2 + y^2 - 1$. The curve where $f(x, y) = 0$ is a circle of radius 1 centered at the origin. There are no places where $f(x, y)$ is not defined. Pick any point inside the circle. We take $(0, 0)$. Since $f(0, 0) = -1$ is less than 0, we have $f(x, y)$ is negative inside the circle. Taking a point outside the circle, we find that $f(x, y)$ is positive outside the circle. Figure 1.4.6 summarizes this information, including the horizontal tangents on the $f(x, y) = 0$ curve. The solutions are sketched in Figure 1.4.7. ◄

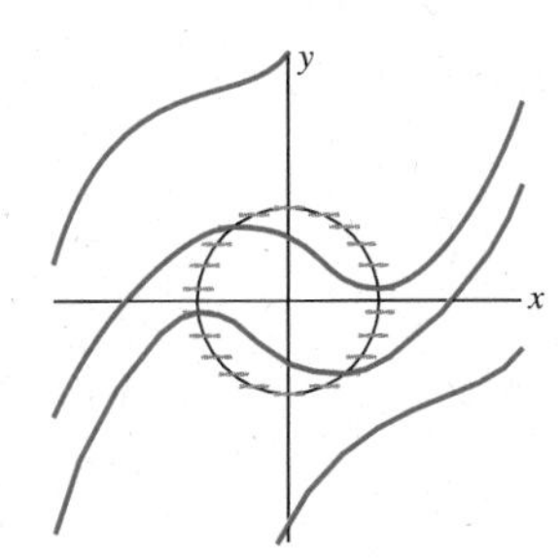

FIGURE 1.4.7

Nullclines and solutions for (11), $dy/dx = x^2 + y^2 - 1$.

EXAMPLE 1.4.6 **Nullclines**

Using nullclines, sketch the solutions of

$$\frac{dy}{dx} = \frac{x - y^2}{y - 1}. \tag{12}$$

Solution Here we have $f(x, y) = (x - y^2)/(y - 1)$. The curve where $f(x, y) = 0$ is a parabola $x - y^2 = 0$. The slope $f(x, y)$ is not defined if $y = 1$. These two curves $x - y^2 = 0$ and $y = 1$ split the xy plane into four regions. Taking a point in each of these four regions, we see whether $f(x, y)$ is positive or negative in that region. Figure 1.4.8 summarizes this information, including the horizontal tangents on the $x - y^2 = 0$ curve. The solutions are sketched in Figure 1.4.9 on the following page. ◄

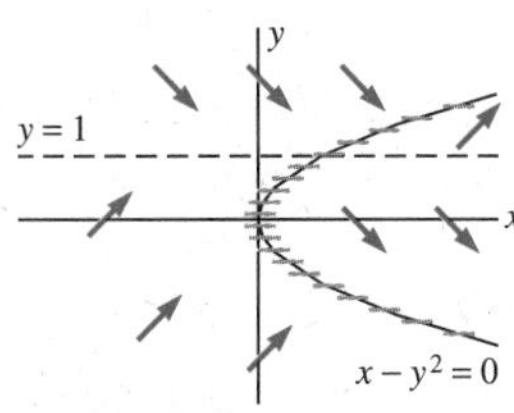

FIGURE 1.4.8

Nullclines for (12), $dy/dx = (x - y^2)/(y - 1)$.

FIGURE 1.4.9

Nullclines and solutions for (12), $dy/dx = (x - y^2)/(y - 1)$.

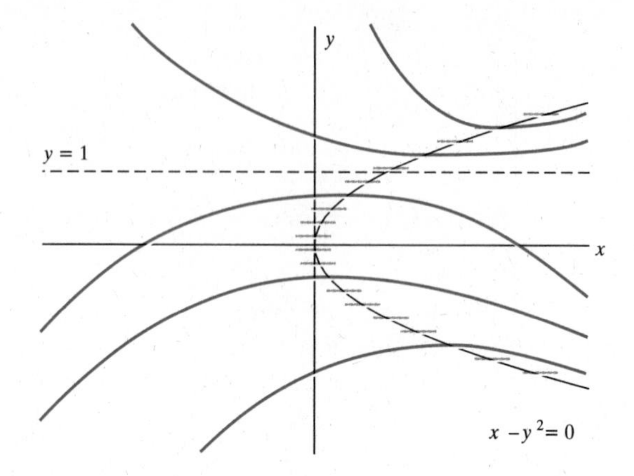

Exercises

In Exercises 1 through 6, the direction field of the given differential equations has been plotted by a computer. Sketch some solutions of the differential equation. (*Hint*: First copy or trace the figure.)

1. $\dfrac{dy}{dx} = y(1 - y^2)$

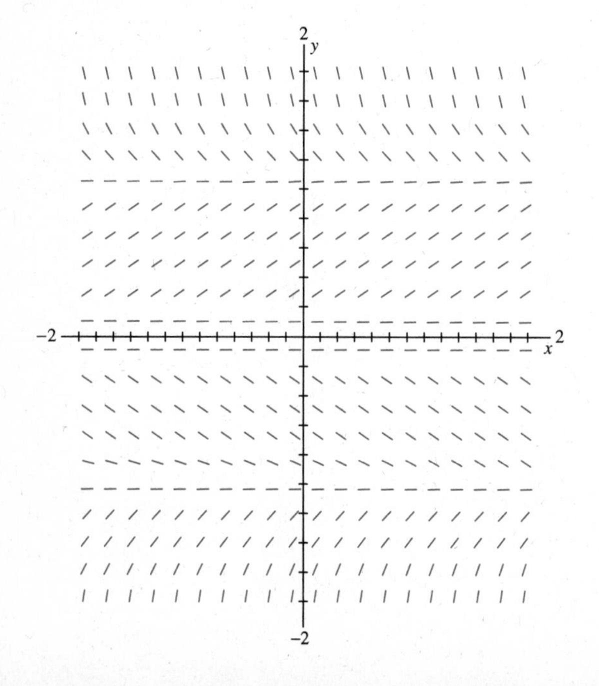

2. $\dfrac{dy}{dx} = x(1 - x^2)$

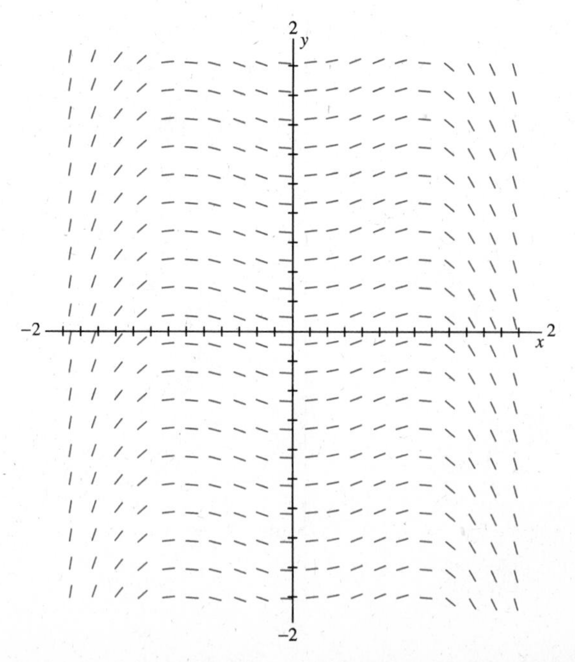

3. $\dfrac{dy}{dx} = y(x - y)$

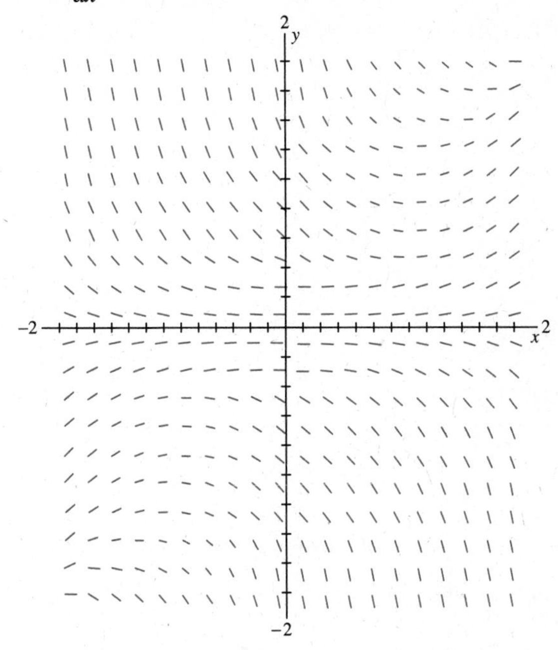

4. $\dfrac{dy}{dx} = x(y - x)$

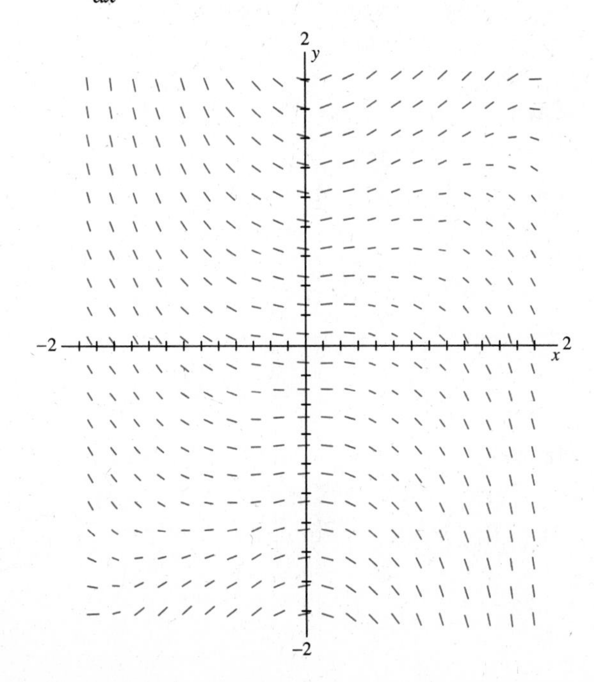

5. $\dfrac{dy}{dx} = x^2 - y^2$

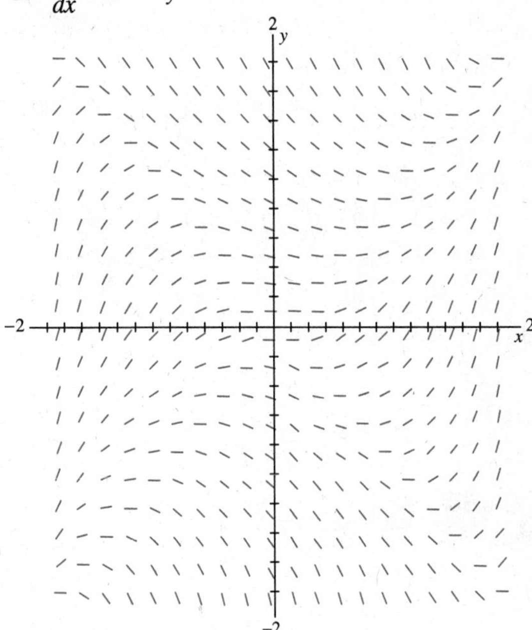

6. $\dfrac{dy}{dx} = x^2 - y$

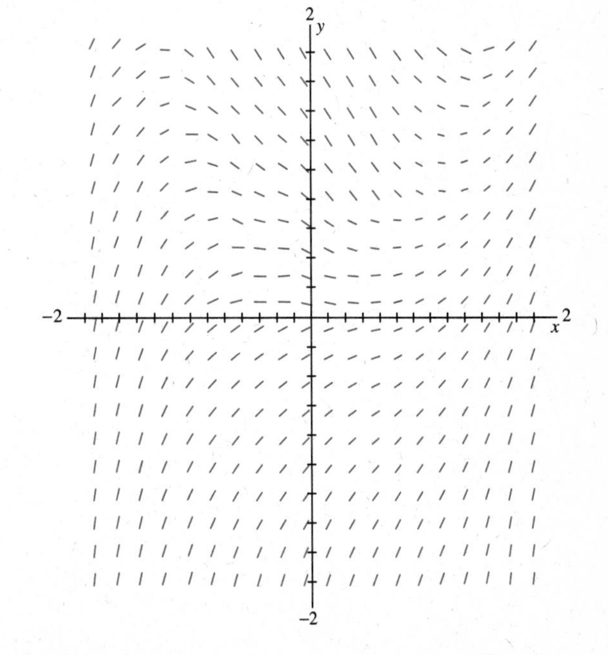

7. Apply the nullcline method to Example 1.4.2 in the text, $dy/dx = x(1 - x)$.

8. Apply the nullcline method to Example 1.4.3 in the text, $dy/dx = y(1 - y)$.

9. Apply the nullcline method to Example 1.4.4 in the text, $dy/dx = y^2 - x$.

10. Apply the nullcline method to Exercise 1.

11. Apply the nullcline method to Exercise 2.

12. Apply the nullcline method to Exercise 3.

13. Apply the nullcline method to Exercise 4.

14. Apply the nullcline method to Exercise 5.

15. Apply the nullcline method to Exercise 6.

In Exercises 16 through 22, apply the nullcline method to the given differential equations.

16. $\dfrac{dy}{dx} = y(1 - x^2 - y^2)$

17. $\dfrac{dy}{dx} = y(y - x)^2$

18. $\dfrac{dy}{dx} = \dfrac{yx - y}{x}$

19. $\dfrac{dy}{dx} = \dfrac{-1 + x^2 + 4y^2}{y - 5x + 10}$

20. $\dfrac{dy}{dx} = \dfrac{y - x^2}{x}$

21. $\dfrac{dy}{dx} = \dfrac{x - y}{2x + y}$

22. $\dfrac{dy}{dx} = \dfrac{y - x^2}{1 + x^2}$

1.5 Existence and Uniqueness

Often the first-order differential equation, $dy/dx = f(x, y)$, cannot be solved by simple integration. It is then important to know when there are solutions. As with the previous elementary examples, the differential equation usually has many solutions, of which only one will satisfy given initial conditions. A more precise statement of the mathematical result that guarantees this is the following.

■ **THEOREM 1.5.1 Basic Existence and Uniqueness Theorem for First-Order Differential Equations**

There exists a unique solution to the differential equation

$$\frac{dy}{dx} = f(x, y) \tag{1}$$

that satisfies the given initial condition

$$y(x_0) = y_0 \tag{2}$$

if both the function $f(x, y)$ and its partial derivative $\partial f/\partial y$ are continuous functions of x and y at and near the initial point $x = x_0$, $y = y_0$. ■

In many cases of practical interest, the continuity conditions are satisfied for all values of (x_0, y_0), so that there exists a unique solution to the initial value problem. An example of an application of this theorem will be given in the next section.

If the conditions of the Basic Existence and Uniqueness Theorem are not met at a point (x_0, y_0), then solutions may not exist at (x_0, y_0), or there may be more than one solution passing through the same point (x_0, y_0), or the solution may not be differentiable at x_0 (Exercise 12 at the end of this section). Such behavior often has a physical interpretation.

EXAMPLE 1.5.1 Existence and Uniqueness

The differential equation

$$\frac{dy}{dx} = xy^2 \tag{3}$$

satisfies the conditions of the uniqueness theorem for all (x, y), since $f = xy^2$ and $\partial f/\partial y = 2xy$ are continuous functions of x and y. Thus, there should be a unique solution of the initial value problem for any (x_0, y_0). For example, there should be a solution at the initial condition $(0, 2)$. If we solve the differential equation (3) by separation,

$$\int \frac{dy}{y^2} = \int x\,dx.$$

After integration and applying the initial condition, we obtain

$$-\frac{1}{y} + \frac{1}{2} = \frac{1}{2}x^2.$$

Solving for y yields the promised unique solution

$$y = \frac{2}{1 - x^2}. \tag{4}$$

However, this solution (4) is not defined for all x. It becomes infinite at $x = \pm 1$. While the Basic Existence and Uniqueness Theorem guarantees a solution of the differential equation that satisfies the initial condition, it does not say that the solution is defined for all x. It is a local existence theorem, guaranteeing only that the solution exists for some x near the initial x_0. In this example, the solution "explodes" at $x = 1$.

The direction field for (3) is graphed in Figure 1.5.1 on page 30. The specific initial condition $y(0) = 2$ and others are shown. One inadequacy of the direction field method is that it may be impossible to tell whether or not solutions become infinite. From the figure it is impossible to tell what occurs outside the plotted region. However, by separation, it can be shown that all solutions of (3) are described by

$$y = \frac{-2}{x^2 + c} \quad \text{and} \quad y = 0.$$

If $c > 0$, solutions exist for all x, while solutions have vertical asymptotes if $c \leq 0$. The case $c = 0$, that is, $y = -2/x^2$ separates solutions that become infinite from those that do not become infinite. This cannot be determined from the direction field. ◄

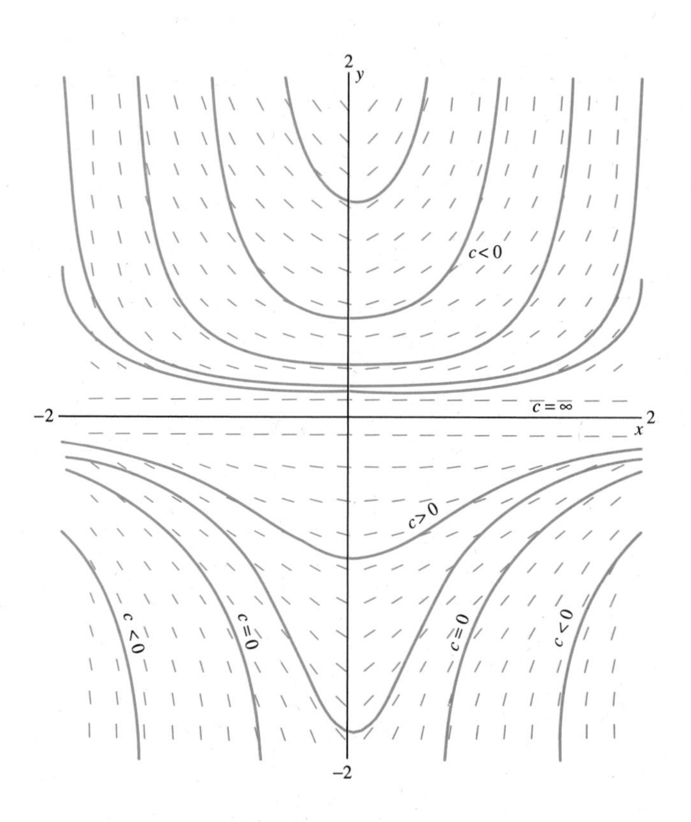

FIGURE 1.5.1

Direction field for (3), $dy/dx = xy^2$.

EXAMPLE 1.5.2 **Initial Condition with No Solution**

The differential equation

$$\frac{dy}{dx} = -\frac{1}{x^2} \tag{5}$$

satisfies the Basic Existence and Uniqueness Theorem everywhere that $x_0 \neq 0$, since $f(x, y) = -1/x^2$ $\partial f/\partial y = 0$ are continuous everywhere that $x \neq 0$. By antidifferentiation, the general solution of (5) is

$$y = \frac{1}{x} + c.$$

Several solutions of (5) are graphed in Figure 1.5.2 on the next page.

If the initial condition is given at $x_0 \neq 0$, then the constant c can be determined uniquely by the initial condition. There is a unique solution to

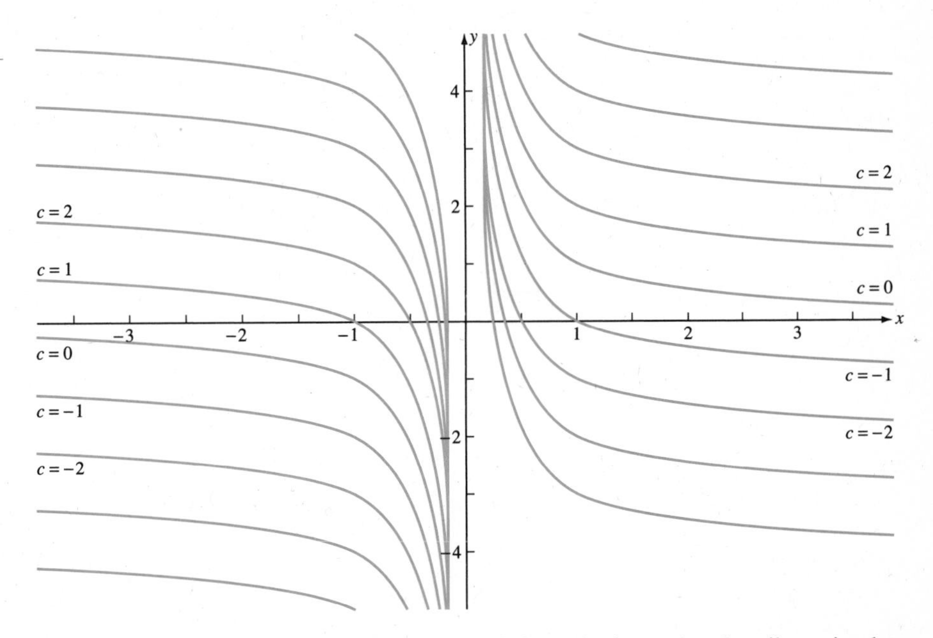

FIGURE 1.5.2

Several solutions of (5), $dy/dx = -1/x^2$.

the initial value problem in this case, and the solution exists locally as in the previous example up to the singularity at $x = 0$. However, if the initial condition is given at $x_0 = 0$, then there are no solutions. For example, there are no solutions of (5) with $y = 1$ at $x = 0$. ◀

EXAMPLE 1.5.3 **Solutions Not Unique**

By separation, we can show that the general solution of

$$\frac{dy}{dx} - 3\frac{y}{x} = 0 \tag{6}$$

is

$$y = cx^3. \tag{7}$$

Several solutions of (7) are plotted in Figure 1.5.3 on the following page.

The functions $f(x, y) = 3y/x$ and $\partial f/\partial y = 3/x$ are continuous everywhere that $x \neq 0$. Thus, if $x_0 \neq 0$, then (7) is the unique solution of the initial value problem ($c = y_0/x_0^3$). However, if $x_0 = 0$, the theorem does not guarantee that there exists a unique solution. Let us see what happens if $x_0 = 0$. From (7) we have $y_0 = c \cdot 0$. If $x_0 = 0$ and $y_0 \neq 0$, there are no solutions. On the other hand, if $x_0 = 0$, $y_0 = 0$, then (7) will be a solution for any value of c and the solution to the initial value problem exists but is not unique. Every solution goes through $(0, 0)$. ◀

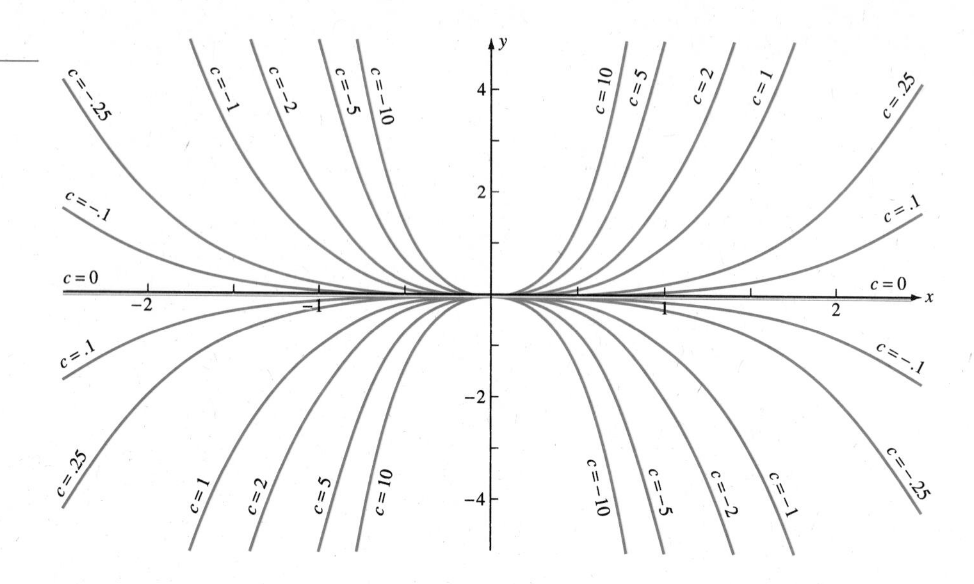

FIGURE 1.5.3

Several solutions $y = cx^3$ of (6), $dy/dx = 3y/x$.

Exercises

For each of the differential equations in Exercises 1 through 10, give those points (x_0, y_0) in the xy plane for which the Basic Existence and Uniqueness Theorem guarantees that a unique solution exists.

1. $\dfrac{dy}{dx} = \dfrac{y}{1 + x^2}$

2. $\dfrac{dy}{dx} = \dfrac{x - y}{3x - 7y}$

3. $\dfrac{dy}{dx} = (1 - x^2 - y^2)^{7/3}$

4. $\dfrac{dy}{dx} = (1 - x^2 - y^2)^{1/3}$

5. $\dfrac{dy}{dx} = (y + x)^{1/5}$

6. $x\dfrac{dy}{dx} = y^2 + 1$

7. $(y - 1)\dfrac{dy}{dx} = \cos x$

8. $\dfrac{dy}{dx} = \dfrac{x^3 + 1}{x(y + 1)}$

9. $\dfrac{dy}{dx} = (1 - x^2 - 2y^2)^{3/2}$

10. $\dfrac{dy}{dx} = y^{1/3}\sqrt{y - 1}$

Exercises 11 and 12 refer to the differential equation

$$\frac{dy}{dx} = x^{-1/3}. \tag{8}$$

11. Show that the Basic Existence and Uniqueness Theorem holds for (8) for all (x_0, y_0) such that $x_0 \neq 0$.

12. Show that, for any constant y_0, there exists a unique function $y(x)$ such that

a) y satisfies (8) if $x \neq 0$,

b) y is continuous for all x,

c) $y(0) = y_0$,

d) y is not differentiable at $x_0 = 0$.

In Exercises 13 through 19, you are given a family of curves $g(x, y) = c$. For each exercise,

a) Verify that the curves implicitly define solutions of the given differential equation.

b) Graph the curves on the same axis for the indicated values of c.

c) Determine for which points (x_0, y_0) the assumptions of the Basic Existence and Uniqueness Theorem fail to hold, and graph these points on the same graph.

13. $x^2 - y^2 = c, \quad \frac{dy}{dx} = \frac{x}{y}, \quad c = 0, \pm 1, \pm 2$

14. $x^2 + y^2 = c, \quad \frac{dy}{dx} = -\frac{x}{y}, \quad c = 1, 2, 3$

15. $x + y^2 = c, \quad 2y\frac{dy}{dx} + 1 = 0, \quad c = 0, \pm 1$

16. $y = cx, \quad \frac{dy}{dx} = \frac{y}{x}, \quad c = 0, \pm 1, \pm 2$

17. $y = c \sin x, \quad \frac{dy}{dx} = y \cot x,$
$c = 0, \pm 1, \pm 2$

18. $y = \tan(x + c), \quad \frac{dy}{dx} = 1 + y^2,$
$c = 0, \pm 1$

19. $y = \frac{1}{x + c}, \quad \frac{dy}{dx} = -y^2, \quad c = 0, \pm 1$

20. **a)** Show that $y = 0$ and $y = (\frac{2}{3}x + c)^{3/2}$ are solutions of

$$y' = y^{1/3}. \tag{9}$$

b) Show that there are at least two solutions of (9) through every point (x_0, y_0) with $y_0 = 0$.

c) Sketch several solutions of (9), including $y = 0$, on the same graph.

d) Note that $f(x, y) = y^{1/3}$ is continuous everywhere. Why aren't this fact and (b) a contradiction of the Basic Existence and Uniqueness Theorem?

21. **a)** Show that $y = (\frac{4}{5}x + c)^{5/4} + 1$ and $y = 1$ are solutions of

$$y' = (y - 1)^{1/5}. \tag{10}$$

b) Show that there are at least two solutions of (10) through every point (x_0, y_0) with $y_0 = 1$.

c) Sketch several solutions of (10), including $y = 1$, on the same graph.

d) Note that $f(x, y) = (y - 1)^{1/5}$ is continuous everywhere. Why aren't this fact and (b) a contradiction of the Basic Existence and Uniqueness Theorem?

22. For what initial conditions $(0, y_0)$ does the solution of $dy/dx = (2x - 2)y^2$ explode at some $x > 0$?

23. Reconsider Example 1.5.1, $dy/dx = xy^2$. For what initial conditions $(0, y_0)$ does the solution explode at some $x > 0$?

24. Solve the differential equation $dy/dx = 2y/x$. Determine for which initial conditions $(0, y_0)$ there are **(a)** no solutions, **(b)** a unique solution, or **(c)** nonunique solutions.

25. Solve the differential equation $dy/dx = -2y/x$. Determine for which initial conditions $(0, y_0)$ there are **(a)** no solutions, **(b)** a unique solution, or **(c)** nonunique solutions.

1.6 First-Order Linear Differential Equations

Linear first-order differential equations,

$$\frac{dy}{dx} + p(x)y = q(x) \tag{1}$$

are comparatively easy to solve, and applications frequently are described by such equations. The function $q(x)$ is called the **forcing function**, the **input**, the **nonhomogeneous term**, or the **right-hand side**, depending upon the application. When $q(x)$ is called the input, the solution of the differential equation $y(x)$ is usually called the **output**. It is important to understand which differential equations are linear and which equations are not linear. A differential equation that is not linear is called a **nonlinear** differential equation. For now, a first-order differential equation $dy/dx = f(x, y)$ is **linear** if $f(x, y)$ is a linear function of y, as written in (1).

Most phenomena in nature do not actually satisfy linear differential equations. However, in many applications, the solutions of interest do not differ too greatly from a function (or operating point). In this case, the differential equation may often be approximated by a linear differential equation. As a consequence, linear differential equations play a fundamental role in many parts of science and engineering.

1.6.1 Form of the General Solution

A **particular** solution y_p of the linear differential equation (1) is any function $y_p(x)$ satisfying the differential equation (1). Thus,

$$\frac{dy_p}{dx} + p(x)y_p = q(x). \tag{2}$$

If $q(x) = 0$, then the equation is called **homogeneous**. If $q(x)$ is not identically zero, then the **associated homogeneous equation** is

$$\frac{dy}{dx} + p(x)y = 0. \tag{3}$$

Thus a solution y_h of the associated homogeneous equation satisfies

$$\frac{dy_h}{dx} + p(x)y_h = 0. \tag{4}$$

We have introduced the ideas of a particular solution and solutions of the associated homogeneous equation because we will see that they play important roles physically and mathematically. We shall show later in this section that

The general solution of any linear first-order differential equation (2) can always be put in the form

$$y = y_p + y_h, \tag{5}$$

where y_p is a particular solution of (2) and y_h is the general solution of the associated homogeneous equation (3).

Furthermore, we will show that

The general solution of a homogeneous first-order linear differential equation is always in the form

$$y_h = cy_1, \tag{6}$$

where y_1 is a nonzero solution of the associated homogeneous equation and c is an arbitrary constant.

Thus the key result is that

For linear first-order equations (2), all the solutions (the general solution) are in the form of a particular solution y_p plus an arbitrary multiple c times one solution y_1 of the associated homogeneous equation.

$$y = y_p + y_h = y_p + cy_1. \tag{7}$$

Equation (7) is fundamentally important. We use the notation y_1 here since in the next chapter, on second-order equations, we will see that the solution of the associated homogeneous equation is made up of two functions y_1 and y_2.

A complete verification of the facts (5) and (6) comes at the end of this section. Here, we will merely verify that $y = y_p + cy_1$ satisfies the first-order linear differential equation

$$\frac{dy}{dx} + p(x)y = q(x). \tag{8}$$

We substitute $y_p + cy_1$ into (8) for y and collect those terms that have c as a multiple and those that do not:

$$\begin{aligned} \frac{dy}{dx} + p(x)y &= \frac{d}{dx}(y_p + cy_1) + p(x)(y_p + cy_1) \\ &= \frac{dy_p}{dx} + p(x)y_p + c\left(\frac{dy_1}{dx} + p(x)y_1\right). \end{aligned} \tag{9}$$

However, y_p satisfies (2) and y_1 satisfies (4), and thus (9) becomes

$$\frac{dy}{dx} + p(x)y = q(x) + c \cdot 0 = q(x).$$

In this way we have verified that $y = y_p + cy_1$ satisfies (8).

We recommend that you memorize that the general solution of linear first-order differential equations (8) is in the form (5), where a particular solution y_p satisfies (2) and y_1 satisfies the associated homogeneous equation (4).

1.6.2 Solutions of Homogeneous First-Order Linear Differential Equations

The general solution of the first-order linear differential equation (8), $dy/dx + p(x)y = q(x)$, is always in the form $y = y_p + cy_1$. However, we have not discussed any methods to obtain the solution y_1 of the associated homogeneous equation or the particular solution y_p. In this subsection we will show that homogeneous solutions of first-order linear differential equations can always be obtained by separation.

Solutions of the related homogeneous equation satisfy

$$\frac{dy}{dx} + p(x)y = 0. \tag{10}$$

Note that $y = 0$ is always a solution of (10). It is called the **trivial solution** of the homogeneous equation. When we say that the general solution of $dy/dx + p(x)y = q(x)$ is always in the form $y = y_p + cy_1$, we mean that y_1 is a nontrivial solution and do not allow $y_1 = 0$.

Nontrivial solutions of the homogeneous equation can always be obtained by separation, since (10) can be written

$$\int \frac{dy}{y} = -\int p(x)\, dx. \tag{11}$$

Indefinite integration yields

$$\ln |y| = -\int p(x)\, dx + c_0,$$

where c_0 is an arbitrary constant of integration. Exponentiating gives

$$|y| = e^{-\int p(x)\, dx + c_0} = e^{c_0} e^{-\int p(x)\, dx}. \tag{12}$$

Note that c_0 is an arbitrary constant, so that e^{c_0} is an arbitrary positive constant. Solving (12) for y gives $y = [\pm e^{c_0}]e^{-\int p(x)\, dx}$. Let $c = \pm e^{c_0}$. Then c is an arbitrary constant, and we have

$$y = ce^{-\int p(x)\, dx}. \tag{13}$$

By separation, we have shown in (13) that the form cy_1 for the solution of a homogeneous first-order linear equation is correct and that we can use

$$y_1 = e^{-\int p(x)\, dx}. \tag{14}$$

If we had used a definite integral instead, which is sometimes preferable if the function $p(x)$ is not simple, then we would obtain an equivalent expression:

$$y_1 = e^{-\int_a^x p(\bar{x})\, d\bar{x}}.$$

The lower limit a can be chosen as any constant. It usually is the initial value x_0 of x.

EXAMPLE 1.6.1 Solution of a Homogeneous Linear Differential Equation

Find the general solution of the homogeneous first-order linear differential equation

$$\frac{dy}{dx} + x^2 y = 0. \tag{15}$$

SOLUTION Since the equation is homogeneous, by separation we obtain

$$\int \frac{dy}{y} = -\int x^2\, dx.$$

Indefinite integration yields

$$\ln|y| = -\frac{1}{3}x^3 + c_0.$$

Exponentiation first yields $|y| = e^{c_0} e^{-\frac{1}{3}x^3}$, which becomes the general solution

$$y = ce^{-\frac{1}{3}x^3}. \tag{16}$$

◄

EXAMPLE 1.6.2 Solution of Homogeneous Equation Requiring a Definite Integral

Find the general solution of the homogeneous first-order linear differential equation

$$\frac{dy}{dx} + \sin(x^2)y = 0. \tag{17}$$

SOLUTION By separation, we obtain

$$\int \frac{dy}{y} = -\int \sin(x^2)\, dx.$$

Integration yields

$$\ln|y| = -\int_0^x \sin(t^2)\, dt + c_0,$$

where we have chosen the lower limit to be zero and the dummy variable of integration to be t. Exponentiation first yields $|y| = e^{c_0} e^{-\int_0^x \sin(t^2)\,dt}$, which becomes the general solution

$$y = ce^{-\int_0^x \sin(t^2)\,dt}. \tag{18}$$

◄

Exercises

In the following exercises, problems where a definite integral is needed or recommended are indicated by an asterisk (*). In Exercises 1 through 10, find the general solution using separation.

1. $\dfrac{dy}{dx} = 3y$

2. $\dfrac{dy}{dx} = -8y$

3. $\dfrac{dy}{dx} = 2xy$

4. $\dfrac{dy}{dx} = -x^3y$

5. $2x\dfrac{dy}{dx} + y = 0$

6. $2x\dfrac{dy}{dx} - y = 0$

7. $\dfrac{dy}{dx} + (\cos x)y = 0$

8. $\dfrac{dy}{dx} + (3 + x^5)y = 0$

9.* $\dfrac{dy}{dx} + \cos(x^{-1/2})y = 0$

10.* $\dfrac{dy}{dx} + \dfrac{\ln x}{1 + e^x}y = 0$

In Exercises 11 through 17, solve the initial value problems using separation.

11. $\dfrac{dy}{dx} = -5y$ with $y(0) = 9$

12. $\dfrac{dy}{dx} = 2y$ with $y(0) = 5$

13. $\dfrac{dy}{dx} = 9y$ with $y(3) = 7$

14. $\dfrac{dy}{dx} = -6y$ with $y(2) = 4$

15.* $\dfrac{dy}{dx} + \dfrac{\sin x}{4 + e^x}y = 0$ with $y(5) = 10$

16.* $\dfrac{dy}{dx} + \ln(7 + xe^x)y = 0$ with $y(0) = 3$.

17. $\dfrac{dy}{dx} + x^{-2}y = 0$ with $y(1) = 3$.

18. Reconsider $\dfrac{dy}{dx} + p(x)y = q(x)$. Suppose y_p is a particular solution and $\tilde{y}$ is any other particular solution.

a) Show that $y = \tilde{y} - y_p$ must be a solution of the associated homogeneous equation $dy/dx + p(x)y = 0$.

b) Using $y_h = cy_1$, the result of this section on solutions of homogeneous equations, show that $y = y_p + cy_1$, as claimed in this section.

1.6.3 Integrating Factors for First-Order Linear Differential Equations

Now that we have a method to obtain the solution for homogeneous first-order linear differential equations, we need a method to obtain a particular solution. We will discuss two methods for obtaining particular solutions. One method uses an integrating factor. This method always works for first-order linear differential equations, but the algebraic steps necessary to produce a satisfactory answer can be unwieldy. We will describe another method which is much easier in the next section. It is called the method of undetermined coefficients. In Chapter 2 the same method is discussed for second-order linear differential equations in Section 9. Unfortunately, this easier method works only on linear differential equations with constant coefficients and will work only on some such problems. However, the technique is very useful in applications.

Although there are methods for deriving this result (see Exercise 55 in this section), it is often easiest for students just to memorize that a particular solution of

$$\frac{dy}{dx} + p(x)y = q(x) \tag{19}$$

may be obtained using the integrating factor

$$u(x) = e^{\int p(x)\,dx}. \tag{20}$$

The expression $u(x)$ is an integrating factor since the differential equation may be solved by an integration if we first multiply the differential equation (19) by the integrating factor (20). Multiplying (19) by (20) gives

$$e^{\int p(x)\,dx}\left(\frac{dy}{dx} + p(x)y\right) = q(x)e^{\int p(x)\,dx}. \tag{21}$$

The left-hand side is an exact derivative of $u(x)y(x)$ [as can be verified using the product rule and the definition of $u(x)$ in (20)]. Thus, (21) is equivalent to

$$\frac{d}{dx}\left(ye^{\int p(x)\,dx}\right) = q(x)e^{\int p(x)\,dx}.$$

Indefinite integration yields

$$ye^{\int p(x)\,dx} = \int q(x)e^{\int p(x)\,dx}\,dx + c, \tag{22}$$

where c is an arbitrary constant of integration. By multiplying both sides of (22) by $e^{-\int p(x)\,dx}$, we obtain the **general solution of any first-order linear differential equation:**

$$y = e^{-\int p(x)\,dx}\int q(x)e^{\int p(x)\,dx}\,dx + ce^{-\int p(x)\,dx}. \tag{23}$$

Although our goal was to just determine a particular solution, the result of using an integrating factor gives the general solution. Formula (23) proves the facts (5), (6), and (7) that the general solution of any first-order linear differential equation is in the form $y = y_p + cy_1$. A solution y_1 of the associated homogeneous equation is seen to be $y_1 = e^{-\int p(x)\,dx}$, whereas a particular solution is possibly more complicated:

$$y_p = e^{-\int p(x)\,dx}\int q(x)e^{\int p(x)\,dx}\,dx.$$

It is not a good idea to memorize this result. Instead, in each example, one usually repeats the following steps.

Summary of the Integrating Factor Method for Solving $\dfrac{dy}{dx} + p(x)y = q(x)$

1. Compute the integrating factor $u = e^{\int p(x)\,dx}$ and simplify.
2. Multiply both sides of the first-order linear differential equation (19) by u to get
$$(uy)' = uq.$$
3. Integrate both sides with respect to x to get $u(x)y(x) = \int u(x)q(x)\,dx + c$.
4. Divide by $u(x)$ to get y.

If the integrals can be evaluated explicitly, then this solution is straightforward. However, often one or more of these integrals cannot be done explicitly, in which case definite integrals should be used. In some of the examples that follow, definite integrals will be used. Also note that the differential equation must be put in the form (19) before beginning this procedure.

The integrating factor can be used even if the linear differential equation is homogeneous. Thus we have two methods to obtain solutions of homogeneous differential equations: separation and the integrating factor.

EXAMPLE 1.6.3 Homogeneous

Find the general solution of

$$\frac{dy}{dx} + x^2 y = 0 \tag{24}$$

using an integrating factor.

Solution From the differential equation we have $p(x) = x^2$. To obtain the solution using an integrating factor, we multiply the linear differential equation (24) by the integrating factor $e^{\int p(x)\,dx} = e^{\frac{1}{3}x^3}$. This gives

$$e^{\frac{1}{3}x^3}\left(\frac{dy}{dx} + x^2 y\right) = 0. \tag{25}$$

The left side is always the derivative of y times the integrating factor. Thus, (25) is equivalent to

$$\frac{d}{dx}\left(e^{\frac{1}{3}x^3} y\right) = 0.$$

Integration gives $e^{\frac{1}{3}x^3}y = c$, and hence

$$y = ce^{-\frac{1}{3}x^3}.$$

This is the same general solution we obtained by separation [see (16)]. ◄

EXAMPLE 1.6.4 **Nonhomogeneous**

The following first-order linear differential equation is not homogeneous:

$$x\frac{dy}{dx} - 3y = x^5.$$

Find the general solution for $x > 0$.

SOLUTION In order to find the integrating factor, $e^{\int p(x)\,dx}$, the differential equation must first be put in the form $dy/dx + p(x)y = q(x)$. This requires dividing the differential equation by x to get

$$\frac{dy}{dx} - \frac{3}{x}y = x^4. \tag{26}$$

Thus, $p(x) = -(3/x)$, and hence the integrating factor is $e^{\int p(x)\,dx} = e^{-3\ln x} = x^{-3}$ (since we assume $x > 0$). Multiplying both sides of (26) by the integrating factor gives

$$\frac{1}{x^3}\left(\frac{dy}{dx} - \frac{3}{x}y\right) = \frac{1}{x^3}x^4 = x.$$

The left-hand side is always an exact derivative of y times the integrating factor:

$$\frac{d}{dx}\left(\frac{1}{x^3}y\right) = x.$$

By indefinite integration, we then obtain

$$\frac{1}{x^3}y = \frac{x^2}{2} + c.$$

Solving for y, we see that the general solution is in the proper form, a particular solution plus an arbitrary multiple of a solution of the associated homogeneous equation:

$$y = \frac{x^5}{2} + cx^3.$$

◀

EXAMPLE 1.6.5 **One Definite Integration**

Solving the following first-order linear differential equation will require a definite integral:

$$\frac{dy}{dx} + x^2y = \cos x.$$

The integrating factor $e^{\int p(x)\,dx} = e^{\frac{1}{3}x^3}$ is straightforward and does not require a definite integral, since $p(x) = x^2$. Multiplying the differential equation by this integrating factor yields

$$e^{\frac{1}{3}x^3}\left(\frac{dy}{dx} + x^2 y\right) = e^{\frac{1}{3}x^3} \cos x.$$

Again, the left-hand side is an exact derivative, so that we get

$$\frac{d}{dx}\left(e^{\frac{1}{3}x^3} y\right) = e^{\frac{1}{3}x^3} \cos x.$$

The right-hand side does not have an elementary integral. Even if the right-hand side has an elementary integral, we could still use a definite integral. We integrate both sides. For convenience we choose the lower limit to be 0:

$$e^{\frac{1}{3}x^3} y = \int_0^x e^{\frac{1}{3}\bar{x}^3} \cos \bar{x}\, d\bar{x} + c.$$

From this, we obtain the general solution as

$$y = e^{-\frac{1}{3}x^3} \int_0^x e^{\frac{1}{3}\bar{x}^3} \cos \bar{x}\, d\bar{x} + ce^{-\frac{1}{3}x^3},$$

a particular solution plus an arbitrary multiple of a solution of the associated homogeneous equation. It is very important to use a dummy variable of integration different from x so that there is no temptation to have $e^{-\frac{1}{3}x^3}$ incorrectly cancel $e^{\frac{1}{3}\bar{x}^3}$. In this example, the integrating factor did not require definite integration. ◄

Examples that require two definite integrals are more complicated. Even a relatively easy-looking linear differential equation may still have a complicated solution.

Exercises

In Exercises 19 through 30, solve the differential equation. If there is no initial condition, give the general solution.

19. $xy' + y = e^x, \quad y(1) = 1$

20. $y' + 2xy = x$

21. $y' = 3e^x$

22. $y' = \dfrac{y + x}{x}$

23. $(x^2 + 1)y' + 2xy = 1$

24. $\dfrac{du}{dt} = 3(u - 1)$

25. $y' + 4y = x, \quad y(0) = 0$

26. $(x + 1)y' + y = x$

27. $xy' = 2y, \quad y(1) = 4$

28. $xy' = -3y + \dfrac{\sin x}{x^2}$

29. $x^2y' + xy = 1$

30. $t^3\dfrac{dy}{dt} + 4t^4 = t^7$

For each of the linear differential equations in Exercises 31 through 33, find the general solution of the differential equation and sketch several solutions ($c = 0$, $c = \pm 1$, $c = \pm 3$ would suffice), and describe the behavior of the solutions near $x = 0$.

31. $xy' + 3y = x$

32. $xy' + 2y = x^{-1}$

33. $xy' - y = x^2$

In Exercises 34 through 45, find the solution of the initial value problem that requires definite integrals. If there are no conditions given, find the general solution.

34. $y' + x^4y = 1$

35. $y' + 2xy = 1$

36. $y' + (\sin x)y = x$

37. $y' + x^2y = x$

38. $y' - 2xy = e^x$

39. $y' + e^xy = 3$

40. $y' - y = e^{x^2}$

41. $\dfrac{dy}{dx} + y = \dfrac{1}{x+1}$ with $y(2) = 1$.

42. $\dfrac{dy}{dx} + \dfrac{3}{x}y = \ln(3 + \cos^2 x)$ with $y(4) = 6$.

43. $3x\dfrac{dy}{dx} - y = x \sin x$ with $y(5) = 0$.

44. $\dfrac{dy}{dx} + y = \sin(e^{4x})$

45. $7x\dfrac{dy}{dx} + y = e^x$

From calculus we know that

$$\frac{dx}{dy} = \frac{1}{dy/dx}.$$

That is, we can think of y as the independent variable and x as the dependent variable provided $dy/dx \neq 0$. In Exercises 46 through 48, the differential equation is nonlinear in the form $dy/dx = f(x, y)$, but linear (after some algebraic manipulations) if written as $dx/dy = 1/f(x, y)$. Find the general solution of the differential equation.

46. $\dfrac{dy}{dx} = \dfrac{1}{y}$

47. $\dfrac{dy}{dx} = \dfrac{1}{y+x}$

48. $\dfrac{dy}{dx} = \dfrac{y}{x+y}$

49. Verify that, if an arbitrary constant c_1 is introduced when $\int p(x)\,dx$ is computed, then the form of the general solution (23) is unchanged. (*Note*: If c is an arbitrary constant, then ce^{-c_1} is still an arbitrary constant.)

50. Show that if $u(x)$ is a function of x, and u satisfies the property

$$u\left(\frac{dy}{dx} + p(x)y\right) = \frac{d(uy)}{dx},$$

then $u = k\exp\left(\int p(x)\,dx\right)$, k a constant.

51. Show that if u is an integrating factor for (19), then so is ku for any constant k.

52. Consider the differential equation $\dfrac{dy}{dx} - xy = 2x - x^3$.

a) Show that $y = x^2$ is a particular solution. Solve the initial value problem $y(0) = 2$.

b) What is the integrating factor? Why is this method difficult?

c) Find the general solution with an integrating factor. Compare to part (a).

53. Consider $dy/dx + [(\sin x)/x]y = q(x)$. Show that the particular solution with $y_p(0) = 0$ can be put into the form $y_p = \int_0^x G(x, \bar{x})q(\bar{x})\,d\bar{x}$ for some function $G(x, \bar{x})$. The function $G(x, \bar{x})$ is called the **influence function** or **Green's function** for the equation.

54. Consider $dy/dx + p(x)y = q(x)$. Show that the particular solution with $y_p(0) = 0$ can be put into the form $y_p = \int_0^x G(x, \bar{x})q(\bar{x})\,d\bar{x}$ for some function $G(x, \bar{x})$. The function $G(x, \bar{x})$ is called the **influence function** or **Green's function** for the differential equation.

55. We briefly present an equivalent method (known as the **method of variation of parameters**) for solving the linear differential equation

$$\frac{dy}{dx} + p(x)y = q(x).$$

Make the change of variables $y = u(x)y_1(x)$, where $y_1(x) = \exp(-\int p(x)\,dx)$ is the solution of the homogeneous equation that was determined previously. Find and solve the simpler first-order linear differential equation for $u(x)$. Compare your solution to the answer obtained by using an integrating factor. (A modification of this method is useful for second-order linear differential equations, since the integrating factor method cannot be applied to second-order linear differential equations.)

1.7 Linear First-Order Differential Equations with Constant Coefficients and Constant Input

Solutions of first-order linear differential equations can always be obtained by the integrating factor method of Subsection 1.6.3. In this section we will describe an easier method for a particular class of first-order linear differential equations. These equations occur quite frequently in applications.

The linear first-order differential equation,

$$\frac{dy}{dx} + p(x)y = q(x),$$

is said to have **constant coefficients** if $p(x)$ is a constant, $p(x) = k$:

$$\frac{dy}{dx} + ky = q(x). \tag{1}$$

We say that the differential equation has constant coefficients even if the forcing function or input $q(x)$ is not constant. It will be easy to obtain a solution y_1 of the associated homogeneous equation for (1). Sometimes the method of undetermined coefficients can be used to obtain a particular solution y_p of a linear differential equation with constant coefficients. Then $y_p + cy_1$ will give the general solution of (1). This section introduces this method by considering the important case when the input $q(x)$ is constant. Other cases are considered in the exercises and in Chapter 2.

1.7.1 Solutions of Homogeneous Linear Differential Equations with Constant Coefficients

The associated homogeneous equation for the linear differential equation with constant coefficients (1) is

$$\frac{dy}{dx} + ky = 0. \tag{2}$$

It is easy to verify that $y = e^{-kx}$ satisfies (2) by substituting this exponential into (2). This solution of the homogeneous equation can be also derived since (2) is separable (Section 1.3) or by the integrating factor method (Subsection 1.6.3). However, it is easier to obtain this solution by just remembering the following simple fact. This fact will be useful in this and later chapters:

> A homogeneous linear constant coefficient differential equation always has at least one solution of the form
> $$y = e^{rx}. \tag{3}$$

For first-order linear equations, r will turn out to be a real number. Statement (3) is still true of second-order equations, but we will see in the next chapter that r can then be a complex number.

We can determine what specific exponential satisfies the homogeneous differential equation with constant coefficients (2) by substituting the unknown exponential $y = e^{rx}$ into the differential equation (2):

$$\frac{d}{dx}e^{rx} + ke^{rx} = 0.$$

Differentiating, we have $re^{rx} + ke^{rx} = 0$. Dividing by e^{rx} then gives that $r = -k$. Since $y_1(x) = e^{-kx}$ is one solution of the homogeneous equation (2), our theory tells us that the general solution of the homogeneous equation will be ce^{-kx}.

> The general solution of
> $$\frac{dy}{dx} + ky = 0$$
> with k constant is
> $$y = ce^{-kx}.$$

This method is illustrated in the next example.

EXAMPLE 1.7.1 **Homogeneous Linear Constant Coefficient**

Determine the general solution of

$$\frac{dy}{dx} + 4y = 0. \tag{4}$$

Solution If we substitute $y = e^{rx}$ into (4) for y, we get

$$\frac{d}{dx}(e^{rx}) + 4e^{rx} = 0.$$

Since $\frac{d}{dx}(e^{rx}) = re^{rx}$, the differential equation (4) is satisfied if

$$re^{rx} + 4e^{rx} = 0,$$

or equivalently,

$$(r + 4)e^{rx} = 0.$$

This equation is to hold for all x. If we cancel the e^{rx} since it never equals zero (or just divide by e^{rx}), we obtain

$$r + 4 = 0, \qquad \text{or equivalently,} \qquad r = -4.$$

Thus, e^{-4x} solves the constant coefficient differential equation (4). Since the differential equation (4) is linear, the general solution is in the form of a particular solution plus an arbitrary multiple of the solution e^{-4x} of the associated homogeneous equation. Equation (4) is homogeneous, so that we can take $y_p = 0$. Thus, the general solution of (4) is

$$y = ce^{-4x}.$$

◄

1.7.2 Constant Coefficient Linear Differential Equations with a Constant Input

Particular solutions of first-order linear differential equations with constant coefficients,

$$\frac{dy}{dx} + ky = q(x), \tag{5}$$

can always be obtained by using the integrating factor method discussed earlier in Section 1.6.3. However, sometimes it is easier to obtain a particular solution by the method of undetermined coefficients. In this subsection, we will discuss only one elementary class of examples. These ideas will be addressed in greater depth in the section on second-order linear differential equations with constant coefficients (Section 2.9).

If the right-hand side of (5) is a constant, $q(x) = I_0$, we refer to the differential equation as having constant input:

$$\frac{dy}{dx} + ky = I_0.$$

EXAMPLE 1.7.2 Constant Coefficient Differential Equation with a Constant Input

Find the general solution of

$$\frac{dy}{dx} + 8y = 9. \tag{6}$$

SOLUTION Since the right-hand side is a constant, we might guess that a particular solution is a constant, $y_p = A$. The constant A is called an **undetermined coefficient**, and this method is called the method of undetermined coefficients. To determine the specific constant A, we substitute

$y_p = A$ into (6) for y to obtain $dA/dx + 8A = 9$. But $dA/dx = 0$, so that

$$8A = 9, \qquad \text{or equivalently,} \qquad A = \frac{9}{8}.$$

Thus, a particular solution of (6) is

$$y_p = A = \frac{9}{8}.$$

The general solution of any linear first-order equation is in the form of a particular solution plus an arbitrary multiple of a solution y_1 of the associated homogeneous equation. Using the techniques of Example 1.7.1, we find that $y_1 = e^{-8x}$. Thus the general solution of (6) is

$$y = \frac{9}{8} + ce^{-8x},$$

where c is an arbitrary constant. ◄

All cases of constant coefficient differential equations with a constant input can be described by

$$\frac{dy}{dx} + ky = I_0, \tag{7}$$

where I_0 is a constant. In the same manner as Example 1.7.2, the general solution of (7) can be obtained as

$$y = \frac{I_0}{k} + ce^{-kx}, \qquad k \neq 0. \tag{8}$$

A particular solution is $y_p = \frac{I_0}{k}$ if $k \neq 0$. More examples of this type of differential equation will appear in the sections on applications of first-order differential equations.

However, if $k = 0$ in (7), then the method of undetermined coefficients as described must be modified. If $k = 0$, (7) is

$$\frac{dy}{dx} = I_0. \tag{9}$$

Any constant is a solution of the homogeneous equation ($dy_h/dx = 0$), and hence a constant cannot be a particular solution. Thus, $y_p \neq A$, contrary to what was assumed earlier. In this case, it is easy to see, by integration of (9), that the guess was wrong and a particular solution is $y_p = I_0 x$. This particular solution is not a constant, but rather is proportional to x. The general solution of (9) by integration is $y = I_0 x + c$. This is again a particular solution plus an arbitrary multiple of a solution, $y_1 = 1$, of the associated homogeneous equation. More will be said about this in our discussion of particular solutions for second-order linear differential equations with constant coefficients. In summary,

If $k \neq 0$, then $\dfrac{dy}{dx} + ky = I_0$ has a constant particular solution.

In applications the constant solutions that result from constant input are variously called **equilibriums**, **steady-state solutions**, and **operating points**.

1.7.3 Constant Coefficient Differential Equations with a Discontinuous Input

Many practical problems involve discontinuous inputs. The discontinuities arise from such physical phenomena as switches, sparks, or digital devices. The method of undetermined coefficients can also be used if the input is a discontinuous input composed of relatively simple functions.

EXAMPLE 1.7.3 Discontinuous Input

Solve the initial value problem

$$\frac{dy}{dx} + 2y = g(x), \qquad \text{with } y(0) = 0, \tag{10}$$

where the input $g(x)$ is the following piecewise constant function:

$$g(x) = \begin{cases} 1, & 0 \leq x < 2, \\ 0, & 2 \leq x. \end{cases}$$

SOLUTION We actually have a two-part problem here. A different differential equation is valid in each region.

For $0 \leq x < 2$, the differential equation (10) is

$$\frac{dy}{dx} + 2y = 1, \tag{11}$$

and must satisfy the given initial condition $y(0) = 0$. The differential equation (11) is linear with constant coefficients and a constant input. A particular solution is a constant, which is easily seen to equal $\frac{1}{2}$. A solution of the associated homogeneous equation is e^{-2x}, so that the general solution is $y = \frac{1}{2} + ce^{-2x}$. The initial condition $y(0) = 0$ implies that $c = -\frac{1}{2}$, so that the solution of the initial value problem for $0 \leq x < 2$ is

$$y(x) = \frac{1}{2} - \frac{1}{2}e^{-2x}. \tag{12}$$

However, the differential equation (10) for $x \geq 2$ is

$$\frac{dy}{dx} + 2y = 0. \tag{13}$$

The general solution of (13) will have one additional arbitrary constant introduced. This constant is not determined from the condition at $x = 0$, but instead is determined from a condition at (13)'s starting value $x = 2$. From physical considerations, we often know that the solution y must be continuous, and thus the initial condition for the differential equation (13) for $x \geq 2$ is

$$y(2) = \lim_{x \to 2^-} y(x). \tag{14}$$

That is, the solutions for $x < 2$ and $x > 2$ should match up, or "meet," when $x = 2$, so that y can be continuous everywhere, including $x = 2$. Since the solution has been determined for $0 < x < 2$, we compute its limit at $x = 2$ from (12) to be

$$y(2) = \lim_{x \to 2^-} y(x) = \frac{1}{2} - \frac{1}{2}e^{-4}. \tag{15}$$

Since (13) is linear and homogeneous (with constant coefficients), its general solution is seen to be

$$y(x) = c_1 e^{-2x}. \tag{16}$$

A different notation for the constant has been introduced so that the constant associated with the differential equation for $x > 2$ is not confused with the constant associated with the differential equation for $x < 2$. The constant c_1 is determined from the initial condition valid at $x = 2$. Letting $x = 2$ in (16) and setting it equal to (15), we get

$$\frac{1}{2} - \frac{1}{2}e^{-4} = c_1 e^{-4}.$$

Thus, $c_1 = \frac{1}{2}(e^4 - 1)$. Hence for $x > 2$ we have

$$y(x) = \frac{1}{2}(e^4 - 1)e^{-2x}. \tag{17}$$

Combining the solutions, (12) and (17), valid in the two regions, we obtain

$$y(x) = \begin{cases} \frac{1}{2}(1 - e^{-2x}), & 0 \leq x < 2, \\ \frac{1}{2}(e^4 - 1)e^{-2x}, & 2 \leq x. \end{cases} \tag{18}$$

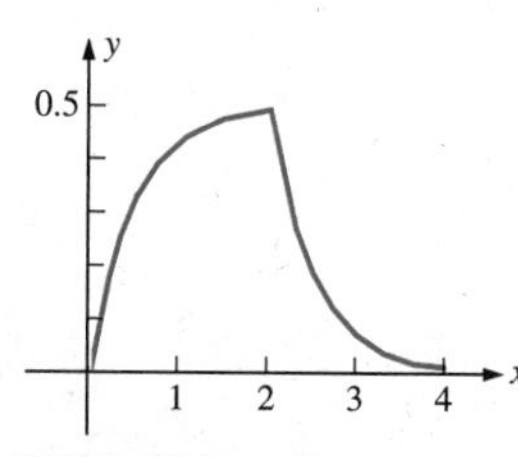

FIGURE 1.7.1

Graph of (18).

The graph of $y(x)$ is given in Figure 1.7.1. Some students may recognize this graph as the charge-discharge curve of a capacitor in an *RC* circuit. This will be discussed more carefully in Section 1.10. ◀

Exercises

In Exercises 1 through 10, find the general solution.

1. $\frac{dy}{dx} = 8y$
2. $\frac{dy}{dx} = 3y$
3. $\frac{dy}{dx} = -2y$
4. $\frac{dy}{dx} = -\frac{1}{3}y$
5. $\frac{dy}{dx} + 7y = 0$
6. $\frac{dy}{dx} - y = 0$
7. $\frac{dy}{dx} + y = 0$
8. $\frac{dy}{dx} - \frac{3}{4}y = 0$
9. $\frac{dy}{dt} = 5y$
10. $\frac{dx}{dt} = -3x$

In Exercises 11 through 22, find the general solution.

11. $\frac{dy}{dx} + 3y = 8$
12. $\frac{dy}{dx} + 6y = 5$
13. $\frac{dy}{dx} - 4y = -9$
14. $\frac{dy}{dx} - y = -\frac{1}{3}$
15. $\frac{dy}{dx} - \frac{4}{5}y = 3$
16. $\frac{dy}{dx} - 8y = 7$
17. $\frac{dy}{dx} + \frac{2}{3}y = -\frac{4}{3}$
18. $\frac{dy}{dx} + \frac{3}{8}y = -4$
19. $\frac{dy}{dx} = 2y + 18$
20. $\frac{dy}{dx} = -8y + 9$
21. $\frac{dy}{dx} = -y - \frac{17}{3}$
22. $\frac{dy}{dx} = 6y - 15$

In Exercises 23 through 30, the input is a constant multiple of an exponential e^{ax}. Suppose that there is a particular solution of the form Ae^{ax} for some constant A. Find the constant A and determine the particular and general solution.

23. $\frac{dy}{dx} + 7y = 8e^{-4x}$
24. $\frac{dy}{dx} - y = 4e^{3x}$
25. $\frac{dy}{dx} - 2y = -3e^{-5x}$
26. $\frac{dy}{dx} + y = -2e^{8x}$
27. $\frac{dy}{dx} + 4y = 3e^{4x}$
28. $\frac{dy}{dx} - y = 5e^{-x}$
29. $\frac{dy}{dx} + 3y = e^{-2x}$
30. $\frac{dy}{dx} - 5y = 6e^{7x}$

In Exercises 31 through 40, obtain just a particular solution (the general solution could always be obtained easily by adding an arbitrary multiple of a solution of the associated homogeneous equation). The forcing function is a polynomial. Guess that a **polynomial of the same degree** is the particular solution and use the **method of undetermined coefficients** to find a particular solution. For example, if you would guess a polynomial of degree 2, substitute $y = Ax^2 + Bx + C$ into the differential equation and determine the constants so that $y = Ax^2 + Bx + C$ satisfies the differential equation.

31. $\frac{dy}{dx} + 7y = 3 + 5x$
32. $\frac{dy}{dx} - 4y = -6 + 9x$
33. $\frac{dy}{dx} + 2y = 14x$
34. $\frac{dy}{dx} - \frac{1}{2}y = -3x$
35. $\frac{dy}{dx} + y = 2x^2 + 5x - 8$

36. $\dfrac{dy}{dx} - 9y = -x^2 + 7x + 2$

37. $\dfrac{dy}{dx} + 3y = x^3$

38. $\dfrac{dy}{dx} - 6y = 4x^2 + 3$

39. $\dfrac{dy}{dx} + 5y = x$

40. $\dfrac{dy}{dx} + 8y = 5x + 11$

In Exercises 41 through 52, obtain just a particular solution (the general solution could always be obtained easily by adding an arbitrary multiple of a solution of the associated homogeneous equation). In these exercises, the forcing function is an elementary sinusoidal function. If the forcing function is $\cos 5x$, then a particular solution of the form $A \cos 5x$ will not work. However, a linear combination of $\cos 5x$ and $\sin 5x$ will work. Thus, the form for y_p is $A \cos 5x + B \sin 5x$, where the constants A and B are determined by substituting the assumed form of the particular solution into the differential equation for y. In Exercises 47–52, the forcing function has several different terms. Use a form for y_p consisting of the sum of the forms for each term.

41. $\dfrac{dy}{dx} + 2y = 3 \sin 6x$

42. $\dfrac{dy}{dx} - 3y = 4 \cos 3x$

43. $\dfrac{dy}{dx} - 5y = 2 \cos x$

44. $\dfrac{dy}{dx} + 8y = \sin x$

45. $\dfrac{dy}{dx} + 4y = 3 \cos 2x + 5 \sin 2x$

46. $\dfrac{dy}{dx} - y = 2 \cos 4x + \sin 4x$

47. $\dfrac{dy}{dx} + 6y = \cos x + \sin 5x$

48. $\dfrac{dy}{dx} + y = \cos 2x + \cos 4x$

49. $\dfrac{dy}{dx} - 9y = 5 + 2 \sin 3x$

50. $\dfrac{dy}{dx} - 2y = -3 + 4 \cos x$

51. $\dfrac{dy}{dx} + y = 2e^{3x} + \sin x$

52. $\dfrac{dy}{dx} + 9y = 4e^{-5x} + \cos 2x$

53. Consider the differential equation $dy/dx + 3y = 8e^{-3x}$. Show that a simple exponential Ae^{-3x} does not work as a particular solution, because e^{-3x} is a solution of the associated homogeneous equation. Instead, solve the differential equation by the integrating factor method. Make a conjecture as to how to modify the method of undetermined coefficients when the simple exponential forcing is a solution of the associated homogeneous equation.

54. Consider the differential equation $dy/dx - ky = P(x)e^{kx}$, where $P(x)$ is a polynomial of degree n. Solve this differential equation by making the change of variables $y = ve^{kx}$. Find and solve the differential equation for v. Show that v is a polynomial of degree $n + 1$. This proves that a particular solution is in the form of an exponential e^{kx} times a polynomial of one higher degree, if the exponential part of the forcing function is a solution of the associated homogeneous equation. A slight generalization of this will be important in Chapter 2.

55. Consider the differential equation $dy/dx - 2y = 7e^{2x}$. Show that a simple exponential Ae^{2x} does not work as a particular solution, because e^{2x} is a solution of the associated homogeneous equation. Instead, solve the differential equation by making a change of variables $y = ve^{2x}$. Find and solve the differential equation for v. Make a conjecture as to how to modify the method of undetermined coefficients when the simple exponential forcing is a solution of the associated homogeneous equation.

In Exercises 56 through 62, a simple exponential will not work as a particular solution, and y_p must have the form Axe^{ax}. Find the general solution.

56. $\dfrac{dy}{dx} - 3y = 8e^{3x}$

57. $\dfrac{dy}{dx} - y = 4e^{x}$

58. $\dfrac{dy}{dx} + 4y = 3e^{-4x}$

59. $\dfrac{dy}{dx} + y = 5e^{-x}$

60. $\dfrac{dy}{dx} - 9y = -2e^{9x}$

61. $\dfrac{dy}{dx} - 7y = 8e^{7x}$

62. $\dfrac{dy}{dx} + 5y = -3e^{-5x}$

In Exercises 63 through 68, find the continuous solution of

$$\frac{dy}{dx} + p(x)y = q(x), \qquad 0 \le x \le 2,$$

and sketch the solution.

63. $p(x) = \begin{cases} 2, & 0 \le x < 1, \\ 1, & 1 \le x \le 2, \end{cases} \quad q(x) = 0,$ $y(0) = 2$

64. $p(x) = 1, \quad q(x) = \begin{cases} 1, & 0 \le x < 1, \\ 0, & 1 \le x \le 2, \end{cases}$ $y(0) = 1$

65. $p(x) = 0, \quad q(x) = \begin{cases} 1, & 0 \le x < 1, \\ -1, & 1 \le x \le 2, \end{cases}$ $y(0) = 0$

66. $p(x) = \begin{cases} 0, & 0 \le x < 1, \\ 1, & 1 \le x \le 2, \end{cases} \quad q(x) = 1,$ $y(0) = 0$

67. $p(x) = \begin{cases} 1, & 0 \le x < 1, \\ 0, & 1 \le x \le 2, \end{cases}$ $q(x) = \begin{cases} 0, & 0 \le x < 1, \\ 1, & 1 \le x \le 2, \end{cases} \quad y(0) = 2$

68. $p(x) = \begin{cases} 1 & 0 \le x < 1, \\ -1 & 1 \le x \le 2, \end{cases} \quad q(x) = 2$ $y(0) = 1$

1.8 Growth and Decay Problems

Differential equations are studied by scientists and engineers because they describe physical problems. Solving a differential equation is only part of applying differential equations. Equally important is showing how the differential equation describes a real-world situation. Experts from a given field will sometimes argue about whether one differential equation or another applies to a given situation. Knowledge about both mathematics and the field of application is required.

This section introduces several applications of first-order equations. In these applications, the reader should pay close attention to how the equations are derived from the given physical problem and should not seek to merely memorize formulas.

1.8.1 A First Model of Population Growth

The first application that we discuss will describe a very idealized situation. More realistic situations can be described, but more complicated mathematical models are required. In this section, one of the simplest models of population growth of a single species will be developed. Let

$$P(t) = \text{population } P \text{ as a function of time } t. \tag{1}$$

The population could be a country's population or the population of the world. Other examples are the number of bacteria in a laboratory experiment or the population of a particularly insidious insect attacking a crop. The derivative of the population with respect to time is important:

$$\frac{dP}{dt} = \text{rate of change of the population.} \tag{2}$$

The derivative is measured in units of the number per unit time. For example, in 1993 the population of the United States was increasing at the rate of 3.2 million per year. However, the **growth rate** is often a more significant measure of growth:

$$\frac{1}{P}\frac{dP}{dt} = \text{growth rate.} \tag{3}$$

The growth rate is the rate of change of the population per individual. Since the population of the United States was 265 million in 1993, the growth rate was $\frac{3.2}{265} = 0.012$, or about 1.2% per year. It is common to measure the growth rate of human populations in percent per year. For other organisms, other time units, such as hours or days, may be used.

To illustrate how a differential equation is used to predict the future, we will make the simplest assumption about growth. We assume the growth rate is a constant k. Then

$$\frac{1}{P}\frac{dP}{dt} = k,$$

or equivalently,

$$\frac{dP}{dt} = kP. \tag{4}$$

Our assumption that the growth rate is a constant is equivalent to assuming that the rate of change of the population is proportional to the population. The differential equation (4) by itself is not enough to predict the future. In addition, we need to know the initial population $P(0)$. The first-order differential equation (4) is linear with constant coefficients. It is also homogeneous, since it has a zero right-hand side ($dP/dt - kP = 0$). We could solve (4) by separation or integrating factors. We shall use the approach of Section 1.7. From Section 1.7 we know that the general solution of (4) will be an arbitrary multiple of a solution of the homogeneous equation. For constant coefficient equations, solutions of homogeneous equations are in the form e^{rx} when x is the independent variable. In many applications including population growth, the independent variable is t, so that solutions of the associated homogeneous equation are sought in the form e^{rt}. If $P = e^{rt}$ is substituted into the differential equation (4), we obtain $r = k$, so that e^{kt} is a solution of the homogeneous equation. Thus, the general solution of (4) is

$$P(t) = ce^{kt}. \tag{5}$$

The arbitrary constant is determined by evaluating the solution (5) at the initial time $t = 0$ using the initial population $P(0)$:

$$P(0) = c.$$

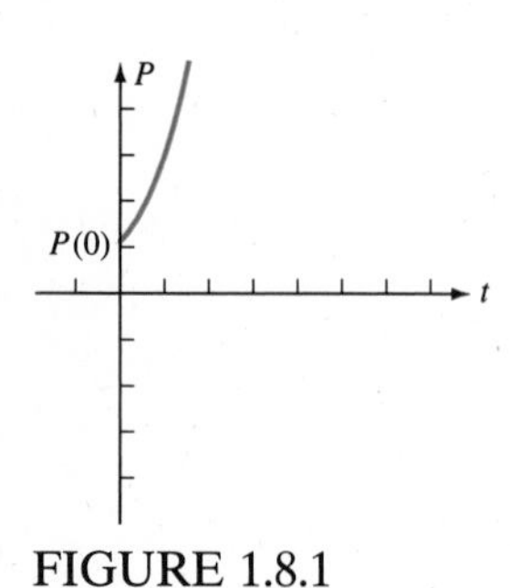

FIGURE 1.8.1

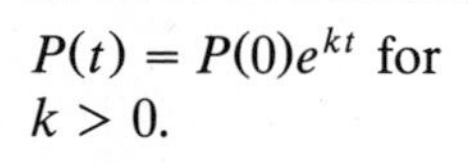

$P(t) = P(0)e^{kt}$ for $k > 0$.

The solution of the initial value problem for the differential equation (4) is then

$$P(t) = P(0)e^{kt}. \tag{6}$$

The solution (6) is an elementary exponential and is sketched for one initial condition in Figure 1.8.1 for the exponential growth case, $k > 0$. Solutions corresponding to other initial conditions are graphed in Figure 1.8.2. Solution curves for $t > 0$ and $t < 0$, $P(0)$ positive and negative, are included in Figure 1.8.3, which we earlier called the direction field. If $k < 0$, the growth rate is negative and the population exponentially decays, as is sketched in Figures 1.8.4 and 1.8.5 on page 55. One reason that exponentials are studied so much in high school mathematics and in calculus is that the solution of the simplest differential equation models involves the exponential. Other continuous growth problems, such as the increase in the amount of money in a bank due to interest and the increase in the cost of living due to inflation, satisfy differential equation (4), and so they also involve exponentials. Examples from physics and chemistry involving exponentials are given in the next sections.

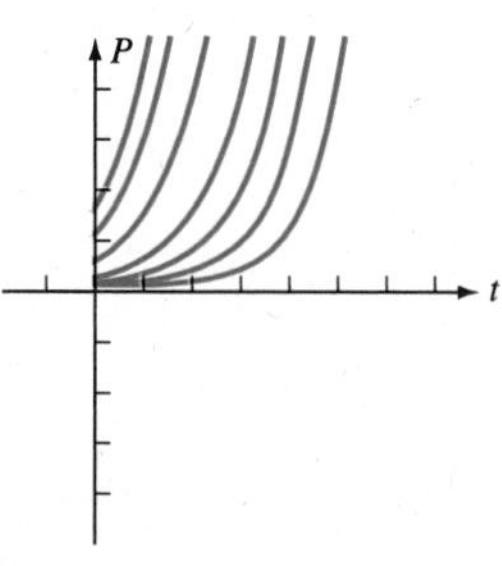

FIGURE 1.8.2

Graphs of $P(t) = P(0)e^{kt}$, $k > 0$, for several $P(0) > 0$.

Doubling Time

An additional way to understand the implications of exponential growth is to investigate how long it takes a population to double, assuming it grows at a constant rate k. We are asking at what time t will the population be twice its initial population $P(0)$. That is, we wish to find t such that $2P(0) = P(t)$. Using (6), the requirement is

$$2P(0) = P(0)e^{kt}. \tag{7}$$

In order to solve (7) for the doubling time t, we first observe that $P(0)$ cancels, so that (7) becomes

$$2 = e^{kt}. \tag{8}$$

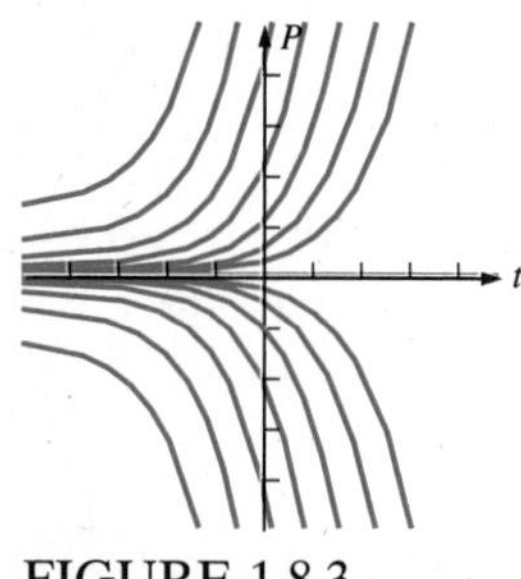

FIGURE 1.8.3

Graphs of $P(t) = P(0)e^{kt}$ for $k > 0$.

Thus, the doubling time does not depend on the initial population. For example, with continuous compounding, your money will double in the same time whether you deposit \$100 or \$1000 (this assumes that the bank doesn't offer a higher rate to the larger depositor and that you do not make additional deposits). To determine the doubling time, we take the natural logarithm of both sides of (8):

$$\ln 2 = kt.$$

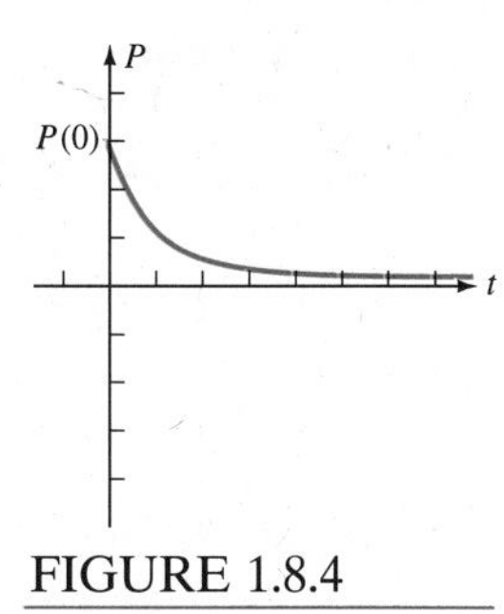

FIGURE 1.8.4

$P(t) = P(0)e^{kt}$ if $k < 0$.

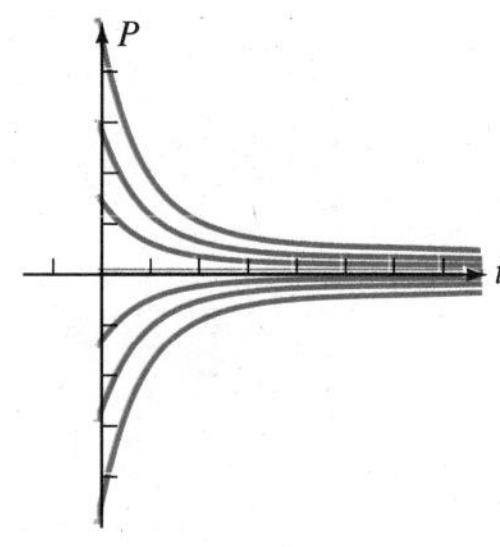

FIGURE 1.8.5

Solutions of (4) with $k < 0$. $dP/dt = kP$.

Thus, the doubling time is given by

$$t_{\text{doubling time}} = t_d = \frac{\ln 2}{k}, \tag{9}$$

where k is the growth rate. Using a calculator or tables, $\ln 2 \approx 0.69315$. This yields a useful observation. From (9), the doubling time for exponential growth is approximately $0.69315/k$. This motivates the **Rule of 70** (some prefer a rule of 72, since 72 is divisible by 3, 6, 9, and 12). As an approximation,

$$t_{\text{doubling time}} \approx \frac{70}{\text{growth rate (measured in \% per year)}}. \tag{10}$$

This is a great formula to memorize, since it communicates so well the implications of the compounding effect of continuous growth.

EXAMPLE 1.8.1 **Using (10)**

If the population of the world is growing at 2% per year, then the world's population is expected to double in approximately $\frac{70}{2} = 35$ years (not 50 years!). If inflation is 5% per year, the cost of living is expected to double in approximately $\frac{70}{5} = 14$ years (not 20 years!). ◄

Sometimes it is advantageous to utilize the doubling time more explicitly in the solution of the differential equation. From (9), the growth rate k may be replaced by the doubling time t_d:

$$k = \frac{\ln 2}{t_d}.$$

The solution (6) of the differential equation (4) becomes

$$P(t) = P(0)e^{kt} = P(0)(e^{\ln 2})^{\frac{t}{t_d}} = P(0)2^{\frac{t}{t_d}}, \tag{11}$$

thus expressing the solution in terms of its doubling time. This shows clearly that the solution doubles after every time t_d.

EXAMPLE 1.8.2 **Using (11)**

Suppose that it is given that some population triples every 7 years. Find the population after 10 years.

Solution Using the ideas in (11), the solution of the differential equation (4) is

$$P(t) = P(0)3^{\frac{t}{7}}.$$

Thus the population could be predicted at $t = 10$ to be $P(0)3^{\frac{10}{7}}$. ◄

Yield

We would like to examine more carefully the effect of compounding when a quantity grows continuously at a constant growth rate k:

$$P(t) = P(0)e^{kt}.$$

When a quantity is growing continuously at the rate of $k\%$ per year, the actual amount at the end of one year ($t = 1$) is

$$P(1) = P(0)e^{k}.$$

In banking, $P(t)$ is interpreted as the amount of money in a fund at time t. The initial amount $P(0)$ is called the **principal**. Exponential growth is referred to as **continuous compounding**. The interest earned after one year is $P(1) - P(0) = P(0)(e^k - 1)$. The **yield (effective interest rate)** is defined to be the actual interest earned over one year divided by the principal:

$$\text{Yield} = \frac{P(1) - P(0)}{P(0)} = e^k - 1. \tag{12}$$

In this analysis, money is being compounded continuously, so that growth can be described by a differential equation.

EXAMPLE 1.8.3 Yield

Suppose a bank has an interest rate of 5% per year that is compounded continually. What will the yield be?

Solution Using (12), we have

$$\text{Yield} = e^{0.05} - 1 \approx 1.0513 - 1 = 0.0513.$$ ◄

This corresponds to the 5.13% yield a bank might advertise when its interest rate is 5% per year. (The real world can be more complicated than this. Banks have, in the past, from the days of hand computation of interest, defined a year to be only 360 days. The consumer would actually benefit by receiving 5 extra days of interest beyond the 5% promised.)

The differential equation $dP/dt = kP$ models the continuous compounding of an initial deposit, where $P(t)$ is the amount of money and k is the growth rate. This model is often sufficiently accurate if money is compounded daily. However, there may be additional deposits or inputs. This leads to a differential equation that is not homogeneous.

EXAMPLE 1.8.4 Continuous Input

Suppose the interest rate on a \$1000 deposit is 7% per year. This means $k = 0.07$. However, suppose that instead of leaving the money to grow, we deposit an additional \$10 per day. Determine the amount of money in the account as a function of time. Assume continuous compounding and continuous deposit.

Solution We must first determine the differential equation that applies. Here, the amount of money increases for two reasons, interest and deposits. In general, we can write a word equation for the rate of change of the amount:

$$\begin{pmatrix}\text{Rate of}\\ \text{change of}\\ \text{money}\end{pmatrix} = \begin{pmatrix}\text{rate of}\\ \text{increase due}\\ \text{to interest}\end{pmatrix} + \begin{pmatrix}\text{rate of}\\ \text{increase due}\\ \text{to deposit}\end{pmatrix}. \tag{13}$$

All three terms in (13) must have the same units, which we take as dollars per year. As discussed before, dP/dt is the rate of change of the amount of money and $kP = 0.07P$ is the amount of interest per year. The deposit rate is a constant. It is \$10 per day. This is equivalent to a deposit rate of \$3650 per year. Thus, the differential equation (13) is

$$\frac{dP}{dt} = 0.07P + 3650.$$

This is a first-order linear differential equation with constant coefficients with a constant input of 3650:

$$\frac{dP}{dt} - 0.07P = 3650. \tag{14}$$

The general solution will be a particular solution plus an arbitrary constant times the solution of the associated homogeneous equation:

$$P = P_p + cP_1.$$

Equation (14) could be solved by separation or integrating factors. We shall use the approach of Section 1.7. Since the input is constant, we look for a particular solution that is a constant $P_p = A$. Substituting $P = A$ into the differential equation, we find that $P_p = A = -\frac{3650}{0.07}$. A solution of the associated homogeneous equation exponentially grows, $P_1 = e^{kt} = e^{0.07t}$. Thus, the general solution of (14) is

$$P(t) = -\frac{3650}{0.07} + ce^{0.07t}. \tag{15}$$

The initial deposit was \$1000, and so the initial condition for the differential equation is $P(0) = 1000$. Thus, the arbitrary constant satisfies $1000 = -\frac{3650}{0.07} + c$. Solving for c, we get the solution to the initial value problem:

$$P(t) = -\frac{3650}{0.07} + \left(1000 + \frac{3650}{0.07}\right)e^{0.07t}. \tag{16}$$

Having solved the differential equation, we can now analyze the effect of making continual deposits at the rate of \$3650 per year. Suppose we were interested in the amount of money that would be accumulated after 10 and 20 years. Without interest, the amount of money at the end of 10 years is \$37,500, and at the end of 20 years it is \$74,000. With interest, the amount of money is determined from (16) and is considerably larger:

$$P(10) = -\frac{3650}{0.07} + \left(1000 + \frac{3650}{0.07}\right)e^{0.7} \approx \$54{,}876$$

$$P(20) = -\frac{3650}{0.07} + \left(1000 + \frac{3650}{0.07}\right)e^{1.4} \approx \$163{,}372$$

We could also make the necessary modifications to compare the effects of various interest rates on the amount of money.

1.8.2 Radioactive Decay

First-order differential equations are very important in understanding radioactive decay. For a given atom of radioactive material, not acted on by radiation or other particles, there is a fixed probability that it will decay in a given time period. For example, suppose the probability is 0.001 that the atom will decay in one year. Now suppose that we have $x(t)$ of these atoms at time t. We would expect at that time to have atoms decaying at a rate of $(0.001)x(t)$ atoms per year. This leads to the following law of radioactive decay:

> The rate of decay for a radioactive material is proportional to the number of atoms present. (17)

Note that the law of decay is not really a law. However, if we have a reasonably large number of atoms and the material is not bombarded by other radiation or particles, then the law of decay is often sufficiently accurate. We do not intend to belabor the point, but it is important to realize that the suitability of a law like (17) depends on the problem and the intended application of the answer.

Since $x(t)$ is the amount of radioactive material at time t, the law (17) may be written as a first-order linear differential equation:

$$\frac{dx}{dt} = -kx. \tag{18}$$

We introduce the minus sign since in radioactive decay the number of atoms diminishes. Here the proportionality constant k, the **decay rate**, will be positive since there is a loss of material.

Equation (18) is homogeneous [this is more easily seen if (18) is rewritten $dx/dt + kx = 0$]. Using the technique of Section 1.7, we note that a solution of the homogeneous equation is in the form $x = e^{rt}$, which when substituted into (18) yields $r = -k$. Thus, e^{-kt} is a solution of the homogeneous equation, and so the general solution of (18) will be an arbitrary multiple of the solution of the homogeneous equation:

$$x(t) = ce^{-kt}.$$

[The differential equation (18) can also be solved by the integrating factor technique or by separation of variables.] The constant c is determined by satisfying the initial condition $x(0) = c$. Thus, we arrive finally at

$$x(t) = x(0)e^{-kt}. \tag{19}$$

Thus, radioactive decay is exponential decay, since $k > 0$ implies that $x(t) \to 0$ at $t \to \infty$, as sketched in Figure 1.8.5.

Half-life

The **half-life** of a radioactive substance is the length of time it takes the material to decay to half its original amount. If T is the half-life, we have by definition

$$x(T) = \frac{1}{2}x(0). \tag{20}$$

To determine the half-life, we substitute the formula (19) for $x(t)$ into (20) to get

$$x(0)e^{-kT} = \frac{1}{2}x(0)$$

or

$$e^{-kT} = \frac{1}{2}.$$

Taking the natural logarithm of both sides and solving for T gives the half-life as

$$T_{\text{half-life}} = T = \frac{\ln 2}{k}. \tag{21}$$

The half-life and the decay rate k are closely related. Determining the half-life for radioactive decay is mathematically analogous to finding the doubling time for population growth.

EXAMPLE 1.8.5 Radioactive Decay

In two years, 3 g of a radioisotope decay to 0.9 g. Determine both the half-life T and the decay rate k.

SOLUTION Let $x(t)$ be the amount of the radioisotope at time t. Measure x in grams and t in years. The basic law is given by (18): $dx/dt = -kx$. Solving this equation gives (19):

$$x(t) = x(0)e^{-kt}. \tag{22}$$

This equation holds for any radioisotope. For the isotope of this example, we are given the initial amount and the amount at $t = 2$:

$$\begin{aligned} 3 &= x(0), \\ 0.9 &= x(2). \end{aligned} \tag{23}$$

Substituting (23) into the general formula (22) with $t = 2$ yields

$$0.9 = 3e^{-2k}.$$

Thus,

$$k = \frac{\ln\left(\frac{1}{0.3}\right)}{2} \approx 0.6.$$

From (21), the half-life is

$$T = \frac{2\ln 2}{-\ln 0.3} \approx 1.2 \text{ years.}$$

See Figure 1.8.6.

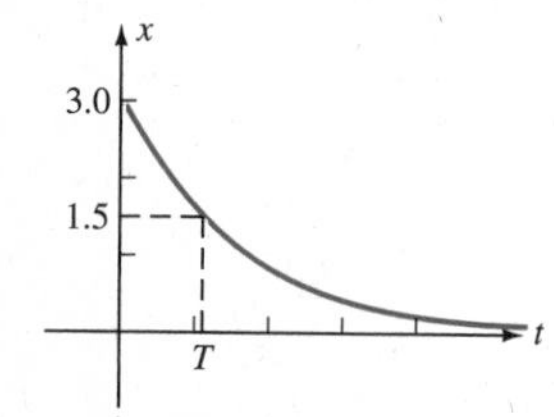

FIGURE 1.8.6

Graph of solution of Example 1.8.5.

1.8.3 Thermal Cooling

Differential equations govern thermal cooling. Ignoring circulation and other effects, **Newton's law of cooling** states that

> The rate of change of the surface temperature of an object is proportional to the difference between the temperature of the object and the temperature of its surroundings (also called the **ambient temperature**) at that time. (24)

If $T(t)$ is the surface temperature of the object at time t and Q_0 is the ambient temperature at time t, then Newton's law of cooling becomes

$$\frac{dT}{dt} = -k(T - Q_0). \tag{25}$$

It is important to understand why we have introduced a minus sign on the right-hand side of (25). If $T > Q_0$, the body's surface temperature is hotter than its surrounding environment, and hence there is a loss of surface temperature. It follows that in this situation, dT/dt must be negative, and consequently k is positive. We have introduced the minus sign so that the proportionality constant k is positive, which is more convenient. On the other hand, if $T < Q_0$, then the surface temperature increases and $dT/dt > 0$. Again the constant k is positive. Note that (24) does not require that the outside temperature Q_0 be constant over time. In using (25) in what follows, we shall assume uniform cooling. That is, we consider the temperature of an object to be the same as its surface temperature.

The differential equation (25) is a first-order linear differential equation with constant coefficients. It is often written in the form

$$\frac{dT}{dt} + kT = kQ_0. \tag{26}$$

Equation (26) can always be solved by the integrating factor technique even if Q_0 depends on t. (Examples with nonconstant Q_0 are in the exercises.)

We will analyze only the simplest case, in which the outside (ambient) temperature Q_0 is a constant. In this case, the constant coefficient differential equation has a constant input. Using the constant input technique, it can be shown that

$$T(t) = Q_0 + ce^{-kt}. \tag{27}$$

We could also solve (26) by separation or by using an integrating factor. We now show how to solve (26) using integrating factors. The integrating factor is $u = e^{\int k\,dt} = e^{kt}$. Multiplying both sides of (26) gives

$$\frac{d}{dt}(e^{kt}T) = kQ_0e^{kt}.$$

Integrating both sides with respect to t, we have

$$e^{kt}T = Q_0e^{kt} + c.$$

Dividing by e^{kt}, we get that the solution is

$$T(t) = Q_0 + ce^{-kt},$$

which is (27).

No matter what the initial condition, we observe that as $t \to \infty$,

$$T(t) \to Q_0.$$

That is, the surface temperature approaches the ambient temperature as time increases. The manner in which the ambient temperature is approached is described by (27). Usually we wish to express the arbitrary constant c in terms of the initial condition $T(0)$ for the temperature. Letting $t = 0$ in (27), we get $T(0) = Q_0 + c$, so that $c = T(0) - Q_0$. Thus, (27) becomes

$$T(t) = Q_0 + [T(0) - Q_0]e^{-kt}. \tag{28}$$

If the initial temperature $T(0)$ equals the ambient temperature Q_0, then (28) shows that the temperature stays a constant, equaling the ambient temperature. Graphs of the temperature as a function of time for various initial conditions are given in Figure 1.8.7. Note that in all cases the solution approaches the ambient temperature Q_0 as $t \to \infty$.

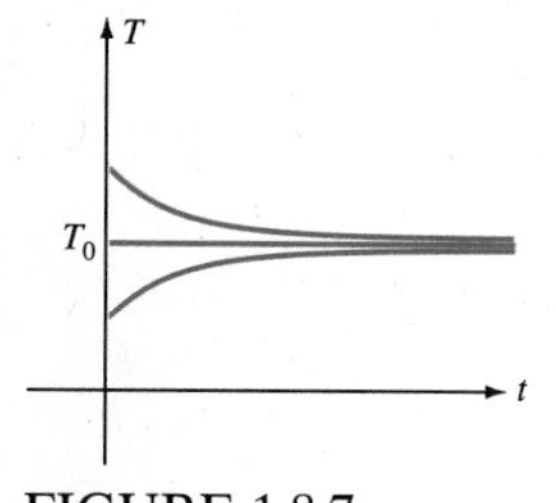

FIGURE 1.8.7

Graph of three solutions to (26).

Equilibrium

An **equilibrium solution** of a differential equation is a solution of the differential equation that is constant in time. For the differential equation (25), we see that $T(t) = Q_0$ is an equilibrium solution. For Newton's law of cooling, we have shown that the temperature approaches this equilibrium as $t \to \infty$ independent of the initial conditions. In Chapter 7, we will discuss equilibrium solutions more fully and introduce the notion of whether an equilibrium solution is stable or unstable.

EXAMPLE 1.8.6 **Cooling**

The room temperature in your office is 70°F. Experience has taught you that the temperature of a cup of coffee brought to your office will drop from 120°F to 100°F in 10 min. What should be the temperature of your cup of coffee when it is brought into the room if you want it to take 20 min before the temperature of the cup drops to 100°F?

SOLUTION First, we define our notation, set up the basic equations, and express the given data in terms of these. Let t be the time in minutes and $T(t)$ the temperature in degrees Fahrenheit at time t. The governing differential equation follows from the cooling law (24):

$$\frac{dT}{dt} = -k(T - Q_0). \tag{29}$$

From the problem description, the ambient temperature is 70°:

$$Q_0 = 70. \tag{30}$$

Not only is the initial temperature prescribed, but we are given the temperature at 10 min:

$$100 = T(10) \qquad \text{when } T(0) = 120. \tag{31}$$

We shall break this problem into two subproblems. The first is to determine k for the flow of thermal energy from our coffee cup to the room. Equation (28) is the solution of the initial value problem for (29):

$$T(t) = Q_0 + [T(0) - Q_0]e^{-kt} = 70 + 50e^{-kt}. \tag{32}$$

We are not given the proportionality constant k and must instead determine it from (32) using the given temperature at 10 min:

$$T(10) = 100 = 70 + 50e^{-10k}.$$

Solving this equation, we obtain

$$e^{-10k} = \frac{3}{5} \qquad \text{so that} \qquad k = \frac{\ln\left(\frac{3}{5}\right)}{-10} \approx 0.05. \tag{33}$$

Now we can solve the second subproblem, which is to determine the initial temperature if instead $T(20) = 100$. From (28) again,

$$T(t) = Q_0 + [T(0) - Q_0]e^{-kt} = 70 + [T(0) - 70]e^{-kt}.$$

However, k is determined in (33). Thus,

$$100 = T(20) = 70 + e^{2 \ln 3/5}[T(0) - 70]$$

or

$$30 = \frac{9}{25}[T(0) - 70].$$

Finally, solving this equation for the desired initial temperature $T(0)$, we obtain

$$T(0) = 70 + \frac{30 \cdot 25}{9} \approx 153.33°\text{F}.$$ ◄

In this example, the coffee-cup surface temperature dropped from 120°F to 100°F in 10 min and from $153\frac{1}{3}$°F to 100°F in 20 min. This shows the impossibility of guessing the answer without using the differential equation and its solution.

Exercises

In all of the following problems assume exponential growth.

1. One estimate of the growth rate of the United States is 1.5% per year. How many years will it take for the population to double?
2. A crystal grows by 5% in one day. When would you expect the crystal to be twice its original size?
3. A certain bacterium is observed to double in 8 h. What is its growth rate?
4. The population of the world is expected to double in the next 30 yr. What is its growth rate?
5. The growth rate of a certain strain of bacteria is unknown, but assumed to be constant. When an experiment started, it was estimated that there were about 1500 bacteria, and an hour later there were 2000. How many bacteria would you predict there are 4 h after the experiment started?
6. A population of bacteria is initially given and grows at a constant rate k_1. Suppose τ hours later the bacteria are put into a different culture such that the population now grows at the constant rate k_2. Determine the population of bacteria for all time.
7. The doubling time for a certain virus is 3 yr. How long will it take for the virus to increase to 10 times its current population level?
8. Initially you have 0.1 g of a bacteria in a large container; 2 h later you have 0.15 g. What is the doubling time for these bacteria?
9. An organism living in a pond reproduces at a rate proportional to the population size. Organisms also die off at a rate proportional to the population size. In addition, organisms are continuously added at a rate of k g/yr. Give the differential equation that models this situation.
10. A bacterium is reproducing in a large vat of nutrients according to an exponential growth law that would cause the population to double in 0.5 h. However, bacteria are continuously siphoned off at a rate of 5 g/h. Initially, there are 10 g of bacteria. How much bacteria are there in the vat after 2 h?
11. The rate of interest at one bank is 3% per year, whereas the yield at another bank is 3% a year. Both offer continuous compounding. Find the two doubling times.
12. The cost of a two-liter bottle of your favorite soft drink was 85 cents two years ago, but it now costs 95 cents. If this rate of increase continues, approximately when will the bottle cost $1.50?
13. The GDP (gross domestic product) of a certain country increased by 6.4% during the last year. If it continued to increase at that rate, approximately how many years would it take for the GDP to double?
14. During one year, food prices increased by 15%. At that rate, in approximately how many years would food prices triple?
15. **Cost of living** is the amount required to purchase a certain fixed list of goods and services in a single year. We assume that it is subject to exponential growth. The growth rate is known as the **rate of inflation**. If the cost of living rose from $10,000 to $11,000 in one year (a 10% net increase), what is the instantaneous rate of increase in the cost of living in that year? Equivalently, what is the rate of inflation?
16. Over a 3-yr period, housing prices increased 15%. At that rate, how many years would it take for housing prices to increase 50%?
17. A radioactive isotope has a half-life of 16 days. You wish to have 30 g at the end of 30 days. How much radioisotope should you start with?
18. A radioisotope is going to be used in an experiment. At the end of 10 days, only 5% is to be left. What should the half-life be?
19. A radioactive isotope sits unused in your laboratory for 10 yr, at which time it is found to

contain only 80% of the original amount of radioactive material.

a) What is the half-life of this isotope?

b) How many *additional* years will it take until only 15% of the original amount is left?

20. At the time an item was produced, 0.01 of the carbon it contained was carbon-14, a radioisotope with a half-life of about 5745 yr.

a) You examine the item and discover that only 0.0001 of the carbon is carbon-14. How old is the object? (This process of determining the age of an object from the amount of carbon-14 it contains is known as carbon-14 dating.)

b) Derive a formula that gives the age A of the object in terms of the fraction of carbon that is carbon-14 at the present time, T.

21. The temperature of an engine at the time it is shut off is 200°C. The surrounding air temperature is 30°C. After 10 min have elapsed, the surface temperature of the engine is 180°C.

a) How long will it take for the surface temperature of the engine to cool to 40°C?

b) For a given temperature T between 200°C and 30°C, let $t(T)$ be the time it takes to cool the engine from 200°C to T. [For example, $t(200) = 0$ and $t(40)$ is the answer to part (a).] Find the formula for $t(T)$ in terms of T and graph the function. (The ambient temperature is still 30°C.)

22. Earlier experiments have shown that a certain component cools in air according to the cooling law (Eq. 25) with constant of proportionality 0.2. At the end of the first processing stage, the temperature of the component is 120°C. The component remains for 10 min in a large room and then enters the next processing stage. At that time the surface temperature is supposed to be 60°C.

a) What must the room temperature be for the desired cooling to take place?

b) Suppose that the entrance and exit temperatures are still set at 120°C and 60°C, respectively, but the length of the wait in the room is w, a constant. Find the desired room temperature as a function of w and graph it.

23. An object at 100°C is to be placed in a 40°C room. What should the constant of proportionality be in Eq. (25) in order that the object be at 60°C after 10 min?

24. The air in a room is cooling. At time t (in hrs) the air temperature is $Q_0(t) = 70 + 20e^{-t/2}$. An object is placed in the room at time $t = 0$. The object is initially at 50°C and changes temperature according to Eq. (25) with $k = \frac{1}{2}$.

a) Find the temperature, $T(t)$, of the object for $0 \leq t \leq 5$.

b) Graph both Q_0 and T on the same axes.

25. An instrument at an initial temperature of 40°C is placed in a room whose temperature is 20°C. For the next 5 h the room temperature $Q_0(t)$ gradually rises and is given by $Q_0(t) = 20 + 10t$, t in hours.

a) Give the form the cooling law (Eq. 25) takes for the instrument.

b) From prior experience, you know that your instrument cools according to Eq. (25) with $k = 1$ if t is measured in hours. If $T(t)$ is the surface temperature at time t, solve the equation in part (a) for $T(t)$.

c) Graph $Q_0(t)$ and $T(t)$ on the same axes for $0 \leq t \leq 5$.

Many banks compound daily. Continuous compounding gives answers that are very close to those given by daily compounding. In Exercises 26 through 31, approximate the effect of daily compounding by assuming continuous compounding.

26. You have $10,000 and intend to invest it for 5 yr in a bank that offers continuous compounding. If you want to have $15,000 in your account at the end of these 5 yr, what annual interest rate do you have to get?

27. You invest $2000 in an account paying 6% a year, compounded daily, on the amount in excess of $500.

a) Express this as a differential equation by assuming continuous compounding.

b) Solve the differential equation and determine the amount in the account after 10 yr.

c) How much more would the account have after 10 yr if the full amount earned interest?

28. An amount of $10,000 is deposited in a bank that pays 9% annual interest compounded daily.

a) If you withdraw $10 a day, how much money do you have after 3 yr?

b) How much can you withdraw each day if the account is to be depleted in exactly 10 yr?

29. A amount of $1000 is deposited in a bank that pays 8% annual interest compounded daily. A deposit of B dollars is made daily.

a) What should B be in order to have $10,000 after 5 yr?

b) Determine the function $B(x)$ that gives the daily deposit needed to have $10,000 after x years. [$B(5)$ is computed in part (a).]

30. An amount of $10,000 is invested at 12% annual interest compounded daily. An additional investment of $\$B$ is made daily. What should B be in order for the investment account to be $100,000 after 10 yr?

31. An amount of $100 is deposited in a foreign bank that pays 20% annual interest compounded daily. Each day you make a transaction of amount $f(t)$ dollars. Part of the year you are able to deposit money [$f(t) > 0$], and part of the year you make withdrawals [$f(t) < 0$]. If t is in years, these transactions occur in the following cyclical yearly pattern:

$$\begin{aligned} f(t) &= \frac{400}{365}\cos(2\pi t)\ \$/\text{day} \\ &= 400\cos(2\pi t)\ \$/\text{year}. \end{aligned}$$

a) Find the amount of money in the account as a function of t.

b) Graph your answer to (a) for 10 yr (a sketch will do if no computer is available).

1.9 Mixture Problems

In this section, a quantity $Q(t)$, such as the amount of some pollutant in a water tank, varies with time. Further amounts of this quantity are being added. The addition will be called inflow. Simultaneously, some of this quantity is being lost. The quantity lost will be called the outflow. In the case of a water tank, the loss could be due to evaporation, overflow, an open valve, or all three. In all cases we assume that the tank is very well mixed (through fast stirring if necessary), so that the concentration of the pollutant will be assumed to be the same throughout all portions of the tank. The beginning idea for analyzing these problems is the fundamental physical principle of conservation of the quantity Q:

$$\begin{bmatrix}\text{Rate of change}\\ \text{of } Q\end{bmatrix} = \frac{dQ}{dt} = \begin{bmatrix}\text{inflow rate}\\ \text{of } Q\end{bmatrix} - \begin{bmatrix}\text{outflow rate}\\ \text{of } Q\end{bmatrix}, \tag{1}$$

where we have noted that dQ/dt, the time derivative of Q, is the rate of change of Q with respect to time t.

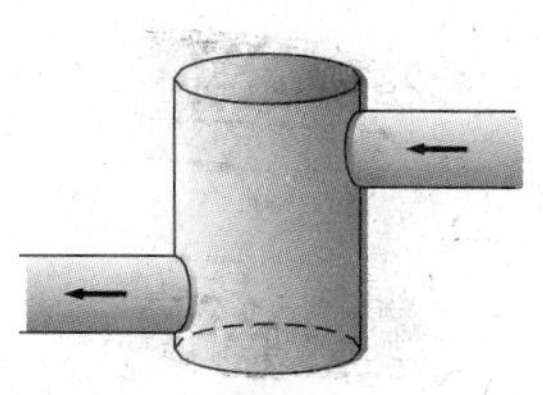

FIGURE 1.9.1

Tank with inflow and outflow.

Procedure for Flow Problems

1. It is helpful to draw a rough sketch of a tank illustrating the inflow and outflow with pipes (see Figure 1.9.1).
2. Label quantities and note the given data.
3. Express inflow rates and outflow rates in terms of the given variables and substitute them into (1).
4. Solve the resulting differential equation.
5. Answer any questions such as "how long?"

We will discuss problems of two different kinds. In one, the volume of water in the tank will be fixed, and the resulting differential equation will be easier to solve. In the other, the volume of water will be changing in time, and the resulting differential equation will be harder to solve. In some of the exercises, the process of setting up the differential equations will be emphasized, and you will not be asked to solve the differential equations.

1.9.1 Mixture Problems with a Fixed Volume

We begin with a fixed volume example.

EXAMPLE 1.9.1 Fixed Volume

Consider a 100-m^3 tank full of water. The water contains a pollutant at a concentration of 0.6 g/m^3. Cleaner water, with a pollutant concentration of 0.15 g/m^3, is pumped into the well-mixed tank at a rate of 5 m^3/s. Water flows out of the tank through an overflow valve at the same rate as it is pumped in.

a. Determine the amount and concentration of the pollutant in the tank as a function of time. Graph the result.
b. At what time will the concentration be 0.3 g/m^3?

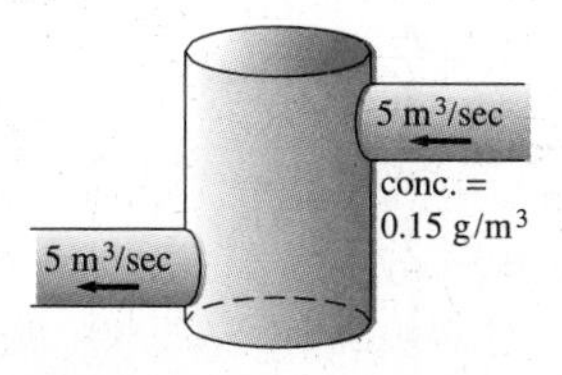

FIGURE 1.9.2

Picture for Example 1.9.1.

Solution In order to illustrate the general principles, and since this is our first mixing problem, we shall include a few more steps than are necessary to solve the particular problem.

In mixture problems, it is best to first draw a rough diagram of a tank indicating the inflow and the outflow (see Figure 1.9.2). Water flows in at the rate of 5 m^3/s, with concentration 0.15 g/m^3, and the mixture flows out at the same rate of 5 m^3/s. Thus, the volume of water in the tank stays the same, equaling 100 m^3. Usually, it is easier to formulate a differential equation for the amount of the pollutant. We let $Q(t)$ be the amount in grams of pollutant in the tank. Q depends on the time t, which we measure in seconds. The amount of pollutant in the tank changes in time as a result of inflow and outflow, so that the rate of change of the amount of pollutant satisfies (1):

$$\frac{dQ}{dt} = \begin{bmatrix}\text{inflow rate}\\ \text{of pollutant}\end{bmatrix} - \begin{bmatrix}\text{outflow rate}\\ \text{of pollutant}\end{bmatrix}. \tag{2}$$

It is given that 5 m^3 of water per second flows in, with the concentration of pollutant being 0.15 $\mathrm{g/m^3}$. Thus, $5 \cdot (0.15)$ g of pollutant per second flows in, since

$$\begin{bmatrix}\text{Inflow rate}\\ \text{of pollutant}\end{bmatrix} = \begin{bmatrix}\text{flow of volume}\\ \text{of water in}\end{bmatrix} \cdot \begin{bmatrix}\text{concentration of}\\ \text{pollutant flowing in}\end{bmatrix}.$$

We now compute the outflow of pollutant. Again 5 m^3 of water per second flows out:

$$\begin{aligned}\begin{bmatrix}\text{Outflow rate}\\ \text{of pollutant}\end{bmatrix} &= \begin{bmatrix}\text{flow of volume}\\ \text{of water out}\end{bmatrix} \cdot \begin{bmatrix}\text{concentration of}\\ \text{pollutant flowing out}\end{bmatrix}\\ &= 5 \cdot \begin{bmatrix}\text{concentration of}\\ \text{pollutant flowing out}\end{bmatrix}.\end{aligned}$$

But we are not given the concentration of the pollutant that flows out. We must compute this concentration. The water in the tank is assumed to be well mixed, so that the concentration that flows out is assumed to be the same as the overall concentration of the pollutant in the tank. To compute the concentration of a pollutant in a tank, we simply divide the amount of pollutant by the total volume:

$$\begin{bmatrix}\text{Concentration of pollutant}\\ \text{in the tank}\end{bmatrix} = \frac{\text{amount of pollutant}}{\text{volume}} = \frac{Q(t)}{100}.$$

In this case, the volume of water is fixed and equals 100 m^3. Here, the amount of pollutant in the tank is unknown, but we called it $Q(t)$. The differential equation for the amount of pollutant follows from (2):

$$\frac{dQ}{dt} = 5 \cdot 0.15 - 5 \cdot \frac{Q}{100} = 0.75 - \frac{1}{20}Q. \tag{3}$$

The differential equation (3) is linear, and we often rewrite it as

$$\frac{dQ}{dt} + \frac{1}{20}Q = 0.75. \tag{4}$$

Since (4) is a linear differential equation with constant coefficients and the input 0.75 is a constant, we can use the method of undetermined coefficients of Section 1.7. (Integrating factors from Section 1.6 also work well.) A particular solution will be a constant, $Q_p = A$. Substituting A for Q in (4), we compute that $A = \frac{3}{4} \cdot 20 = 15$. A solution of the associated homogeneous equation is easily seen to be $Q_1 = e^{rt} = e^{-\frac{1}{20}t}$. Thus the general solution of (4) is

$$Q(t) = 15 + ce^{-\frac{1}{20}t}. \tag{5}$$

The arbitrary constant c is determined from the initial conditions. The word problem states that the 100-m^3 tank initially contains the pollutant at a concentration of 0.6 g/m^3. Thus, initially the amount of pollutant is $Q(0) = 0.6(100) = 60$ g. By substituting this initial condition into the general solution (5), we obtain $c = 45$. Thus, the solution of the initial value problem is

$$Q(t) = 15 + 45e^{-\frac{1}{20}t}, \tag{6}$$

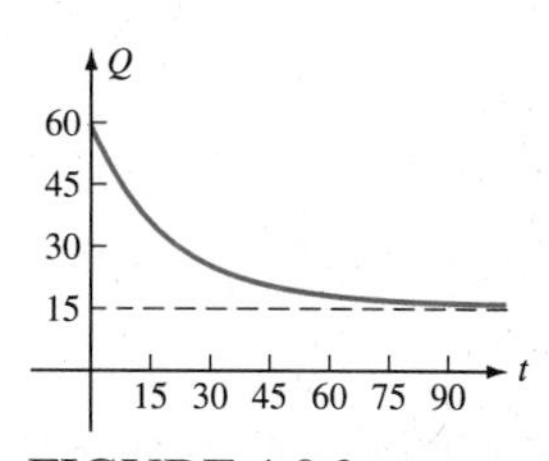

FIGURE 1.9.3

Graph of $15 + 45e^{-t/20}$.

which is graphed in Figure 1.9.3. The amount of pollutant is initially 60, but exponentially decays to 15 as time increases. As $t \to \infty$, $Q(t) \to 15$. It should be physically obvious that the amount of pollutant approaches 15 g as time approaches infinity, because the water entering the tank has a concentration level of 0.15 g/m^3. Eventually the concentration of the pollutant in the tank will approach a level equal to the inflow concentration. Since the tank is 100 m^3, the amount of pollutant in the tank corresponding to the inflow would approach $100 \cdot 0.15 = 15$ g of pollutant.

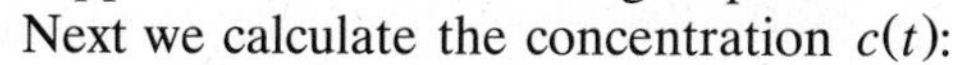

Next we calculate the concentration $c(t)$:

$$c(t) = \frac{Q(t)}{\text{volume}} = \frac{15 + 45e^{-\frac{1}{20}t}}{100} = 0.15 + 0.45e^{-\frac{1}{20}t}. \tag{7}$$

This also shows that the concentration approaches the concentration of the inflow, 0.15, as time increases. We are asked to determine when the concentration equals 0.3, $c(t) = 0.3$. Using (7), we obtain

$$0.3 = 0.15 + 0.45e^{-\frac{1}{20}t}.$$

Solving for t, we find $t = -20\ln(1/3) \approx 21.97$ s. ◄

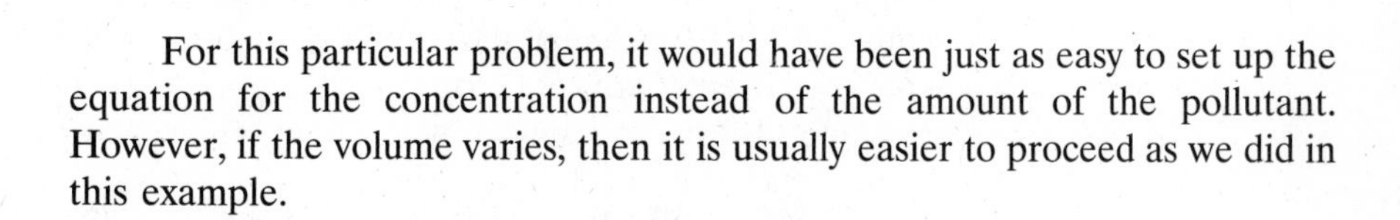

For this particular problem, it would have been just as easy to set up the equation for the concentration instead of the amount of the pollutant. However, if the volume varies, then it is usually easier to proceed as we did in this example.

1.9.2 Mixture Problems with Variable Volumes

Let us reconsider mixture problems in a tank. If the rate of water entering the tank is different from the rate of water leaving the tank, the volume of water in the tank will not be constant as in the previous example. We will show that the resulting differential equation for the amount of a pollutant will still often be linear, but that the coefficients will not be constant. Before we do an example, let us discuss a somewhat general problem.

Suppose we are given the amount of salt initially dissolved in a tank of water. Let us assume that salt water with a concentration of c_i pounds per gallon flows into the tank at the rate of r_i gallons per minute. Here the subscript i stands for the flow in. The tank contents are well mixed, and the salt water mixture is pumped out at the rate of r_o gallons per minute (subscript o for out). This problem is shown diagrammatically in Figure 1.9.4. The volume $V(t)$ of salt water in the tank may vary. Since $V'(t) = r_i - r_o$, we have

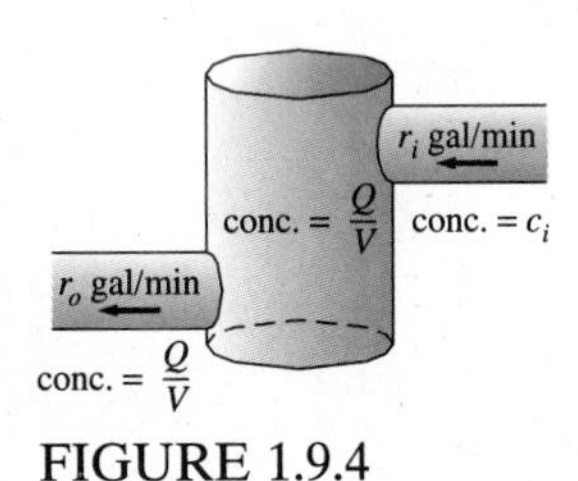

FIGURE 1.9.4

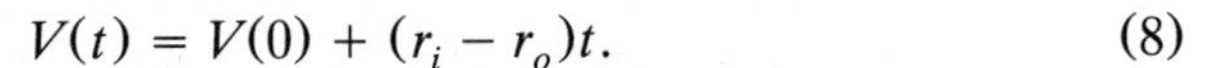

$$V(t) = V(0) + (r_i - r_o)t. \tag{8}$$

Actually this is the somewhat obvious solution of the initial value problem for the differential equation $dV/dt = r_i - r_o$, representing the rate of change of the volume of water in the tank. If the inflow rate of water differs from the outflow rate of water, then the volume varies.

We let $Q(t)$ be the amount of salt in the tank. The inflow rate of salt is $r_i c_i$ (the inflow rate of water r_i times the given concentration of salt c_i in the inflow). The outflow rate of salt is the outflow rate of water r_o times the concentration of salt c_o in the outflow. As before, the concentration of salt in the outflow is the same as the concentration of salt $c(t)$ in the tank:

$$c_o = c(t) = \frac{Q(t)}{\text{volume}} = \frac{Q(t)}{V(0) + (r_i - r_o)t}. \tag{9}$$

The rate of change of the amount of salt is given by the basic relationship (1):

$$\frac{dQ}{dt} = \begin{bmatrix} \text{inflow rate} \\ \text{of salt} \end{bmatrix} - \begin{bmatrix} \text{outflow rate} \\ \text{of salt} \end{bmatrix}.$$

Using the result (9), we have

$$\frac{dQ}{dt} = r_i c_i - r_o c_o = r_i c_i - r_o \frac{Q}{V(0) + (r_i - r_o)t}. \tag{10}$$

Equation (10) is a linear differential equation for $Q(t)$, the amount of salt. The coefficients are not constant if $r_i \neq r_o$. If the coefficients are not constant, (10) may be solved by the integrating factor method.

EXAMPLE 1.9.2 Variable Volume

A 100-gal tank is initially half full of pure water. Then water containing 0.1 lb/gal of salt is added at a rate of 4 gal/min. The well-mixed contents of the tank flow out a pipe at the rate of 2 gal/min. When the tank is full, it overflows. Find the amount and concentration of salt.

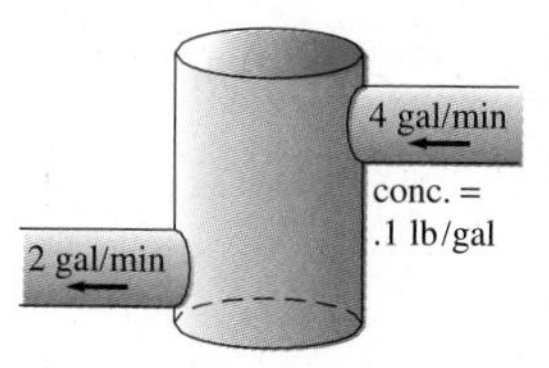

FIGURE 1.9.5

Example 1.9.2.

Solution This problem is sketched in Fig. 1.9.5. The volume of salt water is increasing and is

$$V(t) = 50 + 2t, \tag{11}$$

since $r_i - r_o = 4 - 2 = 2$ and $V(0) = 50$. The 100-gal tank will fill up at $t = 25$ min. First, we solve the problem before the tank is filled.

The First 25 Minutes

We let $Q(t)$ be the amount of salt in the tank, so that the concentration of salt is $c(t) = Q(t)/(50 + 2t)$. The rate of change of the amount of salt satisfies (10):

$$\frac{dQ}{dt} = [4] \cdot [0.1] - [2] \cdot \frac{Q}{50 + 2t}.$$

This can be rewritten as

$$\frac{dQ}{dt} + \frac{2Q}{50 + 2t} = 0.4. \tag{12}$$

We solve this linear differential equation by the integrating factor method. The integrating factor is

$$\exp\left(\int \frac{2}{50 + 2t}\, dt\right) = \exp\left(\ln(50 + 2t)\right) = 50 + 2t.$$

We next multiply (12) by this integrating factor to get

$$\frac{d}{dt}[(50 + 2t)Q] = (50 + 2t)0.4.$$

Antidifferentiating yields

$$(50 + 2t)Q = 0.1(50 + 2t)^2 + c.$$

Solving for Q yields the general solution of the linear differential equation:

$$Q(t) = 0.1(50 + 2t) + c(50 + 2t)^{-1}, \qquad t \leq 25. \tag{13}$$

In this form a particular solution and a solution of the associated homogeneous equation are evident. The concentration can now be determined easily from (11) and (12) as

$$c(t) = \frac{Q(t)}{V(t)} = 0.1 + c(50 + 2t)^{-2}.$$

These formulas are valid until the tank fills, that is, for $t < 25$ min. The constant c can be determined from the initial condition.

$$Q(0) = 0.1 \cdot 50 + c(50)^{-1}, \text{ so that } c = 50Q(0) - 0.1(50)^2$$

In our example, the water is initially pure, so that $Q(0) = 0$. Thus,

$$c = -0.1(50)^2 = -250, \tag{14}$$

which can be substituted into the general solution (13).

After 25 Minutes

After the tank fills up ($t > 25$), the inflow of salt water and the outflow of salt water are the same. They are both 4 gal/min. Thus, the volume is now a constant, 100 gal, and the method of solution is similar to that for the earlier example with constant volume. The differential equation will be

$$\frac{dQ}{dt} = (4) \cdot (0.1) - (4)\frac{Q}{100} = 0.4 - \frac{Q}{25}. \tag{15}$$

A particular solution is easily seen to be $Q_p = A = 0.4 \cdot 25 = 10$, and a solution of the associated homogeneous equation is $Q_1 = e^{rt} = e^{-t/25}$. Thus, the general solution is

$$Q(t) = 10 + c_1 e^{-t/25}, \tag{16}$$

where we have used a different arbitrary constant. This solution (16) is valid for $t > 25$. As $t \to \infty$, the amount of salt $Q(t)$ approaches 10 lb, as expected. The initial condition to determine the constant c_1 occurs at $t = 25$, by assuming the amount of salt Q to vary continuously at $t = 25$. That is, $Q(25) = \lim_{t \to 25^-} Q(t)$. From (13) and (14), we have

$$\lim_{t \to 25^-} Q(t) = 10 + \frac{c}{100} = 10 + \frac{-250}{100} = 7.5. \tag{17}$$

This condition (17) determines the constant c_1 in (16):

$$7.5 = Q(25) = 10 + c_1 e^{-1} \qquad \text{so that} \qquad c_1 = -2.5e.$$

Thus, the solution of the initial value problem (valid for $t > 25$) follows from (16):

$$Q(t) = 10 - (2.5e)e^{-t/25} = 10 - 2.5e^{-t/25}, \qquad t > 25. \tag{18}$$

The solution given by (13), (14), and (18) is shown in Figure 1.9.6. ◄

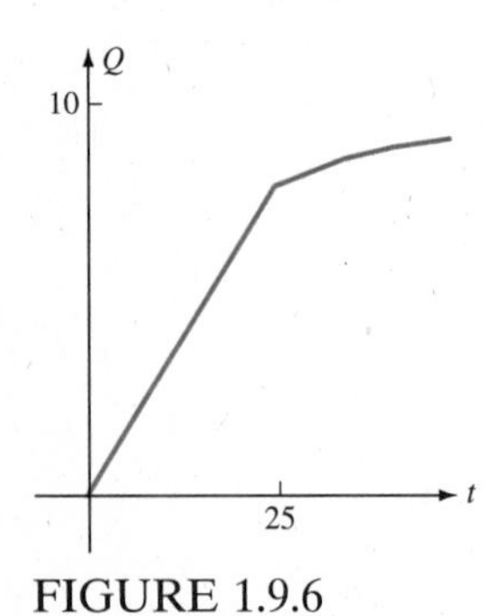

FIGURE 1.9.6

Graph of solution of Example 1.9.2.

Exercises

1. A well-mixed tank contains 300 gal of water with a salt concentration of 0.2 lb/gal. Water containing salt at a concentration of 0.4 lb/gal enters at a rate of 2 gal/min. An open valve allows water to leave the tank at the same rate.

- **a)** Determine the amount and the concentration of salt in the tank as functions of time.
- **b)** How long will it take for the concentration to increase to 0.3 lb/gal?

2. A room has a volume of 800 ft^3. The air in the room contains chlorine at a concentration of 0.1 g/ft^3. Fresh air enters at a rate of 8 ft^3/min. The air in the room is well mixed and flows out of a door at the same rate as the fresh air comes in.

- **a)** Find the chlorine concentration in the room as a function of t.
- **b)** Suppose that the flow rate of fresh air is adjustable. Determine the flow rate required to reduce the chlorine concentration to 0.001 g/ft^3 within 20 min.

3. A well-mixed tank contains 100 L of water with a salt concentration of 0.1 kg/L. Water containing salt at a concentration of 0.2 kg/L enters at a rate of 5 L/h. An open valve allows water to leave at 4 L/h. Water evaporates from the tank at 1 L/h.

- **a)** Determine the amount and concentration of salt as a function of time.
- **b)** Is the limiting concentration the same as that of the inflow?

4. A well-circulated pond contains 1 million L of water that contains a pollutant at a concentration of 0.01 kg/L. Pure water enters from a stream at 100 L/h. Water evaporates from the pond (leaving the pollutant behind) at 50 L/h and flows out an outlet pipe at 50 L/h. How many days will it take for the concentration of pollutant to drop to 0.001 kg/L?

5. A 1000-gal tank is initially half full of water and contains 10 lb of iodine in solution. Pure water enters the tank at a rate of 6 gal/min. An open valve allows water to leave at a rate of 1 gal/min. When full, the tank overflows.

- **a)** Find the amount and concentration of iodine in the tank for the first 100 min.
- **b)** Find the amount and concentration of iodine for the next 100 min.
- **c)** Graph both concentration and amount of iodine.

6. A 100-gal tank is initially full of water containing 10 lb of salt in solution. The tank will overflow whenever additional water is added. A pump is attached to a sensor. It pumps fresh water into the tank at a rate proportional to the concentration of salt in the tank. The constant of proportionality is 10 $(gal)^2/lb \cdot min$. Find the amount and concentration of salt in the tank as functions of t.

7. A well-circulated lake contains 1000 kL of water that is polluted at a concentration of 2 kg/kL. Water from the effluent of a factory enters the lake at the rate of 5 kL/h with a concentration of 7 kg/kL of the pollutant. Water flows out of the lake through an outlet at the rate of 2 kL/h. Determine the amount and concentration of the pollutant as a function of time.

In Exercises 8 through 14, formulate the differential equations (before any tanks empty or overflow) that would be used to solve the problems, but do *not* solve the differential equations. Since Exercises 9–13 involve more than one tank, they require one equation for each tank.

8. A 600-gal tank initially contains 200 gal of brine (salt water) with 25 lb of salt. Brine containing 2 lb of salt per gallon enters the tank at the rate of 13 gal/s. The mixed brine in the tank flows out at the rate of 8 gal/s. How much salt is there in each tank as a function of time?

9. Consider two tanks. Initially, tank 1 contains 150 gal of brine (salt water) with 8 lb of salt, and tank 2 contains 250 gal of brine with 14 lb of salt. Brine containing 3 lb of salt per gallon enters tank 1 at the rate of 13 gal/s. Mixed brine in tank 1 flows out into tank 2 at the rate of 7 gal/s. The mixture in tank 2 flows away at the rate of 28 gal/s. How much salt is there in each tank as a function of time?

10. Consider two tanks. Initially, tank 1 contains 100 gal of brine (salt water) with 35 lb of salt, and tank 2 contains 400 gal of brine with 15 lb of salt. Suppose the mixture flows out of tank 1 into tank 2 at the rate of 17 gal/min and the mixture flows out of tank 2 into tank 1 at the rate of 6 gal/min. How much salt is there in each tank as a function of time?

11. Consider two tanks. Initially, tank 1 contains 230 gal of brine (salt water) with 28 lb of salt, and tank 2 contains 275 gal of brine with 7 lb of salt. Brine containing 5 lb of salt per gallon enters tank 1 at the rate of 21 gal/s. Mixed brine in tank 1 flows out at the rate of 18 gal/s, half flowing into tank 2. The mixture in tank 2 flows away at the rate of 4 gal/s. How much salt is there in each tank as a function of time?

12. Consider three tanks. Initially, tank 1 contains 200 gal of brine (salt water) with 55 lb of salt, tank 2 contains 500 gal of brine with 35 lb of salt, and tank 3 is empty. Suppose the mixture flows out of tank 1 into tank 2 at the rate of 8 gal/min and the mixture flows out tank 2 into tank 3 at the rate of 22 gal/min. How much salt is there in each tank as a function of time?

13. Consider three tanks. Initially, tank 1 contains 100 gal of brine (salt water) with 17 lb of salt, tank 2 contains 200 gal of brine with 19 lb of salt, and tank 3 contains 300 gal of brine with 21 lb of salt. Brine containing 5 lb of salt per gallon enters tank 1 at the rate of 11 gal/s. Mixed brine in tank 1 flows out into tank 2 at the rate of 18 gal/s and the mixture also flows out of tank 2 into tank 3 at the rate of 18 gal/s. How much salt is there in each tank as a function of time?

14. A large tank contains 7 gal of pure water. Polluted water containing 7 g of bacteria per gallon enters at the rate of 14 gal/h. A well-mixed mixture is removed at the rate of 4 gal/h. However, it is also known that the bacteria multiply inside the tank at a growth rate of 2% per hour. Determine the amount of bacteria in the tank as a function of time.

1.10 Electronic Circuits

One of the applications of differential equations that will frequently recur throughout this book is the theory of electronic circuits. There are several reasons for this, among them the importance of circuit theory and the pervasiveness of differential equations in circuit theory. (One of the authors received his first exposure to circuit theory from a highly mathematical electrical engineering professor by the name of Amar Bose. You would be correct if you recognized the significance of his last name.) Also, circuits are one example of what could be called network models. Network models are widely used, for example, in manufacturing and other economic systems.

Circuits will be covered again in greater detail later. In this section we shall introduce the basic circuit concepts we shall use and give some simple examples of circuits that are described by first-order differential equations.

We consider only lumped-parameter circuits. In circuit theory, the word **circuit** means that quantities such as current are determined solely by position along the path. A wire has a finite thickness, and interesting electrical behavior occurs across the wire. We ignore such field effects. **Lumped parameter** means that the effects of the various electrical components may be considered to be concentrated at one point. The converse of a lumped-parameter circuit is a **distributed-parameter** circuit. The analysis of distributed-parameter circuits often involves partial differential equations, which we discuss briefly near the end of this text. Antennas are an example of distributed-parameter systems.

At each point in the circuit there are two quantities of interest: **voltage** (or potential) and **current** (or flow of charge). Current is, by convention, the net flow of positive charge. A **branch** is part of a circuit with two terminals to which connections can be made, a **node** is a point where two or more branches come together (a node is denoted in our sketches —•—), and a **loop** is a closed path formed by connecting branches. Our basic modeling laws are Kirchhoff's circuit laws, the current and voltage laws:

Current law:	The algebraic sum of the currents entering a node at any instant is zero.	(1)
Voltage law:	The algebraic sum of the voltage drops around a loop at any instant is zero.	

The voltage law is equivalent to saying that the voltage drop from one point to another is the same in any direction. We will discuss these laws shortly.

To set up the circuit equations, a current variable is assigned to each branch. One can talk either of the potentials (voltages) at the nodes or of the potential drops across the branches. Kirchhoff's current law may then be applied to each node, and the voltage law to each loop. This procedure exhibits a certain amount of arbitrariness. There is usually some redundancy among the equations, and the determination of a minimal number of equations is generally computationally nontrivial.

In this section we discuss only single- or double-loop circuits. Let us consider the branch containing a two-terminal device shown in Figure 1.10.1. The current is denoted by i. The voltages at the two nodes are denoted v_0 and v_1. The voltage drop is defined to be the difference, $v = v_0 - v_1$. For our purposes, the behavior of the device is completely determined if we know v and i at any time t. The relationship between v and i is called the ***v-i* characteristic** of the particular device.

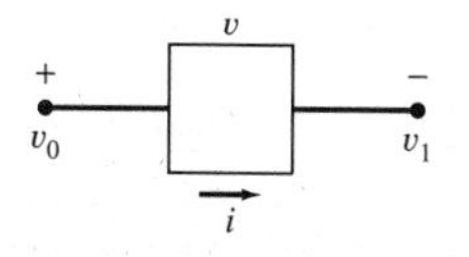

FIGURE 1.10.1

A two-terminal device. The nodal voltages are denoted by v_0 and v_1, the voltage drop $v = v_0 - v_1$, and the current is denoted by i.

We shall consider only the following five basic types of devices.

Resistor If the voltage drop v (measured in volts) is uniquely determined by the current i (measured in amperes) and the time,

$$v = f(i, t),$$

the device is called a current-controlled resistor. If the voltage is proportional to the current,

$$v = iR, \tag{2}$$

and R depends only on t, then the device is a linear resistor and R is called the resistance (usually measured in ohms). One ohm is the resistance that would give a voltage drop of 1 volt if the current were 1 ampere. In many applications R may be approximated by a constant. These resistors will be denoted -\/\/\/-. The coefficient R measures the resistance of the device to the flow of electricity. For a given voltage v, $i = v/R$. Large values of R correspond to small currents, as the device resists the flow of electricity. (Nonlinear resistors will be discussed briefly in Chapters 7 and 9 and denoted -/\/\/\-.)

Capacitor A capacitor stores energy in the form of a charge q (measured in coulombs). The charge q and the voltage drop v across the capacitor are proportional,

$$q = Cv, \tag{3}$$

where C is the capacitance (measured in farads). One farad is the capacitance when 1 coulomb of electricity raises the potential by 1 volt. Charge builds up across a capacitor. The current is due to the flow of electrons. If the charge on a capacitor is constant in time, there is no flow of electrons. The rate of change of the charge is the flow of electrons or the current:

$$i = \frac{dq}{dt}. \tag{4}$$

If the capacitance C is constant, we may differentiate (3), using (4), to obtain the v-i characteristic for a capacitor:

$$i = C\frac{dv}{dt}. \tag{5}$$

The symbol for a capacitor is ·—| |—·. We define only such linear capacitors in this text.

Inductor An inductor stores energy in a magnetic field. The voltage-current relationship for a (linear) inductor is

$$v = L\frac{di}{dt}, \tag{6}$$

where L is called the inductance (and measured in henrys). One henry is the inductance in which one volt is induced by a current varying at the rate of one ampere per second. The symbol for an inductor is -ᘐᘐᘐ-.

TABLE 1.10.1 ELECTRICAL UNITS

Device	Symbol	Unit	v-i Characteristic
Resistor	R	ohm (Ω)	$v = iR$
Capacitor	C	farad (F)	$i = C\, dv/dt$
Inductor	L	henry (H)	$v = L\, di/dt$
Current	i	ampere (A)	

For many devices, such as transistors, we design models by considering them to be made up of linear capacitors, inductors, and current-controlled resistors. Linear resistors will not suffice to fully model a transistor.

No device, of course, is only a resistor or an inductor. However, we can analyze many circuits by considering them to be made up of resistors, inductors, and capacitors. Also, no device is truly linear. However, if restrictions are put on the allowable current and voltage, we can often make the assumption that the device is linear. As with all physical problems, some assumptions are necessary in order for a mathematical model to approximate a physical situation.

Many students will know the fundamental relations for capacitor, inductors, and resistors. For others, we provide the short table, Table 1.10.1.

Voltage Source Some circuits we will consider will also contain a voltage source, which is denoted by $-\bullet\!\!-\!\!(e)\!\!-\!\!\bullet +$. The voltage is higher by e on the $+$ side, so that we say the voltage drop is $-e$ for any current. Batteries are an example of a voltage source.

Current Source Similarly, a current source is denoted by $\bullet\!\!-\!\!(i)\!\!-\!\!\bullet$. In a current source, the current is i for any voltage. Some solar cells are current sources.

Applying Kirchhoff's Laws

In applications of the loop and node laws, it is important to use the correct signs. When we apply the current law at a node, currents entering the node are given the opposite sign to that given the currents leaving the node. Figure 1.10.2 shows several nodes and the corresponding current equations. The voltage drops for resistors, capacitors, and inductors are added if the current

FIGURE 1.10.2

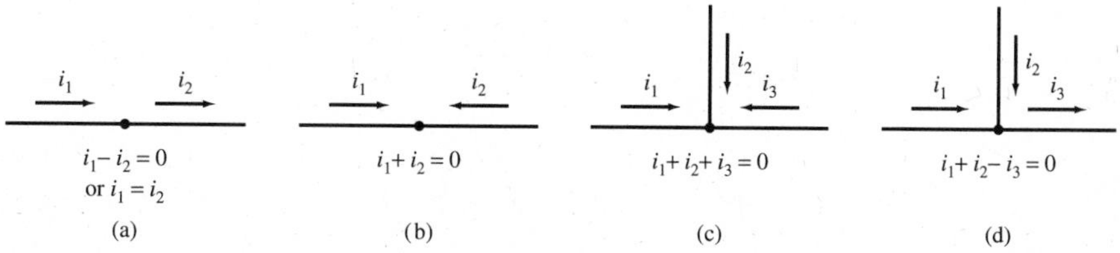

$i_1 - i_2 = 0$ or $i_1 = i_2$ (a)

$i_1 + i_2 = 0$ (b)

$i_1 + i_2 + i_3 = 0$ (c)

$i_1 + i_2 - i_3 = 0$ (d)

variable in that branch is in the same direction as we are moving around the loop. The voltage drops are subtracted if the current variable is in the direction opposite to that in which we are moving around the loop. The converse holds for a voltage source, since in that case we have a voltage gain (negative drop).

EXAMPLE 1.10.1 Circuit with a Resistor, Capacitor, and Voltage Source

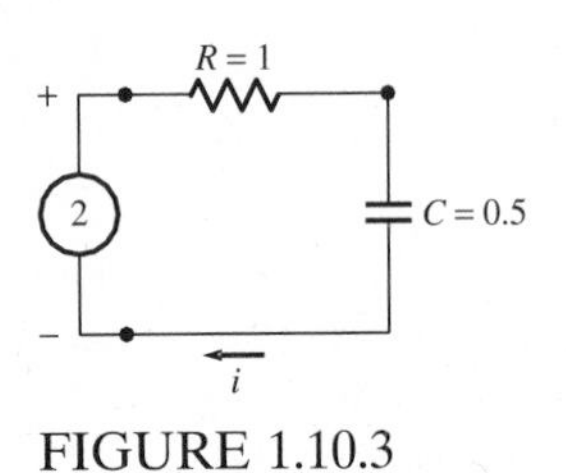

FIGURE 1.10.3

For the linear circuit (with resistance 1 Ω, capacitance 0.5 F, and a 2-V battery) shown in Figure 1.10.3, find the current through the loop and the charge $q(t)$ on the capacitor, given that initially there is no charge on the capacitor, $q(0) = 0$.

SOLUTION Taking the current in each branch to move in a clockwise direction, we see, by the current law (Figure 1.10.2a), that the current is the same in all three branches. We call the current i. By the voltage law, the sum of the voltage drops, starting at the voltage source, is (note the different sign for the voltage gain at the source)

$$-2 + iR + \frac{q}{C} = 0 \qquad \text{or} \qquad -2 + 1i + 2q = 0. \tag{7}$$

The voltage law can be written as a differential equation for the charge or the current by recalling that $i = dq/dt$. In terms of q, we obtain the linear first-order differential equation

$$-2 + \frac{dq}{dt} + 2q = 0 \qquad \text{or} \qquad \frac{dq}{dt} + 2q = 2. \tag{8}$$

Since (8) has constant coefficients and the input (right-hand side) is a constant, we know that a particular solution will be a constant, $q_p = A = 1$. A solution of the associated homogeneous equation is easily seen to be $q = e^{rt} = e^{-2t}$. Thus, the general solution of (8) is

$$q(t) = 1 + ce^{-2t}. \tag{9}$$

In this example, the capacitor is not charged initially, $q(0) = 0$, so that $0 = 1 + c$. Thus, the solution of the initial value problem is

$$q(t) = 1 - e^{-2t}. \tag{10}$$

We see that the capacitor charges and the charge on the capacitor approaches 1 as $t \to \infty$. From the solution (10) we can obtain the current, $i = dq/dt$, by differentiating

$$i(t) = 2e^{-2t}. \tag{11}$$

The current approaches 0 as $t \to \infty$. As $t \to \infty$, there is no voltage drop across the resistor. Thus, as $t \to \infty$, the voltage drop across the capacitor must equal the 2 V of the battery, corresponding to the charge across the capacitor approaching 1 C. ◄

An alternative way to solve this problem is to consider the first-order linear differential equation for the current obtained by differentiating (8) with respect to t:

$$\frac{di}{dt} + 2i = 0. \tag{12}$$

This differential equation is easy to solve, but the initial conditions for the current were not given. Instead we must determine the initial conditions for the current by evaluating the voltage law (7) at $t = 0$. Since $q(0) = 0$, it follows that $i(0) = 2$. In this manner (11) can be obtained from (12). In this alternative method, the charge would be determined by integration, since $dq/dt = i$. Using the initial condition for the charge would give (10).

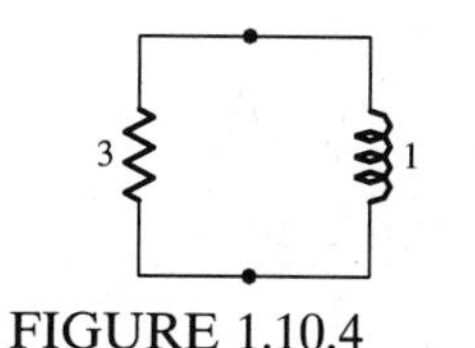

FIGURE 1.10.4

EXAMPLE 1.10.2 Circuit with a Resistor and an Inductor

Consider the circuit in Figure 1.10.4 with a 3-Ω resistor and a 1-H inductor. Determine the current as a function of time in this circuit given that its initial value is 6 A (defined clockwise).

SOLUTION The current i is the same in each branch. In this case, from the voltage law, we directly obtain a differential equation for the current:

$$L\frac{di}{dt} + iR = 0 \quad \text{or} \quad \frac{di}{dt} + 3i = 0. \tag{13}$$

This linear differential equation is homogeneous (there is no input), so that zero is a particular solution, $i_p = 0$. A solution of the homogeneous equation is $i_1 = e^{rt} = e^{-3t}$, so that the general solution of (13) is

$$i(t) = ce^{-3t}. \tag{14}$$

The current is given initially to be 6 A, $i(0) = 6$, so that $c = 6$. The solution of the initial value problem is

$$i(t) = 6e^{-3t}.$$

◄

Exercises

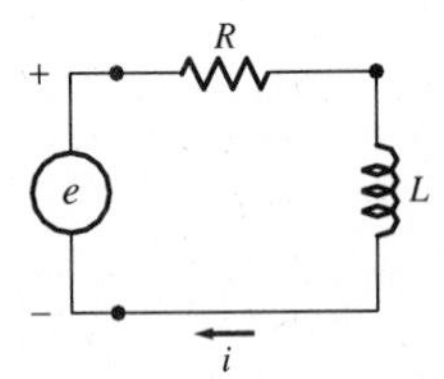

FIGURE 1.10.5

Exercises 1 through 6 refer to the circuit given in Figure 1.10.5. The inductor is linear with inductance of L; the voltage source is $e(t)$.

1. The voltage source is constant, $e = 1$, the resistor is linear with v-i characteristic $v = 2i$, and $L = 1$. Set up the differential equation for the current. Determine the current as a function of time for any initial current $i(0)$. Find any steady-state (equilibrium) solutions.
2. The voltage source is constant, $e = 4$, the resistor is linear with v-i characteristic $v = 6i$, and $L = 2$. Set up the differential equation for the current. Determine the current as a function of time for any initial current $i(0)$. Find any steady-state (equilibrium) solutions.
3. The voltage source is $e(t) = \sin t$, the resistor is linear with v-i characteristic $v = i$, and $L = 1$. Set up the differential equation for the current. Determine the current as a function of time for any initial current $i(0)$.
4. The voltage source is a constant e, the resistor is linear with v-i characteristic $v = iR$, $R > 0$, and the inductance is $L > 0$.
   a) Show that $\lim_{t\to\infty} i(t)$ exists and is the same for any initial current.
   b) If $e = 8.5$, for what values of R and L will $\lim_{t\to\infty} i(t) = 4.2$?
5. The resistor has v-i characteristic $v = i$ and $L = 1$. The voltage source is a 9-V battery that is shorted out after 10 s. That is,
$$e(t) = \begin{cases} 9 & 0 \le t < 10 \\ 0 & t \ge 10. \end{cases}$$
   The current is initially zero. Find $i(t)$ for $0 \le t \le 20$ and graph the result.
6. The resistor has v-i characteristic $v = 2i$ and $L = 1$. The voltage source is a 1.5-V battery that is shorted out for the first 10 s and then unshorted (by a switch) so that
$$e(t) = \begin{cases} 0 & 0 \le t < 10 \\ 1.5 & t \ge 10. \end{cases}$$
   The current is initially zero. Find $i(t)$ for $0 \le t \le 20$ and graph the result.
7. For the circuit in Figure 1.10.6, the capacitance is 0.5, the resistor is linear with v-i characteristic $v = 2i$, and the sinusoidal voltage source is given by $e(t) = 6 \sin t$. Find the charge on the capacitor as a function of t given that it was initially 1.

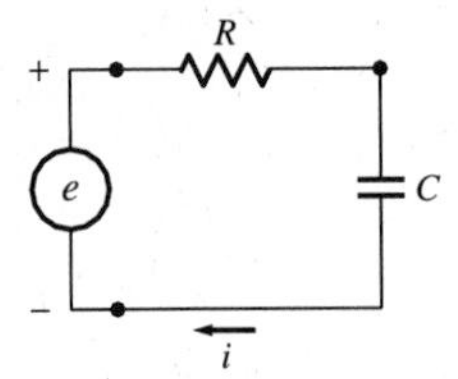

FIGURE 1.10.6

1.11 Mechanics II: Including Air Resistance

The mechanics problems to be considered here are those of straight-line motion of a constant mass m with a resistive force. As stated in Section 1.2.3,

Newton's second law of motion implies that

$$m\frac{d^2y}{dt^2} = F\left(y, \frac{dy}{dt}, t\right), \tag{1}$$

where F is the sum of the applied forces. If F depends on y, dy/dt, and t, then (1) is a second-order differential equation for the position $y(t)$. We have already solved some very elementary problems of this type in Section 1.2.3.

If the force does not depend on the position y,

$$m\frac{d^2y}{dt^2} = F\left(\frac{dy}{dt}, t\right), \tag{2}$$

then this second-order differential equation can be reduced to a first-order equation by solving for the velocity instead,

$$v = \frac{dy}{dt}. \tag{3}$$

In this case, (2) becomes a first-order equation

$$m\frac{dv}{dt} = F(v, t). \tag{4}$$

A common example in which an applied force depends on velocity but not position is air resistance. In general, if an object moves through a fluid (liquid or gas), the fluid exerts a force called the **resistance** on the body. If the fluid has approximately uniform density and the velocity is not too large, then resistance in the real world may often be approximated by the following law:

Resistance is proportional to the magnitude of the velocity and acts in a direction opposite to that of the velocity. (5)

This law is called **linear damping** and it states that the resistive force satisfies

$$\text{Resistive force} = -kv \tag{6}$$

with $k \geq 0$. The constant of proportionality depends on the shape of the body and the nature of the medium the object travels through. The dimensions of k are force/velocity = mass × acceleration/velocity = mass/time.

In the real world, determining the resistance can be quite complicated. A significant amount of effort goes into designing automobiles and airplanes in order to minimize the drag (air resistance). There can be other considerations such as lift (for aircraft and race cars) and cost. Much of the design process in the past relied on experimental wind tunnels, but now mathematical models of the fluid dynamic processes of flow around complicated three-dimensional objects can be analyzed on supercomputers at a tremendous savings in cost over building realistic models and testing them in wind tunnels (although wind tunnel tests are still very important). At high speeds, experiments show that linear damping is not valid, but in some cases the resistive force is instead proportional to the velocity squared.

Here we will analyze problems in which a mass is acted upon only by gravity and a linear resistive force. In this case, the first-order differential equation for the velocity becomes

$$m\frac{dv}{dt} = -mg - kv. \tag{7}$$

We note that (7) is a first-order linear differential equation with constant coefficients with a constant input, $-mg$. The coordinate system has been chosen so that positive y is upward.

EXAMPLE 1.11.1 Air Resistance

Consider a 4-g mass dropped from a height of 6 m. Assume that air resistance acts on the mass with constant of proportionality 12 g/s. Determine the velocity as a function of time.

SOLUTION The differential equation (7) is

$$4\frac{dv}{dt} = -4 \cdot 980 - 12v \qquad \text{or} \qquad \frac{dv}{dt} + 3v = -980. \tag{8}$$

Since this is a linear differential equation, the general solution is in the form of a particular solution plus an arbitrary multiple of a solution of the associated homogeneous equation. Since the input $-mg$ is a constant, we can guess that a particular solution is a constant A. Substituting A into the differential equation (8) for v, we find that $v_p = A = \frac{-980}{3}$. A solution of the associated homogeneous equation is seen to be $v_1 = e^{rt} = e^{-3t}$. Thus, the general solution of (8) is

$$v(t) = \frac{-980}{3} + ce^{-3t}. \tag{9}$$

Formula (9) can also be found by using an integrating factor of $u(t) = e^{3t}$ for (8).

The phrase "dropping an object" means that the initial velocity is zero, $v(0) = 0$. The constant c is then easily determined to be $c = \frac{980}{3}$. Thus,

$$v(t) = \frac{-980}{3} + \frac{980}{3}e^{-3t}. \tag{10}$$

◄

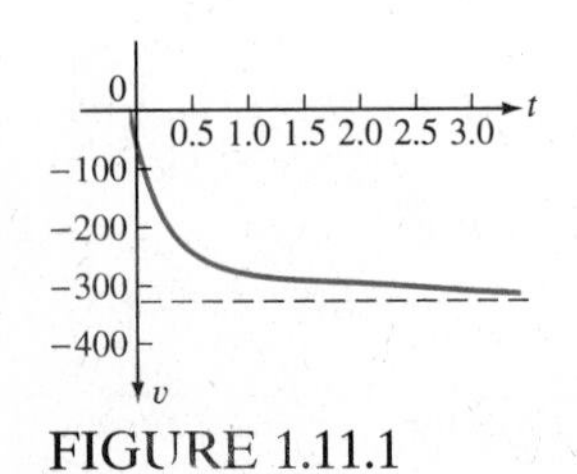

FIGURE 1.11.1

Graph of (10).

The velocity (10) is sketched in Figure 1.11.1. The figure shows that the speed continues to increase as time increases. Note that as $t \to \infty$, $e^{-3t} \to 0$, so that $v(t) \to \frac{-980}{3}$. This limit is referred to as the **terminal velocity**. This is the fastest a body will travel (from rest) if there is a linear resistive force. From (9), we see that the terminal velocity does not depend on the initial condition.

With gravity and no resistance, an object keeps accelerating, and its velocity gets larger and larger. There is no limit to how fast an object will move under gravitational acceleration without air resistance. However, with air resistance, an object has the terminal velocity

$$v_{\text{terminal}} = \frac{-mg}{k}, \tag{11}$$

derived from (7). Heavier objects have a larger terminal velocity. If the resistance were reduced, the terminal velocity would of course be larger. Employing a parachute increases the resistance and hence lowers the terminal velocity.

If the object above was really dropped from 6 m, the question could be how fast the object was going when it hit the ground and how close it was to its terminal velocity when it hit. We leave to the exercises the investigation of these questions.

Exercises

1. A mass of 20 g is dropped from an airplane flying horizontally. Air resistance acts according to (6) with a constant of proportionality of 10 g/s. Considering only vertical motion:

a) Find the velocity as a function of time.

b) Find the velocity after 10 s, assuming that the body has not hit the ground (or a bird).

c) If the gravitational force is assumed constant, what is the limiting velocity?

2. A weight of 64 lb is flung vertically up into the air from the earth's surface. At the instant it leaves the launcher, it has a velocity of 192 ft/s.

a) Ignoring air resistance, determine how long it takes for the object to reach its maximum altitude.

b) If air resistance acts according to (6) with a constant of proportionality of 4 slug/s, how long does it take for the object to reach its maximum altitude?

3. A mass of 70 g is to be ejected downward from a stationary helicopter. Air resistance acts according to (6) with the constant of proportionality of 7 g/s. At what velocity should the mass be ejected if it is to have a velocity of 12,600 cm/s after 5 s?

4. Suppose that an object of mass m is ejected vertically downward with velocity v_0 over the surface of a planet whose gravitational constant is G. The atmosphere exerts resistance on the body according to (6) with constant of proportionality r. (All parameters are in cgs units.)

a) Find the formula for $v(t)$.

b) Find the formula for the limiting velocity.

Assume that the object consists of a payload and a parachute. The payload is 80% of the total mass, and the parachute takes up the balance of the total mass. Assume also that the parachute accounts for essentially all the resistive force.

c) How much would the payload have to be reduced to cut the limiting velocity in half?

5. A 32-lb weight is dropping through a gas near the earth's surface. The resistance is proportional to the square of the velocity, with constant of proportionality 1. At time $t = 0$, the velocity is 1000 ft/s.

a) Find the velocity as a function of time.

b) Is there a limiting velocity? If there is, find it.

1.12 Orthogonal Trajectories

Many physical problems are described using a function $\phi(x, y)$ called a **potential** that represents a quantity such as height, temperature, or pressure. The curves $\phi(x, y) = c$ are called **equipotentials** or **level curves** because the potential is constant along these curves. In the case of height, the equipotentials are the familiar lines connecting points of equal altitude on a topographic map. For temperature, the equipotentials are usually called **isotherms**, and for pressure, they are called **isobars**. (See Figure 1.12.1.)

A potential often causes an action perpendicular to the equipotentials. These curves perpendicular to the equipotentials are usually called **flux** (or **stream**) **lines**. On a topographic map, the flux lines show the direction in which an object would roll downhill (at least initially). In the case of isotherms, the flux lines show the direction of heat flow.

Given the equipotentials, we can often find the flux lines. Since the flux lines are everywhere orthogonal (perpendicular) to the equipotentials, they are sometimes also called an **orthogonal family** of curves.

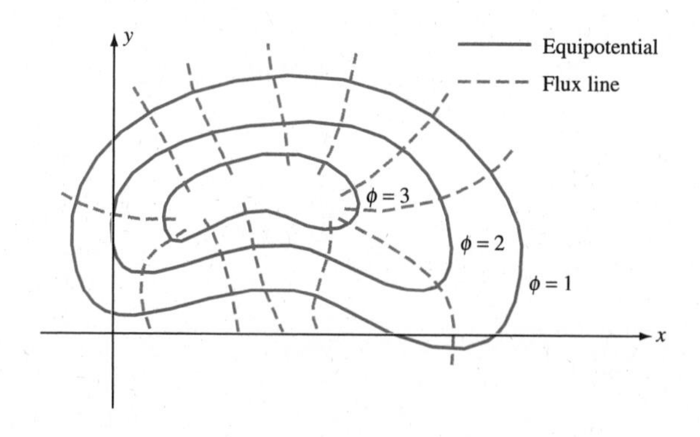

FIGURE 1.12.1

Equipotentials and flux lines.

Procedure for Calculating Flux Lines from Equipotentials

1. First write the family of curves (equipotentials) in the form

$$\phi(x, y) = c.$$

2. Differentiate with respect to x to get

$$\phi_x(x, y) + \phi_y(x, y)\frac{dy}{dx} = 0.$$

3. Solve for dy/dx:

$$\frac{dy}{dx} = -\frac{\phi_x(x, y)}{\phi_y(x, y)}. \tag{1}$$

Equation (1) gives the slope of the equipotentials at (x, y). Since the flux lines are to be orthogonal to the equipotentials, the flux slopes will be the negative reciprocals of the equipotential slopes. Thus the flux lines satisfy the differential equation

$$\frac{dy}{dx} = \frac{-1}{-\phi_x(x, y)/\phi_y(x, y)} = \frac{\phi_y(x, y)}{\phi_x(x, y)}. \tag{2}$$

4. Solve Eq. (2).

EXAMPLE 1.12.1 Orthogonal Family for $\phi(x, y) = c$

The expression $x^2 = c - 2y^2$ defines a family of curves (equipotentials). Find the orthogonal family (flux lines).

Solution First let us write the family of curves in the form $x^2 + 2y^2 = c$. We next differentiate $x^2 + 2y^2 = c$ with respect to x:

$$2x + 4y\frac{dy}{dx} = 0.$$

Solving for dy/dx yields

$$\frac{dy}{dx} = -\frac{x}{2y}.$$

The slope of the orthogonal family at (x, y) is thus given by

$$\frac{dy}{dx} = -\frac{1}{-x/2y} = \frac{2y}{x}.$$

Solving this differential equation by separation of variables, we get

$$\int \frac{1}{y}\, dy = \int \frac{2}{x}\, dx,$$

so that

$$\ln|y| = 2\ln|x| + c_1.$$

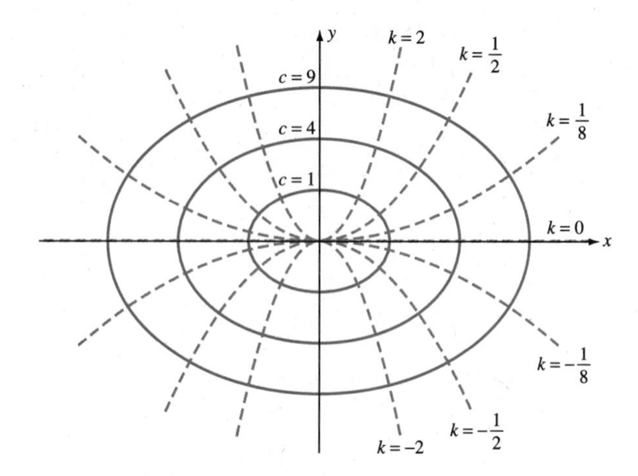

FIGURE 1.12.2

The families of curves $x^2 + 2y^2 = c$ and $y = kx^2$.

We now exponentiate both sides to arrive at the orthogonal family of curves

$$y = kx^2, \qquad k = \pm e^{c_1}.$$

Note that $k = 0$ also gives a solution. Figure 1.12.2 shows both families of curves. ◀

Sometimes the original family is not of the form $\phi(x, y) = c$ but rather $F(x, y, c) = 0$. That is, the parameter c is not solved for. We still need to rewrite the family as $\phi(x, y) = c$. However, if ϕ is complicated to differentiate, we may differentiate $F(x, y, c) = 0$ with respect to x to get

$$F_x(x, y, c) + F_y(x, y, c)y' = 0$$

and then substitute ϕ for c.

EXAMPLE 1.12.2 Orthogonal Family for $F(x, y, c) = 0$

Find the differential equation for the orthogonal family to $x^3 + 3y^2 = x^2c$.

Solution Differentiating gives $3x^2 + 6yy' = 2xc$. But

$$c = \frac{x^3 + 3y^2}{x^2}.$$

Thus

$$y' = \frac{1}{6y}(2xc - 3x^2) = \frac{6y^2 - x^3}{6xy}$$

and the orthogonal family satisfies

$$y' = -\frac{6xy}{6y^2 - x^3}.$$ ◀

Exercises

For each family of curves (equipotentials) in Exercises 1 through 18, compute the orthogonal family (flux lines). Sketch both families for Exercises 1 through 3.

1. $y = x + c$

2. $y = cx$

3. $yx = c$

4. $y^2 = c(x + y)$

5. $y = x^c$

6. $y = (x + c)^3$

7. $y = x^2 + c$

8. $y^3(x + 1) = c$

9. $x = (y - c)^2$

10. $y = e^{cx}$

11. $y = \tan(x + c)$

12. $x^{1/5} + y^{1/5} = c$

13. $y = c \cos x$

14. $y = \cos x + c$

15. $y = cx^2$

16. $e^y - e^x = c$

17. $y^3 + x^2 = c$

18. $\tan y + \tan x = c$

1.13 Exact Equations

Suppose that a first-order differential equation is written in differential form,

$$M(x, y)\,dx + N(x, y)\,dy = 0. \tag{1}$$

For example,

$$\frac{dy}{dx} = -\frac{(y + 2x)}{(x + 4y)}$$

could be written as

$$(y + 2x)\,dx + (x + 4y)\,dy = 0. \tag{2}$$

If there is a function $F(x, y)$ such that

$$dF = M(x, y)\,dx + N(x, y)\,dy,$$

then the differential equation $M\,dx + N\,dy = 0$ would become $dF = 0$, and the general solution would be $F = c$. Such differential equations are called **exact**, since $M\,dx + N\,dy$ is exactly the differential of a function F.

In (2), if $F(x, y) = x^2 + xy + 2y^2$, then

$$dF = \frac{\partial F}{\partial x}\,dx + \frac{\partial F}{\partial y}\,dy = (2x + y)\,dx + (x + 4y)\,dy,$$

so that $(y + 2x)\,dx + (x + 4y)\,dy = 0$ is the same as

$$d(x^2 + xy + 2y^2) = 0.$$

The general solution of (2) is then $F = c$ or

$$x^2 + xy + 2y^2 = c.$$

To develop a method based on this approach, we must address two issues:

1. How do we know when such an $F(x, y)$ exists?
2. If $F(x, y)$ exists, how can we find it?

Notation Using the notation from calculus, we let F_x denote $\partial F/\partial x$, and F_y denote $\partial F/\partial y$. In addition, we let

$$F_{xx} = \frac{\partial^2 F}{\partial x^2}, \qquad F_{yy} = \frac{\partial^2 F}{\partial y^2}, \qquad F_{xy} = \frac{\partial^2 F}{\partial y \partial x}, \qquad F_{yx} = \frac{\partial^2 F}{\partial x \partial y}.$$

Suppose that $dF = M\,dx + N\,dy$. Since from calculus, $dF = F_x\,dx + F_y\,dy$, it follows that

$$F_x = M, \qquad F_y = N. \tag{3}$$

If F has continuous second partials, then from calculus we also know that $F_{xy} = F_{yx}$. But then $(F_x)_y = M_y$ and $(F_y)_x = N_x$ from (3). Thus if an F having continuous second partials exists so that $dF = M\,dx + N\,dy$, then $M_y = N_x$. The fact is that this condition is not only necessary but sufficient.

■ **THEOREM 1.13.1 Conditions for Exactness**

Let $M(x, y)$, $N(x, y)$ and their first partials be continuous on a rectangle $a \le x \le b$, $c \le y \le d$. Then there is an $F(x, y)$ (defined on the same rectangle) such that $dF = M(x, y)\,dx + N(x, y)\,dy$ if and only if $M_y = N_x$. ■

In Eq. (2),

$$\frac{\partial M}{\partial y} = \frac{\partial(y + 2x)}{\partial y} = 1 \quad \text{and} \quad \frac{\partial N}{\partial x} = \frac{\partial(x + 4y)}{\partial x} = 1,$$

so that $M_y = N_x$. Theorem 1.13.1 then guarantees that an $F(x, y)$ exists such that $dF = M\,dx + N\,dy$.

Once we know that there is an $F(x, y)$, we may get it by solving (3). For (2), Eqs. (3) are

$$F_x = M = y + 2x, \tag{4}$$

$$F_y = N = x + 4y. \tag{5}$$

Antidifferentiating (4) with respect to x and thinking of x, y as independent variables give

$$F(x, y) = \int F_x\,dx = \int (y + 2x)\,dx = yx + x^2 + h(y), \tag{6}$$

where $h(y)$ is an unknown function of y. We need to introduce an unknown function of y because $F_x = G_x$ for two functions, $F(x, y)$ and $G(x, y)$, means that they differ by a function that depends only on y. [Similarly, $F_y = G_y$ would mean that $F(x, y)$ and $G(x, y)$ differ by a function of x only.] Substituting expression (6), that is, $F = yx + x^2 + h(y)$, back into Eq. (5), that is, $F_y = x + 4y$, in order to find $h(y)$, we get

$$\frac{\partial F}{\partial y} = \frac{\partial(yx + x^2 + h(y))}{\partial y} = x + 4y$$

or

$$x + h'(y) = x + 4y, \qquad \text{so that} \qquad h'(y) = 4y.$$

Hence $h(y) = 2y^2$ and $F = yx + x^2 + 2y^2$, as noted earlier. We do not need to add a constant c_1 when we find $h(y)$ from $h'(y)$, since the solution of (2) is obtained by setting $F = c$, and $F + c_1 = c$ would be an equivalent answer.

This procedure may be summarized as follows.

Summary of Method for Solution of Exact Equations

1. Write the differential equation in the form $M(x, y)\,dx + N(x, y)\,dy = 0$.
2. Compute M_y and N_x. If $M_y \neq N_x$, the equation is not exact and this technique will not work.
3. Antidifferentiate either $F_x = M$ with respect to x or $F_y = N$ with respect to y. Antidifferentiating will introduce an arbitrary function of the other variable.
4. Take the result for F from step 3 and substitute for F in the other equation from (3) to find the arbitrary function.
5. The solution is $F(x, y) = c$.

EXAMPLE 1.13.1 **Exact Equation**

Solve $(2x + 1 + 2xy)\,dx + (x^2 + 4y^3)\,dy = 0$.

SOLUTION In this problem, $M = 2x + 1 + 2xy$ and $N = x^2 + 4y^3$. Since $M_y = N_x = 2x$, the equation is exact. The partial differential equations for $F(x, y)$ are

$$F_x = M = 2x + 1 + 2xy,$$

$$F_y = N = x^2 + 4y^3.$$

Either equation can be antidifferentiated. We shall antidifferentiate the second one:

$$F = \int F_y\,dy = \int (x^2 + 4y^3)\,dy = x^2y + y^4 + k(x),$$

where $k(x)$ is an unknown function of x. We now substitute this expression for F in the other equation ($F_x = 2x + 1 + 2xy$) in order to find $k(x)$:

$$\frac{\partial[x^2y + y^4 + k(x)]}{\partial x} = 2x + 1 + 2xy.$$

Then

$$2xy + k'(x) = 2x + 1 + 2xy$$

or

$$k'(x) = 2x + 1 \qquad \text{and} \qquad k(x) = x^2 + x.$$

Thus $F(x, y) = x^2y + y^4 + x^2 + x$ and the general solution is

$$F(x, y) = x^2y + y^4 + x^2 + x = c.$$ ◀

A common mistake is to give $F + c$ as an answer instead of $F = c$. Note that $F + c$ does not give any relationship between x and y, and hence does not give us a function of x as an answer.

The above technique usually produces solutions in implicit form. It may not be possible to algebraically solve for y in terms of x.

In some problems the choice of which equation to antidifferentiate can greatly affect the ease of solution.

EXAMPLE 1.13.2 Choice of Integral

Solve the initial value problem

$$(3x^2 + y^3e^y)\,dx + (3xy^2e^y + xy^3e^y + 3y^2)\,dy = 0, \qquad y(2) = 0.$$

Solution In this example, $M = 3x^2 + y^3e^y$ and $N = 3xy^2e^y + xy^3e^y + 3y^2$. Since $M_y = 3y^2e^y + y^3e^y = N_x$, the differential equation is exact. The partial differential equations for $F(x, y)$ are

$$F_x = M = 3x^2 + y^3e^y,$$

$$F_y = N = 3xy^2e^y + xy^3e^y + 3y^2.$$

We have a choice now of evaluating

$$F = \int F_x\,dx = \int (3x^2 + y^3e^y)\,dx$$

or

$$F = \int F_y\,dy = \int (3xy^2e^y + xy^3e^y + 3y^2)\,dy.$$

Note that $\int F_x\,dx$ is quite easy to carry out, whereas $\int F_y\,dy$ is considerably more complicated and requires either integration by parts or the use of tables. Therefore we shall compute the first integral,

$$F = \int F_x\,dx = \int (3x^2 + y^3e^y)\,dx = x^3 + y^3e^yx + h(y),$$

where $h(y)$ is an unknown function of y. Substituting this formula for F into $F_y = N$ in order to find $h(y)$, we have

$$\frac{\partial[x^3 + y^3e^yx + h(y)]}{\partial y} = N = 3xy^2e^y + xy^3e^y + 3y^2,$$

or

$$3y^2e^yx + y^3e^yx + h'(y) = 3xy^2e^y + xy^3e^y + 3y^2,$$

so that

$$h'(y) = 3y^2 \quad \text{and} \quad h(y) = y^3.$$

Thus $F(x, y) = x^3 + y^3e^yx + y^3$, and the general solution is

$$F = x^3 + y^3e^yx + y^3 = c.$$

Applying the initial condition $x = 2$, $y = 0$ gives $c = 8$. ◀

If we start with a first-order equation, for example,

$$\frac{dy}{dx} = \frac{-y^2}{x^2 + 1}, \tag{7}$$

there are many ways of rewriting the equation in the form $M\,dx + N\,dy = 0$. For (7), some possibilities are

$$(x^2 + 1)\,dy + y^2\,dx = 0, \tag{8}$$

$$y^{-2}\,dy + (x^2 + 1)^{-1}\,dx = 0, \tag{9}$$

and

$$dy + \frac{y^2}{x^2 + 1}\,dx = 0. \tag{10}$$

One way of rewriting the equation may be exact and others may not be exact. For example, (9) is exact, whereas (8) and (10) are not. A method of turning some nonexact equations into exact equations is provided in the next section.

That we can treat x and y as independent variables while trying to express y as a function of x may be confusing at first. The point to remember is that

$$M(x, y)\,dx + N(x, y)\,dy = 0$$

is a differential equation. The variables x and y are related. On the other hand,

$$M(x, y)\,dx + N(x, y)\,dy$$

is a differential, and x and y are considered independent. In order to obtain $F(x, y)$, we are using $dF = M\,dx + N\,dy$, not $dF = M\,dx + N\,dy = 0$. Thus, when we compute F, we work with x and y that are independent variables. After we find F, we set $dF = 0$ and get the relationship $F(x, y) = c$ between x and y.

Exercises

In Exercises 1 through 20, determine whether the differential equation is exact. If it is exact, solve it.

1. $(2x + y)\,dx + (2y + x)\,dy = 0$

2. $(x + 3y)\,dx + (y + 2x)\,dy = 0$

3. $x^{-1}y\,dx + (\ln x + 3y^2)\,dy = 0$

4. $(1 + y^2)\,dx + (2xy + 4y)\,dy = 0$

5. $(1 + y^2)\,dx + (3xy + 4y)\,dy = 0$

6. $(3x^2 + 2xy + y^3)\,dx + (x^2 + 3xy^2 + \cos y)\,dy = 0$, $y(0) = 0$

7. $(-2x^{-3} + 2xe^{x^2}y)\,dx + (e^{x^2} + 2y)\,dy = 0$

8. $(2r\theta + 1)\,dr + (r^2 + 1)\,d\theta = 0$

9. $(2u + \theta + 2u\theta^2)\,du + (u + 2u^2\theta + 3\theta^2)\,d\theta = 0$

10. $(2x + y^5e^{3y}\sin y)\,dx + (5xy^4e^{3y}\sin y + 3xy^5e^{3y}\sin y + xy^5e^{3y}\cos y + 2y)\,dy = 0$, $y(0) = 2$

11. $x\,dx + (x + y)\,dy = 0$

12. $x^{-1}\,dx + y^{-1}\,dy = 0$

13. $(2x + 3x^2y^2)\,dx + (2x^3y - 5y^4)\,dy = 0$

14. $(\sin y + e^x)\,dx + (x\cos y - 2y)\,dy = 0$

15. $(2xe^{xy} + x^2ye^{xy})\,dx + (x^3e^{xy} + 3y^2)\,dy = 0$

16. $\dfrac{1}{(x + y)^2 + 1}\,dx + \left[1 + \dfrac{1}{(x + y)^2 + 1}\right]dy = 0$

17. $\dfrac{1}{x + y}\,dx + \left(\dfrac{1}{x + y} + 3y^2\right)dy = 0$

18. $\left[3x^2 + \dfrac{2x}{(x^2 + y^2 + 1)^2}\right]dx + \left[\dfrac{2y}{(x^2 + y^2 + 1)^2} - 4y^3\right]dy = 0$

19. $\cos(x + y)\,dx + [\cos(x + y) - 2y\sin(y^2)]\,dy = 0$

20. $\dfrac{x + 2y}{x + y}\,dy + \left[\ln(x + y) + \dfrac{y}{x + y}\right]dy = 0$

21. Suppose that a differential equation can be solved by separation of variables so that

$$\frac{dy}{dx} = g(x)h(y).$$

This equation is equivalent to

$$\frac{1}{h(y)}\,dy - g(x)\,dx = 0.$$

Show that $[1/h(y)]\,dy - g(x)\,dx = 0$ is an exact differential equation and compare the solution obtained by separation of variables to that arrived at by the exact method.

Exercises 22 through 26 develop one application of exact differential equations. Exercises 22 and 23 review some concepts from calculus.

Exact differentials also arise in calculus when we evaluate line integrals. We recall from calculus that if the differential $M\,dx + N\,dy$ is exact in a rectangular region R and (x_0, y_0) and (x_1, y_1) are two points in R, then

$$\int_{(x_0, y_0)}^{(x_1, y_1)} M(x, y)\,dx + N(x, y)\,dy = \int_{(x_0, y_0)}^{(x_1, y_1)} dF = F(x_1, y_1) - F(x_0, y_0). \qquad (11)$$

One application of line integrals occurs in the computation of work done. For example, let us suppose that at each point (x, y) in R, there is a force that in vector form is written as $[M(x, y), N(x, y)]$. Suppose that $M\,dx + N\,dy$ is an exact differential and M and N have continuous first partials in R. We also assume that the points (x_0, y_0) and (x_1, y_1) are in R.

22. Show that the work in going from (x_0, y_0) to (x_1, y_1) is independent of the path taken as long as the path stays in R and that the work done is given by (11).

23. Show that we can move through R from (x_0, y_0) to (x_1, y_1) without any work being done during any part of the trip if and only if we move along a solution of $M\,dx + N\,dy = 0$.

24. The function $F(x, y)$ is sometimes called a **potential** and the curves $F(x, y) = c$ are called **level curves** or **equipotentials**. In order to have the maximum work done, our motion must be perpendicular to the curves $F(x, y) = c$. Show that at any point (x, y) in R where the solutions of

$$-F_y\,dx + F_x\,dy = 0 \tag{12}$$

exist, they are perpendicular (have perpendicular tangents) to the curves $F = c$. The solutions of (12) and $F = c$ are sometimes called **orthogonal families** (Section 1.12).

25. Show that (12) is exact if and only if F satisfies the partial differential equation

$$F_{xx} + F_{yy} = 0. \tag{13}$$

This is **Laplace's equation**.

26. The solutions of (13) are called **harmonic functions**. Verify that $F(x, y) = x^2 - y^2$ is a harmonic function. Solve (12) for this F and sketch the solutions of (12) and $F = c$ on the same axes.

1.14 Integrating Factors for Exact Equations

As we pointed out in Section 1.13, there are many ways of writing $dy/dx = f(x, y)$ in the form

$$M(x, y)\,dx + N(x, y)\,dy = 0,$$

most of which will not be exact. Since it is impossible to check all these various ways to see which are exact, we shall instead show how to make some nonexact equations exact.

In solving first-order linear equations, we multiplied the equation by a function that turned the left-hand side of $dy/dx + p(x)y = q(x)$ into a derivative. Perhaps the same idea would work here. A function $u(x, y)$ is an **integrating factor** for the differential equation

$$M(x, y)\,dx + N(x, y)\,dy = 0$$

if the equation obtained by multiplying by $u(x, y)$,

$$u(x, y)M(x, y)\,dx + u(x, y)N(x, y)\,dy = 0,$$

is exact. In general, finding an integrating factor may be difficult. However, in the two special cases when the integrating factor depends just on x or just on y, it is possible to compute it with relative ease.

To see how the integrating factor u can be computed in these special cases, let us suppose that

$$M(x, y)\,dx + N(x, y)\,dy = 0$$

has an integrating factor that depends just on x, say $u(x)$. Then

$$u(x)M(x, y)\,dx + u(x)N(x, y)\,dy = 0 \tag{1}$$

is exact. But $u(x)M(x, y)\,dx + u(x)N(x, y)\,dy$ is exact if and only if

$$\frac{\partial[u(x)M(x, y)]}{\partial y} = \frac{\partial[u(x)N(x, y)]}{\partial x},$$

that is,

$$u(x)M_y(x, y) = u'(x)N(x, y) + u(x)N_x(x, y).$$

But then

$$\frac{u'(x)}{u(x)} = \frac{M_y(x, y) - N_x(x, y)}{N(x, y)}. \tag{2}$$

Since the left-hand side of (2) depends only on x, the right-hand side must also depend only on x. Let $Q(x)$ be the right-hand side of (2). Then $u'/u = Q$, so that

$$\ln|u(x)| = \int \frac{u'(x)}{u(x)}\,dx = \int Q(x)\,dx.$$

Thus the integrating factor is

$$u(x) = \exp\left[\int Q(x)\,dx\right].$$

(Since any constant multiple of an integrating factor is an integrating factor, we may omit arbitrary constants and drop the absolute value when finding u.) A similar argument works if there is an integrating factor that is a function of y only. These ideas may be summarized as follows.

Integrating Factor Method

For the differential equation $M(x, y)\,dx + N(x, y)\,dy = 0$, first compute M_y, N_x.

1a. If $(M_y - N_x)/N$ cannot be expressed as a function of x only, then we do not have an integrating factor that is a function of x only. If $(M_y - N_x)/N = Q(x)$ is a function of x, then

$$u(x) = \exp\left[\int Q(x)\,dx\right]$$

is an integrating factor.

1b. If $(N_x - M_y)/M$ cannot be expressed as a function of y only, then we do not have an integrating factor that is a function of y only. If $(N_x - M_y)/M = R(y)$ is a function of y, then

$$u(y) = \exp\left[\int R(y)\,dy\right]$$

is an integrating factor.

2. Multiply $M\,dx + N\,dy = 0$ by the integrating factor.
3. Solve the exact equation $uM\,dx + uN\,dy = 0$.

EXAMPLE 1.14.1 Integrating Factor

Solve the differential equation

$$(3y^2 + 4x)\,dx + (2yx)\,dy = 0. \tag{3}$$

SOLUTION In this example,

$$M = 3y^2 + 4x, \qquad N = 2yx,$$

so that $M_y = 6y$, $N_x = 2y$, and the differential equation is not exact. However,

$$\frac{M_y - N_x}{N} = \frac{6y - 2y}{2yx} = \frac{2}{x}$$

is a function of x. Thus there is an integrating factor

$$u(x) = \exp\left[\int \frac{2}{x}\,dx\right] = \exp\left[2\ln|x|\right] = x^2.$$

Multiplying the differential equation by x^2 gives the new differential equation

$$(3x^2y^2 + 4x^3)\,dx + (2yx^3)\,dy = 0, \tag{4}$$

which is exact. For (4) we have $M = 3x^2y^2 + 4x^3$ and $N = 2yx^3$. Note that $M_y = 6x^2y = N_x$ as expected. The partial differential equations for F are

$$F_x = M = 3x^2y^2 + 4x^3,$$
$$F_y = N = 2yx^3.$$

From the second equation,

$$F = \int F_y\,dy = \int 2yx^3\,dy = y^2x^3 + h(x),$$

where $h(x)$ is an unknown function of x. Substituting $F = y^2x^3 + h(x)$ for F in $F_x = 3x^2y^2 + 4x^3$, we get

$$\frac{\partial[y^2x^3 + h(x)]}{\partial x} = 3x^2y^2 + 4x^3$$

or

$$3y^2x^2 + h'(x) = 3x^2y^2 + 4x^3,$$

so that $h'(x) = 4x^3$ and $h(x) = x^4$. Thus $F = y^2x^3 + x^4$, and the solution of the original differential equation is

$$F = y^2x^3 + x^4 = c.$$

Exercises

In Exercises 1 through 14, solve the differential equation.

1. $(2xy^3 + y^4)\,dx + (xy^3 - 2)\,dy = 0$
2. $(2y^3 + 2)\,dx + (3xy^2)\,dy = 0$
3. $(-2 + x^3y)\,dx + (x^4 + 6yx^3)\,dy = 0$
4. $(3x^2y^{-1} + y)\,dx + (2x + 4y^2)\,dy = 0$
5. $(y + e^{-x})\,dx + dy = 0$
6. $[y\cos(x + y)\sec y - 1]\,dx + \sec y[\sin(x + y) + y\cos(x + y) + x\sin y]\,dy = 0$
7. $2x\,dx + (6ye^y - x^2)\,dy = 0$
8. $(2y^2 + 1 + 2x)\,dx + 2y\,dy = 0$
9. $3x^2y\,dx + (3x^3 + 5y^2)\,dy = 0$
10. $(y + 2e^x)\,dx + (1 + e^{-x})\,dy = 0$
11. $4x^3y\,dx + (5x^4 + 6y)\,dy = 0$
12. $(e^y + e^{-x})\,dx + (e^y + 2ye^{-x})\,dy = 0$
13. $(3x^2y + 2xy^{-2})\,dx + (3x^3 + 2y^{-1})\,dy = 0$
14. $\left[\dfrac{x^2 + x}{x^2 + 2xy + y^2 + 1}\right]dx + \left[\dfrac{x^2 + x}{(x^2 + 2xy + y^2 + 1)} + 3x^2y^2 + 3xy^2\right]dy = 0$
15. Verify that if the linear differential equation $dy/dx + p(x)y = q(x)$ is rewritten as
$$[p(x)y - q(x)]\,dx + dy = 0,$$
then the method of this section and that of Section 1.6.3 produce the same function of x as an integrating factor. Compare the form of the solution obtained by the two methods.
16. Suppose $M\,dx + N\,dy$ is exact, and M and N have continuous first partials and are nonzero. Show that if u is a function of x or y only and $uM\,dx + uN\,dy$ is exact, then u is constant.
17. Show that
$$(y + 2x)\,dx + (x + 2y)\,dy = 0$$
is exact. Also show that if we multiply the given equation by
$$u(x, y) = xy + y^2 + x^2,$$
then the new differential equation is still exact. Compare this example to the conclusion of Exercise 16.
18. Derive the facts about the integrating factor in terms of y in statement 1b of the integrating factor method.

19. Show that

$$(y^2 + 1)\,dx + 2y\,dy = 0$$

has integrating factors $u(x)$ and $\tilde{u}(y)$. Solve the differential equation both ways and show that the answers are equivalent. (Note that this differential equation can also be solved by separation of variables.)

In Exercises 20 through 23, there is an integrating factor of the form $u(x, y) = x^r y^s$. Find r and s and solve the differential equation. [Use $(x^r y^s M)_y = (x^r y^s N)_x$ to find r and s.]

20. $6y^5\,dx + (7xy^4 + 3x^{-5})\,dy = 0$

21. $(2y + 3xy^2)\,dx + (3x + 4x^2y)\,dy = 0$

22. $(3 + 4xy^{-1})\,dx + (-2xy^{-1} - 3x^2y^{-2})\,dy = 0$

23. $(-3x^{-1} - 2y^4)\,dx + (-3y^{-1} + xy^3)\,dy = 0$

CHAPTER

2

Linear Higher-Order Differential Equations

2.1 Introduction and Basic Theory

Chapter 1 discussed first-order equations and gave several examples of their properties, solution, and application. Many applications, however, require second- or higher-order derivatives. Equations involving Newton's law, as discussed in Sections 1.2 and 1.11, may invoke position x, velocity dx/dt, and acceleration d^2x/dt^2. Similarly, as discussed in Section 2.12, the equations describing an RLC circuit may involve charge q, current dq/dt, and the current's rate of change d^2q/dt^2.

Part of this section briefly discusses some of the basic theory for second-order ordinary differential equations. The rest of the chapter will be devoted to second-order (or nth order if Sections 2.4, 2.7, and 2.15 are included) linear equations and their applications.

Both $y = x + 1$ and $y = 1$ are solutions of the second-order differential equation with one initial condition

$$\frac{d^2y}{dx^2} = 0, \qquad y(0) = 1. \tag{1}$$

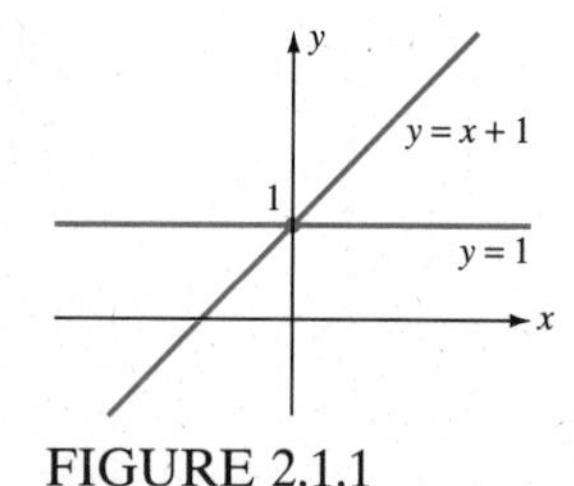

FIGURE 2.1.1

Two solutions of $y'' = 0$ satisfying $y(0) = 1$.

Thus, in contrast to first-order equations, solutions for second-order differential equations are not uniquely determined by their value at a point, and the graphs of distinct solutions may intersect (Figure 2.1.1).

Consider, for a moment, an arrow shot vertically upward. To determine the path of the arrow, we need to know not only the initial position (where the bow is) but also the initial velocity, as discussed in Sections 1.2 and 1.11. The next theorem states that if we have decided on the differential equation (friction, effect of gravity, etc.) to be used and it is a "nice" second-order equation, then initial position and velocity will uniquely determine the solutions. We use the notation

$$y' = \frac{dy}{dx}, \qquad y'' = \frac{d^2y}{dx^2}.$$

■ **THEOREM 2.1.1 Basic Existence and Uniqueness Theorem for Second-Order Ordinary Differential Equations**

Let $f(x, y, z)$ be a function of three variables such that f, f_y, f_z are continuous at and near the point (x_0, y_0, z_0). Then there is a nontrivial interval I containing x_0 such that the differential equation

$$y'' = f(x, y, y') \tag{2a}$$

has a unique solution $y(x)$ defined on I that satisfies the initial conditions

$$y(x_0) = y_0, \qquad y'(x_0) = z_0. \tag{2b}$$

■

EXAMPLE 2.1.1 Existence and Uniqueness

For the initial value problem

$$y'' = xy \sin(yy'), \qquad y(0) = 1, \qquad y'(0) = 2, \tag{3}$$

we have $f(x, y, z) = xy \sin(yz)$. Since f, f_y, f_z are continuous at and near $(0, 1, 2)$, there is a solution to (3) defined on an interval containing $x_0 = 0$, and the solution is unique. ◀

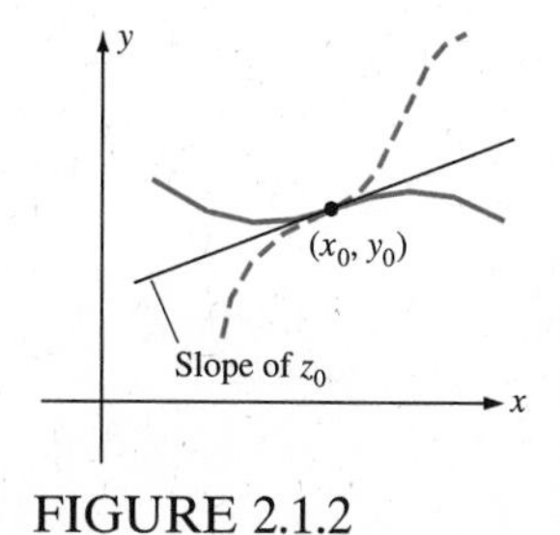

FIGURE 2.1.2

Since the solution of a second-order equation can be found to satisfy two initial conditions (2b), we expect to see two arbitrary constants in solving second-order differential equations.

Graphically, the Existence and Uniqueness Theorem says that if the assumptions hold at (x_0, y_0, z_0), then two solutions can cross at (x_0, y_0) as in Figure 2.1.1, but they cannot also be tangent at (x_0, y_0) as in Figure 2.1.2. The theorem says even more. It says that there will be an infinite number of solutions through (x_0, y_0), one with each possible slope z_0, for which f, f_y, f_z are continuous at (x_0, y_0, z_0). For example, the solutions of Eq. (1) are $y = cx + 1$, and the slope $y' = c$ is arbitrary.

2.1.1 Second-Order Linear Equations

Many physical problems are modeled by second-order differential equations. Some of these equations are linear, but many are not. For many applications involving nonlinear equations, if there is a "moderate" change in position and velocity (or charge and current), the nonlinear equation may be approximated by a linear differential equation. The remainder of this chapter considers only linear differential equations.

In this and the next section, we develop some of the basic theory for second-order linear differential equations. This theory will be extensively used in the solution techniques developed later in this chapter.

The general second-order linear differential equation is of the form

$$\frac{d^2y}{dx^2} + p(x)\frac{dy}{dx} + q(x)y = f(x), \tag{4}$$

where p, q, f are functions of x. The functions p, q are called **coefficients.** The function f is called the **forcing function** or **input**.

The following theorem summarizes several key facts about second-order linear equations.

THEOREM 2.1.2 Basic Existence and Uniqueness Theorem for Second-Order Linear Differential Equations

Suppose that $p(x)$, $q(x)$, and $f(x)$ are continuous functions on the nontrivial interval I. Let x_0 be any number in the interval I, and let y_0, v_0 be any initial conditions. Then there exists a unique solution to the initial value problem

$$y'' + py' + qy = f, \qquad y(x_0) = y_0, \qquad y'(x_0) = v_0, \tag{5}$$

and the solution is defined on all of the interval I. ■

Note that the solutions exist not only near x_0, as promised by Theorem 2.1.1, but over the entire interval I on which p, q, f are continuous. This existence on all of I is a special property of linear equations that not all differential equations share.

EXAMPLE 2.1.2 Continuous Coefficients

The coefficients 1, x^2, x, and the forcing function e^x of

$$y'' + x^2y' + xy = e^x$$

are continuous for all x, so that we take $I = (-\infty, \infty)$. Therefore the solutions will be defined and continuously differentiable for all x. ◄

EXAMPLE 2.1.3 Discontinuous Coefficients

Consider

$$(x-2)y'' + xy' + 6y = (x-2)^3. \tag{6}$$

Dividing by $(x-2)$, we rewrite (6) in the form (2):

$$y'' + \frac{x}{x-2}y' + \frac{6}{x-2}y = (x-2)^2. \tag{7}$$

The coefficients $x/(x-2)$ and $6/(x-2)$ are continuous everywhere except at $x = 2$. The forcing function $(x-2)^2$ is continuous everywhere. Thus we are guaranteed that solutions will exist on $(-\infty, 2)$ and $(2, \infty)$. What happens at $x = 2$ depends on the differential equation and will be discussed in Chapter 4, on series solutions. ◀

Exercises

In Exercises 1 through 12, determine for which points (x_0, y_0) and slopes z_0 the Existence and Uniqueness Theorem (Theorem 2.1.1) guarantees a unique solution through (x_0, y_0) with $y'(x_0) = z_0$.

1. $y'' = (y')^2 + 1$

2. $y'' = (x + y + y')^{1/3}$

3. $y'' = x^{1/3}(y + y')^2$

4. $y'' = y'/y^2$

5. $y'' = \dfrac{y}{(y')^2 + y^2}$

6. $(y' + 1)y'' = 1 + y^2$

7. $y'' = \sin y$

8. $y'' = \tan(y' + xy)$

9. $xy'' = y(y' + 1)^{1/5}$

10. $y'y'' = \sin x$

11. $y(y' + 1)y'' = 1$

12. $y'' = xyy'$

In Exercises 13 through 20, find all solutions of the differential equation and sketch several solutions.

13. $y'' = 2, \quad y(0) = 0$

14. $y'' = x/6, \quad y'(0) = 0$

15. $y'' = 2, \quad y'(0) = 0$

16. $y'' = x/6, \quad y(0) = 0$

17. $y'' = x + 1, \quad y(0) = 1$

18. $y'' = -x^2, \quad y(0) = 1$

19. $y'' = x + 1, \quad y'(0) = -1$

20. $y'' = -x^2, \quad y'(0) = 1$

21. Verify that $y = 0$ and $y = x^2$ are both solutions of the initial value problem

$$yy'' = xy', \qquad y(0) = 0, \qquad y'(0) = 0.$$

Explain why this is not a contradiction of Theorem 2.1.1.

In Chapter 1 we saw that, if f and f_y were continuous in the xy plane, then the graphs of solutions of $y' = f(x, y)$ filled up the xy plane. Exercises 22 and 23 develop the second-order analog of this fact.

22. Suppose that f, f_y, f_z are continuous everywhere in three-dimensional xyz space. Each solution $y(x)$ of $y'' = f(x, y, y')$ defines a curve $(x, y(x), y'(x))$ in three-dimensional space. The curve is parametrized by x. Show that no two of these curves ever intersect.

23. (Continuation of Exercise 22.) Show that these curves fill up three-dimensional space.

2.2 General Solution of Second-Order Linear Differential Equations

In Section 1.6, we analyzed first-order linear differential equations

$$\frac{dy}{dx} + p(x)y = f(x). \tag{1}$$

We showed that the general solution of (1) is always in the form

$$y = y_p + cy_1, \tag{2}$$

where y_p is a particular solution of (1), y_1 is a nonzero solution of the associated homogeneous equation $dy/dx + p(x)y = 0$, and c is an arbitrary constant.

A similar property holds for second-order linear differential equations:

$$\frac{d^2y}{dx^2} + p(x)\frac{dy}{dx} + q(x)y = f(x). \tag{3}$$

If the input $f(x) = 0$, then the linear differential equation is called **homogeneous**. We will show:

Form of Solution of Second-Order Linear Differential Equation

The general solution of

$$\frac{d^2y}{dx^2} + p(x)\frac{dy}{dx} + q(x)y = f(x) \tag{4}$$

is always in the form

$$y = y_p + c_1y_1 + c_2y_2, \tag{5}$$

where

- y_p is a particular solution of (4) and
- y_1 and y_2 are a fundamental set of solutions for the associated homogeneous equation

$$\frac{d^2y}{dx^2} + p(x)\frac{dy}{dx} + q(x)y = 0. \tag{6}$$

The constants c_1 and c_2 are arbitrary constants. The expression $c_1y_1 + c_2y_2$ is called a **linear combination** of y_1 and y_2. We will discuss later in more detail what is meant by a fundamental set of solutions for the associated homogeneous equation. For now,

> If all solutions of (4) can be put in the form (5), then y_1, y_2 are said to be a **fundamental set** of solutions of the associated homogeneous equation (6). For now, we assume that y_1, y_2 is a fundamental set if neither y_1 nor y_2 is a constant multiple of the other. (7)

For a second-order linear equation, there are two solutions of the associated homogeneous equation in the fundamental set [for a first-order linear equation, there is only one solution of the associated homogeneous equation in the general solution; see (2)]. The general solution for a second-order differential equation has two arbitrary constants (first-order equations have one arbitrary constant). For a second-order linear differential equation, the two arbitrary constants will be determined from the two initial conditions (for first-order linear differential equations, the one arbitrary constant is determined from the one initial condition).

Note that $y = 0$ always satisfies the homogeneous differential equation

$$\frac{d^2y}{dx^2} + p(x)\frac{dy}{dx} + q(x)y = 0. \tag{8}$$

Thus, if we are solving (8), we choose $y_p = 0$. It then follows from (5) that

> The general solution of a homogeneous differential equation (8) is a linear combination
>
> $$y = c_1y_1 + c_2y_2 \tag{9}$$
>
> of a fundamental set of solutions y_1, y_2 of the homogeneous equation.

Procedures for determining the y_1, y_2, and y_p will be presented later in this chapter.

EXAMPLE 2.2.1 **Fundamental Set and Initial Value Problem**

a. Verify that $\sin x$, $\cos x$ are a fundamental set of solutions of $y'' + y = 0$.
b. Verify that e^{-3x} is a solution of $y'' + y = 10e^{-3x}$.
c. Give the general solution of

$$y'' + y = 10e^{-3x}. \tag{10}$$

d. Solve the initial value problem

$$y'' + y = 10e^{-3x}, \qquad y(0) = 0, \qquad y'(0) = 1. \tag{11}$$

SOLUTION

a. The verification that $\sin x$, $\cos x$ are solutions of $y'' + y = 0$ is straightforward but provides good practice. Note that $\sin x$ is not a *constant* multiple of $\cos x$ since $\sin x/\cos x$ is not constant. Also, $y'' + y = 0$ is of second order. Thus, by (7), $\{\sin x, \cos x\}$ is a fundamental set of solutions and $c_1 \sin x + c_2 \cos x$ gives all solutions of $y'' + y = 0$.
b. It is straightforward to verify that e^{-3x} satisfies (10). Thus our particular solution of the nonhomogeneous equation is $y_p = e^{-3x}$.
c. Thus, by (5),

$$y = e^{-3x} + c_1 \sin x + c_2 \cos x \tag{12}$$

is the general solution of $y'' + y = 10e^{-3x}$.
d. To solve (11), we apply the initial conditions in (11) to the general solution (12). Evaluating (12) at zero and using the initial condition in (11) gives

$$0 = y(0) = 1 + c_2 \qquad \text{so that } c_2 = -1.$$

Differentiate (12) to get

$$y' = -3e^{-3x} + c_1 \cos x - c_2 \sin x.$$

Evaluate y' at $x = 0$ and use the initial condition $y'(0) = 1$ to get

$$1 = y'(0) = -3 + c_1$$

so that $c_1 = 4$. Thus the solution of (11) is

$$y = e^{-3x} + 4 \sin x - \cos x.$$

◄

Linearity

We have claimed that the form of the general solution of

$$\frac{d^2y}{dx^2} + p(x)\frac{dy}{dx} + q(x)y = f(x) \tag{13}$$

is

$$y = y_p + c_1 y_1 + c_2 y_2. \tag{14}$$

We first proceed to show that $y = y_p + c_1 y_1 + c_2 y_2$ actually solves the differential equation. Later we will discuss the more difficult question of showing that all solutions must be in this form.

We first introduce the notation of a **linear operator**. This notation is not essential for solving simple differential equations, although it can add insight. For more complicated problems, it becomes progressively more important. Our understanding of the properties of the differential equation will be enhanced by the introduction of the **second-order linear differential operator** L:

$$L(y) \equiv \frac{d^2y}{dx^2} + p(x)\frac{dy}{dx} + q(x)y. \tag{15}$$

EXAMPLE 2.2.2 Linear Operator

If $L(y) \equiv xy'' + 3y$, compute $L(\sin x)$.

SOLUTION $L(\sin x) = x(\sin x)'' + 3(\sin x) = -x\sin x + 3\sin x.$ ◄

Linearity Property of Linear Operators (Optional)

It can be shown that the operator

$$L(y) = \frac{d^2y}{dx^2} + p(x)\frac{dy}{dx} + q(x)y \tag{16}$$

satisfies the **linearity property**,

$$L(c_1y_1 + c_2y_2) = c_1L(y_1) + c_2L(y_2) \tag{17}$$

for any twice differentiable functions y_1, y_2.

Using the linear operator notation, the second-order linear differential equation (13) is written

$$L(y) = \frac{d^2y}{dx^2} + p(x)\frac{dy}{dx} + q(x)y = f(x). \tag{18}$$

We have claimed that the general solution appears in the form

$$y = y_p + c_1y_1 + c_2y_2,$$

where y_p is a particular solution satisfying $L(y_p) = f(x)$ and y_1 and y_2 are a fundamental set of homogeneous solutions of the associated homogeneous equation $L(y) = 0$. We now turn to understanding why this holds.

Difference Between Particular Solutions Must Be a Solution of Associated Homogeneous Equation

We first show that

$$\text{If } y \text{ satisfies (13), then } y = y_p + y_h, \tag{19}$$

where y_h is a solution of the associated homogeneous equation

$$\frac{d^2y}{dx^2} + p(x)\frac{dy}{dx} + q(x)y = 0. \tag{20}$$

Verification of (19) Suppose that y and y_p are solutions of (13). Thus, $L(y) = f(x)$ and $L(y_p) = f(x)$. Introduce $y_h = y - y_p$, the difference between y and y_p. Clearly $y = y_p + y_h$. To determine the differential equation satisfied by y_h, we calculate $L(y_h)$ using the linearity property (17):

$$L(y_h) = L(y - y_p) = L(y) - L(y_p) = f - f = 0.$$

Thus y_h is a solution of $L(y_h) = 0$, the associated homogeneous equation, as claimed in (19).

We can also verify (19) without using the linear operator notation. Suppose that y and y_p are solutions of (13). Thus, $y'' + py' + qy = f(x)$ and $y_p'' + py_p' + qy_p = f(x)$. Introduce $y_h = y - y_p$, the difference between y and y_p. Clearly $y = y_p + y_h$. To determine the differential equation satisfied by y_h, we calculate $y_h'' + py_h' + qy_h$ as follows:

$$\begin{aligned} y_h'' + py_h' + qy_h &= (y - y_p)'' + p(y - y_p)' + q(y - y_p) \\ &= y'' - y_p'' + py - py_p' + qy - qy_p \\ &= (y'' + py' + qy) - (y_p'' + py_p' + qy_p) \\ &= f - f = 0. \end{aligned}$$

Thus y_h is a solution of $y_h'' + py_h' + qy_h = 0$, the associated homogeneous equation, as claimed in (19).

Solutions of Homogeneous Equations

In the next section we will show that the general solution of a homogeneous equation must be a linear combination of a fundamental set of solutions. For now, we will be content with just verifying that

> A linear combination
>
> $$c_1y_1 + c_2y_2$$
>
> of two solutions y_1, y_2 of a linear homogeneous differential equation
>
> $$L(y) = \frac{d^2y}{dx^2} + p(x)\frac{dy}{dx} + q(x)y = 0 \tag{21}$$
>
> is also a solution of the homogeneous differential equation.

Verification of (21) Suppose that y_1 and y_2 are solutions of the homogeneous equation $L(y) = 0$ and c_1 and c_2 are constants. Then, by linearity (17),

$$L(c_1y_1 + c_2y_2) = c_1L(y_1) + c_2L(y_2) = c_1 \cdot 0 + c_2 \cdot 0 = 0.$$

Thus, $c_1y_1 + c_2y_2$ is also a solution of $L(y) = 0$, as claimed in (21).

We can also verify (21) without using the linear operator notation. Suppose that y_1 and y_2 are solutions of the homogeneous equation $y'' + py' + qy = 0$ and c_1 and c_2 are constants. We substitute $c_1y_1 + c_2y_2$ for y in $y'' + py' + qy = 0$ to see if it solves that equation:

$$\begin{aligned}(c_1y_1 + c_2y_2)'' &+ p(c_1y_1 + c_2y_2)' + q(c_1y_1 + c_2y_2)\\ &= c_1y_1'' + c_2y_2'' + pc_1y_1' + pc_2y_2' + qc_1y_1 + qc_2y_2\\ &= c_1(y_1'' + py_1' + qy_1) + c_2(y_2'' + py_2' + qy_2)\\ &= c_10 + c_20 = 0.\end{aligned}$$

Thus, $c_1y_1 + c_2y_2$ is also a homogeneous solution of $y'' + py' + qy = 0$, as claimed in (21).

Principle of Superposition

Particular solutions of linear differential equations may be added. The following fact is called the **superposition principle**:

Superposition Principle

If y_p is a particular solution (output) corresponding to one input $f_1(x)$ of

$$y'' + py' + qy = f_1$$

and $\tilde{y}_p$ is a particular solution (output) corresponding to a second input $f_2(x)$ (22)

$$y'' + py' + qy = f_2,$$

then the output corresponding to the sum of the inputs $f + f_2$ is the sum of the outputs $y_p + \tilde{y}_p$. That is, $y_p + \tilde{y}_p$ is a solution of

$$y'' + py' + qy = f_1 + f_2.$$

Verification of the Principle of Superposition (22) Suppose that y_p is a particular solution of $L(y) = f_1$ and $\tilde{y}_p$ is a particular solution of $L(y) = f_2$. We shall show that $y_p + \tilde{y}_p$ is a solution of $L(y) = f_1 + f_2$. Substituting $y_p + \tilde{y}_p$ for y in $y'' + py' + qy$, we obtain, using the linearity property (17),

$$L(y_p + \tilde{y}_p) = L(y_p) + L(\tilde{y}_p) = f_1 + f_2,$$

as claimed in (22).

We can also verify (22) without using the linear operator notation. Suppose that y_p is a particular solution of $y'' + py' + qy = f_1$ and $\tilde{y}_p$ is a particular solution of $y'' + py' + qy = f_2$. We shall show that $y_p + \tilde{y}_p$ is a solution of $y'' + py' + qy = f_1 + f_2$. Substituting $y_p + \tilde{y}_p$ for y in $y'' + py' + qy$, we obtain

$$\begin{aligned} y'' + py' + qy &= (y_p + \tilde{y}_p)'' + p(y_p + \tilde{y}_p)' + q(y_p + \tilde{y}_p) \\ &= y_p'' + \tilde{y}_p'' + py_p' + p\tilde{y}_p' + qy_p + q\tilde{y}_p \\ &= (y_p'' + py_p' + qy_p) + \left(\tilde{y}_p'' + p\tilde{y}_p' + q\tilde{y}_p\right) \\ &= f_1 + f_2, \end{aligned}$$

as claimed in (22).

The superposition principle tells us that, if the forcing function f is a sum of several terms, we may find a particular solution for each term and then add these particular solutions to get a particular solution for the whole forcing function. This fact plays an important role in several of our techniques. The superposition principle may also be expressed as, the response (output) to a sum of two inputs is the sum of the responses (output) to each input. This is an important property of linear devices and systems.

EXAMPLE 2.2.3 Superposition Principle

Suppose you wish to solve

$$y'' - y = e^{2x} - e^{3x}$$

and you determine that

$$y_p(x) = \frac{1}{3}e^{2x} \text{ is a solution of } y'' - y = e^{2x}$$

$$\tilde{y}_p(x) = -\frac{1}{8}e^{3x} \text{ is a solution of } y'' - y = -e^{3x}.$$

The principle of superposition says that

$$y_p(x) + \tilde{y}_p(x) = \frac{1}{3}e^{2x} - \frac{1}{8}e^{3x}$$

is a solution of $y'' - y = e^{2x} - e^{3x}$. ◄

Exercises

For Exercises 1 through 8, verify that the given set is a fundamental set of solutions for the associated homogeneous equation and y_p is a particular solution. Then give the general solution of the differential equation, and solve the initial value problem.

1. $y'' + y = 1$, $y(0) = 0$, $y'(0) = 0$, $\{\sin x, \cos x\}$, $y_p = 1$

2. $y'' - y = e^{3x}$, $y(0) = 0$, $y'(0) = 1$, $\{e^x, e^{-x}\}$, $y_p = \frac{1}{8}e^{3x}$

3. $y'' - 3y' + 2y = 2x$, $y(0) = 1$, $y'(0) = 0$, $\{e^x, e^{2x}\}$, $y_p = x + 3/2$

4. $y'' - 2y' + y = 4e^{2x}$, $y(0) = 0$, $y'(0) = 0$, $\{e^x, xe^x\}$, $y_p = 4e^{2x}$

5. $y'' + 2y' + 2y = 6$, $y(0) = 1$, $y'(0) = 1$, $\{e^{-x}\cos x, e^{-x}\sin x\}$, $y_p = 3$

6. $x^2y'' - 2xy' + 2y = 2x^3$, $y(1) = 2$, $y'(1) = 3$, $\{x, x^2\}$, $y_p = x^3$, $x > 0$

7. $y'' + y = 2\cos x$, $y(0) = 1$, $y'(0) = -1$, $\{\sin x, \cos x\}$, $y_p = x\sin x$

8. $x^2y'' + 4xy' + 2y = 2\ln x + 3$, $y(1) = 0$, $y'(1) = 2$, $\{x^{-1}, x^{-2}\}$, $y_p = \ln x$, $x > 0$

9. a) Verify that both

$$y = c_1e^x + c_2e^{2x} + 2\cosh x \quad (23)$$

and

$$\tilde{y} = \tilde{c}_1e^x + \tilde{c}_2e^{2x} + e^{-x} \quad (24)$$

are a general solution of $y'' - 3y' + 2y = 6e^{-x}$.

b) Verify that, if the initial conditions $y(0) = 4$, $y'(0) = 3$ are applied to (23) and (24), both give the same solution.

10. a) Verify that both

$$y = c_1 + c_2x + x^2 \quad (25)$$

and

$$\tilde{y} = \tilde{c}_1(1 + x) + \tilde{c}_2(1 - x) + x^2 + 3x + 1 \quad (26)$$

are the general solution of $y'' = 2$.

b) Verify that, if the initial conditions $y(0) = 0$, $y'(0) = 1$, are applied to (25) and (26), then both give the same solution.

11. Verify that $-\frac{1}{3}\sin 2x$ is a solution of $y'' + y = \sin 2x$, $-\frac{1}{3}\cos 2x$ is a solution of $y'' + y = \cos 2x$, and $\{\sin x, \cos x\}$ is a fundamental set of solutions for $y'' + y = 0$. Find the general solution of $y'' + y = \sin 2x + \cos 2x$.

12. Verify that e^{2x} is a solution of $y'' - 2y' + y = e^{2x}$, 1 is a solution of $y'' - 2y' + y = 1$, and $\{e^x, xe^x\}$ is a fundamental set of solutions for $y'' - 2y' + y = 0$. Find the general solution of $y'' - 2y' + y = 1 + e^{2x}$.

13. Show that $L(y) = dy/dx$ satisfies the linearity property $L(c_1y_1 + c_2y_2) = c_1L(y_1) + c_2L(y_2)$.

14. Show that $L(y) = d^2y/dx^2$ satisfies the linearity property $L(c_1y_1 + c_2y_2) = c_1L(y_1) + c_2L(y_2)$.

15. Show that $L(y) = d^2y/dx^2 + p(x)(dy/dx) + q(x)y$ satisfies the linearity property $L(c_1y_1 + c_2y_2) = c_1L(y_1) + c_2L(y_2)$.

16. Show that $L(y) = dy/dx + p(x)y$ satisfies the linearity property $L(c_1y_1 + c_2y_2) = c_1L(y_1) + c_2L(y_2)$.

In Exercises 17 through 22, use the linearity property, $L(c_1y_1 + c_2y_2) = c_1L(y_1) + c_2L(y_2)$.

17. Find a particular solution for $L(y) = \cos x$ if it is known that $L(e^x) = 4\cos x$.

18. Find a particular solution for $L(y) = 5e^x$ if it is known that $L(\sin x) = e^x$.

19. Find a particular solution for $L(y) = \tan x - 4\cos(e^x)$ if it is known that $L(\sin x) = 5\tan x$ and $L(x^2) = \cos(e^x)$.

20. Find a particular solution for $L(y) = x^3 + \sin x$ if it is known that $L(e^x) = 5\sin x$ and $L(\cos x) = \frac{1}{4}x^3$.

21. Find some solutions of $L(y) = 0$ if it is known that $L(\cos x) = x^2$ and $L(e^x) = \frac{1}{2}x^2$.

22. Find some solutions of $L(y) = 0$ if it is known that $L(x + e^x) = \sin x$ and $L(e^{-x}) = 4\sin x$.

2.3 Initial Values, Wronskian, and Linear Independence

In this section, we continue to study second-order linear differential equations

$$\frac{d^2y}{dx^2} + p(x)\frac{dy}{dx} + q(x)y = f(x). \tag{1}$$

In the previous section, we have shown that

$$y = y_p + y_h,$$

where y_p is a particular solution of (1) and y_h solves the associated homogeneous equation

$$\frac{d^2y}{dx^2} + p(x)\frac{dy}{dx} + q(x)y = 0. \tag{2}$$

We have claimed that

$$y_h = c_1y_1 + c_2y_2, \tag{3}$$

where y_1, y_2 are a fundamental set of solutions of (2). The proof of (3) will appear in this section. In addition, we will introduce several important concepts that will be used several times in our study of differential equations.

Initial Value Problem

According to the Existence and Uniqueness Theorem (Theorem 2.1.2), there is a unique solution for (2) that satisfies the initial conditions

$$\begin{aligned} y(x_0) &= y_0, \\ y'(x_0) &= v_0 \end{aligned} \tag{4}$$

for *any* numbers y_0, v_0. First, we consider a very practical question. In general, how do we find the two constants c_1 and c_2 in order to satisfy the two initial conditions (4)? We assume (to be proved later) that

$$y = c_1 y_1(x) + c_2 y_2(x). \tag{5}$$

The initial conditions are satisfied if

$$y_0 = c_1 y_1(x_0) + c_2 y_2(x_0) \tag{6}$$

and

$$v_0 = c_1 y_1'(x_0) + c_2 y_2'(x_0). \tag{7}$$

The mathematical problem of solving for the constants consists of solving two linear equations (6) and (7) for the two unknowns c_1 and c_2. Systems of linear equations may have a unique solution, no solution, or an infinite number of solutions, depending on the coefficients of the linear equations. Since the linear equations (6) and (7) each define a line in the c_1, c_2 plane, these three conditions correspond to intersecting nonparallel lines, parallel lines that don't coincide, and parallel lines that do coincide. The result of the analysis to follow is that if the two solutions of a homogeneous linear second-order differential equation are chosen not to be multiples of each other, then no difficulties will arise in solving these linear equations, and a unique solution for c_1 and c_2 will occur.

We may solve the linear system (6) and (7) for c_1 and c_2 in several ways, although all are mathematically equivalent. For example, we can eliminate c_2 by multiplying (6) by $y_2'(x_0)$, multiplying (7) by $y_2(x_0)$, and then subtracting. This calculation gives

$$c_1[y_1(x_0)y_2'(x_0) - y_2(x_0)y_1'(x_0)] = y_2'(x_0)y_0 - y_2(x_0)v_0. \tag{8}$$

If $[y_1(x_0)y_2'(x_0) - y_2(x_0)y_1'(x_0)] = 0$, the coefficient multiplying c_1 is zero in (8), and it turns out that either there will not be a solution for the constants or the solution will not be unique. If $[y_1(x_0)y_2'(x_0) - y_2(x_0)y_1'(x_0)] \neq 0$, we can solve (8) for c_1 to get

$$c_1 = \frac{y_2'(x_0)y_0 - y_2(x_0)v_0}{[y_1(x_0)y_2'(x_0) - y_2(x_0)y_1'(x_0)]}.$$

A similar calculation works for c_2. We see that

There is a unique solution of (6) and (7) precisely when $y_1(x_0)y_2'(x_0) - y_2(x_0)y_1'(x_0) \neq 0$.

We have just rediscovered the fact from the theory of linear equations (see linear algebra review Subsection 2.3.2) that there is a unique solution of (6) and (7) for c_1 and c_2 if and only if the determinant of the coefficients is nonzero, that is,

$$\det \begin{bmatrix} y_1 & y_2 \\ y_1' & y_2' \end{bmatrix} = y_1 y_2' - y_2 y_1' \neq 0. \tag{9}$$

For this reason, we will study a related expression called the Wronskian. We will show that (9) is satisfied (if the two homogeneous solutions are not multiples of each other), so that the constants c_1 and c_2 may be uniquely determined from the initial conditions.

Wronskian

We define the **Wronskian** W of two functions y_1, y_2 to be

$$W[y_1, y_2] = \det\begin{bmatrix} y_1 & y_2 \\ y_1' & y_2' \end{bmatrix} = y_1(x)y_2'(x) - y_2(x)y_1'(x). \tag{10}$$

From our discussion, we know that the Wronskian of two solutions of the homogeneous second-order linear differential equation

$$\frac{d^2y}{dx^2} + p(x)\frac{dy}{dx} + q(x)y = 0 \tag{11}$$

will be important.

We now show that the Wronskian of two solutions y_1, y_2 of (11) itself satisfies a linear first-order differential equation. From that we will analyze the meaning of the Wronskian being zero. We begin by calculating the first derivative of the Wronskian of two solutions y_1, y_2 of (11).

$$\begin{aligned}
\frac{dW}{dx} &= [y_1(x)y_2'(x) - y_2(x)y_1'(x)]' \\
&= y_1(x)y_2''(x) + y_1'(x)y_2'(x) - y_2(x)y_1''(x) - y_2'(x)y_1'(x) \\
&\qquad\qquad\qquad\qquad\qquad\qquad\qquad\qquad \text{(product rule)} \\
&= y_1(x)y_2''(x) - y_2(x)y_1''(x) \qquad \text{(cancellation of two terms)} \\
&= y_1(-py_2' - qy_2) - y_2(-py_1' - qy_1) \\
&\qquad\qquad\qquad\qquad\qquad\qquad [y_1 \text{ and } y_2 \text{ are solutions of (11)}] \\
&= -p[y_1(x)y_2'(x) - y_2(x)y_1'(x)] \quad \text{(cancellation and rearrangement)} \\
&= -pW \qquad [\text{definition of the Wronskian (10)}].
\end{aligned}$$

Thus the Wronskian satisfies the first-order linear differential equation

$$\frac{dW}{dx} = -p(x)W. \tag{12}$$

Here $p(x)$ is the same p that appears as a coefficient in the second-order linear differential equation (11).

The first-order linear differential equation for the Wronskian (12) is separable. It can also be solved using integrating factors. From Section 1.3 or 1.6, the solution of (12) is

$$W(x) = W(x_0)e^{-\int_{x_0}^{x} p(t)\,dt}. \tag{13}$$

The exponential in (13) is never zero. $W(x_0)$ is a constant, the Wronskian evaluated at the initial point. Thus, the Wronskian $W(x)$ is either always zero [if $W(x_0) = 0$] or never zero [if $W(x_0) \neq 0$]. In all problems, the two homogeneous solutions are chosen such that $W(x_0) \neq 0$, and hence in this case $W(x) \neq 0$. The Wronskian arises in many places in the study of linear differential equations. Thus (13) will be quite useful.

Form of Solutions of Homogeneous Equation (Optional)

We can now prove that if two solutions y_1, y_2 of a homogeneous second-order linear differential equation are chosen such that their Wronskian is not zero initially [$W(x_0) \neq 0$], then any homogeneous solution can be written as a linear combination of this y_1 and y_2,

$$y = c_1 y_1 + c_2 y_2. \tag{14}$$

We again consider the initial value problem for the linear homogeneous differential equation

$$\frac{d^2y}{dx^2} + p(x)\frac{dy}{dx} + q(x)y = 0, \tag{15}$$

which satisfies

$$\begin{aligned} y(x_0) &= y_0, \\ y'(x_0) &= v_0. \end{aligned} \tag{16}$$

We do not assume that (14) is valid. By the uniqueness theorem, there is one solution of differential equation (15) that satisfies the initial conditions (16). We call this unique solution $Y(x)$. Its initial value is $y_0 = Y(x_0)$, and its initial derivative is $v_0 = Y'(x_0)$. However, we have shown that it is also possible to obtain one solution of the differential equation in the form

$$y = c_1 y_1 + c_2 y_2 \tag{17}$$

satisfying these initial conditions (if the Wronskian of y_1 and y_2 is initially nonzero). By uniqueness, this must be the same solution as $Y(x)$, so that

$$Y(x) = c_1 y_1 + c_2 y_2. \tag{18}$$

Thus, any homogeneous solution must be a linear combination of y_1 and y_2 (if the Wronskian of y_1 and y_2 is initially nonzero).

Fundamental Solution

We have seen that all solutions of the associated homogeneous equation have been shown to be in the form $c_1 y_1 + c_2 y_2$ if the initial Wronskian of y_1, y_2 is nonzero. Thus two solutions y_1, y_2 of a homogeneous second-order linear differential equation form a **fundamental set** if $W(x_0) \neq 0$. This is an easy test to determine if a set of solutions is fundamental.

Since it can be shown (Exercise 7) that a fundamental set exists, it follows that we have proved that any homogeneous solution can be written as a **linear combination** of that fundamental set

$$y_h = c_1 y_1 + c_2 y_2$$

and that the general solution of the nonhomogeneous differential equation (1) is in the form

$$y = y_p + c_1 y_1 + c_2 y_2.$$

EXAMPLE 2.3.1 Fundamental Set and Initial Value Problem

a. Show that $\{\sin x, \cos x\}$ forms a fundamental set of solutions of the homogeneous linear differential equation

$$y'' + y = 0. \tag{19}$$

b. Find the general solution.
c. Find the solution of the initial value problem $y(0) = 6$, $y'(0) = 3$.

SOLUTION

a. First we should verify that $y_1 = \sin x$ and $y_2 = \cos x$ each satisfy (19). We omit this step. To show that $\{\sin x, \cos x\}$ is a fundamental set of solutions, we then verify that $W[y_1, y_2] \neq 0$ at $x = 0$. Note that $y_1' = \cos x$ and $y_2' = -\sin x$. Thus

$$\begin{aligned} W[y_1, y_2] &= \det \begin{bmatrix} y_1(0) & y_2(0) \\ y_1'(0) & y_2'(0) \end{bmatrix} \\ &= \det \begin{bmatrix} 0 & 1 \\ 1 & 0 \end{bmatrix} = -1 \neq 0, \end{aligned}$$

Hence $\{\sin x, \cos x\}$ forms a fundamental set.

b. Since (19) is homogeneous, its general solution is a linear combination of $\{\sin x \cos x)$,

$$y = c_1 \sin x + c_2 \cos x. \tag{20}$$

c. To satisfy the initial conditions, we first calculate the derivative of (20):

$$y' = c_1 \cos x - c_2 \sin x.$$

The initial conditions are satisfied if

$$\begin{aligned} y(0) &= 6 = c_2, \\ y'(0) &= 3 = c_1. \end{aligned}$$

The solution of the initial value problem is thus

$$y = 3\sin x + 6\cos x.$$

◄

EXAMPLE 2.3.2 **A Fundamental Set**

Determine two solutions of $y'' - 3y' + 2y = 0$ that are of the form e^{rx}, and decide whether they form a fundamental set.

SOLUTION Substituting e^{rx} into $y'' - 3y' + 2y = 0$ gives

$$\begin{aligned}(e^{rx})'' - 3(e^{rx})' + 2e^{rx} &= 0\\ r^2e^{rx} - 3re^{rx} + 2e^{rx} &= 0\\ e^{rx}(r^2 - 3r + 2) &= 0\\ r^2 - 3r + 2 &= 0,\end{aligned}$$

that is, $(r-2)(r-1) = 0$. The roots are $r = 2, 1$, so e^{2x}, e^{1x} are solutions. To determine whether they are a fundamental set of solutions, we compute the Wronskian:

$$\begin{aligned}W[y_1, y_2] &= \det\begin{bmatrix} y_1 & y_2 \\ y_1' & y_2' \end{bmatrix}\\ &= \det\begin{bmatrix} e^{2x} & e^x \\ 2e^{2x} & e^x \end{bmatrix}\\ &= e^{3x} - 2e^{3x} = -e^{3x} \neq 0 \qquad \text{for all } x.\end{aligned}$$

Since the Wronskian is nonzero, $\{e^{2x}, e^x\}$ is a fundamental set of solutions, and every solution of $y'' - 3y' + 2y = 0$ is of the form $c_1e^{2x} + c_2e^x$. ◄

EXAMPLE 2.3.3 **Fundamental Set and Initial Value Problem**

Let y_1 be the solution of

$$y'' + x^2y' + x^3y = 0, \qquad y(1) = 1, \qquad y'(1) = 3, \tag{21}$$

and let y_2 be the solution of

$$y'' + x^2y' + x^3y = 0, \qquad y(1) = -1, \qquad y'(1) = 2. \tag{22}$$

a. Verify $\{y_1, y_2\}$ is a fundamental set of solutions of $y'' + x^2y' + x^3y = 0$.
b. Find constants c_1, c_2 such that $y_3 = c_1y_1 + c_2y_2$ is the solution of

$$y'' + x^2y' + x^3y = 0, \qquad y(1) = 2, \qquad y'(1) = 0. \tag{23}$$

SOLUTION

a. To show that $\{y_1, y_2\}$ is a fundamental set of solutions, it suffices to verify that $W[y_1, y_2] \neq 0$ at any point. We shall verify that $W[y_1, y_2] \neq 0$ at

$x = 1$. Now, at $x = 1$,

$$W[y_1, y_2] = \begin{bmatrix} y_1(1) & y_2(1) \\ y_1'(1) & y_2'(1) \end{bmatrix}$$

$$= \det \begin{bmatrix} 1 & -1 \\ 3 & 2 \end{bmatrix} \qquad \text{[by (21) and (22)]}$$

$$= 5 \neq 0,$$

and hence y_1, y_2 are a fundamental set.

b. We shall use the fact that (3) is equivalent to (4) and (5). Evaluating (4) and (5) at $x_0 = 1$ gives

$$y_3(1) = c_1 y_1(1) + c_2 y_2(1),$$
$$y_3'(1) = c_1 y_1'(1) + c_2 y_2'(1). \tag{24}$$

The initial conditions in (21), (22), and (23) applied to (24) yield two equations in c_1, c_2:

$$2 = c_1 1 + c_2(-1),$$
$$0 = c_1 3 + c_2 2.$$

Solving for c_1, c_2 yields $c_2 = -\frac{6}{5}$, $c_1 = \frac{4}{5}$. Thus $y_3 = \frac{4}{5}y_1 - \frac{6}{5}y_2$. ◄

The next theorem summarizes the key relationships between the Wronskian and the concepts of a fundamental set of solutions.

■ **THEOREM 2.3.1 Fundamental Set Is Equivalent to Nonzero Wronskian**

Suppose that p, q are continuous functions on the interval I and y_1, y_2 are solutions of the linear homogeneous differential equation $y'' + py' + qy = 0$. Then the following are equivalent.

i. $W[y_1, y_2] \neq 0$ for some x in I.
ii. $W[y_1, y_2] \neq 0$ for all x in I.
iii. If y_3 is any solution of

$$y'' + py' + qy = 0. \tag{25}$$

then there exist constants c_1, c_2 such that $y_3 = c_1 y_1 + c_2 y_2$ and the c_1, c_2 are uniquely determined by y_3. That is, $\{y_1, y_2\}$ is a **fundamental set of solutions** for (25) and any other solution y_3 may be written as a **linear combination** of y_1, y_2. ■

Thus, two solutions form a fundamental set if $W[y_1, y_2] \neq 0$.

2.3.1 Linear Independence, Fundamental Set, and the Wronskian

In this subsection we introduce the concept of a linearly independent set of solutions of a homogeneous linear differential equation and show that it is equivalent to being a fundamental set.

Two functions $y_1(x)$, $y_2(x)$ are said to be **linearly independent** if the equation

$$c_1 y_1(x) + c_2 y_2(x) = 0 \tag{26}$$

for all x implies that $c_1 = c_2 = 0$.

For two functions y_1, y_2 this is equivalent to saying that neither function is a constant multiple of the other. (See Exercise 9 at the end of this section.) However, the definition of linear independence (26) extends readily to more than two functions, whereas the concept of being multiples is then no longer appropriate. If functions are not linearly independent, they are called **linearly dependent**.

First, we wish to show that the condition $W[y_1, y_2] = 0$ is equivalent to the functions y_1, y_2 being linearly dependent. If y_1, y_2 are linearly dependent, then either $y_2 = cy_1$ or $y_1 = 0$. In either case, the Wronskian is zero. If $y_2 = cy_1$, then $W[y_1, y_2] = y_1 y_2' - y_2 y_1' = y_1 c y_1' - c y_1 y_1' = 0$. If $y_1 = 0$, then the Wronskian is again zero. This shows that linear dependence implies that the Wronskian is zero. Suppose, on the other hand, that the Wronskian is zero for all x, so that

$$y_1 y_2' - y_2 y_1' = 0, \tag{27}$$

and we want to establish linear dependence in the case when y_1, y_2 are solutions of the associated homogeneous equation. We shall only give the proof in the special case when $y_1 y_2$ is never zero. Dividing both sides of (27) by $y_1 y_2$ gives

$$\frac{y_2'}{y_2} = \frac{y_1'}{y_1}.$$

Integrating with respect to x yields

$$\ln |y_2| = \ln |y_1| + c_0.$$

Exponentiation yields $y_2 = cy_1$. Thus y_1, y_2 are linearly dependent.

Eq. (27) does not in general imply the linear dependence of two functions y_1, y_2. (Take two differentiable functions y_1, y_2, neither of which is identically zero, but with $y_1 y_2 = 0$ for all x. They are linearly independent but $W[y_1, y_2] = 0$.) However, in the special case when y_1, y_2 are solutions of a second order linear homogeneous differential equation, then (27) does imply linear dependence.

We summarize these results in the following theorem.

■ **THEOREM 2.3.2 Linearly Independent and Fundamental Sets**

Suppose that p, q are continuous functions on the interval I and y_1, y_2 are solutions of the linear homogeneous differential equation $y'' + py' + qy = 0$. Then the following four concepts are equivalent:

i. $\{y_1, y_2\}$ is a fundamental set of solutions.
ii. $W[y_1, y_2] \neq 0$.
iii. $\{y_1, y_2\}$ is a linearly independent set of solutions.
iv. The general solution of $y'' + py' + qy = 0$ is $y = c_1 y_1 + c_2 y_2$. ■

2.3.2 Linear Algebra Review

This subsection reviews some of the facts from (linear) algebra used in this section. The exposition is self-contained except that some proofs are deleted. More detail is found in Section 5.8.

For a 2×2 matrix of numbers

$$\mathbf{A} = \begin{bmatrix} a & b \\ c & d \end{bmatrix},$$

recall that the **determinant of A** is the number $ad - bc$. The determinant, denoted $\det(\mathbf{A})$, is closely related to properties of systems of algebraic equations.

Suppose a, b, c, d, e, and f are constants and z, w are variables. Consider the following system of two equations in the two unknowns z, w:

$$\begin{aligned} az + bw &= e, \\ cz + dw &= f. \end{aligned} \tag{28}$$

In elementary algebra the following facts are proved:

Equations (28) have a solution for every e, f if and only if

$$\det \begin{bmatrix} a & b \\ c & d \end{bmatrix} \neq 0. \tag{29}$$

Solutions of (28) are unique for a given e, f if and only if

$$\det \begin{bmatrix} a & b \\ c & d \end{bmatrix} \neq 0. \tag{30}$$

EXAMPLE 2.3.4 **Nonunique Solution**

Consider the system of equations

$$\begin{aligned} z + 2w &= e, \\ 2z + 4w &= f. \end{aligned} \tag{31}$$

Since

$$\det\begin{bmatrix} 1 & 2 \\ 2 & 4 \end{bmatrix} = 4 - 4 = 0,$$

the equations (31) do not have a solution for some e, f, such as $e = 1, f = 7$. When solutions (z, w) do exist, they are not unique. For example, both $(1, 3)$ and $(3, 2)$ are solutions of

$$z + 2w = 7,$$

$$2z + 4w = 14.$$

EXAMPLE 2.3.5 **Unique Solution**

Now consider

$$\begin{aligned} z + 2w &= e, \\ 3z + 4w &= f. \end{aligned}$$

Since

$$\det\begin{bmatrix} 1 & 2 \\ 3 & 4 \end{bmatrix} = 4 - 6 = -2 \neq 0,$$

there is a unique solution for any e, f.

Exercises

In Exercises 1 through 7, use that $W \neq 0$ is equivalent to the solutions of the homogeneous equation forming a fundamental set.

1. Verify that $\sin x$, $\cos x$ are solutions of $y'' + y = 0$. Determine whether they are a fundamental set of solutions.

2. Find all solutions of the form x^r for $x^2y'' - 6y = 0$ on $(0, \infty)$ and determine whether they form a fundamental set of solutions.

3. Find all solutions of the form x^r for $x^2y'' - xy' + y = 0$ on $(0, \infty)$ and determine whether they form a fundamental set of solutions.

4. Find all solutions of the form e^{rx} of $y'' - 4y' + 4y = 0$ on $(-\infty, \infty)$ and determine whether they form a fundamental set of solutions.

5. Let y_1 be the solution on $(0, \infty)$ of
$x^2y'' + y' + xy = 0, \quad y(1) = 1, \quad y'(1) = 1,$
and y_2 the solution on $(0, \infty)$ of
$x^2y'' + y' + xy = 0, \quad y(1) = 0, \quad y'(1) = -1.$

a) Verify that $\{y_1, y_2\}$ is a fundamental set of solutions of $x^2y'' + y' + xy = 0$.

b) Let y_3 be the solution of
$$x^2y'' + y' + xy = 0, \; y(1) = 2, \; y'(1) = 0.$$

Find constants c_1, c_2 such that

$$y_3 = c_1 y_1 + c_2 y_2.$$

6. Let y_1 be the solution of

$$y'' + xy' + y = 0, \quad y(0) = 1, \quad y'(0) = 2,$$

and y_2 the solution of

$$y'' + xy' + y = 0, \quad y(0) = 1, \quad y'(0) = -1.$$

a) Verify that $\{y_1, y_2\}$ is a fundamental set of solutions of $y'' + xy' + y = 0$.

b) Let y_3 be the solution of

$$y'' + xy' + y = 0, \quad y(0) = 2, \quad y'(0) = 2.$$

Find constants c_1, c_2 such that

$$y_3 = c_1 y_1 + c_2 y_2.$$

7. Let y_1 be the solution of

$$y'' + py' + qy = 0, \quad y(x_0) = 1, \quad y'(x_0) = 0,$$

and y_2 the solution of

$$y'' + py' + qy = 0, \quad y(x_0) = 0, \quad y'(x_0) = 1.$$

Verify that y_1, y_2 is a fundamental set of solutions and that the solution y_3 of

$$y'' + py' + qy = 0, \quad y(x_0) = \alpha, \quad y'(x_0) = \beta$$

is $y_3 = \alpha y_1 + \beta y_2$. (The significance of this exercise is that it proves that a fundamental set always exists.)

8. Let $y_1 = x^3 - x, \quad y_2 = x^2 - 1$.

a) Verify that y_1, y_2 are linearly independent functions.

b) Can y_1, y_2 be solutions of a differential equation $y'' + py' + qy = 0$ where p, q are continuous on $(-2, \infty)$?

c) Suppose that y_1, y_2 are solutions of a differential equation $y'' + py' + qy = 0$. Where must p, q have discontinuities?

9. Verify directly that y_1, y_2 are linearly dependent if and only if one of them is a constant multiple of the other.

10. Suppose that y_1, y_2 are differentiable functions. Show that if $y_1 = cy_2$ for a constant c, then $y_1'y_2 - y_1y_2' = 0$ for all x.

11. Suppose that y_1, y_2 are differentiable functions and neither y_1 nor y_2 is ever zero. Show that if $y_1'y_2 - y_1y_2' = 0$ for all x, then $y_1 = cy_2$ for some constant c.

12. Verify that $y = c_1x^2 + c_2x^3$ is the general solution of

$$x^2y'' - 4xy' + 6y = 0 \tag{32}$$

on either $(-\infty, 0)$ or $(0, \infty)$.

13. Verify that $y = c_1x^2 + c_2x^3$ is a solution of (32) on $(-\infty, \infty)$. Show that $\bar{y} = |x|^3$ is a solution of (32) on $(-\infty, \infty)$. Show that there do not exist constants c_1, c_2 such that $|x|^3 = c_1x^2 + c_2x^3$ for $-\infty < x < \infty$.

14. In Section 2.6 we will derive the important **Euler formula**

$$e^{ix} = \cos x + i \sin x, \tag{33}$$

where $i = \sqrt{-1}$. In this exercise, we outline a different derivation of (33). Consider the differential equation

$$y'' + y = 0. \tag{34}$$

a) Show that $\cos x$ and $\sin x$ are solutions of (34).

b) Show that e^{ix} is another solution of (34).

c) Determine the specific linear combination of $\cos x$ and $\sin x$ that will equal e^{ix}. (*Hint*: Use the initial conditions that e^{ix} satisfies.)

15. Show that the Wronskian of two solutions of Airy's differential equation $d^2y/dx^2 + xy = 0$ is a constant.

16. Show that the Wronskian of two solutions for $d^2y/dx^2 + q(x)y = 0$ is a constant.

In Exercises 17 through 21, verify that the given functions are solutions of the given equation, compute the Wronskian, and compare the Wronskian you obtain to (13).

17. $\cos x$ and $\sin x$ for $y'' + y = 0$.

18. e^{2x} and e^{-2x} for $y'' - 4y = 0$.

19. e^x and e^{2x} for $y'' - 3y' + 2y = 0$.

20. x and x^2 for $x^2y'' - 2xy' + 2y = 0$.

21. x^{-1} and x^{-2} for $x^2y'' + 4xy' + 2y = 0$.

22. Verify that $\{\sin 2x, \cos 2x\}$ form a fundamental set of solutions of $y'' + 4y = 0$. Verify that $\tilde{y}(x) = \sin(2x + \pi/4)$ is another solution of $y'' + 4y = 0$. Find c_1, c_2 so that

$$\sin\left(2x + \frac{\pi}{4}\right) = c_1 \sin 2x + c_2 \cos 2x.$$

2.4 *n*th-Order Linear Differential Equations

This section will present the basic theory for nth-order linear differential equations

$$a_n(x)\frac{d^n y}{dx^n} + a_{n-1}(x)\frac{d^{n-1}y}{dx^{n-1}} + \cdots + a_1(x)\frac{dy}{dx} + a_0(x)y = f(x), \tag{1}$$

which is very similar to the theory for second-order equations developed in Sections 2.1 through 2.3. Rather than writing (1) repeatedly, we shall utilize the operator notation of Section 2.2 and write

$$L(y) = f \tag{2}$$

to represent (1). The **associated homogeneous equation** for (1), (2) is then

$$L(y) = 0.$$

Thus $y''' + y = 0$ is the associated homogeneous equation for $y''' + y = \sin x$.

For first-order linear differential equations, we had one arbitrary constant in the general solution, and for second-order equations there were two. Thus we would expect to have n arbitrary constants for an nth-order differential equation.

■ THEOREM 2.4.1 Existence and Uniqueness Theorem

Suppose in (1) that $a_0, \ldots, a_n, f$ are continuous functions on the interval I and $a_n(x)$ is never zero on I. Then for any x_0 in I and any n real numbers $v_0, \ldots, v_{n-1}$, there exists a unique solution to the initial value problem with n initial conditions

$$L(y) = f, \qquad y(x_0) = v_0, y'(x_0) = v_1, \ldots, y^{(n-1)}(x_0) = v_{n-1}. \tag{3}$$

and this unique solution is defined on all of the interval I. ■

The solutions of (1) may be broken into a particular solution and a homogeneous solution, just as for second-order linear differential equations.

We have

THEOREM 2.4.2 General Solution of a Linear Differential Equation

Let $L(y) = f$ be the nth-order linear differential equation (1) and suppose that the assumptions of Theorem 2.4.1 hold. Then

The general solution of $L(y) = f$ is always in the form

$$y = y_p + c_1 y_1 + c_2 y_2 + \cdots + c_n y_n, \tag{4}$$

where

1. y_p is a particular solution of the original equation $L(y) = f$.
2. $\{y_1, y_2, \ldots, y_n\}$ is a fundamental set of solutions of the associated homogeneous equation $L(y) = 0$.

In the special case when $f = 0$, we have

The general solution of $L(y) = 0$ is always in the form

$$y = c_1 y_1 + c_2 y_2 + \cdots + c_n y_n, \tag{5}$$

where $\{y_1, y_2, \ldots, y_n\}$ is a fundamental set of solutions of the associated homogeneous equation $L(y) = 0$.

■

For second-order linear differential equations, we could characterize a fundamental set of solutions as solutions neither of which was a constant multiple of the other. The situation is more complicated with nth-order linear differential equations. We must first discuss the idea of linear independence.

A set of functions $\{y_1, \ldots, y_n\}$ is **linearly independent** if

$$c_1 y_1 + c_2 y_2 + \cdots + c_n y_n = 0 \tag{6}$$

for constants $c_1, \ldots, c_n$ implies that $c_1 = c_2 = \cdots = c_n = 0$. This is equivalent to saying that none of the y_i can be written as a sum of constant multiples of the other y_i's. That is, no y_i is a **linear combination** of the other y_i's. If a set of functions is not linearly independent, it is **linearly dependent**.

EXAMPLE 2.4.1 Linearly Independent Set

Show that the set of functions $\{1, x, x^2\}$ is linearly independent.

SOLUTION We shall verify (6). Suppose, then, that

$$c_1 1 + c_2 x + c_3 x^2 = 0 \qquad \text{for all } x. \tag{7}$$

In general, we could evaluate (7) at several x values, to get equations for the c_i. In this example, we may differentiate (7) with respect to x,

$$c_2 + 2c_3x = 0, \tag{8}$$

and then differentiate again:

$$2c_3 = 0. \tag{9}$$

From (7), (8), and (9) we have $c_3 = 0$, $c_2 = 0$, $c_1 = 0$, and thus $\{1, x, x^2\}$ is linearly independent, by the definition (6). ◄

EXAMPLE 2.4.2 **Linearly Dependent Set**

Show that $\{e^x, e^{-x}, \cosh x\}$ is a linearly dependent set of functions.

SOLUTION Recall that

$$\cosh x = \frac{1}{2}e^x + \frac{1}{2}e^{-x},$$

so that $\cosh x$ is a linear combination of e^x, e^{-x}. Alternatively, by taking $c_1 = \frac{1}{2}$, $c_2 = \frac{1}{2}$, $c_3 = -1$, we get

$$c_1e^x + c_2e^{-x} + c_3 \cosh x = 0,$$

and c_1, c_2, c_3 are not all zero, so, by definition (6), $\{e^x, e^{-x}, \cosh x\}$ is not linearly independent. ◄

A set of solutions $\{y_1, \ldots, y_n\}$ of the nth-order linear homogeneous equation $L(y) = 0$ is a **fundamental set of solutions** on the interval I if

- Every solution $\tilde{y}_h$ of $L(y) = 0$ may be written as a linear combination of $y_1, \ldots, y_n$:

$$\tilde{y}_h = c_1y_1 + c_2y_2 + \cdots + c_ny_n. \tag{10}$$

- The coefficients in (10) are unique. This is equivalent to $\{y_1, \ldots, y_n\}$ being linearly independent.

Provided that $a_n(x)$ is never zero on the interval I, there always exists a fundamental set of solutions. The next example shows one way to construct such a set of solutions.

EXAMPLE 2.4.3 **Fundamental Set of Solutions**

Consider (1) with $a_3 = 1$ and $n = 3$. Then,

$$y''' + a_2(x)y'' + a_1(x)y' + a_0(x)y = 0, \tag{11}$$

or $L(y) = 0$. Assume a_0, a_1, a_2, are continuous on $[-1, 1]$. Let

$$\begin{array}{lll} y_1 \text{ be the solution of } L(y) = 0, & y(0) = 1, & y'(0) = y''(0) = 0; \\ y_2 \text{ be the solution of } L(y) = 0, & y'(0) = 1, & y(0) = y''(0) = 0; \\ y_3 \text{ be the solution of } L(y) = 0, & y''(0) = 1, & y(0) = y'(0) = 0. \end{array} \tag{12}$$

Verify that $\{y_1, y_2, y_3\}$ is a fundamental set of solutions of $L(y) = 0$.

Solution We shall show that $\{y_1, y_2, y_3\}$ is linearly independent by verifying (6). The y_i are differentiable since they are solutions of (11). Assume that

$$c_1 y_1(x) + c_2 y_2(x) + c_3 y_3(x) = 0. \tag{13}$$

Differentiate (13) two times:

$$\begin{aligned} c_1 y_1'(x) + c_2 y_2'(x) + c_3 y_3'(x) &= 0, \\ c_1 y_1''(x) + c_2 y_2''(x) + c_3 y_3''(x) &= 0. \end{aligned} \tag{14}$$

Evaluating (13) and (14) at $x = 0$ and using (12) yields

$$\begin{aligned} c_1 \cdot 1 + c_2 \cdot 0 + c_3 \cdot 0 &= 0, \\ c_1 \cdot 0 + c_2 \cdot 1 + c_3 \cdot 0 &= 0, \\ c_1 \cdot 0 + c_2 \cdot 0 + c_3 \cdot 1 &= 0, \end{aligned}$$

so that $c_1 = c_2 = c_3 = 0$ and $\{y_1, y_2, y_3\}$ is linearly independent by (6).

This way of defining three linearly independent solutions is easy to implement on a computer, since the initial conditions are explicitly known. ◄

EXAMPLE 2.4.4 General Solution

Given that $\{\sin x, \cos x, e^x, e^{-x}\}$ is a fundamental set of solutions of $y'''' - y = 0$ and $e^{3x}/10$ is a solution of $y'''' - y = 8e^{3x}$, find the general solution of $y'''' - y = 8e^{3x}$.

Solution Since $\{\sin x, \cos x, e^x, e^{-x}\}$ is a fundamental set of solutions of the homogeneous equation $y'''' - y = 0$,

$$y_h = c_1 \sin x + c_2 \cos x + c_3 e^x + c_4 e^{-x}$$

is the general solution of $y'''' - y = 0$. We are given that

$$y_p = \frac{e^{3x}}{10}$$

is a particular solution of $y'''' - y = 8e^{3x}$. Thus, by Theorem 2.4.2,

$$y = y_p + y_h = \frac{e^{3x}}{10} + c_1 \sin x + c_2 \cos x + c_3 e^x + c_4 e^{-x}$$

is the general solution of $y'''' - y = 8e^{3x}$. ◄

As with second-order equations, the **Wronskian** may also be utilized. For n functions $\{y_1, \ldots, y_n\}$, the Wronskian is defined as

$$W[y_1, \ldots, y_n] = \det \begin{bmatrix} y_1(x) & y_2(x) & \cdots & y_n(x) \\ y_1'(x) & y_2'(x) & \cdots & y_n'(x) \\ \vdots & \vdots & \vdots & \vdots \\ y_1^{(n-1)}(x) & y_2^{(n-1)}(x) & \cdots & y_n^{(n-1)}(x) \end{bmatrix}, \tag{15}$$

where det again denotes the determinant. In particular, from Section 2.3,

$$W[y_1, y_2] = \det \begin{bmatrix} y_1(x) & y_2(x) \\ y_1'(x) & y_2'(x) \end{bmatrix},$$

and for $n = 3$,

$$W[y_1, y_2, y_3] = \det \begin{bmatrix} y_1(x) & y_2(x) & y_3(x) \\ y_1'(x) & y_2'(x) & y_3'(x) \\ y_1''(x) & y_2''(x) & y_3''(x) \end{bmatrix}.$$

Section 2.3 showed how to evaluate $W[y_1, y_2]$. The evaluation of the determinant (15) is covered in Section 5.8. All of the examples and exercises of this section may be done using the technique of Example 2.4.3 instead of using the Wronskian.

■ **THEOREM 2.4.3 Fundamental Set Is Equivalent to Nonzero Wronskian**

Suppose the nth-order homogeneous linear differential equations (1) and (2) have continuous coefficients and $a_n(x)$ is nonzero on the interval I. Let $\{y_1, \ldots, y_n\}$ be a set of n solutions to $L(y) = 0$. Then

a. $W[y_1, \ldots, y_n]$ is either always zero or never zero on I.
b. $\{y_1, \ldots, y_n\}$ is a fundamental set of solutions if and only if $W[y_1, \ldots, y_n] \neq 0$. ■

EXAMPLE 2.4.5 Fundamental Set of Solutions

a. Verify that $\{\sin 2x, \cos 2x, 1\}$ is a fundamental set of solutions for

$$y''' + 4y' = 0. \tag{16}$$

b. Find the solution of

$$y''' + 4y' = 0, \qquad y(0) = 0, \qquad y'(0) = 0, \qquad y''(0) = 2. \tag{17}$$

SOLUTION

a. First we must verify that $\sin 2x, \cos 2x, 1$ are solutions of (16). This is left to the reader. Since there are *three* solutions and (16) is *third*-order, it suffices to show that $1, \sin 2x, \cos 2x$ are linearly independent. We shall use (6). Suppose

$$c_1 \cdot 1 + c_2 \sin 2x + c_3 \cos 2x = 0, \tag{18}$$

and differentiate this equation twice, to also get

$$\begin{aligned} 2c_2 \cos 2x - 2c_3 \sin 2x &= 0, \\ -4c_2 \sin 2x - 4c_3 \cos 2x &= 0. \end{aligned} \tag{19}$$

Evaluate these equations [(18) and (19)] at $x = 0$, to yield

$$\begin{aligned} c_1 + c_2 \cdot 0 + c_3 \cdot 1 &= 0, \\ 2c_2 \cdot 1 - 2c_3 \cdot 0 &= 0, \\ -4c_2 \cdot 0 - 4c_3 \cdot 1 &= 0. \end{aligned} \tag{20}$$

The last two equations imply $c_2 = c_3 = 0$, and the first then gives $c_1 = 0$. Thus $\sin 2x, \cos 2x, 1$ are linearly independent.

Alternatively (for those using determinants),

$$W[\sin 2x, \cos 2x, 1](0) = \det \begin{bmatrix} 0 & 1 & 1 \\ 2 & 0 & 0 \\ 0 & -4 & 0 \end{bmatrix} = -8 \neq 0,$$

so $\sin 2x, \cos 2x, 1$ are linearly independent by Theorem 2.4.3.

b. Since $\{\sin 2x, \cos 2x, 1\}$ is a fundamental set of solutions for $y''' + 4y' = 0$, the general solution of $y''' + 4y' = 0$ must be a linear combination of $\sin 2x, \cos 2x, 1$:

$$y = c_1 \sin 2x + c_2 \cos 2x + c_3 \cdot 1. \tag{21}$$

Applying the initial conditions in (17) to (21), we get

$$0 = y(0) = c_1 \cdot 0 + c_2 \cdot 1 + c_3,$$
$$0 = y'(0) = c_1 \cdot 2 - c_2 \cdot 0,$$
$$2 = y''(0) = c_1 \cdot 0 - c_2 \cdot 4.$$

Starting with the last equation and working up, we find $c_2 = -\frac{1}{2}$, $c_1 = 0$, $c_3 = \frac{1}{2}$, and the solution of the initial value problem (17) is

$$y = -\frac{1}{2}\cos 2x + \frac{1}{2}.$$

Exercises

1. Is $\{x, x^{-1}, 1\}$ a linearly independent set of functions on $(0, 1]$?

2. Is $\{x + 1, x^2 + x, x^2 - 1\}$ a linearly independent set of functions on $[0, 1]$?

3. For the differential equation $L(y) = y''' - x^2y'' + x^3y = 0$, let y_1 be the solution of

$$L(y) = 0, \quad y(0) = y'(0) = 1, \quad y''(0) = 0;$$

y_2 be the solution of

$$L(y) = 0, \quad y(0) = y''(0) = 1, \quad y'(0) = 0;$$

y_3 be the solution of

$$L(y) = 0, \quad y''(0) = 3, \quad y(0) = y'(0) = 0.$$

a) Verify that $\{y_1, y_2, y_3\}$ is a fundamental set of solutions.

b) Find constants c_1, c_2, c_3 such that $\tilde{y} = c_1y_1 + c_2y_2 + c_3y_3$ is the solution of $L(y) = 0$, $y(0) = y''(0) = y'(0) = 1$.

4. For the differential equation $L(y) = y''' + \cos xy' + y = 0$, let y_1 be the solution of

$$L(y) = 0, \quad y(0) = y'(0) = 1, \quad y''(0) = 0;$$

y_2 be the solution of

$$L(y) = 0, \quad y(0) = y''(0) = 1, \quad y'(0) = 0;$$

y_3 be the solution of

$$L(y) = 0, \quad y'(0) = y''(0) = 1, \quad y(0) = 0.$$

a) Verify that $\{y_1, y_2, y_3\}$ is a fundamental set of solutions.

b) Find constants c_1, c_2, c_3 such that $\tilde{y} = c_1y_1 + c_2y_2 + c_3y_3$ is the solution of $L(y) = 0$, $y(0) = 0$, $y'(0) = 1$, $y''(0) = 2$.

5. a) Verify that $\{1, e^x, xe^x\}$ is a fundamental set of solutions of $y''' - 2y'' + y' = 0$ and that $y_p = x$ is a solution of $y''' - 2y'' + y' = 1$.

b) Give the general solution of $y''' - 2y'' + y' = 1$.

6. a) Verify that $\{\sin x, \cos x, \sin 2x, \cos 2x\}$ is a fundamental set of solutions of $y'''' + 5y'' + 4y = 0$ and that $y_p = 1$ is a solution of $y'''' + 5y'' + 4y = 4$.

b) Give the general solution of $y'''' + 5y'' + 4y = 4$.

7. a) Verify that $\{e^x, xe^x, x^2e^x\}$ is a fundamental set of solutions of $y''' - 3y'' + 3y' - y = 0$ and that $y_p = -e^{-x}$ is a solution of $y''' - 3y'' + 3y' - y = 8e^{-x}$.

b) Give the general solution of $y''' - 3y'' + 3y' - y = 8e^{-x}$.

8. a) Verify that $\{x, x^2, x^3\}$ is a fundamental set of solutions of $x^3y''' - 3x^2y'' + 6xy' - 6y = 0$ on $(0, \infty)$ and that $y_p = 3$ is a solution of $x^3y''' - 3x^2y'' + 6xy' - 6y = -18$.

b) Find the solution of

$$x^3y''' - 3x^2y'' + 6xy' - 6y = -18$$
$$y(1) = 0, \quad y'(1) = 0, \quad y''(1) = 0.$$

2.5 Reduction of Order

In the previous sections we have begun to study the second-order linear differential equation

$$y'' + p(x)y' + q(x)y = f(x).$$

We have shown that the general solution is in the form

$$y = y_p + c_1 y_1 + c_2 y_2.$$

At this point we have not described any methods for obtaining the solutions, y_1, y_2, of the associated homogeneous equation, or a particular solution, y_p. In fact, there are no general methods for obtaining y_1, y_2, y_p for all linear second-order equations (other than numerical methods). This contrasts with our study of linear first-order equations, for which the general solution can always be obtained by an integrating factor. Integrating factors do not exist for general second-order linear differential equations.

The **method of reduction of order** shows that if we have one solution y_1 of a homogeneous equation, then we can always obtain the second solution y_2. (However, there is no general method for obtaining one homogeneous solution in the first place.) We shall discuss the method, work three examples, and then summarize the method.

Suppose we wish to solve the homogeneous equation

$$y'' + p(x)y' + q(x)y = 0, \tag{1}$$

and that we are lucky enough to know one solution $y_1(x)$ of (1). We now show how to reduce (1) to a first-order differential equation. The key is to look for solutions of (1) in the form

$$y = vy_1, \tag{2}$$

where v is an unknown function of x and y_1 is our known solution of the homogeneous equation (1). Substituting $y = vy_1$ into (1) for y gives

$$(vy_1)'' + p(vy_1)' + q(vy_1) = 0.$$

Perform the differentiations,

$$(v''y_1 + 2v'y_1' + vy_1'') + p(v'y_1 + vy_1') + qvy_1 = 0,$$

and regroup by derivatives of v:

$$y_1v'' + (2y_1' + py_1)v' + (y_1'' + py_1' + qy_1)v = 0. \tag{3}$$

But by assumption, y_1 is a solution of the equation $y_1'' + py_1' + qy_1 = 0$. Thus the v terms in (3) always vanish, leaving

$$y_1v'' + (2y_1' + py_1)v' = 0. \tag{4}$$

Now y_1 and $2y_1' + py_1$ are known functions of x. Letting

$$w = v' \tag{5}$$

in (4) gives a first-order homogeneous linear equation in w:

$$y_1 w' + (2y_1' + py_1)w = 0. \tag{6}$$

Equation (6) may be solved by separation (Section 1.3) or by integrating factors (Section 1.6). We now illustrate the method of reduction of order by several examples. This method works whether or not the coefficient in front of the second derivative in (1) is 1.

EXAMPLE 2.5.1 Reduction of Order

First note that $y_1 = x$ is a solution of

$$x^2 y'' + 3xy' - 3y = 0, \qquad x > 0. \tag{7}$$

Find a second solution of (7) and give the general solution.

SOLUTION Let

$$y = vy_1 = vx, \tag{8}$$

and substitute for y in (7):

$$x^2(vx)'' + 3x(vx)' - 3(vx) = 0.$$

Performing the differentiations,

$$x^2(v''x + 2v') + 3x(v'x + v) - 3vx = 0.$$

Regrouping the terms, we see that the v terms cancel as promised, to leave

$$x^3 v'' + 5x^2 v' = 0.$$

Let $w = v'$ and divide by x^3 to get

$$\frac{dw}{dx} + \frac{5}{x}w = 0.$$

This is a first-order homogeneous linear differential equation. It can be solved by separation or integrating factors. By separation,

$$\int \frac{dw}{w} = -\int \frac{5}{x}dx.$$

Integrating (assuming $x > 0$) gives

$$\ln|w| = -5\ln x + c_0.$$

Solving for w yields

$$w = c_1 x^{-5}.$$

But $w = v'$, so that

$$v' = c_1 x^{-5}.$$

Now antidifferentiate to find v:

$$v = \frac{c_1}{-4} x^{-4} + c_2 = \tilde{c}_1 x^{-4} + c_2,$$

where $\tilde{c}_1$ is a new arbitrary constant. (We could just have as well used c_1 again instead of $\tilde{c}_1$.) Finally,

$$y = v y_1 = v x = \tilde{c}_1 x^{-3} + c_2 x \tag{9}$$

is the general solution. The second solution is x^{-3}. ◄

EXAMPLE 2.5.2 Solution with Definite Integral

First note that $y_1 = x$ is a solution of

$$y'' - xy' + y = 0. \tag{10}$$

Find a second solution of (10) and give the general solution.

Solution Let

$$y = v y_1 = v x, \tag{11}$$

and substitute for y in (10):

$$(vx)'' - x(vx)' + (vx) = 0.$$

Performing the differentiation,

$$(v''x + 2v') - x(v'x + v) + vx = 0.$$

Again the v terms cancel, to give

$$xv'' + (2 - x^2)v' = 0.$$

Let $w = v'$ and divide by x to get the first-order linear equation

$$\frac{dw}{dx} + \left(\frac{2}{x} - x\right)w = 0,$$

which can be solved by separation or by integrating factors. By integrating factors, we have

$$w = c_1 e^{-\int (2/x - x)\,dx} = c_1 e^{-2\ln x + \frac{1}{2}x^2} = c_1 x^{-2} e^{\frac{1}{2}x^2}.$$

Since $w = v'$,

$$v' = c_1 x^{-2} e^{\frac{1}{2}x^2}.$$

Antidifferentiation now requires a definite integral to find v:

$$v = c_1 \int_1^x t^{-2} e^{\frac{1}{2}t^2}\, dt + c_2.$$

Finally,

$$y = vy_1 = vx = c_1 x \int_1^x t^{-2} e^{\frac{1}{2}t^2}\, dt + c_2 x \tag{12}$$

is the general solution. It is a linear combination of two solutions of the homogeneous equation (10). The method of reduction of order has been used to obtain the second solution,

$$y_2 = x \int_1^x t^{-2} e^{\frac{1}{2}t^2}\, dt.$$

◄

The next example will be important in Section 2.6.

EXAMPLE 2.5.3 Reduction of Order

i. Find a solution of $y'' - 2y' + y = 0$ of the form e^{rx}.
ii. Use the solution from part (i) to find a fundamental set of solutions for $y'' - 2y' + y = 0$, using reduction of order.

SOLUTION

i. Let $y = e^{rx}$ and substitute into

$$y'' - 2y' + y = 0 \tag{13}$$

to get

$$r^2 e^{rx} - 2re^{rx} + e^{rx} = 0$$

or

$$r^2 - 2r + 1 = 0.$$

Thus $(r - 1)^2 = 0$ and $r = 1$, so e^x is a solution of (13).

ii. We have one solution $y_1 = e^x$ of (13). We shall use reduction of order to find the rest. Let $y = vy_1 = ve^x$, so that (13) becomes

$$(ve^x)'' - 2(ve^x)' + ve^x = 0.$$

Differentiate

$$(v''e^x + 2v'e^x + ve^x) - 2(v'e^x + ve^x) + ve^x = 0$$

and simplify, to get

$$v''e^x = 0, \qquad \text{or} \qquad v'' = 0.$$

Antidifferentiate twice:

$$v' = c_1, \tag{14}$$

$$v = c_1 x + c_2. \tag{15}$$

Thus,

$$y = vy_1 = (c_1 x + c_2)e^x = c_1 xe^x + c_2 e^x$$

is the general solution of (13). A fundamental set of solutions would be $\{e^x, xe^x\}$. The second solution xe^x is obtained by taking $c_1 = 1$ in (14) and $c_2 = 0$ in (15). ◄

Summary of Reduction of Order

Reduction of order can be used to find a fundamental set of solutions of the homogeneous equation $y'' + py' + qy = 0$ given one solution y_1 as follows:

1. Let $y = vy_1$ and substitute into $y'' + py' + qy = 0$.
2. This leads to a second-order linear equation in v with only v'' and v' terms. Let $w = v'$ to get a first-order linear equation (6) in w.
3. Find a nonzero solution for w by either separation or integrating factors.
4. Let v be an antiderivative of w.
5. Let $y_2 = vy_1$. Then $\{y_1, y_2\}$ is a fundamental set of solutions for $y'' + py' + qy = 0$.

Exercises

In Exercises 1 through 8, verify that the given function y_1 is a solution of the given homogeneous differential equation. Then find a fundamental set of solutions and the general solution for the differential equation. Exercises 7 and 8 require definite integrals.

1. $x^2 y'' + 3xy' + y = 0, \quad y_1 = x^{-1}$

2. $x^2 y'' + 5xy' + 3y = 0, \quad y_1 = x^{-1}$

3. $y'' + 10y' + 25y = 0, \quad y_1 = e^{-5x}$

4. $xy'' - (1 + x)y' + y = 0, \quad y_1 = e^x$

5. $xy'' + (x - 1)y' - y = 0, \quad y_1 = e^{-x}$

6. $y'' + 6y' + 9y = 0, \quad y_1 = e^{-3x}$

7. $y'' + xy' - y = 0, \quad y_1 = x$

8. $x^2 y'' + x^2 y' - 2(1 + x)y = 0, \quad y_1 = x^2$

In Exercises 9 through 12, find a solution y_1 of the given differential equations in the form x^r for some constant r. Then find a second solution by reduction of order.

9. $x^2 y'' - 3xy' + 4y = 0$

10. $x^2 y'' + 5xy' + 4y = 0$

11. $x^2 y'' + 7xy' + 9y = 0$

12. $x^2y'' - 5xy' + 9y = 0$

13. Find a solution of $y'' + 4y = 0$ of the form $\sin rx$ for some constant r. Then find the general solution of $y'' + 4y = 0$ by reduction of order.

14. Find a solution of $y'' + 16y = 0$ of the form $\sin rx$ for some constant r. Then find the general solution of $y'' + 16y = 0$ by reduction of order.

In Exercises 15 through 19, find a solution y_1 of the given differential equation in the form e^{rx} for some constant r. Then find a second solution and the general solution by reduction of order.

15. $y'' - 4y' + 4y = 0$

16. $y'' - 6y' + 9y = 0$

17. $y'' + 2y' + y = 0$

18. $y'' - 5y' + 6y = 0$

19. $y'' - 4y = 0$

20. Verify that $y_1(x) = x$ is a solution of (**Legendre equation of order one**) $(1 - x^2)y'' - 2xy' + 2y = 0$ on the interval $(-1, 1)$. Find the general solution on $(-1, 1)$. (*Note*: The integrations are a little more difficult, but can be worked using our techniques.)

21. Verify the statement in the summary of reduction of order that $\{y_1, y_2\}$ form a fundamental set of solutions.

22. Using reduction of order, find a general formula for the second solution of $y'' + p(x)y' + q(x)y = 0$ if y_1 is one solution.

2.6 Homogeneous Linear Constant Coefficient Differential Equations (Second-Order)

Linear constant coefficient differential equations form an important class of differential equations that appear both in physical models and as approximations for more complicated equations. Applications to electric circuits and mechanical systems will be given in Sections 2.8, 2.11, and 2.12.

This section will consider the general linear, second-order homogeneous, **constant coefficient** differential equation,

$$ay'' + by' + cy = 0, \tag{1}$$

where the coefficients a, b, and c are real constants and $a \neq 0$. From Sections 2.2 and 2.3, we know that the general solution of $ay'' + by' + cy = 0$ will be in the form

$$y = c_1y_1 + c_2y_2,$$

where $\{y_1, y_2\}$ is a fundamental set of solutions for $ay'' + by' + cy = 0$.

From linear first-order homogeneous constant coefficient differential equations,

$$\frac{dy}{dx} + ky = 0,$$

we stated in Section 1.7 that

> A homogeneous linear constant coefficient differential equation always has at least one solution of the form $y = e^{rx}$.

This also applies to second- and higher-order homogeneous linear constant coefficient differential equations.

The key to finding $\{y_1, y_2\}$ for linear differential equations with constant coefficients is to look for a solution of the form $y = e^{rx}$, where r is a constant. Substituting $y = e^{rx}$ into $ay'' + by' + cy = 0$ gives

$$a(e^{rx})'' + b(e^{rx})' + ce^{rx} = 0,$$

and, upon differentiation,

$$ar^2e^{rx} + bre^{rx} + ce^{rx} = 0.$$

Finally, divide by e^{rx}, which is always nonzero:

$$ar^2 + br + c = 0.$$

The polynomial $ar^2 + br + c$ is called the **characteristic polynomial** of

$$ay'' + by' + cy = 0.$$

The equation $ar^2 + br + c = 0$ is the **characteristic equation** of

$$ay'' + by' + cy = 0.$$

We have shown that

If r is a root of the characteristic polynomial $ar^2 + br + c$, then e^{rx} is a solution of $ay'' + by' + cy = 0$. (2)

Every second-degree polynomial has two roots. For $ar^2 + br + c = 0$, the roots are given by $(-b \pm \sqrt{b^2 - 4ac})/2a$. There are three cases, depending on whether $b^2 - 4ac > 0$, $b^2 - 4ac = 0$, or $b^2 - 4ac < 0$.

Case 1: Characteristic Polynomial Has Distinct Real Roots ($b^2 - 4ac > 0$)

Suppose the characteristic polynomial $ar^2 + br + c$ has two distinct real roots r_1, r_2. Then by (2), e^{r_1x}, e^{r_2x} are two solutions of $ay'' + by' + cy = 0$. These solutions form a fundamental set. (See Exercise 36.)

Thus

$$y = c_1e^{r_1x} + c_2e^{r_2x}$$

would be the general solution of $ay'' + by' + cy = 0$.

EXAMPLE 2.6.1 Distinct Real Roots

Find the general solution of $y'' - y' - 20y = 0$.

Solution By substituting $y = e^{rx}$ into the differential equation, we derive the characteristic equation $r^2 - r - 20 = 0$. Factoring gives $r^2 - r - 20 = (r - 5)(r + 4) = 0$. There are two distinct roots $r = 5$, $r = -4$. Thus e^{5x}, e^{-4x} are solutions, and the general solution is

$$y = c_1 e^{5x} + c_2 e^{-4x}.$$

◄

EXAMPLE 2.6.2 **Distinct Real Roots**

Find the general solution of

$$y'' + 4y' = 0.$$

Solution By substituting $y = e^{rx}$, we find that the characteristic equation is $r^2 + 4r = r(r + 4) = 0$. There are two distinct real roots $r = 0$, $r = -4$. A fundamental set of solutions would be $\{1, e^{-4x}\}$ since $e^{0x} = 1$. The general solution is

$$y = c_1 \cdot 1 + c_2 e^{-4x} = c_1 + c_2 e^{-4x}.$$

◄

Case 2: Characteristic Polynomial Has Repeated Real Root ($b^2 - 4ac = 0$)

Suppose that the characteristic polynomial has a single repeated root r_1. In this case, there is only one solution of the differential equation of the form e^{rx}. The second solution can always be found by reduction of order (Section 2.5). We shall give an example to illustrate and motivate the general case. However, we shall see that the reduction always produces the same type of second solution, so that it will not be necessary to carry the reduction out each time a differential equation is solved. Students who have omitted reduction of order may skip to the summary of repeated roots.

EXAMPLE 2.6.3 **Reduction of Order**

Using reduction of order, find the general solution of $y'' - 6y' + 9y = 0$.

Solution Substituting $y = e^{rx}$ into $y'' - 6y' + 9y = 0$, we find that the characteristic equation is $r^2 - 6r + 9 = (r - 3)^2 = 0$. Thus $r = 3$ is a repeated root. One solution is e^{3x}. To get the second solution, we will use reduction of order. Let $y = ve^{3x}$ and substitute into the differential equation to get

$$(ve^{3x})'' - 6(ve^{3x})' + 9(ve^{3x}) = 0$$

or, upon differentiation,

$$(v''e^{3x} + 6v'e^{3x} + 9ve^{3x}) - 6(v'e^{3x} + 3ve^{3x}) + 9ve^{3x} = 0.$$

This simplifies to

$$v''e^{3x} = 0 \qquad \text{or} \qquad v'' = 0.$$

Antidifferentiating twice yields

$$v = c_1x + c_2.$$

Thus

$$y = ve^{3x} = c_1xe^{3x} + c_2e^{3x}$$

is the general solution of $y'' - 6y' + 9y = 0$. Note that e^{3x} was one solution, and we have found that a second solution is xe^{3x}. ◄

Summary of Repeated Roots

When $y_1 = e^{r_1x}$ is a solution of (1) corresponding to repeated roots, a second solution is always of the form

$$y_2 = xe^{r_1x}. \tag{3}$$

In the examples, exercises, and text that follow, do not use reduction of order to obtain the second solution unless asked to. Use (3) instead.

EXAMPLE 2.6.4 **Repeated Real Roots**

Find the general solution of $y'' + 2y' + y = 0$.

SOLUTION By substituting $y = e^{rx}$, we find that the characteristic equation is $r^2 + 2r + 1 = (r + 1)^2 = 0$. Thus -1 is a repeated root. One homogeneous solution is e^{-x}. Since the root is repeated, the second solution is xe^{-x} according to (3). Thus the general solution of the homogeneous equation is

$$y = c_1e^{-x} + c_2xe^{-x}.$$ ◄

Case 3: Characteristic Polynomial Has Complex Roots ($b^2 - 4ac < 0$)

Since the differential equation

$$ay'' + by' + cy = 0 \tag{4}$$

has constant coefficients, solutions exist in the form $y = e^{rx}$ if r satisfies the characteristic equation

$$ar^2 + br + c = 0. \tag{5}$$

If $b^2 - 4ac < 0$, then the roots are complex. Suppose that

$$r_1 = \alpha + i\beta \qquad (i^2 = -1),$$

with α, β real numbers, is a complex root of (5). Since a, b, c are real, the other root r_2 must be the **complex conjugate** of r_1. That is,

$$r_2 = \alpha - i\beta.$$

The general solution of (4) can be represented by a linear combination of these two complex exponentials:

$$y = \tilde{c}_1 e^{r_1 x} + \tilde{c}_2 e^{r_2 x}. \tag{6}$$

However, (6) is not particularly useful in this form for some physical problems, since both $e^{r_1 x}$ and $e^{r_2 x}$ involve complex numbers. We shall replace $\{e^{r_1 x}, e^{r_2 x}\}$ by a different fundamental set of solutions. The arbitrary constants in (6) are denoted $\tilde{c}_1$, $\tilde{c}_2$ to distinguish them from the c_1, c_2 we use in the other set of solutions.

Complex exponentials satisfy the usual algebraic properties of exponentials, $e^{r_1 x} = e^{(\alpha + i\beta)x} = e^{\alpha x} e^{i\beta x}$. Thus, the general solution (6) can be written

$$y = \tilde{c}_1 e^{\alpha x} e^{i\beta x} + \tilde{c}_2 e^{\alpha x} e^{-i\beta x}$$

or

$$y = e^{\alpha x}\left(\tilde{c}_1 e^{i\beta x} + \tilde{c}_2 e^{-i\beta x}\right). \tag{7}$$

This solution is still not real. However, because of the Euler formula, $e^{i\beta x} = \cos \beta x + i \sin \beta x$, we will shortly show that an arbitrary linear combination of $e^{\pm i\beta x}$ is equivalent to an arbitrary linear combination of $\cos \beta x$ and $\sin \beta x$. Thus, (7) can be written $y = e^{\alpha x}(c_1 \cos \beta x + c_2 \sin \beta x)$. To summarize:

In the case of complex roots, $\alpha \pm \beta i$, the general solution of

$$ay'' + by' + cy = 0 \text{ is given by}$$

$$\begin{aligned} y &= e^{\alpha x}(c_1 \cos \beta x + c_2 \sin \beta x) \\ &= c_1 e^{\alpha x} \cos \beta x + c_2 e^{\alpha x} \sin \beta x. \end{aligned} \tag{8}$$

To understand this result better requires some further discussion of complex numbers.

Euler's Formula

The relationship between complex exponentials and sines and cosines follows from **Euler's formula,**

$$e^{i\theta} = \cos\theta + i\sin\theta. \tag{9}$$

An additional important relationship follows from replacing θ by $-\theta$ in (9) and using the evenness of cosine, $\cos(-\theta) = \cos\theta$, and the oddness of sine, $\sin(-\theta) = -\sin\theta$, to get

$$e^{-i\theta} = \cos\theta - i\sin\theta. \tag{10}$$

To explain Euler's formula (9), we assume that the Taylor series for the real exponential,

$$e^{x} = 1 + x + \frac{x^2}{2!} + \frac{x^3}{3!} + \frac{x^4}{4!} + \cdots, \tag{11}$$

is valid for the complex exponential. That is, we can let $x = i\theta$ in (11) to get

$$e^{i\theta} = 1 + i\theta - \frac{\theta^2}{2!} - i\frac{\theta^3}{3!} + \frac{\theta^4}{4!} + \cdots.$$

By collecting the real and imaginary parts, we have

$$e^{i\theta} = \left(1 - \frac{\theta^2}{2!} + \frac{\theta^4}{4!} + \cdots\right) + i\left(\theta - \frac{\theta^3}{3!} + \cdots\right).$$

Euler's formula (9) now follows from the Taylor series for $\cos\theta$ and $\sin\theta$:

$$\cos\theta = 1 - \frac{\theta^2}{2!} + \frac{\theta^4}{4!} + \cdots,$$

$$\sin\theta = \theta - \frac{\theta^3}{3!} + \cdots.$$

In our applications to differential equations, $\theta = \beta x$.

Sines and Cosines Are Equivalent to the Complex Exponentials

We have shown that the general solution (7) of the linear differential equation is

$$y = e^{\alpha x}\left(\tilde{c}_1 e^{i\beta x} + \tilde{c}_2 e^{-i\beta x}\right). \tag{12}$$

The linear combination of complex exponentials that appears in (12) may be related to $\cos\beta x$ and $\sin\beta x$ using Euler's formulas [(9) and (10)], as follows:

$$\tilde{c}_1 e^{i\beta x} + \tilde{c}_2 e^{-i\beta x} = \tilde{c}_1(\cos\beta x + i\sin\beta x) + \tilde{c}_2(\cos\beta x - i\sin\beta x).$$

By collecting the $\cos \beta x$ and $\sin \beta x$, we obtain

$$\tilde{c}_1 e^{i\beta x} + \tilde{c}_2 e^{-i\beta x} = c_1 \cos \beta x + c_2 \sin \beta x, \tag{13}$$

where $c_1 = \tilde{c}_1 + \tilde{c}_2$ and $c_2 = i(\tilde{c}_1 - \tilde{c}_2)$. Equation (13) shows that

An arbitrary linear combination of $e^{i\beta x}$ and $e^{-i\beta x}$,

$$\tilde{c}_1 e^{i\beta x} + \tilde{c}_2 e^{-i\beta x}$$

is equivalent to an arbitrary linear combination of $\cos \beta x$ and $\sin \beta x$, (14)

$$c_1 \cos \beta x + c_2 \sin \beta x.$$

It is helpful to memorize (14).

Alternative Derivation

By adding and subtracting Euler's formula [(9) and (10)], we derive fundamental relationships for $\cos \theta$ and $\sin \theta$:

$$\begin{aligned} \cos \theta &= \frac{1}{2} e^{i\theta} + \frac{1}{2} e^{-i\theta} \\ \sin \theta &= \frac{1}{2i} e^{i\theta} - \frac{1}{2i} e^{-i\theta}. \end{aligned} \tag{15}$$

For example, $\cos \theta$ is a specific linear combination of $e^{i\theta}$ and $e^{-i\theta}$. By letting $\theta = \beta x$ in (15) and multiplying by $e^{\alpha x}$, we obtain

$$\begin{aligned} e^{\alpha x} \cos \beta x &= e^{\alpha x}\left(\frac{1}{2} e^{i\beta x} + \frac{1}{2} e^{-i\beta x}\right), \\ e^{\alpha x} \sin \beta x &= e^{\alpha x}\left(\frac{1}{2i} e^{i\beta x} - \frac{1}{2i} e^{-i\beta x}\right). \end{aligned} \tag{16}$$

Since the right-hand sides of (16) are linear combinations of homogeneous solutions, then $e^{\alpha x} \cos \beta x$ and $e^{\alpha x} \sin \beta x$ are also solutions of $ay'' + by' + cy = 0$.

EXAMPLE 2.6.5 **Complex Roots**

Find the general solution of

$$y'' + y' + y = 0.$$

SOLUTION The characteristic polynomial is $r^2 + r + 1 = 0$. By the quadratic formula, the roots are

$$r = \frac{-1 \pm \sqrt{1 - 4}}{2} = -\frac{1}{2} \pm i\frac{\sqrt{3}}{2}.$$

Since $e^{(-\frac{1}{2} \pm i\sqrt{3}/2)x} = e^{-\frac{1}{2}x} e^{\pm i(\sqrt{3}/2)x}$, we have

$$y = c_1 e^{-x/2} \cos\left(\frac{\sqrt{3}}{2}x\right) + c_2 e^{-x/2} \sin\left(\frac{\sqrt{3}}{2}x\right)$$

is the general solution of $y'' + y' + y = 0$.

Note that the root $-(1/2) + i(\sqrt{3}/2)$ does not give the solution $e^{-x/2} \cos[(\sqrt{3}/2)x]$. Rather, the *pair* of solutions $e^{-x/2} \cos[(\sqrt{3}/2)x]$, $e^{-x/2} \sin[(\sqrt{3}/2)x]$ comes from the *pair* of roots

$$-\frac{1}{2} + i\frac{\sqrt{3}}{2}, \qquad -\frac{1}{2} - i\frac{\sqrt{3}}{2}.$$ ◄

EXAMPLE 2.6.6 Pure Imaginary Roots

Find the general solution of $y'' + 4y = 0$.

SOLUTION The characteristic polynomial is $r^2 + 4 = 0$, so that the roots are $r = \alpha \pm i\beta = \pm 2i$. Thus $\alpha = 0$, $\beta = 2$, and a fundamental set of solution is

$$e^{0x} \cos 2x = \cos 2x \qquad \text{and} \qquad e^{0x} \sin 2x = \sin 2x.$$

The general solution is

$$y = c_1 \cos 2x + c_2 \sin 2x.$$ ◄

For convenience, we summarize the three cases.

Solution of $ay'' + by' + cy = 0$ with a, b, c Real Constants

First find the roots r_1, r_2 of the characteristic equation $ar^2 + br + c = 0$. There are three cases:

1. If r_1, r_2 are distinct real roots, then the general solution is $y = c_1 e^{r_1 x} + c_2 e^{r_2 x}$.
2. If $r_1 = r_2$ is a repeated real root, the general solution is $y = c_1 e^{r_1 x} + c_2 x e^{r_1 x}$.
3. If r_1, r_2 are complex roots, they are a conjugate pair:

$$r_1 = \alpha + i\beta, \qquad r_2 = \alpha - i\beta.$$

The general solution =

$$y \text{ is } c_1 e^{\alpha x} \cos \beta x + c_2 e^{\alpha x} \sin \beta x.$$

As will be shown in Sections 2.8 and 2.12, there is a close relationship between electric circuits and mechanical systems and linear differential equations with constant coefficients. In these problems, one often starts knowing the desired response (solution) and wants to design the device (make up the differential equation).

EXAMPLE 2.6.7 Obtaining Differential Equation from Its Solution

Find a second-order linear homogeneous constant coefficient differential equation that has $c_1e^{-3x} + c_2e^{-2x}$ as its general solution.

SOLUTION $c_1e^{-3x} + c_2e^{-2x}$ will be the general solution if -3 and -2 are roots of the characteristic equation. One such characteristic polynomial would be

$$[r - (-3)][r - (-2)] = (r + 3)(r + 2) = r^2 + 5r + 6.$$

A corresponding differential equation is

$$y'' + 5y' + 6y = 0.$$

Note that $2y'' + 10y' + 12y = 0$ would be another correct answer, since the roots determine the polynomial only up to a constant factor. ◄

Exercises

In Exercises 1 through 33, solve the differential equation. Determine the general solution if no initial conditions are given.

1. $y'' + y' - 6y = 0$

2. $y'' - y = 0$

3. $y'' + y = 0$

4. $y'' + 4y' + 4y = 0$

5. $y'' + 4y' + 5y = 0$

6. $y'' - 2y' + y = 0$

7. $y'' - 3y' + 2y = 0$

8. $2y'' - 2y' + y = 0$

9. $y'' - y' = 0$

10. $4y'' + 8y' + 3y = 0$

11. $3y'' = 0$

12. $y'' - 2y' + 2y = 0$

13. $3y'' + 2y' - y = 0$

14. $y'' + 9y = 0$, $y(0) = 1$, $y'(0) = 1$

15. $y'' + y' - 2y = 0$, $y(0) = 0$, $y'(0) = 1$

16. $2y'' + 12y' + 18y = 0$, $y(0) = 1$, $y'(0) = 0$

17. $2y'' + 4y = 0$

18. $3y'' - 24y' + 45y = 0$

19. $2y'' + 8y' + 6y = 0$, $y(0) = 2$, $y'(0) = 0$

20. $y'' - 16y = 0$

21. $y'' + 10y' + 25y = 0$

22. $2y'' + 3y' = 0$

23. $y'' - 14y' + 49y = 0$

24. $y'' + 4y' + 20y = 0$

25. $y'' - 6y' + 25y = 0$

26. $y'' + y' + y = 0$

27. $y'' - 12y = 0$

28. $y'' + 2y' + 8y = 0$

29. $y'' + 4y' + 8y = 0$

30. $y'' + 10y = 0$

31. $y'' + 8y = 0$

32. $y'' + 6y' + 11y = 0$

33. $y'' + 6y' + 7y = 0$

34. Suppose that r is a real number. Verify that e^{rx}, xe^{rx} are linearly independent.

35. Suppose α, β are real numbers and $\beta \neq 0$. Verify that $e^{\alpha x} \cos \beta x, e^{\alpha x} \sin \beta x$ are linearly independent.

36. Suppose that $r_1 \neq r_2$. Verify that $\{e^{r_1 x}, e^{r_2 x}\}$ are linearly independent.

37. Show that if $ar^2 + br + c$ has the repeated root r_1, then $ar^2 + br + c = a(r - r_1)^2$.

38. Using reduction of order and the previous exercise, show that if r_1 is a repeated root of $ar^2 + br + c$, then $xe^{r_1 x}$ is a solution of $ay'' + by' + cy = 0$.

39. From the solutions of Exercises 1 through 33, find an example with repeated roots and *derive* the second solution using reduction of order.

In Exercises 40 through 51, determine a homogeneous second-order linear constant coefficient differential equation with the given expression as its general solution.

40. $y = c_1 e^{3x} + c_2 e^{-4x}$

41. $y = c_1 e^{-x} + c_2 e^{-2x}$

42. $y = c_1 e^{2x} + c_2 x e^{2x}$

43. $y = c_1 e^{3x} + c_2 x e^{3x}$

44. $y = c_1 + c_2 e^{-5x}$

45. $y = c_1 \sin 4x + c_2 \cos 4x$

46. $y = c_1 e^{-x} \sin 2x + c_2 e^{-x} \cos 2x$

47. $y = c_1 \sin 3x + c_2 \cos 3x$

48. $y = c_1 e^{2x} \sin 3x + c_2 e^{2x} \cos 3x$

49. $y = c_1 + c_2 x$

50. $y = c_1 \sin 2x + c_2 \cos 2x$

51. $y = c_1 e^{x} \sin x + c_2 e^{x} \cos x$

2.7 Homogeneous Linear Constant Coefficient Differential Equations (*n*th-Order)

The solution of the nth-order homogeneous linear constant coefficient differential equation

$$a_n y^{(n)} + a_{n-1} y^{(n-1)} + \cdots + a_1 y' + a_0 y = 0 \tag{1}$$

(here $y^{(m)} = d^m y/dx^m$), where the coefficients $a_n, \ldots, a_0$ are constants and $a_n \neq 0$ proceeds almost exactly as for the second-order case ($n = 2$) in Section 2.6. Again, $y = e^{rx}$ will be a solution if r is a root of the **characteristic equation**

$$a_n r^n + a_{n-1} r^{n-1} + \cdots + a_1 r + a_0 = 0. \tag{2}$$

The nth-degree polynomial (2) has n roots, but they need not be distinct. The **multiplicity** of a root is the number of times it is repeated. The sum of the multiplicities of all the distinct roots equals the degree of the polynomial.

EXAMPLE 2.7.1 Multiplicity of a Root

The third-degree polynomial $r^3 - 3r^2 + 3r - 1 = (r - 1)^3$ has 1 as a root of multiplicity 3. ◄

EXAMPLE 2.7.2 Sum of the Multiplicities

The third-degree polynomial $(r - 4)^2(r - 8)$ has two distinct roots; 4 is a root of multiplicity 2, and 8 is a root of multiplicity 1. The sum of the multiplicities $2 + 1 = 3$ equals the degree of the polynomial. ◄

EXAMPLE 2.7.3 Repeated Imaginary Roots

The fourth-degree polynomial $r^4 + 2r^2 + 1 = (r^2 + 1)^2 = (r - i)^2(r + i)^2$ has two complex roots $\pm i$, each of multiplicity 2. ◄

The procedure for the solution of (1) is summarized in the following algorithm.

Procedure for Solution of $a_n y^{(n)} + a_{n-1} y^{(n-1)} + \cdots + a_1 y' + a_0 y = 0$ When $a_n, a_{n-1}, \ldots, a_0$ Are Constants

1. Form the characteristic equation $a_n r^n + a_{n-1} r^{n-1} + \cdots + a_1 r + a_0 = 0$ and determine its roots and their multiplicities. (The characteristic equation can be found by substituting $y = e^{rx}$.)
2. A fundamental set of solutions for (1) is determined as follows:
   a. If r_1 is a real root of multiplicity m, then include the m functions

$$\{e^{r_1 x}, xe^{r_1 x}, \ldots, x^{m-1} e^{r_1 x}\} \tag{3}$$

   in the fundamental set of solutions.

   b. If $\alpha \pm \beta i$ is a pair of complex conjugate roots and they each have multiplicity m, then include the $2m$ functions

$$\{e^{\alpha x} \cos \beta x, e^{\alpha x} \sin \beta x, xe^{\alpha x} \cos \beta x, xe^{\alpha x} \sin \beta x, \ldots,$$
$$x^{m-1} e^{\alpha x} \cos \beta x, x^{m-1} e^{\alpha x} \sin \beta x\} \tag{4}$$

   in the fundamental set of solutions. (If $\alpha + \beta i$ is a root of multiplicity m, then $\alpha - \beta i$ will be a root of multiplicity m also, since $a_n, \ldots, a_o$ are real.)

EXAMPLE 2.7.4 Repeated Imaginary Roots

Find the general solution of $y'''' + 2y'' + y = 0$.

SOLUTION The characteristic equation is $r^4 + 2r^2 + 1 = (r^2 + 1)^2 = 0$, as can be found by substituting $y = e^{rx}$. Thus $\pm i$ are complex conjugate roots of multiplicity 2. In the notation of (4):

$$i = 0 + 1 \cdot i, \qquad \text{so that } \alpha = 0 \qquad \text{and} \qquad \beta = 1.$$

Two solutions are $\cos x, \sin x$ corresponding to the roots $r = \pm i$. Since the complex roots have multiplicity 2, two other independent solutions are $x \sin x$, $x \cos x$. Thus, a fundamental set of solutions is

$$\{\cos x, \sin x, x \cos x, x \sin x\}.$$

The general solution is

$$y = c_1 \cos x + c_2 \sin x + c_3 x \cos x + c_4 x \sin x.$$

◄

EXAMPLE 2.7.5 Repeated Real Roots

Find the general solution of

$$y^{(5)} - 3y^{(4)} + 3y^{(3)} - y^{(2)} = 0.$$

SOLUTION The characteristic equation is

$$\begin{aligned} r^5 - 3r^4 + 3r^3 - r^2 &= r^2(r^3 - 3r^2 + 3r - 1) \\ &= r^2(r - 1)^3 = 0. \end{aligned}$$

There are two distinct roots 0, 1, of multiplicities 2 and 3. Using (3), the root 0 of multiplicity 2 means that we include

$$\{e^{0x}, xe^{0x}\} = \{1, x\}$$

in the fundamental set of solutions. The root 1 of multiplicity 3 means that we include $\{e^x, xe^x, x^2e^x\}$ in the fundamental set of solutions. The general solution is thus

$$y = c_1 + c_2 x + c_3 e^x + c_4 xe^x + c_5 x^2 e^x.$$

◄

EXAMPLE 2.7.6 Repeated Roots

Find the general solution of the linear homogeneous constant coefficient differential equation if the roots of the characteristic equation are known to be $7 \pm 3i$ repeated twice, $\pm 4i$, and 5.

SOLUTION Since there are 7 roots, the general solution is a linear combination of 7 terms using 7 arbitrary constants.

$$y = c_1 e^{7x} \cos 3x + c_2 e^{7x} \sin 3x + c_3 x e^{7x} \cos 3x + c_4 x e^{7x} \sin 3x$$
$$+ c_5 \cos 4x + c_6 \sin 4x + c_7 e^{5x}.$$

In actual applications the roots of the characteristic equation will usually not be integers. They are often found (estimated) using a numerical procedure. In some cases this could be Newton's method from calculus. In this case, care must be taken when determining whether values such as, say, 1.123, 1.124 represent distinct real roots that are close together, or one root of multiplicity 2 whose computation has been influenced by roundoff error or the numerical method.

Exercises

For Exercises 1 through 23, find the general solution.

1. $y''' - 6y'' + 12y' - 8y = 0$

2. $y'''' + 5y'' + 4y = 0$

3. $y'''' - 5y'' + 4y = 0$

4. $y'''' + 8y'' + 16y = 0$

5. $y''' + y'' - 2y' = 0$

6. $y'''' - 2y''' = 0$

7. $y'''' + 4y''' + 6y'' + 4y' + y = 0$

8. $y''' - 2y'' - y' + 2y = 0$

9. $y' - 3y = 0$

10. $y' + 4y = 0$

11. $y''' + y'' - 2y = 0$

12. $y'''' + 4y'' + 4y = 0$

13. $y'''' + 4y''' + 8y'' + 8y' + 4y = 0$

14. $y''' + 3y'' + 3y' + y = 0$

15. $y^{(4)} - 4y^{(3)} + 6y^{(2)} - 4y^{(1)} + y = 0$

16. $y^{(6)} + 3y^{(4)} + 3y^{(2)} + y = 0$

17. $y^{(6)} - 3y^{(4)} + 3y^{(2)} - y = 0$

18. $y^{(4)} - 16y = 0$

19. $y^{(4)} - y = 0$

20. $y'''' + 2y''' + 2y'' = 0$

21. $y'''' + 50y'' + 625y = 0$

22. $3y' + 4y = 0$

23. $y''' + 3y'' + y' - 5y = 0$

For Exercises 24 through 33, you are given the general solution. Write down a homogeneous linear constant coefficient differential equation that has that general solution.

24. $y = c_1 + c_2 x + c_3 x^2 + c_4 x^3$

25. $y = c_1 e^{2x} + c_2 x e^{2x} + c_3$

26. $y = c_1 e^x \sin 2x + c_2 e^x \cos 2x + c_3 e^{-x} + c_4 x e^{-x}$

27. $y = c_1 \sin 5x + c_2 \cos 5x + c_3 x \sin 5x + c_4 x \cos 5x$

28. $y = c_1 e^x + c_2 e^{2x} + c_3 e^{-x}$

29. $y = c_1 e^{-2x} + c_2 x e^{-2x} + c_3 x^2 e^{-2x}$

30. $y = c_1 e^x + c_2 x e^x + c_3 e^{-x} + c_4 x e^{-x}$

31. $y = c_1 e^x \sin x + c_2 e^x \cos x + c_3 x e^x \sin x + c_4 x e^x \cos x$

32. $y = c_1 e^x + c_2 e^{2x} + c_3 e^{3x} + c_4 e^{4x}$

33. $y = c_1 \sin x + c_2 \cos x + c_3 x \sin x + c_4 x \cos x$

34. Find all solutions of $y''' - 3y'' + 3y' - y = 0$ of the form e^{rx}. Then find the general solution by reduction of order (Section 2.5). Compare your answer to that obtained by using (3).

In Exercises 35 through 42, you are given the roots of the characteristic equation of a linear constant coefficient homogeneous differential equation. Find the general solution of the differential equation.

35. 8 repeated four times and $8 \pm 7i$

36. $3 \pm 8i$ repeated two times

37. $4 \pm 5i$, $\pm 5i$ repeated two times, and 0 repeated three times

38. $\pm 5i$ repeated three times, $\pm i$, and 0

39. $2 \pm i$ repeated three times

40. $7 \pm 3i$, 7, -7, $\pm 3i$

41. $2 \pm 6i$, $-2 \pm 6i$, and $\pm 6i$ repeated twice

42. $3 \pm 5i$, $-3 \pm 5i$, and 0 repeated four times

2.8 Mechanical Vibrations I: Formulation and Free Response

2.8.1 Formulation of Equations

This section, like Section 1.11, will study a problem involving the linear motion of a rigid body governed by **Newton's law**,

$$\frac{d(mv)}{dt} = F_T, \tag{1}$$

where m is mass, v is velocity, mv is momentum, t is time, and F_T is the total force acting on the body. The resulting differential equations will be analyzed using the preceding sections on linear constant coefficient differential equations. We shall again use the centimeter-gram-second (cgs) and foot-pound-second (fps) systems of measurement (see Table 2.8.1 at the end of this section).

Horizontal Spring-Mass System

Suppose, then, that we have a horizontal spring connected to a wall with a constant mass m attached (see Figure 2.8.1). We shall assume that the total force F_T is made up of three forces:

F_s = force exerted by the spring
F_r = a friction or resistive force acting on the mass (air resistance, for example)
F_e = any other external forces (such as magnetism)

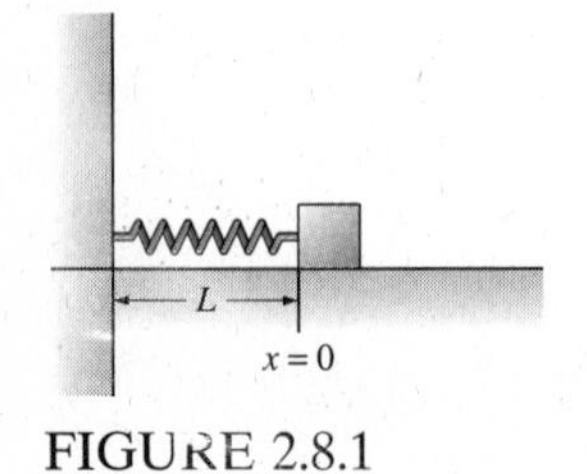

FIGURE 2.8.1

Then Newton's law (1) takes the form

$$m\frac{d^2x}{dt^2} = F_T = F_s + F_r + F_e. \tag{2}$$

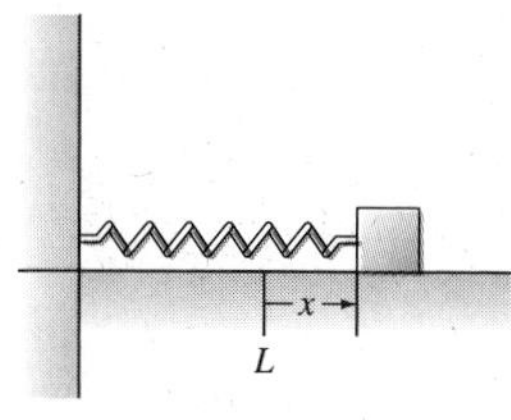

FIGURE 2.8.2

$x > 0$.

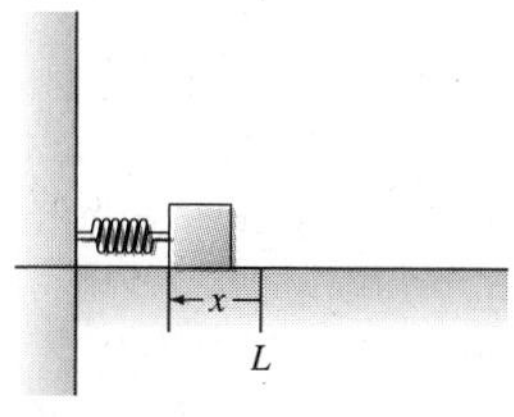

FIGURE 2.8.3

$x < 0$.

Let L be the length of the spring with no mass attached. We introduce a coordinate system in which $x = 0$ is located a distance L from the wall, as illustrated in Figure 2.8.1. Then $x(t)$ is the distance of the mass from this position. If $x > 0$ (as illustrated in Figure 2.8.2), the spring is stretched from its natural length a distance x, and it is known that the spring exerts a force to the left. In Figure 2.8.3, $x < 0$, in which case the spring is compressed and the spring exerts a force to the right.

We can now express the different forces in (2) in terms of x. To begin with, assume that the spring satisfies **Hooke's law**,

> The force exerted by the spring is proportional to the distance the spring is stretched (or compressed).

This is an assumption concerning not only the type of spring but also the mass, external forces, and initial conditions. In particular, we are assuming that the resulting motion is not of too great an amplitude, so that linear equations may be used. Because of the coordinate system we have chosen, the distance the spring is stretched is x, and Hooke's law becomes

$$F_s = -kx \tag{3}$$

where k is a positive constant, called the **spring constant**. k is positive because the spring force acts opposite to the sign of x. That is, if the spring is extended, $x > 0$, then the spring force is negative, and if the spring is compressed, $x < 0$, then the spring force is positive.

Many types of resistive forces at low velocities are (approximately) proportional to the velocity. That is,

$$F_r = -\delta v = -\delta\frac{dx}{dt} \qquad \text{with } \delta > 0. \tag{4}$$

The constant δ is called the **damping constant**. The constant of proportionality, $-\delta$, is negative, since the force of friction acts in the opposite direction to the motion and we want $\delta \geq 0$. Note that (4), like Hooke's law, represents a restriction on the motion to be studied and not just an assumption concerning the type of spring and the type of resistance. Finally, we assume the external forces F_e depend only on time t and not on position x or velocity dx/dt.

Then substituting (3) and (4) into Newton's law (2), we get the linear differential equation

$$m\frac{d^2x}{dt^2} = -kx - \delta\frac{dx}{dt} + f(t),$$

where $f(t) = F_e(t)$. This differential equation is usually written in the form

$$m\frac{d^2x}{dt^2} + \delta\frac{dx}{dt} + kx = f(t). \tag{5}$$

Note that m, δ, and k are all nonnegative constants. The initial conditions usually used with (5) are the initial position $x(0)$ and the initial velocity $v(0) = x'(0)$.

Sometimes it is convenient to think of the resistive force as being due to an attached device, such as a piston moving in a fluid. One such device is called a **dashpot**. Conceptually, this approach is similar to modeling the resistance in wire by the inclusion of a small resistor in the circuit equations instead of thinking of the resistance as distributed throughout the wire.

Spring-Mass System with Gravity

A few additional complications occur for a vertically vibrating spring-mass system in which the force due to gravity must be included. After some effort in choosing an appropriate coordinate system, we will show that (5) is still valid.

Suppose that a spring-mass system is hanging from a fixed support. The force of gravity acting downwards on the mass (the weight) is

$$F_g = mg, \tag{6}$$

where $g = 32\ \text{ft/s}^2$ (fps) or $980\ \text{cm/s}^2$ (cgs). Again let L be the length of the hanging spring with no mass attached. Let y measure a change from this rest length. Thus $y = 0$ is a distance L from the support. We measure y downward so that positive y corresponds to additional spring length. Then Newton's law with the gravitational force (6) becomes

$$m\frac{d^2y}{dt^2} + \delta\frac{dy}{dt} + ky = mg + f(t). \tag{7}$$

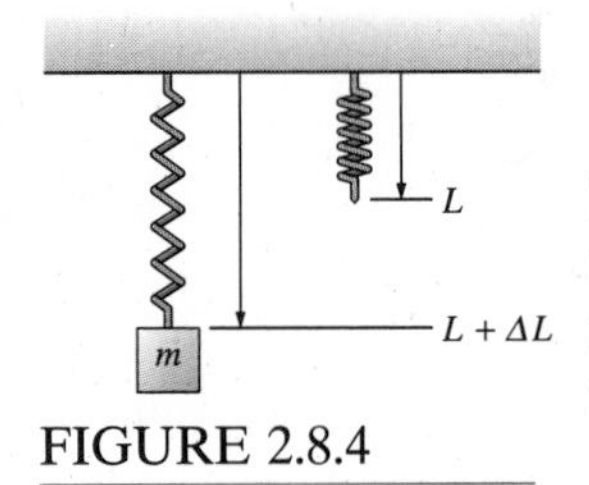

FIGURE 2.8.4

First we analyze what happens without an external force, $f(t) = 0$. From our experience, we know that with gravity the mass will sag (see Figure 2.8.4) and be at rest at $y = \Delta L$ (which we will determine). If the mass is at rest ($y = \Delta L$), then the force of gravity must balance out the spring force, so that

$$k\,\Delta L = mg. \tag{8}$$

The stretching ΔL of the spring with the gravitational force is given by (8). Often in experiments the weight of the object mg is known and ΔL is easily measured. Thus, the spring constant k can be determined from (8).

If we choose (even with an external force) a coordinate system x centered at the sagged rest position (see Figure 2.8.5), then

$$y = x + \Delta L. \tag{9}$$

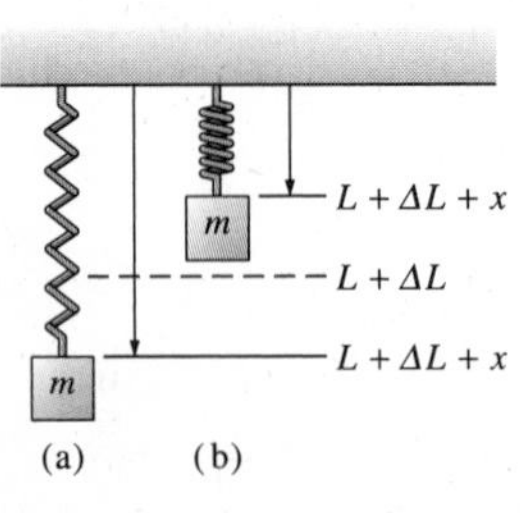

FIGURE 2.8.5

Note that $x = 0$ corresponds to $y = \Delta L$. If (9) is substituted into (7), we obtain

$$m\frac{d^2x}{dt^2} + \delta\frac{dx}{dt} + k(x + \Delta L) = mg + f(t).$$

But $k\,\Delta L = mg$ from (8), and so we have finally

$$m\frac{d^2x}{dt^2} + \delta\frac{dx}{dt} + kx = f(t). \tag{10}$$

Equation (10) is the same as (5), showing that a vertically vibrating spring-mass system with a gravitational force moves the same way as a horizontally vibrating spring-mass system if a coordinate system is chosen that takes into account the sag due to gravity.

Free Response

Suppose that there is no external force other than gravity, so that the external force is $f(t) = 0$. The solutions of the resulting homogeneous differential equation,

$$m\frac{d^2x}{dt^2} + \delta\frac{dx}{dt} + kx = 0, \tag{11}$$

are called the **free** (or **natural**) **response** of the spring-mass system. This is how the system reacts for given initial conditions if it is allowed to proceed without external interference. Note that the free response is the same thing as the solution of the associated homogeneous equation. The resulting behavior is quite different, depending on whether or not there is any damping present.

2.8.2 Simple Harmonic Motion (No Damping, $\delta = 0$)

Suppose there is no damping ($\delta = 0$), so that the free response of a spring-mass system is given by the differential equation

$$m\frac{d^2x}{dt^2} + kx = 0. \tag{12}$$

We can solve (12) using the techniques of Section 2.6. Since the differential equation has constant coefficients (and is homogeneous), we substitute $x = e^{rt}$. The characteristic equation is $mr^2 + k = 0$, with roots $r = \pm i\sqrt{k/m}$, since $m > 0$, $k > 0$. Thus, the general solution of (12) is

$$x = c_1 \cos\sqrt{\frac{k}{m}}\,t + c_2 \sin\sqrt{\frac{k}{m}}\,t, \tag{13}$$

where c_1 and c_2 are arbitrary constants.

It is sometimes convenient to introduce the notation

$$\omega_0 \equiv \sqrt{\frac{k}{m}}\,. \tag{14}$$

This will be explained shortly. With this notation, the differential equation (12) is

$$\frac{d^2x}{dt^2} + \omega_0^2 x = 0, \tag{15}$$

and its general solution (13) is written more easily:

$$x = c_1 \cos \omega_0 t + c_2 \sin \omega_0 t. \tag{16}$$

Note that the number ω_0 that appears in the solution (16) is the square root of the positive constant coefficient in the differential equation (15).

Amplitude-Phase Form

The general solution (13) of the homogeneous equation is a linear combination of $\cos \omega_0 t$ and $\sin \omega_0 t$. We will show, in general, that the addition of these two trigonometric functions yields a simple trigonometric function because

$$c_1 \cos \omega_0 t + c_2 \sin \omega_0 t = R \cos (\omega_0 t - \phi). \tag{17}$$

To show (17), we use the trigonometric addition formula, $\cos (a - b) = \cos a \cos b + \sin a \sin b$, to rewrite the right-hand side of (17) to get

$$c_1 \cos \omega_0 t + c_2 \sin \omega_0 t = R(\cos \omega_0 t \cos \phi + \sin \omega_0 t \sin \phi). \tag{18}$$

Equation (18) holds if

$$c_1 = R \cos \phi,$$
$$c_2 = R \sin \phi. \tag{19}$$

It is seen in Figure 2.8.6 from (19) that R and ϕ are polar coordinates for the point (c_1, c_2). From the Pythagorean theorem,

$$R = \sqrt{c_1^2 + c_2^2}. \tag{20}$$

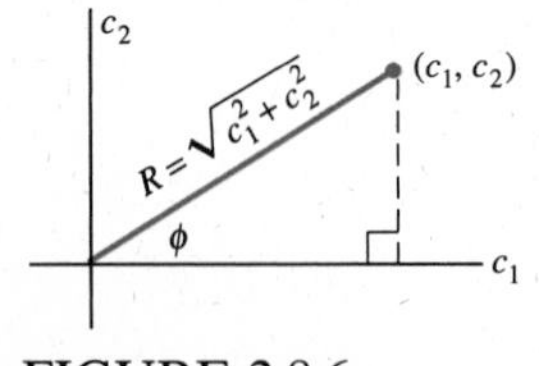

FIGURE 2.8.6

Amplitude R and phase ϕ.

The variable ϕ is called the **phase angle**, and $R = \sqrt{c_1^2 + c_2^2}$ the **amplitude** (or **amplitude of oscillation**). Drawing the triangle and specifying the quadrant are helpful in determining the phase angle. The formula

$$\tan \phi = \frac{c_2}{c_1} \tag{21}$$

is sometimes convenient. However, $\phi = \tan^{-1}(c_2/c_1)$ is correct only if ϕ is in the first or fourth quadrant. If ϕ is in the second or third quadrant, then $\phi = \pi + \tan^{-1}(c_2/c_1)$. Fewer errors are made if the quadrant of the phase angle is stated. In summary:

Amplitude-Phase Form

$$c_1 \cos \omega t + c_2 \sin \omega t = R \cos(\omega t - \phi)$$

$$R = \sqrt{c_1^2 + c_2^2}$$

$$\tan \phi = \frac{c_2}{c_1}$$

$$\phi = \tan^{-1}(c_2/c_1) \quad \text{if } c_1 \geq 0$$

$$\phi = \tan^{-1}(c_2/c_1) + \pi \quad \text{if } c_1 < 0$$

One of the significances of the amplitude and phase form is that it allows a relatively simple graphing of the solution

$$x(t) = R \cos(\omega_0 t - \phi). \tag{22}$$

R is the amplitude, and the solution is a shifted cosine:

$$x(t) = R \cos\left[\omega_0\left(t - \frac{\phi}{\omega_0}\right)\right].$$

One of the times at which the solution is a maximum is $t = \phi/\omega_0$. Formula (22) and Figure 2.8.7 show that the resulting free response, in the absence of friction, is **simple harmonic motion** (given by a cosine function). The mass oscillates sinusoidally around $x = 0$. The solution is periodic in time. Since the trigonometric functions have period 2π, the **period** of motion T is

$$\text{Period} = T = \frac{2\pi}{\omega_0}.$$

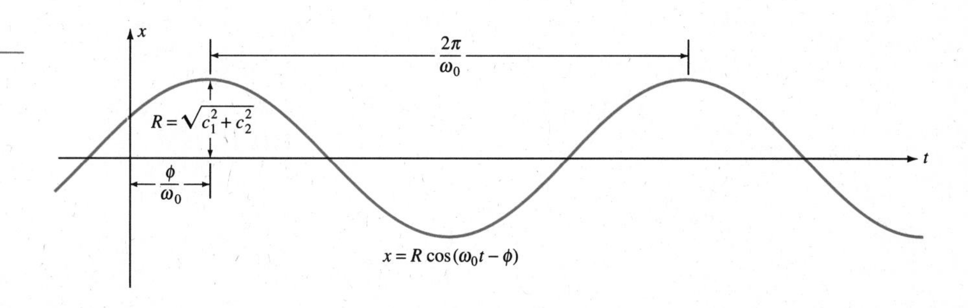

FIGURE 2.8.7
Simple harmonic motion.

TABLE 2.8.1

Units

	cgs	mks	fps
Length	centimeter (cm)	meter (m)	foot (ft)
Mass	gram (g)	kilogram (kg)	slug $= 1\ \text{lb} \cdot \text{s}^2/\text{ft}$
Time	second (s)	second (s)	second (s)
Velocity	cm/s	m/s	ft/s
Acceleration	cm/s^2	m/s^2	ft/s^2
Force ($F = ma$)	dyne $= 1\ \text{g} \cdot \text{cm/s}^2$	newton (N) $= 1\ \text{kg} \cdot \text{m/s}^2$	pound (lb)
Gravity g	$980\ \text{cm/s}^2$	$9.8\ \text{m/s}^2$	$32\ \text{ft/s}^2$

The **frequency** is the number of periods (cycles) per unit time. If the units of time were seconds and the period was $\frac{1}{5}$ s, then the frequency would be 5 cycles per second [or 5 **hertz** (Hz)]. In general, the frequency satisfies

$$\text{Frequency} = \frac{1}{T} = \frac{\omega_0}{2\pi}.$$

This also gives an interpretation of the important parameter ω_0:

$$\omega_0 = 2\pi \cdot \text{frequency}.$$

Since in one period the angle of the cosine changes by 2π radians, ω_0 is sometimes called the **circular frequency**, the number of radians per unit of time.

For the spring-mass system, (12) or (15), the frequency $\omega_0/2\pi = (1/2\pi)\sqrt{k/m}$ is called the **natural frequency** since it is the frequency that arises naturally without any forcing to the system. Stiffer springs (larger k) have larger frequencies.

Table 2.8.1 reviews various units commonly used.

EXAMPLE 2.8.1 **Resulting Motion of Spring-Mass System**

The spring constant of a 24-in. steel spring is measured by hanging a 1-slug (32-lb) mass from the spring and observing that the spring stretches 3 in. Now a $\frac{1}{2}$-slug (16 lb) mass is attached to the spring. The mass is pulled down 3 in. and released with a velocity of 1 ft/s downward. Determine the resulting motion.

SOLUTION First we need to determine the spring constant. From (8),

$$k\,\Delta L = mg.$$

In fps units this is

$$k\left(\frac{1}{4}\ \text{ft}\right) = 1\,\frac{\text{lb} \cdot \text{s}^2}{\text{ft}} \cdot \left(32\ \frac{\text{ft}}{\text{s}^2}\right),$$

or $k = 128$ lb/ft. Since $m = \frac{1}{2}$, the equation of motion is (12),

$$\frac{1}{2}\frac{d^2x}{dt^2} + 128x = 0,$$

or $d^2x/dt^2 + 256x = 0$. The characteristic equation is $r^2 + 256 = 0$, which has roots $r = \pm 16i$. Thus the general solution of the differential equation is

$$x = c_1 \cos 16t + c_2 \sin 16t. \tag{23}$$

The initial conditions are

$$x(0) = \frac{1}{4} \text{ ft}, \qquad x'(0) = 1 \text{ ft/s}.$$

Thus

$$\frac{1}{4} = x(0) = c_1, \qquad 1 = x'(0) = 16c_2$$

and the resulting motion is

$$x = \frac{1}{4} \cos 16t + \frac{1}{16} \sin 16t. \tag{24}$$

In order to more easily visualize this motion, we shall rewrite the solution in the form $x = R\cos(16t - \phi)$. The amplitude is

$$R = \sqrt{c_1^2 + c_2^2} = \sqrt{\left(\frac{1}{4}\right)^2 + \left(\frac{1}{16}\right)^2} = \frac{\sqrt{17}}{16}.$$

The phase angle ϕ is in the first quadrant with

$$\tan\phi = \frac{c_2}{c_1} = \frac{1/16}{1/4} = \frac{1}{4},$$

so that

$$\phi = \tan^{-1}\left(\frac{1}{4}\right) = 0.2498 \text{ radian}.$$

Thus the motion (24) may be rewritten as

$$x = \frac{\sqrt{17}}{16} \cos(16t - 0.2498).$$

The period is $2\pi/16$ s and the frequency is $(16/2\pi)$/s (see Figure 2.8.8). ◄

FIGURE 2.8.8

Graph of $x = \frac{\sqrt{17}}{16} \times \cos(16t - 0.2498)$.

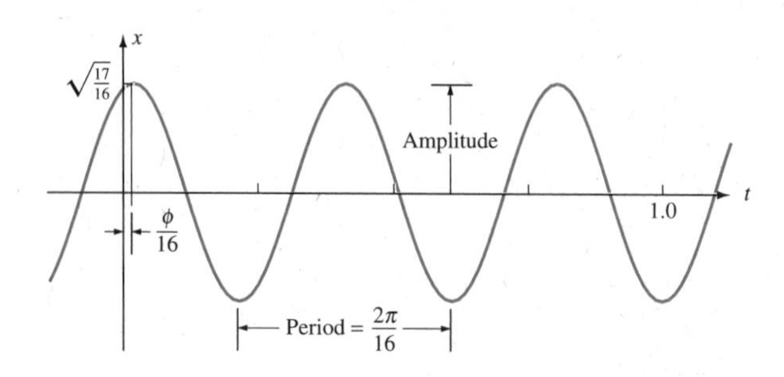

Exercises

In Exercises 1 through 10, put $x(t)$ into the amplitude-phase form $x(t) = R\cos(\omega_0 t - \phi)$:

1. $x(t) = 3\cos 5t - 7\sin 5t$

2. $x(t) = -5\cos 7t + 3\sin 7t$

3. $x(t) = \sqrt{3}\cos 14t + \sin 14t$

4. $x(t) = 5\cos 4t + 5\sqrt{3}\sin 4t$

5. $x(t) = -6\cos 5t + 6\sin 5t$

6. $x(t) = 4\cos 7t - 4\sin 7t$

7. $x(t) = \sqrt{3}\cos 6t - \sin 6t$

8. $x(t) = -\sqrt{3}\cos 9t - \sin 9t$

9. $x(t) = -4\cos 2t + 4\sqrt{3}\sin 2t$

10. $x(t) = 2\cos 2t - 2\sqrt{3}\sin 2t$

Free Response

In Exercises 11 through 24, we assume that Hooke's law is applicable, resistance is negligible, and the mass of the spring is negligible.

11. A mass of 30 g is attached to a spring. At equilibrium the spring has stretched 20 cm. The spring is pulled down another 10 cm and released. Set up the differential equation for the motion, and solve it to determine the resulting motion, ignoring friction.

12. A mass of 400 g is attached to a spring. At equilibrium the point has stretched 245 cm. The spring is pulled down and released. At noon we observe that the mass is 10 cm below equilibrium and traveling upward at $\sqrt{84}$ cm/s. Set up the differential equation for the motion. Solve the differential equation and express the solution in the amplitude-phase form.

13. A mass of 8 slugs ($8 \cdot 32$ lb) is attached to a long spring. The spring stretches 2 ft before coming to rest. The 8-slug mass is removed and a 2-slug mass (weight 64 lb) is attached and placed at equilibrium. The mass is pushed down and released. At the time it is released, the mass is 2 ft below equilibrium and traveling downward at 1 ft/s. Derive the differential equation for the motion of the spring-mass system. Solve the differential equation.

14. A spring-mass system has a spring constant of $k = 5$ g/s^2. What mass should be attached to make the resulting motion have a frequency of 30 Hz?

15. A spring is to be attached to a mass of 10 slugs (weight 320 lb). What should the spring constant be to make the resulting motion have a frequency of 5 Hz?

16. A mass of 6 g is attached to a spring mass system with spring constant 30 g/s^2. What should the initial conditions be to give a response with amplitude 3 and phase angle $\pi/4$?

17. A mass of 16 g is attached to a spring-mass system with spring constant 64 g/s^2. What should the initial conditions be to give a response with amplitude 2 and phase angle $\pi/3$?

18. Suppose you have a spring-mass system as in Figure 2.8.5.

a) What is the effect on the period and frequency of doubling the mass?

b) What is the effect on the period and frequency of doubling the spring constant?

c) What is the effect on the period and frequency of doubling both the spring constant and the mass?

19. A spring-mass system with mass m and spring constant k is subjected to a sudden impulse. The result is that, at time $t = 0$, the mass is at the equilibrium position but has a velocity of 10 cm/s downward.

a) Determine the subsequent motion.

b) Determine the amplitude of the resulting motion as a function of m and k.

c) What is the effect on the amplitude of increasing k?

d) What is the effect on the amplitude of increasing m?

20. (Same situation as Exercise 19.) If the mass is 50 g, what should the spring constant be to give an amplitude of 20 cm?

21. At time $t = 0$, the mass in a spring-mass system with mass m and spring constant k is observed to be 1 ft below equilibrium and traveling downward at 1 ft/s.

a) Determine the subsequent motion.

b) Determine the amplitude as a function of m and k.

c) What is the effect on the amplitude of increasing m or k?

22. (Same situation as Exercise 21.) If the spring constant is $k = 8$ lb/ft, what should the mass be to give an amplitude of 4 ft? At what time will this amplitude be first attained?

23. **(Conservation of mechanical energy)** Suppose that a spring-mass system is modeled by $mx'' + kx = 0$. The quantities $\frac{1}{2}mv^2$ and $\frac{1}{2}kx^2$ are the **kinetic energy** and the **elastic potential energy**. Their sum is the total **mechanical energy**. Multiply $mx'' + kx = 0$ by x' and integrate to show that $\frac{1}{2}mv^2 + \frac{1}{2}kx^2 = \text{constant}$.

24. Let $E = \frac{1}{2}mv^2 + \frac{1}{2}kx^2 = \frac{1}{2}m(x')^2 + \frac{1}{2}kx^2$. Show that if x is a solution of $mx'' + \delta x' + kx = 0$ with $\delta > 0$, then $dE/dt < 0$. Thus E is monotonically decreasing, so that mechanical energy decreases in the presence of resistance.

In Exercises 25 through 30, find the general solution. (If the initial conditions are given, put the solution in amplitude and phase form if possible.)

25. $\dfrac{d^2x}{dt^2} + 7x = 0$

26. $\dfrac{d^2x}{dt^2} + 23x = 0$

27. $\dfrac{d^2x}{dt^2} - 7x = 0$

28. $\dfrac{d^2x}{dt^2} - 23x = 0$

29. $\dfrac{d^2x}{dt^2} + 5x = 0,\ x(0) = 2,\ \dfrac{dx}{dt}(0) = 3$

30. $\dfrac{d^2x}{dt^2} + 9x = 0,\ x(0) = 7,\ \dfrac{dx}{dt}(0) = 5.$

31. Consider an object moving in a circle of radius r, such that the angle θ is steadily increasing, $d\theta/dt = \text{constant}$. How many cycles per second does the object make? Show that the x component of the object satisfies simple harmonic motion (22). Show that $d\theta/dt$ equals the circular frequency.

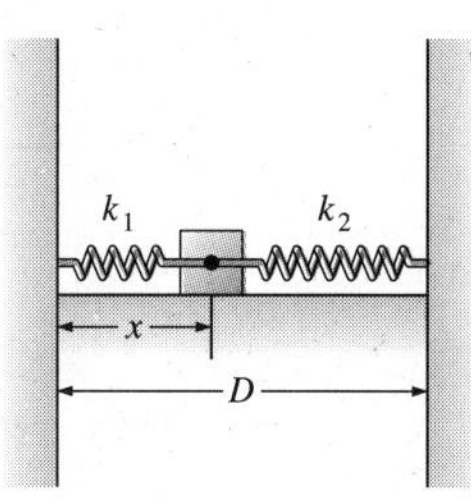

FIGURE 2.8.9

32. Consider a mass m, constrained to move only horizontally, located between two fixed walls a distance D apart and connected to two springs. The left spring has length L_1 and spring constant k_1, while the right spring has length L_2 and spring constant k_2. Let the position of the mass from the left wall be x, as illustrated in Figure 2.8.9 on page 155. Set up the one differential equation for the one mass (acted upon by two forces).

2.8.3 Free Response with Friction ($\delta > 0$)

Suppose now that friction is not negligible, so that the dynamics of the spring-mass system in Figure 2.8.4 are described by the differential equation

$$m\frac{d^2x}{dt^2} + \delta\frac{dx}{dt} + kx = 0. \tag{25}$$

The characteristic equation of (25) is $mr^2 + \delta r + k = 0$, which has the roots

$$r = -\frac{\delta}{2m} \pm \frac{\sqrt{\delta^2 - 4mk}}{2m}. \tag{26}$$

There are three cases to consider, depending on whether $\delta^2 - 4mk$ is negative, zero, or positive. Since we have just discussed the $\delta = 0$ (no friction) case, it is convenient to think of the mass m and spring constant k as fixed, and explain what happens as the friction coefficient δ increases.

Case 1: Underdamping ($0 < \delta < \sqrt{4mk}$)

In this case, $\delta^2 - 4mk < 0$. There is a pair of complex conjugate roots

$$r = -\alpha \pm \beta i, \qquad \alpha = \frac{\delta}{2m}, \qquad \beta = \frac{\sqrt{4mk - \delta^2}}{2m}, \tag{27}$$

so that the solutions of the differential equation $mx'' + \delta x' + kx = 0$ are

$$x = e^{-\alpha t}(c_1 \cos \beta t + c_2 \sin \beta t), \tag{28}$$

or, using (22),

$$x = e^{-\alpha t}\sqrt{c_1^2 + c_2^2}\cos(\beta t - \phi). \tag{29}$$

This is a **damped oscillation**, known as **underdamped** ($\delta < \sqrt{4mk}$). The solution is graphed in Figure 2.8.10. The sinusoidal function $\cos(\beta t - \phi)$ varies between 1 and -1. Thus the solution stays between $\sqrt{c_1^2 + c_2^2}\,e^{-\alpha t}$ and $-\sqrt{c_1^2 + c_2^2}\,e^{-\alpha t}$. To qualitatively sketch (29), first sketch $\sqrt{c_1^2 + c_2^2}\,e^{-\alpha t}$ and its negative $-\sqrt{c_1^2 + c_2^2}\,e^{-\alpha t}$. The solution oscillates between these two curves, as shown in Figure 2.8.10. For underdamped oscillations, the term

$$\sqrt{c_1^2 + c_2^2}\,e^{-\alpha t}$$

is called the **time-varying amplitude** of the oscillation.

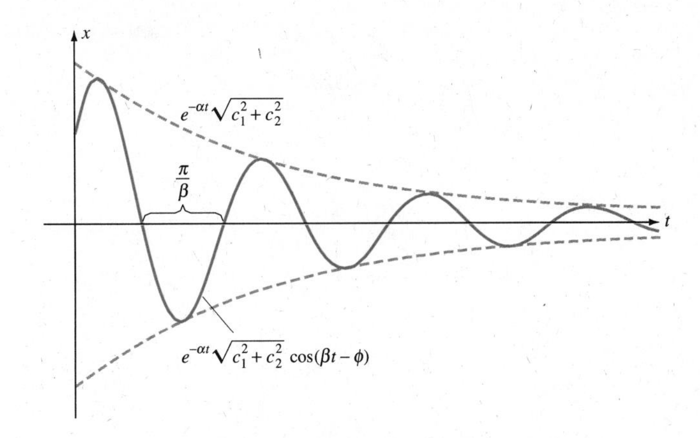

FIGURE 2.8.10

Underdamped oscillation.

With friction the amplitude oscillates with a **pseudo (circular) frequency**

$$\beta = \frac{\sqrt{4mk - \delta^2}}{2m}.$$

This frequency can be related to the (circular) frequency ($\omega_0 = \sqrt{k/m}$) at which the spring-mass system would oscillate without damping:

$$\beta = \sqrt{\frac{k}{m} - \frac{\delta^2}{4m^2}} = \sqrt{\omega_0^2 - \frac{\delta^2}{4m^2}}.$$

The pseudofrequency of an underdamped oscillator is less than the natural frequency. As the resistance δ is increased, the pseudofrequency decreases (the period increases).

One experimental method of determining the damping constant δ is by observing the exponential decay of the amplitude of the oscillation. Another experimental method of determining δ is to compare the frequencies with and without damping.

For a given mass m and spring constant k, increasing the resistance δ (but keeping $\delta < \sqrt{4mk}$) has two effects. First, the oscillation is damped faster; that is, the time-varying amplitude $e^{-\alpha t}\sqrt{c_1^2 + c_2^2}$ decreases faster. Second, the period of the oscillation being damped, $2\pi/\beta$, increases (Figure 2.8.10).

EXAMPLE 2.8.2 **Underdamped Motion**

A mass of 0.5 slug (weight of 16 lb) is suspended on a spring with a spring constant of $k = 2$ lb/ft. The damping is δ lb · s/ft. The mass is pulled down 1 ft and released. Describe and graph the motion for $\delta = 0.1$, 0.25, 1, and 1.75.

SOLUTION The differential equation describing the motion is

$$\frac{1}{2}\frac{d^2x}{dt^2} + \delta\frac{dx}{dt} + 2x = 0, \qquad x(0) = 1, \qquad x'(0) = 0. \tag{30}$$

The characteristic equation of (30) is $1/2r^2 + \delta r + 2 = 0$, which has roots $r = -\delta \pm i\sqrt{4 - \delta^2}$, so that the solution of the differential equation (30) is

$$x = e^{-\delta t}\left[c_1 \cos\left(\sqrt{4 - \delta^2}\,t\right) + c_2 \sin\left(\sqrt{4 - \delta^2}\,t\right)\right], \tag{31}$$

which is underdamped if $\delta < 2$. The initial conditions imply that

$$1 = x(0) = c_1,$$

$$0 = x'(0) = -\delta c_1 + \sqrt{4 - \delta^2}\,c_2.$$

Thus $c_1 = 1$, $c_2 = \delta(4 - \delta^2)^{-1/2}$, and

$$x = e^{-\delta t}\left[\cos\left(\sqrt{4 - \delta^2}\,t\right) + \frac{\delta}{(4 - \delta^2)^{1/2}} \sin\left(\sqrt{4 - \delta^2}\,t\right)\right]. \tag{32}$$

We shall rewrite (32) using (22). Now

$$1 + \left[\frac{\delta}{(4 - \delta^2)^{1/2}}\right]^2 = \frac{4}{4 - \delta^2}.$$

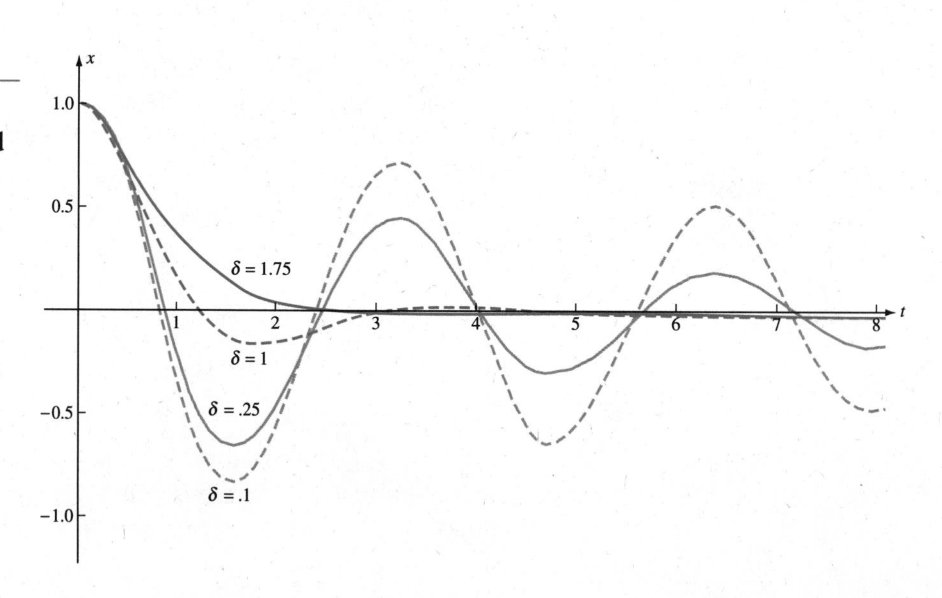

FIGURE 2.8.11

Graph of (33) for $\delta = 0.1$, 0.25, 1, and 1.75.

Thus (32) is

$$x = e^{-\delta t}\frac{2}{\sqrt{4-\delta^2}}\cos\left(\sqrt{4-\delta^2}\,t - \phi\right), \tag{33}$$

where

$$\tan\phi = \delta(4-\delta^2)^{-1/2}.$$

The graphs of (33) for $\delta = 0.1, 0.25, 1$, and 1.75 are shown in Figure 2.8.11. ◄

Case 2: Overdamping ($\delta > \sqrt{4mk}$)

In this case, called **overdamping**, the characteristic equation $mr^2 + \delta r + k = 0$ for the differential equation $mx'' + \delta x' + kx = 0$ has two distinct negative roots:

$$r_1 = -\frac{\delta}{2m} + \frac{\sqrt{\delta^2 - 4mk}}{2m}, \tag{34a}$$

$$r_2 = -\frac{\delta}{2m} - \frac{\sqrt{\delta^2 - 4mk}}{2m}. \tag{34b}$$

[*Note*: $r_1 < 0$ since $\sqrt{\delta^2 - 4mk} < \sqrt{\delta^2} = \delta$.] Thus the solution is

$$x = c_1 e^{r_1 t} + c_2 e^{r_2 t}. \tag{35}$$

Note that $x \to 0$ as $t \to \infty$, since $r_1 < 0$ and $r_2 < 0$.

In general, for overdamped systems, depending on the initial conditions, the solution for x takes one of the three forms shown in Figure 2.8.12, or their negative (see Exercises 41, 42). Figure 2.8.12(a) results if the mass is initially moving away from equilibrium ($x = 0$). The mass slows down and then returns toward the equilibrium. If the mass is initially moving toward the equilibrium, it either approaches the equilibrium [Figure 2.8.12(b)] or, if the

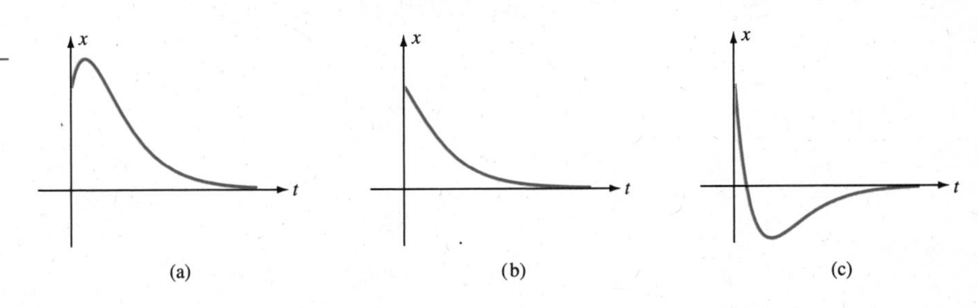

FIGURE 2.8.12

Critically damped or overdamped.

initial velocity is great enough, overshoots the equilibrium and then returns toward the equilibrium position, as in Figure 2.8.12(c).

Case 3: Critical Damping ($\delta = \sqrt{4mk}$)

In this case, $\delta^2 - 4mk = 0$ and (26) shows that there is a repeated real root $r = -\delta/(2m)$. The general solution of the differential equation $mx'' + \delta x' + kx = 0$ is then

$$x = c_1 e^{-(\delta/2m)t} + c_2 t e^{-(\delta/2m)t}. \tag{36}$$

The solution (36) for the **critically damped** case is different from that for the overdamped case (35). However, depending on the values of c_1, c_2, the critically damped solution (36) has a graph that has the same general shape as one of those given for the overdamped case (see Fig. 2.8.12).

EXAMPLE 2.8.3 Critical Damping

A device is being designed that can be modeled as a spring-mass system. The spring constant is $k = 10\ \text{g/s}^2$, and the damping constant is $\delta = 20\ \text{g/s}$.

a. Determine the mass so that the resulting spring-mass system will be critically damped.
b. The mass is pulled down 5 cm from the rest position and released with a downward velocity of 10 cm/s. Determine and solve the equations of motion. Graph the resulting motion.

SOLUTION

a. Critical damping occurs if $\delta^2 - 4mk = 0$. That is, $400 - 40m = 0$. Thus, the desired mass is 10 g.
b. The equation of motion is

$$10x'' + 20x' + 10x = 0, \qquad x(0) = 5, \qquad x'(0) = 10. \tag{37}$$

The characteristic equation for (37) is $10r^2 + 20r + 10 = 0$, or $10(r + 1)^2 = 0$, which has a repeated real root of -1. The general solution of $10x'' + 20x' + 10x = 0$ is thus

$$x = c_1 e^{-t} + c_2 t e^{-t}.$$

Applying the initial conditions in (37) to this general solution gives

$$5 = x(0) = c_1,$$

$$10 = x'(0) = -c_1 + c_2.$$

Solve for c_1, c_2 to get $c_1 = 5$, $c_2 = 15$. Thus

$$x = 5e^{-t} + 15te^{-t},$$

which has a graph like that in Figure 2.8.12(a). ◄

It is important to note that for all three cases in which there is damping ($\delta > 0$), we have

$$\lim_{t \to \infty} x(t) = 0.$$

The importance of this is discussed in Section 2.11.

Exercises

In Exercises 33 through 40, set up the differential equation that describes the motion under the assumptions of this section. Solve the differential equation. State whether the motion of the spring mass system is harmonic, damped oscillation, critically damped, or overdamped. If the motion is a damped oscillation, rewrite in the form (22).

33. The spring-mass system has an attached mass of 10 g. The spring constant is 30 g/s^2. A dashpot mechanism is attached, which has a damping coefficient of 40 g/s. The mass is pulled down and released. At time $t = 0$, the mass is 3 cm below the rest position and moving upward at 5 cm/s.

34. A long spring has a mass of 1 slug attached to it. The spring stretches 16/13 ft and comes to rest. The damping coefficient is 2 slug/s. The mass is subjected to an impulsive force at time $t = 0$, which imparts a velocity of 5 ft/s downward.

35. A mass of 1 g is attached to a spring-mass system for which friction is negligible. The spring stretches 20 cm and comes to rest. The mass is pulled down 1 cm from rest and released with a velocity of 7 cm/s downward.

36. A spring system has an attached mass of 1 g, a spring constant of 5 g/s^2, and a damping coefficient of 4 g/s. The mass is pushed upward 1 cm from rest position and released with a velocity of 3 cm/s downward.

37. A spring with spring constant $k = 12$ slug/s^2 has a mass attached that stretches the spring $2\frac{2}{3}$ ft. The damping coefficient is 7 slug/s. The mass is pushed 1 ft above the rest position and then released with a velocity of 1 ft/s downward.

38. A long spring with spring constant $k = 8$ g/s^2 has a mass attached that stretches the spring 245 cm. The damping coefficient is $\delta = 8$ g/s. At time $t = 0$, the mass is at the equilibrium position and has a velocity of 3 cm/s downward.

39. A spring-mass system has a spring with spring constant $k = 5$ g/s^2, attached mass of $m = 1$ g, and a friction coefficient of $\delta = 4$ g/s. The mass is pulled 2 cm below the equilibrium position and released.

40. A spring-mass system has a spring constant of $k = 1$ lb/ft, a damping coefficient of 2 slug/s, and an attached mass of 1 slug. The mass is pushed upward and released 1 ft above the equilibrium position with a velocity of 1 ft/s upward.

41. Show that in the case of critical damping, $\delta = \sqrt{4mk}$, the mass may change direction at most once [the general solution (36) has at most one horizontal tangent].

42. Show that in the case of overdamping, the mass can change direction at most once.

43. Suppose $\delta > 0$, $k > 0$ are fixed. Describe how varying the mass affects whether the motion is a damped oscillation, overdamped, or critically damped.

44. Suppose a 1-slug mass is attached to a spring with spring constant 400 lb/ft. Determine the damping coefficient if underdamped oscillations are observed with frequency 3 Hz.

45. Suppose a spring-mass system with a 1-slug mass is known to vibrate at 5 Hz without damping but only at 4 Hz when underdamped. Determine the damping coefficient.

For Exercises 46 through 55, solve the differential equation and state whether the corresponding motion is underdamped, overdamped, or critically damped.

46. $2\dfrac{d^2x}{dt^2} + 5\dfrac{dx}{dt} + 3x = 0$

47. $2\dfrac{d^2x}{dt^2} + 4\dfrac{dx}{dt} + 3x = 0$

48. $2\dfrac{d^2x}{dt^2} + \sqrt{48}\dfrac{dx}{dt} + 3x = 0$

49. $3\dfrac{d^2x}{dt^2} + 5\dfrac{dx}{dt} + 4x = 0$

50. $3\dfrac{d^2x}{dt^2} + 7\dfrac{dx}{dt} + 4x = 0$

51. $3\dfrac{d^2x}{dt^2} + 8\dfrac{dx}{dt} + 4x = 0$

52. $9\dfrac{d^2x}{dt^2} + 10\dfrac{dx}{dt} + 4x = 0$

53. $9\dfrac{d^2x}{dt^2} + 12\dfrac{dx}{dt} + 4x = 0$

54. $9\dfrac{d^2x}{dt^2} + 15\dfrac{dx}{dt} + 4x = 0$

55. $3\dfrac{d^2x}{dt^2} + 4\dfrac{dx}{dt} + x = 0$

56. Consider a spring-mass system with critical damping. Suppose that $x(0) = 1$, $x'(0) = v_0$. For what values of v_0 will the solution look like (a), (b), (c) of Figure 2.8.12? (*Hint*: The time of a local maximum or minimum must be positive.)

57. Consider an overdamped spring-mass system. Suppose that $x(0) = 1$, $x'(0) = v_0$. Find a formula for the time of a local maximum or minimum, assuming that one exists.

Exercises 58 through 61 illustrate a general fact but are given only for $m = 1$, $k = 1$. For Exercises 58 and 59, let x_δ be the solution of

$$x'' + \delta x' + x = 0,\ x(0) = 1,\ x'(0) = -1 \quad (38)$$

for a given δ. Note that $\sqrt{4mk} = \sqrt{4} = 2$.

58. Solve (38) for $0 < \delta < 2$ and $\delta = 2$. Show that

$$\lim_{\delta \to 2^-} x_\delta(t) = x_2(t) \qquad \text{for all } t \geq 0.$$

This shows that as the damping coefficient approaches critical damping, the damped oscillation approaches the critical damping solution.

59. Solve (38) for $2 < \delta$ and $\delta = 2$. Show that

$$\lim_{\delta \to 2^+} x_\delta(t) = x_2(t) \qquad \text{for all } t \geq 0.$$

This shows that the overdamped solution approaches that of critical damping.

For Exercises 60 and 61, let x_δ be the solution of

$$x'' + \delta x' + x = 0,\ x(0) = 0,\ x'(0) = 1. \quad (39)$$

60. Solve (39) for $0 < \delta < 2$ and $\delta = 2$. Show that

$$\lim_{\delta \to 2^-} x_\delta(t) = x_2(t) \qquad \text{for all } t.$$

61. Solve (39) for $\delta > 2$ and $\delta = 2$. Show that

$$\lim_{\delta \to 2^+} x_\delta(t) = x_2(t) \qquad \text{for all } t.$$

62. (Requires computer with fairly good graphic capabilities.) Illustrate the limits in Exercises 58 through 61 by graphing x_δ for several values of δ on the same graph. For example, in Exercise 58 you might take $\delta = 2 - (1/n)$ or $\delta = 2 - (1/n^2)$ for $n = 1, 2, 3, \ldots$, and plot on an interval of, say, $0 \leq t \leq 5$. (See Figure 2.8.11.)

2.9 The Method of Undetermined Coefficients

As pointed out in Section 2.2, the general solution of

$$ay''(x) + by'(x) + cy(x) = f(x),$$

where a, b, c are constants and $a \neq 0$, may be written in the form $y = y_p + c_1y_1 + c_2y_2$, where y_1, y_2 are solutions of the associated homogeneous equation $ay'' + by' + cy = 0$ and y_p is a particular solution of $ay'' + by' + cy = f$. The homogeneous equation was solved in Section 2.6. Two methods for finding the particular solution y_p will be given in this chapter. The method of undetermined coefficients will be developed in this section. The method of variation of parameters will be discussed in Sections 2.14 and 2.15.

If we think of the differential equation as modeling a physical system, say a circuit, then f often stands for the input or outside influence, and the solution y is the response. The response is often similar to the input function.

Method of Undetermined Coefficients

The method of undetermined coefficients is used to find a particular solution for a linear differential equation with constant coefficients

$$ay'' + by' + cy = f(x), \tag{1}$$

where a, b, c are constant coefficients and $f(x)$ is one of the following:

1. $f(x)$ = polynomial
2. $f(x)$ = exponential $= e^{\alpha x}$
3. $f(x)$ = sinusoid $= \sin \beta x$ or $\cos \beta x$
4. $f(x)$ = products and linear combinations of 1, 2, 3

Slightly different procedures are used in the method of undetermined coefficients to find a particular solution depending on the type of forcing function. We begin by considering when $f(x)$ involves polynomials or exponentials. Proofs will not be given in this section. The basic understanding of this technique will be gained from examples.

First step The first step using undetermined coefficients is always the same: Analyze solutions of the associated homogeneous equation. Determine the characteristic polynomial and all roots (including the multiplicity of roots).

Second step The second step depends on $f(x)$. For example

If $f(x)$ in (1) is an exponential, $e^{\alpha x}$, times an mth-degree polynomial

$$f(x) = e^{\alpha x} \text{ (polynomial of degree } m),$$

then y_p is of the **form** (2)

$$y_p = e^{\alpha x} x^k (A_0 + A_1 x + \cdots + A_m x^m),$$

where k is the multiplicity of $r = \alpha$ as a root of the characteristic polynomial.

The degree of the polynomial in the particular solution is the same as the degree of the polynomial in the forcing function $f(x)$ if $e^{\alpha x}$ is not a homogeneous solution (that is, $r = \alpha$ is not a root of the characteristic equation). The degree of the polynomial in the particular solution is increased by the number of times $r = \alpha$ is a root of the characteristic polynomial.

Equation (2) gives the form of the particular solution. The A_i are called **undetermined coefficients** and may be determined by substituting the form (2) into the differential equation (1). This method, as described by (2), is valid for first-order equations, and it generalizes the simple example discussed in Section 1.7. The method of undetermined coefficients is also valid with no alterations for finding a particular solution of higher-order constant coefficient linear differential equations. With minor modifications (2) can be used for all seven cases we now consider.

Special Case 1 (Polynomial Forcing)

A special case of (2) occurs if the forcing function $f(x)$ is just a polynomial. This corresponds to $\alpha = 0$. If the forcing function is a polynomial, then a particular solution is also a polynomial (of possibly larger degree). The degree of the particular solution is the same as the degree of the polynomial forcing if $r = 0$ is not a root of the characteristic equation. The degree of the particular solution is increased by the number of times $r = 0$ is a root of the characteristic polynomial. To be more precise

If $f(x)$ in (1) is an mth-degree polynomial, then y_p is of the **form**

$$y_p = x^k (A_0 + A_1 x + \cdots + A_m x^m), \tag{3}$$

where k is the multiplicity of $r = 0$ as a root of the characteristic polynomial.

EXAMPLE 2.9.1 Polynomial Forcing (0 is not a root)

Find the general solution of

$$y'' - 3y' + 2y = 2x^2 + 1. \tag{4}$$

SOLUTION The characteristic equation $r^2 - 3r + 2 = (r-2)(r-1) = 0$ has roots of $r = 1$ and $r = 2$. Thus $y_h = c_1e^x + c_2e^{2x}$. Here $f(x) = 2x^2 + 1$ is a second-degree polynomial. Thus y_p has the form $x^k(A_0 + A_1x + A_2x^2)$. 0 is not a root of the characteristic polynomial, so $k = 0$ and we have

$$y_p = A_0 + A_1x + A_2x^2. \tag{5}$$

Substituting this expression for y in the differential equation (4), we get

$$(A_0 + A_1x + A_2x^2)'' - 3(A_0 + A_1x + A_2x^2)'$$
$$+ 2(A_0 + A_1x + A_2x^2) = 2x^2 + 1$$

or

$$2A_2 - 3A_1 - 6A_2x + 2A_0 + 2A_1x + 2A_2x^2 = 2x^2 + 1.$$

Equating coefficients of like powers of x gives

$$\begin{aligned} 1: &\quad 2A_2 - 3A_1 + 2A_0 = 1 \\ x: &\quad -6A_2 + 2A_1 = 0 \\ x^2: &\quad 2A_2 = 2, \end{aligned}$$

which has the solution $A_0 = 4$, $A_1 = 3$, $A_2 = 1$. Thus the particular solution is $y_p = 4 + 3x + x^2$ and the general solution is

$$y = y_p + y_h = 4 + 3x + x^2 + c_1e^x + c_2e^{2x}.$$ ◄

EXAMPLE 2.9.2 Polynomial Forcing (0 is a root)

Suppose that you were solving the differential equation

$$y'' - 3y' = 2x^2 + 1 \tag{6}$$

using the method of undetermined coefficients. Give the form of the particular solution that you would use.

SOLUTION The characteristic equation $r^2 - 3r = (r-3)r = 0$ has roots of $r = 0$ and $r = 3$. Thus $y_h = c_1 + c_2e^{3x}$. Here $f(x) = 2x^2 + 1$ is a second-degree polynomial. Thus y_p has the form $x^k(A_0 + A_1x + A_2x^2)$. Here 0 is a root of the characteristic polynomial of multiplicity 1, so $k = 1$ and we have

$$y_p = x(A_0 + A_1x + A_2x^2) = A_0x + A_1x^2 + A_2x^3. \tag{7}$$

◄

It is important to note that in both (4) and (6) we had identical right-hand sides $f(x) = 2x^2 + 1$ but different forms (5) and (7) for y_p. To get the correct form for y_p, one has to look at not only the forcing term but also the roots of the characteristic equation. We shall give a physical application of this later in this chapter.

Special Case 2 (Exponential Forcing)

A special case of (2) occurs if the forcing function $f(x)$ is just a constant times $e^{\alpha x}$. The constant corresponds to a zero-degree polynomial, $m = 0$. If the forcing function is this type of exponential, then the form of a particular solution is a similar exponential $y_p = Ae^{\alpha x}$ if $e^{\alpha x}$ is not a homogeneous solution (i.e., if $r = \alpha$ is not a root of the characteristic equation). The form of a particular solution is the same exponential multiplied by a polynomial whose degree is increased (from $m = 0$) by the number of times $r = \alpha$ is a root of the characteristic polynomial. To be more precise,

> If $f(x)$ in (1) is a constant times $e^{\alpha x}$, then y_p has the **form**
>
> $$y_p = Ax^ke^{\alpha x}, \tag{8}$$
>
> where k is the multiplicity of $r = \alpha$ as a root of the characteristic polynomial.

EXAMPLE 2.9.3 Exponential Forcing (α is not a root)

Find a particular solution of

$$y'' + y = 3e^{-x}. \tag{9}$$

Solution Here $f(x) = 3e^{-x}$ is an exponential. By (8), $y_p = Ax^ke^{-x}$, where k is the number of times $\alpha = -1$ is a root of the characteristic polynomial $r^2 + 1$, which has roots $i, -i$. Since -1 is not a root, we have $k = 0$ and

$$y_p = Ae^{-x}.$$

Substituting this form for y_p into (9) gives

$$(Ae^{-x})'' + Ae^{-x} = 3e^{-x}.$$

Differentiating and simplifying gives $2A = 3$, so that $y_p = 3/2e^{-x}$ is a particular solution of (9). ◀

EXAMPLE 2.9.4 Exponential Forcing (α is a root)

Find a particular solution of

$$y'' - y = 3e^{-x}. \tag{10}$$

SOLUTION Here $f(x) = 3e^{-x}$ is an exponential. By (8), $y_p = Ax^k e^{-x}$, where k is the number of times $\alpha = -1$ is a root of the characteristic polynomial $r^2 - 1$, which has roots 1, -1. Since -1 is a root once, we have $k = 1$ and

$$y_p = Axe^{-x}.$$

Substituting this form for y_p into (10) gives

$$(Axe^{-x})'' - Axe^{-x} = 3e^{-x}.$$

Differentiating once,

$$A(e^{-x} - xe^{-x})' - Axe^{-x} = 3e^{-x},$$

and a second time gives

$$A(-2e^{-x} + xe^{-x}) - Axe^{-x} = 3e^{-x}.$$

The xe^{-x} terms cancel. Simplifying gives $-2A = 3$, so that $y_p = -3/2xe^{-x}$ is a particular solution of (10). ◀

Again note that both (9) and (10) had the same forcing function $f(x)$ but different forms for y_p because of the different homogeneous equations. This points out the important fact:

In general, the form for y_p cannot be determined by looking only at the forcing function f. The solution of the homogeneous equation (roots of the characteristic polynomial) must also be considered.

In the examples to follow, we shall make frequent use of the **principle of superposition**

If $f(x) = f_1(x) + \cdots + f_m(x)$, then there is a particular solution of the form obtained by adding up the forms of the particular solutions for each $f_i(x)$ (11)

Also, in several of the examples we shall merely determine the form for y_p and not actually find the constants in the form.

Special Case 3 (Exponential Times Polynomial Forcing)

We give examples of the general case of (2).

EXAMPLE 2.9.5 Superposition

Give the form for y_p if

$$y'' - y' = x^3 + x + e^x - 2xe^x$$

is to be solved by the method of undetermined coefficients.

Solution The characteristic polynomial is $r^2 - r = r(r-1)$, which has roots $r = 0, 1$, so that $y_h = c_1 + c_2 e^x$. The forcing term is

$$f = \underbrace{(x^3 + x)} + \underbrace{(1 - 2x)e^x}.$$

The first term is a third-degree polynomial. Since 0 is a root of multiplicity 1 of the characteristic equation (Special Case 1), y_p must include a term of the form $x^k(A_0 + A_1 x + A_2 x^2 + A_3 x^3)$ with $k = 1$. The second term is of the form $p(x)e^{\alpha x}$, where $p(x) = 1 - 2x$ is a first-degree polynomial and $\alpha = 1$. Since 1 is a root of the characteristic equation, by (2) y_p must include a term of the form $x^k(A_4 + A_5 x)e^x$ with $k = 1$. Thus y_p has the form

$$y_p = x(A_0 + A_1 x + A_2 x^2 + A_3 x^3) + x(A_4 + A_5 x)e^x.$$ ◄

EXAMPLE 2.9.6 Exponential Times Polynomial Forcing (α is a root)

Give the form for y_p if

$$y'' - 2y' + y = 7xe^x$$

is to be solved by the method of undetermined coefficients.

Solution The characteristic polynomial $r^2 - 2r + 1 = (r-1)^2$ has a root 1 of multiplicity 2. Thus $y_h = c_1 e^x + c_2 x e^x$. The forcing term is of the form $p(x)e^{\alpha x}$, where $p(x) = 7x$ is a first-degree polynomial and $\alpha = 1$. Since $\alpha = 1$ is a root of multiplicity 2, by (2) with $k = 2$, $m = 1$, the form for y_p is

$$y_p = x^2(A_0 + A_1 x)e^x.$$ ◄

The method of undetermined coefficients works for any order linear constant coefficient differential equation as long as the forcing term has the correct types of functions.

EXAMPLE 2.9.7 Third-Order Example

Find the form of the general solution of

$$y''' - 6y'' + 12y' - 8y = xe^{2x}. \tag{12}$$

SOLUTION First we need to solve the associated homogeneous equation

$$y''' - 6y'' + 12y' - 8y = 0.$$

The characteristic polynomial is $r^3 - 6r^2 + 12r - 8 = (r-2)^3$, so that 2 is a root of multiplicity 3. A fundamental set of solutions for the associated homogeneous equation is $\{e^{2x}, xe^{2x}, x^2e^{2x}\}$, so that $y_h = c_1e^{2x} + c_2xe^{2x} + c_3x^2e^{2x}$. The forcing term is $f(x) = xe^{2x}$. By (2) with $\alpha = 2$, $m = 1$, y_p must include $x^k[A_0 + A_1x]e^{2x}$, where k is the multiplicity of 2 as a root of the characteristic equation. Thus $k = 3$ and y_p is in the form

$$y_p = x^3[A_0 + A_1x]e^{2x} = A_0x^3e^{2x} + A_1x^4e^{2x}.$$ ◄

The general solution is $y = y_p + y_h$.

Special Case 4 (Sinusoidal Forcing)

If the forcing function is sinusoidal, $f(x) = E_1 \cos \beta x + E_2 \sin \beta x$, we think of the forcing being a linear combination of the complex exponentials $e^{\pm i\beta x}$. Then, according to (8) [which is a special case of (2)], the particular solution should involve the same exponentials $e^{\pm i\beta x}$, which are equivalent to a linear combination of $\cos \beta x$ and $\sin \beta x$. Of course, this should be multiplied by a power of x if $r = \pm i\beta$ are roots of the characteristic equation. To be more precise,

If $f(x) = E_1 \cos \beta x + E_2 \sin \beta x$ in (1), then y_p has the **form**

$$y_p = x^k(A \cos \beta x + B \sin \beta x), \tag{13}$$

where k is the multiplicity of $r = \beta i$ as a root of the characteristic polynomial.

EXAMPLE 2.9.8 **Sinusoidal Forcing ($i\beta$ is not a root)**

Find the general solution of

$$y'' + 2y' + 2y = 3e^{-x} + 4\cos x. \tag{14}$$

Solution First we solve the associated homogeneous equation. The characteristic equation is $r^2 + 2r + 2 = 0$. Its roots are

$$r = -1 \pm i.$$

Thus $y_h = c_1 e^{-x}\cos x + c_2 e^{-x}\sin x$. Now we will determine y_p. Note that f is a sum of two terms, $3e^{-x}$ and $4\cos x$. Consider first $3e^{-x}$. Since -1 is not a root of the characteristic equation, Special Case 2 says that y_p includes a term of the form $A_1 e^{-x}$. Now consider $4\cos x$. Since i is not a root of the characteristic equation, Special Case 4 with $\beta = 1$ says that y_p includes terms of the form $A_2\cos x + B_2 \sin x$. Thus

$$y_p = A_1 e^{-x} + A_2\cos x + B_2\sin x$$

for some constants A_1, A_2, B_2. Substituting this expression into (14) gives

$$[A_1e^{-x} + A_2\cos x + B_2\sin x]'' + 2[A_1e^{-x} + A_2\cos x + B_2\sin x]'$$
$$+ 2[A_1e^{-x} + A_2\cos x + B_2\sin x] = 3e^{-x} + 4\cos x.$$

That is,

$$A_1e^{-x} - A_2\cos x - B_2\sin x - 2A_1e^{-x} - 2A_2\sin x + 2B_2\cos x$$
$$+ 2A_1e^{-x} + 2A_2\cos x + 2B_2\sin x = 3e^{-x} + 4\cos x.$$

There are three functions e^{-x}, $\cos x$, $\sin x$ that appear in this equation. Equating the coefficients of like terms gives

$$\begin{aligned} e^{-x}: &\quad A_1 = 3, \\ \cos x: &\quad A_2 + 2B_2 = 4, \\ \sin x: &\quad B_2 - 2A_2 = 0. \end{aligned} \tag{15}$$

This system of three equations in the three unknowns A_1, A_2, B_2 has the solution

$$A_1 = 3, \qquad A_2 = \frac{4}{5}, \qquad B_2 = \frac{8}{5}.$$

Thus $y_p = 3e^{-x} + \frac{4}{5}\cos x + \frac{8}{5}\sin x$, and, the general solution is

$$y = y_p + y_h = 3e^{-x} + \frac{4}{5}\cos x + \frac{8}{5}\sin x + c_1e^{-x}\cos x + c_2e^{-x}\sin x.$$ ◄

This example emphasizes another common source of errors:

Even though only $\cos \beta x$ appeared in the forcing term f, the form for y_p may require both $x^k A \cos \beta x$ *and* $x^k B \sin \beta x$.

The system of equations (15) consisted of three equations in three unknowns and had a unique solution. It can be proved, using properties of linear independence and the Wronskian, that

If a, b, c are constants and f is the type of function described in the Method of Undetermined Coefficients, then the method always works (perhaps messily). In particular, if the equations for the undetermined constants are not consistent (don't have a solution), then an error has been made.

For example, if one arrives at

$$A_1 \sin x + A_2 e^{-x} = \cos x + 3 \sin x + 5e^{-x},$$

equating the coefficients of *all* the functions that appear gives

$$\begin{aligned} \sin x&: & A_1 &= 3, \\ \cos x&: & 0 &= 1, \\ e^{-x}&: & A_2 &= 5, \end{aligned}$$

which is impossible. Since the method always works for appropriate f, we know that we have made an error. Frequently the error is in finding the form for y_p.

EXAMPLE 2.9.9 Sinusoidal Forcing ($i\beta$ is a root)

Give the form for y_p if

$$y'' + 4y = \sin 2x$$

is to be solved by the method of undetermined coefficients.

SOLUTION The roots of the characteristic polynomial $r^2 + 4$ are $\pm 2i$ and $y_h = c_1 \cos 2x + c_2 \sin 2x$. The forcing term $\sin 2x$ is $\sin \beta x$, where $\beta = 2$. Since βi is a root of the characteristic polynomial of multiplicity 1, we have $k = 1$, and by Special Case 4, the form of the particular solution is

$$y_p = x[A \cos 2x + B \sin 2x].$$

◄

Special Case 5 (Polynomial Times Sinusoidal Forcing)

If the forcing function is a polynomial times a sinusoidal forcing function, the sinusoidal function is treated as a complex exponential as in the previous case. To be more precise,

If $f(x) = p(x) \cos \beta x + q(x) \sin \beta x$ in (1), where $p(x)$ is an mth-degree polynomial and $q(x)$ an nth-degree polynomial, then y_p has the **form**

$$y_p = x^k[(A_0 + A_1 x + \cdots + A_s x^s) \cos \beta x + (B_0 + B_1 x + \cdots + B_s x^s) \sin \beta x], \tag{16}$$

where k is the multiplicity of $r = \beta i$ as a root of the characteristic polynomial and s is the larger of m, n.

EXAMPLE 2.9.10 Polynomial Times Sinusoidal Forcing ($i\beta$ is a root)

Give the form for y_p if

$$y'' + 4y = x^2 \cos 2x - x \sin 2x + \sin 2x = x^2 \cos 2x + (1 - x) \sin 2x$$

is to be solved by the method of undetermined coefficients.

Solution The roots of the characteristic polynomial $r^2 + 4$ are $\pm 2i$ and $y_h = c_1 \cos 2x + c_2 \sin 2x$. The forcing term is of the form

$$p(x) \cos \beta x + q(x) \sin \beta x,$$

where $p(x) = x^2$ is a second-degree polynomial, $q(x) = 1 - x$ is a first-degree polynomial, and $\beta = 2$. Since $\beta i = 2i$ is a root of the characteristic equation of multiplicity 1 by Special Case 5, with $k = 1$, we have

$$y_p = x\big[(A_0 + A_1 x + A_2 x^2) \cos 2x + (B_0 + B_1 x + B_2 x^2) \sin 2x\big]. \blacktriangleleft$$

EXAMPLE 2.9.11 Fourth-Order Example

Give the form for y_p if

$$y'''' + 8y'' + 16y = x \sin x + x^2 \cos 2x \tag{17}$$

is to be solved by the method of undetermined coefficients.

Solution First we solve the associated homogeneous equation

$$y'''' + 8y'' + 16 = 0.$$

The characteristic equation is $r^4 + 8r^2 + 16 = (r^2 + 4)^2 = 0$ and has repeated complex roots $\pm 2i$, $\pm 2i$. A fundamental set of solutions for the associated homogeneous equation is thus

$$\{\sin 2x, \quad \cos 2x, \quad x \sin 2x, \quad x \cos 2x\}.$$

Now we find the form for y_p. Consider the forcing function

$$f = x \sin x + x^2 \cos 2x.$$

By (16), the $x \sin x$ term implies that y_p includes

$$x^{k_1}[(A_0 + A_1 x)\sin x + (B_0 + B_1 x)\cos x] \tag{18}$$

with $k_1 = 0$ since i is not a root of the characteristic polynomial. The $x^2 \cos 2x$ term implies that y_p includes

$$x^{k_2}\left[(A_2 + A_3 x + A_4 x^2)\sin 2x + (B_2 + B_3 x + B_4 x^2)\cos 2x\right] \tag{19}$$

with $k_2 = 2$, since $2i$ has multiplicity 2 as a root of the characteristic equation. In actually using (18) and (19), one may add (18) and (19) to get the form for y_p, substitute into the original differential equation (17), and solve for $A_0, \ldots, A_4, B_0, \ldots, B_4$. Alternatively, one could use (18) to find a particular solution of

$$y'''' + 8y'' + 16y = x \sin x,$$

and then use (19) to find a particular solution of

$$y'''' + 8y'' + 16y = x^2 \cos 2x.$$

Adding these two particular solutions (18) and (19) gives, by the superposition principle, a solution of $y'''' + 8y'' + 16y = x \sin x + x^2 \cos 2x$ as desired. ◀

Special Case 6 (Exponential Times Sinusoidal Forcing)

If the forcing function is an exponential times a sinusoid, $f(x) = E_1 e^{\alpha x} \cos \beta x + E_2 e^{\alpha x} \sin \beta x$, we think of the forcing being a linear combination of the complex exponentials $e^{(\alpha \pm i\beta)x}$. Then, according to (8) [which is a special case of (2)], the particular solution should involve the same exponentials $e^{(\alpha \pm i\beta)x}$, which are equivalent to a linear combination of $e^{\alpha x} \cos \beta x$ and $e^{\alpha x} \sin \beta x$. Of course, this should be multiplied by a power of x if $r = \alpha \pm i\beta$ are roots of the characteristic equation. To be more precise,

If $f(x) = E_1 e^{\alpha x} \cos \beta x + E_2 e^{\alpha x} \sin \beta x$ in (1), then y_p has the form

$$y_p = x^k (A e^{\alpha x} \cos \beta x + B e^{\alpha x} \sin \beta x), \tag{20}$$

where k is the multiplicity of $r = \alpha + \beta i$ as a root of the characteristic polynomial.

EXAMPLE 2.9.12 Exponential Times Sinusoidal Forcing ($\alpha + i\beta$ is a root)

Give the form for y_p if

$$y'' + 2y' + 2y = 5e^{-x}\cos x$$

is to be solved by the method of undetermined coefficients.

Solution The roots of the characteristic polynomial $r^2 + 2r + 2$ are $-1 \pm i$, so that

$$y_h = c_1e^{-x}\cos x + c_2e^{-x}\sin x.$$

The forcing term is of the form $e^{\alpha x}\cos\beta x$, where $\alpha = -1$, $\beta = 1$. Since $-1 + i$ is a root of the characteristic equation of multiplicity 1, by Special Case 6, with $k = 1$,

$$y_p = x(Ae^{-x}\cos x + Be^{-x}\sin x).$$ ◄

EXAMPLE 2.9.13 Superposition

Give the form for y_p if

$$y'' + 2y' + 2y = e^{-x}\cos 2x + e^{-x}\sin 2x + e^{-x} - 3\cos x \tag{21}$$

is to be solved by undetermined coefficients.

Solution Equation (21) has the same characteristic polynomial as Example 2.9.12, so the roots are $-1 \pm i$. The forcing term in (21) is the sum of three groups of terms:

$e^{-x}\cos 2x + e^{-x}\sin 2x$: Since $-1 + 2i$ is not a root, we include $A_0e^{-x}\cos 2x + B_0e^{-x}\sin 2x$, by Special Case 6.

e^{-x}: Since -1 is not a root, we include A_2e^{-x}, by Special Case 2.

$-3\cos x$: Since i is not a root, we include $A_1\cos x + B_1\sin x$, by Special Case 4.

Thus the form for y_p is

$$y_p = A_0e^{-x}\cos 2x + B_0e^{-x}\sin 2x + A_1\cos x + B_1\sin x + A_2e^{-x}.$$ ◄

Case 7 (Polynomial Times Exponential Times Sinusoid Forcing)

If the forcing function is a polynomial times an exponential times a sinusoidal function, then the exponential times the sinusoidal function is treated as a complex exponential as in the previous case. To be more precise,

If $f(x) = p(x)e^{\alpha x}\cos\beta x + q(x)e^{\alpha x}\sin\beta x$ in (1), where $p(x)$ is an mth-degree polynomial and $q(x)$ an nth-degree polynomial, then y_p has the **form**

$$y_p = x^k[(A_0 + A_1x + \cdots + A_sx^s)e^{\alpha x}\cos\beta x + (B_0 + B_1x + \cdots + B_sx^s)e^{\alpha x}\sin\beta x], \tag{22}$$

where k is the multiplicity of $r = \alpha + \beta i$ as a root of the characteristic polynomial and s is the larger of m, n.

This case is actually the general case. It includes all the previous cases for appropriate choices of α, β, m, n.

EXAMPLE 2.9.14 **Polynomial Times Exponential Times Sinusoidal Forcing ($\alpha + i\beta$ is not a root)**

Give the form of y_p if

$$y'' + 3y' + 2y = xe^{-x}\cos 2x$$

is to be solved by the method of undetermined coefficients.

Solution The roots of the characteristic polynomial $r^2 + 3r + 2$ are -1 and -2. Thus, two independent homogeneous solutions are e^{-x} and e^{-2x}. The forcing term is in the form $p(x)e^{\alpha x}\cos\beta x$, where $p(x) = x$ is a first-degree polynomial, and $\alpha + \beta i = -1 + 2i$. Since $-1 + 2i$ is not a root of the characteristic polynomial, from Case 7 (with $k = 0$),

$$y_p = (A_0 + A_1x)e^{-x}\cos 2x + (B_0 + B_1x)e^{-x}\sin 2x.$$ ◄

EXAMPLE 2.9.15 **Polynomial Times Exponential Times Sinusoidal Forcing ($\alpha + i\beta$ is a root)**

Give the form of y_p if

$$y'' + 2y' + 5y = x^3e^{-x}\sin 2x$$

is to be solved by the method of undetermined coefficients.

SOLUTION The roots of the characteristic polynomial $r^2 + 2r + 5$ are $-1 \pm 2i$. Thus two independent homogeneous solutions are $e^{-x} \cos 2x$ and $e^{-x} \sin 2x$. The forcing term is in the form $p(x)e^{\alpha x} \sin \beta x$, where $p(x) = x^3$ is a third-degree polynomial, and $\alpha + \beta i = -1 + 2i$. Since $-1 + 2i$ is a root of the characteristic polynomial of multiplicity 1, from Special Case 7 (with $k = 1$), we have

$$\begin{aligned} y_p &= x(A_0 + A_1 x + A_2 x^2 + A_3 x^3)e^{-x} \cos 2x \\ &\quad + x(B_0 + B_1 x + B_2 x^2 + B_3 x^3)e^{-x} \sin 2x. \end{aligned}$$

When initial conditions are present and the forcing function is nonzero, it is important to remember to apply the initial conditions to the general solution $y_p + c_1 y_1 + c_2 y_2$ and not just to $y_h = c_1 y_1 + c_2 y_2$.

EXAMPLE 2.9.16 **Initial Value Problem**

Solve

$$y'' - y = 3e^{-x}, \qquad y(0) = 0, \qquad y'(0) = 1. \tag{23}$$

SOLUTION First we must find the general solution of $y'' - y = 3e^{-x}$. In Example 2.9.4 we found that $y_p = -\frac{3}{2}xe^{-x}$ and the characteristic polynomial had roots $1, -1$. Thus the general solution is

$$y = -\frac{3}{2}xe^{-x} + c_1 e^x + c_2 e^{-x}.$$

In order to apply the initial conditions, we compute

$$y' = -\frac{3}{2}e^{-x} + \frac{3}{2}xe^{-x} + c_1 e^x - c_2 e^{-x}.$$

The initial conditions then give

$$\begin{aligned} 0 &= y(0) = c_1 + c_2, \\ 1 &= y'(0) = -\frac{3}{2} + c_1 - c_2. \end{aligned}$$

Solving for c_1, c_2 yields $c_1 = \frac{5}{4}$, $c_2 = -\frac{5}{4}$, and the solution of (23) is

$$y = -\frac{3}{2}xe^{-x} + \frac{5}{4}e^x - \frac{5}{4}e^{-x}.$$

Method of Undetermined Coefficients

In summary, the **method of undetermined coefficients** can be used on $ay'' + by' + cy = f$ if a, b, c are constants and f is a linear combination of functions of the form

$$x^m e^{\alpha x} \cos \beta x, \qquad x^m e^{\alpha x} \sin \beta x,$$

where m is a nonnegative integer and α, β are real numbers. Special cases are

$$x^m,\ x^m e^{\alpha x},\ e^{\alpha x},\ e^{\alpha x} \cos \beta x,\ e^{\alpha x} \sin \beta x,\ x^m \cos \beta x,\ x^m \sin \beta x.$$

The method is as follows.

1. First solve the associated homogeneous equation

$$ay'' + by' + cy = 0.$$

2. Determine y_p as a linear combination of functions with unknown coefficients using the following rules.

   Rule 1. If f includes a sum of terms of the form $p(x)e^{\alpha x}$, where $p(x)$ is an mth-degree polynomial, then the form for y_p should include

$$x^k[A_0 + A_1 x + \cdots + A_m x^m]e^{\alpha x}$$

   where k is the multiplicity of α as a root of the characteristic polynomial $ar^2 + br + c$.

   Rule 2. If f includes a sum of terms of the form

$$p(x)e^{\alpha x} \cos \beta x + q(x)e^{\alpha x} \sin \beta x,$$

   where $p(x)$ is an mth-degree polynomial and $q(x)$ is an nth-degree polynomial, then the form for y_p should include

$$x^k[A_0 + A_1 x + \cdots + A_s x^s]e^{\alpha x} \cos \beta x$$
$$+ x^k[B_0 + B_1 x + \cdots + B_s x^s]e^{\alpha x} \sin \beta x,$$

   where s is the larger of m and n and k is the multiplicity of $\alpha + \beta i$ as a root of the characteristic polynomial $ar^2 + br + c$.

3. Substitute the expression for y_p into the differential equation $ay'' + by' + cy = f$ to determine the unknown coefficients A_i, B_i.
4. The general solution of $ay'' + by' + cy = f$ is $y = y_p + y_h$.
5. Apply any initial conditions to $y_p + y_h$ in order to determine arbitrary constants.

Note that if $f(x)$ includes terms like $\ln x$, $x^{1/3} \sin x$, $\tan x$, then undetermined coefficients will not generally work.

In Table 2.9.1, E, F are constants.

TABLE 2.9.1

Case	f includes summands of form	y_p then includes	k is the multiplicity of the root
1.	$p(x)$, mth-degree polynomial	$x^k(A_0 + A_1x + \cdots + A_mx^m)$	0
2.	$Ee^{\alpha x}$	$x^k e^{\alpha x}$	α
3.	$p(x)e^{\alpha x}$ $p(x)$, mth-degree polynomial	$x^k(A_0 + A_1x + \cdots + A_mx^m)e^{\alpha x}$	α
4.	$E_1 \cos \beta x + E_2 \sin \beta x$ (One of E_1, E_2 may be zero)	$x^k(A \cos \beta x + B \sin \beta x)$	βi
5.	$p(x) \cos \beta x + q(x) \sin \beta x$ $p(x)$, mth-degree polynomial $q(x)$, nth-degree polynomial	$x^k[(A_0 + A_1x + \cdots + A_sx^s) \cos \beta x$ $+(B_0 + B_1x + \cdots + B_sx^s) \sin \beta x]$ s = larger of m, n	βi
6.	$E_1e^{\alpha x} \cos \beta x + E_2e^{\alpha x} \sin \beta x$	$x^k(Ae^{\alpha x} \cos \beta x + Be^{\alpha x} \sin \beta x)$	$\alpha + \beta i$
7.	$p(x)e^{\alpha x} \cos \beta x + q(x)e^{\alpha x} \sin \beta x$ $p(x)$, mth-degree polynomial $q(x)$, nth-degree polynomial	$x^k[(A_0 + A_1x + \cdots + A_sx^s)e^{\alpha x} \cos \beta x$ $+(B_0 + B_1x + \cdots + B_sx^s)e^{\alpha x} \sin \beta x]$ s = larger of m, n	$\alpha + \beta i$

The Superposition Principle

The *Superposition Principle* (Section 2.2) for linear differential equations says that

If y_{p1} is a particular solution of $ay'' + by' + cy = f_1$ and y_{p2} is a particular solution of $ay'' + by' + cy = f_2$, then $y_{p1} + y_{p2}$ is a particular solution of $ay'' + by' + cy = f_1 + f_2$. (24)

Intuitively, the superposition principle means that the response (output) that results from the sum (superposition) of two forcing terms (inputs) is the sum of the response from each forcing term. In applications, knowing whether our device or physical problem acts in this manner is a key factor in deciding whether linear equations can be used to analyze the problem.

The superposition principle, combined with the theory of **Fourier series**, enables the method of undetermined coefficients to be used on many additional kinds of forcing functions f. The idea is to write $f(x)$ as an (infinite) linear combination of functions $\cos \beta x$, $\sin \beta x$ for different β, and then use an infinite-series version of the superposition principle. This important idea is discussed in Section 10.4.

Exercises

In Exercises 1 through 12, state whether undetermined coefficients can be applied to the differential equation. If it cannot, explain why not.

1. $y'' + y = x \sin x$
2. $y'' + 3y = x^{1/2} \sin x$
3. $y'' + y = x^2 + x + \ln |x|$
4. $y'' + y = e^{x+1}$
5. $y'' + y = \dfrac{\sin x}{\cos x}$
6. $y'' + y' + y = \cosh x$
7. $y'' + xy' = 3e^{2x}$
8. $y'' + y' = x \sinh 2x$
9. $y'' + y = x^{-1}e^x$
10. $y'' + yy' = e^{2x}$
11. $y' + 3y = e^{-2x} \cos 3x + \sinh 3x$
12. $y'' + 3y' + 4y = \sin^2 x$

In Exercises 13 through 52, solve the differential equation using the method of undetermined coefficients. If no initial conditions are given, give the general solution.

13. $y'' + 9y = x^3 + 6$
14. $y'' - 4y = x^2 + 17x$
15. $y'' + 8y' = 7x + 11$
16. $y'' - 7y' = 5x - 3$
17. $y'' - 7y' + 12y = 5e^{2x}$, $y(0) = 1$, $y'(0) = 0$
18. $y'' + 7y' + 12y = 2e^{5x}$, $y(0) = 0$, $y'(0) = 2$
19. $y'' - 3y' + 2y = 2e^x$
20. $y'' + 4y = 3e^{2x}$
21. $y'' - 2y' + 5y = 4e^x$
22. $y'' + 2y' + 5y = 3e^{-x}$
23. $y'' - 9y = 5e^{-3x}$
24. $y'' - 3y' + 2y = 2e^{-x}$
25. $y'' + 2y' + 5y = 3 \sin x$, $y(0) = 1$, $y'(0) = 1$
26. $y'' + 9y = 5 \cos x$, $y(0) = 0$, $y'(0) = 0$
27. $y'' + 9y = 4 \sin 3x$
28. $y'' + y' + y = 3e^{-x}$
29. $y'' + y = \cos 2x$, $y(0) = 0$, $y'(0) = 2$
30. $y'' + 4y' + 8y = x^2 + 1$, $y(0) = 0$, $y'(0) = 0$
31. $y'' - 4y = xe^{3x}$
32. $y'' - 4y' + 3y = xe^{2x}$
33. $y'' - 4y' + 3y = xe^x$
34. $y'' - 7y' + 12y = (x + 5)e^{4x}$
35. $y'' + 16y = 3 \cos 4x$
36. $y'' + y = \sin x$
37. $y'' - 2y' + 5y = e^x \cos 3x$
38. $y'' + 2y' + y = 3e^x$, $y(0) = 0$, $y'(0) = 2$
39. $y'' + 4y' = 12x^2 + e^x$, $y(0) = 1$, $y'(0) = 1$
40. $y'' + y' + y = \cos 2x$
41. $y'' + 2y' + y = 3e^{-x}$
42. $y'' + 2y' + y = 3xe^{-x} + 2e^{-x}$
43. $y'' - 4y' + 4y = xe^x - e^x + 2e^{3x}$
44. $y'' - 5y' + 4y = 17 \sin x + 3e^{2x}$
45. $y'' + 5y' + 4y = 8x^2 + 3 + 2 \cos 2x$
46. $y' + y = 2e^{-x}$
47. $y' + 3y = x^2 + 1$
48. $y' - y = \sin x$
49. $y'' + 4y = \sin 2x$
50. $2y' + 4y = x$
51. $3y' - 2y = xe^x$
52. $y' - 3y = e^x \sin x$

In Exercises 53 through 96, give the form for y_p if the method of undetermined coefficients were to be used. You need not actually compute y_p.

53. $y'' - 5y' + 6y = x^5 + 7x^3 + 4x$

54. $y'' + 5y' + 6y = x^4 + 3x^2 + 7$

55. $y'' + 9y' = x^3$

56. $y'' - 9y' = 3x^3 + 2x^2 + x + 11$

57. $y'' + 5y' + 6y = 5e^{4x}$

58. $y'' - 5y' + 6y = 6e^{5x}$

59. $y'' - 4y = 5e^{2x}$

60. $y'' + 7y' + 12y = 5e^{-3x}$

61. $y'' + 6y' + 9y = e^{-3x}$

62. $y'' - 8y' + 16y = e^{4x}$

63. $y'' + 2y' + y = x^3e^{-x}$

64. $y'' - 2y' + y = x^2e^x$

65. $y'' - 7y' + 12y = x^5e^{4x}$

66. $y'' - 16y = (x^3 + 3x^2 + 7)e^{-4x}$

67. $y'' - 6y' + 9y = x^4e^{3x}$

68. $y'' + 2y' - 3y = x^3e^x - e^x + e^{-2x} + e^{-3x}$

69. $y'' + 3y' - 10y = x^2e^{2x} + e^{5x}$

70. $y'' - 6y' + 5y = e^{5x} + x^2e^{-5x}$

71. $y'' + 5y' = \cos 5x$

72. $y'' - 25y = \sin 5x$

73. $y'' - 7y' + 12y = x^2 \sin 4x$

74. $y'' + 3y' - 10y = x^2 \cos x$

75. $y'' + 25y = \cos 5x$

76. $y'' + 9y = x^2 \sin 3x + \cos 2x$

77. $y'' - 2y' + 5y = 3e^x \sin 2x$

78. $y'' + 2y' + 5y = 5e^{-x} \cos 3x$

79. $y'' + 9y = xe^{-x} \sin 2x$

80. $y'' - y = x^2e^x \cos x$

81. $y'' + 2y' + 2y = x^2e^x \sin x$

82. $y'' + 2y' + 2y = 3x^2e^{-x} \cos 2x + xe^{-x} \sin 2x$

83. $y'' + 2y' + 2y = e^{-x} \cos x + e^x \sin x$

84. $y'' + 2y' + 2y = e^{-x} \sin x + \cos x$

85. $y'' - y = e^{-x} - e^x + e^x \cos x$

86. $y'' + y = e^{-x} - e^x + e^x \cos x$

87. $y'' + 16y = x \cos 4x + e^{-x} \sin 4x + 3e^{-4x}$

88. $y'' + 2y' + 2y = x^2e^{-x} \cos x$

89. $y'' - 2y' + 2y = x^3e^{-x} \sin x + e^x \cos x$

90. $y'' - 2y' + y = x^2e^x \sin 3x$

91. $y'' + 2y' + y = xe^x \sin 3x + e^x \cos 3x$

92. $y'' + 2y' + 2y = xe^{-x} \sin x$

93. $y'' + 4y' + 8y = x^2e^{-2x} \sin 2x + xe^{-2x} \cos 2x$

94. $y'' + 4y' + 8y = xe^{-2x} \cos x$

95. $y'' + 4y' + 13y = x^3e^{-2x} \sin 3x$

96. $y'' + 4y' + 13y = xe^x \sin 3x$

97. Consider the differential equation
$y'' = x^3 + 7x - 2$
a) Find the form of a particular solution using the method of undetermined coefficients.
b) Find the particular solution by integration.

98. In this exercise, we will show that the method of undetermined coefficients is valid for forcing functions that are an exponential times a polynomial if the method has been proved for polynomials. We assume that if $p(x)$ is an mth-degree polynomial, then there is a particular solution in the form

$$y_p = q(x),$$

where $q(x)$ is a polynomial of degree m if $r = 0$ is not a root of the characteristic equation, $q(x)$ is degree $m + 1$ if $r = 0$ is a root one time, and $q(x)$ has degree $m + 2$ if $r = 0$ is a double root. Suppose

$$ay'' + by' + cy = e^{\alpha x}p(x), \qquad (25)$$

where $p(x)$ is an mth-degree polynomial.
a) Use the change of variables, $y = e^{\alpha x}v$, to show that

$$av'' + (2a\alpha + b)v' + (a\alpha^2 + b\alpha + c)v = p(x).$$

b) Show that there is a particular solution for $v(x)$ that is a polynomial whose degree depends on the number of times $r = \alpha$ is a root of the characteristic equation for (25).

As noted in Examples 2.9.7 and 2.9.11, the method of undetermined coefficients can be applied to equations of order different from two with no changes.

In Exercises 99 through 108, solve the differential equation by the method of undetermined coefficients. If no initial conditions are given, give the general solution.

99. $y''' - 3y'' + 3y' - y = e^{2x}$

100. $y''' - 3y'' + 3y' - y = e^{x}$

101. $y''' - y' = \sin x$

102. $y'''' - 25y'' + 144y = x^2 - 1$

103. $y''' + y' = 3 + 2\cos x$,
$y(0) = y'(0) = y''(0) = 0$

104. $y'''' - y = 2e^{3x} - e^{x}$

105. $y'''' - 16y = 5xe^{x}$

106. $y'''' + 4y'' + 4y = \cos 2x$

107. $y'''' - 5y'' + 4y = e^{2x} - e^{3x}$

108. $y'''' + y''' - 6y'' = 72x + 24$, $y(0) = 0$,
$y'(0) = 0$, $y''(0) = -6$, $y'''(0) = -57$

In Exercises 109 through 116, give the form for y_p that you would use to find a particular solution by the method of undetermined coefficients. Do not actually solve for y_p.

109. $y''' - 3y'' + 3y' - y = x^2e^{x} - 3e^{x}$

110. $y'''' + 2y'' + y = x\sin x$

111. $y'''' - 4y''' + 6y'' - 4y' + y = x^3e^{x} + x^2e^{-x}$

112. $y'''' + 5y'' + 4y = \sin x + \cos 2x + \sin 3x$

113. $y''' + 2y'' + 2y' = 3e^{-x}\cos x$

114. $y''' + 2y'' + 2y' = x^2e^{-x}\cos x - xe^{-x}\sin x$

115. $y'''' + 4y''' + 8y'' + 8y' + 4y = 7e^{-x}\cos x$

116. $y'''' - 2y'' + y = xe^{x} + x^2e^{-x} + e^{2x}$

2.10 Undetermined Coefficients Using Annihilators

As noted in Section 2.9, the method of undetermined coefficients always works if we have a linear constant coefficient differential equation and the forcing function f is of the right kind. This section will provide one explanation of why that is true. It can also be used as an alternative approach for carrying out the method of undetermined coefficients.

Well-designed notation not only helps in the solving of a problem but sometimes suggests new approaches and relationships. The notation of this section is not essential for solving simple differential equations, although it can add insight. For more complicated problems, it becomes progressively more important.

The expression

$$L(y) = a_n(x)\frac{d^ny}{dx^n} + a_{n-1}(x)\frac{d^{n-1}y}{dx^{n-1}} + \cdots + a_1(x)\frac{dy}{dx} + a_0(x)y \tag{1}$$

is called an ***n*th-order linear differential operator**.

Given a function y, $L(y)$ is another function. We already have two examples of operators: differentiation and antidifferentiation. In Chapter 4 the Laplace transform operator is discussed in detail. In this chapter, "operator" will mean linear differential operator.

EXAMPLE 2.10.1 Operator

If $L(y) = xy'' + y$, compute $L(\sin x)$.

Solution $L(\sin x) = x(\sin x)'' + \sin x = -x \sin x + \sin x$. ◄

Just as we sometimes write f for the function whose value at x is $f(x)$, L denotes the operator whose value at y is $L(y)$.

If $L(y)$ is given by (1), then

$$L = a_n(x)\frac{d^n}{dx^n} + a_{n-1}(x)\frac{d^{n-1}}{dx^{n-1}} + \cdots + a_1(x)\frac{d}{dx} + a_0(x). \tag{2}$$

It is important to note that the $a_0(x)$ in (2) stands for multiplication by $a_0(x)$.

EXAMPLE 2.10.2 Operator

Suppose

$$L = x^2\frac{d^2}{dx^2} + x\frac{d}{dx} + x^3.$$

Compute $L(x^4)$.

Solution By (1) with $y = x^4$,

$$\begin{aligned} L(x^4) &= \left(x^2\frac{d^2}{dx^2} + x\frac{d}{dx} + x^3\right)x^4 \\ &= x^2\frac{d^2x^4}{dx^2} + x\frac{dx^4}{dx} + x^3x^4 \\ &= x^2 12x^2 + x4x^3 + x^3x^4 = 16x^4 + x^7. \end{aligned}$$ ◄

■ **Theorem 2.10.1 Linearity**

Differential operators defined by (1) are **linear**. That is, if L is a differential operator, c_1, c_2 are constants, and y_1, y_2 are functions, then

$$L(c_1y_1 + c_2y_2) = c_1L(y_1) + c_2L(y_2). \tag{3}$$

■

(There are nonlinear differential operators, but we shall not discuss them.)

Note that the linear differential equation

$$a_n \frac{d^n y}{dx^n} + a_{n-1} \frac{d^{n-1} y}{dx^{n-1}} + \cdots + a_1 \frac{dy}{dx} + a_0 y = f$$

may be written as

$$L(y) = f,$$

where L is given by (2).

If it is obvious what the independent variable is, D^n is often written for d^n/dx^n, the operation of differentiating n times with respect to the independent variable. For example,

$$x^2 \frac{d^2}{dx^2} + x \frac{d}{dx} + x^3 = x^2 D^2 + xD + x^3.$$

In order to consider a differential operator as a function of functions, we need to specify its domain. We shall always take as the domain those functions that are n times differentiable on a fixed interval.

The product of two differential operators is defined to be their composition as functions:

If L_1, L_2 are differential operators, then $L_1 L_2$ is another differential operator, and it is defined by

$$(L_1 L_2)(y) = L_1(L_2(y)). \tag{4}$$

Note that this means, for example, that $D^2 D^3 = D^5$, since

$$D^2 D^3 y = \frac{d^2}{dx^2}\left(\frac{d^3 y}{dx^3}\right) = \frac{d^5 y}{dx^5} = D^5 y.$$

If the differential operators have constant coefficients, then they may be easily multiplied.

■ **THEOREM 2.10.2 Constant Coefficient Linear Operators Commute**

If L_1, L_2 have **constant coefficients**, then

1. $L_1 L_2 = L_2 L_1$ (commutativity).
2. $L_1 L_2$ can be computed by multiplying out the expressions for L_1, L_2, as if they were ordinary polynomials. ■

The verification of Theorem 2.10.2 is left to the exercises.

EXAMPLE 2.10.3 Commutativity of Operators

Let $L_1 = D + 2$ and $L_2 = D + 1$. Then

$$L_1 L_2 = (D + 2)(D + 1) = D^2 + 3D + 2,$$

and

$$L_2L_1 = (D + 1)(D + 2) = D^2 + 3D + 2.$$

Similarly,

$$L_2^3 = (D + 1)^3 = D^3 + 3D^2 + 3D + 1.$$

2.10.1 Particular Solutions

Suppose that we have the linear differential equation with constant coefficients

$$a_n y^{(n)} + a_{n-1} y^{(n-1)} + \cdots + a_1 y' + a_0 y = f$$

or

$$L_1(y) = f,$$

where $L_1 = a_n D^n + a_{n-1} D^{n-1} + \cdots + a_1 D + a_0$.

If f is a linear combination of terms of the form $x^m e^{\alpha x} \cos \beta x$, $x^m e^{\alpha x} \sin \beta x$, with nonnegative integer m and real α, β, then f is the *solution* of another linear constant coefficient homogeneous differential equation, $L_2(y) = 0$. That is, $L_2(f) = 0$. Thus

If y_p is a particular solution of $L_1(y) = f$, and $L_2(f) = 0$, then y_p is a solution of

$$L_2L_1(y) = L_2(f) = 0.$$

But $L_2L_1(y) = 0$ is a homogeneous equation with constant coefficients that can be solved to give a form for y_p. The operator L_2 is called an **annihilator** since in this way the right hand side becomes zero. L_2 is said to annihilate the right hand side. We may delete the solution of the associated homogeneous equation $L_1(y) = 0$ from the form for y_p.

Rather than give a detailed proof, we shall work an example.

EXAMPLE 2.10.4 **Form of Particular Solution Using Annihilators**

Consider the differential equation

$$y'' - 3y' + 2y = 3xe^x$$

or

$$(D^2 - 3D + 2)y = 3xe^x. \tag{5}$$

We need to find a homogeneous differential equation for which xe^x is a solution. This amounts to applying the rules of Sections 2.6 and 2.7 backwards. Thus xe^x would correspond to a root 1 of the characteristic polynomial of multiplicity 2. Hence $(r-1)^2 = r^2 - 2r + 1$ would be the characteristic polynomial of a linear homogeneous differential equation for which xe^x is a solution. That is,

$$(D^2 - 2D + 1)xe^x = 0.$$

Multiply both sides of the original equation (5) by $(D^2 - 2D + 1)$ to get

$$(D^2 - 2D + 1)(D^2 - 3D + 2)y = 0. \tag{6}$$

The characteristic polynomial of this linear homogeneous equation is

$$(r^2 - 2r + 1)(r^2 - 3r + 2) = (r-1)^3(r-2).$$

Thus the solution of (6) is

$$y = c_1e^x + c_2xe^x + c_3x^2e^x + c_4e^{2x}. \tag{7}$$

Now every solution of the original equation (5) is a solution of (6), so (7) gives a **form** for all solutions of our original equation (5). The terms $c_1e^x + c_4e^{2x}$ in (7) are the general solution of the associated homogeneous equation for our original problem (5). Thus the remaining terms give a form for a particular solution:

$$y_p = c_2xe^x + c_3x^2e^x. \tag{8}$$

The constants c_2, c_3 can be found by substituting into the original equation (5). Note that the form (8) for y_p is precisely that provided by the rules of Section 2.9. ◀

Exercises

In all of the exercises, L denotes a differential operator.

1. If $L = 2D + x$, compute $L(x^3)$.

2. If $L = xD^2 + 1$, compute $L(\sin x)$.

3. If $L = D^2 + xD$, compute $L(3e^{-x})$.

4. If $L = D^2 + 1$, compute $L(\sin x + 3)$.

5. If $L = D^3 - D + 1$, compute $L(e^{-x})$.

6. If $L = xD^2 + x^2D - x + 1$, compute $L(x^4 + 1)$.

For Exercises 7 through 13, compute L_1L_2 and L_2L_1.

7. $L_1 = D$, $L_2 = xD$

8. $L_1 = D + x$, $L_2 = D - x$

9. $L_1 = D + 3$, $L_2 = D^2 + 1$

10. $L_1 = xD$, $L_2 = D^2$

11. $L_1 = D^2 + D$, $L_2 = D - 1$

12. $L_1 = D - 1$, $L_2 = e^xD$

13. $L_1 = D^2 + 1$, $L_2 = \sin xD$

14. Verify that, if y_1 is a solution of $L(y) = e^x$ and y_2 is a solution of $L(y) = \sin x$, then $y_1 + y_2$ is a solution of $L(y) = e^x + \sin x$. (This is sometimes referred to as the **superposition principle.**)

15. Verify that, if y_1 is a solution of $L_1(y) = f$ and f is a solution of $L_2(y) = 0$, then y_1 is a solution of the homogeneous equation $L_2L_1(y) = 0$. (This is the idea behind the method of undetermined coefficients, discussed earlier in this chapter.)

16. Suppose that $L = L_1L_2$. Verify that, if y_1 is a solution of $L_1(y_1) = f$ and y_2 is a solution of $L_2(y_2) = y_1$, then y_2 is a solution of $L(y) = f$.

17. Note that $L = D^2 - 1 = (D + 1)(D - 1) = L_1L_2$. By finding y_1, y_2 as in Exercise 16, solve the second-order differential equation $y'' - y = 3$.

18. Note that $L = D^2 - 3D + 2 = (D - 1)(D - 2) = L_1L_2$. By finding y_1, y_2 as in Exercise 16, solve the second-order linear differential equation $y'' - 3y' + 2y = 2$.

19. Verify Theorem 2.10.2 if L_1, L_2 are second-order constant coefficient differential operators.

20. Suppose that f, g are differentiable functions of x. Show that, if $(D - 1)f = (D - 1)g$, then $f = g + ce^x$ for some constant c.

21. Suppose that f, g are twice-differentiable functions and L is a second-order differential operator. Show that if $L(f) = L(g)$, then $f = g + h$, where h is a solution of $L(y) = 0$.

Exercises 22 through 25 introduce some facts about differential operators that can be used to reduce the amount of computation when using the method of undetermined coefficients (Section 2.9). Here, $p(\lambda) = a_n\lambda^n + a_{n-1}\lambda^{n-1} + \cdots + a_1\lambda + a_0$ is an nth-degree polynomial with constant coefficients, and $p(D) = a_nD^n + a_{n-1}D^{n-1} + \cdots + a_1D + a_0$.

22. **a)** Show that $p(D)e^{\alpha x} = p(\alpha)e^{\alpha x}$ for any scalar α.

b) Compute $(D^3 - 2D^2 + D - 1)e^{3x}$ using part (a).

23. Suppose that $p(\lambda)$, $q(\lambda)$ are polynomials with degree $(p(\lambda)) >$ degree $(q(\lambda))$. Let $r(\lambda)$ be the remainder after dividing $q(\lambda)$ into $p(\lambda)$ using polynomial division. That is, $p(\lambda) = m(\lambda)q(\lambda) + r(\lambda)$ with m, r polynomials and degree $(r(\lambda)) <$ degree $(q(\lambda))$. Suppose that $f(x)$ is a function such that $q(D)f = 0$. Show that $p(D)f = r(D)f$.

24. **a)** Verify that $(D^2 + 2D + 2)e^{-x}\sin x = 0$.

b) Compute $(D^4 + 3D^3 - D)e^{-x}\sin x$ by letting $p(D) = D^4 + 3D^3 - D$, $q(D) = D^2 + 2D + 2$, $f = e^{-x}\sin x$ and using Exercise 23.

25. **a)** Verify that $(D^2 + 2D + 1)xe^{-x} = 0$.

b) Compute $(D^5 + 3D^2 + 1)xe^{-x}$ by letting $p(D) = D^5 + 3D^2 + 1$, $q(D) = D^2 + 2D + 1$, $f = xe^{-x}$ and using Exercise 23.

In Exercises 26 through 33, each of these differential equations is in the form of $L_1(y) = f$, where L_1 has constant coefficients.

i) Find a constant coefficient differential operator L_2 so that $L_2(f) = 0$.

ii) Then find the form for a particular solution of $L_1(y) = f$ by removing terms from the solution of $L_1(y) = 0$ from the solution of $L_2L_1(y) = 0$.

26. $y'' - y = e^{2x}$

27. $y'' - y = e^x$

28. $y'' + y' = x^2 - 1$

29. $y'' + 3y = x^2 + 3x + 1$

30. $y'' - 2y' + y = e^x$

31. $y'' + 4y' + 4y = 3e^{-2x}$

32. $y'' - 5y' + 6y = 2e^{4x} + 6e^{3x}$

33. $y'' - 5y' + 6y = 4e^{2x} + 3e^{3x}$

The exercises of Section 2.9 can also be done with the approach of this section.

2.11 Mechanical Vibrations II: Forced Response

In Section 2.8 we carefully examined the free response of a spring-mass system. In this section we investigate what happens when an additional external force $f(t)$, which depends only on the time t, is applied to the mass.

The equation of motion from Section 2.8 is then

$$m\frac{d^2x}{dt^2} + \delta\frac{dx}{dt} + kx = f(t). \tag{1}$$

The external force $f(t)$ is often referred to as the **forcing function** or the **input**. The solution x is then called the **response** or the **output** of the system. From our theory developed earlier, we know that the solution x of (1) has the form $x = x_p + x_h$, where x_p is a particular solution of (1) and $x_h = c_1x_1 + c_2x_2$ is the general solution of the associated homogeneous equation

$$mx'' + \delta x' + kx = 0. \tag{2}$$

x_h is also called the free response. Note that the arbitrary constants appear only in x_h. We have then

The possible responses to the input $f(t)$ consist of a particular response to $f(t)$ added to the possible free responses of the system.

Most of our discussion will be restricted to sinusoidal forcing, $f(t) = F_0 \cos \omega t$, where F_0 is the **amplitude of the forcing** and ω is the **forcing frequency**. This is an important case to consider for two reasons. First, many forcing functions, such as alternating currents in circuits, are approximately in this form. Also, the theory of Fourier series, discussed in Chapter 10, may be used to express many forcing terms as a sum (series) of terms of the form $\sin \alpha nt$, $\cos \alpha nt$, α a constant.

The study of forced responses will be broken into two cases, depending on whether or not friction is present.

2.11.1 Friction Is Absent ($\delta = 0$)

Suppose that friction is negligible ($\delta = 0$), so that the model for the spring-mass system with periodic forcing is

$$mx'' + kx = F_0 \cos \omega t. \tag{3}$$

The solution of (3) has the form

$$x = x_p + c_1 \cos\left(\sqrt{\frac{k}{m}}\,t\right) + c_2 \sin\left(\sqrt{\frac{k}{m}}\,t\right).$$

Here x_p is a particular solution of (3) and $\cos \omega_0 t$, $\sin \omega_0 t$ are a fundamental set of solutions of $mx'' + kx = 0$ with $\omega_0 = \sqrt{k/m}$ since $\pm i\sqrt{k/m}$ are roots of the characteristic polynomial $mr^2 + k$.

We shall now find a particular solution for (3) by the method of undetermined coefficients (Section 2.9). The characteristic equation

$mr^2 + k = 0$ has roots $r = \pm i\sqrt{k/m} = \pm i\omega_0$. For a spring-mass system (without friction) the natural circular frequency is $\omega_0 = \sqrt{k/m}$. There are two cases for the method of undetermined coefficients depending on whether the forcing function is a solution of the associated homogeneous equation or not. This depends on whether the forcing circular frequency ω equals the natural circular frequency $\omega_0 = \sqrt{k/m}$.

Case 1: ($\omega \neq \sqrt{k/m}$) Forcing Frequency Unequal to the Natural Frequency of Free Response

(Frequency of input unequal to natural frequency of free response) In this case, the form for x_p given by the method of undetermined coefficients is

$$x_p = A \cos \omega t + B \sin \omega t. \tag{4}$$

Substituting this form (4) into the differential equation (3) gives

$$m(A \cos \omega t + B \sin \omega t)'' + k(A \cos \omega t + B \sin \omega t) = F_0 \cos \omega t$$

or

$$-mA\omega^2 \cos \omega t - mB\omega^2 \sin \omega t + kA \cos \omega t + kB \sin \omega t = F_0 \cos \omega t.$$

Equating coefficients of like terms in order to find A, B:

$$\cos \omega t: \qquad -mA\omega^2 + kA = F_0 \Rightarrow A = \frac{F_0}{k - m\omega^2}$$

$$\sin \omega t: \qquad -mB\omega^2 + kB = 0 \Rightarrow B = 0,$$

so that

$$x_p = \frac{F_0}{k - m\omega^2} \cos \omega t. \tag{5}$$

The amplitude of this particular response is $F_0/(k - m\omega^2)$, where F_0 is the amplitude of the input. It is usual to introduce the **frequency-response diagram** in which the response (the ratio of the response amplitude to the input amplitude) is graphed (Figure 2.11.1) as a function of the forcing frequency:

$$\frac{\text{Amplitude of the response}}{\text{Amplitude of the input}} = \frac{1}{k - m\omega^2} = \frac{1}{m(\omega_0^2 - \omega^2)}. \tag{6}$$

The response increases as the forcing frequency approaches the natural frequency. Note that the response diagram has a vertical asymptote when the forcing frequency equals the natural frequency. According to (5) or (6), the amplitude of the response is infinite (undefined) when $\omega = \omega_0 = \sqrt{k/m}$. However, in our analysis we have assumed $\omega \neq \omega_0 = \sqrt{k/m}$. This will be explained in the next subsection. Note from Figure 2.11.1 that the amplitude

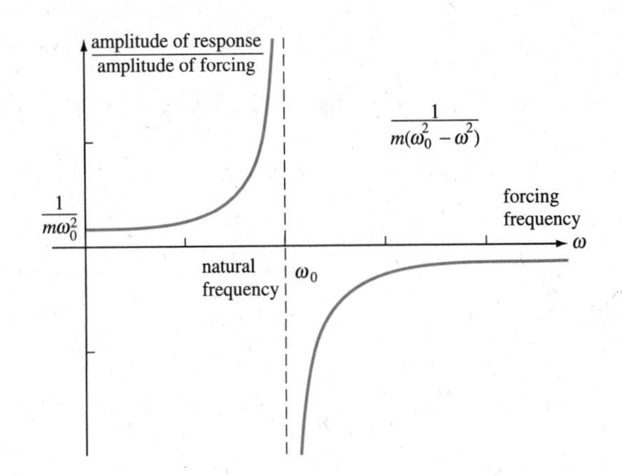

FIGURE 2.11.1

Response diagram (no damping—note resonance).

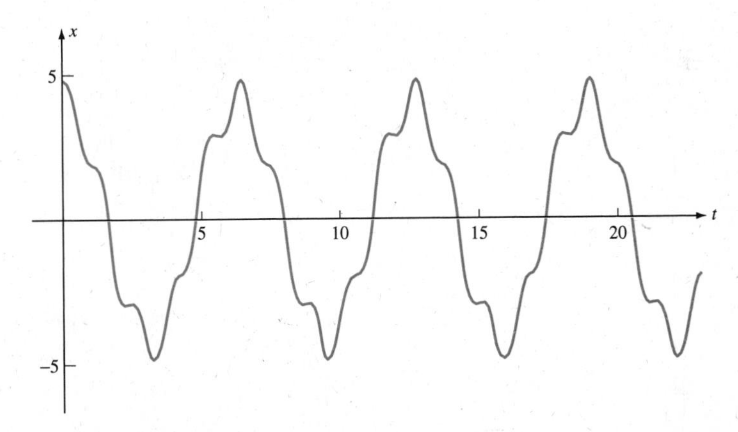

FIGURE 2.11.2

Graph of (7) with $k = 0.25$, $m = 0.01$, $c_1 = c_2 = 0.5$, $\omega = 1$.

is negative if $\omega > \omega_0$. In this case we say the response is 180° out of phase from the input.

The general solution of the differential equation (3) is

$$x = \frac{F_0}{k - m\omega^2} \cos \omega t + c_1 \cos\left(\sqrt{\frac{k}{m}}\, t\right) + c_2 \sin\left(\sqrt{\frac{k}{m}}\, t\right). \tag{7}$$

This is the superposition of two harmonic motions. One is the free response of the system and the other is a periodic forced response with the same period as the forcing function. Figure 2.11.2 gives the graph of (7) for one choice of the parameters m, k, ω, c_1, c_2.

Case 2: ($\omega = \sqrt{k/m}$) Resonance: Forcing Frequency Equals the Natural Frequency of Free Response

If $\omega = \sqrt{k/m}$, then using the method of undetermined coefficients gives the form of the particular solution of (3) as

$$x_p = At \cos \omega t + Bt \sin \omega t. \tag{8}$$

Substituting this form into the differential equation (3) gives

$$m(At \cos \omega t + Bt \sin \omega t)'' + k(At \cos \omega t + Bt \sin \omega t) = F_0 \cos \omega t$$

or

$$m(-2A\omega \sin \omega t - At\omega^2 \cos \omega t + 2B\omega \cos \omega t - Bt\omega^2 \sin \omega t) \\ + kAt \cos \omega t + kBt \sin \omega t = F_0 \cos \omega t.$$

Equating coefficients of like terms gives four equations in A, B:

$$\begin{aligned} \sin \omega t: &\quad -2Am\omega = 0, \\ \cos \omega t: &\quad 2Bm\omega = F_0, \\ t \sin \omega t: &\quad -Bm\omega^2 + kB = 0, \\ t \cos \omega t: &\quad -Am\omega^2 + kA = 0. \end{aligned}$$

Since $\omega^2 = k/m$ by assumption, the last two equations are trivially true, and thus $A = 0$, $B = F_0/(2m\omega)$, from the first two equations. Thus,

$$x = \frac{F_0}{2m\omega} t \sin \omega t + c_1 \cos \omega t + c_2 \sin \omega t \tag{9}$$

is the general solution. The particular solution

$$x_p = \frac{F_0}{2m\omega} t \sin \omega t \tag{10}$$

is of special interest since it illustrates the phenomenon of **resonance**. The amplitude $(F_0/2m\omega)t$ of the oscillation grows proportionally with time. The graph of (10) is given in Figure 2.11.3.

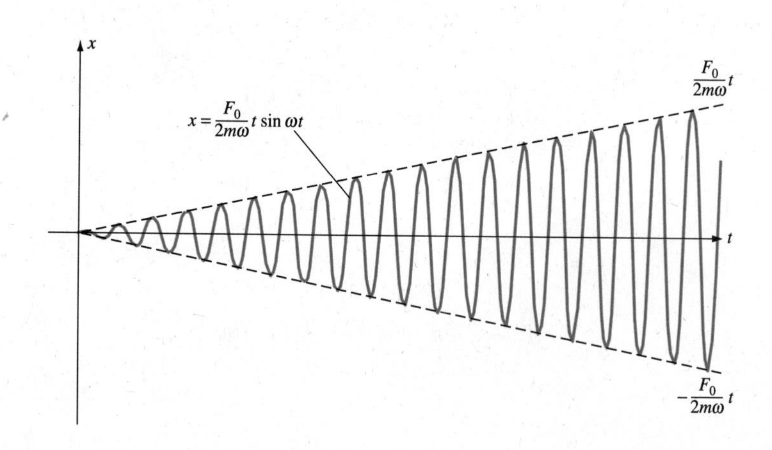

FIGURE 2.11.3
Resonance (forcing frequency = natural frequency).

The forced response is now an unbounded oscillation. Intuitively, the phenomena of resonance may be summarized as follows:

If the period (frequency) of the forcing function is the same as the period (frequency) of the free response of the system, then large-amplitude oscillations may result from forcing terms of small amplitude.

In structures such as bridges or aircraft wings, resonance is to be avoided. Bridges and aircraft wings have both failed on occasion due to resonance. On the other hand, resonance is sometimes desirable in the manipulation of sound waves in musical instruments and detectors of various kinds. Of course, in practice the oscillations do not become arbitrarily large. Rather, either the system is changed (the spring breaks, for example) or the original linear model $mx'' + \delta x' + kx = f$ is no longer a valid model.

Also, "pure resonance" is never observed. In every system there is some friction, so δ, while perhaps very small, is greater than zero. Also, it is almost impossible to make ω *exactly* $\sqrt{k/m}$. However, if δ is close to zero and ω close to $\sqrt{k/m}$, then the forced response may exhibit a large response. The next example illustrates this.

EXAMPLE 2.11.1 Resonance and Near Resonance

Suppose that the model for a spring-mass system with forcing term $\cos \omega t$ is

$$x'' + x = \cos \omega t, \qquad x(0) = 0, \qquad x'(0) = 0. \tag{11}$$

Determine the solution for all ω, and graph the solution for $\omega = 1$ and for several values of ω close to 1.

Solution Using the method of undetermined coefficients [or (7) and (9)], we find that the general solution of $x'' + x = \cos \omega t$ is

$$\omega \neq 1: \qquad x = \frac{1}{1 - \omega^2} \cos \omega t + c_1 \cos t + c_2 \sin t, \tag{12}$$

$$\omega = 1: \qquad x = \frac{1}{2} t \sin t + c_1 \cos t + c_2 \sin t. \tag{13}$$

Applying the initial conditions to (12) gives

$$0 = x(0) = \frac{1}{1 - \omega^2} + c_1$$

$$0 = x'(0) = c_2,$$

so $c_2 = 0$, $c_1 = -1/(1 - \omega^2)$. Thus $x = (1 - \omega^2)^{-1} \cos \omega t - (1 - \omega^2)^{-1} \cos t$. Applying the initial conditions to (13) gives

$$0 = x(0) = c_1,$$

$$0 = x'(0) = c_2.$$

FIGURE 2.11.4

Graph of (14) for $\omega = 0.7, 0.8, 0.9, 1.0$.

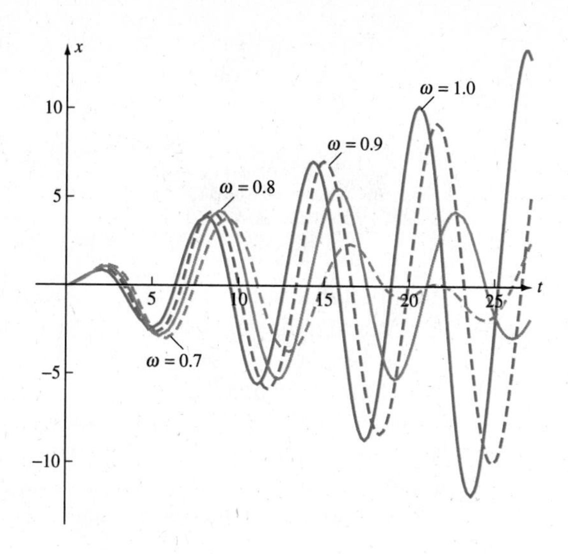

Thus the solution of (11) is

$$x(t) = \begin{cases} \dfrac{1}{1-\omega^2}(\cos \omega t - \cos t) & \text{if } \omega \neq 1, \\ \dfrac{1}{2} t \sin t & \text{if } \omega = 1. \end{cases} \tag{14}$$

The graph of (14) for several values of ω is given in Figure 2.11.4. Notice that for ω close to 1, the solutions exhibit large-amplitude oscillations. ◄

EXAMPLE 2.11.2 Resonance

For what values of the mass m will $mx'' + 25x = 12 \cos \omega t$ exhibit resonance if $12 \cos \omega t$ has a frequency of 28 Hz?

SOLUTION The characteristic polynomial $mr^2 + 25$ has pure imaginary roots $\pm i\sqrt{25/m}$. Resonance occurs if

$$\omega = \sqrt{25/m}. \tag{15}$$

If $\cos \omega t$ has a frequency of 28 Hz, then

$$\frac{\omega}{2\pi} = 28 \quad \text{or} \quad \omega = 56\pi. \tag{16}$$

Combining (15) and (16) gives

$$56\pi = \sqrt{\frac{25}{m}}$$

or $m = 25/(56\pi)^2$. ◄

EXAMPLE 2.11.3 **Road Induced Resonance**

A car is moving along a roadway at a constant horizontal velocity v. Suppose that the vehicle can be modeled by a mass with a vertical spring between the mass and the road. Also suppose that as the car moves along the road, the only forces acting on the spring-mass system are the weight of the car and the changing length of the spring caused by the rising and falling of the road surface. Assume that the road has a surface that is approximately sinusoidal with wavelength L and that the spring-mass system has negligible friction. At what constant horizontal velocity v would resonance be observed?

SOLUTION Let x measure distance horizontally. Thus the varying of the road surface is given by

$$y_r(x) = a \sin \frac{2\pi x}{L}.$$

Let $y(x)$ be the vertical position of the mass at position x. To set up the coordinates, assume that when $x = 0$, the base of the spring is at the average road height, which is $y_r = 0$. Let $y = 0$ correspond to the rest position of the mass there. Thus the spring force and the vehicle weight are balanced out. We measure y in the positive direction. The rest configuration is shown in Figure 2.11.5. Let m be the mass. Since we have measured from the rest position, the gravity term cancels the compression force at rest, and Hooke's law takes the form

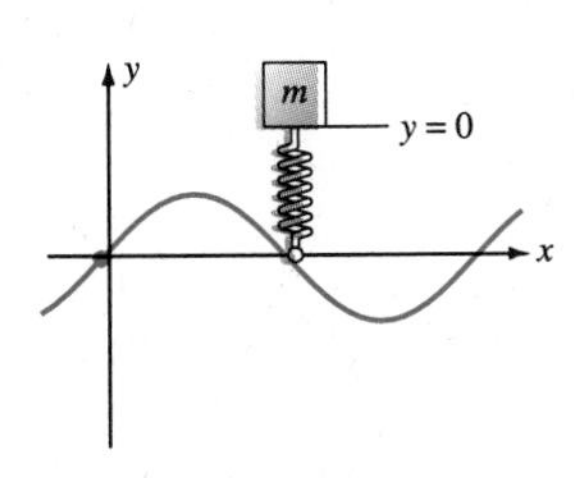

FIGURE 2.11.5

Rest configuration.

$$m\frac{d^2y}{dt^2} = -k(\text{stretching of spring from rest length}). \tag{17}$$

The amount that the spring is stretched from its rest length at position x is $y(x) - y_r(x)$. See Figure 2.11.6. Thus

$$m\frac{d^2y}{dt^2} = -k(y - y_r) = -k\left(y - a\sin\frac{2\pi x}{L}\right). \tag{18}$$

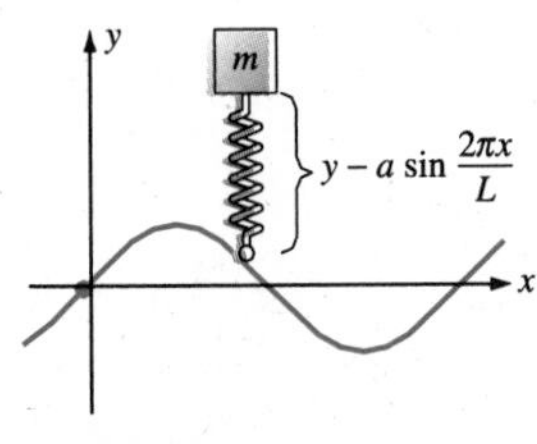

FIGURE 2.11.6

Motion.

The differential equation is usually written in the equivalent form

$$m\frac{d^2y}{dt^2} + ky = ka\sin\frac{2\pi x}{L}.$$

However, the position x is not a constant, but a function of time. If we assume the car is moving at a constant velocity v, then (assuming $x = 0$ is defined to be where the car is at $t = 0$)

$$x = vt.$$

The differential equation of the vertical motion of the car (spring-mass system) is

$$m\frac{d^2y}{dt^2} + ky = ka \sin \frac{2\pi vt}{L}.$$

The spring-mass system is forced at frequency v/L. The spring-mass system resonates if the forcing frequency equals the natural frequency, that is,

$$\frac{2\pi v}{L} = \sqrt{\frac{k}{m}}.$$

Thus, the velocity at which resonance occurs is $v = (L/2\pi)\sqrt{k/m}$. ◄

Exercises

1. For what values of m will $mx'' + 4x = 13 \cos \omega t$ exhibit resonance if $13 \cos \omega t$ has a frequency of 20 Hz?

2. For what values of m will $mx'' + 16x = 12 \cos \omega t$ exhibit resonance if $12 \cos \omega t$ has a frequency of 28 Hz?

3. For what values of k will $36x'' + kx = 4 \cos \omega t$ exhibit resonance if $4 \cos \omega t$ has a frequency of 22 Hz?

4. For what values of k will $25x'' + kx = 2 \cos \omega t$ exhibit resonance if $2 \cos \omega t$ has a frequency of 18 Hz?

5. You know that the forcing function $f(t) = F_0 \cos \omega t$ has a frequency between 10 and 70 Hz. The spring-mass system is $mx'' + 10x = f(t)$. What values of m might lead to resonance?

6. You know that the forcing function $f(t) = F_0 \cos \omega t$ has a frequency between 30 and 50 Hz. The spring-mass system is $mx'' + 9x = f(t)$. What values of m might lead to resonance?

7. If your spring-mass system had $m = 15$, $\delta = 0$, and $k = 8$, what forcing frequency would cause resonance?

8. If your spring-mass system had $m = 5$, $\delta = 0$, and $k = 12$, what forcing frequency would cause resonance?

9. You are building a detector that is to be sensitive (exhibit resonance) to harmonic vibrations at a frequency of 30 Hz. The detector will be a spring-mass system with spring constant $k = 15$ g/s^2. Assuming friction is negligible, what should the attached mass be?

10. A device is being built that can be modeled as a simple spring-mass system with negligible friction. The attached mass is 20 g. The spring constant is adjustable. What values of the spring constant are to be avoided if the device is subjected to external harmonic forces in the *range* of 10 to 50 Hz, and your design goal is to avoid resonance?

11. **(Beats)** Using the trigonometric identities $\cos(\theta \pm \phi) = \cos\theta \cos\phi \mp \sin\phi \sin\theta$, verify that

$$\cos \omega t - \cos \beta t = 2 \sin\left(\frac{(\beta - \omega)}{2}t\right) \sin\left(\frac{(\beta + \omega)}{2}t\right). \quad (19)$$

12. A spring-mass system has an attached mass of 4 g, a spring constant of 16 g/s^2 and negligible friction. It is subjected to a force of $4 \cos(2.2t)$ downward and at time $t = 0$ is initially at rest. Determine the subsequent motion. Using (19) from Exercise 11, rewrite the solution as the

product of two sine functions, and graph the result. The resulting function has a periodic variation in amplitude, or a **beat**.

13. A rectangular closed box is floating in the ocean with its top always parallel to the water's surface. The buoyancy force is proportional to the volume of water displaced, which is also proportional to the depth of the bottom of the box. Let m be the mass of the box and let d be the depth of the bottom of the box when the box is at rest. Let $x(t)$ measure how much deeper than this rest depth the bottom of the box is at time t. Ignore resistance and assume small displacements and velocities. Explain why $mx'' + kx = 0$ is a reasonable model for the vertical movement of the box.

14. Suppose it is known that a 1000-lb weight sags a 3000-lb car by 1 in. What would be a poor wavelength for a road to have if traffic usually moves at 55 mi/h?

15. Consider a mass m attached to a spring with spring constant k without friction ($\delta = 0$). The forcing function is $F_0 \cos \omega t$, where $\omega \neq \sqrt{k/m}$. Determine the solution of the initial value problem if $x(0) = x_0$ and $x'(0) = 0$.

16. Consider a mass m attached to a spring with spring constant k without friction ($\delta = 0$). The forcing function is $F_0 \cos \omega t$, where $\omega \neq \sqrt{k/m}$. Determine the solution of the initial value problem if $x(0) = 0$ and $x'(0) = v_0$.

17. Consider a mass m attached to a spring with spring constant k without friction ($\delta = 0$). The forcing function is $F_0 \sin \omega t$, where $\omega \neq \sqrt{k/m}$. Determine the solution of the initial value problem if $x(0) = x_0$ and $x'(0) = 0$.

18. Consider a mass m attached to a spring with spring constant k without friction ($\delta = 0$). The forcing function is $F_0 \sin \omega t$, where $\omega \neq \sqrt{k/m}$. Determine the solution of the initial value problem if $x(0) = 0$ and $x'(0) = v_0$.

19. Consider a mass m attached to a spring with spring constant k without friction ($\delta = 0$). The forcing function is $F_0 \cos \omega t$, where $\omega = \sqrt{k/m}$. Determine the solution of the initial value problem if $x(0) = 0$ and $x'(0) = 0$.

20. Consider a mass m attached to a spring with spring constant k without friction ($\delta = 0$). The forcing function is $F_0 \sin \omega t$, where $\omega = \sqrt{k/m}$. Determine the solution of the initial value problem if $x(0) = 0$ and $x'(0) = 0$.

In Exercises 21 through 28, find the general solution.

21. $\dfrac{d^2x}{dt^2} + 4x = 8 \cos 5t$

22. $\dfrac{d^2x}{dt^2} + 9x = -5 \sin 5t$

23. $\dfrac{d^2x}{dt^2} - 4x = 8 \cos 5t$

24. $\dfrac{d^2x}{dt^2} - 9x = -5 \sin 5t$

25. $\dfrac{d^2x}{dt^2} + 4x = 3 \cos 2t$

26. $\dfrac{d^2x}{dt^2} + 9x = -4 \sin 3t$

27. $m\dfrac{d^2x}{dt^2} + kx = F_0 \cos \omega t$ with $\omega = \sqrt{\dfrac{k}{m}}$

28. $m\dfrac{d^2x}{dt^2} + kx = F_0 \sin \omega t$ with $\omega = \sqrt{\dfrac{k}{m}}$

2.11.2 Friction Is Present ($\delta > 0$) (Damped Forced Oscillations)

If friction is present, then the solution of $mx'' + \delta x' + kx = f(t)$ takes the form

$$x = x_p + x_h = x_p + c_1x_1 + c_2x_2.$$

As noted earlier there are three possibilities for the fundamental set of solutions $\{x_1, x_2\}$:

$$\{e^{-\alpha t}\cos\beta t, e^{-\alpha t}\sin\beta t\}: \qquad \text{underdamped,}$$
$$\{e^{-\alpha t}, te^{-\alpha t}\}: \qquad \text{critically damped,}$$
$$\{e^{-\alpha_1 t}, e^{-\alpha_2 t}\}; \qquad \text{overdamped.}$$

The initial conditions determine the arbitrary constants c_1, c_2, but $\lim_{t\to\infty} x_1 = 0$, $\lim_{t\to\infty} x_2 = 0$ in all three cases. Thus

If friction is present ($\delta > 0$), the free response x_h is always **transient**. That is, $\lim_{t\to\infty} x_h(t) = 0$.

Another way to say the same thing is that, if friction is present, the effect of the initial conditions dies out (is transient). Thus for many important types of forcing functions, such as constant or periodic, the response is eventually determined almost completely by the forcing function $f(t)$. In many situations, particularly the circuits discussed in the next section, this is a desirable phenomenon.

We consider the differential equation that models a spring-mass system with damping with a periodic forcing function with forcing circular frequency ω of the form $F_0\cos\omega t$;

$$m\frac{d^2x}{dt^2} + \delta\frac{dx}{dt} + kx = F_0\cos\omega t. \tag{20}$$

The forcing function $F_0\cos\omega t$ is not a solution of the associated homogeneous equation $mx'' + \delta x' + kx$, since $\delta > 0$. Thus, according to the method of undetermined coefficients, a particular solution is in the form of a linear combination of sines and cosines of the same frequency as the forcing:

$$x_p = A\cos\omega t + B\sin\omega t. \tag{21}$$

Substituting (21) into (20) yields

$$m\frac{d^2}{dt^2}(A\cos\omega t + B\sin\omega t) + \delta\frac{d}{dt}(A\cos\omega t + B\sin\omega t)$$
$$+ k(A\cos\omega t + B\sin\omega t) = F_0\cos\omega t$$

or

$$-m\omega^2(A\cos\omega t + B\sin\omega t) + \delta\omega(-A\sin\omega t + B\cos\omega t)$$
$$+ k(A\cos\omega t + B\sin\omega t) = F_0\cos\omega t.$$

Equating coefficients of like terms in order to find A and B:

$$\begin{aligned} \cos \omega t: \qquad & (k - m\omega^2)A + \delta\omega B = F_0 \\ \sin \omega t: \qquad & -\delta\omega A + (k - m\omega^2)B = 0. \end{aligned} \tag{22}$$

This system (22) can be solved by elimination. For example, multiplying the first equation by $\delta\omega$, the second by $k - m\omega^2$, and adding yields

$$B\left[(k - m\omega^2)^2 + \delta^2\omega^2\right] = F_0\,\delta\omega.$$

In this way we determine B:

$$B = \frac{F_0\,\delta\omega}{(k - m\omega^2)^2 + \delta^2\omega^2}. \tag{23}$$

The equation for A is best determined using B and the second equation in (22) to get

$$A = \frac{F_0(k - m\omega^2)}{(k - m\omega^2)^2 + \delta^2\omega^2}. \tag{24}$$

A particular solution of (19) is

$$x_p = A\cos\omega t + B\sin\omega t = R\cos(\omega t - \phi), \tag{25}$$

where A and B are given by (22) and (23). The amplitude of this particular response is

$$\begin{aligned} R = \sqrt{A^2 + B^2} &= F_0\frac{\sqrt{(k - m\omega^2)^2 + \delta^2\omega^2}}{(k - m\omega^2)^2 + \delta^2\omega^2} \\ &= \frac{F_0}{\sqrt{(k - m\omega^2)^2 + \delta^2\omega^2}}. \end{aligned} \tag{26}$$

The amplitude-response diagram from damped oscillators follows from (26). The ratio of the amplitude of the response R to the amplitude of the forcing F_0 depends on four parameters: m, k, δ of the spring-mass system and the forcing circular frequency ω. Instead of graphing this relationship as a function of the frequency, it is best to introduce a dimensionless frequency, the ratio of the forcing circular frequency ω to the natural circular frequency $\omega_0 = \sqrt{k/m}$. In this way, it follows from (26) that

$$R = \frac{F_0}{\sqrt{(k - m\omega^2)^2 + \delta^2\omega^2}} = \frac{F_0/k}{\sqrt{(1 - \omega^2/\omega_0^2)^2 + (\delta^2/km)(\omega^2/\omega_0^2)}}, \tag{27}$$

which is graphed in Figure 2.11.7 on the following page.

FIGURE 2.11.7

Response diagram.

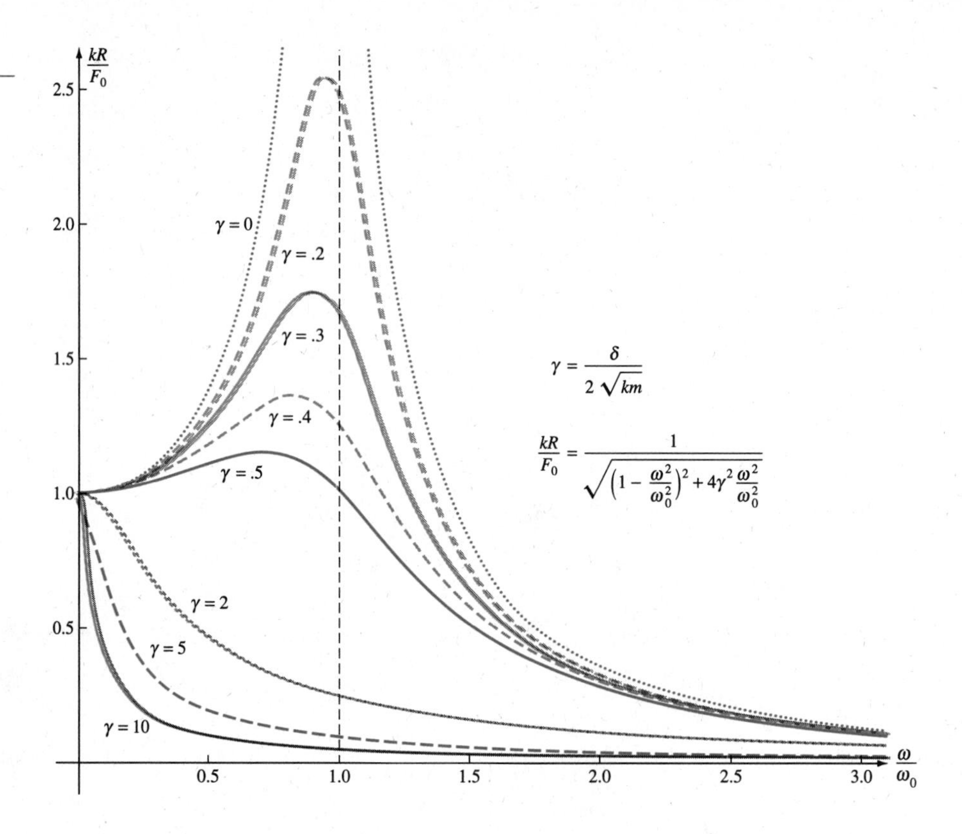

We note that in this way, the graph can be made to depend on only one parameter:

$$\frac{kR}{F_0} = \frac{1}{\sqrt{(1 - \omega^2/\omega_0^2)^2 + 4\gamma^2\omega^2/\omega_0^2}}. \tag{28}$$

The parameter γ is the critical damping ratio

$$\gamma = \frac{\delta}{2\sqrt{km}} = \frac{\delta/2m}{\sqrt{k/m}}. \tag{29}$$

It is also the ratio of the decay rate of underdamped (and critically damped) oscillators to the natural frequency. It can be seen that if $\gamma = 0$, the response curve is the one that we analyzed for forced vibrations without damping. If the forcing frequency equals the natural frequency, $\omega = \omega_0$, then from (28)

$$R = \frac{F_0}{2\gamma k}.$$

This shows that the amplitude of the response is large when the damping is small (γ small). It can be shown that the amplitude response diagram has a maximum for $\omega \neq 0$ if $\gamma < \sqrt{2}/2$. However, if $\gamma \geq \sqrt{2}/2$, then the maximum of the response diagram occurs at $\omega = 0$.

The phase ϕ is probably less important. Recall that $\tan\phi = B/A = \delta\omega/(k - m\omega^2)$.

EXAMPLE 2.11.4 Damped Forced Oscillator

A spring-mass system, with an attached mass of $m = 1$ g, has a spring constant of 50 g/s^2 and a damping coefficient of 2 g/s. At time $t = 0$, the mass is pushed down $\frac{23}{26}$ cm and released with a velocity of $28\frac{2}{13}$ cm/s downward. A force of $41\cos 2t$ dynes acts downward on the mass for $t \geq 0$. Determine the resulting motion and graph it.

SOLUTION The differential equation governing the motion is

$$x'' + 2x' + 50x = 41\cos 2t, \qquad x(0) = \frac{23}{26}, \qquad x'(0) = 28\frac{2}{13}. \tag{30}$$

The characteristic equation $r^2 + 2r + 50 = 0$ has roots $r = -1 \pm 7i$. Thus the solution of the associated homogeneous equation is

$$x_h = c_1 e^{-t}\cos 7t + c_2 e^{-t}\sin 7t.$$

From Section 2.9, the method of undetermined coefficients tells us that there will be a particular solution of the form

$$x_p = A\cos 2t + B\sin 2t.$$

Substituting this form into the differential equation (30) yields

$$(A\cos 2t + B\sin 2t)'' + 2(A\cos 2t + B\sin 2t)'$$
$$+ 50(A\cos 2t + B\sin 2t) = 41\cos 2t$$

or

$$-4A\cos 2t - 4B\sin 2t - 4A\sin 2t + 4B\cos 2t$$
$$+ 50A\cos 2t + 50B\sin 2t = 41\cos 2t.$$

Equating coefficients of $\cos 2t$, $\sin 2t$ gives us equations in A, B;

$$\cos 2t: \quad 46A + 4B = 41,$$
$$\sin 2t: \quad -4A + 46B = 0,$$

so that $B = \frac{1}{13}$, $A = \frac{23}{26}$, and

$$x = x_p + x_h = \frac{23}{26}\cos 2t + \frac{1}{13}\sin 2t + c_1 e^{-t}\cos 7t + c_2 e^{-t}\sin 7t.$$

Applying the initial conditions in (30) in order to determine c_1, c_2, we find

$$\frac{23}{26} = x(0) = \frac{23}{26} + c_1,$$

$$28\frac{2}{13} = x'(0) = \frac{2}{13} - c_1 + 7c_2.$$

Thus $c_1 = 0$, $c_2 = 4$, and

$$x = \frac{23}{26}\cos 2t + \frac{1}{13}\sin 2t + 4e^{-t}\sin 7t.$$

This solution may be simplified, using the form (25) and (26), to give

$$x = \sqrt{\frac{41}{52}}\cos(2t - \phi) + 4e^{-t}\sin 7t, \qquad (31)$$

where $\tan\phi = 2/23$. Note that the long-term or steady response is periodic, with the same period as the input. It is out of phase, however (achieves its maxima at a different time determined by ϕ). As Figure 2.11.8 shows, this motion may be viewed as a damped (transient) oscillation superimposed (added) onto the forced (steady-state) harmonic motion. ◀

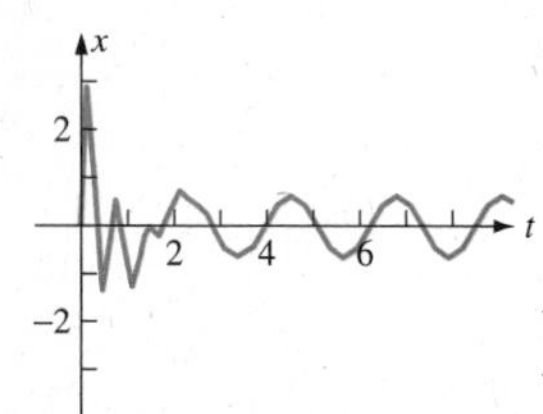

FIGURE 2.11.8

Graph of Eq. (31).

Exercises

In Exercises 29 through 40, find the particular solution with no transient terms and put it into amplitude-phase form.

29. $\dfrac{d^2x}{dt^2} + 6\dfrac{dx}{dt} + 25x = 3\cos 4t$

30. $\dfrac{d^2x}{dt^2} + 6\dfrac{dx}{dt} + 25x = -5\sin 3t$

31. $x'' + 4x' + 13x = 5\cos 2t$

32. $x'' + 4x' + 13x = -2\sin 2t$

33. $x'' + 8x' + 41x = 3\sin t$

34. $x'' + 8x' + 41x = 9\cos t$

35. $x'' + 3x' + 2x = \sin t$

36. $x'' + 7x' + 12x = \sin t$

37. $x'' + 2x' + 2x = \sin t$

38. $x'' + 2x' + 5x = \sin t$

39. $x'' + 2x' + x = \sin t$

40. $x'' + 6x' + 9x = \sin t$

41. Show that the amplitude response curve,

$$y = \frac{1}{\sqrt{(1 - \omega^2/\omega_0^2)^2 + 4\gamma^2\omega^2/\omega_0^2}},$$

graphed in Figure 2.11.7, has a maximum for $\omega \neq 0$ if $\gamma < \sqrt{2}/2$. For what value of ω does a maximum occur? Show that this value of ω approaches ω_0 as $\gamma \to 0$. However, if $\gamma \geq \sqrt{2}/2$, show that the maximum of the response diagram occurs at $\omega = 0$. *Hint*: Let $z = \omega^2/\omega_0^2$.

42. Show that γ defined by (29) is the ratio of the decay rate of underdamped oscillations and the natural frequency. Why does γ have no dimensions?

43. Show that γ defined by (29) is the ratio of the damping coefficient to the damping coefficient at critical damping. That is why it is called the critical damping ratio.

44. Let x_δ be the solution of $x'' + \delta x' + x = \sin t$, $x(0) = 0$, $x'(0) = 1$. Find the formula for x_δ when $\delta = 0$ and $0 < \delta < 2$. Show that, for all $t > 0$, $\lim_{\delta \to 0^+} x_\delta(t) = x_0(t)$. This illustrates the point made earlier in this section that if $\omega = \sqrt{k/m}$ and δ is small, there will be a response similar to resonance.

45. Let $x_\delta(t)$ be the function of t found in Exercise 44. Graph $x_\delta(t)$ for $0 \le t \le 30$ for $\delta = 0.5$, 0.2, 0.1, 0.05 on the same set of coordinates. (Requires a computer with graphics capability. Illustrates the limit of Exercise 44.)

46. Suppose that the top of our vertical spring-mass system is solidly attached to a mechanism that causes the point of attachment to also move in a vertical direction. Let $h(t)$ measure the vertical displacement of the attachment point, with a downward displacement being positive. Explain why a reasonable model for small movements of the mass in the spring-mass system would be $mx'' + \delta x' + kx = kh(t)$, which is in the form (1).

2.11.3 Nonperiodic Forcing

The external force $f(t)$ on a spring-mass system

$$m\frac{d^2x}{dt^2} + \delta\frac{dx}{dt} + kx = f(t)$$

does not have to be periodic. Here we will only consider forcing functions for which a particular solution may be obtained by the method of undetermined coefficients. Other types of forcing functions may be studied using variation of parameters (Section 2.14) or the Laplace transform (Chapter 3).

EXAMPLE 2.11.5 **Constant Force**

A spring-mass system with an attached mass of $m = 1$ g has a spring constant of 10 g/s^2 and a damping coefficient of 2 g/s. The mass is pushed up 1 cm and released. A constant force of 20 dynes acts on the mass in the downward direction. Determine the resulting motion.

SOLUTION The equation of motion is

$$x'' + 2x' + 10x = 20, \qquad x(0) = -1, \qquad x'(0) = 0. \tag{32}$$

First, we find the general free response (solution of the associated homogeneous equation $x' + 2x' + 10x = 0$). The characteristic equation is $r^2 + 2r + 10 = 0$, which has roots $r = -1 \pm 3i$. Thus

$$x_h = c_1e^{-t}\cos 3t + c_2e^{-t}\sin 3t.$$

Using the method of undetermined coefficients (Section 2.9), the form for the particular solution x_p is A, with A a constant. Substituting this into the differential equations (32) gives $10A = 20$, or $A = 2$. Thus

$$x = 2 + c_1e^{-t}\cos 3t + c_2e^{-t}\sin 3t$$

gives the possible (or general) forced response. Applying the initial conditions in (32), we obtain

$$-1 = x(0) = 2 + c_1,$$
$$0 = x'(0) = -c_1 + 3c_2,$$

so that $c_1 = -3$, $c_2 = -1$. Thus

$$x = 2 - 3e^{-t}\cos 3t - e^{-t}\sin 3t,$$

which can be written, using the amplitude phase form (17) of Section 2.8.3, as

$$x = 2 - e^{-t}\sqrt{10}\cos(3t - \phi),$$

where $\phi \approx 0.3218$ radian is the phase angle (Figure 2.11.9).

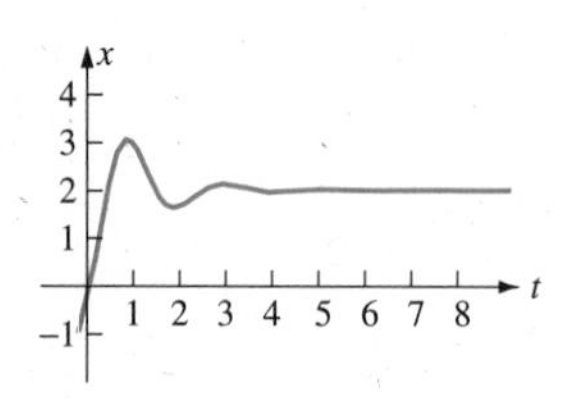

FIGURE 2.11.9

Graph of $x = 2 - e^{-t}\sqrt{10}$ $\cos(3t - 0.3218)$.

In this example, the long-range, equilibrium (steady state) or asymptotic response is $x = 2$, while $-e^{-t}\sqrt{10}\cos(3t - \phi)$ is the transient free response.

Example 2.11.5 illustrates an important fact often used in applications. If a physical system can be described by constant coefficient linear differential equations and the system is stable (in our setting here, this means $\delta > 0$), then a constant forcing function results in a constant output plus transient terms (x_h in our context). The control and regulation of many physical processes utilizes this fact.

Exercises

47. A spring-mass system has mass $m = 4$ g, friction coefficient $\delta = 24$ g/s, and spring constant $k = 52$ g/s^2. The mass is pulled down 3 cm and released. Throughout the subsequent motion, the mass is subjected to a constant external force of 4 dynes downward. Find the subsequent motion. Determine which terms are transient and which are steady-state.

48. Suppose that a spring-mass system has mass m, damping $\delta > 0$, and spring constant k. The system is subjected to a constant force of F dynes downward.

a) Show that the system has an equilibrium and that the equilibrium depends only on k, F and not on m, δ.

b) Show that the solution has the equilibrium as a limit as $t \to \infty$, regardless of the initial condition. Thus, this equilibrium is an attractor.

49. A spring-mass system has an attached mass of 1 g, friction coefficient $\delta = 3$ g/s, and spring constant $k = 2$ g/s^2. The mass is initially at rest. There is an external force of t dynes downward.

a) Determine the subsequent motion.

b) State which terms are transient.

50. A spring-mass system has an attached mass of 2 g, friction coefficient $\delta = 4$ g/s, and spring constant $k = 2$ g/s^2. The mass is initially at rest. There is an external force of $1 - e^{-t}$ dynes downward.

a) Determine the subsequent motion.

b) State which terms are transient.

51. A spring-mass system has an attached mass of 1 g, friction coefficient $\delta = 6$ g/s, and spring constant $k = 25$ g/s^2. The mass is initially 1 cm below the rest position with a velocity of 1 cm/s downward. There is an external force of $3 - e^{-2t}$ dynes downward.

a) Determine the subsequent motion.

b) State which terms are transient.

52. A spring-mass system has an attached mass of 1 g and a spring constant $k = 1$ g/s^2, and friction is negligible. The mass is initially at rest. There is an external force of $t \sin t$. (The forcing function already "looks like" resonance.)

a) Determine the subsequent motion.

b) State which terms are transient.

c) Graph the nontransient terms for $0 \le t \le 5$.

2.12 Linear Electric Circuits

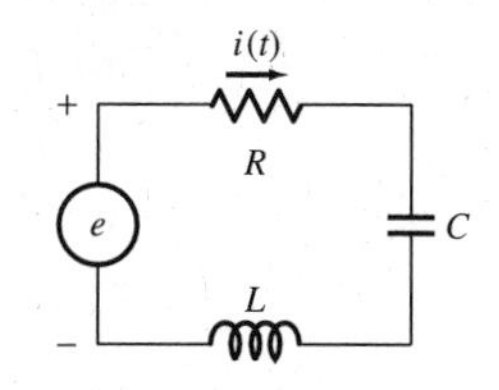

FIGURE 2.12.1

We have already considered circuits in Section 1.10, and will use the units from that section. In this section we shall consider single-loop linear *RLC* circuits, as shown in Figure 2.12.1. The figure shows a linear resistor of resistance R ohms, capacitor of capacitance C farads, inductor of inductance L henries, and voltage source with voltage $e(t)$ volts. Furthermore, we shall assume that R, L, C are nonnegative, which is often the case. (A negative resistance means that we have a device that puts power into the loop rather than dissipating it, as a positive resistor does.)

Recall from Section 1.10 that the **voltage law** says that

The algebraic sum of the voltage drops around a loop at any instant is zero.

Since there is a single loop, the **current law** says that the current is the same in the resistor, inductor, and capacitor. Let this current at time t be $i(t)$ amperes. Let $q(t)$ be the charge in coulombs in the capacitor at time t. From Section 1.10 we know that the voltage drops for each device are:

$$\text{Resistor:} \quad V_R = iR,$$

$$\text{Capacitor:} \quad V_C = \frac{1}{C}q,$$

$$\text{Inductor:} \quad V_I = L\frac{di}{dt},$$

where we assume R, L, C are constants. Thus the voltage law may be written as

$$L\frac{di}{dt} + Ri + \frac{1}{C}q = e.$$

However, current is the time rate of change of charge, so that $i = dq/dt$, and we have the second-order linear differential equation in q

$$L\frac{d^2q}{dt^2} + R\frac{dq}{dt} + \frac{1}{C}q = e(t), \tag{1}$$

or, upon differentiation,

$$L\frac{d^2i}{dt^2} + R\frac{di}{dt} + \frac{1}{C}i = e'(t), \tag{2}$$

which is a second-order linear differential equation in i. Note that this differential equation has exactly the same form as the spring-mass system and L, R, C are also nonnegative by assumption. Thus all of the analysis of the preceding section is still valid, but now mass has become inductance, friction is resistance, and the spring constant is the reciprocal of capacitance. In particular, the discussion of resonance, damping, amplitude, and phase angle is still appropriate.

That many mechanical and nonelectrical problems have the same differential equation (1) as a model is the idea behind the analog computer. To solve the differential equation (1) for a given L, R, C, e with an analog computer, one would build the circuit and then measure the resulting charge (or current) to determine the values of the solution.

Most modern circuits, of course, involve many loops and hence are modeled by systems of differential equations. As circuits have become more and more complex, it has become increasingly expensive and cumbersome to design them by building numerous prototypes. Increasingly, preliminary design work is done by computer simulations, which often involve the solution of systems of differential equations.

Typical Initial Conditions

Mathematically, the appropriate initial conditions for (2) are to be given i and di/dt at $t = 0$. However, often in electrical problems, the current is known initially, but di/dt at $t = 0$ must be calculated. Usually, the initial charge across the capacitor is given. Since $i = dq/dt$, the initial value of di/dt can be determined by evaluating (1) at $t = 0$ to get

$$L\frac{di}{dt}(0) + Ri(0) + \frac{1}{C}q(0) = e(0).$$

Here $e(0)$ is the voltage of the source (battery) at $t = 0$.

Electrical Resonance

We consider an alternating voltage source

$$e(t) = E_0 \cos \omega t.$$

Here, E_0 is the maximum of the voltage source and ω is the (circular) frequency of the voltage source. The differential equation for the current in the RLC circuit follows from (2):

$$L\frac{d^2i}{dt^2} + R\frac{di}{dt} + \frac{1}{C}i = -E_0 \omega \sin \omega t. \tag{3}$$

If the electrical resistance can be neglected, $R = 0$, then the current in the circuit satisfies

$$L\frac{d^2i}{dt^2} + \frac{1}{C}i = -E_0 \omega \sin \omega t. \tag{4}$$

Solutions of the associated homogeneous equation oscillate with natural (circular) frequency $\omega_0 = 1/\sqrt{LC}$. Resonance occurs (without a resistor) if the forcing frequency equals the natural frequency

$$\omega = \omega_0 = \frac{1}{\sqrt{LC}}.$$

Electrical Response

With electrical resistance ($R > 0$), the general solution of (3) is in the form $i = i_p + i_h$ where i_p is a particular solution and i_h corresponds to the solution of the associated homogeneous equation. The solution of the associated homogeneous equation is derived from an exponential e^{rt}, which must correspond to an underdamped, critically damped, or overdamped oscillation (by analogy with the spring-mass system, $m = L$, $\delta = R$, $k = 1/C$). The characteristic equation is

$$Lr^2 + Rr + \frac{1}{C} = 0.$$

The roots of the characteristic equation are $r = (-R \pm \sqrt{R^2 - 4L/C})/2L$. Since L, R, C are positive numbers, in all three cases (underdamped, critically damped, or overdamped), we have $i_h \to 0$ as $t \to \infty$. Thus, after some time, the general solution may be approximated by a particular solution. Using the method of undetermined coefficients, a particular solution may be found in the form

$$i_p = A \cos \omega t + B \sin \omega t.$$

We can obtain formulas for A and B and from them derive formulas for the amplitude of the response. Since we have just calculated this in the section on forced mechanical vibrations, we will just use the previously derived formula (27) of Section 2.11. In addition to $m \Rightarrow L$, $\delta \Rightarrow R$, $k \Rightarrow 1/C$, we note that $F_0 \Rightarrow -E_0\omega$. The extra factor of ω will change some of the properties of the response diagram. Thus, from (25) and (26) of Section 2.11, we obtain

$$i_p = \frac{-E_0\omega}{\sqrt{\left(\frac{1}{C} - L\omega^2\right)^2 + R^2\omega^2}} \sin(\omega t - \phi). \tag{5}$$

We will not be concerned with the phase angle ϕ here.

The amplitude I of the response current satisfies

$$I = \frac{-E_0\omega}{\sqrt{\left(\frac{1}{C} - L\omega^2\right)^2 + R^2\omega^2}}. \tag{6}$$

As with a spring-mass system, the response depends on R, L, C and the forcing circular frequency ω. The response is zero if $\omega = 0$. If $\omega \neq 0$, the response simplifies to

$$I = \frac{-E_0}{\sqrt{\left(\frac{1}{\omega C} - L\omega\right)^2 + R^2}},$$

which is graphed in Figure 2.12.2 on the next page for several values of $\gamma = R/2\sqrt{L/C}$. The maximum of $|I|$ occurs at the minimum of the denominator, which occurs when $1/\omega C - L\omega = 0$ or equivalently $\omega = 1/\sqrt{LC} = \omega_0$. The amplitude response has a maximum when the system is forced at the natural frequency. Smaller values of γ correspond to smaller values of R. The case $R = \gamma = 0$, $\omega = \omega_0$, would be pure resonance.

Radios and Signals

Different radio stations transmit their signals at different frequencies. These signals act as inputs to an electrical device that receives the signals. Primitive radios operated like an *RLC* circuit. The input voltage source would be the sum of the input voltages from the various radio stations. Since the *RLC* circuit is linear, the principle of superposition applies and the response will be the sum of the responses. Each station's signal gets amplified by the amount associated with the electrical response curve. The output will be the sum of signals of different frequencies. The largest amplitude of response will be from the frequency nearest the resonant frequency for the *RLC* circuit. In primitive radios, the capacitance C was adjustable (via a knob). If you wanted to hear your favorite radio station with circular frequency ω^*, you adjusted the dial (changed the capacitance C) so that the resonant circular frequency

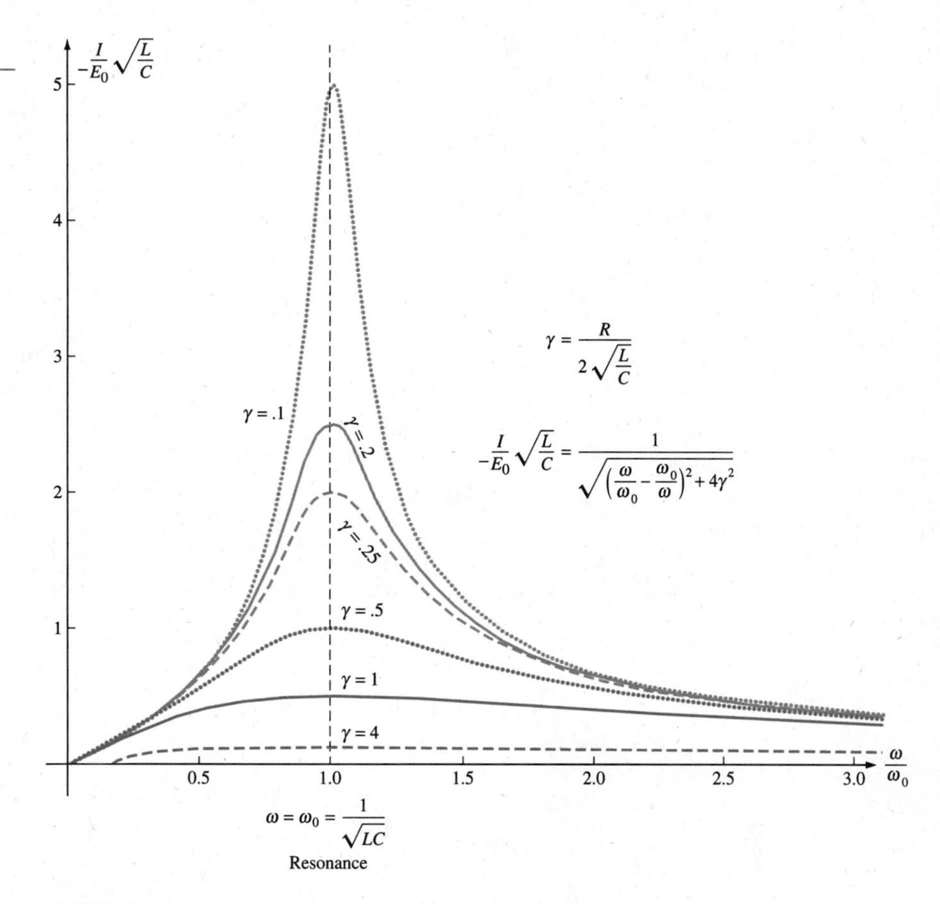

FIGURE 2.12.2

Response diagram.

$\sqrt{1/LC}$ equaled the known radio circular frequency ω^*:

$$\omega^{*2} = \frac{1}{LC}.$$

Actual radio receivers are more sophisticated, but perform essentially in this manner.

Exercises

1. An *RLC* circuit, given by Figure 2.12.1, has a voltage source of $e(t) = 3\cos t$ volts. Values for the components are $R = 3\ \Omega$, $L = 0.5$ H, and $C = 0.4$ F. Initially, the charge on the capacitor is zero and the current in the resistor is 1 A. Find the charge on the capacitor and the current as functions of time.

2. An *RLC* circuit, given by Figure 2.12.1, has a voltage source of $e(t) = 5\cos 2t$ volts. Values for the components are $R = 2\ \Omega$, $L = 1$ H, and $C = \frac{1}{17}$ F. Initially, the charge on the capacitor and the current in the resistor are zero. Find the charge on the capacitor and the current as functions of time.

3. An RLC circuit, given by Figure 2.12.1 on page 203, has a 1.5-V battery as a voltage source $[e(t) = 1.5]$. The values for the components are $R = 1.5\ \Omega$, $L = 1$ H, and $C = 2$ F. First, the charge on the capacitor is 2 C and the current in the resistor is 4 A. Find the charge on the capacitor and the current as functions of time.

4. An RLC circuit, given by Figure 2.12.1, has a 9-V battery as a voltage source $[e(t) = 9]$. Values for the components are $R = 5\ \Omega$, $L = 6$ H, and $C = 1$ F. Initially, the charge on the capacitor is 1 C and the current in the resistor is zero. Find the charge on the capacitor and the current as functions of time.

5. An LC circuit $(R = 0)$, given by Figure 2.12.1, has $C = 0.1$ F and $e(t) = \sin \omega t$. Suppose that ω is a constant such that $e(t)$ has a frequency between 20 and 30 Hz. What values of L will not lead to resonance for any such ω?

6. The RLC circuit shown in Figure 2.12.1 has $R = 20\ \Omega$, $L = 1$ H, and $C = 0.005$ F. The voltage source is shorted out $[e(t) \equiv 0]$. At time $t = 0$, there is a charge of 10 C on the capacitor and no current. Solve the differential equation for the charge, and put in phase amplitude form. How many seconds will it take the variable amplitude to be reduced 99%?

7. The RLC circuit in Figure 2.12.1 has $R = 2\ \Omega$, $L = 1$ H, and $C = 0.5$ F. The initial charge on the capacitor is zero, and the initial current is zero. The voltage source $e(t)$ is a 1-V battery that is shorted out after π seconds. Find the charge on the capacitor and graph for $0 \le t \le 2\pi$ (seconds). (Assume that charge and current are continuous at $t = \pi$.)

8. The LC circuit in Figure 2.12.1 $(R = 0)$ has $L = 4$ H and $C = 0.25$ F. The voltage source is a 2-V battery that is shorted out $(e = 0)$ after 4π seconds. The initial charge is zero and the initial current is 1 A. Find the charge on the capacitor and current in the inductor for $0 \le t \le 8\pi$. (Assume that charge and current are continuous at $t = 4\pi$.)

9. Let $R = 0$ and $e(t) = 0$ in Figure 2.12.1 so that we have an LC circuit with no voltage source. Let $E(i, q) = (L/2)i^2 + (1/2C)q^2$. Show that E is constant by verifying that $dE/dt = 0$. (This is the electrical equivalent of conservation of mechanical energy.)

10. An LC circuit $(R = 0)$ given by Figure 2.12.1 has $L = 8$ H and $C = 2$ F. For what value of ω will the voltage source $e(t) = 2\cos \omega t$ volts create resonance?

11. An LC circuit $(R = 0)$ given by Figure 2.12.1 has $L = 9$ H and $C = 1$ F. The voltage source is $e(t) = 4\cos 2t$. Since $R = 0$, the free response will not be transient. However, there is one choice for $q(0)$, $i(0)$, for which the free response will be absent. What are the values of $q(0)$, $i(0)$?

For the next five exercises, use the correspondence

$$\text{inductance} \leftrightarrow \text{mass},$$

$$\text{resistance} \leftrightarrow \text{friction},$$

and

$$\frac{1}{\text{capacitance}} \leftrightarrow \text{spring constant}$$

between the circuit given by Figure 2.12.1 on page 203 and the spring-mass system in Section 2.8. (A mass of m in slugs or grams becomes an inductance of L henries, etc.)

12. Rewrite Exercise 2.8.18 in terms of this circuit and answer parts (a)–(c).

13. Rewrite Exercise 2.8.19 in terms of this circuit and answer parts (a)–(c).

14. Rewrite Exercise 2.8.20 in terms of this circuit and solve.

15. Rewrite Exercise 2.8.21 in terms of this circuit and solve.

16. Rewrite Exercise 2.8.22 in terms of this circuit and solve.

(Input-Output Voltages) Frequently it is helpful to consider circuits as input-output devices. Suppose for the circuit in Figure 2.12.1 that $R > 0$ so that the free response is transient. We assume that these transient terms are not important and consider the solution $q(t)$ to be given by the terms that are not transient, that is, the forced response. The **input voltage** will be taken as $e(t)$, and the **output voltage** as the voltage across the capacitor, which is $(1/C)q$.

17. Suppose that $E = 1$, $R = 6$, $L = 1$, $C = \frac{1}{13}$, and $\omega = 3$. Find I in (6).

18. Suppose that $E = 1$, $R = 7$, $L = 12$, $C = 1$, and $\omega = 2$. Find I in (6).

2.13 Euler Equation

Linear differential equations with variable coefficients rarely have explicit solutions. Frequently, solutions are obtained numerically (Chapter 6) or by the method of power series (Chapter 4). However, in this section we study linear differential equations with variable coefficients of a specific kind where explicit solutions are not difficult to obtain. The second-order **Euler equation** is

$$ax^2\frac{d^2y}{dx^2} + bx\frac{dy}{dx} + cy = 0, \tag{1}$$

where a, b, c are constants and $a \neq 0$. The third-order Euler equation is

$$a_3x^3\frac{d^3y}{dx^3} + a_2x^2\frac{d^2y}{dx^2} + a_1x\frac{dy}{dx} + a_0y = 0, \tag{2}$$

with a_3, a_2, a_1, a_0 constants and $a_3 \neq 0$, and in general the nth-order Euler's equation is

$$a_nx^n\frac{d^ny}{dx^n} + a_{n-1}x^{n-1}\frac{d^{n-1}y}{dx^{n-1}} + \cdots + a_1x\frac{dy}{dx} + a_0y = 0. \tag{3}$$

There is even a first-order Euler equation,

$$ax\frac{dy}{dx} + by = 0.$$

The Euler equation is also known as the Cauchy-Euler or the equidimensional equation. (See Exercise 15 at the end of this section.) It arises, for example, in some drag problems in uniformly viscous flows.

Euler equations are important not only because they arise in applications, but because in Chapter 4 we will show that the behavior of linear differential equations with singular points is understood better by comparison to the behavior of the solutions of Euler equations.

We consider here the second-order Euler equation,

$$ax^2y'' + bxy' + cy = 0. \tag{4}$$

The variable coefficients of (4) are powers that increase by x for each derivative in (4). This suggests that solutions of (4) are likely to be powers of x,

$$y = x^r, \tag{5}$$

since powers have the property that their power decreases by one each time they are differentiated:

$$\begin{aligned} y' &= rx^{r-1}, \\ y'' &= r(r-1)x^{r-2}. \end{aligned} \tag{6}$$

This is exactly what is needed to counteract the increased power in the variable coefficient of (4).

Substituting $y = x^r$ into $ax^2y'' + bxy' + cy = 0$ gives

$$ax^2r(r-1)x^{r-2} + bxrx^{r-1} + cx^r = 0,$$
$$[ar(r-1) + br + c]x^r = 0,$$

so that

$$\begin{aligned} ar(r-1) + br + c = 0 \\ ar^2 + (b-a)r + c = 0, \end{aligned} \tag{7}$$

which we call the **indicial equation.**

Case 1: Distinct Real Roots

If the polynomial $ar^2 + (b-a)r + c = 0$ has two distinct real roots, r_1, r_2, then x^{r_1}, x^{r_2} provide a fundamental set of solutions. In this case the general solution of (4) is

$$y = c_1x^{r_1} + c_2x^{r_2}.$$

EXAMPLE 2.13.1 Distinct Real Roots

Find the general solution of

$$2x^2\frac{d^2y}{dx^2} + 7x\frac{dy}{dx} - 3y = 0, \qquad x > 0, \tag{8}$$

Solution The student is encouraged to rederive the indicial equation (7) the first few times. Letting $y = x^r$ in (8) and then dividing by x^r yields

$$2r(r-1) + 7r - 3 = 0$$

or

$$2r^2 + 5r - 3 = (2r - 1)(r + 3) = 0.$$

The roots are $r = \frac{1}{2}, -3$ and the general solution is

$$y = c_1 x^{1/2} + c_2 x^{-3}.$$

◄

Case 2: Complex Conjugate Roots

If the indicial equation has complex roots $r_1 = \alpha + i\beta$, $r_2 = \alpha - i\beta$, then the general solution is

$$y = \bar{c}_1 x^{\alpha + i\beta} + \bar{c}_2 x^{\alpha - i\beta} = x^{\alpha}\left(\bar{c}_1 x^{i\beta} + \bar{c}_2 x^{-i\beta}\right). \tag{9}$$

Because of the property of exponentials (for $x > 0$),

$$x^{i\beta} = e^{i\beta \ln x}.$$

Thus, (9) becomes

$$y = x^{\alpha}\left(\bar{c}_1 e^{i\beta \ln x} + \bar{c}_2 e^{-i\beta \ln x}\right).$$

We have shown using Euler's formula in Section 2.6 that a linear combination of $e^{\pm i\theta}$ is equivalent to a linear combination of $\cos\theta$ and $\sin\theta$. In this case, $\theta = \beta \ln x$. Thus, when the roots of the indicial equation for Euler's equation are complex numbers $\alpha \pm i\beta$, the general solution is

$$y = x^{\alpha}[c_1 \cos(\beta \ln x) + c_2 \sin(\beta \ln x)]$$

or equivalently,

$$y = c_1 x^{\alpha} \cos(\beta \ln x) + c_2 x^{\alpha} \sin(\beta \ln x).$$

EXAMPLE 2.13.2 **Complex Conjugate Roots**

Find the general solution of

$$9x^2 y'' + 15xy' + 5y = 0, \qquad x > 0.$$

Solution Letting $y = x^r$ yields the indicial polynomial $9r(r - 1) + 15r + 5 = 9r^2 + 6r + 5 = (3r + 1)^2 + 4$, so the roots are

$$r = -\frac{1}{3} \pm i\frac{2}{3}.$$

Since $x^r = x^{-1/3 \pm i\frac{2}{3}} = x^{-1/3} x^{\pm i\frac{2}{3}}$, the general solution is

$$y = c_1 x^{-1/3} \cos\left(\frac{2}{3} \ln x\right) + c_2 x^{-1/3} \sin\left(\frac{2}{3} \ln x\right).$$

◄

Case 3: Repeated Real Roots

If the indicial equation $ar(r-1)+br+c=0$ has a repeated root $r=r_1$, then only one homogeneous solution is in the assumed form, $y=x^{r_1}$. The second independent homogeneous solution can always be obtained by reduction of order (Section 2.5). In the case of repeated roots for the Euler equation, the second solution is always

$$y = x^{r_1}\ln x. \tag{10}$$

We illustrate this fact with a specific example.

EXAMPLE 2.13.3 Using Reduction of Order

Find the general solution of

$$x^2y'' - 7xy' + 16y = 0, \qquad x > 0. \tag{11}$$

Solution Letting $y=x^r$ yields the indicial equation, $r(r-1)-7r+16 = r^2-8r+16=(r-4)^2=0$, which has repeated root $r=4, 4$. Since we have one solution x^4, the second independent solution can be found by reduction of order. Let

$$y = vx^4.$$

Then (11) becomes

$$x^2(v''x^4 + 2v'4x^3 + v12x^2) - 7x(v'x^4 + v4x^3) + 16vx^4 = 0, \qquad x > 0. \tag{12}$$

The coefficients of v cancel since x^4 is a solution, and (12) reduces to

$$v''x^6 + v'x^5 = 0.$$

Dividing by x^5 yields

$$xv'' + v' = 0. \tag{13}$$

In the method of reduction of order, we let

$$w = v',$$

so that (13) becomes a first-order linear differential equation

$$xw' + w = 0. \tag{14}$$

Separation yields

$$\frac{dw}{w} = -\frac{dx}{x}.$$

Thus (since we assume $x > 0$),

$$\ln|w| = -\ln x + c.$$

After exponentiation,

$$w = \frac{c_1}{x}. \tag{15}$$

Alternatively, (15) may be derived more easily from (14) by noting that (14) is equivalent to

$$\frac{d}{dx}(xw) = 0, \tag{16}$$

from which (15) follows after an integration of (16).

From (15) and $w = v'$, we have

$$v' = \frac{c_1}{x}.$$

Antidifferentiating gives

$$v = c_1 \ln x + c_2,$$

so that

$$y = x^4 v = x^4(c_1 \ln x + c_2) = c_1 x^4 \ln x + c_2 x^4.$$

Thus, the second independent solution is $x^4 \ln x$. ◄

From now on, to obtain the general solution of an Euler equation when there is a repeated root, just note that the second solution is $x^{r_1} \ln x$ and do not use reduction of order unless instructed to do so.

EXAMPLE 2.13.4 Repeated Roots

Find the general solution of

$$x^2 y'' - xy' + y = 0, \qquad x > 0.$$

SOLUTION Letting $y = x^r$ yields the polynomial $r(r - 1) - r + 1 = r^2 - 2r + 1$, which has repeated roots 1, 1. Thus

$$y = c_1 x \ln x + c_2 x, \qquad x > 0.$$ ◄

We summarize these three cases in the following theorem.

■ THEOREM 2.13.1 Euler Equation

The general solution of the second-order Euler equation $ax^2 y'' + bxy' + cy = 0$, $x > 0$, is given as follows: Find the roots r_1, r_2 of the polynomial $ar(r - 1) + br + c$.

1. If $r_1 \neq r_2$ and r_1, r_2 are real, then

$$y = c_1 x^{r_1} + c_2 x^{r_2}. \tag{17}$$

2. If $r_1 = \alpha + \beta i$, $r_2 = \alpha - \beta i$, $\beta \neq 0$, then

$$y = c_1 x^{\alpha} \cos(\beta \ln x) + c_2 x^{\alpha} \sin(\beta \ln x). \tag{18}$$

3. If $r_1 = r_2$, then

$$y = c_1 x^{r_1} + c_2 x^{r_1} \ln x. \tag{19}$$

■

Exercises

In Exercises 1 through 12, solve the second-order Euler equation for $x > 0$. If no initial conditions are given, find the general solution.

1. $x^2y'' + xy' - y = 0$

2. $x^2y'' - 4xy' + 6y = 0$

3. $x^2y'' + 3xy' + y = 0$

4. $x^2y'' - xy' + 2y = 0$,
$y(1) = 0$, $y'(1) = 2$

5. $4x^2y'' + 8xy' + y = 0$

6. $x^2y'' + xy' + y = 0$

7. $x^2y'' + 4xy' + 2y = 0$,
$y(1) = 1$, $y'(1) = 0$

8. $9x^2y'' + 15xy' + 2y = 0$

9. $x^2y'' + xy' + 4y = 0$

10. $x^2y'' + 3xy' + 10y = 0$

11. $x^2y'' + 3xy' + 8y = 0$

12. $x^2y'' - 3xy' + 5y = 0$

13. By looking at the answers to Exercises 1 through 12, find an Euler equation whose indicial equation has repeated roots. For this example, derive the second solution using reduction of order.

14. Find the general solution of $5x(dy/dx) + y = 0$ using separation. Compare your answer to the solution obtained by using the methods of this section.

15. **(Equidimensionality)** Let k be a constant, and perform the change of variables $x = ks$.

i) Show that the Euler equation $ax^2(d^2y/dx^2) + bx(dy/dx) + cy = 0$ becomes

$$as^2\frac{d^2y}{ds^2} + bs\frac{dy}{ds} + cy = 0.$$

That is, the Euler equation is unaltered by a change of scale in the independent variable.

ii) Contrast this with what happens when the same change of scale is performed on the constant coefficient equation $ay'' + by' + cy = 0$.

iii) Verify that, if $k > 0$ is a real constant and r_1, r_2 are distinct real constants, then $x = ks$ changes $c_1x^{r_1} + c_2x^{r_2}$ into $\tilde{c}_1s^{r_1} + \tilde{c}_2s^{r_2}$.

iv) Verify that if $k > 0$ and r_1 is a real constant then $x = ks$ changes $c_1x^{r_1} + c_2x^{r_1}\ln x$ into $\tilde{c}_1s^{r_1} + \tilde{c}_2s^{r_1}\ln s$.

v) Verify that if $k > 0$, and α, β are real constants, $\beta \neq 0$, then $x = ks$ changes $c_1x^{\alpha}\cos(\beta \ln x) + c_2x^{\alpha}\sin(\beta \ln x)$ into $\tilde{c}_1s^{\alpha}\cos(\beta \ln s) + \tilde{c}_2s^{\alpha}\sin(\beta \ln s)$.

16. (Alternative Method) Let $t = \ln x$. ($x = e^t$)

i) Verify that

$$x\frac{dy}{dx} = \frac{dy}{dt}, \tag{20}$$

$$x^2\frac{d^2y}{dx^2} = \frac{d^2y}{dt^2} - \frac{dy}{dt}. \tag{21}$$

ii) Show that the change of variables $t = \ln x$ changes

$$ax^2\frac{d^2y}{dx^2} + bx\frac{dy}{dx} + cy = 0$$

into the constant coefficient equation

$$a\frac{d^2y}{dt^2} + (b - a)\frac{dy}{dt} + cy = 0. \quad (22)$$

Note that the polynomial $ar^2 + (b - a)r + c$ in Theorem 2.13.1 is just the characteristic polynomial of (22).

Euler equations of any order may be solved by either using $y = x^r$ to find the indicial polynomial or performing the change of variables $t = \ln x$ to make the differential equation linear constant coefficient. In Exercises 17 through 20, solve the Euler equations.

17. $x^4y'''' + 6x^3y''' + 7x^2y'' + xy' - y = 0$

18. $x^4y'''' + 6x^3y''' + 9x^2y'' + 3xy' + y = 0$

19. $x^3y''' + xy' - y = 0$

20. $x^3y''' + 4x^2y'' = 0$

2.14 Variation of Parameters (Second-Order)

In Section 2.9 the method of undetermined coefficients was used to find a particular solution of

$$y''(x) + p(x)y'(x) + q(x)y(x) = f(x). \quad (1)$$

There were two major restrictions on the method of undetermined coefficients. First, p, q had to be constants. Second, f had to be in a special form. This section will present a method for finding a particular solution of $y'' + p(x)y' + q(x)y = f$ provided we have first solved the associated homogeneous equation:

$$y'' + p(x)y' + q(x)y = 0. \quad (2)$$

This new method will not require p, q to be constants nor f to be in a special form.

There are no general methods for obtaining a fundamental set of solutions $\{y_1, y_2\}$ of (2). In Section 2.5 we used reduction of order to obtain the second solution y_2 *if* one solution y_1 was known. Unfortunately, there are no general methods for obtaining one solution of (2) if p, q are not constants.

In this section, we will assume we have a fundamental set of solutions $\{y_1, y_2\}$ of (2). We shall first derive the method and then work several examples. We begin by looking for functions v_1 and v_2 so that

$$y = v_1y_1 + v_2y_2, \quad (3)$$

is a solution of (1),

$$y'' + py' + qy = f. \quad (4)$$

This is called the **method of variation of parameters** because the usual constants c_1, c_2 in the solution of the associated homogeneous equation are now varied. When (3) is substituted into (4), one differential equation results, involving two unknowns v_1 and v_2. In the method of variation of parameters, an ingenious (and unmotivated) observation is made. We choose v_1 and v_2 such that the first derivative of y would be the same as if v_1 and v_2 were constants, even though v_1 and v_2 are not constants. That is, we want

$$y' = v_1 y_1' + v_2 y_2'. \tag{5}$$

However, y' should be calculated using the product rule from (3). Thus from (3) we have

$$y' = v_1' y_1 + v_1 y_1' + v_2' y_2 + v_2 y_2'.$$

Equation (5) has only two of these terms. Thus, (5) can be valid only if the sum of the other two terms vanishes:

$$v_1' y_1 + v_2' y_2 = 0. \tag{6}$$

Thus (5) holds if (6) holds. The second derivative of y is calculated by differentiating (5). Using this y'' and y' from (5), we have that (4) is satisfied only if

$$v_1 y_1'' + v_1' y_1' + v_2 y_2'' + v_2' y_2' + p(v_1 y_1' + v_2 y_2') + q(v_1 y_1 + v_2 y_2) = f.$$

Collecting terms yields

$$v_1(y_1'' + p y_1' + q y_1) + v_2(y_2'' + p y_2' + q y_2) + v_1' y_1' + v_2' y_2' = f. \tag{7}$$

Since y_1 and y_2 are homogeneous solutions, satisfying $y'' + py' + qy = 0$, (7) simplifies to

$$v_1' y_1' + v_2' y_2' = f. \tag{8}$$

In summary, we have that if v_1' and v_2' satisfy the linear algebraic equations (6) and (8),

$$\begin{aligned} v_1' y_1 + v_2' y_2 &= 0 \\ v_1' y_1' + v_2' y_2' &= f, \end{aligned} \tag{9}$$

then

$y = v_1 y_1 + v_2 y_2$ is a particular solution of the original differential equation (4).

The linear system (9) can be solved by elimination or Cramer's rule (see the summary at the end of this section) for v_1' and v_2'. By integration (indefinite or definite), we obtain v_1 and v_2. The particular solution is then determined from (3).

We begin by discussing two examples. An additional example appears after the summary. The first example illustrates just the solution technique.

EXAMPLE 2.14.1 Variation of Parameters

$\{x, x^3\}$ is a fundamental set of solutions of $x^2y'' - 3xy' + 3y = 0$. Find the general solution of

$$x^2y'' - 3xy' + 3y = 4x^7.$$

SOLUTION We have $y_1 = x$, $y_2 = x^3$. Dividing by x^2, the differential equation is

$$y'' - \frac{3}{x}y' + \frac{3}{x^2}y = 4x^5.$$

Thus $f(x) = 4x^5$. We must solve (9), which is

$$v_1'x + v_2'x^3 = 0,$$
$$v_1'(x)' + v_2'(x^3)' = 4x^5,$$

or

$$v_1'x + v_2'x^3 = 0,$$
$$v_1' + v_2'3x^2 = 4x^5.$$

Solve the first equation for v_1'; $v_1' = -v_2'x^2$, and substitute into the second equation

$$-v_2'x^2 + v_2'3x^2 = 4x^5,$$

and solve for v_2',

$$v_2' = 2x^3.$$

Then

$$v_1' = -v_2'x^2 = -2x^5.$$

Thus antidifferentiating v_1', v_2' gives

$$v_2 = \frac{x^4}{2}, \qquad v_1 = -\frac{x^6}{3}.$$

A particular solution is

$$y_p = v_1y_1 + v_2y_2 = \left(-\frac{x^6}{3}\right)x + \left(\frac{x^4}{2}\right)x^3 = \frac{x^7}{6},$$

and the general solution is

$$y = \frac{x^7}{6} + c_1x + c_2x^3.$$ ◄

The next example illustrates both the derivation of (9) and the use of a definite integral.

EXAMPLE 2.14.2 Definite Integral

Find the general solution of

$$y'' + y = \frac{1}{x + 1}. \tag{10}$$

Note that (10) is a differential equation with constant coefficients, but $1/(x + 1)$ is not the kind of forcing function to which the method of undetermined coefficients can be applied. We shall solve (10) by the method of variation of parameters since we know how to obtain solutions of the associated homogeneous equation.

SOLUTION First we obtain solutions of the associated homogeneous equation by solving $y'' + y = 0$. By substituting $y = e^{rx}$, we obtain the characteristic equation $r^2 + 1 = 0$, which has roots $\pm i$. Thus, $\cos x$ and $\sin x$ form a fundamental set. According to the method of variation of parameters, we seek a solution of (10) in the form

$$y = v_1 \cos x + v_2 \sin x. \tag{11}$$

The derivative is the same as would occur if v_1 and v_2 were constants,

$$y' = -v_1 \sin x + v_2 \cos x. \tag{12}$$

However, (12) is valid only if

$$v_1' \cos x + v_2' \sin x = 0. \tag{13}$$

Substituting (11) into (10) and taking the derivative of (12) yields

$$-v_1' \sin x - v_1 \cos x + v_2' \cos x - v_2 \sin x + v_1 \cos x + v_2 \sin x = \frac{1}{x + 1}. \tag{14}$$

After canceling the v_1 and v_2 terms, (14) becomes

$$-v_1' \sin x + v_2' \cos x = \frac{1}{x + 1}. \tag{15}$$

The equations for (v_1', v_2') are (13) and (15). For students familiar with solving linear systems of equations, there are several options available, such as using augmented matrices (Section 5.7) or Cramer's rule (see formula (27) at the end of this section). We shall just solve for (v_1', v_2') using elimination. To eliminate v_2' we multiply (13) by $\cos x$, multiply (15) by $-\sin x$, and add the results:

$$v_1'(\cos^2 x + \sin^2 x) = -\frac{\sin x}{x + 1}.$$

Since $\cos^2 x + \sin^2 x = 1$,

$$v_1' = -\frac{\sin x}{x+1}. \tag{16}$$

We can now determine v_2' from (13):

$$v_2' = \frac{\cos x}{x+1}. \tag{17}$$

Since (16) and (17) cannot be antidifferentiated explicitly, we use the definite integral:

$$v_1 = -\int_0^x \frac{\sin \bar{x}}{\bar{x}+1}\, d\bar{x} + c_1$$

$$v_2 = \int_0^x \frac{\cos \bar{x}}{\bar{x}+1}\, d\bar{x} + c_2. \tag{18}$$

For convenience we have chosen the lower limit to be zero.

A solution is formed from (11):

$$y = \sin x \int_0^x \frac{\cos \bar{x}}{\bar{x}+1}\, d\bar{x} - \cos \int_0^x \frac{\sin \bar{x}}{\bar{x}+1}\, d\bar{x}$$

$$+ c_1 \cos x + c_2 \sin x. \tag{19}$$

If the arbitrary constants c_1 and c_2 are kept, the general solution of (10) is obtained. If the constants are zero or any other specific value, then (19) is a particular solution. ◄

Influence Function

Although (19) is a correct and satisfactory general solution of the differential equation (10), some further algebraic manipulations yield an important and interesting result. Since $\bar{x}$ is a dummy variable of integration, the functions of x may be taken inside the integral and the two integrals combined:

$$y = \int_0^x \frac{\sin x \cos \bar{x} - \cos x \sin \bar{x}}{\bar{x}+1}\, d\bar{x} + c_1 \cos x + c_2 \sin x. \tag{20}$$

Furthermore, if we use the trigonometric addition formula, $\sin(a-b) = \sin a \cos b - \cos a \sin b$, then (20) becomes

$$y = \int_0^x \frac{\sin(x-\bar{x})}{\bar{x}+1}\, d\bar{x} + c_1 \cos x + c_2 \sin x. \tag{21}$$

The solution to the differential equation represents the response to the source $1/(x + 1)$. It is seen that the solution at x (called the response location) is the sum (actually an integral) of all sources $\bar{x}$ from $\bar{x} = 0$ (the initial source) to $\bar{x} = x$ (the present source). The function $\sin(x - \bar{x})$ is a weighting function. It is an **influence function** called the **Green's function**, which represents the weight of the contribution to the response at x due to the source at $\bar{x}$:

$$G(x, \bar{x}) \equiv \sin(x - \bar{x}). \tag{22}$$

Using this notation, the solution to the differential equation (10) is

$$y = \int_0^x G(x, \bar{x}) \frac{1}{\bar{x} + 1}\, d\bar{x} + c_1 \cos x + c_2 \sin x. \tag{23}$$

One advantage of solving the differential equation in this way is that if the source changes, then it is easy to change the solution accordingly. In fact, it can be seen that if the source was not described specifically, for example

$$y'' + y = f(x), \tag{24}$$

then the general solution could be written

$$y = \int_0^x G(x, \bar{x}) f(\bar{x})\, d\bar{x} + c_1 \cos x + c_2 \sin x, \tag{25}$$

where the Green's function $G(x, \bar{x})$ is still given by (22).

It is also interesting to note the manner in which initial conditions are satisfied. Evaluating (21) at $x = 0$ yields

$$y(0) = c_1.$$

In this example, it can be shown (but not as easily) that

$$y'(0) = c_2.$$

Summary of Variation of Parameters

Variation of parameters (second-order) is a method of calculating a particular solution of $y'' + py' + qy = f$ given a fundamental set of solutions $\{y_1, y_2\}$ of the associated homogeneous equation $y'' + py' + qy = 0$. The method is as follows:

1. Find a fundamental set of solutions $\{y_1, y_2\}$ of $y'' + py' + qy = 0$.
2. Solve the algebraic system of equations

$$\begin{aligned} v_1' y_1 + v_2' y_2 &= 0, \\ v_1' y_1' + v_2' y_2' &= f \end{aligned} \tag{26}$$

for the functions v_1', v_2'.

3. Antidifferentiate to find v_1, v_2.
4. Then $y = v_1 y_1 + v_2 y_2$ is a particular solution of $y'' + py' + qy = f$.
5. $y = v_1 y_1 + v_2 y_2 + c_1 y_1 + c_2 y_2$ is the general solution of $y'' + py' + qy = f$.

Note If arbitrary constants are introduced in Step 3 when finding v_1, v_2, then $y = v_1 y_1 + v_2 y_2$ will be the general solution.

Note The coefficients of v_1', v_2' in (26) are the entries of the matrix

$$\begin{bmatrix} y_1 & y_2 \\ y_1' & y_2' \end{bmatrix},$$

whose determinant is the Wronskian (see Section 2.3). Since the Wronskian of a fundamental set of solutions is never zero, it follows from matrix theory (Cramer's rule) that (26) can always be solved for v_1', v_2'. Thus the only difficulty is in Step 1.

Two comments are in order. First, a fact from algebra called Cramer's rule can be applied to the system of equations (26) to give formulas for v_1', v_2'. They are

$$v_1' = \frac{\det \begin{bmatrix} 0 & y_2 \\ f & y_2' \end{bmatrix}}{\det \begin{bmatrix} y_1 & y_2 \\ y_1' & y_2' \end{bmatrix}} \quad \text{and} \quad v_2' = \frac{\det \begin{bmatrix} y_1 & 0 \\ y_1' & f \end{bmatrix}}{\det \begin{bmatrix} y_1 & y_2 \\ y_1' & y_2' \end{bmatrix}}.$$

Evaluating the determinants on top gives

$$v_1' = -\frac{f y_2}{W[y_1, y_2]}, \qquad v_2' = \frac{f y_1}{W[y_1, y_2]}, \tag{27}$$

where $W[y_1, y_2]$ is the Wronskian of y_1, y_2. In practice it is probably quicker to use (27) to find v_1', v_2'.

EXAMPLE 2.14.3 Variation of Parameters

Find the general solution of

$$2y'' - 4y' + 2y = x^{-1} e^x, \qquad x > 0. \tag{28}$$

Solution Note that $x^{-1} e^x$ is not the kind of forcing term to which the method of undetermined coefficients can be applied. However, the differential equation has constant coefficients so that we know how to solve the associated homogeneous equation. We shall solve (28) by the method of variation of parameters. We rewrite (28) as

$$y'' - 2y' + y = \frac{1}{2}x^{-1}e^x$$

so that $f(x) = \frac{1}{2}x^{-1}e^x$.

Step 1 We must solve $y'' - 2y' + y = 0$. The characteristic polynomial is $r^2 - 2r + 1 = (r - 1)^2$, which has roots 1, 1. Thus $\{e^x, xe^x\}$ is a fundamental set of solutions. Let $y_1 = e^x$, $y_2 = xe^x$.

Step 2
$y_1 = e^x$, $y_2 = xe^x$,

$$W[e^x, xe^x] = \det\begin{bmatrix} e^x & xe^x \\ e^x & e^x + xe^x \end{bmatrix} = e^{2x},$$

and $f = \frac{1}{2}x^{-1}e^x$. Thus (27) gives

$$v_1' = -\frac{[x^{-1}e^x]\cdot[xe^x]}{2\cdot[e^{2x}]} = -\frac{1}{2}$$

and

$$v_2' = \frac{[x^{-1}e^x]e^x}{2[e^{2x}]} = \frac{x^{-1}}{2}.$$

Step 3

$$v_1 = -\frac{x}{2}$$

$$v_2 = \frac{\ln x}{2}$$

Steps 4 and 5

$$\begin{aligned} y &= v_1y_1 + v_2y_2 + c_1y_1 + c_2y_2 \\ &= \frac{-x}{2}e^x + \frac{\ln x}{2}xe^x + c_1e^x + c_2xe^x \\ &= \frac{xe^x \ln x}{2} + c_1e^x + \tilde{c}_2xe^x \qquad \left(\tilde{c}_2 = c_2 - \frac{1}{2}\right) \end{aligned}$$

is the general solution of $2y'' - 4y' + 2y = x^{-1}e^x$. ◀

Exercises

In Exercises 1 through 14, find the general solution by the method of variation of parameters. Decide whether the method of undetermined coefficients could also have been used.

1. $y'' - y = e^{2x}$

2. $y'' - 4y' + 4y = e^{2x}$

3. $y'' + y = \dfrac{1}{\sin x}$

4. $y'' + 4y = \dfrac{4}{\cos 2x}$

5. $y'' + y = \tan x$

6. $y'' + 2y' + y = \dfrac{e^{-x}}{1 + x^2}$

7. $y'' + 3y' + 2y = \dfrac{1}{1 + e^{2x}}$

8. $4y'' - y = x$

9. $y'' - 6y' + 9y = e^{3x}x^{3/2}, \quad x > 0$

10. $y'' - 4y' + 3y = e^{-x}$

11. $4y'' + 4y' + y = x^{-2}e^{-x/2}$

12. $y'' + 5y' + 4y = e^x$

13. $y'' - y' - 6y = e^{-2x}$

14. $y'' - 3y' + 2y = e^{3x}\cos(e^x)$

In Exercises 15 through 18, a fundamental set of solutions is given for the associated homogeneous equation for $x > 0$. Solve the differential equation, using variation of parameters, and give the general solution. In each case the fundamental set of solutions could be found by the method in Section 2.13 for the Euler equation.

15. $x^2y'' - 2xy' + 2y = x^3, \quad \{x, x^2\}$

16. $x^2y'' + xy' - y = x^{1/2}, \quad \{x, x^{-1}\}$

17. $x^2y'' + 2xy' = x^{-1}, \quad \{1, x^{-1}\}$

18. $x^2y'' - 3xy' + 3y = x, \quad \{x, x^3\}$

19. Show that a particular solution of $y'' + py' + qy = f$ given by variation of parameters may be written as

$$y_p(x) = \int_0^x \frac{y_2(x)y_1(s) - y_1(x)y_2(s)}{W[y_1, y_2](s)} f(s)\, ds. \tag{28}$$

20. Show that (28) gives the unique solution of the initial value problem $y'' + py' + qy = f$, $y(0) = 0$, $y'(0) = 0$.

In Exercises 21 through 25, find the general solution by the method of variation of parameters. In these exercises a definite integral will be necessary as in Example 2.14.2.

21. $y'' - y = e^{-x^2}$

22. $y'' - 4y = \sin(x^{1/3})$

23. $y'' + 5y' + 6y = \dfrac{1}{x + 1}$

24. $y'' - 3y' + 2y = x^{1/5}$

25. $x^2y'' + xy' - y = e^x$

2.15 Variation of Parameters (nth-Order)

The method of variation of parameters given in Section 2.14 may also be used for nth-order linear differential equations.

■ **THEOREM 2.15.1 Variation of Parameters for nth-Order Equations**

Suppose that $\{y_1, \ldots, y_n\}$ is a fundamental set of solutions of the associated homogeneous equation for

$$a_n(x)y^{(n)}(x) + a_{n-1}(x)y^{(n-1)}(x) + \cdots + a_1(x)y'(x) + a_0(x)y(x) = f(x). \tag{1}$$

If $v_1'(x), \ldots, v_n'(x)$ satisfy the system of equations

$$
\begin{aligned}
v_1'(x)y_1(x) &+ v_2'(x)y_2(x) + \cdots + v_n'(x)y_n(x) = 0 \\
v_1'(x)y_1'(x) &+ v_2'(x)y_2'(x) + \cdots + v_n'(x)y_n'(x) = 0 \\
&\vdots \\
v_1'(x)y_1^{(n-1)}(x) &+ v_2'(x)y_2^{(n-1)}(x) + \cdots + v_n'(x)y_n^{(n-1)}(x) = \frac{f(x)}{a_n(x)},
\end{aligned}
\tag{2}
$$

then $y_p = v_1 y_1 + v_2 y_2 + \cdots + v_n y_n$ is a particular solution of (1). ■

In general, solving (2) for $n > 2$ can become quite complicated. For small n, say 3 or 4, the system (2) can be solved by Cramer's rule. Ways to evaluate determinants are covered in Section 5.8. For larger n, computer programs doing symbolic manipulations can be used.

Cramer's rule applied to (2) takes the form

$$
v_1' = \frac{\det\begin{bmatrix} 0 & y_2 & \cdots & y_n \\ \vdots & \vdots & & \vdots \\ 0 & y_2^{(n-2)} & & y_n^{(n-2)} \\ f/a_n & y_2^{(n-1)} & & y_n^{(n-1)} \end{bmatrix}}{W[y_1, \ldots, y_n]}
= (-1)^{n+1} f \frac{\det\begin{bmatrix} y_2 & \cdots & y_n \\ \vdots & & \vdots \\ y_2^{(n-2)} & & y_n^{(n-2)} \end{bmatrix}}{a_n W[y_1, \ldots, y_n]}.
$$

and

$$
v_i' = (-1)^{n+i} \frac{f}{a_n} \frac{\det\begin{bmatrix} y_1 & \cdots & y_{i-1} & y_{i+1} & \cdots & y_n \\ \vdots & & & & & \vdots \\ y_1^{(n-2)} & & y_{i-1}^{(n-2)} & y_{i+1}^{(n-2)} & \cdots & y_n^{(n-2)} \end{bmatrix}}{W[y_1, \ldots, y_n]} \tag{3}
$$

$$
= (-1)^{n+i} \frac{f}{a_n} \frac{W_i}{W[y_1, \ldots, y_n]}. \tag{4}
$$

Here W is the Wronskian of $\{y_1, \ldots, y_n\}$,

$$
W[y_1, \ldots, y_n] = \det\begin{bmatrix} y_1 & y_2 & \cdots & y_n \\ y_1' & y_2' & \cdots & y_n' \\ \vdots & & & \\ y_1^{(n-1)} & y_2^{(n-1)} & \cdots & y_n^{(n-1)} \end{bmatrix},
$$

and W_i is the Wronskian of the $(n-1)$ functions obtained by deleting y_i from the set $\{y_1, \ldots, y_n\}$ that appears in the numerator of (3). The next example could be solved by undetermined coefficients, but we will use it to illustrate variation of parameters.

EXAMPLE 2.15.1 Variation of Parameters

Solve

$$y''' - y'' = e^x, \tag{5}$$

using the method of variation of parameters.

SOLUTION The differential equation (5) has constant coefficients and has characteristic polynomial $r^3 - r^2 = r^2(r-1)$ with roots of 0, 0, 1. Thus $\{1, x, e^x\}$ is a fundamental set of solutions of the associated homogeneous equation $y''' - y'' = 0$. Let $y_1 = 1$, $y_2 = x$, $y_3 = e^x$. Then $a_3 = 1$, $n = 3$, and $f = e^x$. The equations (2) are

$$\begin{aligned} v_1' \cdot 1 + v_2' x + v_3' e^x &= 0, \\ v_1' \cdot 0 + v_2' \cdot 1 + v_3' e^x &= 0, \\ v_1' \cdot 0 + v_2' \cdot 0 + v_3' e^x &= e^x. \end{aligned} \tag{6}$$

This particular example may be easily solved to yield

$$v_3' = 1, \qquad v_2' = -e^x, \qquad v_1' = xe^x - e^x \tag{7}$$

or, upon antidifferentiation,

$$v_3 = x, \qquad v_2 = -e^x, \qquad v_1 = xe^x - 2e^x.$$

Thus

$$y_p = v_1 y_1 + v_2 y_2 + v_3 y_3 = (xe^x - 2e^x)1 + (-e^x)x + x(e^x) = xe^x - 2e^x.$$

The general solution is

$$y = y_p + y_h = xe^x - 2e^x + c_1 + c_2 x + c_3 e^x = xe^x + c_1 + c_2 x + \tilde{c}_3 e^x.$$

Suppose, however, that instead of solving (6) directly, we had used Cramer's rule (4). Then

$$W[1, x, e^x] = \det \begin{bmatrix} 1 & x & e^x \\ 0 & 1 & e^x \\ 0 & 0 & e^x \end{bmatrix} = e^x$$

and

$$v_1' = (-1)^{3+1}\frac{e^x W[x, e^x]}{W[1, x, e^x]} = e^x \frac{\det\begin{bmatrix} x & e^x \\ 1 & e^x \end{bmatrix}}{e^x} = xe^x - e^x,$$

$$v_2' = (-1)^{3+2}\frac{e^x W[1, e^x]}{W[1, x, e^x]} = -e^x \frac{\det\begin{bmatrix} 1 & e^x \\ 0 & e^x \end{bmatrix}}{e^x} = -e^x,$$

$$v_3' = (-1)^{3+3}\frac{e^x W[1, x]}{W[1, x, e^x]} = e^x \frac{\begin{bmatrix} 1 & x \\ 0 & 1 \end{bmatrix}}{e^x} = 1,$$

which agrees with (7). ◄

Exercises

In Exercises 1 through 4, solve the differential equation by the method of variation of parameters and give the general solution.

1. $y''' - y' = e^{2x}$

2. $y''' + y' = \dfrac{1}{\sin x}$

3. $y'''' - y''' = x$

4. $y''' - 6y'' + 11y' - 6y = e^x$

5. Verify that

$$W[e^{ax}, e^{bx}, e^{cx}] = e^{(a+b+c)x}(b-a)(c-a)(c-b).$$

In Exercises 6 through 12, use Exercise 5 and variation of parameters to find the general solution.

6. $y''' - y'' - 4y' + 4y = e^{-x}$

7. $y''' - 2y'' - y' + 2y = e^x$

8. $y''' + y'' - 4y' - 4y = x$

9. $y''' + 2y'' - y' - 2y = 1$

10. $y''' - 3y'' - y' + 3y = \sin x$

11. $y''' + 3y'' - y' - 3y = e^x$

12. $y''' - 3y'' + 2y' = e^{-x}$

13. Verify that

$$W[e^{\alpha x}, xe^{\alpha x}, x^2e^{\alpha x}] = 2e^{3\alpha x}$$

In Exercises 14 through 17, use Exercise 13 and variation of parameters to find the general solution.

14. $y''' - 3y'' + 3y' - y = x^{1/2}e^x$

15. $y''' + 3y'' + 3y' + y = x^{-3}e^{-x}$

16. $y''' + 6y'' + 12y' + 8y = x^{-1}e^{-2x}$

17. $y''' - 6y'' + 12y' - 8y = x^{7/2}e^{2x}$

In Exercises 18 through 21, you are given a fundamental set of solutions for the associated homogeneous equation and f and a_n. Solve the differential equation using variation of parameters.

18. $\{1, x, x^2, x^3\}$, $f = x$, $a_4 = x$

19. $\{e^x, xe^x, x^2e^x\}$, $f = e^x$, $a_3 = 2$

20. $\{1, x^2, x^3\}$, $f = x^{1/2}$, $a_3 = 1$

21. $\{1, x, x^{1/2}\}$, $f = x^3$, $a_3 = x^2$

CHAPTER

3

The Laplace Transform

3.1 Definition and Basic Properties

We have developed several methods for solving differential equations. In this chapter we shall introduce a different type of approach that is very important in many areas of applied mathematics. The idea is to use a *transformation* that changes one set of objects and operations into a different set of objects and operations. Our transformation will be the *Laplace transform*, and it will change a linear differential equation with constant coefficients into a problem in algebra. Many design procedures in such areas as circuit and control theory are based on the algebraic form of the problem provided by the Laplace transform. The Laplace transform is also especially well suited for handling discontinuous forcing functions and impulses. But first we need to develop the basic properties of the Laplace transform. In this chapter we will use t instead of x as the independent variable, since the Laplace transform is most often useful for initial value problems in time t, defined for $t \geq 0$.

Definition of Laplace Transform

Let $f(t)$ be a function defined on the interval $t \geq 0$. The **Laplace transform** of $f(t)$ is obtained by multiplying $f(t)$ by e^{-st} and integrating from $t = 0$ to $t = \infty$. The Laplace transform of $f(t)$ is then a new function of the Laplace transform variable s, and it is given by

$$F(s) = \mathscr{L}[f(t)] = \int_0^\infty e^{-st} f(t)\, dt, \tag{1}$$

provided the improper integral exists. $\mathscr{L}[f(t)]$ is read "the Laplace transform of $f(t)$." The notation $F(s)$ emphasizes that the Laplace transform of $f(t)$ is a function of the variable s. Other notations for the Laplace transform are $\mathscr{L}[f](s)$ and $\mathscr{L}[f]$.

Throughout this chapter lowercase letters will denote the function of t and capital letters will denote its Laplace transform. Thus,

$$\mathscr{L}[g(t)] = G(s) \qquad \text{and} \qquad \mathscr{L}[y(t)] = Y(s).$$

The one exception is $H(t)$ for the Heaviside (unit-step) function in Section 3.4.

EXAMPLE 3.1.1 Calculating Laplace Transform from Definition

Calculate the Laplace transform of e^t.

SOLUTION Recall that, by definition, for any function $h(t)$ defined on $[0, \infty)$,

$$\int_0^\infty h(t)\,dt = \lim_{b\to\infty} \int_0^b h(t)\,dt,$$

and the integral is said to *converge* if this limit exists. If the limit does not exist, the integral is said to *diverge*. Thus, if $f(t) = e^t$, then

$$\mathscr{L}[e^t] = \int_0^\infty e^{-st}e^t\,dt = \lim_{b\to\infty} \int_0^b e^{t(1-s)}\,dt$$

$$= \lim_{b\to\infty} \begin{cases} b & \text{if } s = 1, \\ \dfrac{e^{b(1-s)}}{1-s} - \dfrac{1}{1-s} & \text{if } s \neq 1. \end{cases}$$

If $s < 1$, then $1 - s > 0$ and $\lim_{b\to\infty} e^{b(1-s)} = \infty$, so that the integral diverges. If $s = 1$, then $\lim_{b\to\infty} b = \infty$ and the integral diverges. On the other hand, if $s > 1$, then $1 - s < 0$, and

$$\lim_{b\to\infty} \frac{e^{b(1-s)}}{1-s} - \frac{1}{1-s} = \frac{0}{1-s} - \frac{1}{1-s} = \frac{1}{s-1}.$$

Thus $F(s) = \mathscr{L}[e^t] = 1/(s-1)$, and the domain of the Laplace transform of e^t is $1 < s < \infty$. ◀

EXAMPLE 3.1.2 Calculating Laplace Transform from Definition

Let

$$f(t) = \begin{cases} 0 & \text{if } 0 \le t \le 2, \\ 3 & \text{if } 2 < t \le 4, \\ 0 & \text{if } 4 < t. \end{cases}$$

Calculate $\mathscr{L}[f(t)]$.

SOLUTION

$$
\begin{aligned}
\mathscr{L}[f(t)] &= \int_0^\infty e^{-st} f(t)\,dt \\
&= \int_0^2 e^{-st} f(t)\,dt + \int_2^4 e^{-st} f(t)\,dt + \int_4^\infty e^{-st} f(t)\,dt \\
&= \int_0^2 e^{-st} 0\,dt + \int_2^4 e^{-st} 3\,dt + \int_4^\infty e^{-st} 0\,dt \\
&= 3\int_2^4 e^{-st}\,dt = \left.\frac{3e^{-st}}{-s}\right|_2^4 = \frac{3e^{-4s}}{-s} + \frac{3e^{-2s}}{s}.
\end{aligned}
$$

Obviously, we do not want to have to compute every Laplace transform from the definition. Tables of Laplace transforms have been developed. There is a short table (Table 3.1.1) at the end of this section, and a longer table appears in Section 3.2. But first we need to establish when the Laplace transform exists.

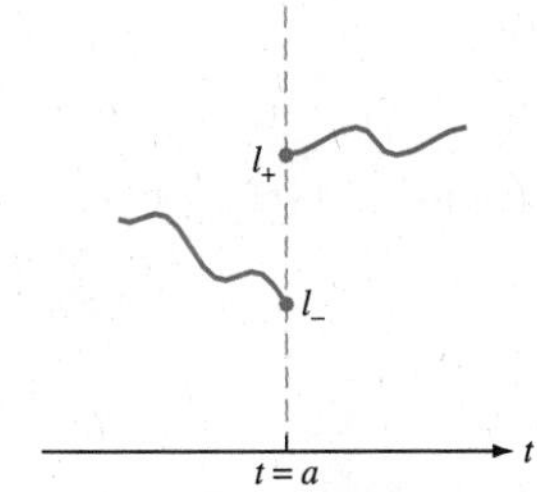

FIGURE 3.1.1

A function $f(t)$ defined on $[0, \infty)$ has a **jump discontinuity** at $a \in [0, \infty)$ if the one-sided limits

$$\lim_{t \to a^+} f(t) = l_+ \qquad \text{and} \qquad \lim_{t \to a^-} f(t) = l_-$$

exist but the function $f(t)$ is not continuous at $t = a$. The most important case is when $l_+ \neq l_-$, as illustrated in Figure 3.1.1.

A function $f(t)$ is **piecewise continuous** on $[0, \infty)$ if, for every number $B > 0$, $f(t)$ is continuous on $[0, B]$ except possibly for a finite number of jump discontinuities. Note that a piecewise continuous function can have an infinite number of discontinuities, but there can be only a finite number of discontinuities on a finite interval. Two piecewise continuous functions are graphed in Figure 3.1.2.

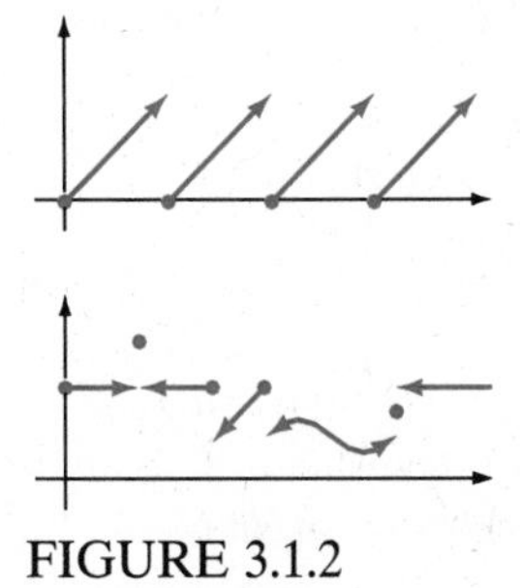
FIGURE 3.1.2

A function $f(t)$ on $[0, \infty)$ is said to be of **exponential order** if there exist constants α, M such that $|f(t)| \leq Me^{\alpha t}$ for $t \geq 0$. That is, as $t \to \infty$, $f(t)$ grows more slowly than a multiple of some exponential. Both of the examples in Figure 3.1.2 are bounded functions and hence of exponential order (take $\alpha = 0$). Note that e^{t^2} is not of exponential order, since t^2 eventually grows faster than αt for any constant α (see Exercise 76 at the end of this section).

THEOREM 3.1.1 Existence of the Laplace Transform

If $f(t)$ is piecewise continuous and of exponential order on $[0, \infty)$, so that $|f(t)| \leq Me^{\alpha t}$, then $\mathscr{L}[f] = F(s)$ exists and is defined at least for $s > \alpha$.

EXAMPLE 3.1.3 Laplace Transform of an Exponential

Compute $\mathscr{L}[e^{at}]$ where a is a constant.

SOLUTION

$$\begin{aligned}
\mathscr{L}[e^{at}] &= \int_0^\infty e^{-st}e^{at}\,dt = \lim_{b\to\infty}\int_0^b e^{t(a-s)}\,dt \\
&= \lim_{b\to\infty} \left.\frac{e^{t(a-s)}}{a-s}\right|_{t=0}^{t=b} \qquad (a \neq s) \\
&= \lim_{b\to\infty} \frac{e^{b(a-s)}}{a-s} - \frac{1}{a-s} \\
&= \begin{cases} -\dfrac{1}{a-s} & \text{if } a-s<0, \\ \text{diverges} & \text{if } a-s>0. \end{cases}
\end{aligned}$$

Thus,

$$\mathscr{L}[e^{at}] = \frac{1}{s-a}, \qquad (s > a). \qquad \text{(T2)} \tag{2}$$

Formulas which appear in the Laplace transform tables will be referred to by T plus the formula number in the table.

An important special case of (2) is when $a = 0$. In this case,

$$\mathscr{L}[1] = \frac{1}{s}, \qquad (s > 0). \qquad \text{(T1)} \tag{3}$$

Inverse Laplace Transform

In the applications to be developed, we shall often have to find a Laplace transform $F(s)$ and then find a function $f(t)$ such that $\mathscr{L}[f(t)] = F(s)$. That is, we will have to invert the Laplace transform. In this situation, it is important to know that $f(t)$ is unique. Since the Laplace transform is given in terms of an integral, changing a few values of $f(t)$ will not change the Laplace transform (see Exercise 75).

THEOREM 3.1.2 Uniqueness of Inverse Laplace Transform

If $F(s)$ is given and there is a continuous function $f(t)$ such that $\mathscr{L}[f(t)] = F(s)$, then $f(t)$ is the *only* continuous function for which $\mathscr{L}[f(t)] = F(s)$. ■

For piecewise continuous functions the result is slightly more technical. If $f(t)$ and $g(t)$ are piecewise continuous and $\mathscr{L}[f] = \mathscr{L}[g]$, then $f(t) = g(t)$ except for a finite number of points in any finite interval.

Since, for all practical purposes (at least ours), $F(s)$ uniquely determines $f(t)$, we may denote this by $f(t) = \mathscr{L}^{-1}[F(s)]$ and call $\mathscr{L}^{-1}$ the **inverse Laplace transform**.

EXAMPLE 3.1.4 Exponential Example

In Example 3.1.3, we showed that $\mathscr{L}[e^{at}] = 1/(s-a)$. Thus,

$$\mathscr{L}^{-1}[1/(s-a)] = e^{at}.$$

It is helpful to memorize this result (and only a few others).

Linearity Property

The Laplace transform has several important properties. One is that it satisfies the linearity property. That is, the Laplace transform of a linear combination of functions equals the same linear combination of their Laplace transforms:

$$\mathscr{L}[c_1 f(t) + c_2 g(t)] = c_1 \mathscr{L}[f(t)] + c_2 \mathscr{L}[g(t)], \tag{4}$$

where c_1, c_2 are constants.

As a corollary, the inverse Laplace transform also satisfies the linearity property:

$$\mathscr{L}^{-1}[c_1 F(s) + c_2 G(s)] = c_1 \mathscr{L}^{-1}[F(s)] + c_2 \mathscr{L}^{-1}[G(s)]. \tag{5}$$

Verification of (4)

$$\begin{aligned}\mathscr{L}[c_1 f(t) + c_2 g(t)] &= \int_0^\infty e^{-st}[c_1 f(t) + c_2 g(t)]\,dt \\ &= c_1 \int_0^\infty e^{-st} f(t)\,dt + c_2 \int_0^\infty e^{-st} g(t)\,dt \\ &= c_1 \mathscr{L}[f(t)] + c_2 \mathscr{L}[g(t)].\end{aligned}$$

EXAMPLE 3.1.5 Using Linearity When Taking Laplace Transforms

Compute $\mathscr{L}[3e^t + 5e^{-2t} + 6]$ using (2), (3), and the linearity property (4).

Solution $\mathscr{L}[3e^t + 5e^{-2t} + 6] = 3\mathscr{L}[e^t] + 5\mathscr{L}[e^{-2t}] + 6\mathscr{L}[1]$
$$= \frac{3}{s-1} + \frac{5}{s+2} + \frac{6}{s}.$$

EXAMPLE 3.1.6 **Using Linearity When Taking Inverse Laplace Transforms**

Given $F(s) = \dfrac{3}{s} + \dfrac{6}{s-3}$, find $f(t) = \mathscr{L}^{-1}[F(s)]$.

Solution From (2) and (3), we have

$$e^{at} = \mathscr{L}^{-1}\left[\frac{1}{s-a}\right] \quad \text{and} \quad 1 = \mathscr{L}^{-1}\left[\frac{1}{s}\right].$$

Thus

$$\mathscr{L}^{-1}\left[\frac{3}{s} + \frac{6}{s-3}\right] = 3\mathscr{L}^{-1}\left[\frac{1}{s}\right] + 6\mathscr{L}^{-1}\left[\frac{1}{s-3}\right]$$
$$= 3 \cdot 1 + 6e^{3t} = 3 + 6e^{3t}.$$

Sinusoidal Functions

From Euler's formula, $e^{ibt} = \cos bt + i \sin bt$, we know that the cosine and sine are linear combinations of imaginary exponentials:

$$\begin{aligned} \cos bt &= \frac{1}{2}e^{ibt} + \frac{1}{2}e^{-ibt} \\ \sin bt &= \frac{1}{2i}e^{ibt} - \frac{1}{2i}e^{-ibt}. \end{aligned} \tag{6}$$

For example, we can derive the Laplace transform of cos bt. Using (6) and the linearity property (4), we get

$$\mathscr{L}[\cos bt] = \frac{1}{2}\mathscr{L}[e^{ibt}] + \frac{1}{2}\mathscr{L}[e^{-ibt}]. \tag{7}$$

We have shown that $\mathscr{L}[e^{at}] = 1/(s-a)$ for $s > a$, which is valid for real exponentials. If a is complex, we can repeat the derivation of its Laplace transform (see Example 3.1.3), and we find that there are no changes except that the restriction $s > a$ should be $s > \text{Re}(a)$. That is, s must be greater than the real part of a. Thus,

$$\mathscr{L}[\cos bt] = \frac{1}{2}\frac{1}{s-ib} + \frac{1}{2}\frac{1}{s+ib} = \frac{1}{2}\frac{s+ib+s-ib}{(s-ib)(s+ib)}$$

so that

$$\mathscr{L}[\cos bt] = \frac{s}{s^2 + b^2}. \qquad \text{(T4)} \tag{8}$$

In a similar manner, we can show that

$$\mathscr{L}[\sin bt] = \frac{b}{s^2 + b^2}. \qquad \text{(T3)} \tag{9}$$

These are very fundamental formulas, and so they are presented in the tables. In Exercises 9 and 10 at the end of this section, these results are derived by explicit integration of the defining Laplace transform (1).

When using (T4) and (T5) and several formulas from the next section to compute inverse Laplace transforms, we will often have to adjust the constants so that the expression corresponds to one in the table.

EXAMPLE 3.1.7 **Inverse Laplace Transform: Denominator Is $s^2 + b^2$**

Given $F(s) = \dfrac{11}{s^2 + 17}$, find $f(t) = \mathscr{L}^{-1}[F(s)]$.

SOLUTION From (9) we have $\mathscr{L}^{-1}\left[\dfrac{b}{s^2 + b^2}\right] = \sin bt$. Then

$$\frac{11}{s^2 + 17} = \frac{11}{s^2 + (\sqrt{17})^2} = \frac{11}{\sqrt{17}} \frac{\sqrt{17}}{s^2 + (\sqrt{17})^2},$$

which involves (9) with $b = \sqrt{17}$. Hence,

$$\mathscr{L}^{-1}\left[\frac{11}{s^2 + 17}\right] = \frac{11}{\sqrt{17}} \sin \sqrt{17}\, t.$$

◄

Frequently, $F(s)$ will have to be expressed as a sum of several terms that can each be evaluated from known formulas.

EXAMPLE 3.1.8 **Inverse Laplace Transform: Denominator Is $s^2 + b^2$**

Given $F(s) = \dfrac{2s + 5}{s^2 + 9}$, find $f(t) = \mathscr{L}^{-1}[F(s)]$.

SOLUTION

$$\begin{aligned}
\mathscr{L}^{-1}\left[\frac{2s + 5}{s^2 + 9}\right] &= \mathscr{L}^{-1}\left[2\frac{s}{s^2 + 9} + \frac{5}{3}\frac{3}{s^2 + 9}\right] \\
&= 2\mathscr{L}^{-1}\left[\frac{s}{s^2 + 9}\right] + \frac{5}{3}\mathscr{L}^{-1}\left[\frac{3}{s^2 + 9}\right] \\
&= 2\cos 3t + \frac{5}{3}\sin 3t \qquad \text{[by (8) and (9)]}.
\end{aligned}$$

◄

Polynomials

Powers and polynomials also have elementary Laplace transforms. There are many ways to derive

$$\mathscr{L}[t^n] = \frac{n!}{s^{n+1}}. \qquad \text{(T9)} \tag{10}$$

The derivation we present requires little knowledge of Laplace transforms. [Further properties of Laplace transforms given in the next section enable one to derive (10) more easily.]

Derivation of (10) From the definition of the Laplace transform,

$$\mathscr{L}[t^n] = \int_0^\infty t^n e^{-st}\,dt.$$

We integrate by parts ($\int_a^b u\,dv = uv|_a^b - \int_a^b v\,du$) one time,

$$u = t^n, \qquad dv = e^{-st}\,dt,$$
$$du = nt^{n-1}\,dt, \qquad v = -\frac{1}{s}e^{-st},$$

so that

$$\mathscr{L}[t^n] = t^n\left(-\frac{1}{s}e^{-st}\right)\Big|_{t=0}^{\infty} + \frac{n}{s}\int_0^\infty t^{n-1}e^{-st}\,dt. \tag{11}$$

If $n \geq 1$, the first term on the right vanishes (assuming $s > 0$), since

$$t^n\left(-\frac{1}{s}e^{-st}\right)\Big|_{t=0}^{\infty} = -\frac{1}{s}\lim_{b\to\infty}[b^n e^{-bs}] - 0 = 0.$$

Thus, (11) becomes a recursion formula for the Laplace transform:

$$\mathscr{L}[t^n] = \frac{n}{s}\mathscr{L}[t^{n-1}]. \tag{12}$$

For example, evaluating (12) for $n = 1$ yields

$$\mathscr{L}[t] = \frac{1}{s}\mathscr{L}[1] = \frac{1}{s^2}, \tag{13}$$

using the fact (3) that $\mathscr{L}[1] = 1/s$. Evaluating (12) at subsequent values of n yields

$$n = 2: \qquad \mathscr{L}[t^2] = \frac{2}{s}\mathscr{L}[t] = \frac{2}{s^3},$$

$$n = 3: \qquad \mathscr{L}[t^3] = \frac{3}{s}\mathscr{L}[t^2] = \frac{3\cdot 2}{s^4}.$$

The pattern has now emerged, and we are able to conclude that (10) is valid.

EXAMPLE 3.1.9 Laplace Transform of a Polynomial

Find the Laplace transform of $f(t) = t^4 + 3t^2 + 5$.

SOLUTION Using the linearity property and (10) yields

$$\mathscr{L}[t^4 + 3t^2 + 5] = \mathscr{L}[t^4] + 3\mathscr{L}[t^2] + 5\mathscr{L}[1] = \frac{4!}{s^5} + 3\frac{2!}{s^3} + 5\frac{1}{s}. \blacktriangleleft$$

EXAMPLE 3.1.10 Inverse Laplace Transform: Denominator Is s^n

If $F(s) = 3/s^5$, find $f(t)$.

SOLUTION Using (10) or (T9) with $n = 4$,

$$f(t) = \mathscr{L}^{-1}[F(s)] = \mathscr{L}^{-1}\left[\frac{3}{s^5}\right] = \frac{3}{4!}\mathscr{L}^{-1}\left[\frac{4!}{s^5}\right] = \frac{3}{4!}t^4. \blacktriangleleft$$

A short table (Table 3.1.1) of elementary Laplace transforms and the corresponding inverse Laplace transform formulas appears later in this section. Some further properties of Laplace transforms that we will shortly discuss also appear in this short table. A more extensive table appears at the end of the next section.

In the tables, the first group of formulas give the Laplace transforms of particular functions. The second group are generally operational formulas. Some of these will be developed shortly. The third group are additional formulas developed in the exercises and other sections. To simplify referencing, formulas have the same numbers in both tables. In these tables, $H(t)$ is a step function, to be discussed in Section 3.4, and $\delta(t)$ is an impulse function, to be discussed in Section 3.6.

3.1.1 The Shifting Theorem (Multiplying by an Exponential)

Many of the more complex Laplace transform formulas are derived from simpler formulas using the Shifting Theorem.

■ **THEOREM 3.1.3 Shifting Theorem**

If $\mathscr{L}[f(t)] = F(s)$, then

$$\mathscr{L}[e^{ct}f(t)] = F(s - c). \quad \text{(T16)} \tag{14}$$

■

This shows that for Laplace transforms, multiplying $f(t)$ by e^{ct} is equivalent to shifting the transform variable from s to $s - c$.

Verification of (14)

$$\begin{aligned}\mathscr{L}[e^{ct}f(t)] &= \int_0^\infty e^{-st}e^{ct}f(t)\,dt \\ &= \int_0^\infty e^{-(s-c)t}f(t)\,dt \\ &= F(s-c).\end{aligned}$$

EXAMPLE 3.1.11 Shifting Theorem

Find the Laplace transform of t^4e^{2t}.

SOLUTION We will apply the Shifting Theorem, (14), with $c = 2$ and $f(t) = t^4$, so that $F(s) = 4!/s^5$:

$$\mathscr{L}[e^{ct}f(t)] = F(s-c)$$

so that

$$\mathscr{L}[e^{2t}t^4] = F(s-2) = \frac{4!}{(s-2)^5}.$$

◄

EXAMPLE 3.1.12 Shifting Theorem

Find the Laplace transform of $e^{-3t}\cos 6t$ using the Laplace transform of $\cos 6t$.

SOLUTION We will apply the Shifting Theorem, (14), with $c = -3$ and $f(t) = \cos 6t$, so that $F(s) = s/(s^2 + 36)$. According to the Shifting Theorem,

$$\mathscr{L}[e^{ct}f(t)] = F(s-c)$$

so that

$$\mathscr{L}[e^{-3t}\cos 6t] = F(s+3) = \frac{s+3}{(s+3)^2 + 36}.$$

◄

Exponential Times Sinusoidal Function

Important formulas for the Laplace transform of simple exponentials times sinusoidal functions may be obtained by generalizing the previous example:

$$\mathscr{L}[e^{at}\cos bt] = \frac{s-a}{(s-a)^2 + b^2} \qquad \text{(T6)} \qquad (15)$$

and

$$\mathscr{L}[e^{at}\sin bt] = \frac{b}{(s-a)^2+b^2}. \qquad \text{(T5)} \qquad (16)$$

Exponential Times Powers

The Shifting Theorem, (14), may also be applied to the Laplace transform formula for powers (10):

$$\mathscr{L}[e^{at}t^n] = \frac{n!}{(s-a)^{n+1}}. \qquad \text{(T11)} \qquad (17)$$

This generalizes Example 3.1.11.

Exercises

1. Sketch the following function and explain why it is not piecewise continuous:

$$f(t) = \begin{cases} t & 0 \le t < 2, \\ 3 & 2 = t, \\ \dfrac{1}{t-2} & 2 < t. \end{cases}$$

In Exercises 2 through 4, sketch the function and explain why it is piecewise continuous.

2. $f(t) = \begin{cases} 1 & 0 \le t \le 1, \\ -1 & 1 < t \le 2, \\ 1 & 2 < t. \end{cases}$

3. $f(t) = \begin{cases} t & 0 \le t < 1, \\ \frac{1}{2} & 1 = t \\ t-1 & 1 < t. \end{cases}$

4. $f(t) = \begin{cases} e^{-t} & 0 \le t < 1, \\ 3 & 1 = t, \\ e^{-t} & 1 < t. \end{cases}$

In Exercises 5 through 7, compute $\mathscr{L}[f(t)]$, using the formula $\mathscr{L}[f] = \int_0^\infty e^{-st}f(t)\,dt$.

5. $f(t) = \begin{cases} 1 & 0 \le t \le 1, \\ 0 & 1 < t. \end{cases}$

6. $f(t) = \begin{cases} 1 & 0 \le t < 1, \\ -1 & 1 \le t < 2, \\ 0 & 2 \le t. \end{cases}$

7. $f(t) = \begin{cases} t & 0 \le t < 1, \\ 2-t & 1 \le t \le 2, \\ 0 & 2 < t. \end{cases}$

8. Verify that $\mathscr{L}[t] = 1/s^2$, using the formula $\mathscr{L}[f] = \int_0^\infty e^{-st}f(t)\,dt$.

9. Verify that $\mathscr{L}[\sin bt] = b/(s^2+b^2)$, using the formula $\mathscr{L}[f] = \int_0^\infty e^{-st}f(t)\,dt$.

10. Verify that $\mathscr{L}[\cos bt] = s/(s^2+b^2)$, using the formula $\mathscr{L}[f] = \int_0^\infty e^{-st}f(t)\,dt$.

11. Using (2), show that $\mathscr{L}[\sinh at] = a/(s^2-a^2)$.

12. Using (2), show that $\mathscr{L}[\cosh at] = s/(s^2-a^2)$.

13. Using (6), show that $\mathscr{L}[\sin bt] = b/(s^2+b^2)$.

In Exercises 14 through 46, use (T1) through (T4) and (T9), [formulas (2), (3), (4), (8), (9) and (10)] to compute $\mathscr{L}[f]$.

14. $f(t) = 3t + 2$

15. $f(t) = 3\cosh 2t$

16. $f(t) = 4e^{3t} + 6e^{-t}$

17. $f(t) = 5\sin 6t$

18. $f(t) = 4 \sin 3t + 5 \cos 7t$

19. $f(t) = -t + 3$

20. $f(t) = e^t - e^{-t} + e^{2t}$

21. $f(t) = 2 + \cos 5t$

22. $f(t) = 2 \sin 3t + 4 \sin 5t$

23. $f(t) = \sinh 3t$

24. $f(t) = t + 3 - e^t$

25. $f(t) = 2e^{-t} + 6e^{3t}$

26. $f(t) = -4 \sin 2t$

27. $f(t) = 3t - 1 + \cosh 2t$

28. $f(t) = 7e^{-5t} - 9e^{3t} - 6$

29. $f(t) = 7t^3 + 11t + 8$

30. $f(t) = 2t^5 + 5t^2$

31. $f(t) = 3t^4 + 4t^3$

32. $f(t) = 2t^6 + 3t^3 + 6t^2$

In Exercises 33 through 46, use the Shifting Theorem (14) or (T16) to compute $\mathscr{L}[f]$.

33. $f(t) = e^{3t} \sin 5t$

34. $f(t) = e^{5t} \sin 3t$

35. $f(t) = e^{4t} \cos 7t$

36. $f(t) = e^{7t} \cos 4t$

37. $f(t) = e^{-3t} \sin 5t$

38. $f(t) = e^{-5t} \sin 3t$

39. $f(t) = e^{-4t} \cos 7t$

40. $f(t) = e^{-7t} \cos 4t$

41. $f(t) = e^{2t}t^6$

42. $f(t) = e^{6t}t^4$

43. $f(t) = e^{-2t}t^6$

44. $f(t) = e^{-6t}t^4$

45. $f(t) = e^{3t}(t^5 + t^3 + 1)$

46. $f(t) = e^{-4t}(t^6 + 6t^2 + 5t + 4)$

In Exercises 47 through 66, $F(s)$ is given. Use formulas (T1) through (T4) and (T9) to compute $f(t) = \mathscr{L}^{-1}[F(s)]$.

47. $\frac{1}{s^2} - \frac{1}{s}$

48. $\frac{3}{s-3} + \frac{4}{s+3}$

49. $\frac{1}{s^2+9}$

50. $\frac{7s+1}{s^2+4}$

51. $\frac{1+s}{s^2}$

52. $\frac{1+s}{s^3}$

53. $\frac{3}{s} - \frac{7}{s^2} + \frac{19}{s^2+1}$

54. $\frac{2}{s} + \frac{3}{s+1} - \frac{7}{s-8}$

55. $\frac{3s+7}{s^2+16}$

56. $\frac{11}{s^2} - \frac{2}{s}$

57. $\frac{5}{s+3} + \frac{7}{s-5}$

58. $\frac{5s+4}{s^2+9}$

59. s^{-6}

60. $4s^{-9}$

61. $\frac{2}{s^2+9}$

62. $\frac{5s}{s^2+13}$

63. $\frac{3}{s^{10}}$

64. $\frac{2}{s^5}$

65. $\frac{3}{2s^2+7}$

66. $\frac{-s+1}{3s^2+11}$

In Exercises 67 through 74, $F(s)$ is given. Use formulas (T1) through (T4) and (T9) to compute $f(t) = \mathscr{L}^{-1}[F(s)]$. The Shifting Theorem (T16) may be required.

67. $\frac{1}{(s-4)^3}$

68. $\dfrac{5}{(s-3)^4}$

69. $\dfrac{(s-5)}{(s-5)^2+9}$

70. $\dfrac{(s+6)}{(s+6)^2+4}$

71. $\dfrac{7}{(s-7)^2+16}$

72. $\dfrac{8}{(s+4)^2+25}$

73. $\dfrac{s+3}{(s+3)^2+5}$

74. $\dfrac{6}{(s-2)^2+3}$

75. Let $f(t) = \begin{cases} 2 & \text{if } t = 1, \\ t & \text{if } t \neq 1. \end{cases}$
Show that $\mathscr{L}[f(t)] = \mathscr{L}[t]$.

76. Show that $\lim_{t\to\infty} e^{-\alpha t}e^{t^2} = \infty$ for any α. Use this to show that e^{t^2} is not of exponential order.

77. Determine if $e^{\sqrt{t}}$ is of exponential order.

78. Show that e^{t^β} is of exponential order if $0 \le \beta \le 1$ and is not of exponential order if $\beta > 1$.

79. Prove Theorem 3.1.1, using the following two facts:

i) If $\int_0^\infty |h(t)|\,dt$ converges, then $\int_0^\infty h(t)\,dt$ converges.

ii) If $0 < g(t) < r(t)$ and $\int_0^\infty r(t)\,dt$ converges, then $\int_0^\infty g(t)\,dt$ converges.

80. Verify by direct integration that $\mathscr{L}[e^{at}] = 1/(s-a)$ if a is a complex number. Note that s must be greater than the real part of a.

3.1.2 Derivative Theorem (Optional)

Some additional results concerning Laplace transforms follow from the definition

$$F(s) = \int_0^\infty e^{-st} f(t)\,dt. \tag{18}$$

If we differentiate (18) with respect to s, we obtain

$$F'(s) = \frac{d}{ds}\int_0^\infty e^{-st} f(t)\,dt = \int_0^\infty \frac{\partial}{\partial s} e^{-st} f(t)\,dt = \int_0^\infty -te^{-st} f(t)\,dt$$

from the laws of differentiating integrals. Thus,

$$\mathscr{L}[tf(t)] = -F'(s). \qquad \text{(T21)} \tag{19}$$

For Laplace transforms, multiplying by t corresponds to taking minus the derivative of the transform.

EXAMPLE 3.1.13 **Using the Derivative Theorem**

Compute the Laplace transform of different powers of t, $\mathscr{L}[t^n]$, using (19).

Solution We already know that if $f(t) = 1$, then $F(s) = 1/s$. Thus,

$$\mathcal{L}[1] = \frac{1}{s},$$

$$\mathcal{L}[t] = \mathcal{L}[t \cdot 1] = -\frac{d}{ds}\mathcal{L}[1] = -\frac{d}{ds}\left(\frac{1}{s}\right) = \frac{1}{s^2},$$

$$\mathcal{L}[t^2] = \mathcal{L}[t \cdot t] = -\frac{d}{ds}\mathcal{L}[t] = -\frac{d}{ds}\left(\frac{1}{s^2}\right) = \frac{2}{s^3},$$

$$\mathcal{L}[t^3] = \mathcal{L}[t \cdot t^2] = -\frac{d}{ds}\mathcal{L}[t^2] = -\frac{d}{ds}\left(\frac{2}{s^3}\right) = \frac{3!}{s^4}.$$

$$\vdots$$

From this we can generalize that

$$\mathcal{L}[t^n] = \frac{n!}{s^{n+1}}. \qquad \text{(T9)} \qquad (20)$$

In this derivation successive usage of (19) has been made. Equation (20) was derived earlier [see (12)] using successive integration by parts. ◄

EXAMPLE 3.1.14 Using the Derivative Theorem

Find $\mathcal{L}[te^{at}]$ using (19).

Solution We use (19) with $f(t) = e^{at}$, so that $F(s) = 1/(s - a)$. Thus,

$$\mathcal{L}[te^{at}] = -\frac{d}{ds}\frac{1}{s-a} = \frac{1}{(s-a)^2}.$$ ◄

Higher powers of t can be obtained in the same way to give

$$\mathcal{L}[t^n e^{at}] = \frac{n!}{(s-a)^{n+1}}. \qquad \text{(T11)} \qquad (21)$$

This can also be derived from (20) using the Shifting Theorem [see (17)].

EXAMPLE 3.1.15 Using the Derivative Theorem

Find $\mathscr{L}[t \sin bt]$.

SOLUTION We use (19) with $f(t) = \sin bt$, so that $F(s) = b/(s^2 + b^2)$. Thus,

$$\mathscr{L}[t \sin bt] = -\frac{d}{ds}\left(\frac{b}{s^2 + b^2}\right) = \frac{2bs}{(s^2 + b^2)^2}, \qquad \text{(T25)} \qquad (22)$$

which we derive later in a different way. The corresponding result for cosines,

$$\mathscr{L}[t \cos bt] = \frac{s^2 - b^2}{(s^2 + b^2)^2}, \qquad \text{(T26)} \qquad (23)$$

can also be derived using (19). ◄

TABLE 3.1.1 SHORT TABLE OF LAPLACE TRANSFORMS

	$f(t)$	$\mathscr{L}[f(t)] = F(s) = \int_0^\infty e^{-st} f(t)\,dt$
(T1)	1	$\frac{1}{s}$
(T2)	e^{at}	$\frac{1}{s-a}$
(T3)	$\sin bt$	$\frac{b}{s^2+b^2}$
(T4)	$\cos bt$	$\frac{s}{s^2+b^2}$
(T9)	t^n	$\frac{n!}{s^{n+1}}$
(T15)	$f(t-c)H(t-c)$	$e^{-cs}F(s) \quad c > 0$
(T16)	$e^{ct}f(t)$	$F(s-c)$
(T18)	$\frac{df}{dt} = f'(t)$	$sF(s) - f(0)$
(T19)	$\frac{d^2f}{dt^2} = f''(t)$	$s^2F(s) - sf(0) - f'(0)$
(T21)	$tf(t)$	$-F'(s)$
(T23)	$\int_0^t f(\tau)\,d\tau$	$\frac{F(s)}{s}$
(T24)	$\int_0^t f(\tau)g(t-\tau)\,d\tau$	$F(s)G(s)$

Exercises

In Exercises 81 through 90, compute the Laplace transform using the Derivative Theorem (19).

81. te^{5t}

82. te^{-5t}

83. $t \cos 5t$

84. $t \sin 5t$

85. $t^2 \cos 3t$

86. $t^2 \sin 3t$

87. $te^{5t} \sin 3t$

88. $te^{5t} \cos 2t$

89. $te^{-4t} \cos 5t$

90. $te^{-4t} \sin 6t$

In Exercises 91 through 100, compute the inverse Laplace transform using the Derivative Theorem (19).

91. $(s + 3)^{-6}$

92. $(s - 2)^{-8}$

93. $\dfrac{5s}{(s^2 + 9)^2}$

94. $\dfrac{2s}{(s^2 + 25)^2}$

95. $\dfrac{s^2 - 9}{(s^2 + 9)^2}$

96. $\dfrac{s^2 - 25}{(s^2 + 25)^2}$

97. $\dfrac{5}{(s^2 + 9)^2}$

98. $\dfrac{3}{(s^2 + 25)^2}$

99. $\dfrac{2s + 1}{(s^2 + 1)^2}$

100. $\dfrac{3s - 5}{(s^2 + 4)^2}$

101. Show that, if $\mathscr{L}[f(t)] = F(s)$, then $\mathscr{L}[f(ct)] = \dfrac{1}{c}F\left(\dfrac{s}{c}\right)$, where c is a positive constant.

102. Using (T21), verify by induction that $\mathscr{L}[t^n f(t)] = (-1)^n F^{(n)}(s)$. This is (T22).

103. Using Exercise 102 and (T2), show that

$$\mathscr{L}[t^n e^{at}] = \frac{n!}{(s - a)^{n+1}}.$$

This is (T11).

Partial fractions results in terms like $(As + B)(s^2 + b^2)^{-2}$, which are not readily related to (22), (23). Thus (T27) is helpful.

104. Using (22), (23), and (T3), show that

$$\mathscr{L}^{-1}\left[\frac{2b^2}{(s^2 + b^2)^2}\right] = \frac{1}{b}\sin bt - t\cos bt. \qquad \text{(T27)}$$

3.2 Inverse Laplace Transforms (Roots, Quadratics, and Partial Fractions)

Many inverse Laplace transforms can be obtained directly from the tables based on the standard Laplace transforms and properties of the previous section. However, in solving differential equations, often we will need the inverse Laplace transform of $F(s)$, where $F(s)$ is a rational function of s—that is, $F(s)$ is a ratio of polynomials in s. In this case partial fractions may be used to express $F(s)$ as a sum of simpler terms.

When the denominator of $F(s)$ has a quadratic factor, $as^2 + bs + c$, we use the discriminant to distinguish between real roots ($b^2 - 4ac > 0$), complex roots ($b^2 - 4ac < 0$), and repeated real roots ($b^2 - 4ac = 0$). We will discuss examples of these types.

Simple Real Roots

When the denominator of $F(s)$ has only simple real roots, then the inverse Laplace transform of the partial fraction decomposition is a straightforward application of the elementary exponential formula $\mathscr{L}[e^{at}] = 1/(s - a)$. Also, the coefficients in the partial fraction expansion are easily determined.

EXAMPLE 3.2.1 **Real Roots in Denominator**

Given $F(s) = \dfrac{3}{s^2 - 4}$, find $f(t) = \mathscr{L}^{-1}[F(s)]$.

SOLUTION Using partial fractions,

$$\frac{3}{s^2 - 4} = \frac{3}{(s - 2)(s + 2)} = \frac{A}{s - 2} + \frac{B}{s + 2}.$$

Multiplying by the denominator gives $3 = A(s + 2) + B(s - 2)$. Note that $s = 2, -2$ are the roots of $s^2 - 4$. Letting $s = 2, -2$ gives $A = \frac{3}{4}$, $B = -\frac{3}{4}$. Thus,

$$\begin{aligned}\mathscr{L}^{-1}\left[\frac{3}{s^2 - 4}\right] &= \frac{3}{4}\mathscr{L}^{-1}\left[\frac{1}{s - 2}\right] - \frac{3}{4}\mathscr{L}^{-1}\left[\frac{1}{s + 2}\right] \\ &= \frac{3}{4}e^{2t} - \frac{3}{4}e^{-2t}.\end{aligned}$$

◄

EXAMPLE 3.2.2 **Real Roots in Denominator**

Given $F(s) = \dfrac{2s + 1}{s^3 - 3s^2 + 2s}$, find $f(t) = \mathscr{L}^{-1}[F(s)]$.

SOLUTION Using partial fractions,

$$\frac{2s + 1}{s^3 - 3s^2 + 2s} = \frac{2s + 1}{s(s - 1)(s - 2)} = \frac{A}{s} + \frac{B}{s - 1} + \frac{C}{s - 2}.$$

Multiplying by the denominator gives $2s + 1 = A(s - 1)(s - 2) + Bs(s - 2) + Cs(s - 1)$. The roots are $s = 0, 1, 2$. Letting $s = 0, 1, 2$ gives $A = \frac{1}{2}$, $B = -3$, $C = \frac{5}{2}$. Thus,

$$\begin{aligned}\mathscr{L}^{-1}\left[\frac{2s + 1}{s^3 - 3s^2 + 2s}\right] &= \frac{1}{2}\mathscr{L}^{-1}\left[\frac{1}{s}\right] - 3\mathscr{L}^{-1}\left[\frac{1}{s - 1}\right] + \frac{5}{2}\mathscr{L}^{-1}\left[\frac{1}{s - 2}\right] \\ &= \frac{1}{2} - 3e^{t} + \frac{5}{2}e^{2t}.\end{aligned}$$

◄

Inverse Laplace Transforms for Quadratics (Completing the Square)

When working with Laplace transforms, we frequently get expressions where the denominator is a quadratic,

$$\frac{As + B}{as^2 + bs + c},$$

that has complex roots. If the denominator has purely imaginary roots (for example, the denominator is $s^2 + 5$), then the formula involving sines and cosines may be used directly, as in Example 3.1.8. If the denominator has roots with nonzero real and imaginary parts ($b^2 - 4ac < 0$), then completing the square enables the inverse Laplace transform to be obtained.

EXAMPLE 3.2.3 Complex Roots: Completing the Square

Given $F(s) = \dfrac{3s - 17}{s^2 - 6s + 13}$, find $f(t) = \mathscr{L}^{-1}[F(s)]$.

Solution Since $b^2 - 4ac = 36 - 52 = -16 < 0$, the denominator does not have real factors. If we were to try to apply partial fractions, we would get the same expression back. This $F(s)$ cannot be simplified further by partial fractions. Thus we must complete the square; $s^2 - 6s + 13 = (s-3)^2 + 4 = (s - 3)^2 + 2^2$. There are two ways to proceed. One is to express the numerator $3s - 17$ in powers of $s - 3$ in order to use the table based on the Shifting Theorem. Alternatively, we may go directly to the answer using the long table. We illustrate the second approach first. There are two formulas (T5), (T6) in Table 3.2.1 with a denominator of $(s - 3)^2 + 2^2$,

$$\frac{s - 3}{(s - 3)^2 + 2^2} \quad \text{and} \quad \frac{2}{(s - 3)^2 + 2^2}.$$

Thus we wish to find constants A, B such that

$$\begin{aligned} \frac{3s - 17}{(s - 3)^2 + 2^2} &= A\frac{(s - 3)}{(s - 3)^2 + 2^2} + B\frac{2}{(s - 3)^2 + 2^2} \\ &= \frac{A(s - 3) + B2}{(s - 3)^2 + 2^2}. \end{aligned} \tag{1}$$

The inverse Laplace transform of (1) is

$$Ae^{3t} \cos 2t + Be^{3t} \sin 2t.$$

From (1), A, B must satisfy

$$3s - 17 = A(s - 3) + B2.$$

Equating powers of s gives $A = 3$. Letting $s = 3$ gives $2B = -8$ or $B = -4$. Thus

$$f(t) = \mathscr{L}^{-1}\left[\frac{3s - 17}{s^2 - 6s + 13}\right] = 3e^{3t}\cos 2t - 4e^{3t}\sin 2t.$$

Alternatively, we can compute $f(t)$ using the short table (Table 3.1.1) and the Shifting Theorem. We express $\dfrac{3s-17}{s^2-6s+13}$ as a function of $s - 3$:

$$\frac{3s - 17}{s^2 - 6s + 13} = \frac{3(s-3) - 8}{(s-3)^2 + 2^2}.$$

But, replacing $s - 3$ by s, we compute that

$$\mathscr{L}^{-1}\left[\frac{3s - 8}{s^2 + 2^2}\right] = \mathscr{L}^{-1}\left[3\frac{s}{s^2 + 2^2} - \frac{8}{2}\frac{2}{s^2 + 2^2}\right] = 3\cos 2t - 4\sin 2t.$$

Thus by the Shifting Theorem we have

$$f(t) = \mathscr{L}^{-1}\left[\frac{3s - 17}{s^2 - 6s + 13}\right] = 3e^{3t}\cos 2t - 4e^{3t}\sin 2t.$$ ◄

The next example has simple real and simple complex roots.

EXAMPLE 3.2.4 Real and Complex Roots

Given $F(s) = \dfrac{s - 8}{(s-5)(s^2 + 2s + 5)}$, find $f(t) = \mathscr{L}^{-1}[F(s)]$.

SOLUTION The quadratic factor has complex roots, since $b^2 - 4ac = -16 < 0$, and so the denominator cannot be factored further. Using partial fractions,

$$\frac{s - 8}{(s-5)(s^2 + 2s + 5)} = \frac{A}{s - 5} + \frac{Bs + C}{(s+1)^2 + 4}. \tag{2}$$

The first term in (2) is easily found in the table. The second term with the quadratic denominator can be determined by either of the techniques used in the last example. Using the first technique, we will get

$$\frac{s - 8}{(s-5)(s^2 + 2s + 5)} = \frac{A}{s - 5} + B\frac{s + 1}{(s+1)^2 + 4} + \tilde{C}\frac{2}{(s+1)^2 + 4}. \tag{3}$$

Finding the constants A, B, $\tilde{C}$ often requires the most work of all the steps. Sometimes it is best to save that work until the end. If A, B, $\tilde{C}$ were known, the inverse transform would be straightforward:

$$\mathscr{L}^{-1}\left[\frac{s-8}{(s-5)(s^2+2s+5)}\right] = Ae^{5t} + Be^{-t}\cos 2t + \tilde{C}e^{-t}\sin 2t.$$

We can either find A, B, C in (2) and then work a problem like the last example or find A, B, $\tilde{C}$ in (3). We shall do the second, since it is quicker. From (3), $s - 8 = A(s^2 + 2s + 5) + (s-5)[B(s+1) + 2\tilde{C}]$. A is easy to obtain by evaluating at the real root $s = 5$, so that $-3 = 40A$ or $A = \frac{-3}{40}$. The remaining coefficients (or all coefficients) may be obtained by equating like powers of s,

$$\begin{aligned} 1: &\quad -8 = 5A - 5B - 10\tilde{C}, \\ s: &\quad 1 = 2A - 5B + B + 2\tilde{C}, \\ s^2: &\quad 0 = A + B. \end{aligned}$$

Since we already know that $A = \frac{-3}{40}$, it follows from the last equation that $B = \frac{3}{40}$ and from the s equation that $2\tilde{C} = 1 - 2A + 4B = 1 + \frac{6}{40} + \frac{12}{40} = 1\frac{18}{40}$. A check on our arithmetic can be done by verifying that the remaining equation, the constant equation, holds. Since $-8 = -\frac{15}{40} - \frac{15}{40} - 5 - \frac{90}{40}$, the check is valid. ◄

Repeated Real Roots

When $s = 0$ is a repeated root, we already know how to obtain the inverse Laplace transform:

$$\mathscr{L}[t^n] = \frac{n!}{s^{n+1}}. \qquad \text{(T9)}$$

A direct application of the Shifting Theorem

$$\mathscr{L}[t^n e^{at}] = \frac{n!}{(s-a)^{n+1}} \qquad \text{(T11)} \qquad (4)$$

enables us to determine the inverse Laplace transform when the repeated root is $s = a$. The inverse Laplace transform of a simple root is the corresponding elementary exponential. Equation (4) shows that the inverse Laplace transform of a multiple root is the corresponding exponential multiplied by one power of t for each time the root is repeated.

EXAMPLE 3.2.5 Repeated Real Roots

If $F(s) = \dfrac{s^2}{(s+1)^3(s+5)}$, find $f(t)$.

Solution Using partial fractions,

$$\frac{s^2}{(s+1)^3(s+5)} = \frac{A}{s+5} + \frac{B}{(s+1)^3} + \frac{C}{(s+1)^2} + \frac{D}{s+1}$$
$$= A\frac{1}{s+5} + \frac{B}{2!}\frac{2!}{(s+1)^3} + C\frac{1}{(s+1)^2} + D\frac{1}{s+1}. \quad (5)$$

Determining the four coefficients can be the hardest part. Once the coefficients A, B, C, D are determined, the inverse Laplace transform can be obtained using the formulas for a simple root and the new formula for repeated roots:

$$\mathscr{L}^{-1}\left[\frac{s^2}{(s+1)^3(s+5)}\right] = Ae^{-5t} + B\frac{1}{2!}t^2e^{-t} + Cte^{-t} + De^{-t}.$$

Determining A, B is not difficult, since $s = -5$ and $s = -1$ are roots. Multiplying (5) by the denominator $(s+5)(s+1)^3$ yields

$$s^2 = A(s+1)^3 + B(s+5) + C(s+1)(s+5) + D(s+5)(s+1)^2.$$

Evaluating at $s = -5$ yields $25 = A(-4)^3$ or $A = 25/(-4)^3 = -\frac{25}{64}$. Evaluating at $s = -1$ yields $B = \frac{1}{4}$. Determining C, D is more difficult, since $s = -1$ is a root of multiplicity 3. There are two choices for finding C, D. The simplest is to equate coefficients of powers of s to get equations to solve for C, D. For example,

$$s^3: \quad 0 = A + D,$$
$$s^2: \quad 1 = 3A + C + 7D.$$

Given A, B, we find that $D = \frac{25}{64}$ and $C = -\frac{9}{16}$. A more elegant way is to multiply (5) by $(s+1)^3$ to get

$$\frac{s^2}{(s+5)} = (s+1)^3\frac{A}{s+5} + B + C(s+1) + D(s+1)^2. \quad (6)$$

Taking the first and second derivatives with respect to s of (6) and evaluating at $s = -1$ yields

$$\left.\frac{d}{ds}\frac{s^2}{(s+5)}\right|_{s=-1} = C$$

$$\left.\frac{d^2}{ds^2}\frac{s^2}{(s+5)}\right|_{s=-1} = 2D.$$

Note that A does not appear in these equations. Carrying out the respective derivatives again gives $C = -\frac{9}{16}$ and $2D = \frac{25}{32}$. Which of these approaches is better depends on what tools you have available, such as calculators or computers that do algebra or calculus, and whether you find calculus or algebra easier. ◄

Repeated Imaginary Roots

Repeated imaginary roots are more difficult than repeated real roots. We will derive only results corresponding to repeated roots of multiplicity 2. Using Euler's formula,

$$\mathscr{L}[t\cos bt] = \tfrac{1}{2}\mathscr{L}[te^{ibt}] + \tfrac{1}{2}\mathscr{L}[te^{-ibt}].$$

Using (4) with $n = 1$, we obtain

$$\mathscr{L}[t\cos bt] = \frac{1}{2}\frac{1}{(s-ib)^2} + \frac{1}{2}\frac{1}{(s+ib)^2} = \frac{1}{2}\frac{(s+ib)^2 + (s-ib)^2}{(s^2+b^2)^2},$$

after using a common denominator. Squaring, adding, and canceling terms yields

$$\mathscr{L}[t\cos bt] = \frac{s^2-b^2}{(s^2+b^2)^2}. \quad \text{(T26)} \tag{7}$$

In a similar way, we obtain

$$\mathscr{L}[t\sin bt] = \frac{2bs}{(s^2+b^2)^2}. \quad \text{(T25)} \tag{8}$$

These results can also be derived using (T21), as outlined in Example 3.1.15 in Section 3.1.2.

A useful inverse transform is obtained directly from (8):

$$\mathscr{L}^{-1}\left[\frac{s}{(s^2+b^2)^2}\right] = \frac{1}{2b}t\sin bt. \tag{9}$$

However, to obtain $\mathscr{L}^{-1}\left[\dfrac{1}{(s^2+b^2)^2}\right]$, we must do some further elementary calculations based on (7). Noting that $s^2 - b^2 = s^2 + b^2 - 2b^2$ in (7) yields

$$\mathscr{L}[t\cos bt] = \frac{-2b^2}{(s^2+b^2)^2} + \frac{1}{s^2+b^2}.$$

Thus,

$$\mathscr{L}^{-1}\left[\frac{-2b^2}{(s^2+b^2)^2}\right] = t\cos bt - \frac{1}{b}\sin bt$$

or

$$\mathscr{L}^{-1}\left[\frac{1}{(s^2+b^2)^2}\right] = \frac{1}{2b^3}\sin bt - \frac{1}{2b^2}t\cos bt. \qquad \text{(T27)} \qquad (10)$$

EXAMPLE 3.2.6 **Repeated Pure Imaginary Roots**

If $F(s) = \dfrac{s^3}{(s^2+9)^2}$, find $f(t)$.

SOLUTION Using partial fractions,

$$\frac{s^3}{(s^2+9)^2} = \frac{As+B}{(s^2+9)^2} + \frac{Cs+D}{s^2+9}$$

$$= A\frac{s}{(s^2+9)^2} + B\frac{1}{(s^2+9)^2} + C\frac{s}{s^2+9} + \frac{D}{3}\left(\frac{3}{s^2+9}\right). \qquad (11)$$

Once the coefficients A, B, C, D are determined, the inverse Laplace transform requires the new formulas (9) and (10):

$$\mathscr{L}^{-1}\left[\frac{s^3}{(s^2+9)^2}\right] = A\frac{1}{2\cdot 3}t\sin 3t + B\left(\frac{1}{2\cdot 3^3}\sin 3t - \frac{1}{2\cdot 3^2}t\cos 3t\right) +$$

$$C\cos 3t + \frac{D}{3}\sin 3t.$$

The coefficients can be determined by multiplying both sides of (11) by $(s^2+9)^2$:

$$s^3 = As + B + (s^2+9)(Cs+D).$$

Equating like powers of s yields

$$\begin{aligned} s^3&:\ 1 = C,\\ s^2&:\ 0 = D,\\ s&:\ 0 = A + 9C,\\ 1 = s^0&:\ 0 = B + 9D. \end{aligned}$$

Thus, $A = -9$, $B = 0$, $C = 1$, $D = 0$. ◄

Repeated Complex Roots

Formulas for repeated complex roots can be obtained directly from the formulas for repeated imaginary roots using the Shifting Theorem:

$$\mathscr{L}^{-1}\left[\frac{s-a}{[(s-a)^2+b^2]^2}\right] = \frac{1}{2b}te^{at}\sin bt, \qquad \text{(T28)}$$

$$\mathscr{L}^{-1}\left[\frac{1}{[(s-a)^2+b^2]^2}\right] = \frac{1}{2b^3}e^{at}\sin bt - \frac{1}{2b^2}te^{at}\cos bt. \qquad \text{(T29)} \tag{12}$$

EXAMPLE 3.2.7 Repeated Complex Roots

If $F(s) = \dfrac{6s+7}{(s^2+6s+25)^2}$, find $f(t)$.

Solution The roots are repeated. By the quadratic formula, the complex roots are $s = -3 \pm 4i$. Completing the square, $s^2+6s+25=(s+3)^2+16$, shows that transforms should be expressed as functions of $s+3$. Since $6s+7=A(s+3)+B$ yields $6s+7=6(s+3)-11$, we have

$$\frac{6s+7}{(s^2+6s+25)^2} = \frac{6(s+3)-11}{[(s+3)^2+4^2]^2}$$

$$= 6\frac{s+3}{[(s+3)^2+4^2]^2} - 11\frac{1}{[(s+3)^2+4^2]^2}.$$

Using (12),

$$\mathscr{L}^{-1}\left[\frac{6s+7}{(s^2+6s+25)^2}\right] = 6e^{-3t}\frac{1}{2\cdot 2}t\sin 2t$$

$$- 11e^{-3t}\left(\frac{1}{2\cdot 2^3}\sin 2t - \frac{1}{2\cdot 2^2}t\cos 2t\right). \quad ◄$$

Many of the basic formulas are used in computing both Laplace and inverse Laplace transforms. However, some formulas tend to be used primarily for taking the Laplace transform, and others are used primarily for finding the inverse Laplace transform. This includes finding the inverse Laplace transform when there are repeated complex roots.

TABLE 3.2.1 LONG TABLE OF LAPLACE TRANSFORMS I

	$f(t)$	$\mathscr{L}[f(t)] = F(s) = \int_0^\infty e^{-st} f(t)\,dt$
(T1)	1	$\frac{1}{s}, \quad s > 0$
(T2)	e^{at}	$\frac{1}{s-a}, \quad s > a$
(T3)	$\sin bt$	$\frac{b}{s^2+b^2}, \quad s > 0$
(T4)	$\cos bt$	$\frac{s}{s^2+b^2}, \quad s > 0$
(T5)	$e^{at} \sin bt$	$\frac{b}{(s-a)^2+b^2}, \quad s > a$
(T6)	$e^{at} \cos bt$	$\frac{s-a}{(s-a)^2+b^2}, \quad s > a$
(T7)	$\sinh bt$	$\frac{b}{s^2-b^2}, \quad s > \lvert b \rvert$
(T8)	$\cosh bt$	$\frac{s}{s^2-b^2}, \quad s > \lvert b \rvert$
(T9)	t^n, n a positive integer	$\frac{n!}{s^{n+1}}, \quad s > 0$
(T10)	t^p, $p > 0$,	$\frac{\Gamma(p+1)}{s^{p+1}}, \quad s > 0$
(T11)	$t^n e^{at}$	$\frac{n!}{(s-a)^{n+1}}, \quad s > a$
(T12)	$\delta(t)$	1
(T13)	$\delta(t-a)$	$e^{-as} \quad a > 0$
(T14)	$H(t-c)$	$\frac{e^{-cs}}{s} \quad c > 0$

TABLE 3.2.2 LONG TABLE OF LAPLACE TRANSFORMS II

	$f(t)$	$\mathscr{L}[f(t)] = F(s) = \int_0^\infty e^{-st} f(t)\,dt$
(T15)	$f(t-c)H(t-c)$	$e^{-cs}F(s), \quad c > 0$
(T16)	$e^{ct}f(t)$	$F(s-c)$
(T17)	$f(ct)$	$\frac{1}{c}F\left(\frac{s}{c}\right), \quad c > 0$
(T18)	$f'(t)$	$sF(s) - f(0)$
(T19)	$f''(t)$	$s^2F(s) - sf(0) - f'(0)$
(T20)	$f^{(n)}(t)$	$s^nF(s) - s^{n-1}f(0) \cdots - sf^{(n-2)}(0) - f^{(n-1)}(0)$
(T21)	$tf(t)$	$-F'(s)$
(T22)	$t^2f(t)$	$F''(s)$
(T22)	$t^nf(t), \quad n > 0$	$(-1)^nF^{(n)}(s)$
(T23)	$\int_0^t f(\tau)\,d\tau$	$\frac{F(s)}{s}$
(T24)	$\int_0^t f(\tau)g(t-\tau)\,d\tau = f * g$	$F(s)G(s)$

TABLE 3.2.3 OTHER FORMULAS INVOLVING THE LAPLACE TRANSFORM

(T25) $\mathscr{L}[t \sin bt] = \dfrac{2bs}{(s^2+b^2)^2}$

(T26) $\mathscr{L}[t \cos bt] = \dfrac{s^2-b^2}{(s^2+b^2)^2}$

(T27) $\mathscr{L}^{-1}\left[\dfrac{2b^2}{(s^2+b^2)^2}\right] = \dfrac{1}{b}\sin bt - t\cos bt$

(T28) $\mathscr{L}[te^{at} \sin bt] = \dfrac{2b(s-a)}{[(s-a)^2+b^2]^2}$

(T29) $\mathscr{L}[te^{at} \cos bt] = \dfrac{(s-a)^2-b^2}{[(s-a)^2+b^2]^2}$

(T30) $\mathscr{L}^{-1}\left[\dfrac{2b^2}{[(s-a)^2+b^2]^2}\right] = \dfrac{1}{b}e^{at}(\sin bt - t\cos bt)$

(T31) $\mathscr{L}[g(t)] = \dfrac{\int_0^T e^{-st}g(t)\,dt}{1-e^{-sT}} = \dfrac{\mathscr{L}[g(t)[1-H(t-T)]]}{1-e^{-sT}}$ g periodic with period T

(T32) $\mathscr{L}^{-1}\left[\dfrac{F(s)}{1-e^{-sT}}\right] = \mathscr{L}^{-1}\left[\sum_{n=0}^{\infty} e^{-nTs}F(s)\right] = \sum_{n=0}^{\infty} f(t-nT)H(t-nT)$

(T33) $\mathscr{L}^{-1}\left[\dfrac{F(s)}{1+e^{-sT}}\right] = \mathscr{L}^{-1}\left[\sum_{n=0}^{\infty} (-1)^n e^{-nTs}F(s)\right] = \sum_{n=0}^{\infty} (-1)^n f(t-nT)H(t-nT)$

Exercises

In Exercises 1 through 68, $F(s)$ is given. Compute $f(t) = \mathscr{L}^{-1}[F(s)]$.

1. $\dfrac{1}{s^2-9}$

2. $\dfrac{2}{s^2-16}$

3. $\dfrac{5s-1}{s^2+7}$

4. $\dfrac{7s-3}{s^2+5}$

5. $\dfrac{s+2}{s^2-1}$

6. $\dfrac{3}{s^2-s}$

7. $\dfrac{2s-1}{s^2-s}$

8. $\dfrac{5}{s^2+5s+6}$

9. $\dfrac{-s-1}{s^2+s-2}$

10. $\dfrac{2s-13}{s^2+8s+15}$

11. $\dfrac{4}{s^2+s-6}$

12. $\dfrac{s-1}{s^2+6s+5}$

13. $\dfrac{s}{s^2-2s+26}$

14. $\dfrac{2s}{s^2+4s+13}$

15. $\dfrac{2s+5}{s^2-6s+18}$

16. $\dfrac{17-3s}{s^2+2s+26}$

17. $\dfrac{3s-2}{s^2+10s+26}$

18. $\dfrac{3s - 1}{s^2 + 2s + 3}$

19. $\dfrac{s}{s^2 - 5s + 6}$

20. $\dfrac{2}{s^2 - 13s + 42}$

21. $\dfrac{1}{s^3 - 3s^2 + 2s}$

22. $\dfrac{2s + 1}{s^3 + 3s^2 + 2s}$

23. $\dfrac{s^2 - 2}{s^3 + 8s^2 + 7s}$

24. $\dfrac{s^2 + s}{s^3 + 4s^2 - 5s}$

25. $\dfrac{s + 3}{s^3 - s}$

26. $\dfrac{3}{s^3 - 4s}$

27. $\dfrac{2s + 1}{s^3 + 4s^2 + 13s}$

28. $\dfrac{2s - 5}{s^3 - 6s^2 + 18s}$

29. $\dfrac{s^2 - 3}{s^3 + 2s^2 + 26s}$

30. $\dfrac{3s - 2}{s^3 + 10s^2 + 26s}$

31. $\dfrac{s - 8}{(s - 5)(s^2 + 4)}$

32. $\dfrac{s - 5}{(s - 8)(s^2 + 9)}$

33. $\dfrac{s^2}{(s - 3)(s^2 - 2s + 26)}$

34. $\dfrac{s + 5}{(s + 3)(s^2 + 2s + 26)}$

35. $\dfrac{s^3 - 1}{s^4 + 10s^2 + 9}$

36. $\dfrac{1}{s^4 + 5s^2 + 4}$

37. $\dfrac{s^3}{s^4 - 5s^2 + 4}$

38. $\dfrac{s^2}{s^4 - 10s^2 + 9}$

39. $\dfrac{s}{s^4 - 16}$

40. $\dfrac{s^2 - 3}{s^4 - 1}$

Repeated roots occur in Exercises 41 through 60.

41. $\dfrac{4}{(s + 3)^6}$

42. $\dfrac{2}{(s - 3)^5}$

43. $\dfrac{4}{(3s + 1)^5}$

44. $\dfrac{s - 2}{(5 - 7s)^9}$

45. $\dfrac{s^2}{(s + 3)^2(s - 3)^2}$

46. $\dfrac{s - 3}{(s + 1)^2(s - 1)^2}$

47. $\dfrac{s}{(s + 1)^2(s^2 + 1)}$

48. $\dfrac{s^3}{(s - 1)^2(s^2 + 4)}$

49. $\dfrac{(s + 5)}{(s + 1)(s - 1)^3}$

50. $\dfrac{s + 1}{(s - 6)^3(s - 4)}$

51. $\dfrac{s^2 - s}{(s^2 + 4)^2}$

52. $\dfrac{2s^3 - 1}{(s^2 + 1)^2}$

53. $\dfrac{s^3 - 1}{(s^2 + 9)^2}$

54. $\dfrac{s^2 + 3s}{(s^2 + 16)^2}$

55. $\dfrac{s^3 - s^2}{(s^2 + 36)^2}$

56. $\dfrac{s^3 + s^2}{(s^2 + 25)^2}$

57. $\dfrac{3s - 1}{(s^2 + 2s + 26)^2}$

58. $\dfrac{s}{(s^2 + 4s + 5)^2}$

59. $\dfrac{s^2}{(s^2+6s+25)^2}$

60. $\dfrac{s^3}{(s^2+6s+25)^2}$

61. $\dfrac{s+5}{(s+1)(s-1)^3}$

62. $\dfrac{s+8}{s(s+1)^3(s-1)}$

63. $\dfrac{12}{s(s-8)^6}$

64. $\dfrac{6s}{(s+1)(s-3)^5}$

65. $\dfrac{s+5}{(s+1)^2(s-1)^2}$

66. $\dfrac{s-1}{s^2(s-3)^2}$

67. $\dfrac{s}{(s^2+1)(s-7)^4}$

68. $\dfrac{5}{(s^2+4)(s-1)^5}$

3.3 Initial Value Problem for Differential Equations

The formulas developed in the previous sections told how to take the Laplace transform (and inverse Laplace transform) of particular functions. This section will begin to show how Laplace transforms may be used in solving differential equations.

The next result is the key to solving linear constant coefficient differential equations using the Laplace transform. For convenience, we will give two special versions and the general version.

THEOREM 3.3.1 Laplace Transforms of Derivatives

First Derivative Case

Suppose that $f(t)$ is continuous and of exponential order on $[0,\infty)$ and $f'(t)$ is piecewise continuous on $[0,\infty)$. Then $\mathscr{L}[f'(t)]$ exists and

$$\mathscr{L}[f'(t)] = s\mathscr{L}[f(t)] - f(0) = sF(s) - f(0). \quad \text{(T18)}$$

Second Derivative Case

Suppose that $f(t)$ and $f'(t)$ are continuous and of exponential order on $[0,\infty)$ and $f''(t)$ is piecewise continuous on $[0,\infty)$. Then $\mathscr{L}[f''(t)]$ exists and

$$\mathscr{L}[f''(t)] = s^2\mathscr{L}[f(t)] - sf(0) - f'(0) = s^2F(s) - sf(0) - f'(0). \quad \text{(T19)}$$

General Case

Suppose that $f(t), f'(t), \ldots, f^{(n-1)}(t)$ are continuous and of exponential order on $[0,\infty)$ and $f^{(n)}(t)$ is piecewise continuous on $[0,\infty)$. Then $\mathscr{L}[f^{(n)}(t)]$ exists and

$$\begin{aligned}\mathscr{L}[f^{(n)}(t)] &= s^n\mathscr{L}[f(t)] - s^{n-1}f(0) - s^{n-2}f'(0) - \cdots - f^{(n-1)}(0)\\ &= s^nF(s) - s^{n-1}f(0) - s^{n-2}f'(0) - \cdots - f^{(n-1)}(0). \quad \text{(T20)}\end{aligned}$$

■

The formulas for the first and second derivatives are included in Table 3.1.1. The formulas for the Laplace transform of time derivatives simplify if all initial conditions are zero.

Verification of the First Derivative Case We calculate the Laplace transform of the first derivative using the definition of the Laplace transform:

$$\mathscr{L}[f'(t)] = \int_0^\infty e^{-st} f'(t)\, dt.$$

Since $f'(t)$ appears in the integrand, we expect that integration by parts ($\int u\, dv = uv - \int v\, du$) may simplify the integral:

$$\begin{aligned} dv &= f'(t)\, dt, & u &= e^{-st} \\ v &= f(t), & du &= -se^{-st}\, dt. \end{aligned}$$

Using integration by parts in this way yields

$$\mathscr{L}[f'(t)] = e^{-st} f(t)\Big|_{t=0}^{\infty} + s\int_0^\infty e^{-st} f(t)\, dt = s\mathscr{L}[f(t)] - f(0), \tag{1}$$

since $e^{-st}f(t)|_{t=0}^{\infty} = \lim_{b\to\infty} e^{-sb}f(b) - f(0) = -f(0)$ if $f(t)$ is of exponential order and s is large enough. The initial condition $f(0)$ is needed for the Laplace transform of the first derivative. The validity of integration by parts here requires that $f(t)$ be continuous. (Note Exercises 29 and 30.)

Verification for Higher Derivatives The Laplace transform of higher derivatives may be done by induction. For example, to show the second-derivative case, compute as follows:

$$\begin{aligned} \mathscr{L}[f''(t)] &= \mathscr{L}[(f'(t))'] = s\mathscr{L}[f'(t)] - f'(0), && \text{using (1),} \\ &= s[s\mathscr{L}[f(t)] - f(0)] - f'(0), && \text{using (1) again,} \\ &= s^2\mathscr{L}[f(t)] - sf(0) - f'(0), && \text{as desired.} \end{aligned}$$

Solving Differential Equations with Laplace Transforms

Linear differential equations with constant coefficients can be solved using Laplace transforms based on the formulas (T18), (T19), and (T20) for the Laplace transform of derivatives. We let $Y(s) = \mathscr{L}[y(t)]$.

EXAMPLE 3.3.1 Solving a Differential Equation

Solve the initial value problem

$$y' - 3y = 4, \qquad y(0) = 7.$$

Solution Take the Laplace transform of both sides of the differential equation, and obtain (using the linearity of the Laplace transform)

$$\mathscr{L}[y'] - 3\mathscr{L}[y] = 4\mathscr{L}[1].$$

From the theorem (T18) on the Laplace transform of the first time derivative,

$$\mathscr{L}[y'] = sY(s) - y(0).$$

Thus, using the initial condition $y(0) = 7$, we have

$$sY(s) - 7 - 3Y(s) = \frac{4}{s}.$$

We solve for $Y(s)$:

$$(s - 3)Y(s) = 7 + \frac{4}{s} = \frac{7s + 4}{s},$$

and hence

$$Y(s) = \frac{7s + 4}{s(s - 3)}.$$

To find $y(t)$, we expand $Y(s)$ using partial fractions:

$$Y(s) = \frac{7s + 4}{s(s - 3)} = \frac{A}{s} + \frac{B}{s - 3}. \tag{2}$$

In terms of the coefficients of the partial fraction expansion, the inverse Laplace transform is

$$y(t) = A + Be^{3t}.$$

To find A, B, multiply both sides of (2) by $s(s - 3)$ and get

$$7s + 4 = A(s - 3) + Bs.$$

Evaluating this at the roots $s = 0, 3$ of the denominator yields $A = -\frac{4}{3}$ and $B = \frac{25}{3}$, so that

$$y(t) = -\frac{4}{3} + \frac{25}{3}e^{3t}.$$ ◄

Since this is a linear first-order equation with constant coefficients, it could also be easily solved by several techniques from Chapters 1 and 2. However, this example motivates the general procedure, which we express for second-order linear differential equations with constant coefficients.

Procedure for Solving Initial Value Problems Using Laplace Transforms

Given the differential equation

$$ay'' + by' + cy = f(t), \tag{3}$$

with a, b, c constants and $f(t)$ a function that has a Laplace transform. Assume the initial conditions $y(0)$ and $y'(0)$ are given.

1. Take the Laplace transform of both sides (using linearity of the Laplace transform):

$$a\mathscr{L}[y''] + b\mathscr{L}[y'] + c\mathscr{L}[y] = \mathscr{L}[f(t)].$$

2. Use the initial conditions on the solution when applying the theorem on the Laplace transforms of derivatives [formulas (T18) and (T19)]:

$$a[s^2Y(s) - sy(0) - y'(0)] + b[sY(s) - y(0)] + cY(s) = F(s).$$

3. Solve for $Y(s) = \mathscr{L}[y(t)]$:

$$Y(s)(as^2 + bs + c) = F(s) + a[sy(0) + y'(0)] + by(0)],$$

or equivalently,

$$Y(s) = \frac{F(s) + a[sy(0) + y'(0)] + by(0)}{as^2 + bs + c}. \tag{4}$$

4. In the specific examples, express $Y(s)$ in terms of functions of s appearing in the Laplace transform column of the Laplace transform table (Tables 3.1.1, 3.2.1, 3.2.2, and 3.2.3).
5. Compute $y(t)$ from $Y(s)$, using these formulas.

We conclude from (4) that when we use Laplace transforms,

> An ordinary differential equation with constant coefficients is always reduced to an algebraic equation for the inverse Laplace transform.

The denominator for the inverse Laplace transform $as^2 + bs + c$ is the same polynomial as the characteristic polynomial $ar^2 + br + c$ for the differential equation (3). Thus, the roots of the denominator are the roots of the characteristic equation.

EXAMPLE 3.3.2 Second-Order Differential Equation

Solve the initial value problem

$$y'' + 3y' + 2y = 1, \qquad y(0) = 0, \qquad y'(0) = 2.$$

Solution Take the Laplace transform of both sides of the differential equation, and obtain (using the linearity of $\mathscr{L}$),

$$\mathscr{L}[y''] + 3\mathscr{L}[y'] + 2\mathscr{L}[y] = \mathscr{L}[1].$$

From Theorem 3.3.1 (or T18 and T19),

$$\mathscr{L}[y''] = s^2Y(s) - sy(0) - y'(0), \qquad \mathscr{L}[y'] = sY(s) - y(0).$$

Thus, using the initial conditions $y(0) = 0$, $y'(0) = 2$, we have

$$s^2Y(s) - 2 + 3sY(s) + 2Y(s) = \frac{1}{s}.$$

Now solve for $Y(s)$:

$$(s^2 + 3s + 2)Y(s) = 2 + \frac{1}{s} = \frac{2s+1}{s}, \quad \text{or} \quad Y(s) = \frac{2s+1}{s(s^2+3s+2)}.$$

To find $y(t)$, expand $Y(s)$ using partial fractions:

$$Y(s) = \frac{2s+1}{s(s+1)(s+2)} = \frac{A}{s} + \frac{B}{s+1} + \frac{C}{s+2}.$$

To find A, B, C, multiply both sides by $s(s+1)(s+2)$ and get

$$2s + 1 = A(s+1)(s+2) + Bs(s+2) + Cs(s+1).$$

Evaluating this expression at the roots $s = 0, -1, -2$ of the denominator yields

$$\begin{aligned} s = 0: &\qquad 1 = A2, \text{ so } A = \frac{1}{2}; \\ s = -1: &\qquad -1 = B(-1)(1), \text{ so } B = 1; \\ s = -2: &\qquad -3 = C(-2)(-1), \text{ so } C = -\frac{3}{2}, \end{aligned}$$

and hence

$$Y(s) = \frac{1}{2}\cdot\frac{1}{s} + \frac{1}{s+1} + \left(-\frac{3}{2}\right)\frac{1}{s+2}. \tag{5}$$

From (T2) and (T1),

$$1 = \mathscr{L}^{-1}\left[\frac{1}{s}\right], \qquad e^{-t} = \mathscr{L}^{-1}\left[\frac{1}{s+1}\right], \qquad e^{-2t} = \mathscr{L}^{-1}\left[\frac{1}{s+2}\right],$$

so that from (5),

$$y(t) = \frac{1}{2} + e^{-t} - \frac{3}{2}e^{-2t}.$$

◄

This problem, of course, could also be done by the methods of undetermined coefficients or variation of parameters discussed in Chapter 2. In the next section we shall consider problems for which the Laplace transform works more easily than either of these techniques.

In concluding this section, three additional examples of solving differential equations using the Laplace transform will be given.

EXAMPLE 3.3.3 **Repeated Roots**

Solve the initial value problem

$$y'' - 2y' + y = 3e^t, \qquad y(0) = 1, \qquad y'(0) = 1.$$

SOLUTION Taking the Laplace transform of both sides gives

$$\mathscr{L}[y''] - 2\mathscr{L}[y'] + \mathscr{L}[y] = 3\mathscr{L}[e^t].$$

Using (T18), (T19), and the initial conditions on the left, and (T2) on the right, we obtain

$$s^2Y(s) - s - 1 - 2[sY(s) - 1] + Y(s) = \frac{3}{s-1}.$$

Solving for $Y(s)$,

$$(s^2 - 2s + 1)Y(s) = \frac{3}{s-1} + s - 1 = \frac{s^2 - 2s + 4}{s-1},$$

$$Y(s) = \frac{s^2 - 2s + 4}{(s^2 - 2s + 1)(s-1)} = \frac{s^2 - 2s + 4}{(s-1)^3}.$$

Expanding $Y(s)$ by partial fractions, we obtain

$$Y(s) = \frac{s^2 - 2s + 4}{(s-1)^3} = \frac{A}{s-1} + \frac{B}{(s-1)^2} + \frac{C}{(s-1)^3}.$$

We multiply by $(s-1)^3$, obtaining

$$\begin{aligned} s^2 - 2s + 4 &= A(s-1)^2 + B(s-1) + C \\ &= A(s^2 - 2s + 1) + B(s-1) + C. \end{aligned}$$

Equating the coefficients of like powers of s,

$$\begin{aligned} 1: &\quad 4 = A - B + C, \\ s: &\quad -2 = -2A + B, \\ s^2: &\quad 1 = A, \end{aligned}$$

and, solving from the last equation up, we get $A = 1$, $B = 0$, $C = 3$. Thus,

$$Y(s) = \frac{1}{s-1} + \frac{3}{(s-1)^3}.$$

Now, from (T2) and (T11),

$$e^t = \mathscr{L}^{-1}\left[\frac{1}{s-1}\right] \quad \text{and} \quad t^2e^t = \mathscr{L}^{-1}\left[\frac{2}{(s-1)^3}\right].$$

Since $\dfrac{2}{(s-1)^3}$ appears in the table, but not $\dfrac{3}{(s-1)^3}$, we need to rewrite $Y(s)$ as

$$Y(s) = \frac{1}{s-1} + \frac{3}{2}\cdot\frac{2}{(s-1)^3}$$

in order to get

$$y(t) = e^t + \frac{3}{2}t^2e^t.$$

EXAMPLE 3.3.4 **Pure Imaginary Roots**

Solve $y'' + 3y = \sin 5t, \qquad y(0) = 0,\ y'(0) = 0.$

SOLUTION Taking the Laplace transform of both sides yields

$$\mathscr{L}[y''] + 3\mathscr{L}[y] = \mathscr{L}[\sin 5t]$$

or, by (T3), (T18), and (T19),

$$s^2Y(s) + 3Y(s) = \frac{5}{s^2+25}.$$

Thus

$$Y(s) = \frac{5}{(s^2+3)(s^2+25)}.$$

By partial fractions,

$$Y(s) = \frac{5}{(s^2+3)(s^2+25)} = \frac{As+B}{s^2+3} + \frac{Cs+D}{s^2+25}.$$

Multiplying both sides by $(s^2+25)(s^2+3)$, we find that

$$5 = (As+B)(s^2+25) + (Cs+D)(s^2+3).$$

Equating like powers of s, we get

$$\begin{aligned} 1&: \quad 5 = 25B + 3D,\\ s&: \quad 0 = 25A + 3C,\\ s^2&: \quad 0 = B + D,\\ s^3&: \quad 0 = A + C. \end{aligned}$$

The second and fourth equations yield $A = C = 0$. The first and third equations give, after a little algebra,

$$B = 5/22, \qquad D = -5/22,$$

so that

$$Y(s) = \frac{5/22}{s^2+3} - \frac{5/22}{s^2+25}.$$

Formula (T3) is that

$$\mathscr{L}^{-1}\left[\frac{\sqrt{3}}{s^2+3}\right] = \sin\sqrt{3}\,t, \qquad \mathscr{L}^{-1}\left[\frac{5}{s^2+25}\right] = \sin 5t.$$

Thus, we need to write $Y(s)$ as

$$Y(s) = \frac{5}{22\sqrt{3}}\cdot\frac{\sqrt{3}}{s^2+3} - \left(\frac{1}{22}\right)\frac{5}{s^2+25}$$

to conclude that

$$y(t) = \frac{5}{22\sqrt{3}}\sin\sqrt{3}\,t - \frac{1}{22}\sin 5t.$$

◄

EXAMPLE 3.3.5 Resonance

Solve $y'' + 4y = 3\sin 2t$ with $y(0) = 0$, $y'(0) = 0$.

SOLUTION This is a resonance problem, since the forcing function $\sin 2t$ is also a solution of the associated homogeneous equation. (The forcing frequency equals the natural frequency.) Taking the Laplace transform of both sides of the differential equation yields

$$\mathscr{L}[y''] + 4\mathscr{L}[y] = 3\frac{2}{s^2+4}.$$

Using the formulas for the Laplace transform of time derivatives (T19) and applying the zero initial conditions gives

$$s^2Y(s) + 4Y(s) = \frac{6}{s^2+4}.$$

Thus,

$$Y(s) = \frac{6}{(s^2+4)^2} = \frac{6}{(s^2+2^2)^2} = \frac{3}{4}\left[\frac{2\cdot 2^2}{(s^2+2^2)^2}\right], \tag{6}$$

which does not require further partial fraction decomposition. Equation (6) may be inverted using Table 3.2.3 or (T27). The solution is

$$y(t) = \frac{3}{4}\left(\frac{1}{2}\sin 2t - t\cos 2t\right).$$

To illustrate the use of some of the other formulas, we shall also find the inverse Laplace transform of (6) using only the short table along with property (T21) for Laplace transforms. We summarize the calculation as follows:

$f(t)$	$F(s)$	Justification
$\cos 2t$	$\dfrac{s}{s^2+4}$	T4
$t\cos 2t$	$-\dfrac{d}{ds}\left(\dfrac{s}{s^2+4}\right) = \dfrac{-(s^2+4)+2s^2}{(s^2+4)^2}$	T21
	$= \dfrac{s^2-4}{(s^2+4)^2} = \dfrac{(s^2+4)-8}{(s^2+4)^2}$	Algebra
	$= \dfrac{-8}{(s^2+4)^2} + \dfrac{1}{s^2+4}$	Algebra
$\frac{1}{2}\sin 2t$	$\dfrac{1}{s^2+4}$	T3
$t\cos 2t - \frac{1}{2}\sin 2t$	$\dfrac{-8}{(s^2+4)^2}$	Addition

Thus,

$$y(t) = \frac{3}{4}\left(\frac{1}{2}\sin 2t - t\cos 2t\right). \blacktriangleleft$$

Computer programs have been written that do symbolic manipulations (they appear to manipulate formulas rather than numbers). This means that Laplace transform techniques and their accompanying algebra involving functions of s may soon be handled as simply as ordinary arithmetic on a pocket calculator.

Exercises

In Exercises 1 through 28, use the Laplace transform to solve the initial value problems.

1. $y'' - 4y = 1, \quad y(0) = 0,\ y'(0) = 1$
2. $y'' - 6y' + 5y = 2, \quad y(0) = 0,\ y'(0) = -1$
3. $y'' + 3y' - 4y = 0, \quad y(0) = 1,\ y'(0) = 1$
4. $y'' - 3y' - 4y = t, \quad y(0) = 0,\ y'(0) = 0$
5. $y'' + 5y' + 6y = 1, \quad y(0) = 1,\ y'(0) = 0$
6. $y'' + y = 1, \quad y(0) = y'(0) = 2$
7. $y'' - 3y' + 2y = 1, \quad y(0) = 0,\ y'(0) = 1$
8. $y'' - y = e^t, \quad y(0) = 1,\ y'(0) = 0$
9. $y'' + 9y = e^{-t}, \quad y(0) = 1,\ y'(0) = 2$
10. $y'' + 4y = 3\cos 5t, \quad y(0) = 0,\ y'(0) = 3$
11. $y'' + 4y' + 13y = 2, \quad y(0) = 1,\ y'(0) = 0$
12. $y'' - 2y' + 5y = 0, \quad y(0) = 2,\ y'(0) = 1$

13. $y^{(4)} - y = 1$,
$y(0) = 3,\ y'(0) = 5,\ y''(0) = 0,\ y'''(0) = 0$

14. $y' + y = e^t \sin t, \quad y(0) = 1$

15. $y' - y = 1 - t, \quad y(0) = -1$

16. $y'' + 2y' + 2y = e^{-t} \sin 2t$,
$y(0) = 0,\ y'(0) = 0$

17. $y'' + 2y' + y = e^{-t}, \quad y(0) = 0,\ y'(0) = 1$

18. $y''' + y'' - y' - y = 0$,
$y(0) = 0,\ y'(0) = 0,\ y''(0) = 1$

19. $y' + 2y = e^{-t} \cos t, \quad y(0) = 0$

20. $y' - 3y = t^2 e^t, \quad y(0) = 1$

Exercises 21 through 28 involve resonance.

21. $y'' + y = \sin t, \quad y(0) = 0,\ y'(0) = 0$

22. $y'' + y = \cos t, \quad y(0) = 0,\ y'(0) = 1$

23. $y'' + 9y = \cos 3t, \quad y(0) = 0,\ y'(0) = 1$

24. $y'' + 9y = \sin 3t, \quad y(0) = 1,\ y'(0) = 0$

25. $y'' + 4y = \cos 2t, \quad y(0) = 0,\ y'(0) = 0$

26. $y'' + 16y = \sin 4t, \quad y(0) = 0,\ y'(0) = 0$

27. $y'' + y = \sin t + \cos t, \quad y(0) = 0,\ y'(0) = 0$

28. $y'' + 4y = \sin 2t + \cos 2t$,
$y(0) = 0,\ y'(0) = 0$

29. Suppose that $f(t)$ and $f'(t)$ are continuous and of exponential order for $t > 0$. Suppose that $\lim_{t \to 0^+} f(t)$ exists and is denoted $f(0^+)$. Show that $\mathscr{L}[f'(t)] = sF(s) - f(0^+)$.

30. Suppose that $f(t)$, $f'(t)$, $f''(t)$ are continuous and of exponential order for $t > 0$. Suppose that $\lim_{t \to 0^+} f(t)$ exists and is denoted $f(0^+)$. Suppose also that $\lim_{t \to 0} f'(t)$ exists and is denoted $f'(0^+)$. Show that $\mathscr{L}[f''(t)] = s^2 F(s) - f'(0^+) - sf(0^+)$.

3.4 Discontinuous Forcing Functions

In the previous section we saw how the Laplace transform could be used to solve linear differential equations with constant coefficients. All of the examples given, however, could easily have been solved by either the method of undetermined coefficients or variation of parameters. In this and the next two sections, we shall examine problems where the Laplace transform works better than either of these two methods.

The key is a notational one. We need a way to write a piecewise continuous function as a simple formula so that it may be handled the way the simpler functions $\cos t$, $\sin 5t$, and $e^{3t} \sin 2t$ were treated.

The needed notation involves the unit-step or Heaviside function. Since manipulating the unit-step function often gives students difficulty, we shall discuss it very carefully.

■ DEFINITION OF HEAVISIDE FUNCTION

The unit step, or Heaviside function $H(t)$, is defined as

$$H(t) = \begin{cases} 1 & \text{if } t \geq 0, \\ 0 & \text{if } t < 0. \end{cases}$$

Its graph is given in Figure 3.4.1. It corresponds to one being turned on at $t = 0$. ■

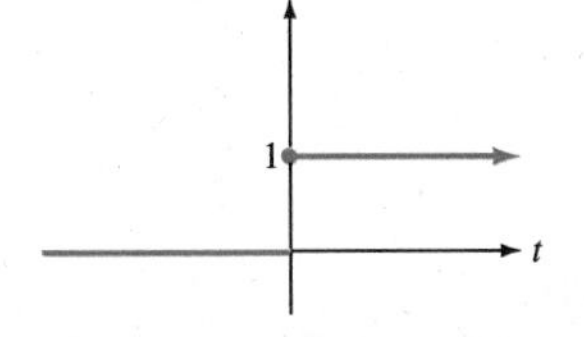

FIGURE 3.4.1

Graph of Heaviside function.

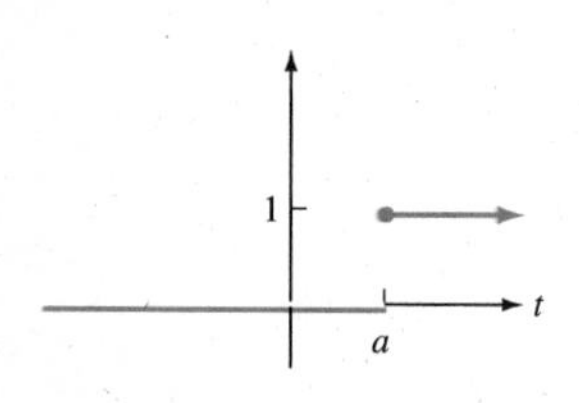

FIGURE 3.4.2

Graph of $H(t-a)$.

The notations $u(t)$, $u_0(t)$, or $1(t)$, are sometimes also used to denote the Heaviside function. Other notations are also used.

The graph of $H(t-a)$, $a > 0$ (Figure 3.4.2) is just a translate of that of $H(t)$. Thus, corresponding to turning one on at $t = a$,

$$H(t-a) = \begin{cases} 0 & \text{if } t < a, \\ 1 & \text{if } t \geq a. \end{cases}$$

EXAMPLE 3.4.1 **Graphing Expressions with Heaviside Functions**

Graph $f(t) = H(t-1) - H(t-3)$.

Solution Since the formula for $H(t-1)$ changes at $t = 1$, and the formula for $H(t-3)$ changes at $t = 3$, the formula for $f(t)$ will undergo changes at $t = 1$ and $t = 3$. Accordingly, we divide the t interval into the intervals $(-\infty, 1)$, $[1, 3)$, and $[3, \infty)$.

If $t < 1$, then $f(t) = H(t-1) - H(t-3) = 0 - 0$, since $H(t-1) = 0$ if $t < 1$ and $H(t-3) = 0$ if $t < 3$.

If $1 \leq t < 3$, then $f(t) = H(t-1) - H(t-3) = 1 - 0 = 1$, since $H(t-1) = 1$ if $t \geq 1$ and $H(t-3) = 0$ if $t < 3$.

If $3 \leq t$, then $f(t) = H(t-1) - H(t-3) = 1 - 1 = 0$, since $H(t-1) = 1$ if $t \geq 1$ and $H(t-3) = 1$ if $t \geq 3$.

Figure 3.4.3 gives the graph of $f(t) = H(t-1) - H(t-3)$. ◄

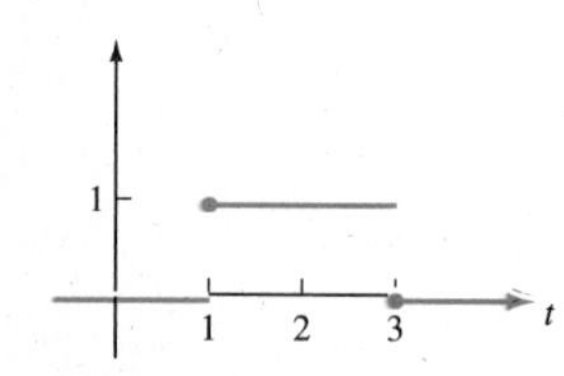

FIGURE 3.4.3

Graph of $H(t-1) - H(t-3)$.

From this example we make the following deduction.

Useful Observation If $0 < a < b$, then

$$H(t-a) - H(t-b) = \begin{cases} 0 & \text{if } t < a, \\ 1 & \text{if } a \leq t < b, \\ 0 & \text{if } b \leq t, \end{cases} \tag{1}$$

and $H(t-a) - H(t-b)$ has the graph in Figure 3.4.4.

Now let $g(t)$ be some function of t. Then if $a < b$,

$$g(t)[H(t-a) - H(t-b)] = \begin{cases} 0 & \text{if } t < a, \\ g(t) & \text{if } a \leq t < b, \\ 0 & \text{if } b \leq t. \end{cases}$$

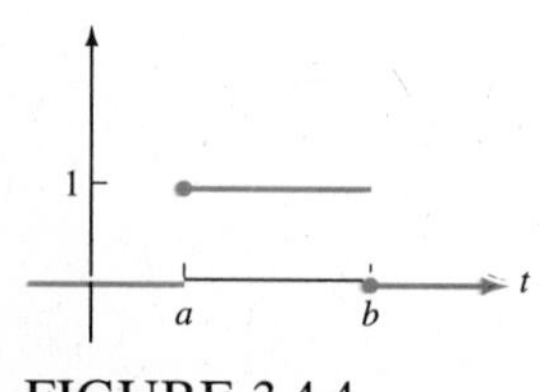

FIGURE 3.4.4

Graph of $H(t-a) - H(t-b)$.

The function $g(t)$ is turned on at $t = a$ and then turned off at $t = b$.

EXAMPLE 3.4.2 Graphing Expressions with Heaviside Functions

Let $f(t) = t^2[H(t-1) - H(t-2)]$. Then

$$t^2[H(t-1) - H(t-2)] = \begin{cases} 0 & \text{if } t < 1, \\ t^2 & \text{if } 1 \le t < 2, \\ 0 & \text{if } 2 \le t, \end{cases}$$

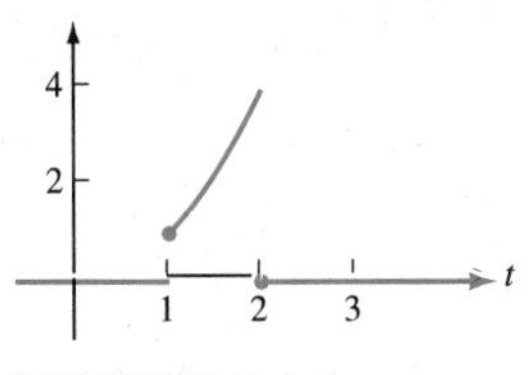

FIGURE 3.4.5

Graph of $t^2[H(t-1) - H(t-2)]$.

and the graph of $t^2[H(t-1) - H(t-2)]$ is given in Figure 3.4.5. The formula $t^2[H(t-1) - H(t-2)]$ picks out that part of the graph of t^2 between 1 and 2, and sets the rest equal to zero. ◄

We are now ready to write a piecewise continuous function in terms of unit-step functions.

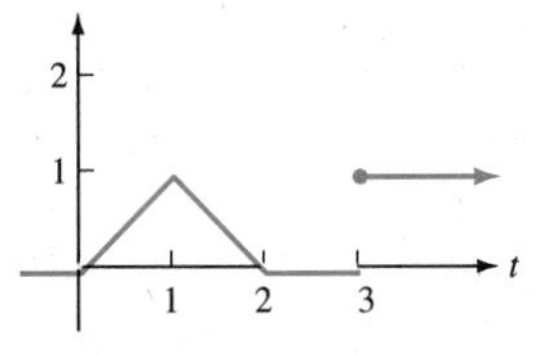

FIGURE 3.4.6

EXAMPLE 3.4.3 Writing in Terms of Heaviside Functions

Write in terms of unit-step functions the function $f(t)$ given by the graph in Figure 3.4.6.

Solution The function $f(t)$ is

$$f(t) = \begin{cases} t & \text{if } 0 \le t < 1, \\ 2 - t & \text{if } 1 \le t < 2, \\ 0 & \text{if } 2 \le t < 3, \\ 1 & \text{if } 3 \le t. \end{cases}$$

Thus

$$f(t) = \underbrace{t[H(t) - H(t-1)]} + \underbrace{(2-t)[H(t-1) - H(t-2)]} + \underbrace{1H(t-3)}. \tag{2}$$

Note that the first term in (2) is zero except when $0 \le t < 1$, and then it takes on the value t. The second term in (2) is zero except when $1 \le t < 2$, and then it takes on the value $2 - t$. The third term in (2) is zero except when $3 \le t$, and then it takes on the value 1. ◄

Note Since we consider only $t \ge 0$, the first term in (2) could have been written $t[1 - H(t-1)]$.

EXAMPLE 3.4.4 Graphing Expressions with Heaviside Functions

Graph $f(t) = 2H(t) + tH(t-1) + (3-t)H(t-2) - 3H(t-4)$ for $t \ge 0$.

TABLE 3.4.1

t interval	$f(t) = 2H(t) + tH(t-1) + (3-t)H(t-2) - 3H(t-4)$
$0 \le t < 1$	$f(t) = 2\cdot 1 + t\cdot 0 + (3-t)\cdot 0 - 3\cdot 0 = 2$
$1 \le t < 2$	$f(t) = 2\cdot 1 + t\cdot 1 + (3-t)\cdot 0 - 3\cdot 0 = 2 + t$
$2 \le t < 4$	$f(t) = 2\cdot 1 + t\cdot 1 + (3-t)\cdot 1 - 3\cdot 0 = 5$
$4 \le t$	$f(t) = 2\cdot 1 + t\cdot 1 + (3-t)\cdot 1 - 3\cdot 1 = 2$

Solution Looking at the Heaviside functions, we see that the formula for $f(t)$ undergoes changes at $t = 1$, $t = 2$, and $t = 4$. Thus we break up the interval $[0, \infty)$ into four subintervals $[0, 1)$, $[1, 2)$, $[2, 4)$, $[4, \infty)$ and see what $f(t)$ looks like on each subinterval. Using the definition of the Heaviside function, we form Table 3.4.1. The graph of this function is given in Figure 3.4.7. ◄

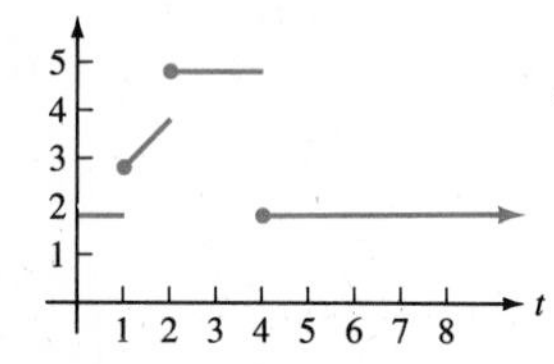

FIGURE 3.4.7

Laplace Transforms of Discontinuous Functions

We have seen that functions that have jump discontinuities can be represented in terms of the unit step function $H(t-c)$. Frequently, the Laplace transform can be calculated using (T15)

$$\mathscr{L}[f(t-c)H(t-c)] = e^{-cs}F(s) \quad \text{for} \quad c > 0. \qquad \text{(T15)} \tag{3}$$

The formula for the Laplace transform of a step function is a special case of (3). We let $f(t-c) = 1$, in which case $f(t) = 1$. Then $F(s) = 1/s$ so that (3) becomes

$$\mathscr{L}[H(t-c)] = \frac{e^{-cs}}{s} \quad \text{for } c > 0. \qquad \text{(T14)} \tag{4}$$

Equation (4) can be independently derived. Formula (3) is trickier to use in other cases, so several examples will be given after its verification.

Verification of (3) for c > 0

$$\begin{aligned}
\mathscr{L}[f(t-c)H(t-c)] &= \int_0^\infty e^{-st}f(t-c)H(t-c)\,dt \\
&= \int_0^c e^{-st}f(t-c)H(t-c)\,dt \\
&\quad + \int_c^\infty e^{-st}f(t-c)H(t-c)\,dt \\
&= \int_0^c e^{-st}f(t-c)\,0\,dt + \int_c^\infty e^{-st}f(t-c)\,1\,dt \\
&= \int_c^\infty e^{-st}f(t-c)\,dt.
\end{aligned}$$

We now make the change of variables $\tau = t - c$ so that $d\tau = dt$:

$$\mathscr{L}[f(t-c)H(t-c)] = \int_0^\infty e^{-s(\tau+c)} f(\tau)\, d\tau$$

$$= e^{-cs}\int_0^\infty e^{-s\tau} f(\tau)\, d\tau = e^{-cs}F(s).$$

Technical Point If $f(t)$ is a piecewise continuous function of exponential order, then $\mathscr{L}[f]$ is the same whether left- or right-hand limits are used for the values of f at the jump discontinuities. This is not really a problem, since whether the value of f at a jump discontinuity is a right- or left-hand limit is not a physically meaningful question. We defined $H(0) = 1$. We could just as well have defined $H(0) = 0$. Since, for our purposes, the value of the function at the jump discontinuity is not important, we will sometimes omit this information from the graphs.

Formula (T15) is tricky to use at first, so several examples will be given.

EXAMPLE 3.4.5 Laplace Transform of Expressions with Heaviside Functions

Compute $\mathscr{L}[t^2H(t-1)]$.

SOLUTION In order to use (3)(T15), we must take

$$t^2H(t-1) = f(t-c)H(t-c).$$

Thus $c = 1$ and $t^2 = f(t-1)$. To compute $e^{-cs}F(s)$ in (3), we need $F(s) = \mathscr{L}[f(t)]$. But we have $f(t-1)$, which is not $f(t)$. To compute $f(t)$, introduce a new variable, say τ, and let $\tau = t - 1$, or $t = \tau + 1$. Then,

$$t^2 = f(t-1),$$

$$(\tau+1)^2 = f(\tau),$$

or

$$\tau^2 + 2\tau + 1 = f(\tau). \tag{5}$$

Now in the formula (5), τ plays the role of a dummy (independent) variable, so that, from (T9),

$$F(s) = \mathscr{L}[f(\tau)] = \mathscr{L}[\tau^2 + 2\tau + 1]$$

$$= \frac{2}{s^3} + \frac{2}{s^2} + \frac{1}{s},$$

and (3) gives that

$$\mathscr{L}[t^2H(t-1)] = e^{-cs}F(s) = e^{-s}\left[\frac{2}{s^3} + \frac{2}{s^2} + \frac{1}{s}\right].$$

EXAMPLE 3.4.6 Laplace Transform of Expressions with Heaviside Functions

Compute $\mathscr{L}[(e^t + 1)H(t-2)]$ using (3).

Solution Taking

$$(e^t + 1)H(t-2) = f(t-2)H(t-2)$$

gives $c = 2$ and $f(t-2) = e^t + 1$. Let $\tau = t - 2$ or $\tau + 2 = t$. Then $f(\tau) = e^{\tau+2} + 1 = e^2 e^\tau + 1$. Thus

$$F(s) = \mathscr{L}[e^2 e^\tau + 1] = e^2 \frac{1}{s-1} + \frac{1}{s}$$

and from (3)

$$\mathscr{L}[(e^t + 1)H(t-2)] = e^{-cs}F(s) = e^{-2s}\left[\frac{e^2}{s-1} + \frac{1}{s}\right].$$ ◀

EXAMPLE 3.4.7 Inverse Laplace Transform Involving e^{-cs}

If $Y(s) = e^{-3s}/s^2$, find $y(t) = \mathscr{L}^{-1}[Y(s)]$.

Solution The presence of e^{-3s} in $Y(s)$ is an indication that we should use (3)(T15). Taking

$$\frac{e^{-3s}}{s^2} = e^{-cs}F(s),$$

we see that $c = 3$ and $F(s) = 1/s^2$. Thus $f(t) = \mathscr{L}^{-1}[F(s)] = t$. The result for $y(t)$ is then

$$y(t) = \mathscr{L}^{-1}[e^{-cs}F(s)] = f(t-c)H(t-c) = (t-3)H(t-3).$$ ◀

EXAMPLE 3.4.8 Inverse Laplace Transform Involving e^{-cs}

If $Y(s) = e^{-2s}\left[\frac{3}{s} + \frac{s}{s^2+4}\right]$, find $y(t) = \mathscr{L}^{-1}[Y(s)]$.

Solution Again the e^{-2s} suggests that we should use formula (T15). Taking

$$e^{-2s}\left[\frac{3}{s} + \frac{s}{s^2+4}\right] = e^{-cs}F(s),$$

we see that $c = 2$ and $F(s) = \dfrac{3}{s} + \dfrac{s}{s^2 + 4}$. Thus $f(t) = 3 + \cos 2t$. Hence,

$$y(t) = \mathscr{L}^{-1}[e^{-cs}F(s)] = f(t-2)H(t-2) = [3 + \cos[2(t-2)]]H(t-2)$$

$$= [3 + \cos(2t-4)]H(t-2).$$

3.4.1 Solution of Differential Equations

We can now use the Laplace transform to solve linear constant coefficient differential equations with piecewise continuous forcing functions of exponential order.

EXAMPLE 3.4.9 **Piecewise Continuous Forcing**

Solve the differential equation

$$y'' + 3y' + 2y = g(t), \qquad y(0) = 0, \qquad y'(0) = 1, \tag{6}$$

using the Laplace transform where $g(t)$ has the graph in Figure 3.4.8.

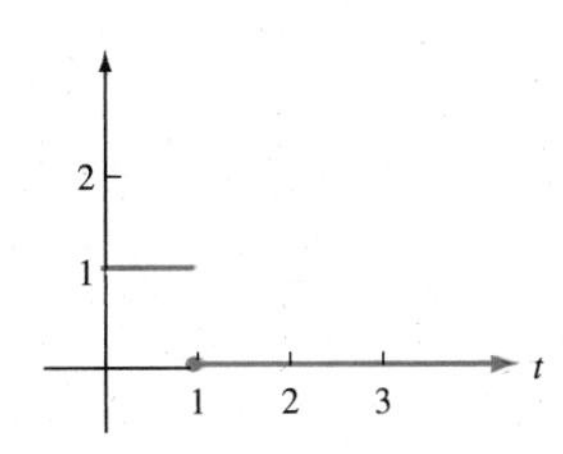

FIGURE 3.4.8

SOLUTION The function $g(t)$ is given by

$$g(t) = \begin{cases} 1 & \text{if } 0 \le t < 1, \\ 0 & \text{if } 1 \le t. \end{cases} \tag{7}$$

Thus $g(t) = [1 - H(t-1)]$ [or equivalently, $H(t) - H(t-1)$], and the differential equation is

$$y'' + 3y' + 2y = 1 - H(t-1), \qquad y(0) = 0, \qquad y'(0) = 1.$$

Taking the Laplace transform of both sides of this differential equation gives

$$s^2Y(s) - sy(0) - y'(0) + 3[sY(s) - y(0)] + 2Y(s) = \frac{1}{s} - \frac{e^{-s}}{s},$$

so that

$$Y(s) = \left[\frac{s+1}{s(s^2 + 3s + 2)}\right] - e^{-s}\left[\frac{1}{s(s^2 + 3s + 2)}\right]. \tag{8}$$

First, we shall find the inverse Laplace transform of the first term in $Y(s)$. Note that

$$\frac{s+1}{s(s^2 + 3s + 2)} = \frac{s+1}{s(s+1)(s+2)} = \frac{1}{s(s+2)}.$$

By partial fractions,

$$\frac{1}{s(s+2)} = \frac{A}{s} + \frac{B}{s+2}$$

so that

$$1 = A(s+2) + Bs.$$

Evaluating at $s = 0$ and $s = -2$ yields $A = \frac{1}{2}$, $B = -\frac{1}{2}$. Thus

$$\mathscr{L}^{-1}\left[\frac{1}{s(s+2)}\right] = \mathscr{L}^{-1}\left[\frac{1/2}{s} + \frac{-1/2}{s+2}\right] = \frac{1}{2} - \frac{1}{2}e^{-2t}. \tag{9}$$

Now the second term of (8) is in the form $e^{-cs}F(s)$, where $c = 1$ and

$$F(s) = \frac{1}{s(s^2+3s+2)} = \frac{1}{s(s+2)(s+1)},$$

so that (3), or (T15), should be used. Again using partial fractions,

$$F(s) = \frac{1}{s(s+2)(s+1)} = \frac{A}{s} + \frac{B}{s+2} + \frac{C}{s+1},$$

so that

$$1 = A(s+2)(s+1) + Bs(s+1) + Cs(s+2).$$

Since the denominator has distinct linear factors, evaluating at their roots $s = 0$, $s = -1$, $s = -2$ yields $A = \frac{1}{2}$, $C = -1$, and $B = \frac{1}{2}$. Hence,

$$F(s) = \frac{1/2}{s} + \frac{1/2}{s+2} + \frac{-1}{s+1}$$

and

$$f(t) = \frac{1}{2} + \frac{1}{2}e^{-2t} - e^{-t}.$$

Thus,

$$\begin{aligned}\mathscr{L}^{-1}[e^{-s}F(s)] &= f(t-1)H(t-1)\\ &= \left[\frac{1}{2} + \frac{1}{2}e^{-2(t-1)} - e^{-(t-1)}\right]H(t-1). \end{aligned} \tag{10}$$

Finally, combining the two terms (9), (10),

$$y(t) = \left[\frac{1}{2} - \frac{1}{2}e^{-2t}\right] - \left[\frac{1}{2} + \frac{1}{2}e^{-2(t-1)} - e^{-(t-1)}\right]H(t-1). \tag{11}$$ ◄

The approach of this section should be contrasted with the examples in Section 1.7.3, where we had to match initial conditions with terminal values at every discontinuity of the "forcing function" $g(t)$.

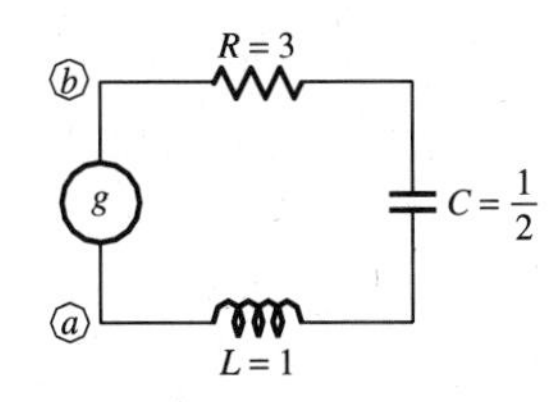

FIGURE 3.4.9

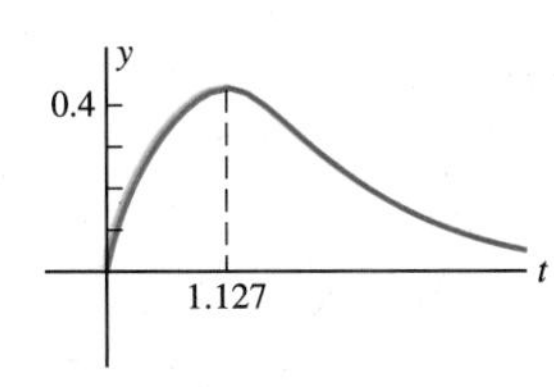

FIGURE 3.4.10

Graph of the solution (11) of (b).

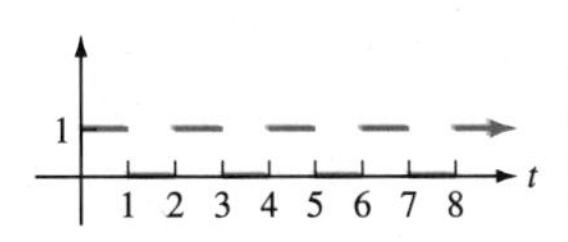

FIGURE 3.4.11

A Physical Interpretation As explained in Section 2.12, under certain operating conditions, equation (6) of the preceding exercise is a mathematical model for the *RLC* circuit in Figure 3.4.9.

In Figure 3.4.9 g is a voltage source that varies with time, $y(0) = 0$ is the initial charge in the capacitor, and $y'(0) = 1$ is the initial current in the capacitor. There are several ways to interpret $g(t)$ as given by (7). One is that it is the voltage across nodes ⓐ and ⓑ in Figure 3.4.9, and these nodes are connected to other circuit elements not shown. Alternatively, g can be thought of as a 1-V battery that is shorted out for $t > 1$. Intuitively, one expects that, because of the relatively large resistance (the circuit is overdamped, in the terminology of Section 2.8), the capacitor will charge for $0 \leq t < 1$ and discharge for $t \geq 1$. Figure 3.4.10 is the graph of $y(t)$. Note that, in fact, the capacitor keeps charging for 1.127 time units before beginning to discharge. That is because the inductor gives the current charging the capacitor the electrical equivalent of momentum. It takes a while for the current charging the capacitor to stop and reverse direction.

EXAMPLE 3.4.10 **Periodic Forcing**

Solve the differential equation

$$y'' + 3y' + 2y = g(t), \qquad y(0) = 0, \qquad y'(0) = 1, \tag{12}$$

using the Laplace transform, where $g(t)$ has the graph shown in Figure 3.4.11.

This is the same as the previous example, except that $g(t)$ is now a periodic function. In the next section, we shall discuss periodic functions in more detail, since they are very important.

SOLUTION First note that

$$\begin{aligned} g(t) &= [1 - H(t-1)] + [H(t-2) - H(t-3)] \\ &\quad + [H(t-4) - H(t-5)] + \cdots \\ &= 1 - H(t-1) + H(t-2) - H(t-3) + H(t-4) - H(t-5) + \cdots \\ &= \sum_{n=0}^{\infty} (-1)^n H(t-n). \end{aligned}$$

While $g(t)$ is given by an infinite series, for any specific value of t only a finite number of terms in the series are nonzero. For example, if $t = 13.7$, then $H(13.7 - n)$ is zero if $n > 13.7$. The differential equation is now written as

$$y'' + 3y' + 2y = \sum_{n=0}^{\infty} (-1)^n H(t-n). \tag{13}$$

It can be shown that it makes sense to take the Laplace transform of the series in (13), and this Laplace transform can be done term by term. Taking the Laplace transform of both sides of (13) gives

$$s^2Y(s) - 1 + 3sY(s) + 2Y(s) = \sum_{n=0}^{\infty} (-1)^n \frac{e^{-ns}}{s}.$$

Solving for $Y(s)$,

$$\begin{aligned} Y(s) &= \frac{1}{s^2+3s+2} + \sum_{n=0}^{\infty} (-1)^n e^{-ns}\left[\frac{1}{s(s^2+3s+2)}\right] \\ &= \frac{1}{s+1} + \frac{-1}{s+2} + \sum_{n=0}^{\infty} (-1)^n e^{-ns}\left[\frac{1}{s(s^2+3s+2)}\right]. \end{aligned} \tag{14}$$

From the preceding example, we know that

$$\mathscr{L}^{-1}\left[e^{-ns}\frac{1}{s(s^2+3s+2)}\right] = f(t-n)H(t-n)$$

and

$$f(t) = \mathscr{L}^{-1}\left[\frac{1}{s(s^2+3s+2)}\right] = \frac{1}{2} + \frac{1}{2}e^{-2t} - e^{-t}.$$

Thus (14) gives that

$$y(t) = e^{-t} - e^{-2t} + \sum_{n=0}^{\infty} (-1)^n \left[\frac{1}{2} + \frac{1}{2}e^{-2(t-n)} - e^{-(t-n)}\right] H(t-n). \tag{15}$$

Again, for any specific value of t, only a finite number of terms in the series in (15) are nonzero. Figure 3.4.12 is the graph of (15). ◀

FIGURE 3.4.12

Graph of (15).

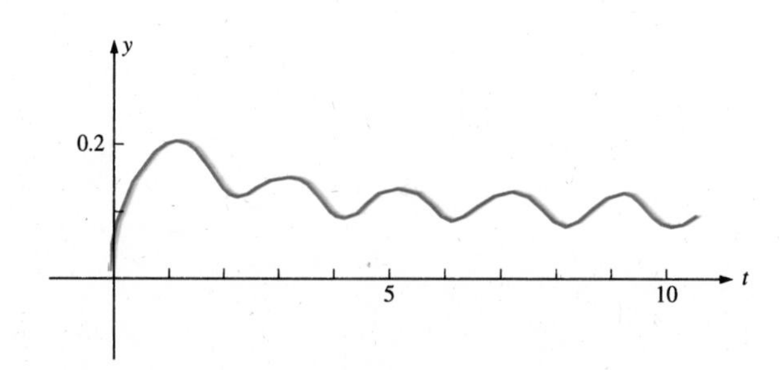

Exercises

In Exercises 1 through 3, write the function defined on $[0, \infty)$ in terms of unit-step functions and graph.

1. $f(t) = \begin{cases} 0 & 0 \le t < 2, \\ 3 & 2 \le t < 5, \\ t & 5 \le t. \end{cases}$

2. $f(t) = \begin{cases} 1 & 0 \le t < 1, \\ -1 & 1 \le t < 2, \\ 1 & 2 \le t < 3, \\ -1 & 3 \le t. \end{cases}$

3. $f(t) = \begin{cases} \sin t & 0 \le t < \pi, \\ 0 & \pi \le t < 2\pi, \\ \sin t & 2\pi \le t < 3\pi, \\ 0 & 3\pi \le t. \end{cases}$

In Exercises 4 through 9, write the function defined on $[0,\infty)$ in terms of unit-step functions.

4.

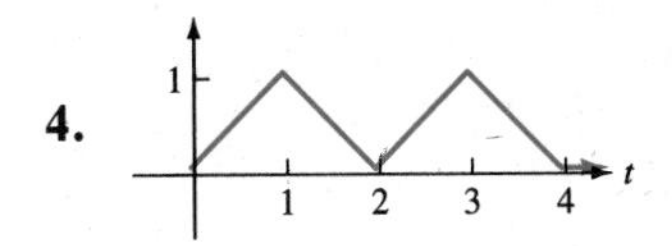

5.

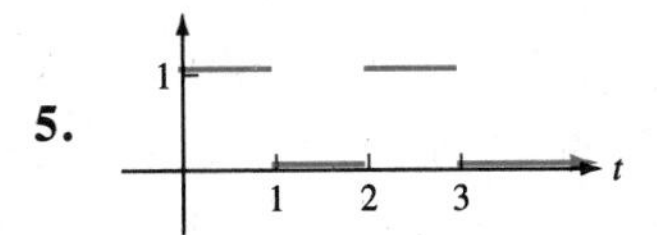

6.

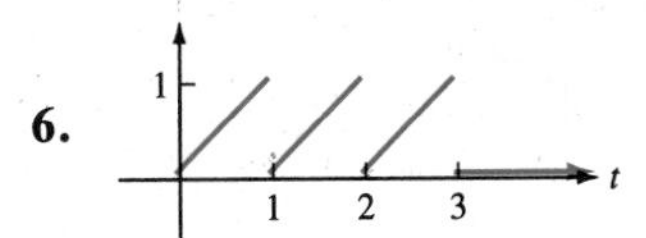

7.

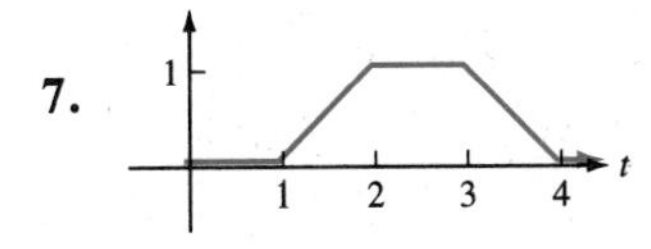

8.

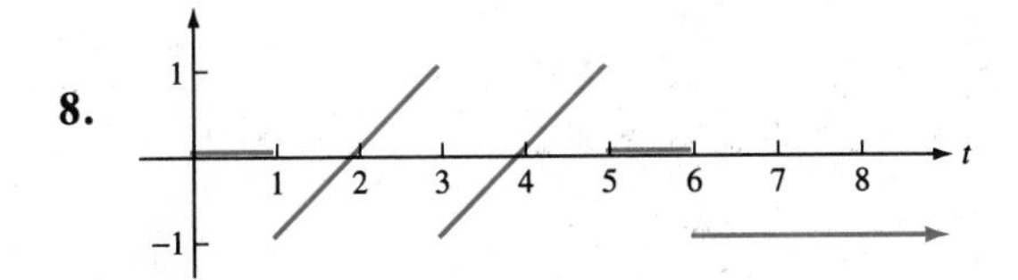

9.

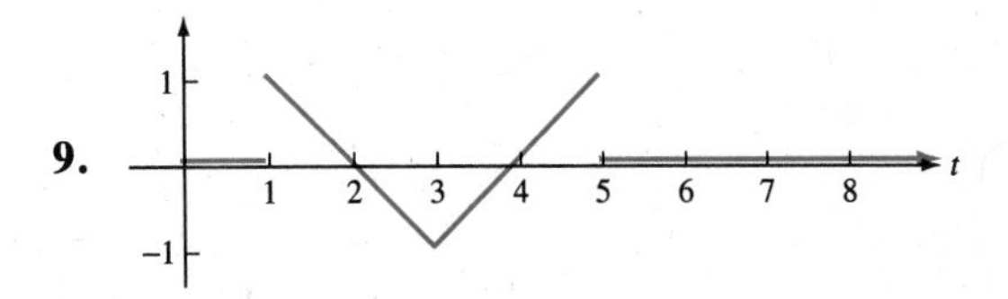

In Exercises 10 through 21, compute the Laplace transform $Y(s)$ of $y(t)$.

10. $y(t) = 2 - 5H(t-1) + 6H(t-3)$

11. $y(t) = tH(t-2)$

12. $y(t) = (t-2)H(t-2)$

13. $y(t) = (t^3 + 1)H(t-1)$

14. $y(t) = e^{3t}H(t-2) + 6H(t-3)$

15. $y(t) = \sin t\, H(t-\pi)$

16. $y(t) = \sin t\, H(t-\pi/2)$

17. $y(t) = \cos t\, H(t-2\pi)$

18. $y(t) = (t^3 + t)H(t-2)$

19. $y(t) = e^{2t}H(t-3)$

20. $y(t) = t^2e^{-3t}H(t-1)$

21. $y(t) = te^{5t}H(t-2)$

In Exercises 22 through 29, compute the inverse Laplace transform $y(t)$ of $Y(s)$.

22. $Y(s) = \dfrac{e^{-3s}}{s^3}$

23. $Y(s) = e^{-2s}\dfrac{1}{s^2+s} + e^{-3s}\dfrac{1}{s^2+s}$

24. $Y(s) = \dfrac{1}{s^2} + \dfrac{e^{-s}}{s^2+4} + e^{-2s}\dfrac{s}{s^2+9}$

25. $Y(s) = \dfrac{e^{-3s}}{s^2+2s+2}$

26. $Y(s) = \dfrac{e^{-2s}}{(s+4)^5}$

27. $Y(s) = \dfrac{e^{-s}}{s^2+1} - \dfrac{e^{-2s}}{s^2+4}$

28. $Y(s) = e^{-3s}\left[\dfrac{1}{s} + \dfrac{1}{s-2}\right]$

29. $Y(s) = e^{-s}\left[\dfrac{4}{s} + \dfrac{6s}{s^2+9}\right]$

30. Using only the definitions of the Laplace transform and the Heaviside function, verify that

$$\mathscr{L}[f(t-c)H(t-c)] = e^{-cs}F(s)$$

if $f(t)$ is piecewise-continuous of exponential order and $\mathscr{L}[f(t)] = F(s)$.

In Exercises 31 through 48, solve the differential equation using the Laplace transform.

31. $y'' + y = g(t), \quad y(0) = y'(0) = 0,$

where $g(t) = \begin{cases} 0, & 0 \le t < 2, \\ 1, & 2 \le t < 5, \\ 0, & 5 \le t. \end{cases}$

32. $y'' - y = g(t), \quad y(0) = y'(0) = 0$

where $g(t) = \begin{cases} 0, & 0 \le t < 1, \\ 1, & 1 \le t < 3, \\ 2, & 3 \le t. \end{cases}$

33. $y' - 3y = g(t)$, $y(0) = 1$, $g(t)$ given by

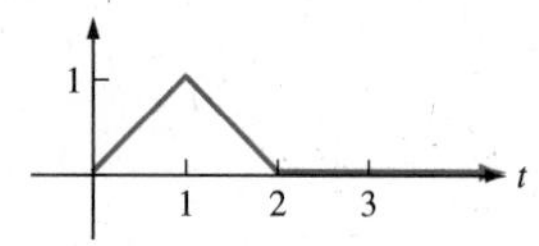

34. $y' + y = g(t)$, $y(0) = 0$, $g(t)$ given by

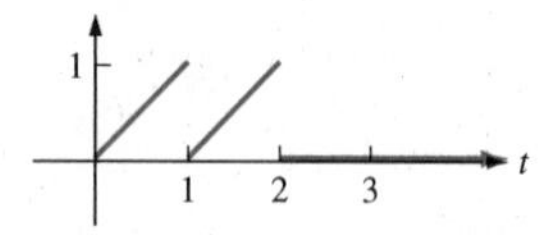

35. $y'' + 2y' + 10y = g(t)$, $y(0) = 1$, $y'(0) = 0$,

where $g(t) = \begin{cases} 1, & 0 \le t < 3, \\ 0, & 3 \le t. \end{cases}$

36. $y'' + 5y' + 6y = g(t)$, $y(0) = y'(0) = 0$,

$g(t) = \begin{cases} 0, & 0 \le t < \pi/2, \\ \cos t, & \pi/2 \le t < 3\pi/2, \\ 0, & 3\pi/2 \le t. \end{cases}$

37. $y'' + 4y = g(t)$, $y(0) = 0$, $y'(0) = 0$, g given by the graph in Exercise 5.

38. $y'' - 2y' + y = g(t)$, $y(0) = 1$, $y'(0) = 1$,

where $g(t) = \begin{cases} t^2, & 0 \le t < 1, \\ 0, & 1 \le t. \end{cases}$

39. $y'' + 9y = g(t)$, $y(0) = 0$, $y'(0) = 0$,

where $g(t) = \begin{cases} 1, & 0 \le t < 2, \\ e^{-t}, & t \ge 2. \end{cases}$

40. $y'' - 9y = g(t)$, $y(0) = 0$, $y'(0) = 0$,

where $g(t) = \begin{cases} e^{-t}, & 0 \le t < 4, \\ 1, & t \ge 4. \end{cases}$

41. $y' - 5y = g(t)$, $y(0) = 0$,

where $g(t) = \begin{cases} e^{3t}, & 0 \le t < 4, \\ 0, & t \ge 4. \end{cases}$

42. $y' + 5y = g(t)$, $y(0) = 0$,

where $g(t) = \begin{cases} 0, & 0 \le t < 4, \\ e^{3t}, & t \ge 4. \end{cases}$

43. $y' + 3y = g(t)$, $y(0) = 0$,

where $g(t) = \begin{cases} 0, & 0 \le t < \dfrac{\pi}{2}, \\ \cos t, & t \ge \dfrac{\pi}{2}. \end{cases}$

44. $y' - 4y = g(t)$, $y(0) = 0$,

where $g(t) = \begin{cases} \cos t, & 0 \le t < \dfrac{\pi}{2}, \\ 0, & t \ge \dfrac{\pi}{2}. \end{cases}$

45. $y' + 2y = g(t)$, $y(0) = 0$,

where $g(t) = \begin{cases} 0, & 0 \le t < 3, \\ \sin t, & t \ge 3. \end{cases}$

46. $y' + 4y = g(t)$, $y(0) = 0$,

where $g(t) = \begin{cases} \sin t, & 0 \le t < 5, \\ 0, & t \ge 5. \end{cases}$

47. $y'' + 9y = g(t)$, $y(0) = 0$, $y'(0) = 1$,

where $g(t) = \begin{cases} \sin t, & 0 \le t < 4, \\ 0, & t \ge 4. \end{cases}$

48. $y'' + 4y = g(t)$, $y(0) = 1$, $y'(0) = 0$,

where $g(t) = \begin{cases} 0, & 0 \le t < 2, \\ \cos t, & t \ge 2. \end{cases}$

In Exercises 49 and 50, use the method given in Example 3.4.10 to solve the differential equation using Laplace transforms.

49. $y'' + 4y = g(t)$, $y(0) = 0$, $y'(0) = 0$, where $g(t)$ is periodic and given by:

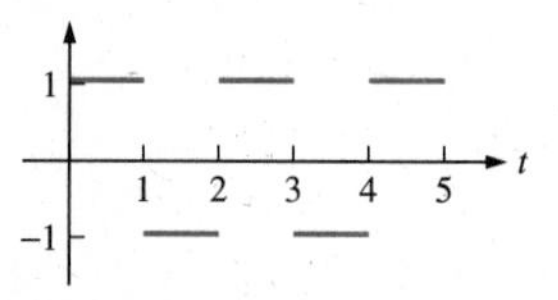

50. $y' + 2y = g(t)$, $y(0) = 1$, where $g(t)$ is periodic and given by:

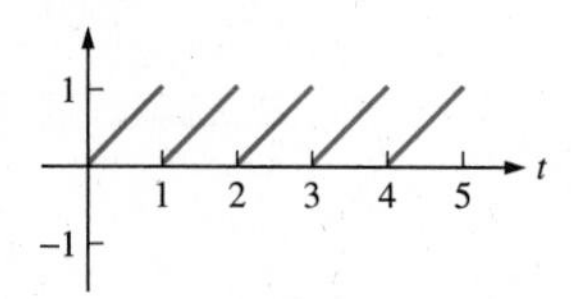

3.5 Periodic Functions

A function $g(t)$ defined on $[0, \infty)$ is **periodic with period** T if $g(t + T) = g(t)$ for all $t > 0$. Periodic functions appear in many applications. We have already seen examples of periodic functions and the Laplace transform. This section will consider them in more detail.

The basic formula is

If $g(t)$ is periodic on $[0, \infty)$ with period T and has a Laplace transform, then:

$$\mathscr{L}[g(t)] = \frac{\int_0^T e^{-st} g(t)\, dt}{1 - e^{-sT}}. \tag{1}$$

Verification Suppose $g(t)$ is periodic with period T and has a Laplace transform. Then

$$\mathscr{L}[g(t)] = \int_0^\infty e^{-st} g(t)\, dt = \int_0^T e^{-st} g(t)\, dt + \int_T^\infty e^{-st} g(t)\, dt$$

(Let $t = T + \tau$ in the second integral.)

$$= \int_0^T e^{-st} g(t)\, dt + \int_0^\infty e^{-s(T+\tau)} g(T + \tau)\, d\tau$$

$$= \int_0^T e^{-st} g(t)\, dt + e^{-sT} \int_0^\infty e^{-s\tau} g(\tau)\, d\tau,$$

[since $g(t)$ has period T]. Thus

$$\mathscr{L}[g(t)] = \int_0^T e^{-st} g(t)\, dt + e^{-sT} \mathscr{L}[g(t)].$$

Solving this equation for $\mathscr{L}[g(t)]$ gives (1).

EXAMPLE 3.5.1 Laplace Transform of Periodic Function

Find $\mathscr{L}[g(t)]$, where

$$g(t) = \begin{cases} 1 & \text{if } 0 \le t < 1, \\ 0 & \text{if } 1 \le t < 2; \end{cases} \qquad \text{period is 2.}$$

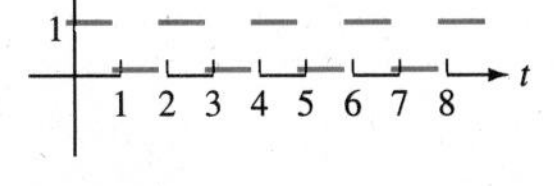

FIGURE 3.5.1

Figure 3.5.1 gives the graph of $g(t)$.

Solution Since the period is 2, formula (1) gives

$$\begin{aligned}\mathscr{L}[g(t)] &= \frac{1}{1-e^{-2s}}\int_0^2 e^{-st}g(t)\,dt \\ &= \frac{1}{1-e^{-2s}}\int_0^1 e^{-st}\,dt \\ &= \frac{1}{1-e^{-2s}}\left(\frac{e^{-st}}{-s}\bigg|_{t=0}^{t=1}\right) \\ &= \frac{1}{1-e^{-2s}}\left(\frac{e^{-s}}{-s}+\frac{1}{s}\right);\end{aligned}$$

which can be simplified to

$$\frac{1}{s(1+e^{-s})}.$$

The most difficult part of (1) is computing the integral $\int_0^T e^{-st}g(t)\,dt$. This can be done using Laplace transform tables, as follows. If

$$\hat{g}(t) = \begin{cases} g(t) & \text{for } 0 \le t \le T, \\ 0 & \text{for } t > T, \end{cases}$$

then

$$\int_0^T e^{-st}g(t)\,dt = \mathscr{L}[\hat{g}(t)], \tag{2}$$

since

$$\begin{aligned}\int_0^T e^{-st}g(t)\,dt &= \int_0^T e^{-st}\hat{g}(t)\,dt \\ &= \int_0^\infty e^{-st}\hat{g}(t)\,dt = \mathscr{L}[\hat{g}(t)].\end{aligned}$$

Note that $\hat{g}(t) = g(t)[1 - H(t - T)]$. This approach is (T31).

EXAMPLE 3.5.2 **Laplace Transform of Periodic Function**

Let $g(t)$ have period 2, where

$$g(t) = \begin{cases} t & \text{for } 0 \le t \le 1, \\ 2 - t & \text{for } 1 \le t \le 2. \end{cases}$$

Find $G(s)$.

FIGURE 3.5.2

g from Example 3.5.2
$\hat{g}$ from Example 3.5.2

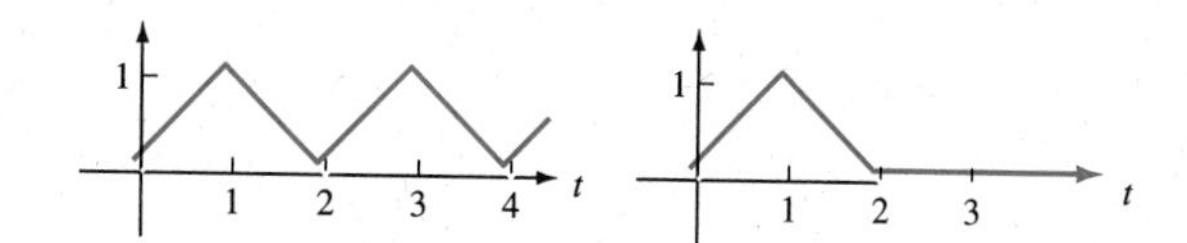

SOLUTION Let

$$\begin{aligned}\hat{g}(t) &= t[1 - H(t-1)] + (2-t)[H(t-1) - H(t-2)] \\ &= t + 2(1-t)H(t-1) + (t-2)H(t-2).\end{aligned}$$

Then $\hat{g}(t) = g(t)$ for $0 \le t \le 2$ and $\hat{g}(t) = 0$ for $t > 2$. Graphs of g and $\hat{g}$ are given in Figure 3.5.2. Thus by (1), (2),

$$\begin{aligned}\mathscr{L}[g(t)] &= \frac{1}{1-e^{-2s}}\int_0^2 e^{-st}g(t)\,dt \\ &= \frac{1}{1-e^{-2s}}\mathscr{L}[\hat{g}(t)] \\ &= \frac{1}{1-e^{-2s}}\mathscr{L}[t + 2(1-t)H(t-1) + (t-2)H(t-2)] \\ &= \frac{1}{1-e^{-2s}}\left[\frac{1}{s^2} - 2\frac{e^{-s}}{s^2} + \frac{e^{-2s}}{s^2}\right].\end{aligned}$$

Inverse Laplace transforms are not usually computed using (1). One method to compute inverse Laplace transformations is by the use of series. Recall that, for $|x| < 1$,

$$\frac{1}{1-x} = 1 + x + x^2 + \cdots = \sum_{n=0}^{\infty} x^n. \tag{3}$$

If $s > 0$, $T > 0$, then $e^{-sT} < 1$. Letting $x = e^{-sT}$, (3) becomes

$$\frac{1}{1-e^{-sT}} = 1 + e^{-sT} + e^{-2sT} + \cdots = \sum_{n=0}^{\infty} e^{-nsT}. \tag{4}$$

Thus, if $\mathscr{L}[f(t)] = F(s)$,

$$\frac{1}{1-e^{-sT}}F(s) = \sum_{n=0}^{\infty} e^{-nsT}F(s).$$

Taking the inverse Laplace transform term by term using (T15) with $c = nT$ gives

$$\mathscr{L}^{-1}\left[\frac{1}{1-e^{-sT}}F(s)\right] = \sum_{n=0}^{\infty} f(t-nT)H(t-nT). \qquad \text{(T32)} \qquad (5)$$

Similarly, letting $x = -e^{-sT}$ in (3) gives

$$\frac{1}{1+e^{-sT}} = \sum_{n=0}^{\infty} (-1)^n e^{-nsT},$$

and

$$\mathscr{L}^{-1}\left[\frac{1}{1+e^{-sT}}F(s)\right] = \sum_{n=0}^{\infty} (-1)^n f(t-nT)H(t-nT). \qquad \text{(T33)} \quad (6)$$

EXAMPLE 3.5.3 **Inverse Laplace Transform with $1 - e^{-sT}$ in the Denominator**

Let $G(s) = \dfrac{e^{-2s}-1}{s^2(1-e^{-3s})}$. Find $g(t) = \mathscr{L}^{-1}[G(s)]$.

Solution Using (4), we get

$$\begin{aligned} G(s) &= \frac{1}{s^2}(e^{-2s}-1)\sum_{n=0}^{\infty} e^{-3ns} \\ &= \sum_{n=0}^{\infty} \frac{1}{s^2}e^{-(2+3n)s} - \sum_{n=0}^{\infty} \frac{1}{s^2}e^{-3ns}. \end{aligned}$$

Now, by (T15), $\mathscr{L}^{-1}\left[\frac{1}{s^2}e^{-cs}\right] = (t-c)H(t-c)$. Thus,

$$g(t) = \sum_{n=0}^{\infty} (t-2-3n)H(t-2-3n) - \sum_{n=0}^{\infty} (t-3n)H(t-3n).$$

This example also illustrates that, if $G(s)$ has a $1/[1-e^{-sT}]$-factor, it is not necessarily the case that $g(t)$ is periodic. ◄

Alternative Solution We could also do Example 3.5.3 using formula (5) with $T = 3$ and $F(s) = -\dfrac{1}{s^2} + \dfrac{1}{s^2}e^{-2s}$. Then $f(t) = -t + (t-2)H(t-2)$, and

(5) is

$$\sum_{n=0}^{\infty} f(t-3n)H(t-3n)$$

$$= \sum_{n=0}^{\infty} [-(t-3n) + (t-3n-2)H(t-3n-2)]H(t-3n)$$

$$= \sum_{n=0}^{\infty} -(t-3n)H(t-3n) + (t-3n-2)H(t-3n-2)H(t-3n)$$

$$= \sum_{n=0}^{\infty} -(t-3n)H(t-3n) + (t-3n-2)H(t-3n-2).$$

The last equality follows, since

$$H(t-a)H(t-b) = H(t-c).$$

where c is the larger of a, b.

Exercises

In Exercises 1 through 8, sketch $g(t)$ and find $G(s)$ using formula (1), or (1) and (2).

1. $g(t) = |\sin t|$ (rectified sine)

2. $g(t) = |\cos t|$ (rectified cosine)

3. $g(t) = \begin{cases} t^2 & \text{for } 0 \le t < 1, \\ 2 - t^2 & \text{for } 1 \le t < 2, \end{cases}$ period 2.

4. $g(t) = t$ for $0 \le t < 1$, period 1.

5. $g(t) = e^t$ for $0 \le t < 1$, period 1.

6. $g(t) = t^3$ for $0 \le t < 1$, period 1.

7. $g(t) = \begin{cases} 1 & \text{for } 0 \le t < 1, \\ -1 & 1 \le t < 2, \end{cases}$ period 2.

8. $g(t) = \begin{cases} t & \text{for } 0 \le t < 1, \\ 1 & 1 \le t < 2, \\ -1 & \text{for } 2 \le t < 3, \end{cases}$ period 3.

In Exercises 9 through 20, find $g(t)$, given $G(s)$.

9. $G(s) = \dfrac{1}{1-e^{-s}}\left[\dfrac{1}{s^2} + \dfrac{1}{s^3}\right]$

10. $G(s) = \dfrac{1}{1+e^{-3s}}\left[\dfrac{2s}{s^2+6s+13}\right]$

11. $G(s) = \dfrac{1}{1-e^{-s}}\left[\dfrac{s}{s^2+4}\right]$

12. $G(s) = \dfrac{1}{s}\tanh\dfrac{s}{2}$

13. $G(s) = \dfrac{1}{1-e^{-2s}}\left[\dfrac{1}{s^2} + \dfrac{e^{-s}}{s^3}\right]$

14. $G(s) = \dfrac{1}{s}\operatorname{sech} s$

15. $G(s) = \dfrac{1}{1-e^{-\pi s}}\left[\dfrac{1}{s} + \dfrac{e^{-\pi s/2}}{s^2+1}\right]$

16. $G(s) = \dfrac{1}{1-e^{-3s}}\left[\dfrac{1}{s} + \dfrac{e^{-s}}{s^2} + \dfrac{e^{-2s}}{s^3}\right]$

17. $G(s) = \dfrac{1}{1+e^{-5s}}\left[\dfrac{1}{s^3} + \dfrac{e^{-2s}}{s^4}\right]$

18. $G(s) = \dfrac{1}{1+e^{-s}}\left[\dfrac{1}{s^2-1} + \dfrac{e^{-2s}}{s^2-4}\right]$

19. $G(s) = \dfrac{1}{s^2}\operatorname{csch} s$

20. $G(s) = \dfrac{1}{1+e^{-3s}}\left[\dfrac{1}{s+2} + \dfrac{e^{-2s}}{s-2}\right]$

21. Verify that $H(t-a)H(t-b) = H(t-c)$, where c is the larger of a, b.

In Exercises 22 through 28, solve the initial value problem, using the Laplace transform.

22. $y' + y = g(t)$, $y(0) = 0$, $g(t)$ from Exercise 1.

23. $y'' + y = |\sin 2t|$, $y(0) = 0$, $y'(0) = 0$

24. $y'' - y = \sin t - |\sin t|, \quad y(0) = 0, y'(0) = 1$

25. $y' + y = g(t)$, $y(0) = 1$, $g(t)$ as in Exercise 5.

26. $y' - 2y = g(t)$, $y(0) = 0$, $g(t)$ as in Exercise 4.

27. $y'' - 4y = g(t)$, $y(0) = ky'(0) = 0$, $g(t)$ as in Exercise 7.

28. $y'' + 2y' + y = g(t), \quad y(0) = y'(0) = -1$, $g(t)$ as in Exercise 8.

29. Suppose that the series $\Sigma_{n=0}^{\infty} t^n/n!$ for e^t can be Laplace-transformed term by term. Show that the resulting series gives $\mathscr{L}[e^t]$, which is $1/(s-1)$.

3.6 Integrals and the Convolution Theorem

We have seen that, intuitively, the Laplace transform of the derivative is obtained by multiplying by s since

$$\mathscr{L}[g'(t)] = sG(s) - g(0). \qquad \text{(T18)} \tag{1}$$

Thus, it should not be a surprise that the Laplace transform of an integral is obtained by dividing by s. To be precise,

$$\mathscr{L}\left[\int_0^t f(\tau)\,d\tau\right] = \frac{F(s)}{s}. \qquad \text{(T23)} \tag{2}$$

Derivation of (2) We use (1) with $g(t) = \int_0^t f(\tau)\,d\tau$, so that $g'(t) = f(t)$, $g(0) = 0$:

$$\mathscr{L}[g'(t)] = F(s) = sG(s) - g(0) = s\mathscr{L}\left[\int_0^t f(\tau)\,d\tau\right], \tag{3}$$

since $g(0) = 0$. Equation (2) follows from (3) by algebraically solving for $\mathscr{L}[\int_0^t f(\tau)\,d\tau]$.

EXAMPLE 3.6.1 **Inverse Laplace Transform Using (T23)**

Although it is better to use partial fractions, calculate $\mathscr{L}^{-1}\left[\dfrac{3}{s(s+5)}\right]$ using (2).

SOLUTION We let $F(s) = \dfrac{3}{s+5}$, so that $f(t) = 3e^{-5t}$. According to (2)

$$\mathscr{L}^{-1}\left[\frac{F(s)}{s}\right] = \mathscr{L}^{-1}\left[\frac{3}{s(s+5)}\right] = \int_0^t f(\tau)\,d\tau$$

$$= \int_0^t 3e^{-5\tau}\,d\tau = 3\left.\frac{e^{-5\tau}}{-5}\right|_0^t = \frac{3}{5}(1 - e^{-5t}).$$

Finding the inverse Laplace transform of a product of transforms is a question that arises frequently. The following theorem is quite useful. The integral theorem (2) for Laplace transforms is a special case of this theorem.

■ **THEOREM 3.6.1 Convolution**

Suppose that $f(t)$ and $g(t)$ are functions with Laplace transforms $\mathscr{L}[f(t)] = F(s)$ and $\mathscr{L}[g(t)] = G(s)$. Then

$$\mathscr{L}\left[\int_0^t f(\tau)g(t-\tau)\,d\tau\right] = F(s)G(s) \qquad \text{(T24)} \tag{4}$$

and

$$\mathscr{L}^{-1}[F(s)G(s)] = \int_0^t f(\tau)g(t-\tau)\,d\tau. \tag{5}$$

■

The inverse of a product of transforms equals $\int_0^t f(\tau)g(t-\tau)\,d\tau$, which is a new function of t called the **convolution of f and g**. A convenient notation for the convolution of f and g is $f * g$,

$$f * g = \int_0^t f(\tau)g(t-\tau)\,d\tau. \tag{6}$$

Fortunately it does not matter whether we have the convolution of f and g or the convolution of g and f, since they are the same:

$$f * g = g * f. \tag{7}$$

Verification of (7) By the definition of the convolution of f and g (6),

$$f * g = \int_0^t f(\tau)g(t-\tau)\,d\tau. \tag{8}$$

We make a change of variables (t fixed) in the integral in (8), $s = t - \tau$. Then $ds = -d\tau$ and (8) becomes

$$f * g = \int_t^0 f(t-s)g(s)(-ds) = \int_0^t f(t-s)g(s)\,ds = g * f.$$

Thus in the key convolution theorems (4) and (5) it does not matter whether $\int_0^t f(\tau)g(t-\tau)\,d\tau$ or $\int_0^t g(\tau)f(t-\tau)\,d\tau$ is used, since they are equal.

EXAMPLE 3.6.2 **Convolution**

Suppose $f(t) = e^t$ and $g(t) = e^{4t}$. Compute $f * g$.

Solution Using (6), the convolution $f * g$ is just another function of t:

$$\begin{aligned} e^t * e^{4t} &= \int_0^t e^{\tau} e^{4(t-\tau)}\,d\tau \\ &= e^{4t}\int_0^t e^{-3\tau}\,d\tau \\ &= e^{4t}\frac{e^{-3\tau}}{-3}\bigg|_{\tau=0}^{\tau=t} = \frac{1}{3}e^{4t}(1-e^{-3t}) = \frac{1}{3}(e^{4t}-e^t). \end{aligned}$$

EXAMPLE 3.6.3 **Inverse Laplace Transform Using the Convolution**

Use (5) to evaluate $\mathscr{L}^{-1}\left[\dfrac{1}{(s+1)(s+2)}\right]$.

Solution Let $F(s) = \dfrac{1}{s+1}$, $G(s) = \dfrac{1}{s+2}$. Then

$$f(t) = \mathscr{L}^{-1}[F(s)] = \mathscr{L}^{-1}\left[\frac{1}{s+1}\right] = e^{-t},$$

$$g(t) = \mathscr{L}^{-1}\left[\frac{1}{s+2}\right] = e^{-2t}.$$

Formula (5) now gives

$$\begin{aligned} \mathscr{L}^{-1}\left[\frac{1}{(s+1)(s+2)}\right] &= e^{-t} * e^{-2t} = \int_0^t e^{-\tau}e^{-2(t-\tau)}\,d\tau \\ &= e^{-2t}\int_0^t e^{\tau}\,d\tau = e^{-2t}(e^t-1) = e^{-t}-e^{-2t}. \end{aligned}$$

Of course, this example could also have been done using a partial fractions expansion.

Application of the Convolution Theorem to Differential Equations

Consider the initial value problem

$$y'' + b^2 y = f(t), \tag{9}$$

with $y(0) = 0$ and $y'(0) = 0$.

Taking the Laplace transform of both sides of the differential equation yields

$$s^2Y(s) + b^2Y(s) = F(s),$$

since $y(0) = 0$ and $y'(0) = 0$. Here $Y(s)$ is the Laplace transform of $y(t)$ and $F(s)$ is the Laplace transform of $f(t)$. Solving for the Laplace transform of the solution gives

$$Y(s) = \frac{F(s)}{s^2 + b^2}.$$

We note that we can represent $Y(s)$ as a product of transforms of known functions

$$Y(s) = F(s)G(s),$$

where

$$G(s) = \frac{1}{s^2 + b^2}.$$

We have introduced $G(s)$ whose inverse transform can be obtained easily (if necessary using the table)

$$g(t) = \frac{1}{b} \sin bt.$$

Since $Y(s)$ equals a product of transforms $F(s)G(s)$, $y(t)$ may be obtained using the convolution theorem (5)

$$y(t) = \int_0^t f(\tau)g(t - \tau)\, d\tau = \int_0^t f(\tau)\frac{1}{b} \sin b(t - \tau)\, d\tau. \tag{10}$$

The function $(1/b)\sin b(t - \tau)$ is an **influence function** expressing how much the input at τ influences the output at t. The relatively simple solution (10) of (9) was obtained using the convolution theorem for Laplace transforms.

Without using Laplace transforms, the usual method to solve (9) is the method of variation of parameters. To show how (10) could have been obtained using variation of parameters, we use the trigonometric addition formula $\sin(a - b) = \sin a \cos b - \cos a \sin b$, so that (10) becomes

$$y = v_1 \cos bt + v_2 \sin bt, \tag{11}$$

where $v_1 = -\int_0^t f(\tau)(1/b) \sin b\tau\, d\tau$ and $v_2 = \int_0^t f(\tau)(1/b) \cos b\tau\, d\tau$. Equation (11) is in the form assumed in the method for variation of parameters.

3.6.1 Derivation of the Convolution Theorem (Optional)

The convolution theorem (T24)

$$\mathscr{L}^{-1}[F(s)G(s)] = \int_0^t f(\tau)g(t - \tau)\, d\tau \tag{12}$$

is very important, but not easy to verify. We will be able to recognize the inverse transform of a product of transforms $F(s)G(s)$ if we consider different dummy variables of integration for the defining Laplace transforms, $F(s) = \int_0^\infty e^{-s\tau} f(\tau)\, d\tau$ and $G(s) = \int_0^\infty e^{-su} g(u)\, du$, so that

$$F(s)G(s) = \int_0^\infty e^{-s\tau} f(\tau)\, d\tau \int_0^\infty e^{-su} g(u)\, du$$

$$= \int_0^\infty \int_0^\infty e^{-s(u+\tau)} f(\tau) g(u)\, du\, d\tau. \tag{13}$$

We make a change of variables in the u integral, and let $t = u + \tau$, as motivated by the exponent in (13). We integrate over t (rather than u) for fixed τ, so that $dt = du$:

$$F(s)G(s) = \int_{\tau=0}^\infty \int_{t=\tau}^\infty e^{-st} f(\tau) g(t-\tau)\, dt\, d\tau. \tag{14}$$

In the Laplace transforms, the functions $f(t)$ and $g(t)$ are only defined for $t \geq 0$. If we define

$$g(Q) = 0 \qquad \text{for } Q < 0, \tag{15}$$

then we can extend the t integral in (14) to begin from $t = 0$ instead of $t = \tau$ since this only adds 0 to the integral. Thus,

$$F(s)G(s) = \int_{\tau=0}^\infty \int_{t=0}^\infty e^{-st} f(\tau) g(t-\tau)\, dt\, d\tau$$

$$= \int_{t=0}^\infty \int_{\tau=0}^\infty e^{-st} f(\tau) g(t-\tau)\, d\tau\, dt$$

$$= \int_{t=0}^\infty e^{-st} \left[\int_{\tau=0}^\infty f(\tau) g(t-\tau)\, d\tau \right] dt.$$

We now recognize that the function in brackets above must be the function whose Laplace transform is $F(s)G(s)$. Thus,

$$\mathscr{L}\left[\int_{\tau=0}^\infty f(\tau) g(t-\tau)\, d\tau \right] = F(s)G(s). \tag{16}$$

However, $g(Q)$ is zero [see (15)] for negative arguments, which occur for $\tau > t$. Thus, there are no contributions to the integral in (16) for $\tau > t$, so that

$$\mathscr{L}\left[\int_{\tau=0}^t f(\tau) g(t-\tau)\, d\tau \right] = F(s)G(s). \tag{17}$$

Equation (17) is the convolution theorem for Laplace transforms, and it is equivalent to (12).

Exercises

1. If $f(t) = t$, $g(t) = e^t$, find $f * g$ from the definition (1).
2. If $f(t) = e^t$, $g(t) = \cos t$, find $f * g$ from the definition (1).
3. Compute $1 * 1$.
4. Compute $1 * t$.
5. Using formula (5), compute
$$\mathscr{L}^{-1}\left[\frac{1}{s^2} \cdot \frac{1}{(s^2+1)}\right].$$
6. Using formula (5), compute
$$\mathscr{L}^{-1}\left[\frac{s}{(s^2+1)^2}\right].$$

In Exercises 7 through 16, solve the initial value problem using the convolution theorem of Laplace transforms. Here $f(t)$ is an unspecified function that has a Laplace transform.

7. $y'' - 5y' + 4y = f(t)$, $y(0) = 0$, $y'(0) = 0$
8. $y'' + 7y' + 10y = f(t)$, $y(0) = 0$, $y'(0) = 0$
9. $y'' - 9y = f(t)$, $y(0) = 0$, $y'(0) = 0$
10. $y'' - 16y = f(t)$, $y(0) = 0$, $y'(0) = 0$
11. $y'' + 4y = f(t)$, $y(0) = 0$, $y'(0) = 7$
12. $y'' + 25y = f(t)$, $y(0) = 3$, $y'(0) = 0$
13. $y' + 7y = f(t)$, $y(0) = 2$
14. $y' - 3y = f(t)$, $y(0) = -6$
15. $y'' - 2y' + 10y = f(t)$, $y(0) = 0$, $y'(0) = 0$
16. $y'' + 6y' + 25y = f(t)$, $y(0) = 0$, $y'(0) = 0$
17. $y'' + 4y' + 13y = f(t)$, $y(0) = 0$, $y'(0) = 0$

An example of an **integral equation** is

$$x(t) = \int_0^t h(t-\tau)x(\tau)\,d\tau + g(t)$$
$$= (h * x)(t) + g(t),$$

where h, g are known functions and $x(t)$ is an unknown function. This integral equation can be solved by taking the Laplace transform, solving for $X(s)$, and then computing $\mathscr{L}^{-1}[X(s)]$.

In Exercises 18 through 25, solve the given integral equation using the Laplace transform.

18. $x(t) = \frac{1}{2}\int_0^t x(s)\,ds + 1$
19. $x(t) = \int_0^t \cos(t-\tau)x(\tau)\,d\tau + \sin t$
20. $x(t) = \int_0^t (t-\tau)x(\tau)\,d\tau + 1$
21. $x(t) = \int_0^t e^{-t+\tau}x(\tau)\,d\tau + 2$
22. $6x(t) = \int_0^t (t-\tau)^3 x(\tau)\,d\tau + t$
23. $x(t) = \int_0^t 2\sin(2t-2\tau)x(\tau)\,d\tau + \sin t$
24. $x(t) = \int_0^t \cos(2t-2\tau)x(\tau)\,d\tau + e^{3t}$
25. $x(t) = -\int_0^t \sinh(t-\tau)x(\tau)\,d\tau + 3$
26. Suppose that f, g, h are piecewise continuous functions on $[0,\infty)$ and α, β are constants. Using the definition (1) of convolution, verify that

    a) $(\alpha f + \beta g) * h = \alpha(f * h) + \beta(g * h)$.

    b) $f * (\alpha g + \beta h) = \alpha(f * g) + \beta(f * h)$.

    Using (5), verify that

    c) $f * (g * h) = (f * g) * h$.
27. Suppose that f is piecewise continuous, of exponential order ($|f(t)| \le Me^{\alpha t}$). Show that, for $s > \alpha$,
$$|F(s)| \le \frac{M}{s-\alpha}.$$
28. Using Exercise 27 and the Laplace transform, show that there do not exist f, g both piecewise continuous, of exponential order, such that $f * g = 1$.

29. The following example illustrates one way in which integral equations are "nicer" than differential equations. Let

$$g(t) = \begin{cases} e^t & \text{for } 0 \le t < 1, \\ -1 + e^t + e^{t-1} & \text{for } 1 \le t. \end{cases}$$

i) Show that $g(t)$ is continuous for $t \ge 0$, but $g'(1)$ is not defined. Also show that $g(t)$ is a solution of $y' - y = H(t-1)$, $y(0) = 1$ for $t \ge 0$, $t \ne 1$.

ii) Show that $g(t)$ satisfies the integral equation $y(t) = \int_0^t [y(\tau) + H(\tau - 1)]\, d\tau + 1$ for *all* $t \ge 0$, including $t = 1$.

30. Suppose that $x(t)$ is a piecewise continuous function for $-\infty < t < \infty$. Think of t as being the location along the real axis. That is, t measures a position rather than time. Let $x(t)$ be a quantity that varies along the t-axis, such as light intensity or frequency. A measurement of $x(t)$ would actually be a (possibly weighted) average of x over an interval, say $[t-a, t+a]$, which contains t. The *unweighted average* over $[t-a, t+a]$ would be

$$\frac{1}{2a} \int_{-a}^{a} x(t + \tau)\, d\tau. \tag{18}$$

Let $(1 * x)(z)$ be the value of $(1 * x)$ evaluated at $t = z$. Show that (18) can be written as

$$\frac{1}{2a} [(1 * x)(t + a) - (1 * x)(t - a)].$$

3.7 Impulses and Distributions

Suppose that a capacitor having charge Q_0 for time $t \le t_0$ is suddenly partially discharged (think of the spark in a spark plug) to a lower charge Q_1 at time t_1. The capacitor then maintains the charge Q_1 for $t \ge t_1$. The amount of charge lost is $Q_1 - Q_0$. If this discharge happens quickly, then, for a very brief period of time, the current $i(t)$, which is the rate of change of the charge $q(t)$, must be very large. The faster the discharge, the higher the current gets. The relationship between change in charge and current is

$$\Delta Q = Q_1 - Q_0 = \int_{t_0}^{t_1} \frac{dq}{dt}\, dt = \int_{t_0}^{t_1} i(t)\, dt. \tag{1}$$

Typical graphs of $q(t)$ and $i(t)$ are given in Figure 3.7.1. If the time interval is shortened, we have the graphs shown in Figure 3.7.2.

Since $i(t)$ is zero outside of $[t_0, t_1]$, (1) may be replaced by

$$\Delta Q = \int_{-\infty}^{\infty} i(t)\, dt.$$

Similarly, one may think of a mass m being acted on by a large force for a brief period of time, leading to a change in momentum. In this case the momentum mv plays the role of the charge, and the instantaneous force $(mv)'$ plays the role of the current.

Suppose that we take the limit as the length of the time interval $[t_0, t_1]$ goes to zero by letting $t_1 \to t_0^+$. The charge $q(t)$ approaches the function (graphed in Figure 3.7.3):

$$\tilde{q}(t) = Q_0 - (Q_0 - Q_1)H(t - t_0).$$

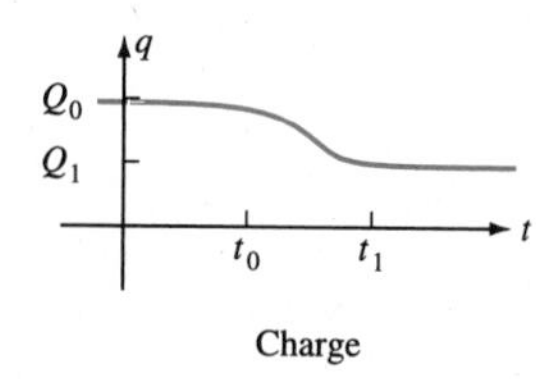

Charge

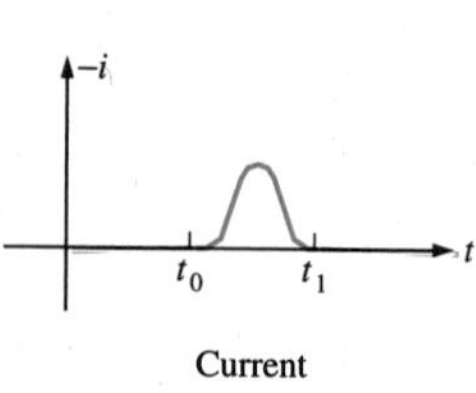

Current

FIGURE 3.7.1

Let $\tilde{i}(t)$ be the limit of $i(t)$ as $t_1 \to t_0^+$. But $i(t)$ goes to zero for $t \neq t_0$ and $i(t_0) \to \infty$. On the other hand, we still want (1) to hold:

$$\Delta Q = Q_1 - Q_0 = \int_{-\infty}^{\infty} \tilde{i}(t)\, dt,$$

which means $\tilde{i}(t)$ should have a finite area. Clearly, this limiting current $\tilde{i}(t)$ is not a function in the sense we are familiar with. It is an example of a **distribution** or **impulse function**. These generalized functions are a very convenient mathematical notation.

Charge

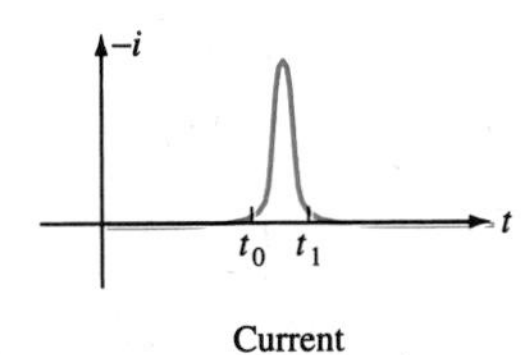

Current

FIGURE 3.7.2

■ DEFINITION OF DELTA FUNCTION

$\delta(t)$ is a mathematical object known as the **delta function**. It is an example of a **distribution** or **generalized function**. It has the following properties:

i) $\delta(t) = 0$ if $t \neq 0$.
ii) $\delta(0)$ is not defined.
iii) $\int_{-\infty}^{\infty} \delta(t)\, dt = 1$.
iv) If $g(t)$ is a continuous function on $(-\infty, \infty)$, then

$$\int_{-\infty}^{\infty} g(t)\delta(t)\, dt = g(0).$$

■

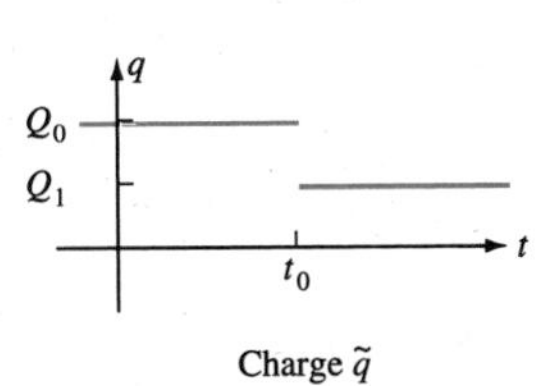

Charge $\tilde{q}$

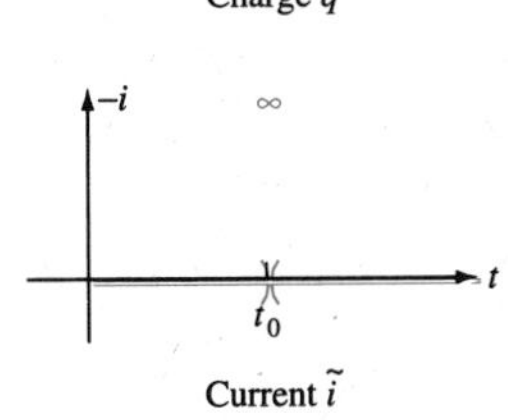

Current $\tilde{i}$

FIGURE 3.7.3

It is possible to build a logical, rigorous definition of $\delta(t)$, but we shall not do so. Intuitively, we may think of $\delta(t)$ as an approximation of a physical impulse of magnitude 1 at time $t = 0$. For example, it could be the rapid transfer of one unit of charge at time zero. It can be shown that if a is a constant, then

v) $\delta(t - a) = 0$ if $t \neq a$.
vi) $\int_{-\infty}^{\infty} \delta(t - a)\, dt = 1$.
vii) If $g(t)$ is a continuous function on $(-\infty, \infty)$, then

$$\int_{-\infty}^{\infty} g(t)\delta(t - a)\, dt = g(a).$$

From these formulas, we get that

$$\int_{-\infty}^{t} \delta(t)\, dt = \begin{cases} 1 & \text{if } t > 0, \\ 0 & \text{if } t < 0. \end{cases}$$

Thus, formally,

$$\int_{-\infty}^{t} \delta(t)\, dt = H(t),$$

and $\delta(t)$ may be considered, in some sense, to be the derivative of the Heaviside function. One nice property of the Laplace transform is that it works almost as easily for distributions and impulses as it does for ordinary functions.

Proceeding formally,

$$\mathscr{L}[\delta(t-a)] = \int_0^\infty e^{-st}\delta(t-a)\,dt = e^{-sa}\,[\text{by (vii) above}].$$

Thus,

$$\mathscr{L}[\delta(t-a)] = e^{-as} \tag{T13}$$

In particular,

$$\mathscr{L}[\delta(t)] = 1,$$

which is (T12). There are many other distributions, for example, "derivatives" of $\delta(t)$, but they will not be considered here (note Exercise 11). ■

EXAMPLE 3.7.1 Impulsive Forcing Function

Solve $y' + y = \delta(t-1)$, $y(0) = 1$.

SOLUTION Taking the Laplace transform of both sides gives

$$sY(s) - y(0) + Y(s) = e^{-s}.$$

Solving for $Y(s)$,

$$Y(s) = \frac{1}{s+1} + \frac{e^{-s}}{s+1},$$

so that, by (T2) and (T15),

$$y(t) = \mathscr{L}^{-1}\left[\frac{1}{s+1}\right] + \mathscr{L}^{-1}\left[e^{-s}\frac{1}{s+1}\right] = e^{-t} + e^{-(t-1)}H(t-1). \tag{2}$$

◄

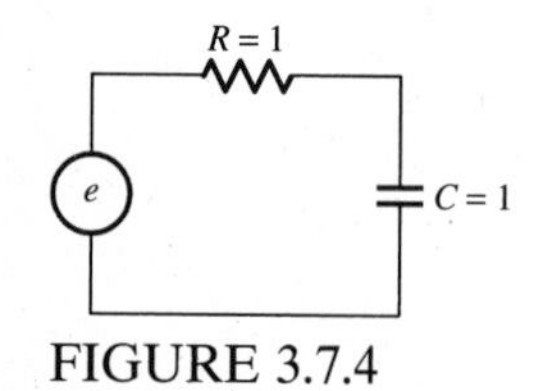

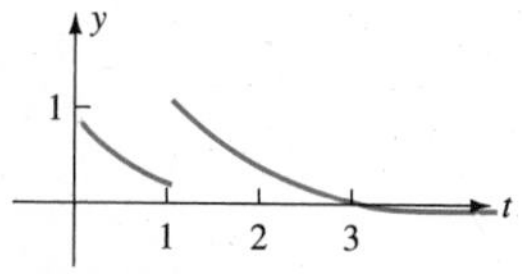

FIGURE 3.7.4

FIGURE 3.7.5
Graph of (2)

Physically, this example could be viewed as the simple linear *RC* circuit in Figure 3.7.4, where y is the charge on the capacitor at time t, and there is an initial charge of one on the capacitor. For $0 \le t < 1$, the voltage e is zero and the capacitor is discharging. At $t = 1$, there is a voltage impulse, that is, a very large voltage is applied for a brief period, which recharges the capacitor. Then the voltage is again zero, and the capacitor resumes discharging.

The graph of (2) is given in Figure 3.7.5. This graph should be interpreted as meaning that, in a real problem, $y(t)$ would be given by a function like that in Figure 3.7.6.

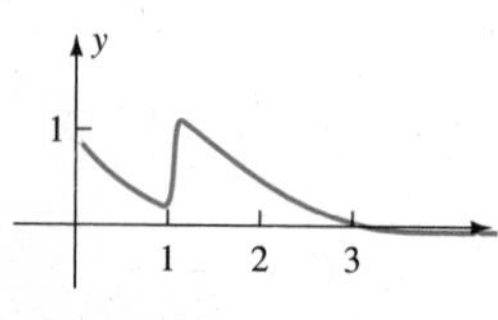

FIGURE 3.7.6

Needless to say, in real problems involving impulses, some care should be taken to ensure that the equations being used still provide accurate models in the presence of the large, but brief, values given by the impulse.

Exercises

In Exercises 1 through 8, solve the differential equation.

1. $y' + 8y = \delta(t-1) + \delta(t-2), \quad y(0) = 0$

2. $y'' + 2y' + 2y = \delta(t-5)$,
$y(0) = 1,\ y'(0) = 0$

3. $y'' + 6y' + 109y = \delta(t-1) - \delta(t-7)$,
$y(0) = 0,\ y'(0) = 0$

4. $y'' + 3y' + 2y = -2\delta(t-1)$,
$y(0) = 1,\ y'(0) = 0$

5. $y'' + 4y' + 3y = 1 + \delta(t-3)$,
$y(0) = 0,\ y'(0) = 1$

6. $y' + 4y = 1 - \delta(t-4), \quad y(0) = 1$

7. $y'' + y = 1 + \delta(t-2\pi)$,
$y(0) = 1,\ y'(0) = 0$

8. $y'' + 4y = t + 4\delta(t-4\pi),\ y(0) = y'(0) = 1$

9.* (Requires personal computer or access to computer facilities for Exercises 2 through 4.) For Exercises 1 through 4, above:

a) Sketch the solution of the differential equation.

b) Obtain the graph of the solution, and compare it to your sketch.

10.* For Exercises 5 through 8, above:

a) Sketch the solution of the differential equation.

b) Obtain the graph of the solution, and compare to your sketch.

11. Let $\delta^{(n)}(t)$, $n \geq 0$, n an integer, have the properties that

i) $\delta^{(n)}(t) = 0$ if $t \neq 0$,

ii) $\int_{-\infty}^{\infty} \delta^{(n)}(t-a)g(t)\,dt = (-1)^n g^{(n)}(a)$ for continuous g.

Show formally that $\mathcal{L}[\delta^{(n)}(t-a)] = s^n e^{-as}$.

CHAPTER

4

Variable Coefficient Linear Equations and Series Solutions

4.1 Introduction to Ordinary Differential Equations

We have developed methods in Chapters 2 and 3 for solving linear differential equations with constant coefficients:

$$ay'' + by' + cy = f(x). \tag{1}$$

Obtaining two linearly independent homogeneous solutions of (1) is straightforward because the coefficients a, b, c are constant. If $f(x)$ is a linear combination of the "right sort" of functions, the particular solution can be obtained by the method of undetermined coefficients. If $f(x)$ is the wrong kind of function for the method of undetermined coefficients, then the method of variation of parameters (Section 2.14) may be used to obtain a particular solution. Alternatively, one can use Laplace transforms (Chapter 3).

However, in many applications the coefficients a, b, c are not all constants, but can depend on x:

$$a(x)y'' + b(x)y' + c(x)y = f(x).$$

We have not discussed variable coefficient linear differential equations extensively. All we have noted is that the Euler equation (Section 2.13) is easily solvable, but the variable coefficients for that equation are very special.

However, the form of the solution of Euler's equation will be used to motivate some of the results on singular points in Sections 4.4 and 4.5. We recommend reading Section 2.13 before Sections 4.4 and 4.5.

The most important goal of this chapter is to discuss some qualitative properties of linear differential equations with variable coefficients. We will show that near most points, the solution to the differential equation can be approximated by a polynomial. However, at some special points, the solution to the differential equation can behave in a more complicated way. The solutions to differential equations can often be represented by infinite series whose first few terms represent interesting and easily accessible approximations to these solutions.

There are many different ways of presenting this material. If a short version of this chapter is preferred, we wish to suggest a brief discussion of the theory (Section 4.3 and the theory portion of 4.6) without the longer discussion of the series methods, since the theory is very important and can be understood without the calculations. (This advice is not the usual one the authors give. Usually we suggest that the differential equations are best understood through explicit computed examples.)

Historically, series methods were also an important numerical method of solving differential equations. Today, there are generally better and more accessible numerical methods. Hence some instructors may wish to deemphasize the series method. We wish to provide this option to the instructor.

Before discussing differential equations, we shall review some key facts about power series.

4.2 Review of Power Series

This section will review from calculus some of the basic facts about series and power series.

If $\{a_0, a_1, \ldots\} = \{a_n\}$ is a **sequence** of numbers, then the expression

$$\sum_{n=0}^{\infty} a_n = a_0 + a_1 + a_2 + \cdots$$

is called an **infinite series**. The mth **partial sum** s_m is given by

$$s_m = \sum_{n=0}^{m} a_n = a_0 + a_1 + \cdots + a_m.$$

The series $\sum_{n=0}^{\infty} a_n$ is said to **converge** if the partial sums have a limit. If L is that limit, that is,

$$\lim_{m \to \infty} s_m = L,$$

then L is called the **sum** of the series, and we write

$$\sum_{n=0}^{\infty} a_n = L.$$

Thus, by definition,

$$\sum_{n=0}^{\infty} a_n = \lim_{m\to\infty} \sum_{n=0}^{m} a_n = \lim_{m\to\infty} s_m. \tag{1}$$

If the series does not converge, it is called **divergent**. Note that there are two sequences here: the sequences being added up, $\{a_n\}$, and the sequence of partial sums $\{s_m\}$.

$$\text{If the series } \sum_{n=0}^{\infty} a_n \text{ converges, } \quad \text{then } \lim_{n\to\infty} a_n = 0. \tag{2}$$

Essentially, (2) says that if we can add up an infinite number of terms $\{a_n\}$ and get a finite sum L, then the terms a_n must become smaller and smaller, eventually approaching zero.

EXAMPLE 4.2.1 Geometric Series

Verify that $\sum_{n=0}^{\infty} \frac{1}{2^n} = 2.$

SOLUTION In this example, $a_n = 1/2^n$, and

$$s_m = \sum_{n=0}^{m} \frac{1}{2^n} = 1 + \frac{1}{2} + \cdots + \frac{1}{2^m}. \tag{3}$$

The series is a **geometric series**, and we use the algebraic technique of multiplying both sides of (3) by $(1 - \frac{1}{2})$ to get

$$\left(1 - \frac{1}{2}\right)s_m = 1 + \frac{1}{2} + \cdots + \frac{1}{2^m} - \frac{1}{2} - \frac{1}{2^2} - \cdots - \frac{1}{2^{m+1}} = 1 - \frac{1}{2^{m+1}}.$$

Solving for s_m,

$$s_m = \frac{1 - \dfrac{1}{2^{m+1}}}{\dfrac{1}{2}} = 2 - \frac{1}{2^m}.$$

Since $\lim_{m\to\infty} s_m = 2 - 0 = 2$, we have finished. Note also that $\lim_{n\to\infty} a_n = 0$. ◄

This example illustrates the concept of convergence. We will not actually need to sum series in this manner.

Note the similarity between the series $\sum_{n=0}^{\infty} a_n$ and the improper integral $\int_0^{\infty} f(x)\,dx$. Whether they converge depends on $\{a_n\}$ or $f(x)$. Both are defined as the limit of a sum (integral) over a finite interval:

$$\sum_{n=0}^{\infty} a_n = \lim_{m\to\infty} \sum_{n=0}^{m} a_n, \qquad \int_0^{\infty} f(x)\,dx = \lim_{m\to\infty} \int_0^{m} f(x)\,dx.$$

While $\lim_{n\to\infty} a_n = 0$ if $\sum_{n=0}^{\infty} a_n$ converges, the converse is not true. It is possible to have $\lim_{n\to\infty} a_n = 0$, but $\sum_{n=0}^{\infty} a_n$ does not converge.

EXAMPLE 4.2.2 Divergent Series

Consider the series $\sum_{n=0}^{\infty}(n+1)^{-1} = 1 + \frac{1}{2} + \frac{1}{3} + \cdots$. We have $a_n = 1/(n+1)$, so that $\lim_{n\to\infty} a_n = 0$. However, the series diverges, since it is known that $\lim_{m\to\infty} s_m = \infty$. ◄

A series is called **absolutely convergent** if $\sum_{n=0}^{\infty} a_n$ and $\sum_{n=0}^{\infty} |a_n|$ both converge. Absolutely convergent series have the nice property that the terms can be arranged in any order and the series still has the same sum. For example, we could add up all the odd-numbered terms and then add up all the even-numbered terms.

A series $\sum_{n=0}^{\infty} a_n$ is **conditionally convergent** if $\sum_{n=0}^{\infty} a_n$ converges but $\sum_{n=0}^{\infty}|a_n|$ is divergent. As noted earlier, $\lim_{n\to\infty} a_n = 0$ is not usually enough to guarantee convergence of $\sum_{n=0}^{\infty} a_n$. There is an important exception.

$$\begin{gathered}\text{If } a_n > 0,\ a_n \geq a_{n+1} \text{ for all } n > 0 \text{ and, } \lim_{n\to\infty} a_n = 0, \\ \text{then } \sum_{n=0}^{\infty}(-1)^n a_n \text{ is a convergent } (\textbf{alternating}) \text{ series.}\end{gathered} \tag{4}$$

EXAMPLE 4.2.3 Conditionally Convergent Series

Consider the series $\displaystyle\sum_{n=0}^{\infty} \frac{(-1)^n}{n+1} = 1 - \frac{1}{2} + \frac{1}{3} - \cdots$. Since

$$\sum_{n=0}^{\infty} \left|(-1)^n \frac{1}{n+1}\right| = \sum_{n=0}^{\infty} \frac{1}{n+1},$$

we know from Example 4.2.2 that the series does not converge absolutely. However,

$$\frac{1}{n+1} > 0, \qquad \frac{1}{n+1} \geq \frac{1}{n+2}, \qquad \text{and} \qquad \lim_{n\to\infty} \frac{1}{n+1} = 0,$$

so that, by (4), the series does converge. Since the convergence of the series is not absolute, it is conditional. ◄

In general, testing for conditional convergence can be difficult. Fortunately, the following test for absolute convergence will be sufficient for most of our examples.

■ **THEOREM 4.2.1 Ratio Test**

Suppose $a_n \neq 0$ for all n.

1. If $\displaystyle\lim_{n\to\infty} \frac{|a_{n+1}|}{|a_n|} = L < 1$, then $\displaystyle\sum_{n=0}^{\infty} a_n$ converges absolutely.

2. If $\displaystyle\lim_{n\to\infty} \frac{|a_{n+1}|}{|a_n|} = L > 1$, then $\displaystyle\sum_{n=0}^{\infty} a_n$ diverges.

3. If $\displaystyle\lim_{n\to\infty} \frac{|a_{n+1}|}{|a_n|} = 1$, then this test does not tell whether the series converges or diverges. ■

Power Series

The expression

$$\sum_{n=0}^{\infty} a_n(x-b)^n \tag{5}$$

is called a **power series centered at *b***. It is also sometimes called a **Taylor series centered at *b*** or **Maclaurin's series** if $b = 0$. A power series defines a function of x for those values of x where the series converges (the region of convergence), and we often write

$$f(x) = \sum_{n=0}^{\infty} a_n(x-b)^n. \tag{6}$$

Region of Convergence for a Power Series

One way to determine the region of convergence of a power series is the ratio test. According to the ratio test, the power series will converge if

$$\lim_{n\to\infty} \left| \frac{a_{n+1}(x-b)^{n+1}}{a_n(x-b)^n} \right| < 1.$$

Thus, the power series

$$\text{converges if} \quad |x-b| < r, \tag{7}$$

and

$$\text{diverges if} \quad |x-b| > r, \tag{8}$$

where r is called the **radius of convergence** and

$$r = \frac{1}{\lim_{n\to\infty} \left|\frac{a_{n+1}}{a_n}\right|} = \lim_{n\to\infty} \left|\frac{a_n}{a_{n+1}}\right|,$$

provided the limit exists. It turns out that for a power series, there is always a number r such that (7) and (8) hold even if the ratio test does not apply. In working with differential equations we will not be calculating the radius of convergence in this way, but the concept is important. In summary,

> The region of convergence of a power series is always:
> An interval of some radius r centered at b.
> This includes two special cases:
> When $r = 0$, the power series converges only at $x = b$.
> When $r = \infty$, the power series converges for all x.

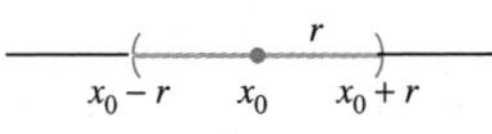

FIGURE 4.2.1

We graph the region of convergence in Figure 4.2.1.

Facts (7) and (8) are still valid for complex x, b, with $|\cdot|$ meaning the magnitude of a complex number. Thus the region of convergence in the complex plane is a disc of radius r centered at b. This region of convergence is graphed in Figure 4.2.2.

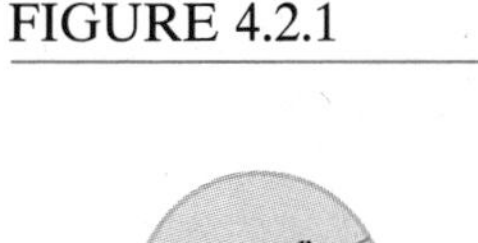

FIGURE 4.2.2

EXAMPLE 4.2.4 **Ratio Test**

Determine the convergence set of

$$\sum_{n=0}^{\infty} n!x^n.$$

Solution By the ratio test (Theorem 4.2.1),

$$\lim_{n\to\infty} \frac{|a_{n+1}|}{|a_n|} = \lim_{n\to\infty} \frac{(n+1)!}{n!} = \lim_{n\to\infty} (n+1) = +\infty.$$

Thus $1/r = \infty$ or (by convention) $r = 0$, and the convergence set is just the center $b = 0$. Alternatively we could use Theorem 1:

$$\lim_{n\to\infty} \frac{(n+1)!|x|^{n+1}}{n!|x|^n} = \lim_{n\to\infty} (n+1)|x| = \begin{cases} 0 & \text{if } x = 0, \\ \infty & \text{if } x \neq 0. \end{cases}$$

Thus the series converges only for $x = 0$. ◀

EXAMPLE 4.2.5 **Ratio Test**

Determine the convergence set of

$$\sum_{n=0}^{\infty} \frac{1}{n+1}(x-2)^n.$$

SOLUTION Using the power series version of the ratio test (Theorem 4.2.1), we obtain

$$\frac{1}{r} = \lim_{n \to \infty} \frac{1/(n+2)}{1/(n+1)} = \lim_{n \to \infty} \frac{n+1}{n+2} = 1,$$

so that $r = 1$. Thus the series converges for $|x - 2| < 1$ and diverges for $|x - 2| > 1$. The endpoints of this interval are

$$|x - 2| = 1 \qquad \text{or} \qquad x = 1, 3.$$

To determine whether they are included in the convergence set, we must examine them separately. If $x = 1$, then the series is

$$\sum_{n=0}^{\infty} \frac{1}{n+1}(1-2)^n = \sum_{n=0}^{\infty} \frac{(-1)^n}{n+1},$$

which converges, from Example 4.2.3. On the other hand, if $x = 3$, then the series is

$$\sum_{n=0}^{\infty} \frac{1}{n+1}(3-2)^n = \sum_{n=0}^{\infty} \frac{1}{n+1},$$

which diverges, from Example 4.2.2. Thus the convergence set is the interval $[1, 3)$. ◀

EXAMPLE 4.2.6 **Ratio Test**

Find the convergence set of $\sum_{n=0}^{\infty} \frac{x^n}{n!}$. (*Note*: $0! = 1$.)

SOLUTION Again using the ratio test (Theorem 4.2.1),

$$\frac{1}{r} = \lim_{n \to \infty} \frac{1/(n+1)!}{1/(n!)} = \lim_{n \to \infty} \frac{n!}{(n+1)!} = \lim_{n \to \infty} \frac{1}{n+1} = 0.$$

Thus, by convention, $r = \infty$ and the series converges for all x. That is, the convergence set is the real line $(-\infty, \infty)$. ◀

The properties of power series that we shall need in order to solve differential equations derive from the following fundamental fact, which is not necessarily true for other types of series.

4.2.1 Analytic Functions and Taylor Series

If a function f can be written as a power series at b (which converges in some region around b), then f is said to be **analytic** at b, and

$$f(x) = \sum_{n=0}^{\infty} a_n(x-b)^n.$$

Letting $x = b$, we see that $a_0 = f(b)$. If f is analytic, f' exists and can be obtained by term-by-term differentiation, $f'(x) = \sum_{n=1}^{\infty} na_n(x-b)^{n-1}$. Letting $x = b$, we get that $a_1 = f'(b)$. These first two terms correspond to the linearization or tangent line approximation for $f(x)$ around $x = b$. If f is analytic, this process of differentiation can be continued indefinitely. Thus,

If a function is an analytic function
(given by a power series with a nonzero radius of convergence),
then the function is infinitely differentiable
and all its derivatives are also analytic.

In general, if f is analytic,

$$f(x) = \sum_{n=0}^{\infty} a_n(x-b)^n,$$
$$a_n = \frac{f^{(n)}(b)}{n!}. \tag{9}$$

In this case the series in (9) is called the **Taylor series** representation of the analytic function $f(x)$. The series is said to be **centered** at b.

Many of the functions we discuss in differential equations are analytic for most values of x. This will include most of our standard functions, such as polynomials, exponentials, trigonometric functions, and fractions, compositions, and inverses of these functions.

EXAMPLE 4.2.7 Taylor Series

Verify that the Taylor series around $x = 0$ for $f(x) = \dfrac{1}{1-x}$ is the well known geometric series,

$$\frac{1}{1-x} = 1 + x + x^2 + \cdots = \sum_{n=0}^{\infty} x^n, \text{ for } |x| < 1. \tag{10}$$

SOLUTION We could verify this as in Example 4.2.1. We shall use (9) instead. Let

$$f(x) = (1-x)^{-1}.$$

Then

$$f'(x) = (1-x)^{-2},$$
$$f''(x) = 2(1-x)^{-3},$$

and, in general,

$$f^{(n)}(x) = n!(1-x)^{-n-1}. \tag{11}$$

Thus, $f(0) = 1$, $f'(0) = 1$, and, in general, $f^{(n)}(0) = n!$, so that, by (9),

$$\frac{1}{1-x} = \sum_{n=0}^{\infty} \frac{f^{(n)}(0)}{n!} x^n = \sum_{n=0}^{\infty} \frac{n!}{n!} x^n = \sum_{n=0}^{\infty} x^n.$$

The ratio test can be used to show that $\sum_{n=0}^{\infty} x^n$ converges for $|x| < 1$. However, $1/(1-x)$ makes sense for all $x \neq 1$. To write $1/(1-x)$ as a power series for $x > 1$ or for $x < -1$, we must pick a different center. For example, if we take $b = 3$, then from (11), $f^{(n)}(3) = n!(1-3)^{-n-1}$, and

$$\frac{1}{1-x} = \sum_{n=0}^{\infty} \frac{f^{(n)}(3)}{n!}(x-3)^n = \sum_{n=0}^{\infty} \frac{(-1)^{n+1}}{2^{n+1}}(x-3)^n,$$

which converges for $|x-3| < 2$, which is the interval $(1, 5)$. ◄

Technical Point What we have just shown is that if $1/(1-x)$ can be written as a power series centered at zero, then it is given by (10). We have not shown that $1/(1-x)$ can be written as a power series centered at zero! There do exist infinitely differentiable functions that are not analytic. However, we will have a theorem in the next section that ensures that solutions of the differential equations we consider can be written as power series.

4.2.2 Taylor Polynomials

If $f(x)$ is l times continuously differentiable in an open interval containing b, then, for $m \leq l$,

$$\rho_m(x) = \sum_{n=0}^{m} \frac{f^{(n)}(b)}{n!}(x-b)^n \tag{12}$$

is called the mth-degree **Taylor polynomial** for $f(x)$, centered at b. $\rho_1(x) = f(b) + f'(b)(x-b)$ is the linearization of f (the tangent line approximation to f) at b. If $f(x)$ has a power-series expansion (9), then $\rho_m(x)$ is just the mth partial sum of the Taylor series for $f(x)$, and

$$\lim_{m \to \infty} \rho_m(x) = f(x)$$

for all x in the convergence set of the power series. Thus the Taylor polynomials can be used to approximate f. Even if f does not have a power series, we have, from Taylor's theorem in calculus, that, if $f(x)$ is l times continuously differentiable in the interval $(b - x, b + x)$ and $m < l$, then

$$f(x) = f(b) + f'(b)(x - b) + \cdots + \frac{f^{(m)}(b)}{m!}(x - b)^m + R_m(x) \quad (13)$$

or

$$f(x) = p_m(x) + R_m(x),$$

where $R_m(x)$ is the remainder. We may estimate $R_m(x)$ by using the fact that

$$R_m(x) = \frac{f^{(m+1)}(\xi)}{(m + 1)!}(x - b)^{m+1}, \quad (14)$$

where ξ is some number between b and x.

EXAMPLE 4.2.8 **Taylor Polynomial**

From Example 4.2.7, we have that if $f(x) = 1/(1 - x)$, then the mth-degree Taylor polynomial for $f(x)$ centered at zero is

$$p_m(x) = 1 + x + \cdots + x^m.$$

Exercises

For Exercises 1 through 13, determine which of the following expressions are power series.

1. $\sum_{n=0}^{\infty} x^n$

2. $\sum_{n=0}^{\infty} n^x$

3. $\sum_{n=0}^{\infty} \sin(n)x^n$

4. $\sum_{n=0}^{\infty} \sin(nx)$

5. 3

6. $x^{1/2}$

7. $\sum_{n=-2}^{\infty} 3x^n$

8. $\sum_{n=1}^{\infty} 2^n x^{2n}$

9. $\sum_{n=0}^{\infty} nx^{3n+1}$

10. $8 - x^2$

11. $\dfrac{1}{x}$

12. $x - 3^{1/2}$

13. $e^{-3}x$

For Exercises 14 through 16, use the ratio test (Theorem 4.2.1) to determine whether or not the series converges. If the ratio test fails, so state.

14. $\sum_{n=1}^{\infty} \dfrac{n}{2^n}$

15. $\sum_{n=0}^{\infty} \dfrac{1}{(n + 1)^2}$

16. $\sum_{n=1}^{\infty} \dfrac{1}{3n}$

For Exercises 17 through 19, determine the radius of convergence and the convergence set for the power series. Determine, if possible, whether the endpoints of the interval are in the convergence set.

17. $\sum_{n=0}^{\infty} \dfrac{x^n}{\sqrt{n!}}$

18. $\sum_{n=0}^{\infty} 3^n(x-2)^n$

19. $\sum_{n=1}^{\infty} n!\frac{x^n}{2^n}$

For Exercises 20 through 28, write the indicated function as a Taylor series with the given center b.

20. $f(x) = e^x, \quad b = 0$

21. $f(x) = e^{2x}, \quad b = 1$

22. $f(x) = \sin x, \quad b = 0$

23. $f(x) = \cos x, \quad b = 0$

24. $f(x) = \sin x, \quad b = \pi/2$

25. $f(x) = x^3, \quad b = 0$

26. $f(x) = x^3, \quad b = 1$

27. $f(x) = x, \quad b = 0$

28. $f(x) = x, \quad b = -3$

In Exercises 29 through 34, use the following method to obtain the power series of the given function around the given point. Geometric series (10) can be used to obtain related geometric series. For example, to obtain the power series of $1/(1-x)$ centered at $x = 3$, we introduce $z = x - 3$. In this case,

$$\frac{1}{1-x} = \frac{1}{1-(z+3)} = \frac{1}{-2-z} = \frac{1}{-2\left(1+\frac{z}{2}\right)}.$$

Using (10), we obtain

$$\frac{1}{1-x} = \frac{1}{-2\left(1+\frac{z}{2}\right)} = -\frac{1}{2}\sum_{n=0}^{\infty}\left(-\frac{z}{2}\right)^n$$

$$= -\frac{1}{2}\sum_{n=0}^{\infty}\left(-\frac{1}{2}\right)^n(x-3)^n,$$

which is valid for $|z/2| < 1$ or $|x-3| < 2$.

29. $f(x) = \frac{1}{1-x}, \quad b = 5$

30. $f(x) = \frac{1}{1-x}, \quad b = -7$

31. $f(x) = \frac{1}{4-x}, \quad b = 0$

32. $f(x) = \frac{1}{4-x}, \quad b = 1$

33. $f(x) = \frac{1}{4-x}, \quad b = 6$

34. $f(x) = \frac{1}{4-x}, \quad b = -4$

Since the power series for a function at a given center is unique, it is often possible to obtain the series by manipulating simpler series.

35. Take the series for $1/(1-x)$ in Example 4.2.7. Make the substitution $-x^2$ for x to obtain the series for $1/(1+x^2)$ centered at zero.

36. Antidifferentiate the result of Exercise 35 termwise to get a series for $\tan^{-1} x$.

37. Use the answer for Exercise 20 to obtain a Taylor series for $f(x) = e^{x^3}$. Note that obtaining this series, using (9), becomes progressively more complicated.

38. In Example 4.2.3, it was noted that $\sum_{n=0}^{\infty}(-1)^n/(n+1)$ converges. It can be shown that

$$\sum_{n=0}^{\infty}(-1)^n\frac{1}{n+1} = \ln 2. \qquad (15)$$

Starting with $s_0 = 1$, compute

$$s_m = \sum_{n=0}^{m}(-1)^n\frac{1}{n+1}$$

for larger and larger m, until s_m gives the correct answer to three decimal places. (This requires access to a computer or programmable calculator and a simple loop program. It illustrates how convergence can be slow.) Now repeat the experiment until s_m is accurate to five places.

39. Can the interval $[0, \infty)$ be the convergence set for a power series? Explain your answer.

In Exercises 40 through 45, find the mth-degree Taylor polynomial of $f(x)$ centered at b.

40. $m = 3$ and $f(x)$, b as in Exercise 20.

41. $m = 2$ and $f(x)$, b as in Exercise 21.

42. $m = 5$ and $f(x)$, b as in Exercise 22.

43. $m = 5$ and $f(x)$, b as in Exercise 24.

44. $m = 2$ and $f(x)$, b as in Exercise 25.

45. $m = 1$ and $f(x)$, b as in Exercise 27.

46. Define the function $f(x)$ by

$$f(x) = \begin{cases} e^{-1/x^2} & \text{if } x \neq 0, \\ 0 & \text{if } x = 0. \end{cases}$$

Show that $f(x)$ is infinitely differentiable everywhere, including zero, and that $f^{(n)}(0) = 0$ for all n. Thus

$$\sum_{n=0}^{\infty} f^{(n)}(0)\frac{x^n}{n!} = 0 \qquad \text{for all } x,$$

which does not agree with f anywhere except at $x = 0$.

4.3 Solution at an Ordinary Point (Theory)

In this chapter we consider second-order linear differential equations that can be put in the form

$$a(x)y''(x) + b(x)y'(x) + c(x)y(x) = h(x). \tag{1}$$

We will assume that the possibly variable coefficients $a(x)$, $b(x)$, $c(x)$ and the forcing function $h(x)$ are extremely nice functions, analytic for all x including complex x. A function $g(x)$ is **analytic** at x_0 if the function has a power-series expansion centered there and the power series converges to $g(x)$ for x near x_0. Not only is an analytic function at x_0 continuous at x_0, but all its derivatives are continuous there as well. Functions that are analytic for all real and complex x include any polynomial, $\sin x$, $\cos x$, e^x, and many other functions that arise in applications as coefficients of linear differential equations. Any sum, product, or composition of everywhere analytic functions is analytic everywhere. (The theory is simplest if a, b, c, h are analytic for all x, and many important examples can be put in this form.)

It is frequently more convenient to state some of the results if both sides of (1) are divided by $a(x)$ to give

$$y''(x) + p(x)y'(x) + q(x)y(x) = f(x), \tag{2}$$

where

$$p(x) \equiv \frac{b(x)}{a(x)}, \qquad q(x) \equiv \frac{c(x)}{a(x)}, \qquad f(x) \equiv \frac{h(x)}{a(x)}.$$

Definitions of an Ordinary Point and a Singular Point

If the functions a, b, c, and h are analytic for all x and $a(x_0) \neq 0$, then x_0 is called an **ordinary point** of the differential equation (1) or (2). If $a(x_0) = 0$, then x_0 is called a **singular point**, since p, q, and f may be undefined there.

EXAMPLE 4.3.1 **Ordinary and Singular Points**

For the differential equation $xy'' + 4xy' + 3y = 0$, determine which points are ordinary points and which points are singular points.

SOLUTION All x are ordinary points except $x = 0$, which is a singular point since $a(x) = x$ is zero there. ◄

EXAMPLE 4.3.2 Ordinary and Singular Points

For the differential equation $(x+1)(x+2)y'' + 4y = 0$, determine which points are ordinary points and which points are singular points.

SOLUTION All x are ordinary points except $x = -1$ and $x = -2$, which are singular points since $a(x) = (x+1)(x+2)$ is zero there. ◄

EXAMPLE 4.3.3 Ordinary and Singular Points

For the differential equation $y'' + \sin xy = 0$, determine which points are ordinary points and which points are singular points.

SOLUTION We have $a(x) = 1$, $b(x) = 0$, $c(x) = \sin x$. a, b, and c are analytic for all x, and $a(x)$ is never zero. All x are ordinary points. There are no singular points. ◄

The reason we have carefully defined an ordinary point is that there are many important properties of differential equations that are valid at ordinary points.

■ **THEOREM 4.3.1 Solutions at an Ordinary Point**

If x_0 is an ordinary point, then

1. All solutions of the linear differential equation (1) are continuous and all derivatives are continuous at and near $x = x_0$. (3)
2. Every solution y of (1) can be written as a power series centered at $x = x_0$ that converges to $y(x)$,
$$y = \sum_{n=0}^{\infty} c_n (x - x_0)^n. \tag{4}$$
3. All solutions to the linear differential equation can be approximated near $x = x_0$ by the polynomials obtained by truncating the series in (4). (5)
4. The region of convergence for this infinite power series is an interval around x_0 whose radius is at least the distance from x_0 to the nearest singular point (real or complex) such that $a(x) = 0$. This gives a guaranteed region of convergence in which the series converges. The actual region of convergence may be larger. (6) ■

A similar result holds for the first-order equation

$$a(x)y'(x) + b(x)y(x) = h(x)$$

and its equivalent form

$$y'(x) + p(x)y(x) = f(x).$$

EXAMPLE 4.3.4 Ordinary Point

Consider the differential equation

$$y'' + \frac{1}{x-1}y' + xy = e^x.$$

The coefficient $1/(x-1)$ is not continuous at $x = 1$, so that $1/(x-1)$ is not analytic at $x = 1$. To put the equation in the form (1), we multiply by $x - 1$:

$$(x-1)y'' + y' + x(x-1)y = (x-1)e^x. \tag{7}$$

Now all coefficients are analytic, since they are polynomials or polynomials times simple exponentials. But $a(x) = x - 1$ is zero at $x = 1$. Thus, all values of x are ordinary points except $x = 1$, which is a singular point. According to the theorem, all solutions to the differential equation (7) are guaranteed to be well behaved (analytic) everywhere except at $x = 1$. (What happens near the singular point $x = 1$ will be explained in a later section devoted to singular points.) If your friend in the computer center solves the differential equation (7) numerically as an initial value problem starting at $x = 0$ and obtains a solution that has bizarre behavior at $x = \frac{1}{4}$, this theorem tells you to question the validity of your friend's computation, since you know that the solution must be well behaved there.

Suppose $x_0 = 3$ is a special point of interest. Since $x = 3$ is an ordinary point, all solutions of (7) can be written as a power series centered at $x = 3$,

$$y = \sum_{n=0}^{\infty} c_n(x-3)^n.$$

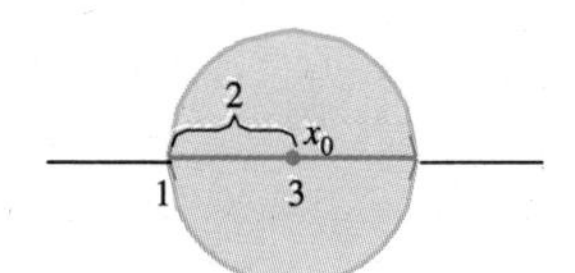

FIGURE 4.3.1

Furthermore, the region of convergence will be an interval centered at $x = 3$ at least up to the nearest singular point, which is $x = 1$. Thus, from this theorem it is guaranteed that the series converges for $|x - 3| < 2$. The region of convergence may be larger. See Figure 4.3.1. ◄

EXAMPLE 4.3.5 Singular Point at a Complex Value

Consider the differential equation

$$(x^2 + 1)y'' + y' + xy = 0. \tag{8}$$

The coefficients are all analytic, since they are polynomials, and $a(x) = x^2 + 1$ is never zero for real x. Thus, all real values of x are ordinary points. According to the theorem, all solutions to the differential equation (8) are guaranteed to be well behaved (analytic) for all real x.

Suppose $x = 0$ is a special point of interest. Since $x = 0$ is an ordinary point, all solutions of (8) have a power series centered at $x = 0$,

$$y = \sum_{n=0}^{\infty} c_n x^n. \tag{9}$$

To analyze the region where this series representation is convergent, we first locate all singular points $[a(x) = 0]$, including any in the complex plane. Singular points for (8) satisfy $x^2 + 1 = 0$. In this case there are no real singular points, but $x = \pm i$ are singular points in the complex plane and are sketched in Figure 4.3.2. The region of convergence is at least a circle in the complex plane centered at $x = 0$ with radius of convergence up to the nearest singular point, which is located at $x = \pm i$. Thus, the radius of convergence will be at least 1. It turns out that the radius of convergence of (9) is 1. In this example, the series representation of the solution converges for $|x| < 1$, even though the actual solution to the differential equation must be well behaved (analytic) everywhere on the real axis. ◄

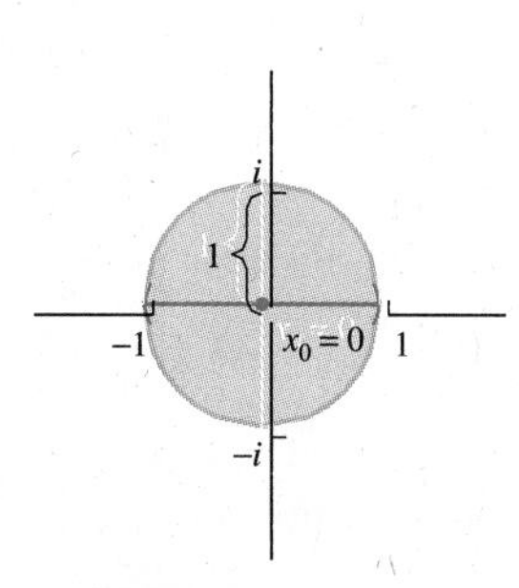

FIGURE 4.3.2

If the power-series representation of a solution to a differential equation does not converge at some point, this does not necessarily mean that the actual solution of the differential equation is not well behaved there. This is subtle. It is the same distinction that occurs between a function $1/(x^2 + 1)$, which is very nice everywhere along the real axis, and its Taylor series centered at 0 (which follows from the corresponding geometric series)

$$\frac{1}{1 + x^2} = 1 - x^2 + x^4 - x^6 + \cdots,$$

which converges only for $|x| < 1$.

In "solving" a differential equation by power series, there are three levels of solution. Suppose the solution is

$$y(x) = \sum_{n=0}^{\infty} c_n (x - b)^n, \tag{10}$$

where the c_n are constants.

Level I

Given a number, say 5, we can determine the first five coefficients $\{c_0, c_1, c_2, c_3, c_4\}$ of (10). The Taylor polynomial $\Sigma_{n=0}^{4} c_n (x - b)^n$ then produces an approximation to the solution. The method of Section 4.4 produces this level of solution.

Level II
We have a recursion relationship (also known as a difference equation; see Chapter 8) for the coefficients. For example, it could take the form

$$c_{n+1} = nc_n + c_{n-1} + n^2 c_{n-2}. \tag{11}$$

With this relationship we can automatically generate as many coefficients as needed since, given $\{c_0, \ldots, c_n\}$, formula (11) gives c_{n+1}. Then it will give c_{n+2}, etc. The method of Section 4.5 produces this level of solution.

Level III
We have a formula for the coefficients themselves, for example, $c_n = n^2 + 1$. Occasionally, Level III can be reached by inspection of a Level I or Level II solution. Sometimes the expression for c_n can be obtained by solving the recursion relationship found in a Level II solution. We will not emphasize Level III solutions further in this text, but their importance should not be overlooked.

Exercises

For Exercises 1 through 6, state at which points Theorem 4.3.1 will guarantee the existence of a power-series solution centered at that point.

1. $y'' + xy = e^x$

2. $y'' + xy = \tan x$

3. $xy'' + e^x y = \dfrac{1}{x-1}$

4. $(x^2 - 2x)y'' + xy' + y = \cos x$

5. $(\sin x)y'' + \dfrac{1}{x-2}y' + y = |x - 3|$

6. $e^x y'' + (\cosh x)y' - y = \sinh x$

For Exercises 7 through 14, give the guaranteed region of convergence of the power-series solution (by Theorem 4.3.1).

7. $y'' + \dfrac{1}{x-1}y = \sin x, \qquad x_0 = 0$

8. $(x^2 + 4)y'' + y = e^x, \qquad x_0 = 0$

9. $(x + 1)y' + e^x y = \cos x, \qquad x_0 = 2$

10. $e^{3x}y' + (\sinh x)y = \cosh x, \qquad x_0 = 6$

11. $(x^2 + 1)y'' + xy' = e^x - \cos x, \qquad x_0 = 0$

12. $y'' + \dfrac{1}{x+1}y' + \dfrac{1}{x-1}y = x, \qquad x_0 = \frac{1}{2}$

13. $y'' + \dfrac{1}{x-2}y' + \dfrac{3}{x-5}y = e^x, \qquad x_0 = 3.3$

14. $xy'' + \dfrac{1}{x-1}y' = e^x, \qquad x_0 = 0.2$

For Exercises 15 through 20, determine which real points are ordinary points and which points (including complex ones) are singular points.

15. $y'' + (x + 1)(x + 2)y' + 4y = 0$

16. $y'' + (\cos x)y = 0$

17. $(x^2 - 1)y'' + xy' + 6y = 0$

18. $x(x^2 + 4)y'' + xy' - 5y = 0$

19. $(x^2 + 9)y'' + (x - 1)(x - 2)y' + xy = 0$

20. $x(x^2 - 16)y'' + (x^2 + 1)y' + 4xy = 0$

In Exercises 21 through 24, determine the values of x for which all solutions are guaranteed to be continuous. Also, where is the power-series solution (in powers of x) guaranteed to converge?

21. $(x^2 + 9)y'' + xy' + 7y = 0$

22. $(x - 16)(x^2 + 1)y'' + x^2y' - 16y = 0$

23. $(x^2 - 9)y'' + (x^2 + 1)y' + xy = 0$

24. $(x^2 + 16)y'' + (x^2 - 1)y' - xy = 0$

4.4 Solution at an Ordinary Point (Taylor-Series Method)

The first method we present for solving a differential equation using series has certain advantages and disadvantages. They will be discussed at the end of this section.

The Taylor-series method is based on two ideas. In order to get a series for $y(x)$ centered at x_0, we need only compute the derivatives of y at x_0, $y^{(n)}(x_0)$. Also, the differential equation of which y is a solution can be used to express the higher derivatives in terms of lower derivatives.

EXAMPLE 4.4.1 Taylor Series

Find the first five terms of the Taylor-series expansion for the solution of

$$y'' + 3y' + xy = \sin x, \qquad y(0) = 0, \qquad y'(0) = 2, \tag{1}$$

centered at $x_0 = 0$, using the Taylor-series method.

SOLUTION Note that, by Theorem 4.3.1, (1) has a power series solution around $x_0 = 0$ since $x_0 = 0$ is an ordinary point. For convenience, rewrite the differential equation (1) as

$$y'' = \sin x - 3y' - xy. \tag{2}$$

The initial conditions (1) give

$$y(0) = 0, \qquad y'(0) = 2. \tag{3}$$

Evaluate (2) at $x = 0$, to get

$$y''(0) = \sin(0) - 3y'(0) - 0y(0) = -3 \cdot 2 = -6. \tag{4}$$

Now differentiate (2):

$$y''' = \cos x - 3y'' - y - xy'. \tag{5}$$

Evaluate (5) at $x = 0$ to find $y'''(0)$, using (3) and (4):

$$y'''(0) = 1 - 3y''(0) - y(0) = 1 - 3(-6) - 0 = 19. \tag{6}$$

Now differentiate (5):

$$y^{(4)} = -\sin x - 3y''' - y' - xy'' - y'. \tag{7}$$

Evaluate (7) at $x = 0$ to find $y^{(4)}(0)$ using (3), (4), and (6):

$$y^{(4)}(0) = -0 - 3y'''(0) - 2y'(0) = -3(19) - 2(2) = -61. \tag{8}$$

Thus the solution of the initial value problem (1) (to five terms) is

$$y(x) = y(0) + y'(0)x + \frac{y''(0)}{2}x^2 + \frac{y'''(0)}{3!}x^3 + \frac{y^{(4)}(0)}{4!}x^4 + \cdots$$
$$= 2x - 3x^2 + \frac{19}{6}x^3 - \frac{61}{24}x^4 + \cdots .$$

If our purpose is to get an estimate for the solution $y(x)$ near $x = 0$, then

$$2x - 3x^2 + \frac{19}{6}x^3 - \frac{61}{24}x^4$$

may be sufficient. We will not address the important problem of determining the number of needed terms.

Summary of Taylor-Series Method

To solve

$$a(x)y''(x) + b(x)y'(x) + c(x)y(x) = f(x), \tag{9}$$

1. Choose the center x_0 and be sure a, b, c, f can be written as power series centered at x_0 with radii of convergence greater than zero, and $a(x_0) \neq 0$. (In our problems, x_0 will be given.)
2. Determine how many terms of the expansion

$$y(x) = \sum_{n=0}^{\infty} c_n(x - x_0)^n = \sum_{n=0}^{\infty} \frac{y^{(n)}(x_0)}{n!}(x - x_0)^n$$

   are needed. Note that the constants $\{c_n\}$ are the coefficients of the Taylor-series expansion of y and are given by

$$c_n = \frac{y^{(n)}(x_0)}{n!}.$$

3. If initial conditions for y are given at x_0, use them to determine c_0 and c_1. If no initial conditions are given for (9), take c_0, c_1 as arbitrary.
4. Evaluate (9) at x_0. Then solve to find $y''(x_0)$. Set $c_2 = y''(x_0)/2$.
5. Differentiate (9) to get

$$a'(x)y''(x) + a(x)y'''(x) + b'(x)y'(x) + b(x)y''(x) + c'(x)y(x)$$
$$+ c(x)y'(x) = f'(x). \tag{10}$$

6. Evaluate (10) at x_0 and use the known $y(x_0)$, $y'(x_0)$, $y''(x_0)$ to find $y'''(x_0)$. Let $c_3 = y'''(x_0)/3!$.
7. Continue in this manner by repeatedly differentiating (10) and evaluating at x_0 to find $y^{(n)}(x_0)$ for as many n as needed.

Modifications for First-Order Equation

For

$$a(x)y'(x) + b(x)y(x) = f(x),$$

the procedure is essentially the same, except that if no initial condition is given, only c_0 is taken to be arbitrary.

There are several advantages to this method.

Advantages of Method

i. Done just by evaluation and differentiation.
ii. Does not require knowing series for $a(x)$, $b(x)$, $c(x)$, $f(x)$.
iii. Does not require manipulation of series or summation notation.
iv. Can be used, in principle, to find any needed number of terms.

There are also disadvantages.

Disadvantages of Method

i. If $a(x)$, $b(x)$, $c(x)$, $f(x)$ are at all complicated, the repeated differentiations may become quite messy.
ii. If the expression for a general term is required, then it is more difficult to find this way, and requires some "insight."
iii. It may sometimes be difficult to determine a recursion relationship for coefficients that would enable one to easily generate additionally needed terms from known terms.

If one has access to a symbolic differentiation program, then the first disadvantage is easily surmounted, since the computer does all the differentiation. The second problem still remains. The method of the next section will address these problems.

In concluding this section, an example of computing the general solution of a linear second-order differential equation will be given.

EXAMPLE 4.4.2 **Taylor Series**

Find the first five terms of the Taylor-series expansion for the general solution of

$$y'' + xy' + x^2y = 3 + x, \tag{11}$$

centered at $x_0 = 0$, using the Taylor-series method.

Solution Since no initial conditions are given and (11) is a second-order linear differential equation, we know that there are two arbitrary constants, and

$$y(0) = c_0, \qquad y'(0) = c_1$$

may be taken as arbitrary. Solve (11) for y'':

$$y'' = 3 + x - xy' - x^2y. \tag{12}$$

Evaluate (12) at $x = 0$ to find $y''(0)$:

$$y''(0) = 3 + 0 - 0 - 0 = 3. \tag{13}$$

Differentiate (12):

$$y''' = 1 - y' - xy'' - 2xy - x^2y'. \tag{14}$$

Evaluate (14) at $x = 0$:

$$y'''(0) = 1 - y'(0) = 1 - c_1. \tag{15}$$

Differentiate (14):

$$y^{(4)} = -2y'' - xy''' - 2y - 2xy' - 2xy' - x^2y''. \tag{16}$$

Evaluate (16) at $x = 0$:

$$y^{(4)}(0) = -2y''(0) - 2y(0) = -6 - 2c_0. \tag{17}$$

Thus,

$$y(x) = c_0 + c_1x + \frac{3}{2!}x^2 + \frac{(1 - c_1)}{3!}x^3 + \frac{(-6 - 2c_0)}{4!}x^4 + \cdots \tag{18}$$

Recall that, from the theory of linear differential equations (Section 2.2), the general solution $y(x)$ of (11) may be written as

$$y(x) = y_p(x) + \tilde{c}_1y_1(x) + \tilde{c}_2y_2(x),$$

where $y_p(x)$ is a particular solution of (11), $y_1(x)$, $y_2(x)$ are solutions of the associated homogeneous equation, and $\tilde{c}_1$, $\tilde{c}_2$ are arbitrary constants. Note that the general solution (18) can be rewritten as

$$y(x) = \left[\frac{3}{2}x^2 + \frac{x^3}{3!} - \frac{x^4}{4} + \cdots\right] + c_0\left[1 - \frac{x^4}{12} + \cdots\right] + c_1\left[x - \frac{x^3}{3!} + \cdots\right].$$

That is,

$$\frac{3}{2}x^2 + \frac{x^3}{3!} - \frac{x^4}{4} + \cdots \qquad \text{are the first five terms}$$

(the first two are zero) of a power series around $x_0 = 0$ for a particular solution y_p, and

$$\left.\begin{array}{l} 1 - \dfrac{x^4}{12} + \cdots \\ x - \dfrac{x^3}{3!} + \cdots \end{array}\right. \quad \text{are the first five terms}$$

of power series for y_1 and y_2, respectively, which are solutions of the associated homogeneous equation $y'' + xy' + x^2y = 0$. ◄

Note In Examples 4.4.1 and 4.4.2, we first rewrote the differential equation by solving for y''. If $a(x)$ is not constant, it may be simpler to omit this step and follow the general procedure of repeatedly differentiating the differential equation.

Exercises

In Exercises 1 through 6, use the method of Section 4.4 to find the indicated number of terms (m) of the Taylor-series expansion of the solution of the given differential equation. The center x_0 is where the initial conditions are applied.

1. $y'' + y = x, \quad y(0) = 0, \quad y'(0) = 2, \quad m = 5$

2. $y'' + x^2y' + y = \sin x,$
$y(0) = 1, \quad y'(0) = 1, \quad m = 5$

3. $y' + x^2y = 1, \quad y(0) = 2, \quad m = 4$

4. $y' + x^2y = 1, \quad y(1) = 2, \quad m = 4$

5. $xy' + y = x^2, \quad y(1) = 0, \quad m = 4$

6. $y'' + xy = e^x, \quad y(-1) = 0, \quad y'(-1) = 1,$
$m = 5$

In Exercises 7 through 16, use the method of Section 4.4 to find the first five terms of the general solution of the indicated differential equation. Center the series at x_0 and express your answer in the form $y_p + y_h$ (see Example 4.4.2).

7. $y'' + y = 0, \quad x_0 = 0$

8. $y' + xy = \sin x, \quad x_0 = 0$

9. $y'' + x^2y = e^x, \quad x_0 = 0$

10. $xy' + y = \ln x, \quad x_0 = 1$

11. $y'' - (\sin x)y = \cos x, \quad x_0 = \pi/2$

12. $y'' + y' + y = x^{-1}, \quad x_0 = 1$

13. $xy' + 2y = x, \quad x_0 = 1$

14. $y'' - xy = 0, \quad x_0 = 1$

15. $y'' + \dfrac{1}{x-1}y = x^2, \quad x_0 = 0$

16. $y'' + xy' + xy = \sin x, \quad x_0 = 0$

An alternative method to find the series solution at a center $x_0 \neq 0$ is to do the change of variables $z = x - x_0$, and find the series centered at $z_0 = 0$.

17. Verify that the change of variables $z = x - 1$ changes

$$\frac{dy}{dx} + x^2y = 1, \qquad y(1) = 2,$$

to

$$\frac{dy}{dz} + (z+1)^2y = 1, \qquad y(0) = 2.$$

Find the first five terms of the power-series solution for $y(z)$, centered at $z_0 = 0$, and compare your answer to that of Exercise 4.

18. Verify that the change of variables $z = x + 1$ changes $d^2y/dx^2 + xy = e^x$, $y(-1) = 0$, $y'(-1) = 1$, to

$$\frac{d^2y}{dz^2} + (z - 1)y = e^{z-1},$$
$$y(0) = 0, \qquad y'(0) = 1. \tag{19}$$

Find the first five terms of the power-series solution to (19), centered at $z = 0$, and compare your answer to that of Exercise 6.

19. Let $z = x - 1$, and rewrite the differential equation in Exercise 10 as a differential equation in y and z. Find the first five terms of a power-series solution of y, centered at $z_0 = 0$, and compare your answer to that of Exercise 10.

20. Let $z = x - \pi/2$, and rewrite the differential equation in Exercise 11 as a differential equation in y and z. Find the first five terms of a power-series solution of y, centered at $z_0 = 0$, and compare your answer to that of Exercise 11.

21. Let $z = x - 1$, and rewrite the differential equation in Exercise 12 as a differential equation in y and z. Find the first five terms of a power-series solution of y, centered at $z_0 = 0$, and compare your answer to that of Exercise 12.

4.5 Series Solution at an Ordinary Point (Undetermined Coefficients)

This section will present an alternative method to that of Section 4.4 for the determination of a series solution of

$$a(x)y''(x) + b(x)y'(x) + c(x)y(x) = f(x) \tag{1}$$

at a point x_0, where a, b, c, f all have power-series expansions and $a \neq 0$—that is, at an **ordinary point** of the differential equation. The idea behind the method is much like the method of undetermined coefficients in Chapter 2. Given a form for the solution, in this case a Taylor series, we substitute the form into the differential equation. An algebraic problem in the unknown coefficients of the series then must be solved.

Shifting Summation Indices for Series

In solving the power-series examples and exercises of this and later sections, an elementary shifting of summation indices is usually required. Consider the power series around $x = 0$,

$$\sum_{n=0}^{\infty} c_n x^n = c_0 + c_1 x + c_2 x^2 + \cdots. \tag{2}$$

Frequently, we will want to consider the same series, but for example in powers of $n + 2$ instead of n. We claim

> The series is the same if n is increased by 2 in every term as long as the series starts 2 sooner.

This is valid because

$$\sum_{n=-2}^{\infty} c_{n+2}x^{n+2} = c_0 + c_1x + c_2x^2 + \cdots. \tag{3}$$

Comparing (2) and (3), we see that

$$\sum_{n=0}^{\infty} c_n x^n = \sum_{n=-2}^{\infty} c_{n+2}x^{n+2}. \tag{4}$$

This result is also valid if the index of the series is increased by any constant, not just 2. The index can also be decreased if the starting value is increased. One can also establish formula (4) by doing a change of variables. Letting $m = n - 2$ in (2) and noting that $m = -2$ when $n = 0$, we get

$$\sum_{n=0}^{\infty} c_n x^n = \sum_{m=-2}^{\infty} c_{m+2}x^{m+2},$$

which is equivalent to (4). Students who have trouble using (4) may find using a substitution like $m = n - 2$ helpful at first.

This type of shifting of the index greatly simplifies finding the series representation of the solution of linear differential equations.

EXAMPLE 4.5.1 Series Solution at an Ordinary Point $x = 0$

Find the general solution of

$$y'' + x^2y = 0, \tag{5}$$

using power-series solutions centered at $x_0 = 0$.

SOLUTION There are no singular points for (5). Thus, the solution will be well behaved (analytic) for all x. Since $x = 0$ is an ordinary point, a power-series solution exists:

$$y = \sum_{n=0}^{\infty} c_n x^n = c_0 + c_1x + c_2x^2 + c_3x^3 + \cdots. \tag{6}$$

The differential equation (5) is second-order, so that the initial conditions $y(0)$, $y'(0)$ can be taken as arbitrary. But from (6) we see that $y(0) = c_0$ and $y'(0) = c_1$. Thus the constants c_0 and c_1 can be taken as arbitrary. To determine the other coefficients c_n, we substitute (6) into (5). We need the second derivative of y:

$$y' = \sum_{n=0}^{\infty} c_n n x^{n-1},$$

$$y'' = \sum_{n=0}^{\infty} c_n n(n-1)x^{n-2}.$$

Now substituting these series for y and y'' into (5) yields

$$\sum_{n=0}^{\infty} c_n n(n-1)x^{n-2} + x^2 \sum_{n=0}^{\infty} c_n x^n$$

$$= \sum_{n=0}^{\infty} c_n n(n-1)x^{n-2} + \sum_{n=0}^{\infty} c_n x^{n+2} = 0. \tag{7}$$

The algebra will simplify if we can add these two series. To do so, we get all series in terms of the same powers of x. There are several possibilities. We usually change the notation of the series with the smaller powers. We wish to change the leftmost series in (7) from $n-2$ to $n+2$. That is, n must be increased by 4. According to our theorem concerning the manipulation of infinite series, we must be consistent. Every n in the first sum can be increased by 4 if the sum starts 4 sooner:

$$\sum_{n=0}^{\infty} c_n n(n-1)x^{n-2} = \sum_{n=-4}^{\infty} c_{n+4}(n+4)\,[(n+4)-1]x^{(n+4)-2}$$

$$= \sum_{n=-4}^{\infty} c_{n+4}(n+4)(n+3)x^{n+2}.$$

Using this result, (7) becomes

$$\sum_{n=-4}^{\infty} c_{n+4}(n+4)(n+3)x^{n+2} + \sum_{n=0}^{\infty} c_n x^{n+2} = 0. \tag{8}$$

We cannot combine the series yet, since they do not begin with the same value of n. This is taken care of by writing the extra terms ($n = -4, -3, -2, -1$) of the first series separately:

$$0x^{-2} + 0x^{-1} + 2c_2x^0 + 6c_3x^1$$

$$+ \sum_{n=0}^{\infty} c_{n+4}(n+4)(n+3)\,x^{n+2} + \sum_{n=0}^{\infty} c_n x^{n+2} = 0.$$

Now combine the two series to get

$$2c_2x^0 + 6c_3x^1 + \sum_{n=0}^{\infty} [c_{n+4}(n+4)(n+3) + c_n]x^{n+2} = 0. \tag{9}$$

On each side of the equal sign in (9) we have a power series. (The one on the right is $0 = 0 + 0x + 0x^2 + 0x^3 + \cdots$). Since the power-series expansion at a given point is unique, the corresponding coefficients must be equal. Thus, equating coefficients of x^n gives

$$x^0: \quad 2c_2 = 0 \quad \text{or} \quad c_2 = 0,$$

$$x^1: \quad 6c_3 = 0 \quad \text{or} \quad c_3 = 0,$$

and

$$x^{n+2} \text{ for } n \geq 0\text{: } c_{n+4}(n+4)(n+3) + c_n = 0. \tag{10}$$

Equation (10) is called a **recursion** formula. It can be solved for subsequent values of the coefficients:

$$c_{n+4} = -\frac{c_n}{(n+4)(n+3)} \qquad \text{for } n \geq 0. \tag{11}$$

We can take c_0, c_1 arbitrary, and we know that $c_2 = c_3 = 0$. This formula enables us to find the remaining coefficients recursively in terms of earlier values of the coefficients. Thus, for example, (11) tells us that

$$\begin{aligned}
n = 0: &\quad c_4 = -\frac{1}{4 \cdot 3}c_0, \\
n = 1: &\quad c_5 = -\frac{1}{5 \cdot 4}c_1, \\
n = 2: &\quad c_6 = -\frac{1}{6 \cdot 5}c_2 = 0, \\
n = 3: &\quad c_7 = -\frac{1}{7 \cdot 6}c_3 = 0, \\
n = 4: &\quad c_8 = -\frac{1}{8 \cdot 7}c_4 = \frac{1}{8 \cdot 7 \cdot 4 \cdot 3}c_0.
\end{aligned}$$

Therefore,

$$\begin{aligned}
y &= \sum_{n=0}^{\infty} c_n x^n = c_0 + c_1 x + c_2 x^2 + c_3 x^3 + \cdots \\
&= c_0 + c_1 x - \frac{1}{4 \cdot 3}c_0 x^4 - \frac{1}{5 \cdot 4}c_1 x^5 + \frac{1}{8 \cdot 7 \cdot 4 \cdot 3}\, c_0 x^8 + \cdots \\
&= c_0\left(1 - \frac{1}{4 \cdot 3}x^4 + \frac{1}{8 \cdot 7 \cdot 4 \cdot 3}x^8 + \cdots\right) + c_1\left(x - \frac{1}{5 \cdot 4}x^5 + \cdots\right).
\end{aligned} \tag{12}$$

The theory of homogeneous linear differential equations developed in Chapter 2 says that $y = \tilde{c}_1 y_1 + \tilde{c}_2 y_2$. From (12) we know that the series representation of two linearly independent homogeneous solutions of the original differential equation (5) is

$$y_1 = 1 - \frac{1}{4 \cdot 3}x^4 + \frac{1}{8 \cdot 7 \cdot 4 \cdot 3}x^8 + \cdots \qquad \text{and} \qquad y_2 = x - \frac{1}{5 \cdot 4}x^5 + \cdots.$$

These series solutions are most useful as approximations to the solutions of the differential equation near the ordinary point $x = 0$. If solutions are desired away from $x = 0$, there are generally two good choices: Either obtain a different series solution in the neighborhood of the desired point or use numerical solutions instead of series solutions. ◄

EXAMPLE 4.5.2 **Series Solution at an Ordinary Point $x = 0$**

Find the general solution of

$$y'' + x^2 y' - \frac{3}{2}xy = 4 + 7x, \tag{13}$$

using power-series solutions centered at $x_0 = 0$.

Solution There are no singular points for (13). Thus, the solution will be well behaved (analytic) for all x. Since $x = 0$ is an ordinary point, a power-series solution exists:

$$y = \sum_{n=0}^{\infty} c_n x^n. \tag{14}$$

To determine the coefficients c_n, we substitute (14) into (13). We need the first and second derivatives of y:

$$y' = \sum_{n=0}^{\infty} c_n n x^{n-1},$$

$$y'' = \sum_{n=0}^{\infty} c_n n(n-1)x^{n-2}.$$

Substituting these series for y, y', y'' into (13) yields

$$\sum_{n=0}^{\infty} c_n n(n-1)x^{n-2} + x^2 \sum_{n=0}^{\infty} c_n n x^{n-1} - \frac{3}{2}x \sum_{n=0}^{\infty} c_n x^n$$

$$= \sum_{n=0}^{\infty} c_n n(n-1)x^{n-2} + \sum_{n=0}^{\infty} c_n n x^{n+1} - \frac{3}{2} \sum_{n=0}^{\infty} c_n x^{n+1} = 4 + 7x. \tag{15}$$

We wish to change the first series from $n - 2$ to $n + 1$. That is, n must be increased by 3. According to our theorem concerning the manipulation of infinite series, we must be consistent. Every n in the first sum can be increased by 3 if the sum starts 3 sooner. Thus (15) becomes

$$\sum_{n=-3}^{\infty} c_{n+3}(n+3)(n+2)x^{n+1} + \sum_{n=0}^{\infty} \left(n - \frac{3}{2}\right)c_n x^{n+1} = 4 + 7x, \tag{16}$$

where the second and third series in (15) have been combined. We cannot combine the two series in (16) yet, since they do not begin with the same value of n. This is taken care of by writing the extra terms ($n = -3, -2, -1$) separately and combining the two infinite series:

$$0x^{-2} + 0x^{-1} + 2c_2x^0 + \sum_{n=0}^{\infty} \left[c_{n+3}(n+3)(n+2) + \left(n - \frac{3}{2}\right)c_n\right]x^{n+1} = 4 + 7x.$$

Equating coefficients of powers of x gives

$$\begin{aligned} x^0: &\quad 2c_2 = 4, \\ (n=0)x^1: &\quad 6c_3 - \frac{3}{2}c_0 = 7, \end{aligned} \tag{17}$$

and

$$n \geq 1: \quad c_{n+3}(n+3)(n+2) + \left(n - \frac{3}{2}\right)c_n = 0. \tag{18}$$

Thus,

$$c_{n+3} = -\frac{(n - \frac{3}{2})c_n}{(n+3)(n+2)} \qquad \text{for } n \geq 1. \tag{19}$$

From (17), we have $c_2 = 2$ and $c_3 = \frac{7}{6} + \frac{1}{4}c_0$. Then (19) gives

$$n = 1: \quad c_4 = \frac{\frac{1}{2}}{4 \cdot 3}c_1$$

$$n = 2: \quad c_5 = -\frac{\frac{1}{2}}{5 \cdot 4}c_2 = -\frac{1}{20}.$$

Therefore,

$$\begin{aligned} y &= \sum_{n=0}^{\infty} c_n x^n = c_0 + c_1 x + c_2 x^2 + c_3 x^3 + \cdots \\ &= c_0 + c_1 x + 2x^2 + \left(\frac{7}{6} + \frac{1}{4}c_0\right)x^3 + \frac{1}{24}c_1 x^4 - \frac{1}{20}x^5 + \cdots \\ &= 2x^2 + \frac{7}{6}x^3 - \frac{1}{20}x^5 + \cdots + c_0\left(1 + \frac{1}{4}x^3 + \cdots\right) + c_1\left(x + \frac{1}{24}x^4 + \cdots\right). \end{aligned}$$

From the theory of linear differential equations developed in Chapter 2, which says that $y = y_p + \tilde{c}_1 y_1 + \tilde{c}_2 y_2$, we know that

$$y_p = 2x^2 + \frac{7}{6}x^3 - \frac{1}{20}x^5 + \cdots \tag{20}$$

is a series representation around $x = 0$ of a particular solution of (13) and that

$$y_1 = 1 + \frac{1}{4}x^3 + \cdots \qquad \text{and} \qquad y_2 = x + \frac{1}{24}x^4 + \cdots$$

are series representations of two linearly independent homogeneous solutions of the original differential equation (13). It can be shown that (20) is the unique particular solution corresponding to the initial conditions $y(0) = 0$ and $y'(0) = 0$. ◄

EXAMPLE 4.5.3 Series Solution at an Ordinary Point $x \neq 0$

For the linear differential equation

$$y'' + xy' + 2y = 0, \tag{21}$$

find the series solution centered at $x_0 = 1$, using the method of this section. Find the first four terms of the solution and the recursion relationship for finding additional terms. (The interest in $x_0 = 1$ could come from initial conditions being specified at $x = 1$.)

SOLUTION Since $x = 1$ is an ordinary point, we let

$$y = \sum_{n=0}^{\infty} c_n(x-1)^n. \tag{22}$$

The coefficient x of y' needs to be written as a power series centered at $x_0 = 1$, that is, in terms of powers of $(x - 1)$. This can be done using the Taylor-series formulas of Section 4.2 or by inspection:

$$x = 1 + (x - 1).$$

Alternatively, we introduce a new variable

$$z = x - 1$$

and consider obtaining a series solution for the differential equation (21) centered at $z = 0$ (which is equivalent to $x = 1$). We shall use this latter approach, since it reduces the amount of writing and simplifies the notation. If we let $z = x - 1$, then $x = z + 1$, and the differential equation (21) becomes

$$\frac{d^2y}{dz^2} + (1 + z)\frac{dy}{dz} + 2y = 0 \tag{23}$$

because of the chain rule $[dy/dx = (dy/dz) \cdot (dz/dx) = dy/dz]$. The series (22) for y and its derivatives are

$$\begin{aligned} y &= \sum_{n=0}^{\infty} c_n z^n, \\ \frac{dy}{dz} &= \sum_{n=0}^{\infty} c_n n z^{n-1}, \\ \frac{d^2y}{dz^2} &= \sum_{n=0}^{\infty} c_n n(n-1) z^{n-2}. \end{aligned} \tag{24}$$

Substituting (24) into (23) and multiplying out the term with $(1 + z)$ yields

$$\sum_{n=0}^{\infty} c_n n(n-1)z^{n-2} + \sum_{n=0}^{\infty} c_n n z^{n-1} + \sum_{n=0}^{\infty} c_n n z^n + 2\sum_{n=0}^{\infty} c_n z^n = 0. \tag{25}$$

We now rewrite the first two series of (25) in terms of z^n by shifting the summation index by 2 and 1, respectively:

$$\sum_{n=-2}^{\infty} c_{n+2}(n+2)(n+1)z^n + \sum_{n=-1}^{\infty} c_{n+1}(n+1)z^n + \sum_{n=0}^{\infty} c_n n z^n + 2\sum_{n=0}^{\infty} c_n z^n = 0. \tag{26}$$

However, in the first sum there is no contribution from $n = -2$ and $n = -1$, and in the second sum there is no contribution from $n = -1$. Thus, these series can be combined as

$$\sum_{n=0}^{\infty} [c_{n+2}(n+2)(n+1) + c_{n+1}(n+1) + c_n(n+2)]z^n = 0. \tag{27}$$

Now equate the coefficients of powers of z on both sides of the equal sign. We obtain the recursion relationship

$$n \geq 0: \qquad c_{n+2}(n+2)(n+1) + c_{n+1}(n+1) + c_n(n+2) = 0$$

or

$$n \geq 0: \qquad c_{n+2} = -\frac{1}{n+2}c_{n+1} - \frac{1}{n+1}c_n. \tag{28}$$

From (28),

$$n = 0: \qquad c_2 = -\frac{1}{2}c_1 - c_0$$

$$n = 1: \qquad c_3 = -\frac{1}{3}c_2 - \frac{1}{2}c_1 = -\frac{1}{3}\left(-\frac{1}{2}c_1 - c_0\right) - \frac{1}{2}c_1 = -\frac{1}{3}c_1 + \frac{1}{3}c_0$$

Therefore,

$$\begin{aligned} y &= \sum_{n=0}^{\infty} c_n z^n = \sum_{n=0}^{\infty} c_n(x-1)^n \\ &= c_0 + c_1(x-1) + c_2(x-1)^2 + c_3(x-1)^3 + \cdots \\ &= c_0\left[1 - (x-1)^2 + \frac{1}{3}(x-1)^3 + \cdots\right] \\ &\quad + c_1\left[x - \frac{1}{2}(x-1)^2 - \frac{1}{3}(x-1)^3 + \cdots\right] \\ &= c_0 y_1 + c_1 y_2. \end{aligned}$$

This is a linear combination of two linearly independent homogeneous solutions, both of which are well behaved (analytic) at $x = 1$ since $x = 1$ is an ordinary point. ◀

We summarize the method discussed in this section as follows:

Method of Undetermined Series Coefficients

1. Select the center x_0 and make sure that $a(x)$, $b(x)$, $c(x)$, $f(x)$ all have power-series expansions centered at x_0 and $a(x_0) \neq 0$. (In our problems the center will be given.)
2. Let
$$y = \sum_{n=0}^{\infty} c_n (x - x_0)^n, \tag{29}$$
where the c_n are the unknown coefficients of the solution.
3. Write $a(x)$, $b(x)$, $c(x)$, $f(x)$ as Taylor series centered at x_0.
4. Substitute the unknown expansion (29) for y and the series for a, b, c, f into the differential equation (1).
5. Algebraically manipulate the result so as to obtain a single series on each side of the equal sign. This procedure usually takes the following order:
   i. Multiply out any products between series for the coefficients and the series for y.
   ii. By shifting summation variables, get all series in terms of $(x - x_0)^n$.
   iii. By possibly removing the first few terms of some of the series and adding them as separate terms, get all series summed over the same values of n.
   iv. Combine all the series on each side of the equal sign into a single series.
6. Equate coefficients of corresponding powers of $x - x_0$ to obtain equations for the coefficients.

Exercises

In Exercises 1 through 14, find the series solution, centered at x_0, of the differential equation, using the method of this section. Give the recursion relationship for the coefficients and also find the first five coefficients.

1. $y' + 3y = 0, \quad y(0) = 2, \quad x_0 = 0$

2. $y'' + y = 1, \quad y(0) = 0, \quad y'(0) = 1, \quad x_0 = 0$

3. $xy' + y = 0, \quad y(1) = 2, \quad x_0 = 1$

4. $y'' + xy = 0, \quad y(0) = 1, \quad y'(0) = 1, \quad x_0 = 0$

5. $y'' + xy = \dfrac{1}{1-x}, \quad y(0) = 0, \quad y'(0) = 0, \quad x_0 = 0$

6. $y'' - 2xy' + 4y = 0, \quad y(0) = 1, \quad y'(0) = 0, \quad x_0 = 0$

7. $y'' - 2xy' + 6y = 0, \quad y(0) = 0, \quad y'(0) = 1, \quad x_0 = 0$

8. $x^2y' - 2y = 1, \quad y(1) = 0, \quad x_0 = 1$

9. $y' + 3x^2y = e^x, \quad y(0) = 0, \quad x_0 = 0$

10. $y' - 2x^2y = xe^x, \quad y(0) = 1, \quad x_0 = 0$

11. $y'' + x^2y = \sin x, \quad y(0) = 0, \quad y'(0) = 0, \quad x_0 = 0$

12. $y'' + xy' + x^3y = \cos x, \quad y(0) = 1, \quad y'(0) = 1, \quad x_0 = 0$

13. $y' + (x^2 + 1)y = x^2$, $\quad y(0) = 2$, $\quad x_0 = 0$

14. $y' + (x^3 - x)y = e^x$, $\quad y(0) = 1$, $\quad x_0 = 0$

In Exercises 15 through 22, express the general solution as a linear combination of a linearly independent set of solutions, using the method of this section. Obtain recursion formulas and determine the first few terms. Take all power series centered at $x_0 = 0$.

15. $y'' - y = 0$

16. $y'' - x^3y = 0$

17. $(1 + x^2)y'' + y = 0$

18. $(1 - x^2)y'' + xy' + 3y = 0$

19. $y'' + 2xy = 0$

20. $y' - 2xy = 0$

21. $y' + 3x^5y = 0$

22. $y'' + 5x^2y' + (1 + x)y = 0$

23. Suppose that $a(x)$, $b(x)$, $c(x)$, in (1) have power-series expansions centered at $x_0 = 0$ and that $f \equiv 0$ and $a(0) \neq 0$.

a) Show that (1) has solutions with power-series expansions

$$g(x) = x + \sum_{n=2}^{\infty} \alpha_n x^n, \quad h(x) = 1 + \sum_{n=2}^{\infty} \beta_n x^n,$$

for unique constants α_n, β_n.

b) Show that g, h are a fundamental set of solutions for (1).

c) Show that if $y(x) = \sum_{n=0}^{\infty} c_n x^n$ is the solution of (1) with $y(0) = c_0$, $y'(0) = c_1$, then $y = c_0 h + c_1 g$ and $c_n = \beta_n c_0 + \alpha_n c_1$ for all $n \geq 2$.

In Exercises 24 through 31, express the general solution as a linear combination of linearly independent homogeneous solutions, using the method of this section. Obtain a recursion formula and determine the first few terms. Take the power series centered at the given x_0.

24. $y'' - x^2y = 0$ around $x_0 = 3$

25. $y'' - x^2y = 0$ around $x_0 = 4$

26. $y'' + xy = 0$ around $x_0 = 1$

27. $y'' + xy = 0$ around $x_0 = 2$

28. $y'' + xy' + 5y = 0$ around $x_0 = 3$

29. $y'' - 2xy' + 3y = 0$ around $x_0 = 1$

30. $y'' + 4y' - (1 + 3x)y = 0$ around $x_0 = -1$

31. $y'' - (4 + x)y' + 3xy = 0$ around $x_0 = 2$

4.6 Approximate Solution at a Regular Singular Point (Theory)

The preceding sections have discussed

$$a(x)y'' + b(x)y' + c(x)y = h(x),$$

or equivalently,

$$y'' + p(x)y' + q(x)y = f(x).$$

In summary, if p, q, f have Taylor series around $x = x_0$ (in which case $x = x_0$ is called an ordinary point), all solutions are well behaved (analytic) at $x = x_0$ and may be represented by a Taylor series around $x = x_0$:

$$y = \sum_{n=0}^{\infty} c_n (x - x_0)^n.$$

In many applications, either $p(x)$ or $q(x)$ does not have a Taylor series around $x = x_0$ [because of a zero of $a(x)$]. In this case, we say that $x = x_0$ is a singular point. What can we say about the solution of the differential equation near $x = x_0$ if $x = x_0$ is a singular point? If $x = x_0$ is a singular point, then it is not guaranteed that all solutions are well behaved (analytic) at $x = x_0$. If $x = x_0$ is a singular point, then solutions may or may not be well behaved (analytic) at $x = x_0$.

EXAMPLE 4.6.1 Singular Point with Solution Not Analytic

We begin with an elementary example of a first-order equation:

$$xy' + y = 0. \tag{1}$$

All points are ordinary points except $x = 0$, which is a singular point. All solutions are guaranteed to be well behaved everywhere except possibly at the singular point $x = 0$.

Equation (1) can be solved by separation, since it is first-order linear and homogeneous. However, (1) may be solved even more quickly because it is a first-order Euler equation (Section 2.13). A homogeneous solution may be obtained by substituting $y = x^r$ in (1). This immediately yields $r + 1 = 0$ or $r = -1$. Thus, the general solution of (1) is

$$y = cx^{-1}.$$

This solution is well behaved everywhere except at the singular point $x = 0$, where the solution has a vertical asymptote if $c \neq 0$. ◄

From this example, we see that the solution can be infinite or undefined at the singular point. However, the next example shows that the solution to the differential equation might be well behaved at a singular point.

EXAMPLE 4.6.2 Singular Point with Analytic Solution

Consider the first-order Euler equation

$$xy' - 2y = 0. \tag{2}$$

Again all points are ordinary points except $x = 0$, which is a singular point. The solutions are not guaranteed to be well behaved at $x = 0$. A homogeneous solution may be obtained by substituting $y = x^r$, in which case we obtain $r - 2 = 0$ or $r = 2$. Thus, the general solution of (2) is

$$y = cx^2. \tag{3}$$

Even though $x = 0$ is a singular point, in this example, the solution is well behaved at the singular point. Note, however, from (3) that $y(0) = 0$ for all c.

Thus, while the solution is well behaved at the singular point, we cannot pick the initial condition $y(0)$ to be arbitrary at the singular point. ◄

Second-Order Euler Equation

The behavior of solutions to second-order Euler equations,

$$ax^2y'' + bxy' + cy = 0, \tag{4}$$

with a, b, c constant, will indicate some of the wide variety of phenomena that can occur at a singular point (here $x = 0$) for second-order equations. At least one solution is always of the form

$$y = x^r,$$

where r could be any number. The exponent r was found as the root of the quadratic equation we called the indicial equation. For Euler equations, negative powers are possible. Even for positive powers, the power may be fractional (in which case some derivative does not exist at the singular point). The oscillations associated with a linear combination of $x^\alpha \sin(\beta \ln x)$ and $x^\alpha \cos(\beta \ln x)$ are also possible if the roots are complex, $r = \alpha \pm i\beta$.

Analysis of Regular Singular Points

Here we will only analyze the homogeneous differential equation,

$$a(x)y'' + b(x)y' + c(x)y = 0, \tag{5}$$

or equivalently,

$$y'' + p(x)y' + q(x)y = 0. \tag{6}$$

Our goal is twofold: to determine the behavior in the neighborhood of the singular point and to obtain a series representation of the solution that is valid near, but not necessarily at, the singular point. A singular point is a place where $a(x) = 0$. Usually we analyze differential equations in the form (6). A singular point is characterized by either $p(x)$ or $q(x)$ being undefined at the singular point because of division by zero. Frequently $p(x)$ or $q(x)$ will be a ratio of functions whose denominator is zero at the singular point.

We begin by assuming that $x_0 = 0$ is the singular point. If the singular point x_0 is not equal to zero, then the change of independent variables $z = x - x_0$ gives an equation in z for which $z = 0$ is a singular point. Thus, there is no loss of generality by discussing only the case in which $x_0 = 0$ is a singular point.

First we need to specify the types of equations our method will work on. The method we develop in the next few sections is not valid for all singular points; it is valid only for singular points of a special kind, called regular singular points. First we need the definition of a pole.

Definition of a Pole The function $g(x)$ is said to have a **pole** at x_0 if

$$g(x) = \frac{G(x)}{(x - x_0)^m},$$

where m is a positive integer and $G(x)$ is analytic at x_0 with $G(x_0) \neq 0$. The pole is called single (simple), double, or triple depending on whether $m = 1$, 2, or 3. A pole is a special kind of singularity.

Definition of a Regular Singular Point The point $x_0 = 0$ is a **regular singular point** of

$$y'' + p(x)y' + q(x)y = 0 \tag{7}$$

if the singularity of the coefficient $p(x)$ is no worse than a **simple pole** and the singularity of the coefficient $q(x)$ is no worse than a **double pole**. The precise definitions of these types of singularities are

$$\begin{aligned} p(x) &= \frac{P(x)}{x}, \\ q(x) &= \frac{Q(x)}{x^2}, \end{aligned} \tag{8}$$

where $P(x)$ and $Q(x)$ are analytic at $x = 0$ (that is, P, Q both have Taylor-series expansions centered at $x = 0$).

If the singularity is at $x = x_0$ instead of $x = 0$, then

$$\begin{aligned} p(x) &= \frac{P(x)}{x - x_0}, \\ q(x) &= \frac{Q(x)}{(x - x_0)^2}, \end{aligned} \tag{9}$$

where $P(x)$ and $Q(x)$ are analytic at $x = x_0$.

We believe (7) and (8) is the simplest way to understand and memorize the definition of a regular singular point at $x = 0$. An alternative expression for this definition follows by using (8) to replace $p(x)$ and $q(x)$ in the differential equation (7),

$$y'' + \frac{P(x)}{x}y' + \frac{Q(x)}{x^2}y = 0,$$

and multiplying by x^2:

$$x^2y'' + xP(x)y' + Q(x)y = 0. \tag{10}$$

Again $P(x)$ and $Q(x)$ are analytic at $x = 0$ (have Taylor-series expansions centered at $x = 0$).

If a singular point is not regular, we call it **irregular**. Similar questions could be asked about the behavior of solutions of differential equations in the neighborhood of an irregular singular point.

Corresponding Euler (Equidimensional) Equation

We are interested in the approximate behavior of the solutions of the differential equation (7) or (10) in the neighborhood of a regular singular point $x = 0$. We should not be surprised by the importance of the **corresponding Euler (equidimensional) equation** obtained by replacing in (10) the analytic functions $P(x)$ and $Q(x)$ by their constant values at $x = 0$:

$$x^2y'' + xP(0)y' + Q(0)y = 0. \tag{11}$$

This can also be obtained in the same way directly from (7) and (8).

We claim that in some sense (to be described more precisely later in this section) the solution of (7) or (10) may be approximated in the neighborhood of a regular singular point by the solution of the corresponding Euler equation (11). Since (11) is an Euler equation, some solutions of (11) may be easily obtained in the form $y = x^r$, where r satisfies the **indicial equation**

$$r(r-1) + P(0)r + Q(0) = 0. \tag{12}$$

In some situations, the solutions of (7) or (10) can be approximated by the exact solutions

$$y \approx c_1x^{r_1} + c_2x^{r_2}$$

of the corresponding Euler equation (11). We will learn in what sense the approximation is accurate, and in what examples it must be modified.

EXAMPLE 4.6.3 Regular Single Point, Corresponding Euler Equation, and the Indicial Equation

For the linear differential equation

$$x(x+4)y'' + y' + \frac{5x}{(x+4)(x-2)^3}y = 0: \tag{13}$$

a. Find all real singular points. Classify the real singular points as to those that are regular and those that are irregular.
b. For each real regular singular point, determine the corresponding Euler equation.
c. Derive the indicial equation for each regular singular point from part (b).

SOLUTION

a. To determine and classify the singular points, it is best to make the coefficient of y'' to be 1 by dividing (13) by $x(x+4)$:

$$y'' + \frac{1}{x(x+4)}y' + \frac{5}{(x+4)^2(x-2)^3}y = 0. \tag{14}$$

It can be seen from (14) that all x are ordinary points except

$x = 0$, which is a regular singular point,
$x = -4$, which is a regular singular point,
$x = 2$, which is an irregular singular point.

We discuss the regular singular points in the following paragraphs.
$x = 2$ is an irregular singular point because the coefficient $q(x) = 5/(x+4)^2(x-2)^3$ has a singularity at $x = 2$ worse than a double pole. (It has a triple pole at $x = 2$.)

b. To analyze the singular point $x = 0$, we put (14) in the appropriate form:

$$p(x) = \frac{P(x)}{x} = \frac{1}{x(x+4)} \qquad \text{so that} \quad P(x) = \frac{1}{x+4}$$

$$q(x) = \frac{Q(x)}{x^2} = \frac{5}{(x+4)^2(x-2)^3} \qquad \text{so that} \quad Q(x) = \frac{5x^2}{(x+4)^2(x-2)^3}.$$

$x = 0$ is a regular singular point because $P(x)$ and $Q(x)$ are analytic at $x = 0$. For $x = 0$ to be a regular singular point, $p(x)$ can have at worst a simple pole (which it does) and $q(x)$ can have at worst a double pole [actually $q(x)$ is well behaved at $x = 0$, which is certainly nicer than an allowable double pole]. The corresponding Euler equation at $x = 0$ is

$$x^2y'' + \frac{1}{4}xy' + 0y = 0,$$

since $P(0) = \frac{1}{4}$ and $Q(0) = 0$.

c. The indicial equation is derived by letting $y = x^r$:

$$r(r-1) + \frac{1}{4}r = r\left(r - \frac{3}{4}\right) = 0.$$

The roots are $r = 0$ and $r = \frac{3}{4}$. Near $x = 0$, we expect one solution of the differential equation to behave like $y \approx x^0$ and a second independent solution to behave like $y \approx x^{\frac{3}{4}}$.

b. To analyze the singular point $x = -4$, we put (14) in the appropriate form:

$$p(x) = \frac{P(x)}{x+4} = \frac{1}{x(x+4)} \qquad \text{so that} \quad P(x) = \frac{1}{x}$$

$$q(x) = \frac{Q(x)}{(x+4)^2} = \frac{5}{(x+4)^2(x-2)^3} \qquad \text{so that} \quad Q(x) = \frac{5}{(x-2)^3}.$$

$x = -4$ is a regular singular point because $P(x)$ and $Q(x)$ are analytic at $x = -4$. For $x = -4$ to be a regular singular point, $p(x)$ can have at worst a simple pole (which it does) and $q(x)$ can have at worst a double pole (which it does). The corresponding Euler equation is

$$(x+4)^2 y'' - \frac{1}{4}(x+4)y' - \frac{5}{6^3}y = 0,$$

since $P(-4) = -\frac{1}{4}$ and $Q(-4) = -(5/6^3)$.

c. The indicial equation is derived by letting $y = (x+4)^r$:

$$r(r-1) - \frac{1}{4}r - \frac{5}{6^3} = 0.$$

Its real roots could be determined from the quadratic formula. ◄

Summary

From these examples we see how to obtain the indicial equation at a regular singular point directly from the corresponding Euler equation. In the next section we will obtain a series representation of the solution in the neighborhood of the regular singular point and show that the solutions of the corresponding Euler equation approximate in some sense the solution of the differential equation in the neighborhood of the regular singular point.

Exercises

In Exercises 1 through 13:

a) Find all real singular points. Classify the real singular points as to those that are regular and those that are irregular.

b) For each regular singular point, determine the corresponding Euler (equidimensional) equation.

c) Derive the indicial equation for each from part (b).

1. $(x+1)^2(x+2)^2y'' + 4x^2y = 0$

2. $y'' + \dfrac{1}{x(x-3)}y' + \dfrac{6}{x^2(x-3)^2}y = 0$

3. $y'' + \dfrac{1}{x(x^2+9)}y' + \dfrac{7}{x^2(x+3)^2}y = 0$

4. $y'' + \dfrac{1}{x^2(x+5)}y' + \dfrac{8}{x(x+5)^2}y = 0.$

5. $y'' + \dfrac{1}{x(x+2)^2}y' + \dfrac{4(x+1)}{x^2(x+2)^2}y = 0$

6. $y'' + \dfrac{(x+3)}{x(x-2)}y' + \dfrac{5}{(x+3)^2(x-2)^2}y = 0$

7. $y'' + \dfrac{x}{(x+3)(x+5)}y' + \dfrac{2x^2}{(x+3)^2(x+5)^2}y = 0$

8. $y'' + \dfrac{5}{x^3(x-1)}y' + \dfrac{1}{x(x+4)^2(x-1)^2}y = 0$

9. $y'' + \dfrac{1}{(x^2-1)(x+4)}y' + \dfrac{(x+3)}{(x+4)^2(x-1)^2}y = 0$

10. $y'' + \dfrac{(x^2-1)}{x(x+4)(x-2)}y' + \dfrac{(x^2-1)^2}{x^2(x+4)^2(x-2)^2}y = 0$

11. $y'' + \dfrac{1}{x(x^2+4)}y' + \dfrac{(x+1)}{x^2(x-2)^2}y = 0$

12. $y'' + \dfrac{1}{x(x-1)}y' + \dfrac{5}{x^2(x^2+4)^2(x-1)^3}y = 0$

13. $y'' + \dfrac{1}{x(x+4)(x-5)}y' + \dfrac{5}{x^2(x+4)(x-5)^2}y = 0$

4.7 Series Solution at a Regular Singular Point (Method of Frobenius)

If a point is a regular singular point of a differential equation, then all solutions of the differential equation may be represented by an infinite series in the neighborhood of the regular singular point. The **method of Frobenius** is used to obtain these solutions and is summarized in the next theorem.

■ **THEOREM 4.7.1 Frobenius**

Suppose that $x = 0$ is a regular singular point of $y'' + p(x)y' + q(x)y = 0$ and that $y'' + p(x)y' + q(x)y = 0$ is written as

$$x^2y'' + xP(x)y + Q(x)y = 0, \tag{1}$$

where P, Q are analytic at $x = 0$ (have Taylor-series expansions centered at $x = 0$). The **indicial equation**, derived from the corresponding Euler equation, is

$$r(r-1) + P(0)r + Q(0) = 0. \tag{2}$$

The indicial equation has roots r_1, r_2. We take $r_1 \geq r_2$ if the roots are real. There are three possibilities.

a. If the roots of the indicial equation differ by a noninteger ($r_1 \neq r_2$ and $r_1 - r_2$ is not an integer), then there are two independent solutions of (1) in the form

$$y_1(x) = |x|^{r_1} \sum_{n=0}^{\infty} c_n x^n, \qquad y_2(x) = |x|^{r_2} \sum_{n=0}^{\infty} \tilde{c}_n x^n, \tag{3}$$

where c_n, $\tilde{c}_n$ are constants with $c_0 \neq 0$ and $\tilde{c}_0 \neq 0$.

b. If the roots of the indicial equation are repeated ($r_1 = r_2$), then one solution is again of the form

$$y_1(x) = |x|^{r_1} \sum_{n=0}^{\infty} c_n x^n, \tag{4}$$

with $c_0 \neq 0$. The form of the second independent solution involves logarithms:

$$y_2(x) = y_1(x) \ln |x| + |x|^{r_1} \sum_{n=0}^{\infty} b_n x^n. \tag{5}$$

c. If the roots of the indicial equation differ by an integer ($r_1 - r_2 = N > 0$, N an integer), then the solution corresponding to the largest root is always of the simpler form

$$y_1(x) = |x|^{r_1} \sum_{n=0}^{\infty} c_n x^n, \tag{6}$$

with $c_0 \neq 0$. However, the form of the second independent solution may or may not involve logarithms, and is of the form

$$y_2(x) = a y_1(x) \ln |x| + |x|^{r_2} \sum_{n=0}^{\infty} b_n x^n, \tag{7}$$

with $b_0 \neq 0$. In this case, "usually" $a \neq 0$, in which case logarithms exist and we may take $a = 1$. However, it is possible that in some examples (after some computation) $a = 0$, in which case both solutions have the simpler form even though the roots differ by an integer. ■

The simplest examples of regular singular points will be case (a), with real distinct roots differing by a noninteger. If the roots are complex, then they differ by a noninteger and case (a) still applies. It can be shown that the real and imaginary part of $y_1(x)$ will give two real independent solutions of (1). However, we shall not work any examples with complex roots.

If we only consider $x > 0$, the absolute values may be dropped. In all cases, the coefficients of the series solutions are determined by substitution.

Comparison of Method of Frobenius to Corresponding Euler Equation

In case (a), where the roots differ by a noninteger, the approximation of the solution to the differential equation based on taking the first term in each infinite series is the same as the solution of the corresponding Euler equation. Thus, in this case the Euler equation predicts the correct behavior of the solution in the neighborhood of the regular singular point.

In case (b), where the roots are equal, the second solution involves logarithms. This is not very surprising, since the corresponding Euler equation with repeated roots has a similar logarithmic structure. When solving the Euler equation, if only the first term is taken for $y_1(x)$, i.e., $y_1(x) \approx c_0 x^{r_1}$, then $y_1 \ln x \approx c_1 x^{r_1} \ln x$ is the same as the second solution of the corresponding Euler equation, since it has equal roots. The other part of the second solution is smaller near the regular singular point $x = 0$. Thus, in this case the Euler equation predicts the correct behavior of the solution in the neighborhood of the regular singular point.

In case (c), where the roots differ by a positive integer, there are two cases. If $a = 0$, then the solutions behave like the solutions do when the roots do not differ by an integer and have the same approximate behavior in the neighborhood of the regular singular point as the solution of the corresponding Euler equation. In the "usual" case where $a \neq 0$, the second solution has logarithmic behavior, while there is no logarithmic behavior for the solutions of the corresponding Euler equation. Even in this case, the solution near the regular singular point $x = 0$ is approximately the same as that predicted by the corresponding Euler equation:

$$y_2 \approx a c_0 x^{r_1} \ln x + \tilde{c}_0 x^{r_2} \approx \tilde{c}_0 x^{r_2},$$

using (6) and (7), since r_1 is the larger root, $r_1 \geq r_2$.

In all situations, the corresponding Euler equation predicts the correct approximate behavior of the solution in the neighborhood of a regular singular point.

4.7.1 Examples with Roots Not Differing by an Integer

If $x = 0$ is a regular singular point and the roots of the indicial equation differ by a noninteger, then two solutions exist in the form (assuming $x > 0$)

$$y(x) = x^r \sum_{n=0}^{\infty} c_n x^n = \sum_{n=0}^{\infty} c_n x^{n+r}, \tag{8}$$

where $r = r_1$ and $r = r_2$ are the two roots of the indicial equation. Both solutions can be obtained by substituting (8) into the differential equation. The method will work only if r satisfies the indicial equation. It is usually necessary to calculate y' and y'' from (8):

$$y' = \sum_{n=0}^{\infty} (n + r)c_n x^{n+r-1},$$
$$y'' = \sum_{n=0}^{\infty} (n + r)(n + r - 1)c_n x^{n+r-2}. \tag{9}$$

EXAMPLE 4.7.1 Method of Frobenius with Roots Not Differing by an Integer

Find two linearly independent solutions that are series in powers of x for the second-order linear differential equation

$$y'' + \frac{1}{x}y' + \frac{x - \frac{1}{9}}{x^2}y = 0. \tag{10}$$

Solution Here $p(x) = 1/x$ has a simple pole and $q(x) = (x - \frac{1}{9})/x^2$ has a double pole, so that $x = 0$ is a regular singular point. To determine the corresponding Euler equation, we want $p(x) = 1/x = P(x)/x$. Thus $P(x) = 1$. Also, $q(x) = (x - \frac{1}{9})/x^2 = Q(x)/x^2$, so that $Q(x) = x - \frac{1}{9}$. This may be obvious directly from (10). The method of Frobenius is best implemented if we multiply (10) by x^2 to get

$$x^2y'' + xy' + \left(x - \frac{1}{9}\right)y = 0. \tag{11}$$

In general, $x^2y'' + xP(x)y' + Q(x)y = 0$. The corresponding Euler equation is $x^2y'' + xy' - \frac{1}{9}y = 0$, since $P(0) = 1$ and $Q(0) = -\frac{1}{9}$. Thus, the indicial equation is

$$r(r - 1) + r - \frac{1}{9} = r^2 - \frac{1}{9} = 0,$$

which has roots $r_1 = \frac{1}{3}$ and $r_2 = -\frac{1}{3}$. Since the roots are distinct and do not differ by an integer, there will be two solutions of the form

$$y = \sum_{n=0}^{\infty} c_n x^{n+r}. \tag{12}$$

It is quicker to initially consider both values of r simultaneously. Substituting (12) into (11) using (9) and multiplying the coefficients times the series gives

$$\sum_{n=0}^{\infty} (n + r)(n + r - 1)c_n x^{n+r} + \sum_{n=0}^{\infty} (n + r)c_n x^{n+r} + \sum_{n=0}^{\infty} c_n x^{n+r+1} - \frac{1}{9}\sum_{n=0}^{\infty} c_n x^{n+r} = 0.$$

Divide through by x^r to get

$$\sum_{n=0}^{\infty} (n + r)(n + r - 1)c_n x^n + \sum_{n=0}^{\infty} (n + r)c_n x^n + \sum_{n=0}^{\infty} c_n x^{n+1} - \frac{1}{9} \sum_{n=0}^{\infty} c_n x^n = 0. \tag{13}$$

The index $n + 1$ in the third sum needs to be reduced by 1, which can be done using the shift of the summation index we used for ordinary points (see Section 4.5):

$$\sum_{n=0}^{\infty} (n + r)(n + r - 1)c_n x^n + \sum_{n=0}^{\infty} (n + r)c_n x^n + \sum_{n=1}^{\infty} c_{n-1} x^n - \frac{1}{9} \sum_{n=0}^{\infty} c_n x^n = 0.$$

There are three terms corresponding to $n = 0$ and four terms for $n \geq 1$. Combining coefficients of like powers of x^n and setting them equal to zero yields

$$n = 0: \qquad \left[r(r - 1) + r - \frac{1}{9}\right]c_0 = 0, \tag{14}$$

$$n \geq 1: \qquad \left[(n + r)(n + r - 1) + (n + r) - \frac{1}{9}\right]c_n + c_{n-1} = 0. \tag{15}$$

The terms corresponding to the smallest value of n, (14), will always correspond to the indicial equation. In this example, $r(r - 1) + r - \frac{1}{9} = 0$ is the indicial equation. From (14), we have $0 \cdot c_0 = 0$, so that c_0 is arbitrary if r satisfies the indicial equation ($r = \frac{1}{3}$ or $r = -\frac{1}{3}$).

Equation (15) is the recursion formula. Solving for c_n yields

$$n \geq 1: \qquad c_n = -\frac{c_{n-1}}{(n + r)(n + r - 1) + (n + r) - \frac{1}{9}}. \tag{16}$$

The denominator can always be related to the indicial polynomial $r(r - 1) + r - \frac{1}{9}$ by replacing r by $n + r$. (That turns out to be the reason why it is very important that the roots of the indicial equation differ by a noninteger for this method to work in a relatively straightforward way.) Since the indicial equation $r(r - 1) + r - \frac{1}{9} = 0$ factors as $(r - \frac{1}{3})(r + \frac{1}{3}) = 0$, the denominator of (16) also factors:

$$n \geq 1: \qquad c_n = -\frac{c_{n-1}}{(n + r - \frac{1}{3})(n + r + \frac{1}{3})}. \tag{17}$$

This can be checked by multiplying out the denominator. If you are confused by this step, it is OK just to use (16) instead of its equivalent (17). We must now consider the recursion formula (17) for each root separately.

Root $\boldsymbol{r_1 = \frac{1}{3}}$
Substituting $r = \frac{1}{3}$ into (17) gives

$$n \geq 1: \qquad c_n = -\frac{c_{n-1}}{n(n+\frac{2}{3})}. \tag{18}$$

It is important to note that the denominator is zero when $n = 0$ and $n = -\frac{2}{3}$, which are prohibited by the constraint $n \geq 1$. Solving the first term

$$n = 1: \qquad c_1 = -\frac{1}{\frac{5}{3}}c_0 = -\frac{3}{5}c_0$$

$$\cdots$$

There is one constant, c_0, that is arbitrary. All the other constants are determined in terms of c_0. Therefore since $r = \frac{1}{3}$,

$$y = \sum_{n=0}^{\infty} c_n x^{n+r} = x^{\frac{1}{3}} \sum_{n=0}^{\infty} c_n x^n = c_0 x^{\frac{1}{3}}\left(1 - \frac{3}{5}x + \cdots\right). \tag{19}$$

Equation (19) represents one linearly independent homogeneous solution. This solution has a cube root singularity at the regular singular point $x = 0$. This behavior near the regular singular point was predicted by the corresponding Euler equation. The second independent solution corresponds to $r = -\frac{1}{3}$.

Root $\boldsymbol{r_2 = -\frac{1}{3}}$
Substituting $r = -\frac{1}{3}$ into (17) gives

$$n \geq 1: \qquad c_n = -\frac{c_{n-1}}{n(n-\frac{2}{3})}. \tag{20}$$

It is again important to note that the denominator cannot be zero because of the constraint $n \geq 1$. (In other problems, the constraint that n is an integer may prevent the denominator from being zero when the roots differ by a noninteger.) Solving the first term,

$$n = 1: \qquad c_1 = -\frac{1}{\frac{1}{3}}c_0 = -3c_0.$$

$$\cdots$$

There is one constant, c_0, that is arbitrary. It is a different constant from the one corresponding to $r = \frac{1}{3}$. We shall denote it by $\bar{c}_0$. All the other constants are determined in terms of $\bar{c}_0$. Therefore, since $r = -\frac{1}{3}$,

$$y = \sum_{n=0}^{\infty} c_n x^{n+r} = x^{-\frac{1}{3}} \sum_{n=0}^{\infty} c_n x^n = \bar{c}_0 x^{-\frac{1}{3}}(1 - 3x + \cdots). \tag{21}$$

Equation (21) represents the second linearly independent homogeneous solution. This solution is infinite at the regular singular point, since it is proportional to $x^{-\frac{1}{3}}$ there. Its behavior near the regular singular point was predicted by the corresponding Euler equation.

The general solution of (11) is a linear combination of the two solutions (19) and (21):

$$y = c_0 x^{\frac{1}{3}}\left(1 - \frac{3}{5}x + \cdots\right) + \bar{c}_0 x^{-\frac{1}{3}}(1 - 3x + \cdots), \tag{22}$$

where c_0 and $\bar{c}_0$ are arbitrary constants. The basic singular behavior that occurs at the regular singular point $x = 0$ is predicted from the corresponding Euler equation. ◀

4.7.2 Examples with Roots Differing by an Integer

EXAMPLE 4.7.2 Repeated Root

Find two linearly independent solutions using the method of Frobenius expanded around $x = 0$ for the second-order linear differential equation

$$y'' + \frac{1}{x}y' + y = 0. \tag{23}$$

Solution Here, $p(x) = 1/x$ has a simple pole at $x = 0$ and $q(x) = 1$ is analytic (which is less singular than the allowable double pole at $x = 0$). Thus, $x = 0$ is a regular singular point. To determine the corresponding Euler equation, we want $p(x) = 1/x = P(x)/x$, so that $P(x) = 1$, and $q(x) = 1 = Q(x)/x^2$, so that $Q(x) = x^2$. This may be obvious directly from (23). The method of Frobenius is best implemented if we multiply (23) by x^2:

$$x^2y'' + xy' + x^2y = 0. \tag{24}$$

In general, $x^2y'' + xP(x)y' + Q(x)y = 0$. The corresponding Euler equation is $x^2y'' + xy' + 0y = 0$, since $P(0) = 1$ and $Q(0) = 0$. Thus, the indicial equation

$$r(r - 1) + r = r^2 = 0$$

has $r_1 = r_2 = 0$ as a repeated root. From Theorem 4.7.1, we have that (24) has one solution of the form

$$y = \sum_{n=0}^{\infty} c_n x^{n+r} = \sum_{n=0}^{\infty} c_n x^n, \tag{25}$$

since $r = 0$. Substituting (25) into (24) and performing some algebra gives

$$\sum_{n=0}^{\infty} n(n-1)c_n x^n + \sum_{n=0}^{\infty} nc_n x^n + \sum_{n=0}^{\infty} c_n x^{n+2} = 0.$$

The index $n + 2$ in the last sum needs to be reduced by 2 to be the same as the others. This can be done by shifting the summation index to get

$$\sum_{n=0}^{\infty} n(n-1)c_n x^n + \sum_{n=0}^{\infty} nc_n x^n + \sum_{n=2}^{\infty} c_{n-2} x^n = 0.$$

Collecting coefficients of x^n and setting them equal to zero then yields

$$\begin{aligned} n &= 0: & 0 &= 0, \\ n &= 1: & c_1 &= 0, \\ n &\geq 2: & n^2 c_n + c_{n-2} &= 0. \end{aligned} \tag{26}$$

Note that (26) implies that

$$0 = c_1 = c_3 = c_5 = \cdots .$$

Only even powers occur. We see that c_0 is arbitrary (corresponding to the arbitrary multiple of one solution). All other coefficients will be proportional to c_0. We can let $c_0 = 1$. From (26),

$$\begin{aligned} c_2 &= -\frac{1}{2^2} c_0 = -\frac{1}{2^2}, \\ c_4 &= -\frac{1}{4^2} c_2 = \frac{1}{8^2}, \\ &\cdots \end{aligned}$$

so that one solution is

$$y_1(x) = 1 - \frac{1}{4}x^2 + \frac{1}{64}x^4 \cdots . \tag{27}$$

Higher-order terms can be obtained if needed. It can be shown by seeing a pattern in the solution of (26) that $c_{2n} = (-1)^n / 2^{2n}(n!)^2$, but we will not use this result.

There are several ways to find the second solution y_2, none of which are simple. We shall just substitute the form (5) from the Frobenius Theorem into the differential equation (24). That is,

$$y_2(x) = y_1(x) \ln x + w(x),$$

where

$$w(x) = \sum_{n=0}^{\infty} b_n x^n, \qquad \text{since } r = 0. \tag{28}$$

Here the b_n are new constants to be determined. We first find a differential equation for $w(x)$. Substituting (28) into (24) and performing the differentiation yields, after regrouping,

$$(x^2 y_1'' + x y_1' + x^2 y_1) \ln x + x^2 \left(2y_1' \frac{1}{x} - y_1 \frac{1}{x^2} \right) + x y_1 \frac{1}{x} + x^2 w'' + x w' + x^2 w = 0. \tag{29}$$

Since y_1 is a homogeneous solution, the $\ln x$ terms in (29) vanish and $w(x)$ solves the following nonhomogeneous differential equation:

$$x^2 w'' + x w' + x^2 w = -2 x y_1' = -2 \sum_{n=0}^{\infty} c_n n x^n, \tag{30}$$

where we have used the series (25) for $y_1(x)$. We have $c_0 = 1$, and other values of c_n can be determined from the recursion formula (26). The differential equation (30) for $w(x)$ is the same differential equation as the one for $y(x)$ except for the nonhomogeneous term $-2xy_1'$.

We substitute the series expansion (28) for $w(x)$ (and do the same arithmetic on the sequence b_n as we previously did on the sequence c_n):

$$\sum_{n=0}^{\infty} n(n-1)b_n x^n + \sum_{n=0}^{\infty} nb_n x^n + \sum_{n=2}^{\infty} b_{n-2}x^n = -2\sum_{n=0}^{\infty} c_n nx^n. \tag{31}$$

Equating coefficients of x^n yields

$$n = 0: \qquad 0 = 0$$

$$n = 1: \qquad b_1 = -2c_1 = 0$$

and

$$n \geq 2: \qquad n^2 b_n + b_{n-2} = -2nc_n$$

or

$$b_n = -\frac{1}{n^2}(b_{n-2} + 2nc_n). \tag{32}$$

Solving the first few terms,

$$n = 2: \qquad b_2 = -\frac{1}{4}(b_0 + 4c_2) = -\frac{1}{4}(b_0 - 1). \tag{33}$$

In (33) b_0 is arbitrary, since it corresponds to a multiple of the previously determined homogeneous solution $y_1(x)$. One possibility (not the only possibility) is to take $b_0 = 0$, in which case

$$b_2 = \frac{1}{4}.$$

With this choice, the second solution is

$$y_2(x) = \left(1 - \frac{1}{4}x^2 + \cdots\right)\ln x + \left(\frac{1}{4}x^2 + \cdots\right).$$

There are many different second solutions. ◄

In some applications, however, one wants to use only those solutions that are bounded, or even continuous at the singularity. In these cases, only $y_1(x)$ would be used if it can be shown that the other solution has a ln x term.

EXAMPLE 4.7.3 Roots Differing by a Positive Integer

Find two linearly independent solutions using the method of Frobenius centered at $x = 0$ for the second-order linear differential equation

$$y'' - \frac{2}{x}y' + \frac{1}{x}y = 0. \tag{34}$$

Solution Here $p(x) = -2/x$ has a simple pole at $x = 0$, and $q(x) = 1/x$ also has a simple pole at $x = 0$ (which is not as singular as the double pole that is allowable). Thus, $x = 0$ is a regular singular point. To determine the corresponding Euler equation, we want $p(x) = -2/x = P(x)/x$, so that $P(x) = -2$, and $q(x) = 1/x = Q(x)/x^2$, so that $Q(x) = x$. This may be obvious directly from (34). The method of Frobenius is best implemented if we multiply (34) by x^2:

$$x^2y'' - 2xy' + xy = 0. \tag{35}$$

In general, $x^2y'' + xP(x)y' + Q(x)y = 0$. The corresponding Euler equation is $x^2y'' - 2xy' + 0y = 0$, since $P(0) = -2$ and $Q(0) = 0$. Thus, the indicial equation,

$$r(r-1) - 2r = r(r-3) = 0,$$

has roots $r_1 = 3$ and $r_2 = 0$. The roots are distinct but differ by an integer. By the method of Frobenius (Theorem 4.7.1), (35) has solutions of the form

$$y_1 = x^{r_1} \sum_{n=0}^{\infty} c_n x^n, \tag{36}$$

$$y_2 = ay_1 \ln x + x^{r_2} \sum_{n=0}^{\infty} b_n x^n. \tag{37}$$

If $a = 0$, then both series have the same (simple) form with different r's. If $a \neq 0$, then one solution will still be in this form. Thus, we begin as we have done in the other cases, by substituting

$$y = x^r \sum_{n=0}^{\infty} c_n x^n = \sum_{n=0}^{\infty} c_n x^{n+r} \tag{38}$$

into (35). There will always be a solution for the larger root. If we can get a second linearly independent solution for the smaller root, then $a = 0$. If, however, the method fails to find a second linearly independent solution of the form (38) for the smaller root, then we must return to (37) and use $a \neq 0$.

Substituting (38) into (35) and multiplying the coefficients times the series yields

$$\sum_{n=0}^{\infty} (n+r)(n+r-1)c_n x^{n+r} - 2\sum_{n=0}^{\infty} (n+r)c_n x^{n+r} + \sum_{n=0}^{\infty} c_n x^{n+r+1} = 0.$$

Dividing through x^r and shifting the summation index in the third series gives

$$\sum_{n=0}^{\infty} (n+r)(n+r-1)c_n x^n - 2\sum_{n=0}^{\infty} (n+r)c_n x^n + \sum_{n=1}^{\infty} c_{n-1} x^n = 0. \tag{39}$$

Equating coefficients then yields

$$n = 0: \qquad [r(r-1) - 2r]c_0 = 0, \tag{40}$$

$$n \geq 1: \qquad [(n+r)(n+r-1) - 2(n+r)]c_n + c_{n-1} = 0. \tag{41}$$

Equation (40) holds if r is either root of the indicial equation. Then (41) is

$$n \geq 1: \qquad (n+r)(n+r-3)c_n + c_{n-1} = 0. \tag{42}$$

Now we consider the two roots r_1, r_2. We first consider $r_1 = 3$, since there is always a solution for the larger root.

Root $r_1 = 3$
Let $r = r_1 = 3$. Then (42) is

$$n \geq 1: \qquad (n+3)nc_n + c_{n-1} = 0,$$

so that

$$n \geq 1: \qquad c_n = (-1)\frac{c_{n-1}}{n(n+3)}. \tag{43}$$

It is important to note that the denominator is zero when $n = 0$ and $n = -3$, which are prohibited by the constraint $n \geq 1$. Thus, taking $c_0 = 1$,

$$n = 1: \qquad c_1 = -\frac{1}{4}c_0 = -\frac{1}{4},$$

$$n = 2: \qquad c_2 = -\frac{1}{10}c_1 = \frac{1}{40},$$

$$\cdots$$

It can be shown that $c_n = (-1)^n 3!/n!(n+3)!$, but we do not use this fact. Since the larger root is $r = 3$, one solution is

$$y_1 = x^3 \sum_{n=0}^{\infty} c_n x^n = x^3\left(1 - \frac{1}{4}x + \frac{1}{40}x^2 + \cdots\right). \tag{44}$$

This behavior near the regular singular point $x = 0$ was predicted by the corresponding Euler equation.

Root $r_2 = 0$
We first try $a = 0$. Let $r = r_2 = 0$. Then (42) becomes

$$n \geq 1: \qquad n(n-3)b_n + b_{n-1} = 0. \tag{45}$$

The usual method of solving for b_n may have difficulty at $n = 3$, since one would incorrectly divide by zero when solving for b_3. This problem occurs because the roots of the indicial equation differ by an integer. If $n = 3$, then (45) implies that $b_2 = 0$ (and b_3 is arbitrary). Letting $n = 2$ and $n = 1$ in

(45), we find that $b_1 = 0$ and $b_0 = 0$. But then we have

$$y_2 = \sum_{n=0}^{\infty} b_n x^n = \sum_{n=3}^{\infty} b_n x^n = x^3 \sum_{n=0}^{\infty} b_{n+3} x^n,$$

which is the same as (38) with $b_{n+3} = c_n$.

Second Independent Solution

Since taking $a = 0$ does not give a second solution, we take $a = 1$ in (37). Thus

$$y_2(x) = y_1(x) \ln x + w(x)$$

where

$$w(x) = \sum_{n=0}^{\infty} b_n x^{n+r_2} \qquad \text{(In this example, } r_2 = 0.) \tag{46}$$

The b_n are new constants to be determined. By substituting (46) into (35), differentiating, and regrouping, we find a differential equation for $w(x)$:

$$\begin{gathered}(x^2 y_1'' - 2xy_1' + xy_1) \ln x + x^2\left(2y_1'\frac{1}{x} - y_1\frac{1}{x^2}\right) - 2xy_1\frac{1}{x} + \\ x^2 w'' - 2xw' + xw = 0.\end{gathered} \tag{47}$$

Since y_1 is a homogeneous solution of (34), the coefficient of $\ln x$ in (47) is zero. Thus, $w(x)$ solves the following nonhomogeneous differential equation:

$$x^2 w'' - 2xw' + xw = -2xy_1' + 3y_1 = \sum_{n=0}^{\infty} c_n[3 - 2(n+3)]x^{n+3}. \tag{48}$$

The series (44) for $y_1(x)$ is used to get the series on the right in (48). We have $c_0 = 1$. The other values for the c_n are determined from the recursion formula from (43). The differential equation (48) for $w(x)$ is the same differential equation as the one for $y(x)$ except for the nonhomogeneous term $-2xy_1' + 3y_1$.

We substitute the series expansion (46) for $w(x)$ (and do the same arithmetic on the sequence b_n as we previously did on the sequence c_n) to get

$$\sum_{n=0}^{\infty} [n(n-1) - 2n]b_n x^n + \sum_{n=1}^{\infty} b_{n-1} x^n = \sum_{n=3}^{\infty} c_{n-3}(3 - 2n)x^n. \tag{49}$$

Equating coefficients of x^n yields

$$\begin{aligned} n = 0&: & 0 &= 0, \\ n = 1&: & -2b_1 + b_0 &= 0, \\ n = 2&: & -2b_2 + b_1 &= 0, \\ n \geq 3&: & n(n-3)b_n + b_{n-1} &= c_{n-3}(3 - 2n). \end{aligned} \tag{50}$$

The $n = 3$ equation is the only term where b_n cannot be determined from the earlier values. For $n = 3$, (50) implies that b_3 is arbitrary and

$$b_2 = -3c_0 = -3.$$

Thus, from (50),

$$b_0 = 2b_1 = 4b_2 = -12.$$

The remaining b_n may be found recursively from (50) for $n \geq 4$. However, b_3 is arbitrary. (This happens because we can add a multiple of the homogeneous solution y_1 to y_2 and still have a linearly independent homogeneous solution.) The simplest choice for b_3 is

$$b_3 = 0.$$

We then determine b_n for $n \geq 4$ from (50). Using the terms we have found so far, we have

$$y_2(x) = x^3\left(1 - \frac{1}{4}x + \cdots\right)\ln x - 12 - 6x - 3x^2 + \cdots. \tag{51}$$

Equation (51) represents the second linearly independent homogeneous solution. This solution is not analytic at the regular singular point $x = 0$ because of the $x^3 \ln x$ singularity there. In this case, y_2 exists at $x = 0$, since $\lim_{x \to 0} x^3 \ln x = 0$, but a sufficiently high derivative (y''') does not exist at $x = 0$. When the roots of the indicial equation differ by an integer, the singularity of the solution cannot be predicted from the behavior of the corresponding Euler equation. However, the leading-order approximate behavior of $y_2(x)$, namely that $y_2(x)$ can be approximated by a constant near $x = 0$, was predicted by the corresponding Euler equation (for $x > 0$). ◄

Exercises

In Exercises 1 through 9, $x = 0$ is a regular singular point. First, find the indicial equation from the corresponding Euler equation and show that the roots differ by a noninteger. Then, find two linearly independent solutions using the method of Frobenius. Show that the same indicial equation is valid. Determine the recursion formula and solve for the first few terms.

1. $36x^2y'' + x^2y' + 5y = 0$

2. $3xy'' + 2y' + y = 0$

3. $6x(1 - x)y'' + (2 - 15x)y' - 3y = 0$

4. $2x^2y'' - 3xy' + (3 + x)y = 0$

5. $2x^2y'' - 5xy' + (5 + 4x)y = 0$

6. $2x^2y'' - 5xy' + (5 + 3x^2)y = 0$

7. $2x^2y'' - 3xy' + (3 + 2x^2)y = 0$

8. $36x^2y'' + (x^2 + 5x^3)y' + 5y = 0$

9. $3xy'' + 2y' + (1 + 4x)y = 0$

In Exercises 10 through 17, $x = 0$ is a regular singular point. First, find the indicial equation from the corresponding Euler equation and show that the roots are the same or differ by an integer. Then, find two linearly independent solutions using the method of Frobenius. Show that the same indicial equation is valid. Determine the recursion formula and solve for the first few terms.

10. $xy'' - y' - y = 0$

11. $xy'' - y = 0$

12. $xy'' + y' + y = 0$

13. $x^2y'' - 5xy' + (9 - 4x)y = 0$

14. $x^2y'' - 4xy' + (6 + x)y = 0$

15. $x^2y'' - 6xy' + (12 + x)y = 0$

16. $x^2y'' - 6xy' + (12 + 3x^2)y = 0$

17. $x^2y'' - 4xy' + (6 + x^2)y = 0$

Sometimes the infinite series obtained by the method of Frobenius can be summed and the solution of the differential equation recognized as an elementary function. In Exercises 18 and 19, $x = 0$ is a regular singular point. First, find the indicial equation from the corresponding Euler equation and show that the roots differ by a noninteger. Then, find two linearly independent solutions using the method of Frobenius. Show that the same indicial equation is valid. Determine and solve the recursion formula. Show that the solution obtained by series methods is equivalent to the given elementary functions.

18. $4x^2y'' + 2xy' + xy = 0$, $\{y_1 = \sin(x^{\frac{1}{2}}), y_2 = \cos(x^{\frac{1}{2}})\}$

19. $2x(1 - x)y'' + (1 - 5x)y' - y = 0$, $\left\{y_1 = \dfrac{x^{\frac{1}{2}}}{1 - x}, y_2 = \dfrac{1}{1 - x}\right\}$

Sometimes the infinite series obtained by the method of Frobenius can be summed and the solution of the differential equation recognized as an elementary function. In Exercises 20 through 23, $x = 0$ is a regular singular point. First, find the indicial equation from the corresponding Euler equation and show that the roots are the same or differ by an integer. Then find two linearly independent solutions using the method of Frobenius. Show that the same indicial equation is valid. Determine and solve the recursion formula. Show that the solution obtained by series methods is equivalent to the given elementary functions.

20. $x(1 - x)y'' + (1 - 2x)y' = 0$, $\{y_1 = 1, y_2 = \ln x - \ln(1 - x)\}$

21. $x(1 - x)y'' + (1 - 3x)y' - y = 0$, $\left\{y_1 = \dfrac{1}{1 - x}, y_2 = \dfrac{\ln x}{1 - x}\right\}$

22. $x(1 - x)y'' + (2 - 3x)y' - y = 0$, $\left\{y_1 = -\dfrac{1 \ln(1 - x)}{x}, y_2 = \dfrac{1}{x}\right\}$

23. $x(1 - x)y'' + (3 - 2x)y' = 0$, $\{y_1 = 1, y_2 = x^{-2}(1 - 2x)\}$

4.8 Bessel Functions

Many vibrating strings are analyzed using the wave equation, which is a partial differential equation. The solutions for vibration problems take the form of series involving terms like $\sin \alpha nx$, $\cos \alpha nx$. Such Fourier series are interpreted as the sum of the basic harmonic motions. For example, for a violin string of length π with fixed ends, the first four basic harmonic motions are constant multiples of the functions shown in Figure 4.8.1. Rectangular membranes also involve trigonometric functions.

However, if we have a round membrane, fixed at the edges, such as a drum, the shape of the fundamental harmonic motions is different. There is still a function, call it $J(x)$, which has an infinite number of zeros, and the basic harmonic motions take the form $J(\alpha_n x)$ for constants α_n. However, this function is no longer the sine or cosine. The function $J(x)$ will be a **Bessel function**. Bessel functions appear in many problems involving partial differential equations. Since these functions play such an important role in applied

FIGURE 4.8.1

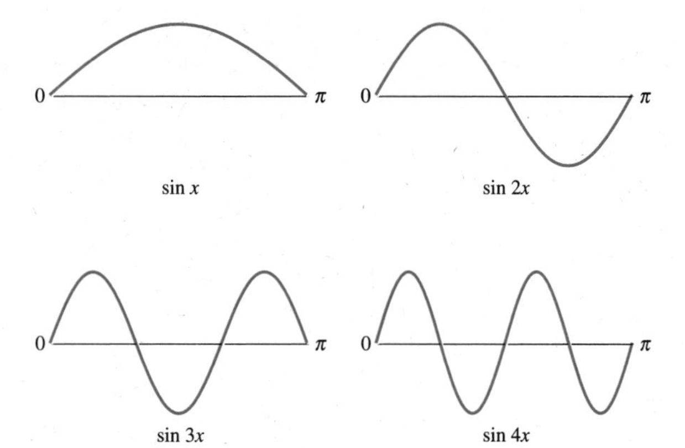

mathematics and because our analysis of the preceding section can be used to study them, this section will briefly examine Bessel functions.

The differential equation

$$x^2\frac{d^2y}{dx^2} + x\frac{dy}{dx} + (x^2 - \nu^2)y = 0 \tag{1}$$

with $\nu \geq 0$ a parameter is called **Bessel's equation**. Note that $x_0 = 0$ is a regular singular point for (1), so that the solutions of (1) near $x = 0$ can be approximated by the solutions of the corresponding Euler equation, $x^2y'' + xy' - \nu^2 y = 0$, whose indicial equation is

$$r(r-1) + r - \nu^2 = r^2 - \nu^2 = 0, \tag{2}$$

which has roots $r = \nu, -\nu$. Thus, if $\nu > 0$, one solution, which we denote $J_\nu(x)$, can be approximated using x^ν near $x = 0$, and a second linearly independent solution, denoted $Y_\nu(x)$, can be approximated using $x^{-\nu}$ near $x = 0$. If $\nu = 0$, then the roots of the indicial equation are repeated 0, 0, and only one independent solution [which we denote $J_0(x)$] can be approximated using x^0 near $x = 0$. If $\nu = 0$, the second independent solution [denoted $Y_0(x)$] can be approximated using $x^0 \ln x$ near $x = 0$. The general solution of (1) can be written as a linear combination of these two linearly independent solutions (whether $\nu > 0$ or $\nu = 0$):

$$y = \alpha J_\nu(x) + \beta Y_\nu(x), \tag{3}$$

where α, β are arbitrary constants. The function (to be described more specifically later) $J_\nu(x)$ is called the **Bessel function of the first kind of order ν**. It has the property that it is bounded near $x = 0$ and can be approximated (in some sense) using x^ν near $x = 0$. The function $Y_\nu(x)$ is called the **Bessel function of the second kind of order ν**, and it is unbounded near $x = 0$. In particular, near $x = 0$, $Y_\nu(x)$ can be approximated using $x^{-\nu}$ if $\nu > 0$ and approximated using $x^0 \ln x$ if $\nu = 0$.

In some applications, solutions to Bessel's differential equation (1) are needed that are bounded at $x = 0$. From the properties of the Bessel functions near $x = 0$, it follows that the solution given by (2) will be bounded near $x = 0$ only if $\beta = 0$, in which case $y = \alpha J_\nu(x)$.

In many applications, ν will not be an integer. In this case, we will need to use the **gamma function** $\Gamma(x)$. The gamma function extends the idea of a factorial to numbers other than integers. We shall not rigorously define the gamma function. For our purposes, it suffices to think of it as "like a factorial" and note two facts. First, if x is a positive integer, then $\Gamma(x) = (x-1)!$. Thus, for example, $\Gamma(5) = 4! = 24$. Second, the gamma function acts just like a factorial in that $\Gamma(x) = (x-1)\Gamma(x-1)$. Thus, for example, $\Gamma(\frac{7}{2}) = \frac{5}{2}\Gamma(\frac{5}{2})$.

We obtain series solutions of Bessel's differential equation (1) by the method of Frobenius. Following the last section, we look for solutions in the form

$$y = x^r \sum_{n=0}^{\infty} c_n x^n. \tag{4}$$

Substituting (4) into (1), we obtain

$$x^2 \sum_{n=0}^{\infty} (n+r)(n+r-1)c_n x^{n+r-2} + x \sum_{n=0}^{\infty} (n+r)c_n x^{n+r-1} + (x^2 - \nu^2) \sum_{n=0}^{\infty} c_n x^{n+r} = 0.$$

Divide by x^r and multiply the series by the coefficients:

$$\sum_{n=0}^{\infty} (n+r)(n+r-1)c_n x^n + \sum_{n=0}^{\infty} (n+r)c_n x^n - \nu^2 \sum_{n=0}^{\infty} c_n x^n + \sum_{n=0}^{\infty} c_n x^{n+2} = 0.$$

The last term $\sum_{n=0}^{\infty} c_n x^{n+2}$ equals $\sum_{n=2}^{\infty} c_{n-2} x^n$.

Equating coefficients of powers of x to zero gives the equations:

$$n = 0: \quad [r(r-1) + r - \nu^2]c_0 = 0, \tag{5}$$

$$n = 1: \quad [(1+r)r + (r+1) - \nu^2]c_1 = 0, \tag{6}$$

$$n \geq 2: \quad [(n+r)(n+r-1) + (n+r) - \nu^2]c_n + c_{n-2} = 0, \tag{7}$$

and, after simplification,

$$n = 0: \quad [r^2 - \nu^2]c_0 = 0, \tag{8}$$

$$n = 1: \quad [r + 1 - \nu][r + 1 + \nu]c_1 = 0, \tag{9}$$

$$n \geq 2: \quad [n + r - \nu][n + r + \nu]c_n + c_{n-2} = 0. \tag{10}$$

From (8) the roots of the indicial equation are $r = \nu$ and $r = -\nu$.

Bessel Function of the First Kind of Order ν

We can always obtain a solution corresponding to the larger root $r = \nu$.

For $r = \nu$, we have from (9) and (10)

$$[1 + 2\nu]c_1 = 0, \tag{11}$$

$$n[n + 2\nu]c_n + c_{n-2} = 0. \tag{12}$$

Since $\nu \geq 0$, (11) implies that $c_1 = 0$, and hence, by (12),

$$c_3 = c_5 = \cdots = 0.$$

To find the c_n with n even, note that, from (12),

$$c_n = (-1)\frac{1}{n(n + 2\nu)}c_{n-2}. \tag{13}$$

Let $n = 2m$. Then (13) is

$$\begin{aligned}
c_{2m} &= (-1)\frac{1}{2m(2m + 2\nu)}c_{2(m-1)} = \frac{(-1)}{2^2 m(m + \nu)}c_{2(m-1)} \\
&= \frac{(-1)}{2^2 m(m + \nu)}\,\frac{(-1)}{2^2(m - 1)(m - 1 + \nu)}c_{2(m-2)} \\
&\quad \vdots \\
&= \frac{(-1)^m}{2^{2m} m(m - 1)(m - 2)\cdots 1(m + \nu)(m - 1 + \nu)\cdots(1 + \nu)}c_0 \\
&= \frac{(-1)^m}{2^{2m} m!(m + \nu)(m - 1 + \nu)\cdots(1 + \nu)}c_0 \\
&= (-1)^m \frac{\Gamma(\nu + 1)}{2^{2m} m!\Gamma(\nu + m + 1)}c_0.
\end{aligned} \tag{14}$$

By convention, we take $c_0 = 1/[2^\nu \Gamma(\nu + 1)]$ and the resulting solution is denoted by $J_\nu(x)$;

$$J_\nu(x) = x^\nu \sum_{m=0}^{\infty} \frac{(-1)^m x^{2m}}{2^{2m+\nu} m!\Gamma(\nu + m + 1)}. \tag{15}$$

The function $J_\nu(x)$ is called the **Bessel function of the first kind of order ν.**

Second Independent Solution

When 2ν is not an integer, then the roots of the indicial equation differ by a noninteger. The second linearly independent solution can be obtained by the method of Frobenius with $r = -\nu$. When 2ν is an integer, then the roots of the indicial equation differ by an integer and the method may need to be modified. In any case, we first attempt to find a second solution in the simpler form.

We will need a formula to extend the validity of the factorial (and gamma) functions to negative numbers.

If $\tau < 0$ is a negative number that is not a negative integer, then there exists an integer k such that $\tau + k + 1 > 0$. Then we can define $\Gamma(\tau)$ by

$$\Gamma(\tau) = \frac{\Gamma(\tau + k + 1)}{\tau(\tau + 1) \cdots (\tau + k)}. \tag{16}$$

With this convention, we can try to solve (8), (9), and (10) with $r = -\nu$. The equations (8), (9), and (10) are then

$$n = 0: \qquad 0 = 0, \tag{17}$$

$$n = 1: \qquad [1 - 2\nu]\tilde{c}_1 = 0, \tag{18}$$

$$n \geq 2: \qquad [n - 2\nu]n\tilde{c}_n + \tilde{c}_{n-2} = 0. \tag{19}$$

If 2ν is not an integer, then (18) and (19) again imply $0 = \tilde{c}_1 = \tilde{c}_3 = \tilde{c}_5 = \cdots$ and

$$\tilde{c}_{2m} = (-1)\frac{1}{2m(2m - 2\nu)}\tilde{c}_{2(m-1)} = \frac{-1}{2^2 m(m - \nu)}\tilde{c}_{2(m-1)}. \tag{20}$$

Thus,

$$\begin{aligned}\tilde{c}_{2m} &= \frac{(-1)}{2^2 m(m - \nu)} \frac{(-1)}{2^2(m - 1)(m - 1 - \nu)}\tilde{c}_{2(m-2)} \\ &= \frac{(-1)}{2^2 m(m - \nu)} \frac{(-1)}{2^2(m - 1)(m - 1 - \nu)} \\ &\quad \times \frac{(-1)}{2^2(m - 2)(m - 2 - \nu)}\tilde{c}_{2(m-3)},\end{aligned}$$

so that

$$\begin{aligned}\tilde{c}_{2m} &= \frac{(-1)^m}{2^{2m} m(m - 1) \cdots 1(m - \nu)(m - 1 - \nu) \cdots (1 - \nu)}\tilde{c}_0 \\ &= (-1)^m \frac{\Gamma(-\nu + 1)}{2^{2m} m!\Gamma(-\nu + m + 1)}\tilde{c}_0.\end{aligned}$$

Letting $\tilde{c}_0 = 1/[2^{-\nu}\Gamma(-\nu + 1)]$ gives the Bessel function

$$J_{-\nu}(x) = x^{-\nu} \sum_{m=0}^{\infty} \frac{(-1)^m x^{2m}}{2^{2m-\nu} m!\Gamma(-\nu + m + 1)}. \tag{21}$$

Note that the series (21) for $J_{-\nu}$ comes by replacing ν by $-\nu$ in the series (15) for J_ν.

If $\nu = (2p + 1)/2$ for an integer p, then 2ν is an *odd* integer. Thus we can again solve (8), (9), and (10) by taking $0 = \tilde{c}_1 = \tilde{c}_3 = \tilde{c}_5 = \cdots$ and $\tilde{c}_{2m}$ given by (14), so that (21) also holds if 2ν is an odd integer.

The only remaining possibility is that ν is an integer and hence 2ν is an even integer. Then (18) and (19) imply that $\tilde{c}_1 = \tilde{c}_3 = \tilde{c}_5 = \cdots = 0$. For $n = 2\nu$, in (19), we get $\tilde{c}_{n-2} = 0$, and hence, from (19), that

$$\tilde{c}_{n-2} = \tilde{c}_{n-4} = \cdots = \tilde{c}_0 = 0.$$

Thus we do not get a second solution in the form (4) and, according to Theorem 4.7.1, must look for a solution in the form

$$y_2(x) = J_\nu(x) \ln x + x^{-\nu} \sum_{n=0}^{\infty} b_n x^n. \tag{22}$$

■ **THEOREM 4.8.1 Solutions of Bessel's Equation**

If ν is not an integer, then $\{J_\nu(x), J_{-\nu}(x)\}$, given by (15), (21), is a fundamental set of solutions of Bessel's equation (1). If $\nu = n$ is an integer, then a second solution $Y_n(x)$ can be found that involves a $\ln x$ term, and $\{J_n(x), Y_n(x)\}$ will be a fundamental set of solutions. ■

There are several ways to define Y_n. We omit the details but, if $n = 0$, then it can be shown that the Frobenius method gives

$$y_2 = J_0(x) \ln x + \sum_{n=1}^{\infty} \frac{(-1)^{n-1} h_n}{2^{2n}(n!)^2} x^{2n},$$

where $h_n = 1 + \frac{1}{2} + \cdots + \frac{1}{n}$.

Rather than use y_2, we define

$$Y_0(x) = \frac{2}{\pi}(\gamma - \ln 2) J_0(x) + \frac{2}{\pi} y_2(x), \tag{23}$$

where γ is the **Euler constant** ($\gamma \approx 0.5772\ldots$). Then, since $Y_0(x)$ is a linear combination of the solutions J_0, y_2, it will also be a solution of (1) for $\nu = 0$. Y_0 is called the **Bessel function of the second kind of order 0**. For integers $n > 0$, one can derive a similar expansion. (See Exercise 9 at the end of this section.) Finally, for ν not an integer, define

$$Y_\nu(x) = \frac{1}{\sin \nu\pi}[J_\nu(x) \cos \nu\pi - J_{-\nu}(x)], \tag{24}$$

which is a linear combination of J_ν, $J_{-\nu}$. Then Y_ν is called the **Bessel function of the second kind of order ν**. It can be shown that, for an integer $n \geq 0$,

$$\lim_{\nu \to n} Y_\nu(x) = Y_n(x).$$

THEOREM 4.8.2 Fundamental Set for Bessel's Differential Equation

For any $\nu \geq 0$, $\{J_\nu(x), Y_\nu(x)\}$ is a fundamental set of solutions of Bessel's equation of order ν. Furthermore, the only bounded solutions of Bessel's equation are multiples of $J_\nu(x)$. ■

We now wish to consider what the graphs of the Bessel functions look like. We begin with the Bessel functions of the first kind $J_\nu(x)$. The important points are covered in the next theorem.

THEOREM 4.8.3 Properties of Bessel Functions

The function $J_\nu(x)$ is continuous and bounded on the interval $0 \leq x < \infty$. The series in (15) converges for all x. Furthermore, $J_\nu(x)$ has an infinite number of real zeros denoted $\lambda_{\nu s}$; $s = 0, 1, 2, \ldots$ and $\lim_{s \to \infty} \lambda_{\nu s} = \infty$. ■

This theorem makes the Bessel functions sound somewhat like the sine and cosine. However, there are important differences. To illustrate, we consider $\nu = \frac{1}{2}$.

If $\nu = \frac{1}{2}$, then it turns out that $J_\nu(x), J_{-\nu}(x)$ can be expressed in **closed form**. Returning to the derivation of $J_\nu(x)$, we see from (13) that if $\nu = \frac{1}{2}$, then

$$c_{2m} = (-1)^m[(2m+1)!]^{-1}c_0.$$

Thus,

$$J_{1/2}(x) = x^{1/2} \sum_{m=0}^{\infty} \frac{(-1)^m x^{2m}}{(2m+1)!} \frac{1}{2^{1/2}\Gamma(3/2)}. \tag{25}$$

However,

$$\Gamma\left(\frac{1}{2}\right) = \sqrt{\pi},$$

so that (25) becomes

$$J_{1/2}(x) = \sqrt{\frac{2}{\pi x}} \sin x, \tag{26}$$

and, similarly,

$$J_{-1/2}(x) = \sqrt{\frac{2}{\pi x}} \cos x. \tag{27}$$

$J_{1/2}$ and $J_{-1/2}$ are graphed in Figure 4.8.2.

The formulas for $J_{1/2}(x), J_{-1/2}(x)$ illustrate the important fact that, if $y(x)$ is any solution of Bessel's equation (for any $\nu \geq 0$), then

$$\lim_{x \to \infty} y(x) = 0. \tag{28}$$

FIGURE 4.8.2

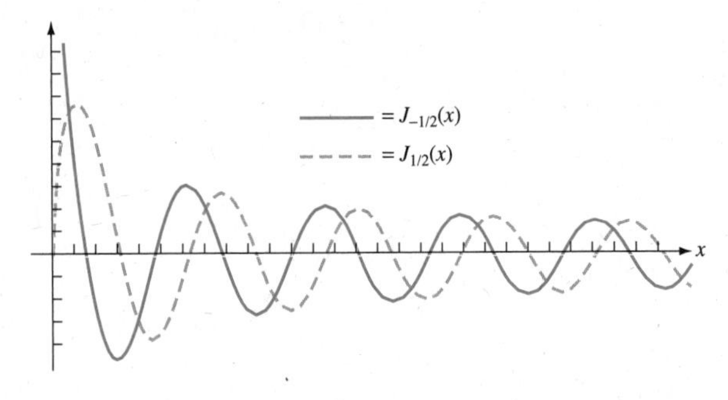

The formulas (27) and (28), however, are somewhat misleading, in that the zeros of $J_{1/2}(x)$, $\lambda_{1/2s} = \pi s$ are evenly spaced. This is not true in general. (Note also Exercise 8.) For example, the first four zeros of $J_0(x)$ are

$$\lambda_{01} = 2.40482,$$

$$\lambda_{02} = 5.52008,$$

$$\lambda_{03} = 8.65373,$$

$$\lambda_{04} = 11.7915.$$

Other than the fact that $J_0(0) = 1$ and $J_\nu(0) = 0$ if $\nu > 0$, all of the Bessel functions have graphs similar in general appearance to those in Figure 4.8.2, the major differences being the rate at which $\lim_{x \to 0^+} J_\nu(x) = 0$ for $\nu > 0$, and the rate at which $\lim_{x \to 0^+} Y_\nu(x) = -\infty$.

Additional properties of the Bessel functions appear in the exercises.

Exercises

1. Verify that $J_{-1/2}(x) = \sqrt{2/(\pi x)} \cos x$.

2. Using the series (15), verify that, for $\nu \geq 1$,

$$J_{\nu-1}(x) + J_{\nu+1}(x) = \frac{2\nu}{x} J_\nu(x). \tag{29}$$

3. Using the series (15), verify that, for $\nu \geq 1$,

$$J_{\nu-1}(x) - J_{\nu+1}(x) = 2J_\nu'(x), \tag{30}$$

where $J_\nu'(x)$ is the derivative with respect to x of $J_\nu(x)$.

4. Using (29) and (30), verify that, for $\nu \geq 1$,

$$J_{\nu+1}(x) = \frac{\nu}{x} J_\nu(x) - J_\nu'(x). \tag{31}$$

5. Assuming that (31) holds for $\nu \geq 0$ (which is true), use formula (26) for $J_{1/2}(x)$ to derive a formula for $J_{3/2}(x)$.

6. Assuming that (31) holds for $\nu \geq 0$, use Exercise 5 and (31) to derive a formula for $J_{5/2}(x)$.

7. Assuming that (31) holds for $\nu \geq 0$, use (31) to conclude that $J_1(x) = -J_0'(x)$. Assume that J_ν has the property that the only places where $J_\nu'(x) = 0$ are at a max or min, and that all maxima are positive and all minima are negative. Conclude that

$$\lambda_{00} < \lambda_{10} < \lambda_{01} < \lambda_{11} < \lambda_{02} < \lambda_{12} < \cdots.$$

That is, the zeros of $J_0(x), J_1(x)$ have the **interlacing property**.

8. Using the formula for $J_{3/2}(x)$ found in Exercise 5, show that the zeros of $J_{3/2}(x)$ are the solutions of $\tan\lambda = \lambda$. By graphing $y = \tan\lambda$, $y = \lambda$, conclude that

$$\lim_{s\to\infty}\left|\lambda_{(3/2)s} - \frac{\pi}{2} - \pi s\right| = 0.$$

so that, for large s, the zeros of $J_{3/2}(x)$ are about π units apart.

9. Verify that, for $n > 0$, n an integer, and $h_m = 1 + \frac{1}{2} + \cdots + \frac{1}{m}$,

$$\pi Y_n(x) = 2J_n(x)\left(\ln\frac{x}{2} + \gamma\right)$$

$$+ x^n\sum_{m=0}^{\infty}\frac{(-1)^{m-1}(h_m + h_{m+n})}{2^{2m+n}m!(m+n)!}x^{2m}$$

$$- x^{-n}\sum_{m=0}^{n-1}\frac{(n-m-1)!}{2^{2m-n}m!}x^{2m}$$

is a solution of Bessel's equation (1) with $\nu = n$.

10. Graph the functions

$$f_n(x) = J_{1/2}(nx)$$

on the interval $0 \le x \le \pi$ for $n = 1, 2, 3, 4$ and compare them to the curves in Figure 4.8.1.

11. Euler's constant γ can be defined as

$$\gamma = \lim_{n\to\infty} h_n - \ln n$$

where $h_n = 1 + (1/2) + \cdots (1/n)$. Compute the value of $q_n = h_n - \ln n$ for several values of n, and observe the apparent convergence.

CHAPTER

5

Linear Systems of Differential Equations

5.1 Introduction

5.1.1 General Linear Systems

Up to this point we have considered only single differential equations with one dependent variable. In many applications, however, one is led to simultaneously consider several ordinary differential equations with several dependent variables and one independent variable. Such **systems of differential equations** may be linear or nonlinear. Nonlinear systems are studied in Chapter 9. If all of the differential equations are linear in the dependent variables, the resulting linear systems of differential equations are most naturally studied using vector notation and matrix theory, which we will cover beginning in Section 5.7. The remainder of this section introduces some basic concepts for linear systems. Sections 5.2 and 5.3 discuss methods that are helpful with linear systems consisting of a few equations. Sections 5.4, 5.5, and 5.6 cover three applications in which systems arise. The applications may also be covered after Section 5.11 if Sections 5.2 and 5.3 are omitted.

A simple example of a linear system of differential equations is

$$\begin{aligned} x'(t) &= 2x(t) + 3y(t) + \sin t, \\ y'(t) &= 4x(t) - 5y(t) + \cos(t^2) + t. \end{aligned} \tag{1}$$

In the system (1), there are two unknown functions $x(t)$, $y(t)$. A pair of functions $x(t)$, $y(t)$ would be a **solution** of (1) if they satisfied both equations.

EXAMPLE 5.1.1 **Solution of System**

Verify that $x = e^t$, $y = e^{2t}$ is a solution of the system

$$x' = x + 3y - 3e^{2t}, \tag{2a}$$

$$y' = 4x + 2y - 4e^t. \tag{2b}$$

SOLUTION We must show that $x = e^t$, $y = e^{2t}$ satisfies both (2a) and (2b). Substituting $x = e^t$, $y = e^{2t}$ into (2a) gives

$$(e^t)' = e^t + 3e^{2t} - 3e^{2t} \quad \text{or} \quad e^t = e^t,$$

and substituting $x = e^t$, $y = e^{2t}$ into (2b) gives

$$(e^{2t})' = 4e^t + 2e^{2t} - 4e^t \quad \text{or} \quad 2e^{2t} = 2e^{2t}.$$

Thus $x = e^t$, $y = e^{2t}$ is a solution of the system (2). ◀

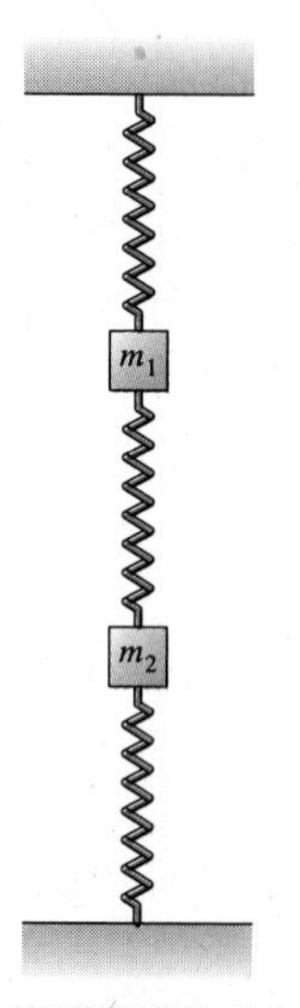

FIGURE 5.1.1

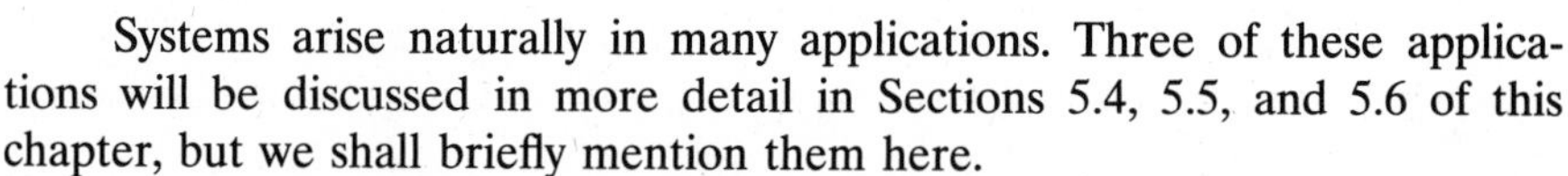

Systems arise naturally in many applications. Three of these applications will be discussed in more detail in Sections 5.4, 5.5, and 5.6 of this chapter, but we shall briefly mention them here.

One application is mechanical systems. For example, the linear motion of several interconnected masses as in Figure 5.1.1 (a flexible structure) would have a dependent variable for each mass. If other than straight-line motion is considered, each mass would have the x, y, z coordinates of its position as dependent variables.

A second application involves multiloop circuits. Consider, for example, the circuit in Figure 5.1.2. The voltage law applied to the top and bottom loops of this circuit yields

$$\frac{di_1}{dt} + 3i_1 - i_2 = e_1(t), \tag{3}$$

$$2\frac{di_3}{dt} + 5i_3 + i_2 = e_2(t), \tag{4}$$

respectively. The current law at node (a) is

$$i_1 + i_2 - i_3 = 0.$$

Using this equation to eliminate i_2 from (3) and (4) gives the system

$$\begin{aligned} \frac{di_1}{dt} + 4i_1 - i_3 &= e_1(t), \\ 2\frac{di_3}{dt} - i_1 + 6i_3 &= e_2(t). \end{aligned} \tag{5}$$

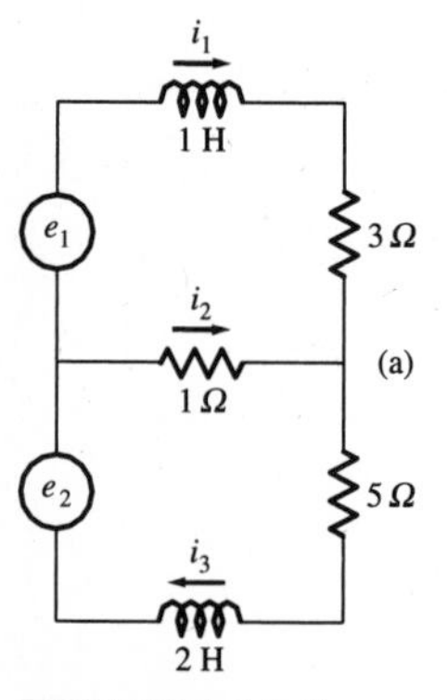

FIGURE 5.1.2

The final application we shall consider will be mixing problems, in which there will be several interrelated concentrations and volumes of interest.

In this chapter we will consider only systems of linear differential equations, especially those with constant coefficients. Nonlinear systems, with a few simple exceptions, are either solved numerically (Chapter 6) or analyzed qualitatively (Chapter 9).

The existence and uniqueness properties of general systems of differential equations are not as straightforward as for single equations. In particular, determining the expected number of initial conditions or arbitrary constants is a little trickier. One cannot tell how many constants there will be merely by looking at the system.

EXAMPLE 5.1.2 Three Derivatives, Two Arbitrary Constants

Consider the linear system of differential equations with constant coefficients

$$x''(t) + y'(t) + x(t) + 3y(t) = t, \tag{6a}$$

$$2x''(t) + 2y'(t) + 2x(t) + 7y(t) = 3. \tag{6b}$$

Since this system involves second derivatives of x and first derivatives of y, one might expect the answer to involve three arbitrary constants. Note, however, that subtracting twice equation (6a) from (6b) gives the new **equivalent system** (one that has the same solutions)

$$x''(t) + y'(t) + x(t) + 3y(t) = t, \tag{7a}$$

$$y(t) = 3 - 2t; \tag{7b}$$

or, using (7b) to eliminate $y(t)$ in (7a),

$$x''(t) + x(t) = 7t - 7, \tag{8a}$$

$$y(t) = 3 - 2t. \tag{8b}$$

Using the method of undetermined coefficients, we can solve (8a) for x. The result is

$$x(t) = 7t - 7 + c_1 \cos t + c_2 \sin t, \tag{9a}$$

$$y(t) = 3 - 2t. \tag{9b}$$

Note that there are only two arbitrary constants in this solution, and there are no arbitrary constants in the y portion of the solution. ◄

There is a general procedure for determining how many arbitrary constants there will be in the general solution of a linear system of differential equations with constant coefficients. Its use is mostly restricted to small-sized systems, since it uses determinants, and determinants are "expensive" to compute for systems with many dependent variables. We shall illustrate each step by applying it to system (6) of Example 5.1.2.

Procedure for Determining the Number of Arbitrary Constants

Procedure for determining the number of arbitrary constants in a system of linear differential equations with constant coefficients:

1. **First rewrite the system in operator notation** (see Section 2.10). Let D stand for the operation of differentiation with respect to the independent variable, and define $D^n = d^n/dt^n$. Thus, for example, system (6) of Example 5.1.2 would be written as

$$\begin{aligned} (D^2+1)x + (D+3)y &= t, \\ (2D^2+2)x + (2D+7)y &= 3. \end{aligned} \tag{10}$$

2. **Take the determinant of the coefficient matrix**, thinking of D as a variable. For (10) this would be

$$\begin{aligned} \det\begin{bmatrix} D^2+1 & D+3 \\ 2D^2+2 & 2D+7 \end{bmatrix} &= (D^2+1)(2D+7) - (D+3)(2D^2+2) \\ &= D^2+1. \end{aligned}$$

3. **The degree of the resulting polynomial in D will be the number of arbitrary constants.** In our example, D^2+1 is a second-degree polynomial in D, and so the system (6) will have two arbitrary constants in its general solution, which is what was observed in (9).

If the determinant is zero, the system is called **degenerate**. As the next two examples show, degenerate systems can have properties quite different from those we would expect from our study (in Chapter 2) of linear differential equations with constant coefficients.

EXAMPLE 5.1.3 **Inconsistent Degenerate System**

Consider the system of differential equations

$$x' + y' + x + y = e^t, \tag{11a}$$

$$x + y = e^t, \tag{11b}$$

or, equivalently,

$$\begin{aligned} (D+1)x + (D+1)y &= e^t, \\ x + y &= e^t. \end{aligned}$$

Since

$$\det\begin{bmatrix} D+1 & D+1 \\ 1 & 1 \end{bmatrix} = 0,$$

this system is degenerate. It has no solution. If equation (11b) is substituted into (11a), we get the contradiction that $(x+y)' = 0$ so that $x+y$ is a constant, but, from (11b), $(x + y) = e^t$. A system with no solutions is called **inconsistent**. ◄

EXAMPLE 5.1.4 Consistent Degenerate System

Consider the system

$$x' + y = 0,$$
$$2x' + 2y = 0,$$

or

$$Dx + y = 0,$$
$$2Dx + 2y = 0.$$

Since

$$\det \begin{bmatrix} D & 1 \\ 2D & 2 \end{bmatrix} = 0,$$

this system is also degenerate. Note that the second equation, $2x' + 2y = 0$, is just a constant multiple of the first equation, $x' + y = 0$. In this example, one could take $x(t)$ to be an **arbitrary differentiable function**, and then $y(t)$ would be $-x'(t)$. Having an arbitrary function in the solution is quite different from having an arbitrary constant. One difference is that no finite set of initial conditions of the types we have considered can determine an arbitrary function. ◄

The problem of determining what initial conditions are possible for a system of differential equations, and what equations can be simplified or eliminated is an important problem. In many fields, such as electrical engineering, procedures have been developed that can be implemented on a computer and are based on the area of mathematics known as graph theory.

5.1.2 First-Order Systems

The situation is simpler for some **first-order** linear systems. A system is first-order if it involves only first derivatives.

A system of first-order linear differential equations is said to be **explicit** if it is in the form

$$\begin{aligned} x_1'(t) &= a_{11}(t)x_1(t) + a_{12}(t)x_2(t) + \cdots + a_{1n}(t)x_n(t) + f_1(t), \\ x_2'(t) &= a_{21}(t)x_1(t) + a_{22}(t)x_2(t) + \cdots + a_{2n}(t)x_n(t) + f_2(t), \\ &\vdots \\ x_n'(t) &= a_{n1}(t)x_1(t) + a_{n2}(t)x_2(t) + \cdots + a_{nn}(t)x_n(t) + f_n(t). \end{aligned} \tag{12}$$

Here the dependent variables are $x_1, \ldots, x_n$; the functions a_{ij}, with $1 \le i \le n$, $1 \le j \le n$, are the **coefficients**; and $f_1, \ldots, f_n$ are the **input** or **forcing functions**.

We have seen several examples of such systems already—for example, systems (1) and (2), and the circuit example (5). Another example is

$$\begin{aligned} x_1'(t) &= 3x_1(t) - 5x_2(t) + \cos t, \\ x_2'(t) &= tx_1(t) + 3x_2(t) + 6x_3(t) + \sin t, \\ x_3'(t) &= x_1(t) - e^t x_3(t). \end{aligned} \tag{13}$$

In this example, $a_{11} = 3$, $a_{12} = -5$, $a_{13} = 0$, $f_1(t) = \cos t$; $a_{21} = t$, $a_{22} = 3$, $a_{23} = 6$, $f_2(t) = \sin t$; $a_{31} = 1$, $a_{32} = 0$, $a_{33} = -e^t$, and $f_3(t) = 0$.

While we will not go into matrix theory until Section 5.7, note that (12) may be simplified by using matrix notation. Let

$$\mathbf{x}(t) = \begin{bmatrix} x_1(t) \\ x_2(t) \\ \vdots \\ x_n(t) \end{bmatrix}, \quad \mathbf{A}(t) = \begin{bmatrix} a_{11}(t) & \cdots & a_{1n}(t) \\ a_{21}(t) & \cdots & a_{2n}(t) \\ \vdots & & \vdots \\ a_{n1}(t) & \cdots & a_{nn}(t) \end{bmatrix}, \quad \mathbf{f}(t) = \begin{bmatrix} f_1(t) \\ f_2(t) \\ \vdots \\ f_n(t) \end{bmatrix},$$

and recall from calculus that

$$\mathbf{x}'(t) = \begin{bmatrix} x_1'(t) \\ x_2'(t) \\ \vdots \\ x_n'(t) \end{bmatrix}.$$

Then (12) can be written as

$$\mathbf{x}'(t) = \mathbf{A}(t)\mathbf{x}(t) + \mathbf{f}(t), \tag{14}$$

or simply

$$\mathbf{x}' = \mathbf{A}\mathbf{x} + \mathbf{f}. \tag{15}$$

In particular, the example (13) would be

$$\begin{bmatrix} x_1'(t) \\ x_2'(t) \\ x_3'(t) \end{bmatrix} = \begin{bmatrix} 3 & -5 & 0 \\ t & 3 & 6 \\ 1 & 0 & -e^t \end{bmatrix} \begin{bmatrix} x_1(t) \\ x_2(t) \\ x_3(t) \end{bmatrix} + \begin{bmatrix} \cos t \\ \sin t \\ 0 \end{bmatrix}. \tag{16}$$

For a first-order linear system, the existence and uniqueness result is what we would expect.

THEOREM 5.1.1 Existence and Uniqueness for First-Order Linear Systems

Suppose that, in the explicit system (12), $f_1(t), \ldots, f_n(t)$ and the $a_{ij}(t)$ are continuous functions on the interval I. Let t_0 be in the interval I, and let $x_{10}, \ldots, x_{n0}$ be n real numbers. Then there exists a unique solution $x_1(t), \ldots, x_n(t)$ to (12), defined for all t in the interval I, such that

$$x_1(t_0) = x_{10}, \qquad x_2(t_0) = x_{20}, \qquad \ldots, \qquad x_n(t_0) = x_{n0}.$$

This theorem will be proved for the constant coefficient case in Section 5.9.

EXAMPLE 5.1.5 Existence and Uniqueness

By Theorem 5.1.1, the system

$$x' = 3x - 2y + e^t, \qquad x(0) = 4,$$

$$y' = -2x + 5y - e^{3t}, \qquad y(0) = 5$$

has a unique solution $x(t)$, $y(t)$ defined for all t.

The explicit first-order form (12) is actually fairly general, as the next two facts show. When we talk of rewriting a system, it is understood that the new system is equivalent to the old one (has the same solutions).

THEOREM 5.1.2 Explicit Form and Initial Conditions

If a system of n linear, constant coefficient, first-order differential equations with n dependent variables $x_1, \ldots, x_n$ has n arbitrary initial conditions in its general solution, then the system can be rewritten in the explicit form (12) by adding and subtracting constant multiples of the equations. In particular, $x_1(t_0), \ldots, x_n(t_0)$ may be specified arbitrarily at any fixed time t_0. Conversely, if a system of n linear, constant coefficient differential equations has less than n arbitrary constants in its general solution, it cannot be written in the explicit form (12).

EXAMPLE 5.1.6 Initial Conditions for First-Order Systems

Consider the system of two first-order linear, differential equations

$$x'(t) + 3y'(t) = x(t) - 5y(t) + t, \tag{17a}$$

$$x'(t) - 2y'(t) = 3x(t) + y(t) + e^t. \tag{17b}$$

Since

$$\det\begin{bmatrix} D-1 & 3D+5 \\ D-3 & -2D-1 \end{bmatrix} = -5D^2 + 5D + 16,$$

we have, from Theorem 5.1.1, that the system (17) will have two arbitrary constants in its general solution. Theorem 5.1.2 then tells us that they may be taken as $x(t_0)$, $y(t_0)$. Note that adding -1 times equation (17a) to (17b) gives

$$x' + 3y' = x - 5y + t, \tag{18a}$$

$$-5y' = 2x + 6y + e^t - t. \tag{18b}$$

If we then add $\frac{3}{5}$ of equation (18b) to (18a) and then multiply (18b) by $-\frac{1}{5}$, we get the explicit system

$$x' = \frac{11}{5}x - \frac{7}{5}y + \frac{2}{5}t + \frac{3}{5}e^t,$$

$$y' = -\frac{2}{5}x - \frac{6}{5}y - \frac{1}{5}(e^t - t),$$

as promised by Theorem 5.1.2. ◄

The consideration of first-order systems may seem to be restrictive, since we know, from Sections 2.8, 2.11, and 2.12, that many electric circuits and mechanical systems will lead to systems of second-order linear differential equations. However,

> Every system of differential equations can be rewritten as a first-order system of differential equations.

The easiest way to illustrate this fact is by examples.

EXAMPLE 5.1.7 Rewriting as a First-Order System

Rewrite

$$\begin{aligned} x'' + 3x' - y'' + y &= \sin t, \\ x' - 4x + 5y'' - 6y' &= \cos t, \end{aligned} \tag{19}$$

as a first-order system.

SOLUTION We introduce the two new variables $z = x'$, $w = y'$. Then (19) may be written as a first-order system in x, y, z, w:

$$\begin{aligned} z' + 3z - w' + y &= \sin t, \\ z - 4x + 5w' - 6w &= \cos t, \\ x' - z &= 0, \\ y' - w &= 0. \end{aligned} \tag{20}$$

The resulting system (20) is not unique. For example, one could also use $x' - 4x + 5w' - 6y' = \cos t$ for the second equation in (20). ◄

In the case of a single-loop *RLC* circuit, this procedure amounts to thinking of the circuit as a first-order system in the charge and current rather than as a second-order equation in the charge. For a mechanical system, we have a first-order system in velocity and position rather than a second-order system in position.

EXAMPLE 5.1.8 **Rewriting as a First-Order System**

Rewrite

$$\begin{aligned} x''' + x' + y'' + y &= t, \\ x + y'' - y &= 1, \end{aligned} \tag{21}$$

as a first-order system.

SOLUTION The system (21) involves third-order derivatives of x and second-order derivatives of y. Accordingly, we introduce the variables $u = x'$, $v = x'' = u'$, $w = y'$. One way to rewrite (21) would be

$$\begin{aligned} v' + u + w' + y &= t, \\ x + w' - y &= 1, \\ x' - u &= 0, \\ u' - v &= 0, \\ y' - w &= 0. \end{aligned}$$

◄

Exercises

In Exercises 1 through 4, verify that the indicated functions form a solution of the system of differential equations.

1. $x' + 2x + y = 3e^t + e^{2t}$, $x = e^t$,
$x' + y' + 3x = 4e^t + 2e^{2t}$, $y = e^{2t}$

2. $x_1'' + x_1 + x_2 = \cos 2t$, $x_1 = \sin t$,
$x_1' + x_2'' + 4x_2 = \cos t$, $x_2 = \cos 2t$

3. $x_1' = x_1 + x_2 + x_3 - 3t + 1$,
$x_2' = x_1 - x_3$,
$x_3' = x_1 + x_2 - 2t$,
$x_1 = t$, $x_2 = t + 1$, $x_3 = t - 1$

4. $x'' - x + y' + y - z = 3e^{2t} - e^{3t}$,
$x' + x + y'' - z' = 2e^t + 4e^{2t} - 3e^{3t}$,
$x' + y' - 2y + z' - 3z = e^t$,
$x = e^t$, $y = e^{2t}$, $z = e^{3t}$

In Exercises 5 through 10, determine whether or not the system is degenerate. If it is not degenerate, determine the number of arbitrary constants in the general solution.

5. $x' + x + y' - y = t$,
$x' - x + 2y' + y = \sin t$

6. $x_1'' + x_1 + x_2'' = \sin 3t + 1$,
$x_1' + x_2' = t^2$

7. $x'' + y'' + y = t$,
$3x' + x + 4y' - y = \sin t$

8. $x_1'' + x_1 + x_2'' - x_2' = \cos t$,
$x_1'' - x_1' - 2x_2'' + x_2 = t^3 - 2t$

9. $x' + x + y' = t$,
$x'' + x' + y'' = t - 1$

10. $x'' - 2x + y'' - y = 0$,
$2x' + x + 2y' + 3y = t$

In Exercises 11 through 14, determine whether or not the system is degenerate. If it is not degenerate, determine the number of arbitrary constants in the general solution. [*Note*: These exercises require a knowledge of determinants of 3×3 matrices (see Section 5.8).]

11. $x' + x + y + z = t$,
$2x + y' = 2t$,
$y + 3z' = t$

12. $x'' + x + y + z' = 1$,
$x' + y' + z'' = t$,
$x'' - x' + x - y' + y - z'' + z' = e^t$

13. $x_1' + x_1 + x_2' + x_2 + x_3 = t$,
$x_1' + x_2' + 3x_3 = e^t$,
$x_2' + x_3' = t$

14. $x'' + y'' + z'' = 0$,
$x' + y' + y + z' + z = 1$,
$x' + 3z = t$

In Exercises 15 through 20, rewrite the given system of differential equations as a first-order system of differential equations.

15. $x'' - x + y = t$,
$x' + y' = 0$

16. $x''' + 3x' - 5x = \cos t$

17. $x_1'' + x_2'' = \sin t$,
$x_1''' - x_1' + x_2 = t$

18. $x'''' + x'' + x = e^t$

19. $x_1'' = 3x_1' + 4x_2' - x_1$,
$x_2'' = x_1' - x_2' + x_2$

20. $x_1'' + x_2'' + x_3 = t$,
$x_3''' + x_1 - 6x_2 + x_1' = e^t$,
$x_2' + x_1' + x_3 + x_3' = t^2$

In Exercises 21 through 24, determine (using Theorem 5.1.2) whether or not the system of differential equations can be put in explicit form using the same number of dependent variables by adding constant multiples of the equations. If it can be put in explicit form, do so.

21. $2x' + 3x + 5y' = t$,
$x' - x + 2y' + y = \cos t$

22. $x' + x + 2y' - y = 0$,
$2x' - 3x + 4y' - 7y = t$

23. $x' + y' = 6y + t$,
$-3x' + 7y' = 3y - x$

24. $x' + x - 3y = t$,
$6x' - 5y' - x + y = \cos t$

5.2 Elimination Methods

The algebraic system of equations

$$2x + 3y = 7, \tag{1a}$$

$$4x - 5y = 13, \tag{1b}$$

can be solved by adding constant multiples of one equation to another to obtain an equation in just one variable or, equivalently, by solving one equation, say (1a), for one of the variables, say x, and substituting this expression into the other equation to obtain an equation in just one variable. A similar approach will be developed in this section for linear systems of differential equations with constant coefficients. This method is usually best suited for systems involving a small number of dependent variables, and will enable us to solve the differential equations that arise in the applications of Sections 5.4, 5.5, and 5.6. It is also possible to program this method on a computer that does symbolic calculus and algebra. However, a more common way to analyze and solve more complicated systems of linear differential equations is either numerically (Chapter 6) or by the matrix techniques developed later in this chapter.

To begin with, suppose we have two linear, constant coefficient differential equations involving two dependent variables $x(t)$, $y(t)$. This system may be written as

$$L_1x + L_2y = f_1, \tag{2a}$$

$$L_3x + L_4y = f_2, \tag{2b}$$

where L_1, L_2, L_3, L_4 are differential operators (Section 2.10) and f_1, f_2 are known functions of t. We assume that f_1, f_2 are functions for which the method of undetermined coefficients (Section 2.9) is appropriate.

EXAMPLE 5.2.1 Operator Notation

The system of differential equations

$$x'' + 3x' + y'' = t, \tag{3a}$$

$$x' + 3x + y'' + y' = e^{2t}, \tag{3b}$$

may be written in the form (2) as follows:

$$(D^2 + 3D)x + D^2y = t, \tag{4a}$$

$$(D + 3)x + (D^2 + D)y = e^{2t}. \tag{4b}$$

Here

$$L_1 = D^2 + 3D, \qquad L_2 = D^2, \qquad L_3 = D + 3, \qquad L_4 = D^2 + D,$$

and

$$f_1 = t, \qquad f_2 = e^{2t}. \quad ◀$$

In order to solve the system (2), we will derive a single equation in just one of the dependent variables. One way to do this is to "multiply" (2a) by L_3 and (2b) by L_1, to get

$$L_3L_1x + L_3L_2y = L_3f_1, \tag{5a}$$

$$L_1L_3x + L_1L_4y = L_1f_2. \tag{5b}$$

But $L_1L_3 = L_3L_1$, since L_3, L_1 have constant coefficients. Thus, upon subtracting (5b) from (5a), we get a differential equation solely in terms of y:

$$(L_3L_2 - L_1L_4)y = L_3f_1 - L_1f_2.$$

To apply this process to (4), we "multiply" (4b) by $L_1 = D^2 + 3D$, and (4a) by $L_3 = D + 3$, to get

$$\begin{aligned}(D+3)(D^2+3D)x + (D+3)D^2y &= (D+3)t \\ &= 1 + 3t, \\ (D^2+3D)(D+3)x + (D^2+3D)(D^2+D)y &= (D^2+3D)e^{2t} \\ &= 4e^{2t} + 6e^{2t} \\ &= 10e^{2t}.\end{aligned}$$

Subtracting the second equation from the first yields

$$-(D^4 + 3D^3)y = 1 + 3t - 10e^{2t},$$

that is,

$$y'''' + 3y''' = 10e^{2t} - 1 - 3t.$$

This differential equation can be solved for y, for example, by the method of undetermined coefficients, to give

$$y = \frac{e^{2t}}{4} - \frac{t^4}{24} + c_1 + c_2t + c_3t^2 + c_4e^{-3t}, \tag{6}$$

which has four arbitrary constants. This formula for y can be substituted into either of the original differential equations (3a) or (3b) to get a differential equation in x only. If (6) is substituted into (3b) for y, we get a first-order differential equation in x which, upon solution, provides one additional arbitrary constant, for a total of five arbitrary constants. On the other hand, if formula (6) is substituted for y in (3a), a second-order differential equation in x results, which, upon solution, gives us *two* additional arbitrary constants, for a total of six arbitrary constants. Thus, depending on how these calculations are done, we get either five or six arbitrary constants. But from Section 5.1, the correct total number of arbitrary constants is four, since $L_3L_2 - L_1L_4 = -D^4 - 3D^3$ is fourth-order.

To see how we could have avoided this problem in this particular example, return to the system (4) and note that $D^2 + 3D = D(D + 3)$. Thus, to get the same coefficient in front of x in both equations, we could have merely multiplied (4b) by D to give the new system

$$(D^2 + 3D)x + D^2y = t, \tag{7a}$$

$$D(D + 3)x + D(D^2 + D)y = De^{2t} = 2e^{2t}. \tag{7b}$$

Now if we subtract the first equation (7a) from (7b), we get

$$D^3y = 2e^{2t} - t,$$

which can be solved by antidifferentiating three times to yield

$$y = \frac{e^{2t}}{4} - \frac{t^4}{24} + c_1t^2 + c_2t + c_3, \tag{8}$$

where c_1, c_2, c_3 are arbitrary constants. Substituting (8) into equation (3b) gives the first-order differential equation in x

$$x' + 3x = \frac{-e^{2t}}{2} + \frac{t^3}{6} + \frac{t^2}{2} - 2c_1t - (c_2 + 2c_1). \tag{9}$$

This differential equation (9) can be solved by the integrating-factor method, but it is probably quicker to use undetermined coefficients, as follows. The characteristic polynomial is $r + 3$, with a single root $r = -3$, so that

$$x_h = c_4e^{-3t},$$

and x_p is of the form

$$x_p = A + Bt + Ct^2 + Et^3 + Fe^{2t}.$$

Substitute this form into (9) for x, to get

$$(B + 2Ct + 3Et^2 + 2Fe^{2t}) + 3(A + Bt + Ct^2 + Et^3 + Fe^{2t})$$
$$= -\frac{e^{2t}}{2} + \frac{t^3}{6} + \frac{t^2}{2} - 2c_1t - (c_2 + 2c_1).$$

Equating the corresponding coefficients, we find

$$\begin{aligned}
1: &\qquad B + 3A = -(c_2 + 2c_1),\\
t: &\qquad 2C + 3B = -2c_1,\\
t^2: &\qquad 3E + 3C = \frac{1}{2},\\
t^3: &\qquad 3E = \frac{1}{6},\\
e^{2t}: &\qquad 5F = -\frac{1}{2}.
\end{aligned}$$

In solving these equations, we treat A, B, C, E, F as the unknowns, and the c_1, c_2, c_3 as fixed, but known, constants. Thus, starting with the last equation and back-substituting yields

$$\begin{aligned}
F &= -\frac{1}{10},\\
E &= \frac{1}{18},\\
C &= \frac{1}{6} - E = \frac{1}{9},\\
B &= -\frac{2}{3}C - \frac{2}{3}c_1 = -\frac{2}{27} - \frac{2}{3}c_1,\\
A &= -\frac{1}{3}B - \frac{1}{3}(c_2 + 2c_1) = \frac{2}{81} + \frac{2}{9}c_1 - \frac{1}{3}(c_2 + 2c_1)\\
&= \frac{2}{81} - \frac{4}{9}c_1 - \frac{1}{3}c_2.
\end{aligned}$$

The solution of (4) is thus

$$y = \frac{e^{2t}}{4} - \frac{t^4}{24} + c_1t^2 + c_2t + c_3, \tag{10a}$$

$$\begin{aligned}
x &= \left(\frac{2}{81} - \frac{4}{9}c_1 - \frac{1}{3}c_2\right) - \left(\frac{2}{27} + \frac{2}{3}c_1\right)t\\
&\quad + \frac{t^2}{9} + \frac{t^3}{18} - \frac{e^{2t}}{10} + c_4e^{-3t},
\end{aligned} \tag{10b}$$

which has the correct number of arbitrary constants, four.

Caution If an arbitrary constant appears several places in the solution, as c_1 does in (10), we cannot change it in one location without changing it everywhere. For example, if we were to try to simplify (10b) by letting $\tilde{c}_1 = c_1/9$, we would have to replace c_1 by $9\tilde{c}_1$ everywhere in (10a) and (10b).

We shall summarize these observations and then carefully work additional examples.

Elimination Method

Elimination method for a system of two linear, constant coefficient, differential equations in two unknowns:

1. First write the system in the form (2):

$$L_1x + L_2y = f_1, \tag{11a}$$

$$L_3x + L_4y = f_2. \tag{11b}$$

2. Multiply (11a) and/or (11b) by constant coefficient differential operators, so that both equations have the same x or y coefficient. [One can always multiply (11a) by L_3 and (11b) by L_1, or (11a) by L_4 and (11b) by L_2.]
3. Subtract one equation from the other to get a differential equation in just one dependent variable, and solve this differential equation.
4. Substitute the answer for part 3 into one of the equations in the original system, to get a differential equation in the other dependent variable. Solve this differential equation.
5. If the number of arbitrary constants is the same as the order of $L_1L_4 - L_2L_3$, we have the general solution.
6. If the number of arbitrary constants is greater than the order of $L_1L_4 - L_2L_3$, substitute the formulas for x, y from parts 3 and 4 into the original differential equation not used in part 4, in order to eliminate any extra constants.
7. If initial conditions are given, use them to eliminate the remaining arbitrary constants.

EXAMPLE 5.2.2 **Elimination Method**

Using elimination, solve the system

$$4x' + 4x + y' = 1, \tag{12a}$$

$$3x' + y' - y = t. \tag{12b}$$

SOLUTION

Step 1
First rewrite the system as

$$4(D + 1)x + Dy = 1, \tag{13a}$$

$$3Dx + (D - 1)y = t. \tag{13b}$$

Step 2
We shall eliminate the y term. Multiply (13a) by $(D-1)$ and (13b) by D, to get

$$4(D^2-1)x+(D-1)Dy=(D-1)1=-1, \tag{14a}$$
$$3D^2x+D(D-1)y=Dt=1. \tag{14b}$$

Step 3
Subtract (14b) from (14a), to give

$$(D^2-4)x=-2$$

or

$$x''-4x=-2. \tag{15}$$

We shall solve this differential equation by the method of undetermined coefficients. The characteristic equation is $r^2-4=0$ with roots $r=\pm 2$. Thus $x_h=c_1e^{-2t}+c_2e^{2t}$. The particular solution has the form $x_p=A$, A a constant. Substituting this form into (15) gives $A''-4A=-2$ or $-4A=-2$ and $A=\frac{1}{2}$. Thus

$$x=\frac{1}{2}+c_1e^{-2t}+c_2e^{2t}. \tag{16}$$

Step 4
This expression for x may be substituted into either (12a) or (12b), to find y. Since (12a) will give a slightly simpler equation for y, we substitute the formula (16) for x into (12a), to get

$$y'=1-4x'-4x$$

or

$$y'=-1+4c_1e^{-2t}-12c_2e^{2t}.$$

Antidifferentiate both sides with respect to t, to find

$$y=-t-2c_1e^{-2t}-6c_2e^{2t}+c_3. \tag{17}$$

Step 5
To check whether the number of arbitrary constants in (16) and (17) is correct, compute

$$\det\begin{bmatrix}4D+4 & D\\ 3D & D-1\end{bmatrix}=4D^2-4-3D^2=D^2-4.$$

This is *second*-order, and so our answer (16) and (17) should have *two* arbitrary constants. Since we have three arbitrary constants, we proceed to Step 6.

Step 6
Since (12a) was used to find y from x, we know that (12a) holds. Thus, we use the other equation, (12b), or, equivalently, (13b), to eliminate the extra

constant. Substitute (16) and (17) into (12b), $3x' + y' - y = t$, to find

$$3(\tfrac{1}{2} + c_1e^{-2t} + c_2e^{2t})' + (-t - 2c_1e^{-2t} - 6c_2e^{2t} + c_3)'$$
$$-(-t - 2c_1e^{-2t} - 6c_2e^{2t} + c_3) = t,$$

or

$$-6c_1e^{-2t} + 6c_2e^{2t} - 1 + 4c_1e^{-2t} - 12c_2e^{2t} + t$$
$$+ 2c_1e^{-2t} + 6c_2e^{2t} - c_3 = t,$$

or, upon simplification,

$$-1 - c_3 = 0 \quad \text{and} \quad c_3 = -1.$$

Thus the general solution of (12) is

$$x = \frac{1}{2} + c_1e^{-2t} + c_2e^{2t},$$

$$y = -1 - t - 2c_1e^{-2t} - 6c_2e^{2t}.$$

◀

EXAMPLE 5.2.3 **Elimination Method**

Using the elimination method, solve the initial value problem

$$x' = x + 5y, \qquad x(0) = 0, \tag{18a}$$
$$y' = -x - y, \qquad y(0) = 1. \tag{18b}$$

Solution Since x appears only once in (18b) and is undifferentiated, we could simply solve (18b) for $x = -y' - y$ and then substitute into (18a) to get $(-y' - y)' = (-y' - y) + 5y$. However we shall follow the general procedure. First, rewrite (18) as

$$(D - 1)x - 5y = 0, \tag{19a}$$
$$x + (D + 1)y = 0. \tag{19b}$$

Multiply the second equation by $(D - 1)$:

$$(D - 1)x - 5y = 0,$$
$$(D - 1)x + (D^2 - 1)y = 0,$$

and subtract the first from the second, to give an equation in only y:

$$(D^2 + 4)y = 0,$$

or $y'' + 4y = 0$. The characteristic polynomial is $r^2 + 4$ with roots of $\pm 2i$. Thus,

$$y = c_1 \cos 2t + c_2 \sin 2t. \tag{20a}$$

If this formula for y is substituted into (18a), a first-order differential equation in x results, whereas substituting into (18b) gives an algebraic equation for x. We substitute, then, into (18b):

$$x = -y' - y = -(c_1 \cos 2t + c_2 \sin 2t)' - (c_1 \cos 2t + c_2 \sin 2t),$$

so that

$$x = (-2c_2 - c_1) \cos 2t + (2c_1 - c_2) \sin 2t. \tag{20b}$$

The solution (20a), (20b) has the correct number of constants, since $L_1L_4 - L_2L_3 = D^2 + 4$ is second-order. The initial conditions $x(0) = 0$, $y(0) = 1$ can now be applied to the general solution (20a), (20b):

$$0 = x(0) = -2c_2 - c_1,$$
$$1 = y(0) = c_1.$$

Thus $c_1 = 1$, $c_2 = -\frac{1}{2}$, and the solution of the original initial value problem (18) is

$$x = \frac{5}{2} \sin 2t$$

$$y = \cos 2t - \frac{1}{2} \sin 2t.$$ ◄

The elimination method may also be used to solve systems with more than two equations.

Later in this chapter we shall solve (18) using matrix theory. Even for the relatively simple problem (18), the added simplicity of the matrix method will be apparent.

Exercises

In Exercises 1 through 8, use the method of elimination to solve the differential equation. Find the general solution if no initial conditions are given.

1. $x' = -x + 3y$,
$y' = -2x + 4y$

2. $x' = -x - y$,
$y' = 2x + y$

3. $x' = -2x - 3y + t$,
$y' = x + 2y + 3$

4. $x' = x + y + 1$, $\quad x(0) = 1$,
$y' = x + y + e^t$, $\quad y(0) = 1$

5. $x' = x - 2y$, $\quad x(0) = 0$,
$y' = 2x + y$, $\quad y(0) = 1$

6. $x' = 2x - y + t^2$,
$y' = x$

7. $x' = 2x - y + \sin t$,
$y' = 2x + 4y + \cos t$

8. $x' = x - y + 2t$, $\quad x(0) = 3$,
$y' = x - y - t$, $\quad y(0) = 0$

In solving the algebraic equations (1), we could divide the first equation by 2 if we wanted to. However, we have carefully avoided dividing by a polynomial in $D = d/dt$. Exercises 9 through 11 examine why we have not done this.

9. If $(D - 1)x = 0$ for a function $x(t)$, then x need not be equal to zero. Find all x for which $(D - 1)x = 0$.

10. Suppose that $x(t)$, $y(t)$ are functions such that

$$(D^2 - 1)x + (D - 1)y = 0.$$

Note that $D^2 - 1$ and $D - 1$ are "divisible" by $D - 1$. Show that x, y are solutions of

$$(D + 1)x + y = c_1 e^t$$

for some constant c_1.

11. Suppose that L_1, L_2, L_3 are constant coefficient differential operators. Show that if x, y are functions of t such that

$$L_1 L_2 x + L_1 L_3 y = 0,$$

then

$$L_2 x + L_3 y = f,$$

where f is a solution of $L_1 f = 0$.

In Exercises 12 through 25, solve the system of differential equations by the method of elimination. Find the general solution if no initial conditions are given.

12. $x'' + y' - y = 0,$
$x' + y' + y = 0$

13. $x' + x + y' - y = 0,$
$x' - x + 2y' + y = 0$

14. $x'' + x + y'' = 0,$
$x' + y' = t$

15. $x'' - 2x + y'' - y = 1,$
$2x' + x + 2y' + 2y = 2$

16. $x' = y + z,$
$y' = 2x + y + 2z,$
$z' = 3x + 3y + 2z$

17. $x' = 2x - y + z,$ $\quad x(0) = 0,$
$y' = -x + 2y + z,$ $\quad y(0) = 1,$
$z' = x + y,$ $\quad z(0) = 0$

18. $x' = 2x - y + z + 1,$
$y' = 2x - y + 2z,$
$z' = x - y + 2z + 2$

19. $x' = 3x - y + z,$
$y' = -x + 3y + z,$
$z' = x + y + z$

20. $x'' + 2x - y' - y = 0,$
$-x' + x + y'' - y = 0$

21. $x'' = 8x - y - y',$
$y'' = x - x' + y$

22. $x'' + 2x' + x - 3y' - 3y = 0,$
$2x + y' - 4y = 0$

23. $x'' + 2x' + 3y' = 1,$
$x' + 2y' - y'' = 0$

24. $x'' - x' + 2y' = 0,$
$-2x' + y'' - y' = 0$

25. $x'' - x + y'' - 2y' + y = e^{2t},$
$x' - x + 2y' + y = 0$

5.3 Solution by Laplace Transform

An alternative to the elimination method for the solution of systems of linear differential equations with constant coefficients is to use the Laplace transform. This approach, sometimes in a somewhat disguised format, is especially popular in the electrical engineering and system and control literature. This method has the advantage that one can divide freely and not have to worry about having too many arbitrary constants. In order to show the similarities and differences of the two methods, we shall illustrate the method by reworking one of the examples from Section 5.2.

Note that the system to be solved for X, Y is very similar to the system of Section 5.2, except that now we have an s in place of D. We can, if we want to, safely divide an equation by, say, $s - 1$, whereas "dividing" by $(D - 1)$ is to be avoided unless you have a very good understanding of what the operator notation means. (See Exercises 9 through 11 of Section 5.2.)

EXAMPLE 5.3.1 Solution Using Laplace Transforms

Solve the system of differential equations

$$4x' + 4x + y' = 1, \qquad x(0) = 2, \tag{1a}$$

$$3x' + y' - y = t, \qquad y(0) = -6, \tag{1b}$$

using Laplace transforms (this is Example 5.2.2 with initial conditions added).

Solution Taking the Laplace transform of both sides of equations (1a), (1b) gives

$$\begin{aligned} 4[sX - x(0)] + 4X + sY - y(0) &= \frac{1}{s}, \\ 3[sX - x(0)] + [sY - y(0)] - Y &= \frac{1}{s^2}. \end{aligned} \tag{2}$$

Using the initial conditions $x(0) = 2$, $y(0) = -6$, and combining terms, (2) becomes

$$4(s + 1)X + sY = \frac{1}{s} + 2 = \frac{1 + 2s}{s}, \tag{3a}$$

$$3sX + (s - 1)Y = \frac{1}{s^2}. \tag{3b}$$

Solve (3a) for Y,

$$Y = \frac{1 + 2s}{s^2} - \frac{4(s + 1)}{s}X, \tag{4}$$

and substitute into (3b) to yield an equation in X:

$$3sX + (s - 1)\left[\frac{1 + 2s}{s^2} - 4\frac{(s + 1)}{s}X\right] = \frac{1}{s^2}. \tag{5}$$

Simplifying (5), we obtain

$$(-s^2 + 4)X = \frac{1}{s}(2 + s - 2s^2),$$

and, solving for X, we get

$$X = \frac{2s^2 - s - 2}{s(s^2 - 4)} = \frac{2s^2 - s - 2}{s(s - 2)(s + 2)}. \tag{6}$$

By partial fractions,

$$X = \frac{A}{s} + \frac{B}{s - 2} + \frac{C}{s + 2},$$

where

$$2s^2 - s - 2 = A(s-2)(s+2) + Bs(s+2) + Cs(s-2). \qquad (7)$$

Since the three factors of the denominator are all linear factors, the simplest way to find A, B, C is to evaluate (7) at the roots $s = 0$, $s = 2$, $s = -2$, of the factors in the denominator. This gives $A = \frac{1}{2}$, $B = \frac{1}{2}$, $C = 1$. Thus,

$$X(s) = \frac{1/2}{s} + \frac{1/2}{s-2} + \frac{1}{s+2},$$

so that

$$x(t) = \frac{1}{2} + \frac{1}{2}e^{2t} + e^{-2t}. \qquad (8)$$

To find $y(t)$, we could substitute $X(s)$ from (6) into (4), get $Y(s)$, and calculate the inverse Laplace transform of $Y(s)$. It is somewhat quicker to substitute formula (8) for $x(t)$ into (1a) and then antidifferentiate to find y. We shall follow the second approach. From (1a),

$$y' = 1 - 4x' - 4x = -1 - 6e^{2t} + 4e^{-2t}.$$

Antidifferentiating gives

$$y = -t - 3e^{2t} - 2e^{-2t} + c_1.$$

The initial condition $y(0) = -6$ gives an equation for c_1:

$$-6 = y(0) = -3 - 2 + c_1,$$

so that $c_1 = -1$ and the solution of the differential equation (1) is

$$x = \frac{1}{2} + \frac{1}{2}e^{2t} + e^{-2t},$$

$$y = -1 - t - 3e^{2t} - 2e^{-2t}.$$

◄

Exercises

In Exercises 1 through 18, solve the initial value problem, using the Laplace transform. Except for the addition of initial conditions, Exercises 1 through 12 are Exercises 1 through 8 and 16 through 19 of Section 5.2.

1. $x' = -x + 3y, \quad x(0) = 2,$
$y' = -2x + 4y, \quad y(0) = 1$

2. $x' = -x - y, \quad x(0) = 1,$
$y' = 2x + y, \quad y(0) = -2$

3. $x' = -2x - 3y + t, \quad x(0) = 6,$
$y' = x + 2y + 3, \quad y(0) = -5$

4. $x' = x + y + 1, \quad x(0) = -\frac{1}{2},$
$y' = x + y + e^t, \quad y(0) = 1$

5. $x' = x - 2y, \quad x(0) = 0,$
$y' = 2x + y, \quad y(0) = 1$

6. $x' = 2x - y + t^2, \quad x(0) = 4,$
$y' = x, \quad y(0) = 5$

7. $x' = 2x - y + \sin t, \quad x(0) = \frac{23}{78}$
$y' = 2x + 4y + \cos t, \quad y(0) = -\frac{67}{39}$

8. $x' = x - y + 2t, \quad x(0) = 3,$
$y' = x - y - t, \quad y(0) = 0$

9. $x' = y + z, \quad x(0) = 2,$
$y' = 2x + y + 2z, \quad y(0) = 0,$
$z' = 3x + 3y + 2z, \quad z(0) = 4$

10. $x' = 2x - y + z, \quad x(0) = 0,$
$y' = -x + 2y + z, \quad y(0) = -2,$
$z' = x + y, \quad z(0) = -7$

11. $x' = 2x - y + z + 1, \quad x(0) = 3,$
$y' = 2x - y + 2z, \quad y(0) = 6,$
$z' = x - y + 2z + 2, \quad z(0) = 1$

12. $x' = 3x - y + z, \quad x(0) = 1,$
$y' = -x + 3y + z, \quad y(0) = 3,$
$z' = x + y + z, \quad z(0) = 5$

13. $x'' = -2x + y + y' + t, \quad x(0) = x'(0) = 0,$
$y'' = -x + x' + y + 1, \quad y(0) = y'(0) = 0$

14. $x'' = 8x - y - y', \quad x(0) = 1, \quad x'(0) = -5,$
$y'' = x - x' + y, \quad y(0) = -\frac{9}{4}, \quad y'(0) = \frac{61}{4}$

15. $x'' + 2x' + 3y' = 1, \quad x(0) = 1, \quad x'(0) = 6,$
$x' + 2y' - y'' = 0, \quad y(0) = 6, \quad y'(0) = -3$

16. $x'' + 2x' + x - 3y' - 3y = 0,$
$2x + y' - 4y = 0,$
$y(0) = 0, \quad y'(0) = 0, \quad y''(0) = 1$
[*Hint*: Compute $x(0)$, $x'(0)$.]

17. $x'' - x + y'' - 2y' + y = -e^{2t}, \quad x(0) = -\frac{47}{14},$
$x' - x + 2y' + y = 0, \quad y(0) = \frac{43}{14}, \quad y'(0) = -\frac{69}{7}$
[*Hint*: Compute $x'(0)$.]

18. $x'' - x' + 2y' = 0, \quad x(0) = 0, \quad x'(0) = 1,$
$-2x' + y'' - y' = 0, \quad y(0) = 1, \quad y'(0) = 0$

5.4 Mixing Problems

In this and the next two sections we will give several applications of systems of differential equations. Mixing problems will be discussed in this section. It is assumed that the reader has read Section 1.11. At the end of this section, we shall briefly discuss how a wide variety of problems can be viewed as mixing/flow problems.

In our problems we will have two or more tanks interconnected with pipes. The tanks contain a chemical dissolved in a fluid, which we take as salt (NaCl) and water, respectively. At a given time t, we are interested in the amount of salt in each tank.

EXAMPLE 5.4.1 Two-Tank Mixing Problem (Fixed Volume)

We have two 100-gal tanks. Tank A is initially full of salt water with a salt concentration of 0.5 lb/gal. Tank B is full of pure water. Pure water enters tank A at 2 gal/min. Water exits the well-mixed tank A at 2 gal/min and flows into tank B. Water also exits tank B at 2 gal/min and flows out of the system (down the drain?). Pictorially, we have Figure 5.4.1. Determine the differential equation for the amount of salt in each tank. Solve the differential equations.

SOLUTION We assume that the amount of water in the pipe between the tanks is negligible. Let $x(t)$ be the amount of salt in pounds in tank A at time

FIGURE 5.4.1

Picture for Example 5.4.1.

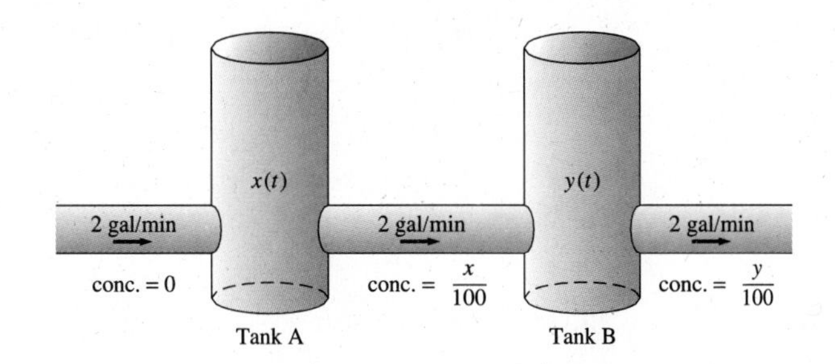

t, and $y(t)$ the amount of salt in pounds in tank B at time t. The inflow and outflow rates of water in each tank balance, so the volumes are constant. Thus the concentrations in tanks A and B are $x/100$ and $y/100$, respectively (amount/volume). The equation describing the rate of change of the amount of salt in tank A is

$$\frac{dx}{dt} = -\begin{bmatrix}\text{outflow rate}\\ \text{of salt}\\ \text{from A}\end{bmatrix} = -\begin{bmatrix}\text{flow rate}\\ \text{of}\\ \text{water}\end{bmatrix}\cdot\begin{bmatrix}\text{conc. of}\\ \text{salt in}\\ \text{water}\end{bmatrix}$$

$$= -2\frac{x}{100}. \tag{1}$$

For tank B we have

$$\frac{dy}{dt} = \begin{bmatrix}\text{inflow rate}\\ \text{of salt}\\ \text{from A}\end{bmatrix} - \begin{bmatrix}\text{outflow rate}\\ \text{of salt}\\ \text{from B}\end{bmatrix}$$

$$= 2\left(\frac{x}{100}\right) - 2\left(\frac{y}{100}\right). \tag{2}$$

Combining (1) and (2) and using our initial conditions, we have the linear system of differential equations

$$x' = -\frac{1}{50}x, \qquad x(0) = 50, \tag{3a}$$

$$y' = \frac{1}{50}x - \frac{1}{50}y, \qquad y(0) = 0. \tag{3b}$$

Since (3a) involves only x, we do not need to use elimination. Equation (3a) can be solved several ways. The quickest is to observe that the characteristic polynomial $r + 1/50$ has only the single root $-1/50$. Thus, $x = c_1e^{-t/50}$. The initial condition in (3a) implies that $c_1 = 50$ and

$$x = 50e^{-t/50}. \tag{4}$$

Substitute (4) for x in (3b) to get

$$y' + \frac{1}{50}y = e^{-t/50}. \tag{5}$$

This could be solved by undetermined coefficients. We shall use the integrating factor $e^{\int p(t)dt} = e^{t/50}$ (Section 1.6.3),

$$(e^{t/50}y)' = e^{-t/50}e^{t/50} = 1,$$

and antidifferentiate:

$$e^{t/50}y = t + c_2.$$

Solving for y and using the initial condition in (3b) to find $c_2 = 0$ gives

$$y = te^{-t/50}; \tag{6}$$

thus the solution of the system of differential equations (3) is

$$x = 50e^{-t/50}, \qquad y = te^{-t/50}. \tag{7}$$

◀

The solution has the expected physical behavior. The amount of salt in tank A exponentially decays. In tank B there is an increase at first because of the inflow of salt from tank A, and then the amount of salt in tank B decreases. This much we could have guessed. However, our solution (7) makes it possible to answer questions whose answer is not obvious. For example, is there ever more salt in tank B than in tank A? The answer is yes, as the graph of (7) in Figure 5.4.2 shows.

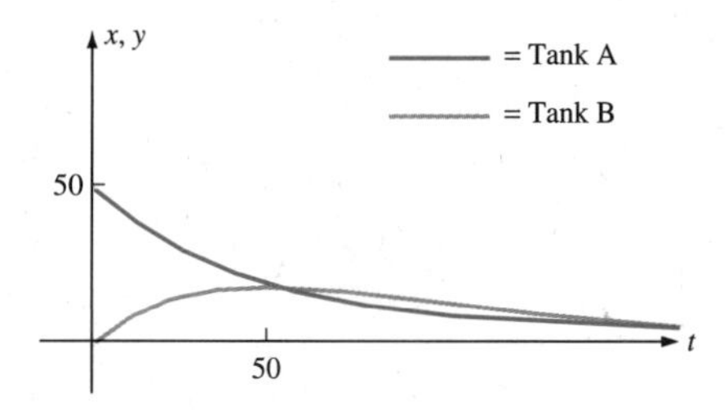

FIGURE 5.4.2

Graphs of (7).

EXAMPLE 5.4.2 Two Tanks with Evaporation

Two well-mixed 100-L tanks are initially full of pure water. Water containing salt at a concentration of 30 g/L flows into tank A at a rate of 4 L/min. Water flows from tank A to tank B at 4 L/min. Water evaporates from tank B at 2 L/min and flows out at 2 L/min. We want to find the amount of salt in both tanks as a function of time. Pictorially, we have Figure 5.4.3.

Solution Let $x(t)$ be the amount of salt in tank A at time t and $y(t)$ the amount of salt in tank B at time t; t is measured in minutes, and x and y are

FIGURE 5.4.3

Picture for Example 5.4.2.

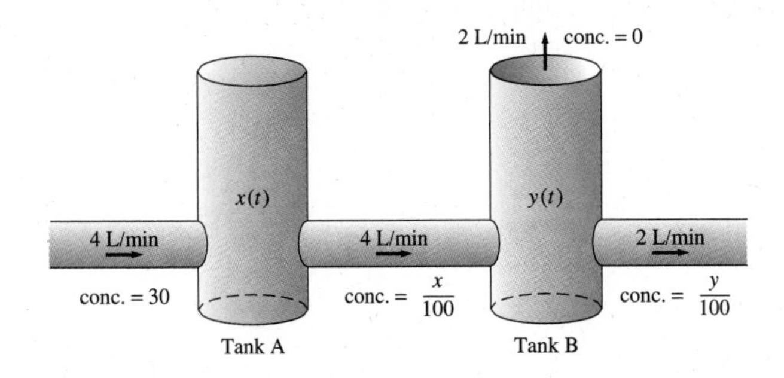

in grams. Note that the volume of both tanks is constant. The equations describing the change in the amounts of salt are

$$\frac{dx}{dt} = \begin{bmatrix}\text{inflow rate of}\\ \text{water into}\\ \text{tank A}\end{bmatrix}\begin{bmatrix}\text{conc. of}\\ \text{NaCl}\\ \text{in inflow}\end{bmatrix} - \begin{bmatrix}\text{outflow rate of}\\ \text{water from}\\ \text{tank A}\end{bmatrix}\begin{bmatrix}\text{conc. of}\\ \text{NaCl}\\ \text{in outflow}\end{bmatrix},$$

$$\frac{dy}{dt} = \begin{bmatrix}\text{inflow rate of}\\ \text{water into}\\ \text{tank B}\end{bmatrix}\begin{bmatrix}\text{conc. of}\\ \text{NaCl}\\ \text{in inflow}\end{bmatrix} - \begin{bmatrix}\text{outflow rate of}\\ \text{water excluding}\\ \text{evaporation from}\\ \text{tank B}\end{bmatrix}\begin{bmatrix}\text{conc. of}\\ \text{NaCl}\\ \text{in outflow}\end{bmatrix},$$

since no salt is lost due to evaporation. Thus,

$$\frac{dx}{dt} = 4(30) - 4\frac{x}{100}, \qquad x(0) = 0,$$

$$\frac{dy}{dt} = 4\frac{x}{100} - 2\frac{y}{100}, \qquad y(0) = 0.$$

That is,

$$\frac{dx}{dt} + \frac{x}{25} = 120, \qquad x(0) = 0, \tag{8a}$$

$$\frac{dy}{dt} + \frac{y}{50} - \frac{x}{25} = 0, \qquad y(0) = 0. \tag{8b}$$

We could solve (8) in the same way we solved (3). Instead, we shall use the Laplace transform. Taking the Laplace transform of (8a), (8b), and using the initial conditions gives

$$sX + \frac{1}{25}X = \frac{120}{s}, \tag{9a}$$

$$sY + \frac{1}{50}Y - \frac{1}{25}X = 0. \tag{9b}$$

Solve (9a) for X:

$$X(s) = \frac{120}{s(s + 1/25)}. \tag{10}$$

Using partial fractions,

$$X = \frac{A}{s} + \frac{B}{s + 1/25},$$

where $120 = A(s + 1/25) + Bs$, so that $A = 25 \cdot 120 = 3000$ and $B = -3000$. Thus the inverse Laplace transform of $X(s)$ gives

$$x(t) = 3000 - 3000e^{-t/25}. \tag{11}$$

From (9b) and (10),

$$Y = \frac{1}{25(s + 1/50)}X = \frac{120}{25(s + 1/50)s(s + 1/25)}.$$

By partial fractions,

$$Y = \frac{A}{s} + \frac{B}{s + 1/50} + \frac{C}{s + 1/25}, \tag{12}$$

where

$$\frac{120}{25} = A\left(s + \frac{1}{50}\right)\left(s + \frac{1}{25}\right) + Bs\left(s + \frac{1}{25}\right) + Cs\left(s + \frac{1}{50}\right).$$

Evaluating at $s = 0$, $s = -1/25$, $s = -1/50$ yields

$$A = 6000, \qquad B = -12{,}000, \qquad C = 6000.$$

Taking the inverse Laplace transform in (12) and recalling (11) gives the solution of (8) as

$$\begin{aligned} x &= 3000 - 3000e^{-t/25}, \\ y &= 6000e^{-t/25} - 12{,}000e^{-t/50} + 6000. \end{aligned} \tag{13}$$

The graphs of (13) are given in Figure 5.4.4. ◄

Both tanks approach an equilibrium, or fixed amount of salt. Initially the salt accumulates faster in tank A, but then it increases more rapidly in tank B as more is passed on. If, instead of tanks, we think of a series of reservoirs in an arid region, each flowing into the next, this solution has interesting consequences. What do you think they are?

FIGURE 5.4.4

Graph of (13).

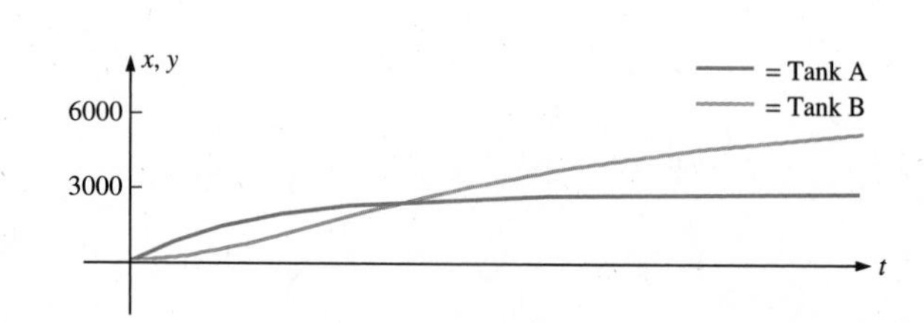

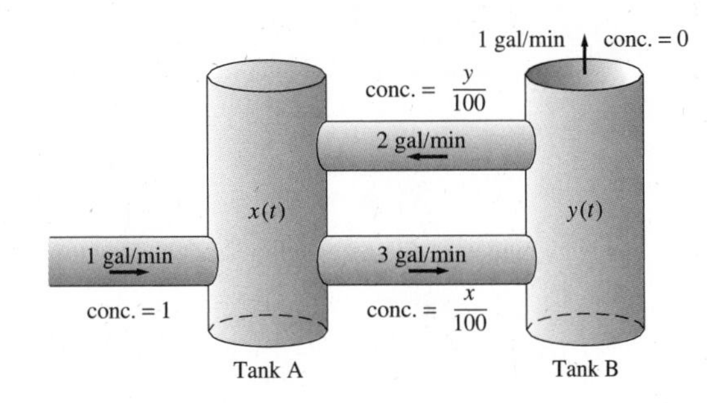

FIGURE 5.4.5

Picture for Example 5.4.3.

EXAMPLE 5.4.3 **Salt Lake**

Two 100-gal tanks are both initially full of pure water. Water containing salt at a concentration of 1 lb/gal flows into tank A from an outside source at 1 gal/min. Water flows from tank A to tank B at 3 gal/min. Water evaporates from tank B at 1 gal/min and is also piped back to tank A at 2 gal/min. Find the amount of salt in each tank as a function of time. Pictorially, we have Figure 5.4.5.

SOLUTION Note that the amount of water in each tank is constant. Let x be the amount of salt in tank A at time t, and y the amount of salt in tank B at time t. The units are gal-lb-min. Then,

$$\frac{dx}{dt} = \begin{bmatrix} \text{inflow rate} \\ \text{NaCl from} \\ \text{outside} \end{bmatrix} + \begin{bmatrix} \text{inflow rate} \\ \text{NaCl from} \\ \text{tank B} \end{bmatrix} - \begin{bmatrix} \text{outflow rate} \\ \text{NaCl from} \\ \text{tank A} \end{bmatrix},$$

$$\frac{dy}{dt} = \begin{bmatrix} \text{inflow rate} \\ \text{NaCl from} \\ \text{tank A} \end{bmatrix} - \begin{bmatrix} \text{outflow rate} \\ \text{NaCl from} \\ \text{tank B} \end{bmatrix},$$

or

$$\frac{dx}{dt} = 1 \cdot 1 + 2\frac{y}{100} - 3\frac{x}{100}, \qquad x(0) = 0,$$

$$\frac{dy}{dt} = 3\frac{x}{100} - 2\frac{y}{100}, \qquad y(0) = 0.$$

We have, then, the linear system of differential equations

$$\frac{dx}{dt} = -\frac{3}{100}x + \frac{y}{50} + 1, \qquad x(0) = 0, \tag{14a}$$

$$\frac{dy}{dt} = \frac{3}{100}x - \frac{y}{50}, \qquad y(0) = 0. \tag{14b}$$

We shall solve (14) by elimination. First, we rewrite (14) in operator notation as

$$(100D + 3)x - 2y = 100, \tag{15a}$$

$$-3x + (100D + 2)y = 0. \tag{15b}$$

Multiply (15b) by $\frac{1}{3}(100D + 3)$ and add to (15a) to get a differential equation just in terms of y:

$$\left[\frac{1}{3}(100D + 3)(100D + 2) - 2\right]y = 100,$$

or

$$(100D^2 + 5D)y = 3; \tag{16}$$

that is, $100y'' + 5y' = 3$. The characteristic polynomial $100r^2 + 5r$ of (16) has roots $r = 0, -\frac{1}{20}$. Since 0 is a root, the method of undetermined coefficients says that there will be a particular solution to (16) of the form $y_p = Ct$. Substituting this form for y_p into (16), we get

$$100(Ct)'' + 5(Ct)' = 3,$$

which implies that $5C = 3$ or $C = \frac{3}{5}$. Thus the general solution of (16) is

$$y = \frac{3}{5}t + c_1 + c_2e^{-t/20}. \tag{17}$$

Then, from (14b), $x = \frac{1}{3}(100y' + 2y)$, so that

$$x = 20 + \frac{2}{3}c_1 + \frac{2}{5}t - c_2e^{-t/20}. \tag{18}$$

Equations (17), (18) give the general solution. The initial conditions $x(0) = 0$, $y(0) = 0$ applied to (17), (18) give

$$x(0) = 0 = 20 + \frac{2}{3}c_1 - c_2,$$

$$y(0) = 0 = c_1 + c_2,$$

and thus $c_1 = -12$, $c_2 = 12$. The solution of (14) is then

$$x = 12 + \frac{2}{5}t - 12e^{-t/20},$$

$$y = -12 + \frac{3}{5}t + 12e^{-t/20}, \tag{19}$$

which is graphed in Figure 5.4.6. ◄

Both x and y have asymptotes, and for large t, $x \approx 12 + \frac{2}{5}t$, $y \approx -12 + \frac{3}{5}t$, $y/x \approx \frac{3}{2}$. Thus while both tanks get steadily more "salty," tank B will approach being 1.5 times as "salty" as tank A.

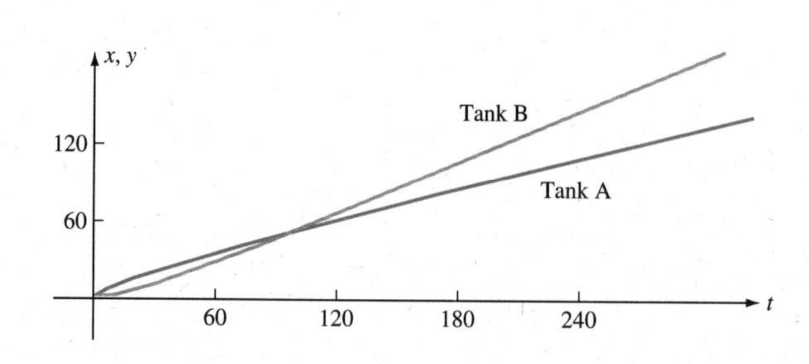

FIGURE 5.4.6

Graph of (19).

Comment While we have talked only of water tanks and salt, the modeling process used in our mixing problems is frequently used whenever the following situation occurs:

- There are several quantities of interest.
- The rate at which one quantity is changed into another quantity is proportional to a linear combination of the quantities.

In our examples, the quantities were the amount of salt in each tank. However, there are numerous other possibilities. Some of the more widely used ones are

i. A region is divided into several areas. The quantities of interest are the populations in each area. The populations change by birth, death, and migration. Birth and death rates are proportional to the number present. Migration rates may be proportional to the population of an area or to population differences.
ii. A species of animal is divided into several age groups. The quantities of interest are the number in each age group. Individuals die, or change age groups, at rates proportional to the number in the group. They are born at rates proportional to the numbers in the fertile age groups.
iii. An economy is divided into different sectors. The quantities of interest are the amount of goods of each sector. Goods from some sectors either are consumed, use goods from other sectors, or produce goods for other sectors at rates that are linear combinations of the amount of goods in each sector.
iv. A lake is divided into regions on the basis of stable circulation patterns. The quantities of interest are the amount of a pollutant in the different regions.

Examples (i) and (ii) are sometimes referred to as **Leslie population models**, whereas (iii) is referred to in economics as a **Leontief model** of a multisectored economy.

Exercises

1. Two jars are initially full of pure water. Jar A is 0.5 L and jar B is 0.25 L. Water containing salt at a concentration of 100 g/L is pumped into jar A at 0.5 L/h. Water flows from jar A to jar B at 0.5 L/h. Water flows out of jar B and down a drain at 0.5 L/h. Find the amounts of salt in each jar as a function of time t. Graph your solutions. (Figure 5.4.1 is applicable.)

2. There are two laboratory beakers. Beaker A contains 0.15 L of pure water, and beaker B contains 20 g of salt dissolved in 0.1 L of water. Water containing salt at a concentration of 100 g/L flows into beaker A at a rate of 0.3 L/h. Water flows from beaker A to beaker B at 0.3 L/h. Water flows from beaker B at 0.3 L/h and goes down the drain. Find the amount of salt in each beaker as a function of time, and graph your solution.

3. Water containing salt at a concentration of 1 lb/gal flows into a 2-gal tank at 0.2 gal/min. Water flows out of the first tank and into a second 3-gal tank at 0.2 gal/min. Both tanks are initially full of pure water. Pure water from a tap flows directly into the second tank at 0.1 gal/min. Water is piped out of the second tank and down a drain at 0.3 gal/min. Find the amount of salt in each tank as a function of time and graph your solution.

4. A tank initially contains 2 L of water and 5 g of salt. Water containing salt at a concentration of 5 g/L flows into this tank at 2 L/h. Water flows from this tank into a second tank at 2 L/h. The second tank contains 1 L of fluid and initially contains 10 g of salt. Water evaporates from the second tank at 1 L/h and flows from the second tank and down a drain at 1 L/h. Find the amount of salt in each tank as a function of time, and graph the solutions.

5. There are two tanks. Tank A is a 100-gal tank, initially full of water containing salt at a concentration of 0.5 lb/gal. Tank B is a 200-gal tank, initially full of water containing salt at a concentration of 0.1 lb/gal. Starting at time $t = 0$, water is pumped from tank A to tank B at 2 gal/min and from tank B to tank A at 2 gal/min. Find the amount of salt in each tank as a function of time, and graph the solutions.

6. There are two 100-gal tanks full of water. Tank A contains salt at a concentration of 0.4 lb/gal, and tank B contains pure water. Pure water flows into tank A from an outside source at 2 gal/min. Water is pumped from tank A to tank B at 3 gal/min. Water evaporates from tank B at a rate of 2 gal/min. Water is also pumped from tank B to tank A at 1 gal/min. Find the amount of salt in each tank as a function of time, and graph your solution.

7. Two 2-L jars are initially full of pure water. Water containing salt at a concentration of 10 g/L is pumped into jar A at a rate of 2 L/h and into jar B at a rate of 2 L/h. Water is piped from jar A to jar B at 3 L/h and from jar B to jar A at 1 L/h. In addition, water flows from jar B down the drain at 4 L/h. Find the amount of salt in the jars as a function of time, and graph the solutions.

8. There are two tanks. Tank A contains 100 gal of pure water, and tank B contains 10 lb of salt dissolved in 200 gal of water. Pure water enters tank A at 5 gal/min. Water is pumped from tank A to tank B at 8 gal/min and from tank B to tank A at 3 gal/min. In addition, 5 gal/min of water is pumped out of tank B and sent out of the system. Find the amount of salt in each tank as a function of time, and graph the amounts.

9. Two large tanks initially each contain 100 gal of pure water. Water containing salt at a concentration of 0.5 lb/gal is pumped into tank A at 14 gal/min. Water is pumped from tank A to tank B at 9 gal/min. Water is pumped from tank B and sent down a drain at 6 gal/min. Find a differential equation for the amount of salt in each tank that is valid as long as the tanks are not full. You do not need to solve the differential equation. (*Note*: The volumes are not constant.)

10. Each of two large tanks in the desert initially contains 500 gal of water with a salt concentration of 0.01 lb/gal. Water containing salt at a concentration of 0.2 lb/gal is pumped into tank A at a rate of 10 gal/h. Water evaporates from tank A at a rate of 7 gal/h and is pumped from tank A into tank B at 5 gal/h. Water evaporates from tank B at a rate of 8 gal/h. Water is also pumped out of tank B at 3 gal/h. Write the differential equation for the amount of salt in each tank, which is valid until one of the tanks goes dry. You do not need to solve the differential equation.

11. There are three tanks, each initially containing 100 gal of pure water. Water containing salt at a concentration of 2 lb/gal flows into tank A at a rate of 5 gal/min. Water is pumped from tank A to tank B at 2 gal/min and from tank A to tank C at 3 gal/min. Water is pumped from tank C to tank B at 3 gal/min. Water is pumped out of tank B and then down a drain at a rate of 5 gal/min. Derive the differential equation for the amount of salt in each tank as a function of time. You need not actually solve the differential equation.

12. Three shallow ponds are out in the sun. Each is initially full of 5000 gal of pure water. Water containing salt at a concentration of 0.2 lb/gal flows into pond A at a rate of 10 gal/h. Water evaporates from pond A at 2 gal/h. Water flows from pond A to pond B at 8 gal/h. Water evaporates from pond B at 3 gal/h. Water flows from pond B to pond C at 5 gal/h. In pond C, water evaporates at 4 gal/h and flows out at 1 gal/h. Derive the differential equation for the amount of salt in each pond. You do not need to solve the differential equation.

13. Two tanks contain V_1 and V_2 gallons of water, respectively. Water containing salt at a concentration of δ lb/gal flows into the first tank at α gal/min, and β gal/min $(0 < \beta \leq \alpha)$ of water is pumped from the first tank into the second. Finally, γ gal/min of water is pumped out of the second tank $(0 \leq \gamma \leq \beta)$. Assume that evaporation rates for the tanks are such that V_1, V_2 are constant.

i) Set up the differential equation for the amount of salt in each tank.

ii) Using elimination, find a differential equation for the amount of salt in the second tank.

iii) Find the roots of the characteristic equation from part (ii).

iv) Determine for what values of the parameters there will be equilibrium solutions and determine the equilibrium solution if there is one.

(Population Models) Exercises 14 through 19 deal with the following situation. A species is divided into m groups, which we take to be age groups. Suppose that w is the size of a group at time t. For that group we assume

i) There is a loss due to death that is proportional to group size, $-\delta w$, with $\delta > 0$. (This could also represent harvesting for species such as trees.)

ii) Individuals "graduate" from one group to the next at a rate proportional to group size, $-gw$, with $g > 0$.

iii) Fertile groups give rise to offspring at a rate proportional to group size, βw, with $\beta > 0$.

14. Suppose that a population consists of two groups: adults and children. Let x be the number of children and y the number of adults. Assume that children cannot have offspring. Explain why the model

$$\begin{aligned} \frac{dx}{dt} &= -(\delta_1 + g)x + \beta y, \quad x(t_0) = x_0 \geq 0, \\ \frac{dy}{dt} &= -\delta_2 y + gx, \quad y(t_0) = y_0 \geq 0 \end{aligned} \tag{20}$$

with δ_1, δ_2, g, β positive constants might be reasonable given assumptions (i) through (iii).

Exercises 15 through 17 illustrate the types of behavior possible for (20).

15. Suppose that (20) holds and $\delta_1 = g = 1$, $\delta_2 = 2$, and $\beta = 1$. Show that $\lim_{t\to\infty} x(t) = 0$, $\lim_{t\to\infty} y(t) = 0$, so that the species dies out.

16. Suppose that (20) holds and $\delta_1 = g = 1$, $\delta_2 = 2$, and $\beta = 4$.

a) Show that (20) has nonzero equilibrium solutions $(\bar{x}, \bar{y})$.

b) Show that every solution $x(t)$, $y(t)$ converges to one of these equilibrium solutions.

17. Suppose that (20) holds and $\delta_1 = g = 1$, $\delta_2 = 2$, and $\beta = 9$. Show that for initial conditions $x(0) > 0$, $y(0) > 0$ we have $\lim_{t\to\infty} x(t) = \infty$, $\lim_{t\to\infty} y(t) = \infty$, and $\lim_{t\to\infty} \dfrac{x(t)}{y(t)}$ exists and is finite.

18. Suppose that (20) holds. Show that

a) If $\delta_2(\delta_1 + g) - \beta g > 0$, then $\lim_{t\to\infty} x(t) = \lim_{t\to\infty} y(t) = 0$.

b) If $\delta_2(\delta_1 + g) - \beta g = 0$, then there are nonzero equilibriums and every solution of (20) converges to one of these equilibriums as $t \to \infty$.

c) If $\delta_2(\delta_1 + g) - \beta g < 0$, then $\lim_{t\to\infty} x(t) = \lim_{t\to\infty} y(t) = \infty$ for initial conditions $x(0) > 0$, $y(0) > 0$.

d) Give a biological interpretation of parts (a) through (c) in terms of the effect of the birth rate constant β.

19. Suppose that the population consists of three stages, which we shall call larva (x), pupa (y), and adult (z). Only adults can produce larva.

a) Explain why a reasonable model might be

$$\begin{aligned} x' &= -(\delta_1 + g_1)x + \beta z \\ y' &= -(\delta_2 + g_2)y + g_1 x \\ z' &= -\delta_3 z + g_2 y \end{aligned} \tag{21}$$

with all constants positive and the initial conditions nonnegative.

b) Discuss what other assumptions would probably need to be made for (21) to be an accurate model.

20. **(Interest)** Two investment accounts are set up with \$1000 initially in account A and \$2000 initially in account B. Account A, the long-term account, earns 10% a year compounded daily. Account B earns 5% a year compounded daily. Deposits are made into B at the rate of \$10 a day. Every day the bank transfers money from B to A at an annual rate of 20% of the difference between B and \$2000. Set up the differential equations that model this situation. (Interest is first discussed in Section 1.8.)

5.5 Mechanical Systems

In this section we will discuss how some mechanical systems can be modeled by linear systems of differential equations. This is a continuation of Sections 1.3, 1.13, 2.8, and 2.11. In order to avoid nonlinear problems, we shall consider point masses, connected by springs in a linear array, undergoing small oscillations. Larger three-dimensional arrays can be used to model many physical structures and mechanical devices. We consider two configurations.

A Horizontal Array of Springs and Masses

Suppose we have two point masses of mass m_1, m_2 and three springs of lengths L_1, L_2, L_3 and spring constants k_1, k_2, k_3. The springs and masses are arranged as in Figure 5.5.1.

The left end of spring 1 and the right end of spring 3 are attached to immovable surfaces. The masses are in contact with a surface whose coefficient of friction (damping constant) is δ. We assume that the mass of the springs is negligible, friction does not affect the springs, Hooke's law is applicable to the springs, motion occurs only horizontally, and resistance is proportional to the velocity. As noted in the earlier sections, this will give us linear equations, and assumes small velocities, small displacements, and small masses.

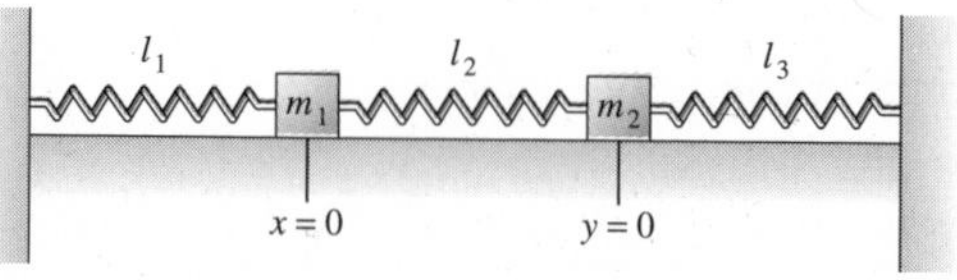

FIGURE 5.5.1

Spring–mass system at rest.

l_1 m_1 l_2 m_2 l_3

$x = 0$ $y = 0$

It is not difficult to show that the configuration of Figure 5.5.1 can be at equilibrium (at rest). Let l_1, l_2, l_3 be the lengths of the springs when the system is at rest (equilibrium). Let

$$x = \text{distance of mass } m_1 \text{ from its rest position,}$$
$$y = \text{distance of mass } m_2 \text{ from its rest position.}$$

A positive value indicates displacement to the right; a negative value is displacement to the left. Also let the difference between a spring's length at rest and its length be ΔL, so that

$$\Delta L_1 = l_1 - L_1,$$
$$\Delta L_2 = l_2 - L_2,$$
$$\Delta L_3 = l_3 - L_3.$$

We now use Newton's law to describe the motion of the masses m_1, m_2. Let F_{T_i} be the total force on mass i. Then

$$F_{T_1} = \begin{bmatrix}\text{force from}\\ \text{spring 1}\end{bmatrix} + \begin{bmatrix}\text{force from}\\ \text{spring 2}\end{bmatrix} + \begin{bmatrix}\text{resistance}\\ \text{on } m_1\end{bmatrix} \tag{1}$$

and

$$F_{T_2} = \begin{bmatrix}\text{force from}\\ \text{spring 2}\end{bmatrix} + \begin{bmatrix}\text{force from}\\ \text{spring 3}\end{bmatrix} + \begin{bmatrix}\text{resistance}\\ \text{on } m_2\end{bmatrix}. \tag{2}$$

The lengths of the three springs are

$$\text{Spring 1:} \qquad L_1 + \Delta L_1 + x = l_1 + x,$$
$$\text{Spring 2:} \qquad L_2 + \Delta L_2 + y - x = l_2 + y - x,$$
$$\text{Spring 3:} \qquad L_3 + \Delta L_3 - y = l_3 - y.$$

Thus $\Delta L_1 + x$, $\Delta L_2 + y - x$, $\Delta L_3 - y$ measure the amount of extension or compression of each spring. Keeping in mind that spring 2 exerts an equal but opposite force on both m_1 and m_2, and that the force exerted by a spring is proportional to the amount by which its length differs from the rest length, we get that (1) and (2) are

$$(m_1 x')' = -k_1(\Delta L_1 + x) + k_2(\Delta L_2 + y - x) - \delta x' \tag{3a}$$

and

$$(m_2 y')' = -k_2(\Delta L_2 + y - x) + k_3(\Delta L_3 - y) - \delta y'. \tag{3b}$$

Since l_1, l_2, l_3 are the spring lengths at rest, the spring forces must balance at these lengths. Thus,

$$k_1 \Delta L_1 = k_2 \Delta L_2 = k_3 \Delta L_3,$$

and (3) simplifies to the system of two homogeneous, second-order linear differential equations,

$$m_1x'' + \delta x' + (k_1 + k_2)x - k_2y = 0, \tag{4a}$$

$$m_2y'' + \delta y' + (k_2 + k_3)y - k_2x = 0. \tag{4b}$$

If outside forces $f_1(t)$, $f_2(t)$ act horizontally on the masses m_1, m_2, we get the nonhomogeneous system

$$m_1x'' + \delta x' + (k_1 + k_2)x - k_2y = f_1, \tag{5a}$$

$$m_2y'' + \delta y' + (k_2 + k_3)y - k_2x = f_2. \tag{5b}$$

It is possible that there are different damping coefficients for m_1 and m_2. This could be caused, for example, by differences in the surfaces on which m_1, m_2 sit, or by properties of a mechanical device being modeled by this spring–mass system. In this event, $\delta x'$ is replaced by $\delta_1 x'$ in (5a) and $\delta y'$ is replaced by $\delta_2 y'$ in (5b). In our examples we assume that $\delta_1 = \delta_2$.

In order to determine the number of arbitrary constants in the solution of (4), we rewrite (4) in operator notation, as

$$[m_1D^2 + \delta D + k_1 + k_2]x - k_2y = 0, \tag{6a}$$

$$-k_2x + [m_2D^2 + \delta D + k_2 + k_3]y = 0, \tag{6b}$$

and take the determinant of the coefficients to get

$$[m_1D^2 + \delta D + k_1 + k_2][m_2D^2 + \delta D + k_2 + k_3] - k_2^2, \tag{7}$$

which is a fourth-order polynomial in D. Thus, by Section 5.1, the system (4) has four arbitrary constants in its general solution. In fact, one can arbitrarily specify $x(0)$, $x'(0)$, $y(0)$, $y'(0)$ [although large values may not make physical sense, since the differential equation (4) is then no longer appropriate].

We shall refer to $(m_1r^2 + \delta r + k_1 + k_2)(m_2r^2 + \delta r + k_2 + k_3) - k_2^2$ as the **characteristic equation** of (6) or (7).

EXAMPLE 5.5.1 Spring–Mass System without Friction

Suppose that the spring–mass system of Figure 5.5.1 has two 1-g masses and the spring constants are all 1 g/s^2. Find the resulting motion if the first mass is initially at its equilibrium position with a velocity of 16 cm/s to the right, while the second mass is initially at rest at its equilibrium position. Friction is assumed negligible.

SOLUTION We have $m_1 = m_2 = 1$, $k_1 = k_2 = k_3 = 1$, and $\delta = 0$. The system (4) is thus

$$x'' + 2x - y = 0, \qquad x(0) = 0,\ x'(0) = 16, \tag{8a}$$

$$y'' + 2y - x = 0, \qquad y(0) = y'(0) = 0. \tag{8b}$$

We shall solve (8) by elimination. Rewriting the system (8) in operator notation, we obtain

$$(D^2 + 2)x - y = 0, \tag{9a}$$

$$-x + (D^2 + 2)y = 0. \tag{9b}$$

Add $(D^2 + 2)$ times the second equation (9b) to the first equation (9a) to give

$$[(D^2 + 2)(D^2 + 2) - 1]y = 0$$

or

$$(D^2 + 3)(D^2 + 1)y = 0. \tag{10}$$

Then the characteristic polynomial of this differential equation is $p(r) = (r^2 + 3)(r^2 + 1)$, which has roots $r = \pm\sqrt{3}i, \pm i$. Thus

$$y = c_1 \cos t + c_2 \sin t + c_3 \cos\sqrt{3}t + c_4 \sin\sqrt{3}t. \tag{11a}$$

Since $x = y'' + 2y$ from (9b) we use (11a) to give

$$x = c_1 \cos t + c_2 \sin t - c_3 \cos\sqrt{3}t - c_4 \sin\sqrt{3}t. \tag{11b}$$

The initial conditions of (8) applied to the general solution (11) yields

$$0 = x(0) = c_1 - c_3,$$

$$16 = x'(0) = c_2 - c_4\sqrt{3},$$

$$0 = y(0) = c_1 + c_3,$$

$$0 = y'(0) = c_2 + c_4\sqrt{3}.$$

Thus $c_1 = c_3 = 0$, and $c_2 = 8$, $c_4 = -8/\sqrt{3}$. The solution of the differential equation (8) is

$$x = 8 \sin t + \frac{8}{\sqrt{3}} \sin\sqrt{3}t, \tag{12a}$$

$$y = 8 \sin t - \frac{8}{\sqrt{3}} \sin\sqrt{3}t. \tag{12b}$$

Figure 5.5.2 shows a graph of y. ◄

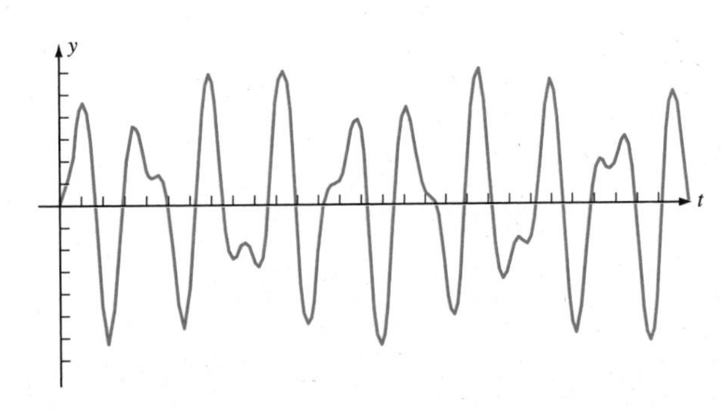

FIGURE 5.5.2

Graph of (12b).

This example is typical in the following respects. If there is no friction, ($\delta = 0$), then the polynomial (7) always has two distinct pairs of purely imaginary conjugate roots. The free response will then consist of the superposition of two harmonic motions of different frequencies (periods). If the ratio of the frequencies is not a rational number, the free response will not be periodic except for very special initial conditions (see Exercises 12 through 14 at the end of the section). The functions (12) are not periodic.

Physical Implications Recall from Section 2.11 that resonance can occur if the forcing term has the same frequency as the free response. For Example 5.5.1, we would expect resonance with forcing terms of frequencies $1/(2\pi)$ or $\sqrt{3}/(2\pi)$ cycles/s. Suppose now we had 50 masses arranged linearly, as in Figure 5.5.3. If friction were negligible, we might expect (and in fact it's approximately true) that the free response would contain harmonic functions of 50 different frequencies. Thus, a forcing function containing terms at *any* of these 50 frequencies would cause resonance. Now, imagine a large, highly Øexible structure to be made of lightweight materials (for example, a solar collector or large antenna in space). The previous discussion suggests that such a structure might have a large number of frequencies at which it would exhibit resonance. The problem, then, is to build the structure so that these frequencies are not within the range of frequencies of the disturbances expected to act on the structure. Alternatively, we would need to design devices to counteract the disturbances at certain frequencies (active or passive controllers).

FIGURE 5.5.3

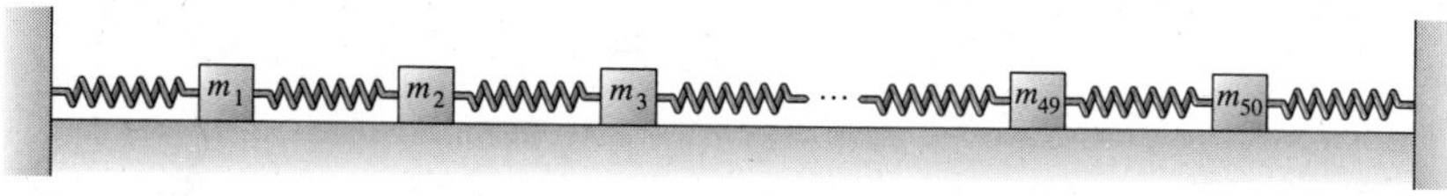

EXAMPLE 5.5.2 Spring–Mass System with Friction

Again we consider a spring–mass system arranged as in Figure 5.5.1 with $m_1 = m_2 = 1$, $k_1 = k_2 = k_3 = 1$, but now we consider friction with a damping coefficient of $\delta = 2$ g/s. Mass m_1 is displaced 1 cm to the left and released with a velocity of 11 cm/s to the right, while at the same instant m_2 is displaced 1 cm to the right and released with a velocity of 9 cm/s to the right.

Solution The system of differential equations (4) is

$$x'' + 2x' + 2x - y = 0, \qquad x(0) = -1, x'(0) = 11, \tag{13a}$$

$$-x + y'' + 2y' + 2y = 0, \qquad y(0) = 1, y'(0) = 9, \tag{13b}$$

or, in operator notation,

$$(D^2 + 2D + 2)x - y = 0, \tag{14a}$$

$$-x + (D^2 + 2D + 2)y = 0. \tag{14b}$$

Using elimination to solve (14), we add $D^2 + 2D + 2$ times (14b) to (14a) and get

$$[(D^2 + 2D + 2)(D^2 + 2D + 2) - 1]y = 0. \tag{15}$$

Now, in general, we have a fourth-degree polynomial. The roots will usually be found numerically on a computer, although formulas exist. However, because of the choice of coefficients, we may solve the characteristic equation

$$(r^2 + 2r + 2)(r^2 + 2r + 2) - 1 = 0 \tag{16}$$

as follows. Equation (16) is

$$(r^2 + 2r + 2)^2 = 1.$$

Thus, taking the square root of both sides yields

$$r^2 + 2r + 2 = 1 \qquad \text{or} \qquad r^2 + 2r + 2 = -1,$$

that is,

$$r^2 + 2r + 1 = 0 \qquad \text{or} \qquad r^2 + 2r + 3 = 0.$$

Solving these quadratics gives the roots of (16) as

$$r = -1, -1 \qquad \text{or} \qquad r = -1 \pm i\sqrt{2}.$$

We have then that

$$y = c_1 e^{-t} + c_2 t e^{-t} + c_3 e^{-t} \cos\sqrt{2}\,t + c_4 e^{-t} \sin\sqrt{2}\,t. \tag{17a}$$

We may find x from (14b) as $x = y'' + 2y' + 2y$, so that (17a) gives (note shortcut below):

$$x = c_1 e^{-t} + c_2 t e^{-t} - c_3 e^{-t} \cos\sqrt{2}\,t - c_4 e^{-t} \sin\sqrt{2}\,t. \tag{17b}$$

Using the initial conditions from (13),

$$\begin{aligned} -1 &= x(0) = c_1 - c_3, \\ 11 &= x'(0) = -c_1 + c_2 + c_3 - \sqrt{2}\,c_4, \\ 1 &= y(0) = c_1 + c_3, \\ 9 &= y'(0) = -c_1 + c_2 - c_3 + \sqrt{2}\,c_4, \end{aligned}$$

we find that $c_1 = 0$, $c_3 = 1$, $c_2 = 10$, $c_4 = 0$, and the solution of (13) is

$$\begin{aligned} x &= 10te^{-t} - e^{-t} \cos\sqrt{2}\,t, \\ y &= 10te^{-t} + e^{-t} \cos\sqrt{2}\,t. \end{aligned}$$

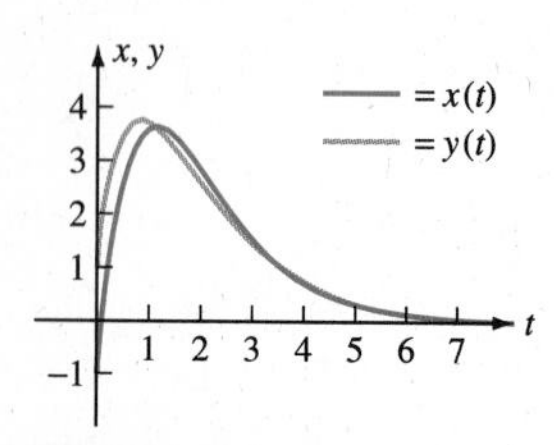

FIGURE 5.5.4

Solution of Example 5.5.2.

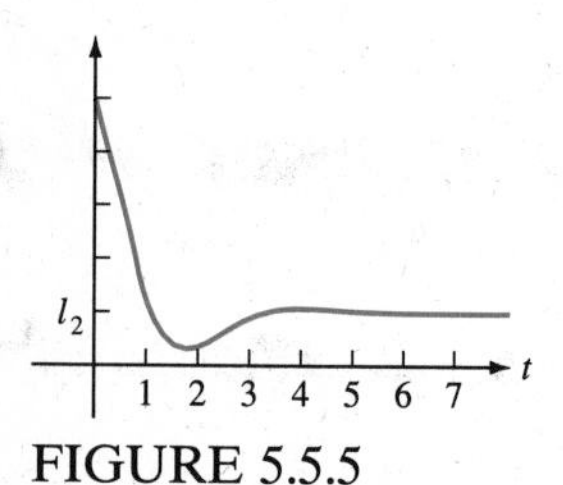

FIGURE 5.5.5

Distance between masses in Example 5.5.2.

This solution is graphed in Figure 5.5.4. The distance between the two masses is graphed in Figure 5.5.5. ◄

In this example we had two real roots and a complex conjugate pair. By changing δ, one may get several other types of solutions, ranging from four complex numbers to four distinct real numbers (see Exercise 9).

A Shortcut

It is possible to sometimes reduce the amount of work in finding the second half of the solution when doing elimination. To see how this works, return to Example 5.5.2 and the derivation of y in (16) and (17). Since -1 is a repeated root of $r^2 + 2r + 1$, we know that (note Exercises 2.10.22 through 2.10.25)

$$(D^2 + 2D + 1)e^{-t} = 0, \qquad (D^2 + 2D + 1)te^{-t} = 0. \tag{18}$$

Similarly, since $-1 \pm i\sqrt{2}$ are the roots of $r^2 + 2r + 3$, we have

$$(D^2 + 2D + 3)e^{-t}\cos\sqrt{2}t = 0, \qquad (D^2 + 2D + 3)e^{-t}\sin\sqrt{2}t = 0. \tag{19}$$

Now, to find x from (14b) and (17a), we have, from (14b), that $x = (D^2 + 2D + 2)y$. Now substitute in y, from (17a):

$$\begin{aligned} x = {} & c_1(D^2 + 2D + 2)e^{-t} + c_2(D^2 + 2D + 2)te^{-t} \\ & + c_3(D^2 + 2D + 2)e^{-t}\cos\sqrt{2}t \\ & + c_4(D^2 + 2D + 2)e^{-t}\sin\sqrt{2}t. \end{aligned} \tag{20}$$

But $D^2 + 2D + 2 = [D^2 + 2D + 1] + 1$ and $D^2 + 2D + 2 = [D^2 + 2D + 3] - 1$, so that, using (18) and (19) gives (20) as

$$\begin{aligned} x = {} & c_1(0 + 1e^{-t}) + c_2(0 + 1te^{-t}) + c_3(0 - 1e^{-t}\cos\sqrt{2}t) \\ & + c_4(0 - 1e^{-t}\sin\sqrt{2}t), \end{aligned}$$

which is (17b). If the solution involves two or four complex roots, this method can be much quicker than actually differentiating the expression for y twice and then combining terms. When real exponentials are involved,

$$p(D)e^{\alpha t} = p(\alpha)e^{\alpha t},$$

for any polynomial $p(\lambda)$. For example,

$$(D^3 - 3D^2 + D - 6)e^{5t} = (5^3 - 3 \cdot 5^2 + 5 - 6)e^{5t} = 49e^{5t}.$$

A Vertical Array of Springs and Masses

Now consider two masses hanging on springs in the configuration of Figure 5.5.6. Again, let L_1, L_2 be the lengths of the springs, k_1, k_2 the spring

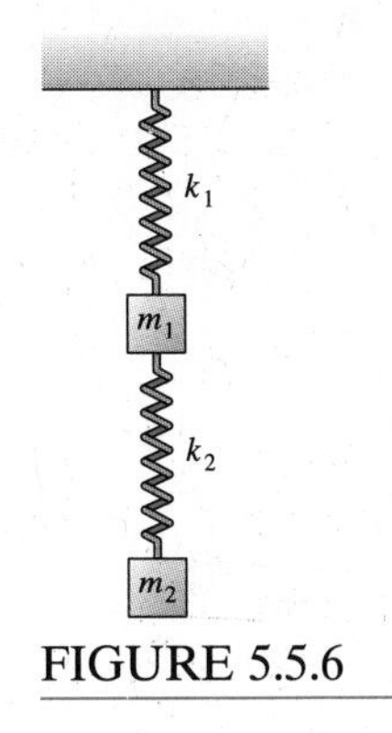

FIGURE 5.5.6

constants, and l_1, l_2 the lengths of the springs at rest with the masses attached. Set $\Delta L_i = l_i - L_i$ for $i = 1, 2$. The resistive force acting on each mass will have damping coefficient δ_i. We may have $\delta_1 \neq \delta_2$ if the masses move in different media or if the spring–mass system is a model of a different physical structure and the δ_i represent internal damping. Let x, y measure the displacement of masses m_1, m_2 from equilibrium with the downward direction as positive. Then, from Newton's laws, ignoring external forces other than gravity,

$$(m_1x')' = \begin{bmatrix}\text{force from}\\ \text{spring 1}\end{bmatrix} + \begin{bmatrix}\text{force from}\\ \text{spring 2}\end{bmatrix} + \begin{bmatrix}\text{gravity on}\\ \text{mass 1}\end{bmatrix} + \begin{bmatrix}\text{friction on}\\ \text{mass 1}\end{bmatrix},$$

$$(m_2y')' = \begin{bmatrix}\text{force from}\\ \text{spring 2}\end{bmatrix} + \begin{bmatrix}\text{gravity on}\\ \text{mass 2}\end{bmatrix} + \begin{bmatrix}\text{friction on}\\ \text{mass 2}\end{bmatrix},$$

or

$$\begin{aligned} m_1x'' &= -k_1(\Delta L_1 + x) + k_2(\Delta L_2 + y - x) + m_1g - \delta_1x', \\ m_2y'' &= -k_2(\Delta L_2 + y - x) + m_2g - \delta_2y'. \end{aligned}$$

At equilibrium $k_1\Delta L_1 = k_2\Delta L_2 + m_1g$ and $k_2\Delta L_2 = m_2g$, so that our system of differential equations is

$$\begin{aligned} m_1x'' + \delta_1x' + (k_1 + k_2)x - k_2y &= 0, \\ m_2y'' + \delta_2y' + k_2y - k_2x &= 0. \end{aligned} \tag{21}$$

Since (21) is so similar to (4), we shall not consider the configuration of Figure 5.5.6 in the exercises.

Exercises

In Exercises 1 through 7, we have a spring–mass system arranged as in Figure 5.5.1. Set up the differential equation that describes the dynamics, and solve the resulting differential equation. If friction is absent, tell for which frequencies a forcing term $\sin \alpha t$ would cause resonance. If you have access to a computer, you may wish to graph some of the solutions.

1. Two 1-g masses are arranged in a spring–mass system, as in Figure 5.5.1. The spring constants are $k_1 = 1$, $k_2 = 4$, and $k_3 = 1$ g/s^2. Resistance is negligible. Initially, both masses are at their equilibrium positions and the first mass is at rest while the second mass is moving 1 cm/s to the right.

2. Two 2-g masses are arranged in a spring–mass system, as in Figure 5.5.1. The spring constants are $k_1 = 4$, $k_2 = 6$, $k_3 = 4$ g/s^2. Resistance is negligible. At time $t = 0$, both masses are released. Mass one is 1 cm to the left of its rest position; mass two is 1 cm to the right of its rest position.

3. Two 2-g masses are arranged in a spring–mass system as in Figure 5.5.1. The spring constants are $k_1 = 6$, $k_2 = 1$, $k_3 = 6$ g/s^2. The same resistance acts on both masses. The damping constant is 8 g/s. At time $t = 0$, both masses are at their rest positions. Mass one is initially moving at 1 cm/s to the right; mass two is initially moving at 1 cm/s to the left.

4. Two 1-g masses are arranged in a spring–mass system as in Figure 5.5.1. The spring constants are $k_1 = 2$, $k_2 = 1$, $k_3 = 2$ g/s^2. The same resistance acts on both masses. The damping constant is 2 g/s. Initially, both masses are at their rest positions but moving to the right at 3 cm/s and 1 cm/s, respectively.

5. Two 1-g masses are arranged in a spring–mass system, as in Figure 5.5.1. The spring constants are $k_1 = 4$, $k_2 = 6$, $k_3 = 4$ g/s^2. Resistance is negligible. Initially, mass one is at its rest position and moving 2 cm/s to the left while mass two is initially 2 cm to the right of its rest position and moving 6 cm/s to the right.

6. Two 1-g masses are arranged in a spring–mass system as in Figure 5.5.1. The spring constants are $k_1 = 1$, $k_2 = 12$, $k_3 = 1$ g/s^2. Resistance is negligible. Initially, the first mass is at its rest (equilibrium) position and moving to the right at 5 cm/s. The second mass is initially 2 cm to the right of its rest position and moving 5 cm/s to the left.

7. Two 2-g masses are arranged in a spring–mass system as in Figure 5.5.1. The spring constants are $k_1 = 10$, $k_2 = 3$, $k_3 = 10$ g/s^2. The damping coefficient is $\delta = 12$ g/s. Initially the first mass is 1 cm to the left of its rest position and moving to the right at 4 cm/s. The second mass is initially 5 cm to the right of its rest position and moving to the left at 16 cm/s.

8. Show that, if $\delta = 0$, then the characteristic polynomial for (7) always has two distinct pairs of purely imaginary conjugate complex roots.

9. Two 1-g masses are arranged in a spring–mass system as in Figure 5.5.1. The spring constants are $k_1 = k_3 = k$ and $k_2 = l$. The damping coefficient is δ. Thinking of k, l as unknown but fixed, we examine the effect of increasing the resistance.

a) For what values of δ are all roots of the characteristic equation for (7) complex?

b) For what values of δ will there be two complex and two real roots for the characteristic equation of (7)?

c) For what values of δ will there be four distinct real roots for the characteristic equation of (7)?

10. Derive the system of differential equations that models the spring–mass system in Figure 5.5.7 under the physical assumptions of this section. Include friction.

FIGURE 5.5.7

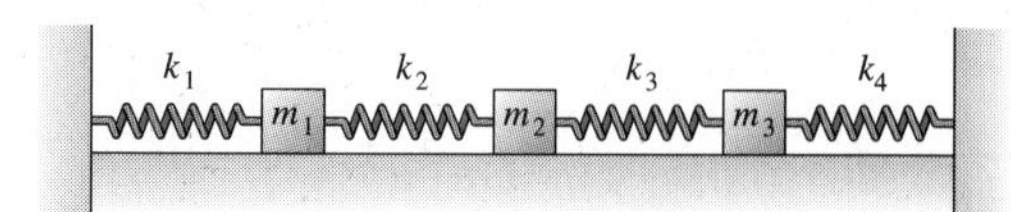

11. Derive the system of differential equations that models the spring–mass system with n masses and $(n + 1)$ springs in Figure 5.5.8. Include friction.

FIGURE 5.5.8

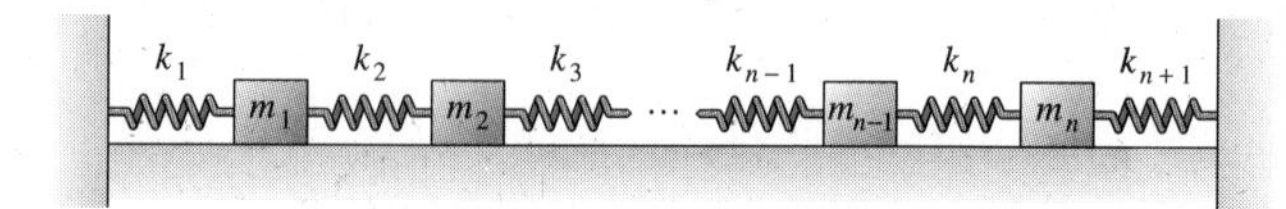

12. Show that, if $A \neq 0$, $B \neq 0$, and $\alpha \neq 0$, $\beta \neq 0$, then

$$f(t) = A \sin \alpha t + B \sin \beta t$$

is periodic if and only if α/β is a rational number. (*Hint*: Use linear combinations of $f(t)$, $f''(t)$ to argue that $\sin \alpha t$, $\sin \beta t$ both have period τ if f has period τ.)

13. (Continuation of Exercise 12.) Show that if $\beta/\alpha = m/n$, where m, n are integers, then $A \sin \alpha t + B \sin \beta t$ has period $\tau = 2\pi n/\alpha = 2\pi m/\beta$.

14. (Continuation of Exercise 12.) Show that if $\beta/\alpha = m/n$, where m, n are integers, then $\tau = 2\pi n/\alpha$ is the smallest period of $A \sin \alpha t + B \sin \beta t$ with $A \neq 0$, $B \neq 0$, if and only if the rational number m/n is in reduced form (m, n have no common integer factors except 1 and hence are **relatively prime**).

Exercises 15 through 17 require a knowledge of matrix multiplication and differentiation. They may be assigned after Section 5.9 if the reader has not seen these concepts before.

15. Write the system (4) in the form

$$\mathbf{M}\mathbf{x}'' + \delta \mathbf{x}' + \mathbf{K}\mathbf{x} = \mathbf{0} \tag{22}$$

where **M**, **K** are 2×2 matrices and $\mathbf{x} = \begin{bmatrix} x \\ y \end{bmatrix}$. Observe that (22) looks like the scalar equation $mx'' + \delta x' + kx = 0$ from Section 2.8 for a single mass and spring.

16. A 2×2 matrix is called **symmetric** if it can be written as

$$\begin{bmatrix} a & b \\ b & c \end{bmatrix}$$

for some scalars a, b, c. Symmetric matrices have several useful properties and play an important role in many applications. Show that the **M**, **K** from Exercise 15 are symmetric matrices.

17. A 3×3 matrix is called **symmetric** if its entries have the pattern

$$\begin{bmatrix} a & b & c \\ b & d & e \\ c & e & f \end{bmatrix}$$

for scalars a, b, c, d, e, f. Write the system of differential equations from Exercise 10 in the form (22), where **M**, **K** are 3×3 matrices. Verify that **M**, **K** are symmetric matrices.

18. Verify that, if $p(r)$ is a polynomial, then

$$p(D)e^{\alpha t} = p(\alpha)e^{\alpha t}.$$

5.6 Multiloop Circuits

In Section 2.12 we discussed simple circuits. However, most circuits have more than one loop and may often be modeled by a system of differential equations. Unfortunately, the application of Kirchhoff's voltage and current laws usually leads to a system with too many variables. Algorithms based on graph theory exist for finding a reduced set of variables. However, we shall consider only two-loop circuits of the general form shown in (1).

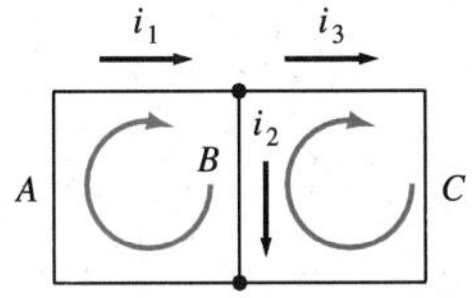

i_1 i_3 i_2 A B C (1)

This circuit has three **branches** denoted A, B, C with currents i_1, i_2, and i_3, respectively. On each branch we will allow some combination of linear resistors, inductors, capacitors, and independent voltage sources in series. Units are ohms, amps, coulombs, volts, etc., as discussed in Sections 1.10 and 2.12.

The dynamics of (1) will be determined by three equations:

The voltage law applied to loop AB; (2)

The voltage law applied to loop BC; (3)

The current law at the top node, which is

$$i_1 - i_2 - i_3 = 0. \tag{4}$$

We also assume that at least one branch does not have a capacitor in it. Our procedure is then as follows:

1. Write down the voltage law for loops AB and BC, using the charge as the dependent variable if there is a capacitor. Otherwise use the current in the branch. This gives two differential equations in three unknowns.
2. Use the current law (4) to eliminate a current variable corresponding to a branch without a capacitor.

The result is a system of differential equations solvable by the methods of this chapter.

EXAMPLE 5.6.1 **Two-Loop Circuit**

Using this procedure, write the differential equation for (5).

(5)

Solution Let i_1, i_2, i_3 be the currents in each branch and q_1, q_2 the charges in the capacitors in branch A and branch B. Then the voltage law applied to loops AB and BC gives

$$i_1 + 3i_2 + \frac{1}{4}q_2 + \frac{1}{6}q_1 + 5i_1' = e_1 \tag{6a}$$

and

$$2i_3 + 7i_3' - \frac{1}{4}q_2 - 3i_2 = e_2, \tag{6b}$$

respectively. Note that, since branch B is traversed in the opposite direction in loop BC, we reverse the signs of the corresponding voltages in (6b). Since there is no capacitor in branch C, use the current law $i_1 - i_2 - i_3 = 0$ to solve for i_3:

$$i_3 = i_1 - i_2,$$

which, upon substitution into (6), gives

$$i_1 + 3i_2 + \frac{1}{4}q_2 + \frac{1}{6}q_1 + 5i_1' = e_1, \tag{7a}$$

$$2i_1 - 2i_2 + 7i_1' - 7i_2' - \frac{1}{4}q_2 - 3i_2 = e_2. \tag{7b}$$

Now use the fact that $i_2 = q_2'$, $i_1 = q_1'$, to finally get

$$\begin{aligned} \left[5q_1'' + q_1' + \frac{1}{6}q_1\right] + \left[3q_2' + \frac{1}{4}q_2\right] &= e_1(t), \\ [7q_1'' + 2q_1'] + \left[-7q_2'' - 5q_2' - \frac{1}{4}q_2\right] &= e_2(t), \end{aligned} \tag{8}$$

or, in the notation of Section 5.2,

$$\begin{aligned} \left[5D^2 + D + \frac{1}{6}\right]q_1 + \left[3D + \frac{1}{4}\right]q_2 &= e_1(t), \\ [7D^2 + 2D]q_1 - \left[7D^2 + 5D + \frac{1}{4}\right]q_2 &= e_2(t). \end{aligned} \tag{9}$$

◄

In multiloop circuits with several voltage sources, such as (5), we need to be careful about their polarity (which node is +). If the + and − were reversed on the right side of (5), then we would have $-e_2$ instead of e_2 on the right side of (6b).

Note that this procedure always leads to a system of the form

$$\begin{aligned} L_1x + L_2y &= e_1(t), \\ L_3x + L_4y &= e_2(t), \end{aligned} \tag{10}$$

where L_1, L_2, L_3, L_4 are constant coefficient linear differential operators of order at most two, and x, y are either a charge or current from two different branches.

Exercises

In Exercises 1 through 4, find the system of differential equations that model the circuit. Note that, if more than one branch is missing a capacitor, then there is more than one possible differential equation. Do not solve the resulting differential equations.

1.

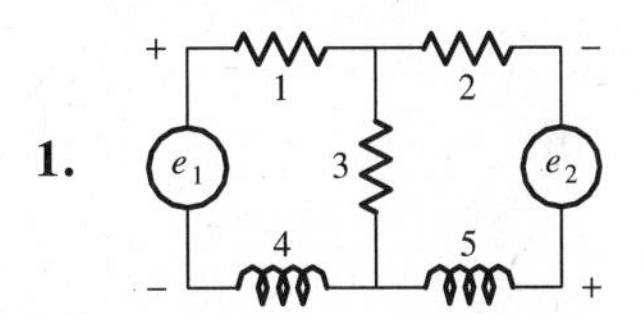

2.

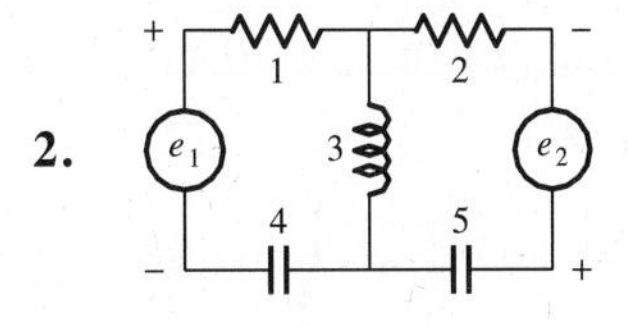

3.

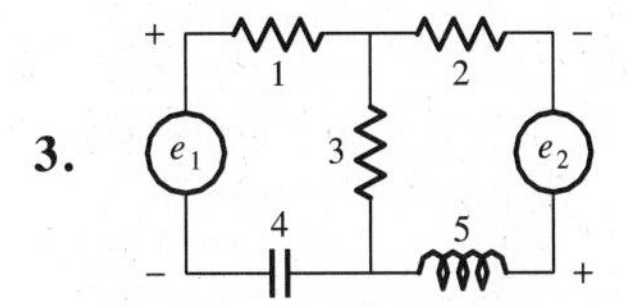

4.

Exercises 5 through 8 refer to the circuit in (11).

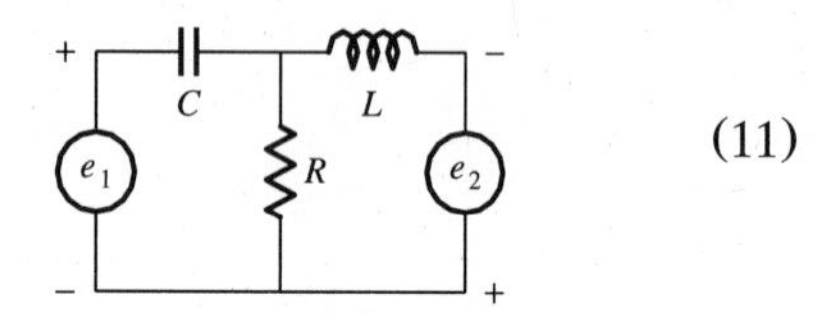

(11)

5. Suppose that $e_1 = e_2 = 0$, $L = 2$, $C = 2$, and $R = \frac{1}{2}$. Determine the free response of (11).

6. Take the same values of e_1, e_2, L, and C as in Exercise 5, but raise the resistance R to 1. Determine the free response to (11).

7. Take the same values of e_1, e_2, L, and C as in Exercise 5, but assume $R > \frac{1}{2}$. Show that the free response is made up of terms $e^{\alpha t}\cos\beta t$, $e^{\alpha t}\sin\beta t$, where $\alpha = -(4R)^{-1}$, $\beta = [4R^2 - 1]^{1/2}/(4R)$.

8. For large values of R, the circuit (11) would be expected to act like the circuit (12).

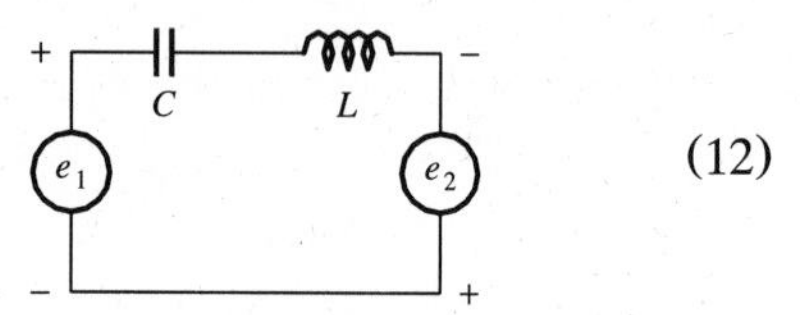

(12)

Determine the free response of (12) and compare to the results of Exercises 5, 6, and 7. In particular, show that as $R \to \infty$, the free response of (11) converges to that of (12).

Exercises 9 and 10 refer to the circuit (13). This example shows how the number of initial conditions can vary.

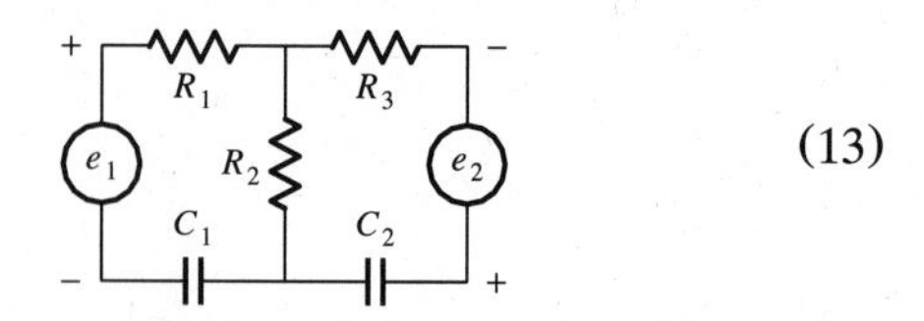

(13)

9. Take $e_1 = e_2 = 0$, $R_1 = R_3 = 0$, $R_2 = 1$, and $C_1 = C_2 = 1$ in (13). Derive the differential equation and determine the free response.

10. Take $e_1 = e_2 = 0$ in (13) and let $R_1 = 1$, $R_2 = 1$, $R_3 = 1$, $C_2 = 1$, $C_1 = 1$. Derive the differential equation and determine the free response.

Exercises 11 and 12 refer to the circuit (14).

(14)

11. Show that the differential equation for (14) may be written as

$$\left(L_1D^2 + R_2D + \frac{1}{C_1}\right)q_1 - R_2Dq_2 = e_1,$$
$$-R_2Dq_1 + \left(L_2D^2 + R_2D + \frac{1}{C_2}\right)q_2 = e_2, \tag{15}$$

where q_1, q_2 are the charges on the capacitors.

12. Let $C_1 = C_2 = 1$, $R_2 = 1$, $L_1 = L_2 = 1$, $e_1 = e_2 = 0$, and (i) determine the free response; (ii) observe that a harmonic oscillation is present even though there is a resistor; (iii) determine a set of initial conditions so that $q_1(t) = \sin t$.

5.7 Matrices and Vectors

5.7.1 Introduction

In the preceding sections, we have come to realize that the modeling of many important problems naturally leads to systems of differential equations.

Frequently, these systems either are linear or may be approximated by linear equations. In Sections 5.1, 5.2 and 5.3, we solved some fairly simple systems. The methods of those sections do not lend themselves well to many of the problems encountered in practice, where there are many more dependent variables. Neither do those techniques provide us with a sufficient theoretical understanding.

What is needed are some ideas and techniques from matrix theory and linear algebra. The remainder of this chapter is devoted to developing some of these ideas and applying them to systems of differential equations. Most students have probably seen matrices and vectors before. However, our presentation is self-contained and contains only that material that is needed for solving systems of differential equations of the types we shall study. The applications of matrix theory to differential equations begin in Section 5.9.

There is a fundamental pedagogical problem at this point. The ways in which these ideas are most easily presented are not the way they are usually computed in practice. Throughout the remainder of this chapter, we have tried to be as honest as possible. Where the method presented is primarily a textbook technique, we will so state. Unfortunately, to discuss how eigenvalues, eigenvectors, etc., are really found is well beyond the range of this book, and requires a separate course in numerical linear algebra. However, software such as MATLAB or MAPLE can be used to easily compute these quantities.

5.7.2 Basic Terminology and Matrix Operations

A rectangular array is called a **matrix**. The **entries** of a matrix may be numbers, functions, or even other matrices. Unless stated otherwise, the entries in this chapter will be either **scalars** (real or complex numbers) or scalar-valued functions. If the matrix has m **rows** and n **columns**, it is called an $m \times n$ matrix (pronounced "m by n") and $m \times n$ is referred to as the **size** of the matrix.

EXAMPLE 5.7.1 **Size of a Matrix**

For example,

$$\begin{bmatrix} 3 \\ 1 \\ 0 \end{bmatrix} \text{ is } 3 \times 1, \qquad [t, t^2, 1] \text{ is } 1 \times 3,$$

$$\begin{bmatrix} i & -i \\ 2 & 0 \\ 1 & 4 \end{bmatrix} \text{ is } 3 \times 2, \qquad \text{and} \qquad [5] \text{ is } 1 \times 1.$$

◄

An $m \times 1$ matrix is called an m-dimensional **(column) vector**, and a $1 \times n$ matrix is called an n-dimensional **(row) vector**. The first matrix in Example 1 is a column vector, and the second is a row vector. Unless stated otherwise, **vector** will mean a column vector. Matrices will be denoted by boldface capitals such as **A**, **B**, and row and column vectors will be denoted by boldface lowercase letters such as **u**, **v**, **x**.

The entry (or element) of **A** in the ith row and jth column is called the i, j entry of **A** and is denoted a_{ij}. Thus, if **A** is 3×4, we would have

$$\mathbf{A} = \begin{bmatrix} a_{11} & a_{12} & a_{13} & a_{14} \\ a_{21} & a_{22} & a_{23} & a_{24} \\ a_{31} & a_{32} & a_{33} & a_{34} \end{bmatrix}.$$

If the number of rows equals the number of columns, **A** is called a **square matrix**. Two matrices **A**, **B** are **equal** if they are the same size and corresponding entries are equal. That is, $a_{ij} = b_{ij}$ for all values of i, j.

In order to add or subtract two matrices, they must be the same size. If **A**, **B** are both $m \times n$ matrices, then $\mathbf{A} + \mathbf{B}$ and $\mathbf{A} - \mathbf{B}$ are also $m \times n$ matrices. The i, j entry of $\mathbf{A} + \mathbf{B}$ is $a_{ij} + b_{ij}$. That is, they are added entrywise. Similarly, the i, j entry of $\mathbf{A} - \mathbf{B}$ is $a_{ij} - b_{ij}$.

EXAMPLE 5.7.2 Matrix Addition and Subtraction

For example,

$$[1 \quad 2 \quad 0] + [3 \quad 4 \quad \pi] = [4 \quad 6 \quad \pi],$$

and

$$\begin{bmatrix} 3 \\ 6 \end{bmatrix} - \begin{bmatrix} 2 \\ 7 \end{bmatrix} = \begin{bmatrix} 1 \\ -1 \end{bmatrix},$$

while

$$\begin{bmatrix} t & t^2 \\ 1 & 0 \end{bmatrix} + \begin{bmatrix} 1 & -t \\ 0 & t \end{bmatrix} = \begin{bmatrix} t+1 & t^2 - t \\ 1 & t \end{bmatrix}.$$

If α is a number (also called a **scalar**), and **A** is an $m \times n$ matrix, then $\alpha\mathbf{A}$ is also an $m \times n$ matrix. The i, j entry of $\alpha\mathbf{A}$ is αa_{ij}. For example,

$$3\begin{bmatrix} 1 & 2 \\ t & \pi \end{bmatrix} = \begin{bmatrix} 3 & 6 \\ 3t & 3\pi \end{bmatrix}.$$

Matrix Multiplication

If $\mathbf{r}$ is a $1 \times n$ row vector, and $\mathbf{c}$ is an $n \times 1$ column vector,

$$\mathbf{r} = [r_1, \ldots, r_n], \qquad \mathbf{c} = \begin{bmatrix} c_1 \\ \vdots \\ c_n \end{bmatrix},$$

then their **product** is the 1×1 matrix

$$\mathbf{rc} = r_1 c_1 + r_2 c_2 + \cdots + r_n c_n = \sum_{i=1}^{n} r_i c_i.$$

For example,

$$[1 \quad 2 \quad 3]\begin{bmatrix} 4 \\ 5 \\ 6 \end{bmatrix} = 1 \cdot 4 + 2 \cdot 5 + 3 \cdot 6 = 32.$$

The expression $\sum_{i=1}^{n} r_i c_i$ is also called a **dot** or **inner product**.

Suppose, now, that $\mathbf{A}$ is $m \times n$ and $\mathbf{B}$ is $n \times p$. If $\mathbf{r}_1, \ldots, \mathbf{r}_m$ are the rows of $\mathbf{A}$ and $\mathbf{b}_1, \ldots, \mathbf{b}_p$ are the columns of $\mathbf{B}$, then the **product** of $\mathbf{A}$ and $\mathbf{B}$, denoted $\mathbf{AB}$, is an $m \times p$ matrix. Its i, j entry is $\mathbf{r}_i \mathbf{b}_j$. That is, $\sum_{k=1}^{n} a_{ik} b_{kj}$. Thus,

$$\mathbf{AB} = \begin{bmatrix} \mathbf{r}_1 \\ \mathbf{r}_2 \\ \vdots \\ \mathbf{r}_m \end{bmatrix} [\mathbf{b}_1, \ldots, \mathbf{b}_p] = \begin{bmatrix} \mathbf{r}_1\mathbf{b}_1 & \mathbf{r}_1\mathbf{b}_2 & \cdots & \mathbf{r}_1\mathbf{b}_p \\ \mathbf{r}_2\mathbf{b}_1 & \mathbf{r}_2\mathbf{b}_2 & \cdots & \mathbf{r}_2\mathbf{b}_p \\ \vdots & & & \vdots \\ \mathbf{r}_m\mathbf{b}_1 & \mathbf{r}_m\mathbf{b}_2 & \cdots & \mathbf{r}_m\mathbf{b}_p \end{bmatrix}. \tag{1}$$

Note the pattern of the sizes

$$(m \times n)(n \times p) = m \times p,$$

and that $\mathbf{A}$ must have the same number of columns as $\mathbf{B}$ has rows.

EXAMPLE 5.7.3 Matrix Multiplication

Find $\mathbf{AB}$ if

$$\mathbf{A} = \begin{bmatrix} 1 & 2 \\ 3 & 0 \\ 0 & 1 \end{bmatrix}, \qquad \mathbf{B} = \begin{bmatrix} 1 & 0 & 1 & 2 \\ -1 & 1 & 3 & 0 \end{bmatrix}.$$

Solution Note that $\mathbf{A}$ is 3×2 and $\mathbf{B}$ is 2×4, so that it is possible to form the product and $\mathbf{AB}$ will be 3×4. Then from (1),

$$\mathbf{AB} = \begin{bmatrix} [1 \quad 2]\begin{bmatrix} 1 \\ -1 \end{bmatrix} & [1 \quad 2]\begin{bmatrix} 0 \\ 1 \end{bmatrix} & [1 \quad 2]\begin{bmatrix} 1 \\ 3 \end{bmatrix} & [1 \quad 2]\begin{bmatrix} 2 \\ 0 \end{bmatrix} \\ [3 \quad 0]\begin{bmatrix} 1 \\ -1 \end{bmatrix} & [3 \quad 0]\begin{bmatrix} 0 \\ 1 \end{bmatrix} & [3 \quad 0]\begin{bmatrix} 1 \\ 3 \end{bmatrix} & [3 \quad 0]\begin{bmatrix} 2 \\ 0 \end{bmatrix} \\ [0 \quad 1]\begin{bmatrix} 1 \\ -1 \end{bmatrix} & [0 \quad 1]\begin{bmatrix} 0 \\ 1 \end{bmatrix} & [0 \quad 1]\begin{bmatrix} 1 \\ 3 \end{bmatrix} & [0 \quad 1]\begin{bmatrix} 2 \\ 0 \end{bmatrix} \end{bmatrix}$$

$$= \begin{bmatrix} -1 & 2 & 7 & 2 \\ 3 & 0 & 3 & 6 \\ -1 & 1 & 3 & 0 \end{bmatrix}.$$ ◄

Comment Note that, to multiply two $n \times n$ matrices takes n^3 multiplications and $n^2(n-1)$ additions. For even moderately large n, this can become time-consuming. Many applied problems involve matrices where m, n are in the thousands. For this reason, in actually solving problems, one often tries to avoid computing matrix products whenever possible.

A very important special case of matrix multiplication is when $\mathbf{A}$ is a column vector and $\mathbf{B}$ is a row vector.

EXAMPLE 5.7.4 Matrix Multiplication

Compute $\mathbf{uv}$ if

$$\mathbf{u} = \begin{bmatrix} 1 \\ 0 \\ 3 \end{bmatrix}, \qquad \mathbf{v} = [1 \quad 0 \quad \pi \quad 2].$$

Solution Since $\mathbf{u}$ is 3×1 and $\mathbf{v}$ is 1×4, the product will be 3×4, and it is

$$\mathbf{uv} = \begin{bmatrix} 1 \cdot 1 & 1 \cdot 0 & 1 \cdot \pi & 1 \cdot 2 \\ 0 \cdot 1 & 0 \cdot 0 & 0 \cdot \pi & 0 \cdot 2 \\ 3 \cdot 1 & 3 \cdot 0 & 3 \cdot \pi & 3 \cdot 2 \end{bmatrix} = \begin{bmatrix} 1 & 0 & \pi & 2 \\ 0 & 0 & 0 & 0 \\ 3 & 0 & 3\pi & 6 \end{bmatrix}.$$ ◄

It is also helpful to note that, if $\mathbf{A}$ is $m \times n$ and $\mathbf{B}$ is $n \times p$ and $\mathbf{b}_1, \ldots, \mathbf{b}_p$ are the columns of $\mathbf{B}$, then

$$\mathbf{AB} = \mathbf{A}[\mathbf{b}_1, \ldots, \mathbf{b}_p] = [\mathbf{Ab}_1, \mathbf{Ab}_2, \ldots, \mathbf{Ab}_p]. \tag{2}$$

That is,

> Multiplying the matrix **A** times **B** is the same as multiplying each column of **B** by **A**. (3)

Matrix multiplication is associative, $\mathbf{A}(\mathbf{BC}) = (\mathbf{AB})\mathbf{C}$, and distributive, $\mathbf{A}(\mathbf{B} + \mathbf{C}) = \mathbf{AB} + \mathbf{AC}$. However, in working with matrix products, it is important to keep in mind that, even if **AB** and **BA** are both defined, **AB** is usually *not* the same as **BA**. If $\mathbf{AB} = \mathbf{BA}$, we say they **commute**.

EXAMPLE 5.7.5 Matrices Which Do Not Commute

Let

$$\mathbf{A} = \begin{bmatrix} 0 & 1 \\ 0 & 0 \end{bmatrix}, \qquad \mathbf{B} = \begin{bmatrix} 1 & 0 \\ 0 & 0 \end{bmatrix}.$$

Then

$$\mathbf{AB} = \begin{bmatrix} 0 & 0 \\ 0 & 0 \end{bmatrix}, \qquad \mathbf{BA} = \begin{bmatrix} 0 & 1 \\ 0 & 0 \end{bmatrix},$$

so that $\mathbf{AB} \neq \mathbf{BA}$. ◄

Inverse of a Matrix

A matrix (vector) with all entries zero is called a **zero matrix (zero vector)** and is denoted **0**. Example 5.7.5 also shows that

$$\mathbf{AB} = \mathbf{0} \qquad \text{need not imply that } \mathbf{B} = \mathbf{0} \text{ or } \mathbf{A} = \mathbf{0}. \tag{4}$$

Similarly,

$$\mathbf{AB} = \mathbf{AC} \qquad \text{need not imply that} \qquad \mathbf{B} = \mathbf{C}. \tag{5}$$

In order to understand (4) and (5) better, we define the **identity matrix I** to be a square matrix with its i, i entries equal to one, and all the other entries zero. Examples of identity matrices are

$$[1], \qquad \begin{bmatrix} 1 & 0 \\ 0 & 1 \end{bmatrix}, \qquad \begin{bmatrix} 1 & 0 & 0 \\ 0 & 1 & 0 \\ 0 & 0 & 1 \end{bmatrix}.$$

Note that, for a given n, there is only one $n \times n$ identity matrix. If **I** is $m \times m$ and **A** is $m \times n$, then

$$\mathbf{IA} = \mathbf{A},$$

while

$$\mathbf{AI} = \mathbf{A},$$

if $\mathbf{I}$ is $n \times n$. The identity matrix is a special case of a diagonal matrix. An $n \times n$ matrix is a **diagonal matrix** if $a_{ij} = 0$ for $i \neq j$. Examples of diagonal matrices are

$$\mathbf{I}, \quad \alpha\mathbf{I}, \quad \begin{bmatrix} 1 & 0 \\ 0 & 2 \end{bmatrix}, \quad \begin{bmatrix} 1 & 0 & 0 \\ 0 & -3 & 0 \\ 0 & 0 & 1 \end{bmatrix}, \quad \begin{bmatrix} 2 & 0 & 0 \\ 0 & 0 & 0 \\ 0 & 0 & 0 \end{bmatrix}.$$

If $\mathbf{B}$ is a matrix such that

$$\mathbf{AB} = \mathbf{BA} = \mathbf{I},$$

then $\mathbf{B}$ is called the **inverse** of $\mathbf{A}$ and is denoted $\mathbf{A}^{-1}$. Only a square matrix can have an inverse, although not all square matrices are **invertible** (have an inverse). The inverse, when it exists, is unique (Exercise 5.7.38). For example,

$$\begin{bmatrix} -2 & 1 \\ \frac{3}{2} & -\frac{1}{2} \end{bmatrix}\begin{bmatrix} 1 & 2 \\ 3 & 4 \end{bmatrix} = \begin{bmatrix} 1 & 0 \\ 0 & 1 \end{bmatrix} = \mathbf{I},$$

so that

$$\begin{bmatrix} -2 & 1 \\ \frac{3}{2} & -\frac{1}{2} \end{bmatrix} = \begin{bmatrix} 1 & 2 \\ 3 & 4 \end{bmatrix}^{-1} \quad \text{and} \quad \begin{bmatrix} 1 & 2 \\ 3 & 4 \end{bmatrix} = \begin{bmatrix} -2 & 1 \\ \frac{3}{2} & -\frac{1}{2} \end{bmatrix}^{-1}.$$

When matrices have inverses, we do not get the "unusual" behavior in (4) and (5).

EXAMPLE 5.7.6 Matrix Algebra

Show that if $\mathbf{A}$ has an inverse and $\mathbf{AB} = \mathbf{AC}$, then $\mathbf{B} = \mathbf{C}$.

SOLUTION Suppose $\mathbf{A}$ has an inverse $\mathbf{A}^{-1}$ and

$$\mathbf{AB} = \mathbf{AC}.$$

Multiply both sides by $\mathbf{A}^{-1}$ on the left,

$$\mathbf{A}^{-1}\mathbf{AB} = \mathbf{A}^{-1}\mathbf{AC}.$$

Since $\mathbf{A}^{-1}\mathbf{A} = \mathbf{I}$ and matrix multiplication is associative, we have

$$\mathbf{IB} = \mathbf{IC}$$

or

$$\mathbf{B} = \mathbf{C}.$$

◄

We shall return to the inverse at several points in this chapter and develop additional properties as needed.

5.7.3 Systems of Linear Equations

One of the most common problems in applications is to solve a system of m equations in n unknowns $x_1, \ldots, x_n$,

$$\begin{aligned} a_{11}x_1 + a_{12}x_2 + \cdots + a_{1n}x_n &= b_1, \\ a_{21}x_1 + a_{22}x_2 + \cdots + a_{2n}x_n &= b_2, \\ \vdots \quad \vdots \quad & \\ a_{m1}x_1 + a_{m2}x_2 + \cdots + a_{mn}x_n &= b_m. \end{aligned} \tag{6}$$

If we let

$$\mathbf{A} = \begin{bmatrix} a_{11} & a_{12} & \cdots & a_{1n} \\ \vdots & & & \vdots \\ a_{m1} & a_{m2} & \cdots & a_{mn} \end{bmatrix}, \quad \mathbf{x} = \begin{bmatrix} x_1 \\ \vdots \\ x_n \end{bmatrix}, \quad \mathbf{b} = \begin{bmatrix} b_1 \\ \vdots \\ b_m \end{bmatrix}, \tag{7}$$

then (6) may be written as

$$\mathbf{Ax} = \mathbf{b}. \tag{8}$$

The notation of (6), (7), and (8), including the size $m \times n$, will be used for the remainder of this chapter. Since we shall frequently need to solve systems such as (6) in the remainder of this chapter, the solution of (6) will be discussed fairly carefully. To solve $\mathbf{Ax} = \mathbf{b}$ for the unknowns $\mathbf{x}$, we must solve for variables, and substitute. However, it turns out to be much easier to set up an algorithm if we manipulate the equations by performing one of three operations:

i) Multiply an equation by a nonzero scalar (9a)

ii) Exchange two equations (9b)

iii) Add a multiple of one equation to another (9c)

EXAMPLE 5.7.7 **Solution without Matrix Notation**

Solve the system of equations

$$\begin{aligned} y + 2z &= 1, \\ x + y + z &= 2, \\ 2x + 4y + 8z &= 3. \end{aligned} \tag{10}$$

SOLUTION For reasons that will become clear shortly, we shall perform what appears to be a couple of unnecessary operations. Exchange Eqs. (1) and (2) in (10) to get

$$\begin{aligned} x + y + z &= 2, \\ y + 2z &= 1, \\ 2x + 4y + 8z &= 3. \end{aligned} \tag{11}$$

Add -2 times Eq. (1) to Eq. (3) in (11) to eliminate the x from the third equation:

$$\begin{aligned} x + y + z &= 2, \\ y + 2z &= 1, \\ 2y + 6z &= -1. \end{aligned} \tag{12}$$

The last two equations involve only y and z. Now add -2 times the second equation to the third, to get

$$\begin{aligned} x + y + z &= 2, \\ y + 2z &= 1, \\ 2z &= -3. \end{aligned} \tag{13}$$

The last equation now involves only z. Now perform the two operations of adding -1 times Eq. (3) to Eq. (2) in (13), and adding $-\frac{1}{2}$ times Eq. (3) to Eq. (1):

$$\begin{aligned} x + y &= 3\frac{1}{2}, \\ y &= 4, \\ 2z &= -3, \end{aligned} \tag{14}$$

which eliminates z from Eqs. (1) and (2). Finally, add -1 times Eq. (2) to Eq. (1) and multiply Eq. (3) by $\frac{1}{2}$, to get:

$$\begin{aligned} x &= -\frac{1}{2}, \\ y &= 4, \\ z &= -\frac{3}{2}. \end{aligned} \tag{15}$$

◄

Looking back over this example, we see that, in the calculations, we need not write down the variables x, y, z and the equal sign. Instead, we can proceed as follows:

■ ALGORITHM 5.7.1 Solution of $\mathbf{Ax} = \mathbf{b}$ by Gaussian Elimination

Write down the **augmented matrix** $[\mathbf{A} \mid \mathbf{b}]$. This is $\mathbf{A}$ with an extra column $\mathbf{b}$ added. Perform the **elementary row operations** of

Multiplying a row by a nonzero scalar, (16a)

Exchanging two rows, (16b)

Adding a multiple of one row to another row, (16c)

[which correspond to the operations (9) on the equations] according to the following pattern:

Part 1

a. Find a nonzero entry in the first column. If there is none, move to the next column. Repeat until a nonzero entry is found.
b. If this nonzero entry is not in the first row, exchange the row this entry is in with the first row.
c. Add multiples of this row to all t he entries below it, so that all other entries in that column are zeroed out (made zero).

- Now ignore the first row and repeat (a), (b), (c) on the matrix formed by the rest of the rows.
- Now ignore the first two rows and repeat (a), (b), (c) on the matrix formed by the rest of the rows.
- Continue this pattern.

Part 1 terminates when we run out of rows; all the rest of the rows are identically zero, or there is a row whose only nonzero entry is the last one:

If there is a row whose only nonzero entry is the last one, the system of equations is **inconsistent**. That is, there is no solution.

If the system of equations is **consistent** (not inconsistent), then proceed to Part 2.

Part 2

Starting with the last nonzero row, find its first nonzero entry, and zero out all other entries in the column above. Then take the second to last nonzero row, find its first nonzero entry, and zero out all other entries in the column above. Continue until reaching the first row. Often one also makes the leading nonzero entries in each nonzero row equal to one. The resulting matrix is called the **row echelon form** of the original matrix.

The columns that have the first nonzero entry of some row in them are called the **distinguished columns**:

If every column is distinguished except the last one, we have found a unique solution. (17)

If some columns other than the last one are not distinguished, then there are infinitely many solutions, and the variables corresponding to all undistinguished columns (except the last one) may be taken as arbitrary. The variables corresponding to the distinguished columns are found in terms of the arbitrary variables. (18)

■

EXAMPLE 5.7.8 Solution Using Gaussian Elimination and Augmented Matrix

Solve the system of Example 5.7.7 using Gaussian elimination:

$$\begin{aligned} y + 2z &= 1, \\ x + y + z &= 2, \\ 2x + 4y + 8z &= 3. \end{aligned} \tag{19}$$

SOLUTION Here

$$\mathbf{A} = \begin{bmatrix} 0 & 1 & 2 \\ 1 & 1 & 1 \\ 2 & 4 & 8 \end{bmatrix}, \qquad \mathbf{x} = \begin{bmatrix} x \\ y \\ z \end{bmatrix}, \qquad \mathbf{b} = \begin{bmatrix} 1 \\ 2 \\ 3 \end{bmatrix},$$

and the augmented matrix is

$$[\mathbf{A} \,|\, \mathbf{b}] = \left[\begin{array}{ccc|c} 0 & 1 & 2 & 1 \\ 1 & 1 & 1 & 2 \\ 2 & 4 & 8 & 3 \end{array}\right]. \tag{20}$$

We shall now apply the method of Gaussian elimination to (20) as described in Algorithm 5.7.1. By referring back to Example 5.7.7, the reader will see that the elementary row operations (16) on the augmented matrix (20) are equivalent to the earlier operations (9) on the system (10).

There are two nonzero entries in the first column of (20). To get a nonzero entry in the 1,1 position, we exchange rows 1 and 2 to get

$$\left[\begin{array}{ccc|c} 1 & 1 & 1 & 2 \\ 0 & 1 & 2 & 1 \\ 2 & 4 & 8 & 3 \end{array}\right]. \tag{21}$$

(We could have exchanged rows 1 and 3, instead.)

Note that (21) is the augmented matrix of (11). To eliminate the rest of the entries in the first column, add -2 times the first row to the third row, to get

$$\left[\begin{array}{ccc|c} 1 & 1 & 1 & 2 \\ 0 & 1 & 2 & 1 \\ 0 & 2 & 6 & -1 \end{array}\right], \tag{22}$$

which is the augmented matrix of (12). Leaving row 1 alone, we look at the remaining two rows. The 2,2 entry is already nonzero, so we need to zero out all entries in the second column below it. To do this, add -2 times the second row to the third row:

$$\left[\begin{array}{ccc|c} 1 & 1 & 1 & 2 \\ 0 & 1 & 2 & 1 \\ 0 & 0 & 2 & -3 \end{array}\right], \tag{23}$$

which is the augmented matrix of (13). This completes Part 1. For Part 2, we use the 3,3 entry to zero out all of the entries above it in the third column. This is done by adding -1 times the third row to the second row, and $-\frac{1}{2}$ times the third row to the first row:

$$\left[\begin{array}{ccc|c} 1 & 1 & 0 & \frac{7}{2} \\ 0 & 1 & 0 & 4 \\ 0 & 0 & 2 & -3 \end{array}\right], \tag{24}$$

which is the augmented matrix of (14). Next, we go to the second row and use the 2,2 entry to zero out all of the entries above it. This is done by adding -1 times row 2 to row 1:

$$\left[\begin{array}{ccc|c} 1 & 0 & 0 & -\frac{1}{2} \\ 0 & 1 & 0 & 4 \\ 0 & 0 & 2 & -3 \end{array}\right]. \tag{25}$$

Multiply row 3 by $\frac{1}{2}$ to get

$$\left[\begin{array}{ccc|c} 1 & 0 & 0 & -\frac{1}{2} \\ 0 & 1 & 0 & 4 \\ 0 & 0 & 1 & -\frac{3}{2} \end{array}\right]. \tag{26}$$

This is the augmented matrix of

$$x = -\frac{1}{2},$$
$$y = 4,$$
$$z = -\frac{3}{2},$$

which gives the solution of (19) as desired. ◄

We now give two more examples illustrating Gaussian elimination (Algorithm 5.7.1).

EXAMPLE 5.7.9 **Inconsistent System**

Find all solutions of

$$\begin{aligned} x_1 + 2x_2 &= 3, \\ x_1 - x_2 + x_3 &= 1, \\ x_1 + 5x_2 - x_3 &= 10. \end{aligned} \tag{27}$$

Solution The augmented matrix is

$$\left[\begin{array}{ccc|c} 1 & 2 & 0 & 3 \\ 1 & -1 & 1 & 1 \\ 1 & 5 & -1 & 10 \end{array}\right]. \tag{28}$$

Since the 1,1 entry is already nonzero, add -1 times row one to both rows 2 and 3 to zero the rest of the first column:

$$\left[\begin{array}{ccc|c} 1 & 2 & 0 & 3 \\ 0 & -3 & 1 & -2 \\ 0 & 3 & -1 & 7 \end{array}\right]. \tag{29}$$

The 2,2 entry is already nonzero. Zero the entry below it by adding row 2 to row 3,

$$\left[\begin{array}{ccc|c} 1 & 2 & 0 & 3 \\ 0 & -3 & 1 & -2 \\ 0 & 0 & 0 & 5 \end{array}\right]. \tag{30}$$

The last row is $[0 \quad 0 \quad 0 \,|\, 5]$, which is the equation

$$0x_1 + 0x_2 + 0x_3 = 5,$$

which has no solution. Thus the system (27) is inconsistent. There are no solutions. ◄

EXAMPLE 5.7.10 **Arbitrary Constants and Vector Form of Solution**

Find all the solutions of

$$\begin{aligned} x + y + z + 3w &= 2, \\ x + y + w &= 1, \\ -2x - 2y + z &= -1, \\ -x - y + 2z + 3w &= 1. \end{aligned} \tag{31}$$

Solution The augmented matrix is

$$\left[\begin{array}{cccc|c} 1 & 1 & 1 & 3 & 2 \\ 1 & 1 & 0 & 1 & 1 \\ -2 & -2 & 1 & 0 & -1 \\ -1 & -1 & 2 & 3 & 1 \end{array}\right]. \tag{32}$$

The 1,1 entry is already nonzero. Use it to zero the rest of column 1 by adding -1 times row 1 to row 2, adding twice row 1 to row 3, and adding row

1 to row 4:

$$\left[\begin{array}{cccc|c} 1 & 1 & 1 & 3 & 2 \\ 0 & 0 & -1 & -2 & -1 \\ 0 & 0 & 3 & 6 & 3 \\ 0 & 0 & 3 & 6 & 3 \end{array}\right]. \tag{33}$$

Ignoring the first row, the first nonzero entry appears in column 3. Since the 2,3 entry is nonzero, we do not need to exchange rows. Now eliminate the entries below the 2,3 entry by adding 3 times row 2 to rows 3 and 4:

$$\left[\begin{array}{cccc|c} 1 & 1 & 1 & 3 & 2 \\ 0 & 0 & -1 & -2 & -1 \\ 0 & 0 & 0 & 0 & 0 \\ 0 & 0 & 0 & 0 & 0 \end{array}\right]. \tag{34}$$

This completes Part 1. Eliminate the entry above the 2,3 entry by adding row 2 to row 1. Multiply row 2 by -1 to get each leading entry positive:

$$\left[\begin{array}{cccc|c} 1 & 1 & 0 & 1 & 1 \\ 0 & 0 & 1 & 2 & 1 \\ 0 & 0 & 0 & 0 & 0 \\ 0 & 0 & 0 & 0 & 0 \end{array}\right]. \tag{35}$$

The matrix (35) is the row echelon form of (32). The distinguished columns are the first and third. Thus, by (18), the second and fourth variables are arbitrary:

$$y, w \qquad \text{are arbitrary}. \tag{36}$$

From (35),

$$\begin{aligned} x + y + w &= 1, \\ z + 2w &= 1, \end{aligned} \tag{37}$$

or

$$\begin{aligned} x &= 1 - y - w, \\ z &= 1 - 2w, \end{aligned} \tag{38}$$

gives the other variables. It is very convenient for our later applications to write this solution (36) and (38) in vector notation:

$$\mathbf{x} = \begin{bmatrix} x \\ y \\ z \\ w \end{bmatrix} = \begin{bmatrix} 1 - y - w \\ y \\ 1 - 2w \\ w \end{bmatrix} = \begin{bmatrix} 1 \\ 0 \\ 1 \\ 0 \end{bmatrix} + y \begin{bmatrix} -1 \\ 1 \\ 0 \\ 0 \end{bmatrix} + w \begin{bmatrix} -1 \\ 0 \\ -2 \\ 1 \end{bmatrix}, \tag{39}$$

where y, w are arbitrary constants. ◄

Two different people applying Algorithm 5.7.1 to a system of algebraic equations will always get the same answer, no matter in what order they do the calculations. If there is more than one solution and a choice of arbitrary variables different from that suggested in Algorithm 5.7.1 is made, the answer will appear different.

For example, in the previous problem, the original system (31) has been reduced to the equivalent one (37). In (37) one could take x, w arbitrary and then find y, z in terms of x and w,

$$y = 1 - x - w, \qquad z = 1 - 2w,$$

so that the solution vectors have the form

$$\mathbf{x} = \begin{bmatrix} x \\ y \\ z \\ w \end{bmatrix} = \begin{bmatrix} x \\ 1 - x - w \\ 1 - 2w \\ w \end{bmatrix} = \begin{bmatrix} 0 \\ 1 \\ 1 \\ 0 \end{bmatrix} + x \begin{bmatrix} 1 \\ -1 \\ 0 \\ 0 \end{bmatrix} + w \begin{bmatrix} 0 \\ -1 \\ -2 \\ 1 \end{bmatrix}. \tag{40}$$

While (39) and (40) appear different, they actually describe the same set of vectors.

Not just any pair of variables can be taken arbitrary in (37). For example, one cannot take z, w both arbitrary since $z + 2w = 1$. The beginning student is advised to follow Algorithm 5.7.1.

Algorithm 5.7.1 can be modified to find $\mathbf{A}^{-1}$ if it exists. The key is the observation that the **matrix equation** $\mathbf{AX} = \mathbf{B}$ can be solved by working with the augmented matrix $[\mathbf{A} \mid \mathbf{B}]$ and $\mathbf{A}^{-1}$ is the solution of $\mathbf{AX} = \mathbf{I}$.

■ ALGORITHM 5.7.2 For Computation of $\mathbf{A}^{-1}$, if A Is Square and $\mathbf{A}^{-1}$ Exists

1. Write down the augmented matrix $[\mathbf{A} \mid \mathbf{I}]$.
2. Perform elementary row operations to convert **A** to the identity, to give $[\mathbf{I} \mid \mathbf{C}]$.
3. Then $\mathbf{C} = \mathbf{A}^{-1}$.
4. If at any point in the calculations, the left side has a zero row, then $\mathbf{A}^{-1}$ does not exist. ■

EXAMPLE 5.7.11 Computing the Inverse of a Matrix

Find $\mathbf{A}^{-1}$, if possible, for

$$\mathbf{A} = \begin{bmatrix} 1 & 2 \\ 3 & 4 \end{bmatrix}.$$

SOLUTION Following Algorithm 5.7.2, we write down the augmented matrix

$$\left[\begin{array}{cc|cc} 1 & 2 & 1 & 0 \\ 3 & 4 & 0 & 1 \end{array}\right].$$

Adding -3 times row 1 to row 2 gives

$$\left[\begin{array}{rr|rr} 1 & 2 & 1 & 0 \\ 0 & -2 & -3 & 1 \end{array}\right].$$

Add row 2 to row 1,

$$\left[\begin{array}{rr|rr} 1 & 0 & -2 & 1 \\ 0 & -2 & -3 & 1 \end{array}\right],$$

and multiply row 2 by $-\frac{1}{2}$,

$$\left[\begin{array}{rr|rr} 1 & 0 & -2 & 1 \\ 0 & 1 & \frac{3}{2} & -\frac{1}{2} \end{array}\right].$$

Since the left side is the identity, we have

$$\mathbf{A}^{-1} = \begin{bmatrix} -2 & 1 \\ \frac{3}{2} & -\frac{1}{2} \end{bmatrix}.$$

If the left side had a row of zeros, there would be no inverse. ◄

Comments If $\mathbf{Ax} = \mathbf{b}$ and $\mathbf{A}$ has an inverse, then multiplying both sides of $\mathbf{Ax} = \mathbf{b}$ by $\mathbf{A}^{-1}$ gives $\mathbf{x} = \mathbf{A}^{-1}\mathbf{b}$. Thus one could solve $\mathbf{Ax} = \mathbf{b}$ by computing $\mathbf{A}^{-1}$ (if it exists) and using $\mathbf{x} = \mathbf{A}^{-1}\mathbf{b}$. However, this is substantially more work than working directly with $[\mathbf{A} \mid \mathbf{b}]$. Inverses are usually not computed in practice. Also, while it can be shown that Algorithm 5.7.1 involves the least work of any general algorithm for solving $\mathbf{Ax} = \mathbf{b}$, it is not numerically implemented as stated, to avoid dividing by small numbers. Care must be used in choosing which rows to use to eliminate others (choice of pivots). For large systems and certain others that are numerically delicate, totally different techniques must be used.

Exercises

For Exercises 1 through 15, let

$$\mathbf{A} = \begin{bmatrix} 1 & 2 & 0 \\ 1 & -1 & 1 \end{bmatrix}, \quad \mathbf{B} = \begin{bmatrix} 1 & 3 \\ -1 & 4 \end{bmatrix},$$

$$\mathbf{u} = \begin{bmatrix} 1 \\ 2 \end{bmatrix}, \quad \mathbf{v} = \begin{bmatrix} 1 \\ 0 \\ 1 \end{bmatrix}, \quad \mathbf{w} = [3 \quad -4].$$

In each of Exercises 1 through 15, state whether or not it is possible to compute the given expression. Compute if possible.

1. $\mathbf{AB}$

2. $\mathbf{BA}$

3. $\mathbf{uv}$

4. $\mathbf{uw}$

5. $\mathbf{wu}$

6. $\mathbf{wv}$

7. $2\mathbf{A}$

8. $\mathbf{wAv}$

9. $\mathbf{A} + \mathbf{B}$

10. $3\mathbf{I}$ ($\mathbf{I}$ is 3×3)

11. $\mathbf{B} - 2\mathbf{I}$ ($\mathbf{I}$ is 2×2)

12. $3\mathbf{AI}$ ($\mathbf{I}$ is 3×3)

13. $\mathbf{B} - 2\mathbf{A}$

14. $\mathbf{B}^2$ ($\mathbf{B}^2 = \mathbf{BB}$)

15. $\mathbf{A} + \mathbf{0}$ ($\mathbf{0}$ is 2×3)

16. Let $\mathbf{A} = \begin{bmatrix} 1 & -1 \\ 1 & -1 \end{bmatrix}$ and compute $\mathbf{A}^2 = \mathbf{AA}$.

17. Let $\mathbf{A} = \begin{bmatrix} 1 & -1 \\ 0 & -1 \end{bmatrix}$ and compute $\mathbf{A}^2 = \mathbf{AA}$.

For Exercises 18 through 29, solve the indicated system of algebraic equations, using Gaussian elimination via augmented matrices and the method of Algorithm 5.7.1. If the system is inconsistent, so state. If there are infinitely many solutions, express them in vector form like (39) of Example 5.7.10.

18. $$\begin{aligned} x + y &= 1, \\ x + 2y &= 2 \end{aligned}$$

19. $$\begin{aligned} 2x + 3y &= 2, \\ 4x + 6y &= 4 \end{aligned}$$

20. $$\begin{aligned} 2x + 3y &= 2, \\ 4x + 6y &= 3 \end{aligned}$$

21. $$\begin{aligned} 2x - y &= 0, \\ -6x + 3y &= 0 \end{aligned}$$

22. $$\begin{aligned} x - y &= 1, \\ -2x + 2y &= 3 \end{aligned}$$

23. $$\begin{aligned} x + y + z &= 1, \\ y + 2z &= 0, \\ 2x + 3y + 4z &= 0 \end{aligned}$$

24. $$\begin{aligned} x + 2y + 3z &= 0, \\ 2x + 4y + 6z &= 0, \\ x + 2y + 3z &= 0 \end{aligned}$$

25. $$\begin{aligned} x + y - z &= 1, \\ x - y - z &= 2, \\ 2x + 3y - z &= 0 \end{aligned}$$

26. $$\begin{aligned} x + y &= 0, \\ y + z &= 0, \\ x - z &= 0 \end{aligned}$$

27. $$\begin{aligned} x + y &= 1, \\ y - z &= 2, \\ x - z &= 3 \end{aligned}$$

28. $$\begin{aligned} x + y + z + w &= 0, \\ -x - y - z - w &= 0, \\ 2x + 2y + 2z + 2w &= 0, \\ x + y + z + w &= 0 \end{aligned}$$

29. $$\begin{aligned} x - y + z - w &= 1, \\ x - y + z - 3w &= 7, \\ 2x - 2y + 2z + 2w &= -10, \\ -x + y + z + w &= -4 \end{aligned}$$

Exercises 30 through 38 provide some practice in manipulating matrix expressions.

30. Is $(\mathbf{A} + 2\mathbf{I})^2 = \mathbf{A}^2 + 4\mathbf{A} + 4\mathbf{I}$ for any square matrix $\mathbf{A}$ if $\mathbf{A}$, $\mathbf{I}$ are the same size?

31. If $\mathbf{A}$, $\mathbf{B}$ are square matrices of the same size, is $(\mathbf{A} + \mathbf{B})^2 = \mathbf{A}^2 + 2\mathbf{AB} + \mathbf{B}^2$?

32. Let $\mathbf{u}$ be $n \times 1$ and $\mathbf{v}$ be $1 \times n$ matrices such that $\mathbf{vu} = \mathbf{0}$.

a) Must $\mathbf{uv} = \mathbf{0}$?

b) Let $\mathbf{N} = \mathbf{uv}$. Show that $\mathbf{N}^2 = \mathbf{0}$.

33. Suppose $\mathbf{x}_1$, $\mathbf{x}_2$ are two solutions of $\mathbf{Ax} = \mathbf{0}$. Show $c_1\mathbf{x}_1 + c_2\mathbf{x}_2$ is also a solution of $\mathbf{Ax} = \mathbf{0}$ for any scalars c_1, c_2.

34. If $\mathbf{x}_1$ is a solution of $\mathbf{Ax} = \mathbf{b}$ and $\mathbf{x}_2$ is a solution of $\mathbf{Ax} = \mathbf{0}$, show that $\mathbf{x}_1 + c\mathbf{x}_2$ is a solution of $\mathbf{Ax} = \mathbf{b}$ (for any constant c).

35. If $\mathbf{x}_1$ is a solution of $\mathbf{Ax} = \mathbf{b}_1$ and $\mathbf{x}_2$ is a solution of $\mathbf{Ax} = \mathbf{b}_2$, show that $\mathbf{x}_1 + \mathbf{x}_2$ is a solution of $\mathbf{Ax} = \mathbf{b}_1 + \mathbf{b}_2$.

36. Suppose that $\mathbf{A}$ is $n \times n$ and $\mathbf{u}$ is $n \times 1$. Show that if $\mathbf{u} \neq \mathbf{0}$ and $\mathbf{Au} = \mathbf{0}$, then $\mathbf{A}$ cannot be invertible.

37. Suppose that $\mathbf{A}$, $\mathbf{B}$ are $n \times n$ invertible matrices. Show that $\mathbf{AB}$ is invertible and $(\mathbf{AB})^{-1} = \mathbf{B}^{-1}\mathbf{A}^{-1}$.

38. Suppose that $\mathbf{A}$ is invertible and $\mathbf{B}$, $\mathbf{C}$ are both inverses of $\mathbf{A}$. Show that $\mathbf{B} = \mathbf{C}$; that is, the inverse is unique.

In Exercises 39 through 42, compute $\mathbf{A}^{-1}$ using Algorithm 5.7.2.

39. $\mathbf{A} = \begin{bmatrix} 1 & 2 \\ 1 & 1 \end{bmatrix}$

40. $\mathbf{A} = \begin{bmatrix} 1 & 3 \\ 2 & 1 \end{bmatrix}$

41. $\mathbf{A} = \begin{bmatrix} 0 & 1 \\ 1 & 2 \end{bmatrix}$

42. $\mathbf{A} = \begin{bmatrix} 2 & 1 \\ -1 & 0 \end{bmatrix}$

In Exercises 43 through 54, you are given the reduced row echelon form of the augmented matrix of the linear system $\mathbf{Ax} = \mathbf{b}$. If the system is not consistent, state that fact. If the system is consistent, write out the solution in vector form.

43. $\left[\begin{array}{ccccc|c} 1 & 0 & 0 & 1 & 0 & 1 \\ 0 & 1 & 0 & 0 & 1 & 2 \\ 0 & 0 & 1 & 1 & 1 & 3 \\ 0 & 0 & 0 & 0 & 0 & 0 \end{array}\right]$

44. $\left[\begin{array}{ccccc|c} 1 & 0 & 0 & 1 & 1 & 3 \\ 0 & 1 & 0 & 2 & 1 & 2 \\ 0 & 0 & 1 & 0 & 0 & 1 \\ 0 & 0 & 0 & 0 & 0 & 0 \end{array}\right]$

45. $\left[\begin{array}{cccc|c} 0 & 1 & 0 & 2 & 1 \\ 0 & 0 & 1 & 3 & 0 \\ 0 & 0 & 0 & 0 & 0 \\ 0 & 0 & 0 & 0 & 0 \end{array}\right]$

46. $\left[\begin{array}{cccc|c} 1 & 1 & 1 & 2 & 1 \\ 0 & 0 & 0 & 0 & 0 \\ 0 & 0 & 0 & 0 & 0 \end{array}\right]$

47. $\left[\begin{array}{cccc|c} 0 & 1 & 2 & 0 & 2 \\ 0 & 0 & 0 & 1 & 1 \\ 0 & 0 & 0 & 0 & 0 \\ 0 & 0 & 0 & 0 & 0 \end{array}\right]$

48. $\left[\begin{array}{cccc|c} 0 & 1 & 0 & 1 & 2 \\ 0 & 0 & 1 & 3 & 2 \\ 0 & 0 & 0 & 0 & 1 \\ 0 & 0 & 0 & 0 & 0 \end{array}\right]$

49. $\left[\begin{array}{cccc|c} 1 & 1 & 2 & -1 & 0 \\ 0 & 0 & 0 & 0 & 1 \\ 0 & 0 & 0 & 0 & 0 \end{array}\right]$

50. $\left[\begin{array}{ccc|c} 1 & 0 & 0 & 2 \\ 0 & 1 & 0 & 3 \\ 0 & 0 & 1 & 5 \\ 0 & 0 & 0 & 0 \end{array}\right]$

51. $\left[\begin{array}{ccccc|c} 0 & 0 & 0 & 0 & 1 & 3 \\ 0 & 0 & 0 & 0 & 0 & 0 \\ 0 & 0 & 0 & 0 & 0 & 0 \end{array}\right]$

52. $\left[\begin{array}{ccccc|c} 0 & 0 & 0 & 1 & 0 & 2 \\ 0 & 0 & 0 & 0 & 0 & 0 \\ 0 & 0 & 0 & 0 & 0 & 0 \end{array}\right]$

53. $\left[\begin{array}{ccccc|c} 1 & 1 & 2 & 0 & 1 & 2 \\ 0 & 0 & 0 & 1 & 1 & 5 \\ 0 & 0 & 0 & 0 & 0 & 0 \\ 0 & 0 & 0 & 0 & 0 & 0 \end{array}\right]$

54. $\left[\begin{array}{ccccc|c} 1 & 0 & 0 & 0 & 1 & 0 \\ 0 & 0 & 1 & 1 & 1 & 5 \\ 0 & 0 & 0 & 0 & 0 & 0 \\ 0 & 0 & 0 & 0 & 0 & 0 \end{array}\right]$

5.8 Determinants and Linear Independence

If $\mathbf{A}$ is a square matrix, then the **determinant** of $\mathbf{A}$, denoted $\det \mathbf{A}$, is a number (scalar). There are several ways to define the determinant. We shall give two ways to compute it (in lieu of a definition).

a. If $\mathbf{A}$ is 1×1, then

$$\det [a_{11}] = a_{11}. \tag{1}$$

b. If $\mathbf{A}$ is 2×2, then

$$\det \begin{bmatrix} a_{11} & a_{12} \\ a_{21} & a_{22} \end{bmatrix} = a_{11}a_{22} - a_{21}a_{12}. \tag{2}$$

In general, if $\mathbf{A}$ is $n \times n$, then $\det \mathbf{A}$ can be expressed in terms of n smaller $(n-1) \times (n-1)$ determinants, as follows:

Take any row or column of $\mathbf{A}$. Suppose the ith row is selected; its entries are $a_{i1}, a_{i2}, \ldots, a_{in}$. Multiply each a_{ij} by $(-1)^{i+j} M_{ij}$, where M_{ij} is the determinant of the matrix obtained by deleting the ith row and jth column of $\mathbf{A}$. Adding the $a_{ij}(-1)^{i+j} M_{ij}$ gives $\det \mathbf{A}$:

$$\det \mathbf{A} = \sum_{j=1}^{n} a_{ij}(-1)^{i+j} M_{ij}. \tag{3}$$

Thus, if $\mathbf{A}$ were, say, 5×5, formula (3) could be used to write $\det \mathbf{A}$ as a sum of five 4×4 determinants. Then (3) could be used to express each of these 4×4 determinants as the sum of four 3×3 determinants each of which could then be expressed as three 2×2 determinants. The M_{ij} are called **minors**, and $A_{ij} = (-1)^{i+j} M_{ij}$ are called **cofactors**.

EXAMPLE 5.8.1 **Determinants Using Cofactors**

Compute $\det \mathbf{A}$, where

$$\mathbf{A} = \begin{bmatrix} 1 & 1 & 2 \\ 1 & 0 & -1 \\ 2 & 1 & 4 \end{bmatrix}. \tag{4}$$

SOLUTION Since we can pick any row or column, it is a little less work to use one with a zero in it. If we select row 2, (3) becomes

$$\begin{aligned} \det \mathbf{A} &= a_{21}(-1)^{2+1} M_{21} + a_{22}(-1)^{2+2} M_{22} + a_{23}(-1)^{2+3} M_{23} \\ &= 1(-1) \det \begin{bmatrix} 1 & 2 \\ 1 & 4 \end{bmatrix} + 0(1) \det \begin{bmatrix} 1 & 2 \\ 2 & 4 \end{bmatrix} + (-1)(-1) \det \begin{bmatrix} 1 & 1 \\ 2 & 1 \end{bmatrix}. \end{aligned}$$

The 2×2 determinants are evaluated using (2), to give

$$\det \mathbf{A} = -1(4-2) + 0 + 1(1-2) = -3.$$

This is obviously a lot of work for even fairly small matrices. In fact, one can show that, if $\mathbf{A}$ is $n \times n$, the number of multiplications is approximately $n!$. For this reason, evaluation of the determinant is usually avoided or computed by alternative means.* However, we will have no alternative to this method when we need to compute the characteristic polynomial two sections from now.

* *All of our discussion of numerical considerations assumes that the calculations are done* **serially**. *Different considerations apply on* **parallel computers**.

There are two ways to make the determinants in our problems easier to compute. The first is an alternative method for the 3×3 case.

Alternative Method for the 3 × 3 Determinant Take the 3×3 matrix

$$\mathbf{A} = \begin{bmatrix} a & b & c \\ d & e & f \\ g & h & i \end{bmatrix},$$

and repeat the first two columns. Then multiply the elements down each of the six diagonals, as shown in (5).

$$\left[\begin{array}{ccc|cc} a & b & c & a & b \\ d & e & f & d & e \\ g & h & i & g & h \end{array}\right]. \tag{5}$$

$$-gec - hfa - idb + aei + bfg + cdh = \det \mathbf{A}.$$

Subtract the first three products from the sum of the last three products to get $\det \mathbf{A}$. This formula may be used even if the entries depend on another variable.

EXAMPLE 5.8.2 Special 3 × 3 Method

Consider Example 5.8.1,

$$\mathbf{A} = \begin{bmatrix} 1 & 1 & 2 \\ 1 & 0 & -1 \\ 2 & 1 & 4 \end{bmatrix},$$

and compute $\det \mathbf{A}$ using (5).

SOLUTION From (5) we have

$$\left[\begin{array}{ccc|cc} 1 & 1 & 2 & 1 & 1 \\ 1 & 0 & -1 & 1 & 0 \\ 2 & 1 & 4 & 2 & 1 \end{array}\right].$$

$$-(0) - (-1) - (4) + 0 + (-2) + 2 = -3.$$ ◄

However, probably the quickest way to compute $\det \mathbf{A}$ is based on the following result.

■ **THEOREM 5.8.1 Determinant Properties**

Suppose $\mathbf{A}$ is an $n \times n$ matrix. Then,

1. Exchanging two rows or columns of $\mathbf{A}$ changes the sign of $\det \mathbf{A}$.

2. Adding a multiple of one row (column) to another row (column) does not change $\det \mathbf{A}$.
3. If $\mathbf{A}$ is **upper triangular** (all entries below the principal diagonal are zero), then $\det \mathbf{A}$ is the product of the diagonal entries. ■

This theorem can be used two ways. One can use parts 1 and 2 to get a row (or column) with only one nonzero entry, and then expand the determinant along that row or column, using (3). Alternatively, row operations may be performed on $\mathbf{A}$ following the Gaussian elimination pattern of the last section, and the determinant taken of the resulting upper triangular matrix.

EXAMPLE 5.8.3 Determinant Using Row Operations

Find

$$\det \begin{bmatrix} 1 & 1 & 2 \\ 1 & 0 & -1 \\ 2 & 1 & 4 \end{bmatrix}.$$

SOLUTION

$$\det \begin{bmatrix} 1 & 1 & 2 \\ 1 & 0 & -1 \\ 2 & 1 & 4 \end{bmatrix} \quad \text{adding } -1 \text{ times row 1 to row 2, adding } -2 \text{ times row 1 to row 3,}$$

$$= \det \begin{bmatrix} 1 & 1 & 2 \\ 0 & -1 & -3 \\ 0 & -1 & 0 \end{bmatrix} \quad \text{adding } -1 \text{ times row 2 to row 3,}$$

$$= \det \begin{bmatrix} 1 & 1 & 2 \\ 0 & -1 & -3 \\ 0 & 0 & 3 \end{bmatrix}$$

$$= 1(-1)(3) = -3 \quad \text{by part 3 of Theorem 5.8.1.}$$ ◄

For even small matrices such as 4×4 or 5×5, this method is much quicker than the expansion formula (3). Unfortunately, if the entries of $\mathbf{A}$ depend on a variable, then using row operations may become impractical or at least more difficult.

EXAMPLE 5.8.4 Determinant of Matrix with Parameters

Find $\det \mathbf{A}$ if

$$\mathbf{A}(\lambda) = \begin{bmatrix} \lambda & 2 & 1 \\ 1 & \lambda - 1 & 1 \\ 1 & 2 & \lambda \end{bmatrix}. \tag{6}$$

Solution We could evaluate this determinant by using the 3,1 entry to zero both entries above, and then expanding along the first column by (3) (Exercise 21). Instead, we shall use (3) directly and expand along the first row:

$$\begin{aligned}\det \mathbf{A}(\lambda) &= \lambda(-1)^{1+1}\det\begin{bmatrix}\lambda-1 & 1\\ 2 & \lambda\end{bmatrix} + 2(-1)^{1+2}\det\begin{bmatrix}1 & 1\\ 1 & \lambda\end{bmatrix}\\ &\quad +1(-1)^{1+3}\det\begin{bmatrix}1 & \lambda-1\\ 1 & 2\end{bmatrix}\\ &= \lambda(\lambda^2-\lambda-2)-2(\lambda-1)+(3-\lambda)\\ &= \lambda^3-\lambda^2-5\lambda+5.\end{aligned}$$

5.8.1 Linear Independence

A set of r vectors $\{\mathbf{b}_1,\ldots,\mathbf{b}_r\}$ all the same size are **linearly independent** if

$c_1\mathbf{b}_1 + c_2\mathbf{b}_2 + \cdots + c_r\mathbf{b}_r = \mathbf{0}$ for constants $c_1, c_2, \ldots, c_r$ implies that the only solution is $c_1 = c_2 = \cdots = c_r = 0$. (7)

The expression $c_1\mathbf{b}_1 + c_2\mathbf{b}_2 + \cdots + c_r\mathbf{b}_r$ is a **linear combination**. Intuitively, a set of vectors is linearly independent if no one of the vectors can be written as a linear combination of the others. If a set of vectors is not linearly independent, it is called **linearly dependent**.

Two vectors are linearly independent if neither is a multiple of the other. Three vectors are linearly independent if they do not lie in the same plane.

EXAMPLE 5.8.5 **Linear Independence**

Is the set of vectors

$$\left\{\begin{bmatrix}1\\1\\0\end{bmatrix},\begin{bmatrix}1\\0\\1\end{bmatrix},\begin{bmatrix}5\\2\\3\end{bmatrix}\right\}$$

linearly independent?

Solution From (7), this is equivalent to asking whether

$$c_1\begin{bmatrix}1\\1\\0\end{bmatrix} + c_2\begin{bmatrix}1\\0\\1\end{bmatrix} + c_3\begin{bmatrix}5\\2\\3\end{bmatrix} = \begin{bmatrix}0\\0\\0\end{bmatrix} \tag{8}$$

has $c_1 = c_2 = c_3 = 0$ as its only solution. The equation (8) is equivalent to

$$\begin{aligned} c_1 + c_2 + 5c_3 &= 0,\\ c_1 + 2c_3 &= 0,\\ c_2 + 3c_3 &= 0.\end{aligned} \tag{9}$$

If we apply Gaussian elimination to (9), we finally get an augmented matrix of

$$\left[\begin{array}{ccc|c} 1 & 0 & 2 & 0 \\ 0 & 1 & 3 & 0 \\ 0 & 0 & 0 & 0 \end{array}\right].$$

Thus we can take c_3 arbitrary and $c_1 = -2c_3$, $c_2 = -3c_3$. In particular, c_3 need not be zero, so the vectors are not linearly independent. ◄

EXAMPLE 5.8.6 Linear Independence

Is the set of vectors

$$\left\{\begin{bmatrix} 1 \\ 2 \end{bmatrix}, \begin{bmatrix} 2 \\ 1 \end{bmatrix}\right\}$$

linearly independent?

Solution From (7), this is equivalent to asking whether

$$c_1\begin{bmatrix} 1 \\ 2 \end{bmatrix} + c_2\begin{bmatrix} 2 \\ 1 \end{bmatrix} = \begin{bmatrix} 0 \\ 0 \end{bmatrix} \tag{10}$$

has as the only solution $c_1 = c_2 = 0$. Equation (10) is equivalent to

$$\begin{aligned} c_1 + 2c_2 &= 0, \\ 2c_1 + c_2 &= 0. \end{aligned} \tag{11}$$

Solving the system, we find $c_1 = 0$, $c_2 = 0$. Thus

$$\left\{\begin{bmatrix} 1 \\ 2 \end{bmatrix}, \begin{bmatrix} 2 \\ 1 \end{bmatrix}\right\}$$

is a linearly independent set of vectors. ◄

The following theorem relates several of our concepts.

■ **Theorem 5.8.2 Invertibility and Linear Independence**

Suppose $\mathbf{A}$ is an $n \times n$ matrix. Then the following are equivalent:

The rows of $\mathbf{A}$ are linearly independent. (12)

The columns of $\mathbf{A}$ are linearly independent. (13)

$\det \mathbf{A} \neq 0$. (14)

$\mathbf{A}$ has an inverse. (15)

This theorem gives us a way to check linear independence. ■

■ **ALGORITHM 5.8.1 Checking Linear Independence**

To check whether n vectors that are $n \times 1$ (or $1 \times n$) are linearly independent, make a matrix **A** with the vectors as either rows or columns. The vectors are linearly independent if and only if $\det \mathbf{A} \neq 0$. ■

Exercises

In Exercises 1 through 7, compute $\det \mathbf{A}$ if

1. $\mathbf{A} = \begin{bmatrix} 1 & 2 \\ 3 & 4 \end{bmatrix}$

2. $\mathbf{A} = \begin{bmatrix} 1 & -1 \\ 2 & -6 \end{bmatrix}$

3. $\mathbf{A} = \begin{bmatrix} 1 & 1 & 2 \\ 0 & 1 & 0 \\ 2 & 1 & -1 \end{bmatrix}$

4. $\mathbf{A} = \begin{bmatrix} 1 & 1 & 1 \\ 2 & 2 & 2 \\ 3 & 3 & 3 \end{bmatrix}$

5. $\mathbf{A} = \begin{bmatrix} 1 & 1 & 0 \\ 0 & 1 & 1 \\ 1 & 0 & 1 \end{bmatrix}$

6. $\mathbf{A} = \begin{bmatrix} 1 & 2 & 3 & 4 \\ 0 & -1 & 2 & 6 \\ 0 & 0 & 3 & 5 \\ 0 & 0 & 0 & 9 \end{bmatrix}$

7. $\mathbf{A} = \begin{bmatrix} 1 & 2 & 0 \\ -1 & 1 & 1 \\ 1 & 2 & 3 \end{bmatrix}$

In Exercises 8 through 12, determine whether the given set of vectors is linearly independent.

8. $\left\{ \begin{bmatrix} 1 \\ 1 \end{bmatrix}, \begin{bmatrix} 1 \\ -1 \end{bmatrix} \right\}$

9. $\left\{ \begin{bmatrix} 1 \\ -1 \end{bmatrix}, \begin{bmatrix} -1 \\ 1 \end{bmatrix} \right\}$

10. $\left\{ \begin{bmatrix} 1 \\ 2 \\ 1 \\ 0 \end{bmatrix}, \begin{bmatrix} 0 \\ 1 \\ 1 \\ 1 \end{bmatrix}, \begin{bmatrix} 1 \\ 0 \\ 0 \\ 1 \end{bmatrix} \right\}$

11. $\left\{ \begin{bmatrix} 1 \\ 1 \\ 1 \end{bmatrix}, \begin{bmatrix} 1 \\ -1 \\ 1 \end{bmatrix}, \begin{bmatrix} -1 \\ 0 \\ 1 \end{bmatrix} \right\}$

12. $\left\{ \begin{bmatrix} 1 \\ 1 \\ 0 \end{bmatrix}, \begin{bmatrix} 0 \\ 1 \\ 1 \end{bmatrix}, \begin{bmatrix} 1 \\ 0 \\ -1 \end{bmatrix} \right\}$

13. a) Let **I** be the 2×2 identity. Verify that

$$\det(5\mathbf{I}) = 25.$$

b) Let **I** be the 3×3 identity. Verify that

$$\det(5\mathbf{I}) = 125.$$

c) Let **I** be the 4×4 identity. Verify that

$$\det(5\mathbf{I}) = 625.$$

14. (Note Exercise 13.) Show that if **I** is $n \times n$ and α is a scalar, then $\det(\alpha \mathbf{I}) = \alpha^n$.

In Exercises 15 through 17, compute $\det(\lambda \mathbf{I} - \mathbf{A})$, where **I** is 2×2 and λ is a parameter (variable) and:

15. $\mathbf{A} = \begin{bmatrix} 1 & 0 \\ 0 & 3 \end{bmatrix}$

16. $\mathbf{A} = \begin{bmatrix} 1 & 2 \\ 3 & 0 \end{bmatrix}$

17. $\mathbf{A} = \begin{bmatrix} 0 & 3 \\ 6 & 0 \end{bmatrix}$

In Exercises 18 through 20, compute $\det(\lambda \mathbf{I} - \mathbf{A})$, where **I** is 3×3 and λ is a parameter, and:

18. $\mathbf{A} = \begin{bmatrix} 3 & 6 & 7 \\ 0 & 0 & 5 \\ 0 & 0 & 2 \end{bmatrix}$

19. $\mathbf{A} = \begin{bmatrix} 1 & 0 & 1 \\ 0 & 1 & 0 \\ 1 & 0 & 1 \end{bmatrix}$

20. $\mathbf{A} = \begin{bmatrix} 0 & 0 & 1 \\ -1 & 3 & 2 \\ 2 & 0 & -1 \end{bmatrix}$

21. Let

$$\mathbf{A}(\lambda) = \begin{bmatrix} \lambda & 2 & 1 \\ 1 & \lambda - 1 & 1 \\ 1 & 2 & \lambda \end{bmatrix}.$$

Compute $\det \mathbf{A}(\lambda)$ by adding -1 times row 3 to row 2, and $-\lambda$ times row 3 to row 1, and then expanding down the first column, using (3). Compare your answer to that of Example 5.8.4.

22. If $\mathbf{A}$, $\mathbf{B}$ are $n \times n$ matrices, then it can be shown that $\det(\mathbf{AB}) = \det \mathbf{A} \det \mathbf{B}$. Use this to show that if $\mathbf{A}$ is invertible, then $\det(\mathbf{A}^{-1}) = (\det \mathbf{A})^{-1}$.

5.9 Differential Equations: Basic Theory

In this section we will begin to apply matrix theory to linear systems of differential equations in the explicit form

$$\begin{aligned} x_1'(t) &= a_{11}(t)x_1(t) + \cdots + a_{1n}(t)x_n(t) + f_1(t), \\ x_2'(t) &= a_{21}(t)x_1(t) + \cdots + a_{2n}(t)x_n(t) + f_2(t), \\ &\vdots \\ x_n'(t) &= a_{n1}(t)x_1(t) + \cdots + a_{nn}(t)x_n(t) + f_n(t), \end{aligned} \tag{1}$$

with initial conditions

$$x_1(t_0) = a_1, \ldots, x_n(t_0) = a_n. \tag{2}$$

In order to do so, we must briefly discuss matrix-valued functions (or equivalently, matrices whose entries are functions):

If $\mathbf{A}(t)$ is an $m \times n$ matrix whose entries depend on a variable t, then

$$\frac{d\mathbf{A}}{dt} = \mathbf{A}'(t) \tag{3}$$

is an $m \times n$ matrix whose i,j entry is $a_{ij}'(t)$; that is, matrices are differentiated element-wise.

Of course, (3) requires the entries of $\mathbf{A}(t)$ to be differentiable functions.

EXAMPLE 5.9.1 **Derivative of a Matrix**

If

$$\mathbf{A}(t) = \begin{bmatrix} t & 1 & \sin t \\ 0 & 2 & \cos t \end{bmatrix}, \quad \text{then } \mathbf{A}'(t) = \begin{bmatrix} 1 & 0 & \cos t \\ 0 & 0 & -\sin t \end{bmatrix}.$$ ◄

Notational Comment The prime is also sometimes used in matrix theory to denote something called the transpose, so be careful when consulting other texts.

Differentiating matrix-valued functions works almost exactly like differentiating scalar functions:

$$\frac{d(\mathbf{A}(t) + \mathbf{B}(t))}{dt} = \frac{d\mathbf{A}(t)}{dt} + \frac{d\mathbf{B}(t)}{dt}; \tag{4}$$

$$\frac{d\mathbf{A}(t)}{dt} = \mathbf{0} \text{ for all } t \text{ implies that } \mathbf{A}(t) \text{ is a constant matrix;} \tag{5}$$

$$\frac{d\mathbf{A}}{dt} = \frac{d\mathbf{B}}{dt} \text{ implies that } \mathbf{A}(t) = \mathbf{B}(t) + \mathbf{C}, \tag{6}$$

where $\mathbf{C}$ is a constant matrix of the same size as $\mathbf{A}$ and $\mathbf{B}$, and

$$\frac{d(\mathbf{A}(t)\mathbf{B}(t))}{dt} = \frac{d\mathbf{A}(t)}{dt}\mathbf{B}(t) + \mathbf{A}(t)\frac{d\mathbf{B}(t)}{dt}, \tag{7}$$

which is the **product rule.**

The only thing one has to be careful with is that the order of the products in the product rule cannot be changed. Thus, we cannot write $(\mathbf{AB})' = \mathbf{A}'\mathbf{B} + \mathbf{B}'\mathbf{A}$, which would be true with scalar functions.

If we introduce the notation

$$\mathbf{x}(t) = \begin{bmatrix} x_1(t) \\ \vdots \\ x_n(t) \end{bmatrix}, \qquad \mathbf{A}(t) = \begin{bmatrix} a_{11}(t) & \cdots & a_{1n}(t) \\ \vdots & & \vdots \\ a_{n1}(t) & \cdots & a_{nn}(t) \end{bmatrix},$$

$$\mathbf{f}(t) = \begin{bmatrix} f_1(t) \\ \vdots \\ f_n(t) \end{bmatrix}, \qquad \mathbf{a} = \begin{bmatrix} a_1 \\ \vdots \\ a_n \end{bmatrix},$$

the nth-order linear system of differential equations (1) may be written as

$$\mathbf{x}'(t) = \mathbf{A}(t)\mathbf{x}(t) + \mathbf{f}(t), \qquad \mathbf{x}(t_0) = \mathbf{a},$$

or, more simply, as

$$\mathbf{x}' = \mathbf{A}\mathbf{x} + \mathbf{f}, \qquad \mathbf{x}(t_0) = \mathbf{a}. \tag{8}$$

The vector $\mathbf{a}$ is called the **initial condition**, $\mathbf{A}$ is the **coefficient matrix**, $\mathbf{f}$ is the **input** or **forcing term**, and $\mathbf{x}$ is the **output** or **response**. A solution is a vector of functions $\mathbf{x}(t)$ (equivalently, a vector-valued function) that satisfies (8). We shall consider only systems where $\mathbf{A}$ is square. Theorem 5.1.1 can now be restated in matrix notation.

■ **THEOREM 5.9.1 Fundamental Theorem of Linear Systems**

Suppose that the entries of **A** and **f** are continuous on an interval I containing t_0. Let **a** be an arbitrary $n \times 1$ vector of scalars. Then

> there exists a unique solution to $\mathbf{x}' = \mathbf{A}\mathbf{x} + \mathbf{f}$, $\mathbf{x}(t_0) = \mathbf{a}$, and this solution is defined on all of the interval I. ■

Just as in the scalar case, we get a breakdown of the solution into free response and forced response.

■ **THEOREM 5.9.2 Particular and Homogeneous**

Suppose that the entries of **A** and **f** are continuous on the interval I. Then

> Every solution of $\mathbf{x}' = \mathbf{A}\mathbf{x} + \mathbf{f}$ is of the form $\mathbf{x} = \mathbf{x}_p + \mathbf{x}_h$, where $\mathbf{x}_p$ is a **particular solution** of $\mathbf{x}' = \mathbf{A}\mathbf{x} + \mathbf{f}$, and $\mathbf{x}_h$ is a solution of the **associated homogeneous equation** $\mathbf{x}' = \mathbf{A}\mathbf{x}$.

■

The particular solution $\mathbf{x}_p$ is sometimes called the **forced response** to the input **f**, and $\mathbf{x}_h$ is then called the **free response.**

Thus, solving linear systems can again be broken down into two problems: solving the associated homogeneous equations and finding a particular solution. In the next two sections, we shall show how to do both when **A** is constant. The key facts about the solution of the associated homogeneous equation are covered in the next result.

■ **THEOREM 5.9.3 Form of Solutions of Homogeneous Equations**

Suppose that the entries of the $n \times n$ matrix $\mathbf{A}(t)$ are continuous on the interval $t_0 \leq t \leq t_1$. Then the **general solution** of $\mathbf{x}'(t) = \mathbf{A}(t)\mathbf{x}(t)$ may be written as

$$\mathbf{x} = c_1\mathbf{x}_1(t) + c_2\mathbf{x}_2(t) + \cdots + c_n\mathbf{x}_n(t), \tag{9}$$

where $\{\mathbf{x}_1, \ldots, \mathbf{x}_n\}$ is any set of n solutions of $\mathbf{x}'(t) = \mathbf{A}(t)\mathbf{x}(t)$, such that the vectors $\{\mathbf{x}_1(t_0), \mathbf{x}_2(t_0), \ldots, \mathbf{x}_n(t_0)\}$ are linearly independent and the $c_1, \ldots, c_n$ are arbitrary constants. The set $\{\mathbf{x}_1(t), \ldots, \mathbf{x}_n(t)\}$ is called a **fundamental set of solutions** for $\mathbf{x}' = \mathbf{A}\mathbf{x}$. ■

A consequence of the Fundamental Theorem 5.9.1 on existence and uniqueness is that, if $\{\mathbf{x}_1, \ldots, \mathbf{x}_n\}$ are solutions of $\mathbf{x}' = \mathbf{A}\mathbf{x}$ and the set of vectors $\{\mathbf{x}_1(t), \ldots, \mathbf{x}_n(t)\}$ is linearly independent for one value of t, then they are linearly independent for all t.

Another way to express this is as follows:

Suppose $\mathbf{A}(t)$ is $n \times n$. Let $\{\mathbf{x}_1, \ldots, \mathbf{x}_n\}$ be n solutions of $\mathbf{x}' = \mathbf{A}\mathbf{x}$ on the interval I. Let $\mathbf{X}(t)$ be the matrix with the $\mathbf{x}_i$ as columns. Let $\mathbf{c}$ be an $n \times 1$ vector of arbitrary constants. Then the following are equivalent:

$$\{\mathbf{x}_1, \ldots, \mathbf{x}_n\} \text{ is a fundamental set of solutions of } \mathbf{x}' = \mathbf{A}\mathbf{x}. \tag{10}$$

$$\det \mathbf{X}(t) \neq 0 \quad \text{for some value of } t \text{ in } I. \tag{11}$$

$$\det \mathbf{X}(t) \neq 0 \quad \text{for all values of } t \text{ in } I. \tag{12}$$

$$\mathbf{X}(t)\mathbf{c} \text{ is the general solution of } \mathbf{x}' = \mathbf{A}\mathbf{x}. \tag{13}$$

EXAMPLE 5.9.2 Fundamental Set of Solutions

Verify that

$$\left\{ \begin{bmatrix} e^{2t} \\ e^{2t} \end{bmatrix}, \begin{bmatrix} 1 \\ -1 \end{bmatrix} \right\}$$

is a fundamental set of solutions of

$$\mathbf{x}' = \begin{bmatrix} 1 & 1 \\ 1 & 1 \end{bmatrix} \mathbf{x}. \tag{14}$$

SOLUTION We need to show that

$$\{\mathbf{x}_1, \mathbf{x}_2\} = \left\{ \begin{bmatrix} e^{2t} \\ e^{2t} \end{bmatrix}, \begin{bmatrix} 1 \\ -1 \end{bmatrix} \right\}$$

are both solutions and that they are linearly independent. First, we verify they are solutions: Substituting $\mathbf{x}_1$ into (14) gives

$$\begin{bmatrix} e^{2t} \\ e^{2t} \end{bmatrix}' \stackrel{?}{=} \begin{bmatrix} 1 & 1 \\ 1 & 1 \end{bmatrix} \begin{bmatrix} e^{2t} \\ e^{2t} \end{bmatrix}$$

or

$$\begin{bmatrix} 2e^{2t} \\ 2e^{2t} \end{bmatrix} = \begin{bmatrix} 2e^{2t} \\ 2e^{2t} \end{bmatrix},$$

so that $\mathbf{x}_1$ is a solution. Similarly,

$$\begin{bmatrix} 1 \\ -1 \end{bmatrix}' \stackrel{?}{=} \begin{bmatrix} 1 & 1 \\ 1 & 1 \end{bmatrix} \begin{bmatrix} 1 \\ -1 \end{bmatrix}$$

or

$$\begin{bmatrix} 0 \\ 0 \end{bmatrix} = \begin{bmatrix} 0 \\ 0 \end{bmatrix},$$

so that both $\mathbf{x}_1$, $\mathbf{x}_2$ are solutions. To verify that they are linearly independent, we compute that

$$\det[\mathbf{x}_1, \mathbf{x}_2] = \det\begin{bmatrix} e^{2t} & 1 \\ e^{2t} & -1 \end{bmatrix} = -2e^{2t} \neq 0.$$ ◄

EXAMPLE 5.9.3 Fundamental Set, General Solution, and Initial Conditions

Given that we have found that

$$\{\mathbf{x}_1, \mathbf{x}_2\} = \left\{\begin{bmatrix} e^{2t} \\ e^{2t} \end{bmatrix}, \begin{bmatrix} 1 \\ -1 \end{bmatrix}\right\}$$

is a fundamental set of solutions of (14),

a. Find the general solution of (14);
b. Find the solution of the initial value problem

$$\mathbf{x}' = \begin{bmatrix} 1 & 1 \\ 1 & 1 \end{bmatrix}\mathbf{x}, \qquad \mathbf{x}(0) = \begin{bmatrix} 2 \\ 3 \end{bmatrix}. \tag{15}$$

SOLUTION

a. From Theorem 5.9.3, the general solution will be

$$\mathbf{x} = \begin{bmatrix} x_1 \\ x_2 \end{bmatrix} = c_1\begin{bmatrix} e^{2t} \\ e^{2t} \end{bmatrix} + c_2\begin{bmatrix} 1 \\ -1 \end{bmatrix}, \tag{16}$$

or, equivalently,

$$x_1 = c_1 e^{2t} + c_2, \qquad x_2 = c_1 e^{2t} - c_2.$$

b. Applying the initial condition in (15) to the general solution in (16) gives

$$\mathbf{x}(0) = c_1\begin{bmatrix} 1 \\ 1 \end{bmatrix} + c_2\begin{bmatrix} 1 \\ -1 \end{bmatrix} = \begin{bmatrix} 2 \\ 3 \end{bmatrix}. \tag{17}$$

This is two equations in two unknowns,

$$c_1 + c_2 = 2,$$
$$c_1 - c_2 = 3.$$

The solution is $c_1 = \frac{5}{2}$, $c_2 = -\frac{1}{2}$, so that

$$\mathbf{x} = \frac{5}{2}\begin{bmatrix} e^{2t} \\ e^{2t} \end{bmatrix} - \frac{1}{2}\begin{bmatrix} 1 \\ -1 \end{bmatrix} = \begin{bmatrix} \frac{5}{2}e^{2t} - \frac{1}{2} \\ \frac{5}{2}e^{2t} + \frac{1}{2} \end{bmatrix},$$

is the solution of (15). ◄

Exercises

1. If $\mathbf{A}(t) = \begin{bmatrix} 1 & t \\ t^2 & 1 \end{bmatrix}$, compute $\mathbf{A}'(t)$.

2. Verify that $(\mathbf{ABC})' = \mathbf{A}'\mathbf{BC} + \mathbf{AB}'\mathbf{C} + \mathbf{ABC}'$.

3. a) Verify that

$$\left\{\begin{bmatrix} e^{-t} \\ -e^{-t} \end{bmatrix}, \begin{bmatrix} e^{3t} \\ e^{3t} \end{bmatrix}\right\}$$

is a fundamental set of solutions of

$$\mathbf{x}' = \begin{bmatrix} 1 & 2 \\ 2 & 1 \end{bmatrix}\mathbf{x}. \tag{18}$$

b) Find the general solution of (18).

c) Find the solution of (18) that satisfies the initial condition

$$\mathbf{x}(0) = \begin{bmatrix} 2 \\ -1 \end{bmatrix}.$$

4. a) Verify that $\left\{\begin{bmatrix} \cos t \\ \sin t \end{bmatrix}, \begin{bmatrix} -\sin t \\ \cos t \end{bmatrix}\right\}$ is a fundamental set of solutions for

$$\mathbf{x}' = \begin{bmatrix} 0 & -1 \\ 1 & 0 \end{bmatrix}\mathbf{x}. \tag{19}$$

b) Find the general solution of (19).

c) Find the solution of (19) for which

$$\mathbf{x}(0) = \begin{bmatrix} 5 \\ 7 \end{bmatrix}.$$

5. a) Verify that $\left\{\begin{bmatrix} e^{3t} \\ e^{3t} \end{bmatrix}, \begin{bmatrix} 1 \\ -2 \end{bmatrix}\right\}$ is a fundamental set of solutions for

$$\mathbf{x}' = \begin{bmatrix} 2 & 1 \\ 2 & 1 \end{bmatrix}\mathbf{x}.$$

b) Find the general solution of this differential equation.

c) Find the solution for which

$$\mathbf{x}(0) = \begin{bmatrix} 7 \\ -7 \end{bmatrix}.$$

6. a) Verify that $\left\{\begin{bmatrix} e^{4t} \\ e^{4t} \end{bmatrix}, \begin{bmatrix} 3e^{-t} \\ -2e^{-t} \end{bmatrix}\right\}$ is a fundamental set of solutions for

$$\mathbf{x}' = \begin{bmatrix} 1 & 3 \\ 2 & 2 \end{bmatrix}\mathbf{x}.$$

b) Find the general solution of this differential equation.

c) Find the solution for which

$$\mathbf{x}(0) = \begin{bmatrix} -11 \\ 4 \end{bmatrix}.$$

7. a) Verify that $\left\{\begin{bmatrix} \cos 2t \\ \sin 2t \end{bmatrix}, \begin{bmatrix} -\sin 2t \\ \cos 2t \end{bmatrix}\right\}$ is a fundamental set of solutions for

$$\mathbf{x}' = \begin{bmatrix} 0 & -2 \\ 2 & 0 \end{bmatrix}\mathbf{x}.$$

b) Find the general solution of this differential equation.

c) Find the solution for which

$$\mathbf{x}(0) = \begin{bmatrix} 19 \\ -37 \end{bmatrix}.$$

8. a) Verify that $\left\{ \begin{bmatrix} e^{2t} \\ e^{2t} \end{bmatrix}, \begin{bmatrix} 3e^{t} \\ 2e^{t} \end{bmatrix} \right\}$ is a fundamental set of solutions of

$$\mathbf{x}' = \begin{bmatrix} -1 & 3 \\ -2 & 4 \end{bmatrix}\mathbf{x}.$$

b) Find the general solution of this differential equation.

c) Find the solution for which

$$\mathbf{x}(0) = \begin{bmatrix} 2 \\ \frac{3}{2} \end{bmatrix}.$$

9. Verify that a matrix $\mathbf{X}(t)$ satisfies the matrix differential equation

$$\mathbf{X}'(t) = \mathbf{A}(t)\mathbf{X}(t)$$

if and only if each column of $\mathbf{X}(t)$ satisfies $\mathbf{x}' = \mathbf{A}(t)\mathbf{x}$.

10. Suppose that $\mathbf{X}(t)$ is a matrix that satisfies $\mathbf{X}'(t) = \mathbf{A}(t)\mathbf{X}(t)$. Let $\mathbf{a}$ be a constant vector. Verify that $\mathbf{X}(t)\mathbf{a}$ is a solution of the vector differential equation $\mathbf{x}' = \mathbf{A}(t)\mathbf{x}$.

The theory of this section not only resembles that of Chapter 2; it includes it as a special case. To see this, consider the scalar linear homogeneous differential equation:

$$y''(t) + b(t)y'(t) + c(t)y(t) = 0, \qquad (20)$$

$$y(t_0) = a_0, \qquad y'(t_0) = a_1, \qquad (21)$$

where a_0, a_1 are constants.

11. Rewrite (20), (21) as a system in the form (recall Example 5.1.7) $(z = y')$

$$\mathbf{x}'(t) = \mathbf{A}(t)\mathbf{x}(t), \qquad (22)$$

$$\mathbf{x}(t_0) = \mathbf{a}, \qquad (23)$$

with

$$\mathbf{x}(t) = \begin{bmatrix} y(t) \\ z(t) \end{bmatrix}.$$

12. (Same notation as Exercise 11.) Show that two scalar functions $y_1(t)$, $y_2(t)$ are a fundamental set of solutions for the scalar equation $y'' + by' + cy = 0$ if and only if the vectors

$$\mathbf{x}_1(t) = \begin{bmatrix} y_1(t) \\ z_1(t) \end{bmatrix}, \qquad \mathbf{x}_2(t) = \begin{bmatrix} y_2(t) \\ z_2(t) \end{bmatrix}$$

are a fundamental set of solutions to the system (22).

13. (Same notation as Exercise 11, 12.) Show that the $\det \mathbf{X}(t)$ in (12) for the system (22) is just the *Wronskian* of the solution $\{y_1, y_2\}$ of (20).

Usually when problems are modeled, for example, circuits, they are not initially in the *explicit* form

$$\mathbf{x}'(t) = \mathbf{A}(t)\mathbf{x}(t) + \mathbf{f}(t)$$

but rather in the *implicit* form

$$\mathbf{E}(t)\mathbf{x}'(t) + \mathbf{F}(t)\mathbf{x}(t) = \mathbf{g}(t). \qquad (24)$$

Several examples appear in Sections 5.1, 5.2, 5.3. Exercises 14 through 16 elaborate on some of the ideas behind Theorem 5.1.2.

14. Suppose that $\mathbf{E}(t)$, $\mathbf{F}(t)$ are $n \times n$ matrices, and $\mathbf{E}(t)$ is invertible for each value of t. Show that (24) can be written in the form

$$\mathbf{x}'(t) = \mathbf{A}(t)\mathbf{x}(t) + \mathbf{f}(t) \qquad (25)$$

using the matrix $\mathbf{E}(t)^{-1}$.

15. (Uses Exercise 14.)

a) Rewrite the system

$$\begin{aligned} x' + y' + 2x - 3y &= t, \\ x' - y' + 2x - y &= 1, \end{aligned} \qquad (26)$$

in the form $\mathbf{Ex}' + \mathbf{Fx} = \mathbf{g}(t)$.

b) Compute $\mathbf{E}^{-1}$ and use it to rewrite the system (26) in the form

$$\mathbf{x}'(t) = \mathbf{Ax}(t) + \mathbf{f}(t).$$

16. **a)** Rewrite the system

$$\begin{aligned} x' + y' + x - y + z &= \cos t, \\ y' - z' - x + 2y - 3z &= 0, \\ x' + 2z' + 2x - 3y - z &= 1 \end{aligned} \tag{27}$$

in the form $\mathbf{E}\mathbf{x}'(t) + \mathbf{F}\mathbf{x}(t) = \mathbf{g}(t)$.

b) Compute $\mathbf{E}^{-1}$ and use it to rewrite (27) in the form (25).

For Exercises 17 and 18, suppose that $\mathbf{A}(t)$ is 2×2 and continuous for all t. A solution of $\mathbf{x}'(t) = \mathbf{A}(t)\mathbf{x}(t)$ can be thought of as a point

$$\mathbf{x}(t) = \begin{bmatrix} x_1(t) \\ x_2(t) \end{bmatrix}$$

in the x_1, x_2 plane tracing out a curve (or **trajectory**).

17. Let

$$\mathbf{A}(t) = \frac{1}{1+t^2}\begin{bmatrix} t & 1 \\ -1 & t \end{bmatrix}$$

for $-\infty < t < \infty$.

a) Verify that

$$\mathbf{x}_1(t) = \begin{bmatrix} 1 \\ -t \end{bmatrix}, \quad \mathbf{x}_2(t) = \begin{bmatrix} t \\ 1 \end{bmatrix}$$

is a fundamental set of solutions for $\mathbf{x}' = \mathbf{A}\mathbf{x}$.

b) Graph the trajectories $\mathbf{x}_1(t)$, $\mathbf{x}_2(t)$ for all t in the plane, and observe that the trajectories cross, since

$$\mathbf{x}_1(-1) = \mathbf{x}_2(1) = \begin{bmatrix} 1 \\ 1 \end{bmatrix}.$$

(This does not violate uniqueness, since $\mathbf{x}_1$, $\mathbf{x}_2$ never are at the same place at the same time.)

18. Suppose that $\mathbf{A}$ is a constant matrix and α is a scalar.

a) Show that, if $\mathbf{z}(t)$ is a solution of $\mathbf{x}'(t) = \mathbf{A}\mathbf{x}(t)$, then the function $\mathbf{w}(t) = \mathbf{z}(t - \alpha)$ is also a solution of $\mathbf{x}'(t) = \mathbf{A}\mathbf{x}(t)$.

b) Using the existence and uniqueness Theorem 5.9.1 and part (a), show that if $\mathbf{A}$ is a 2×2 constant matrix and the trajectories in the plane of two solutions $\mathbf{x}_1$, $\mathbf{x}_2$ ever intersect, then the solutions give the same trajectory and, in fact, there exists a scalar α so that $\mathbf{x}_1(t) = \mathbf{x}_2(t - \alpha)$ for all t. (*Hint*: Start by assuming $\mathbf{x}_1(t_0) = \mathbf{x}_2(t_1)$ for some fixed t_0, t_1.)

19. Verify that, if $\mathbf{x}_1$, $\mathbf{x}_2$ are solutions of $\mathbf{x}' = \mathbf{A}(t)\mathbf{x}$, then $c_1\mathbf{x}_1 + c_2\mathbf{x}_2$ with c_1, c_2 constant is also a solution of $\mathbf{x}' = \mathbf{A}(t)\mathbf{x}$.

20. Suppose that $\mathbf{x}_p$ is a solution of $\mathbf{x}' = \mathbf{A}\mathbf{x} + \mathbf{f}$. Show that any other solution $\mathbf{x}$ of $\mathbf{x}' = \mathbf{A}\mathbf{x} + \mathbf{f}$ may be written as $\mathbf{x} = \mathbf{x}_p + \mathbf{x}_h$, where $\mathbf{x}_h$ is a solution of $\mathbf{x}' = \mathbf{A}\mathbf{x}$.

21. Show that, if $\mathbf{A}(t)$ is invertible for each t, then $(\mathbf{A}^{-1})' = -\mathbf{A}^{-1}\mathbf{A}'\mathbf{A}^{-1}$.

22. Show that (13) follows from (9).

23. Using (13), show that $\mathbf{X}(t)\mathbf{X}^{-1}(0)\mathbf{a}$ is the solution of $\mathbf{x}' = \mathbf{A}\mathbf{x}$ which satisfies the initial condition $\mathbf{x}(0) = \mathbf{a}$.

5.10 Homogeneous Systems of Differential Equations with Constant Coefficients Using Eigenvectors

5.10.1 Introduction and Real Eigenvalues

This section will show how to solve

$$\mathbf{x}' = \mathbf{A}\mathbf{x}$$

when $\mathbf{A}$ is a constant $n \times n$ matrix. One solution is $\mathbf{x}(t) = \mathbf{0}$. We wish to find nontrivial (nonzero) solutions. If $\mathbf{A}$ were a scalar, we know that a solution would be $x = e^{\lambda t}$ with $\lambda = \mathbf{A}$. This suggests looking for solutions of the form

$$\mathbf{x} = e^{\lambda t}\mathbf{u}, \tag{1}$$

where **u** is a constant nonzero vector and λ is a scalar. Substituting $e^{\lambda t}\mathbf{u}$ for **x** in $\mathbf{x}' = \mathbf{Ax}$, we get

$$(e^{\lambda t}\mathbf{u})' = \mathbf{A}(e^{\lambda t}\mathbf{u})$$

or

$$\lambda e^{\lambda t}\mathbf{u} = e^{\lambda t}\mathbf{Au},$$

and, upon division by $e^{\lambda t}$,

$$\lambda\mathbf{u} = \mathbf{Au}. \tag{2}$$

In summary,

> If λ is a scalar and **u** is a **nonzero vector** such that $\mathbf{Au} = \lambda\mathbf{u}$, then $e^{\lambda t}\mathbf{u}$ is a nontrivial solution of $\mathbf{x}' = \mathbf{Ax}$. (3)

The scalar λ in (2) is called an **eigenvalue** of **A** and the vector **u** is called an **eigenvector** of **A** (**associated** with the eigenvalue λ). It is not obvious at first that such a λ, **u** need exist. However, by the end of this section we shall see that eigenvalues and eigenvectors exist and are very helpful in solving linear systems of differential equations with constant coefficients.

EXAMPLE 5.10.1 Eigenvalue and Eigenvector

Let

$$\mathbf{A} = \begin{bmatrix} 1 & 2 \\ 2 & 1 \end{bmatrix}, \qquad \mathbf{u} = \begin{bmatrix} 1 \\ 1 \end{bmatrix}, \qquad \lambda = 3.$$

Then

$$\mathbf{Au} = 3\mathbf{u} = \begin{bmatrix} 3 \\ 3 \end{bmatrix}.$$

Thus **u** is an eigenvector of **A** associated with the eigenvalue 3. ◄

We now turn to the problem of computing eigenvalues and eigenvectors. We cannot, without a major digression, present any of the methods that are often used in practice. Instead, we shall present a method that, while generally applicable in principle, is actually used only for small examples like the ones we shall work in this section.

Notational Comment If **z** is an $n \times 1$ vector of variables, say in the equation $\mathbf{Az} = \mathbf{0}$, we shall usually denote these variables by $\{z_1, \ldots, z_n\}$, so that

$$\mathbf{z} = \begin{bmatrix} z_1 \\ z_2 \\ \vdots \\ z_n \end{bmatrix}.$$

On the other hand, a collection of r vectors, for example, r solutions of $\mathbf{Az} = \mathbf{0}$, would be denoted $\{\mathbf{z}_1, \mathbf{z}_2, \ldots, \mathbf{z}_r\}$. We shall not need a notation to refer to the individual entries of a collection of vectors.

Note that $\mathbf{Au} = \lambda\mathbf{u}$ may be rewritten as

$$(\lambda\mathbf{I} - \mathbf{A})\mathbf{u} = \mathbf{0}. \tag{4}$$

If $\lambda\mathbf{I} - \mathbf{A}$ were invertible, we could multiply (4) by $(\lambda\mathbf{I} - \mathbf{A})^{-1}$ and conclude that $\mathbf{u} = \mathbf{0}$. But eigenvectors are always nonzero. Thus $\lambda\mathbf{I} - \mathbf{A}$ cannot be invertible if λ is an eigenvalue. But, from Theorem 5.8.2, $\lambda\mathbf{I} - \mathbf{A}$ is not invertible if and only if $\det(\lambda\mathbf{I} - \mathbf{A}) = 0$.

THEOREM 5.10.1 Eigenvalues and the Characteristic Polynomial

Let $p(\lambda) = \det(\lambda\mathbf{I} - \mathbf{A})$, where $\mathbf{A}$, $\mathbf{I}$ are $n \times n$ matrices. Then $p(\lambda)$ is an nth-degree polynomial called the **characteristic** (or **auxiliary**) **polynomial** of $\mathbf{A}$. The eigenvalues of $\mathbf{A}$ are precisely the roots of the characteristic polynomial $p(\lambda)$. ■

EXAMPLE 5.10.2 Characteristic Polynomial

Let

$$\mathbf{A} = \begin{bmatrix} 1 & 2 \\ 2 & 1 \end{bmatrix}.$$

Find the characteristic polynomial and eigenvalues of $\mathbf{A}$.

SOLUTION The characteristic polynomial is

$$\begin{aligned} p(\lambda) &= \det(\lambda\mathbf{I} - \mathbf{A}) = \det\left(\begin{bmatrix} \lambda & 0 \\ 0 & \lambda \end{bmatrix} - \begin{bmatrix} 1 & 2 \\ 2 & 1 \end{bmatrix}\right) \\ &= \det\begin{bmatrix} \lambda - 1 & -2 \\ -2 & \lambda - 1 \end{bmatrix} \\ &= (\lambda - 1)^2 - 4 = \lambda^2 - 2\lambda - 3 = (\lambda - 3)(\lambda + 1). \end{aligned}$$

The eigenvalues of $\mathbf{A}$ are the roots of the characteristic polynomial which, in this example, are $\lambda_1 = 3$, $\lambda_2 = -1$. ◀

Eigenvectors

Once the eigenvalues of $\mathbf{A}$ are found, we proceed as follows. To find the eigenvectors of $\mathbf{A}$, solve the system

$$(\lambda\mathbf{I} - \mathbf{A})\mathbf{u} = \mathbf{0} \tag{5}$$

for each eigenvalue. Equation (5) does not uniquely determine the eigenvectors. There will be one or more arbitrary constants in the solution. Once (5) is solved for a given eigenvalue, call it λ_0, we expand the solution in vector form, as in Example 5.7.10. Taking the vector times each arbitrary constant gives a set of linearly independent eigenvectors corresponding to the eigenvalue λ_0. The number of linearly independent eigenvectors is the same as the number of arbitrary constants in the solution of (5).

EXAMPLE 5.10.3 Solutions of $\mathbf{x}' = \mathbf{A}\mathbf{x}$ from an Eigenvalue and Eigenvector

Suppose that we solve (5) for a given eigenvalue, say $\lambda = 2$, and find that the solution is

$$\mathbf{u} = \begin{bmatrix} 3u_2 \\ -u_2 \\ u_2 \end{bmatrix},$$

or upon factoring,

$$\mathbf{u} = u_2 \begin{bmatrix} 3 \\ -1 \\ 1 \end{bmatrix}.$$

Then we would get

$$e^{2t} \begin{bmatrix} 3 \\ -1 \\ 1 \end{bmatrix}, \tag{6}$$

as a solution of $\mathbf{x}' = \mathbf{A}\mathbf{x}$. ◄

EXAMPLE 5.10.4 Solutions of $\mathbf{x}' = \mathbf{A}\mathbf{x}$ from Eigenvalue with Several Eigenvectors

Suppose that we solve (5) for a given eigenvalue, say $\lambda = -3$, and find that the solution of $(\lambda \mathbf{I} - \mathbf{A})\mathbf{u} = \mathbf{0}$ is:

$$\mathbf{u} = \begin{bmatrix} u_2 - u_3 \\ 2u_2 \\ 3u_3 - u_2 \end{bmatrix} = u_2 \begin{bmatrix} 1 \\ 2 \\ -1 \end{bmatrix} + u_3 \begin{bmatrix} -1 \\ 0 \\ 3 \end{bmatrix}.$$

Then

$$e^{-3t} \begin{bmatrix} 1 \\ 2 \\ -1 \end{bmatrix}, \quad e^{-3t} \begin{bmatrix} -1 \\ 0 \\ 3 \end{bmatrix} \tag{7}$$

will both be solutions of $\mathbf{x}' = \mathbf{A}\mathbf{x}$. ◄

Any nonzero multiple of an eigenvector may be used. For example, we could use

$$e^{-3t}\begin{bmatrix} 2 \\ 4 \\ -2 \end{bmatrix} \quad \text{instead of} \quad e^{-3t}\begin{bmatrix} 1 \\ 2 \\ -1 \end{bmatrix} \quad \text{in (7).}$$

It turns out that the solutions of $\mathbf{x}' = \mathbf{A}\mathbf{x}$ generated in this manner are always linearly independent.

■ **THEOREM 5.10.2 Linear Independence of Eigenvalue-Eigenvector Solutions**

If $\mathbf{A}$ is an $n \times n$ matrix and $\mathbf{A}$ has n *distinct* eigenvalues $\lambda_1, \ldots, \lambda_n$, then for each i, the solution of $(\lambda_i \mathbf{I} - \mathbf{A})\mathbf{u} = \mathbf{0}$ will have a solution of the form $c\mathbf{u}_i$, where c is an arbitrary constant. The set of vector functions

$$\{e^{\lambda_1 t}\mathbf{u}_1, e^{\lambda_2 t}\mathbf{u}_2, \ldots, e^{\lambda_n t}\mathbf{u}_n\}$$

is a fundamental set of solutions for $\mathbf{x}' = \mathbf{A}\mathbf{x}$. Here $\mathbf{u}_i$ is an eigenvector corresponding to the eigenvalue λ_i. In this case the general solution of $\mathbf{x}' = \mathbf{A}\mathbf{x}$ is

$$\mathbf{x} = c_1 e^{\lambda_1 t}\mathbf{u}_1 + \cdots + c_n e^{\lambda_n t}\mathbf{u}_n.$$

■

EXAMPLE 5.10.5 Solution of $\mathbf{x}' = \mathbf{A}\mathbf{x}$; Distinct Eigenvalues

Find a fundamental set of solutions and use it to find the general solution

$$\begin{aligned} x' &= x + 2y, \\ y' &= 2x + y. \end{aligned} \tag{8}$$

SOLUTION Let

$$\mathbf{A} = \begin{bmatrix} 1 & 2 \\ 2 & 1 \end{bmatrix}, \quad \mathbf{x} = \begin{bmatrix} x \\ y \end{bmatrix},$$

so that (8) can be written as $\mathbf{x}' = \mathbf{A}\mathbf{x}$. From Example 5.10.2 we know that $\mathbf{A}$ has characteristic polynomial

$$p(\lambda) = \det(\lambda\mathbf{I} - \mathbf{A}) = \lambda^2 - 2\lambda - 3 = (\lambda - 3)(\lambda + 1),$$

so that the eigenvalues are $\lambda_1 = 3$, $\lambda_2 = -1$. We now need to compute the eigenvectors using (5). For $\lambda_1 = 3$, $(\lambda\mathbf{I} - \mathbf{A})\mathbf{u} = \mathbf{0}$ becomes

$$\left(\begin{bmatrix} 3 & 0 \\ 0 & 3 \end{bmatrix} - \begin{bmatrix} 1 & 2 \\ 2 & 1 \end{bmatrix}\right)\mathbf{u} = \mathbf{0}$$

or

$$\begin{bmatrix} 2 & -2 \\ -2 & 2 \end{bmatrix}\begin{bmatrix} u_1 \\ u_2 \end{bmatrix} = \begin{bmatrix} 0 \\ 0 \end{bmatrix}.$$

Performing two row operations on the augmented matrix

$$\left[\begin{array}{rr|r} 2 & -2 & 0 \\ -2 & 2 & 0 \end{array}\right]$$

yields

$$\left[\begin{array}{rr|r} 1 & -1 & 0 \\ 0 & 0 & 0 \end{array}\right].$$

Thus the eigenvectors are u_2 arbitrary, $u_1 - u_2 = 0$, or $u_1 = u_2$, that is,

$$\mathbf{u} = \begin{bmatrix} u_1 \\ u_2 \end{bmatrix} = \begin{bmatrix} u_2 \\ u_2 \end{bmatrix} = u_2 \begin{bmatrix} 1 \\ 1 \end{bmatrix}.$$

We take

$$\begin{bmatrix} 1 \\ 1 \end{bmatrix}$$

as our eigenvector (any nonzero multiple of it would do) corresponding to the eigenvalue $\lambda_1 = 3$ and

$$e^{3t} \begin{bmatrix} 1 \\ 1 \end{bmatrix}$$

is one solution of (8). Similarly for the other eigenvalue $\lambda_2 = -1$, $(\lambda\mathbf{I} - \mathbf{A})\mathbf{u} = \mathbf{0}$ becomes

$$\left(\begin{bmatrix} -1 & 0 \\ 0 & -1 \end{bmatrix} - \begin{bmatrix} 1 & 2 \\ 2 & 1 \end{bmatrix} \right) \mathbf{u} = \mathbf{0}$$

or

$$\begin{bmatrix} -2 & -2 \\ -2 & -2 \end{bmatrix} \mathbf{u} = \mathbf{0}.$$

The solutions are $u_1 = -u_2$, u_2 arbitrary, or

$$\mathbf{u} = \begin{bmatrix} -u_2 \\ u_2 \end{bmatrix} = u_2 \begin{bmatrix} -1 \\ 1 \end{bmatrix}.$$

We may take

$$\begin{bmatrix} -1 \\ 1 \end{bmatrix}$$

as our eigenvector corresponding to the eigenvalue $\lambda_2 = -1$ and

$$e^{-t} \begin{bmatrix} -1 \\ 1 \end{bmatrix}$$

is a second solution of (8). Since $\mathbf{A}$ is 2×2 and we have the two linearly independent solutions of (8),

$$\left\{ e^{3t} \begin{bmatrix} 1 \\ 1 \end{bmatrix},\ e^{-t} \begin{bmatrix} -1 \\ 1 \end{bmatrix} \right\}$$

is a fundamental set of solutions. The general solution of (8) would be

$$\mathbf{x} = c_1 e^{3t}\begin{bmatrix} 1 \\ 1 \end{bmatrix} + c_2 e^{-t}\begin{bmatrix} -1 \\ 1 \end{bmatrix}$$

or

$$\begin{aligned} x &= c_1 e^{3t} - c_2 e^{-t}, \\ y &= c_1 e^{3t} + c_2 e^{-t}. \end{aligned} \tag{9}$$

◀

If some eigenvalue is a repeated root of the characteristic polynomial, then the method we have just described may not find all solutions. How to find the other solutions in this case will be discussed in Section 5.10.3. In Section 5.10.2 we will discuss how to handle complex eigenvalues.

EXAMPLE 5.10.6 **Solution of $\mathbf{x}' = \mathbf{A}\mathbf{x}$; Repeated Eigenvalues**

Solve the system of differential equations

$$\begin{aligned} x_1' &= -2x_1 + x_2 + x_3, \\ x_2' &= x_1 - 2x_2 + x_3, \\ x_3' &= x_1 + x_2 - 2x_3. \end{aligned} \tag{10}$$

SOLUTION Letting

$$\mathbf{x} = \begin{bmatrix} x_1 \\ x_2 \\ x_3 \end{bmatrix}, \qquad \mathbf{A} = \begin{bmatrix} -2 & 1 & 1 \\ 1 & -2 & 1 \\ 1 & 1 & -2 \end{bmatrix},$$

we may rewrite (10) as $\mathbf{x}' = \mathbf{A}\mathbf{x}$. First, we find the characteristic polynomial of $\mathbf{A}$,

$$p(\lambda) = \det(\lambda\mathbf{I} - \mathbf{A}) = \det\begin{bmatrix} \lambda+2 & -1 & -1 \\ -1 & \lambda+2 & -1 \\ -1 & -1 & \lambda+2 \end{bmatrix}.$$

Expanding along the first row, we obtain

$$\begin{aligned} p(\lambda) &= (\lambda+2)(-1)^{1+1}\det\begin{bmatrix} \lambda+2 & -1 \\ -1 & \lambda+2 \end{bmatrix} \\ &\quad +(-1)(-1)^{1+2}\det\begin{bmatrix} -1 & -1 \\ -1 & \lambda+2 \end{bmatrix} \\ &\quad +(-1)(-1)^{1+3}\det\begin{bmatrix} -1 & \lambda+2 \\ -1 & -1 \end{bmatrix} \\ &= (\lambda+2)[(\lambda+2)^2 - 1] + (-\lambda - 2 - 1) - (1 + \lambda + 2) \\ &= \lambda^3 + 6\lambda^2 + 9\lambda = \lambda(\lambda+3)^2. \end{aligned}$$

The roots of $p(\lambda)$ are $\lambda_1 = 0$ and $\lambda_2 = -3$. The **multiplicity** of an eigenvalue is its multiplicity as a root of the characteristic polynomial. In this example, $\lambda_1 = 0$ is an eigenvalue of multiplicity one, whereas λ_2 is an eigenvalue of multiplicity two. We now need to find the corresponding eigenvectors.

For $\lambda_1 = 0$, $(\lambda_1\mathbf{I} - \mathbf{A})\mathbf{u} = \mathbf{0}$ is

$$\begin{bmatrix} 2 & -1 & -1 \\ -1 & 2 & -1 \\ -1 & -1 & 2 \end{bmatrix}\mathbf{u} = \mathbf{0}. \tag{11}$$

To solve (11) using Gaussian elimination, Algorithm 5.7.1, we consider the augmented matrix

$$\left[\begin{array}{rrr|r} 2 & -1 & -1 & 0 \\ -1 & 2 & -1 & 0 \\ -1 & -1 & 2 & 0 \end{array}\right]. \tag{12}$$

Adding $\frac{1}{2}$ times row 1 to rows 2 and 3 yields

$$\left[\begin{array}{rrr|r} 2 & -1 & -1 & 0 \\ 0 & \frac{3}{2} & -\frac{3}{2} & 0 \\ 0 & -\frac{3}{2} & \frac{3}{2} & 0 \end{array}\right].$$

Now add row 2 to row 3 and then multiply row 2 by $\frac{2}{3}$:

$$\left[\begin{array}{rrr|r} 2 & -1 & -1 & 0 \\ 0 & 1 & -1 & 0 \\ 0 & 0 & 0 & 0 \end{array}\right].$$

This completes Part 1 of Algorithm 5.7.1. Now add row 2 to row 1 and multiply row 1 by $\frac{1}{2}$, to get the row echelon form

$$\left[\begin{array}{rrr|r} 1 & 0 & -1 & 0 \\ 0 & 1 & -1 & 0 \\ 0 & 0 & 0 & 0 \end{array}\right].$$

The distinguished columns are 1 and 2. Thus we may take u_3 arbitrary and

$$u_1 - u_3 = 0, \qquad u_2 - u_3 = 0 \qquad \text{gives us} \qquad u_1 = u_3, \qquad u_2 = u_3.$$

Thus the eigenvectors for $\lambda_1 = 0$ are

$$\mathbf{u} = \begin{bmatrix} u_3 \\ u_3 \\ u_3 \end{bmatrix} = u_3\begin{bmatrix} 1 \\ 1 \\ 1 \end{bmatrix}.$$

We choose

$$\begin{bmatrix} 1 \\ 1 \\ 1 \end{bmatrix}$$

as our eigenvector corresponding to $\lambda = 0$, so that

$$\mathbf{x}_1 = e^{0t}\begin{bmatrix}1\\1\\1\end{bmatrix} = \begin{bmatrix}1\\1\\1\end{bmatrix} \tag{13}$$

is one solution of $\mathbf{x}' = \mathbf{A}\mathbf{x}$.

Now we must find the eigenvectors for $\lambda_2 = -3$. The eigenvectors $\mathbf{u}$ satisfy $(\lambda_2\mathbf{I} - \mathbf{A})\mathbf{u} = \mathbf{0}$, or

$$\begin{bmatrix}-1 & -1 & -1\\-1 & -1 & -1\\-1 & -1 & -1\end{bmatrix}\mathbf{u} = \mathbf{0}. \tag{14}$$

The augmented matrix for the system of equations (14) is

$$\left[\begin{array}{ccc|c}-1 & -1 & -1 & 0\\-1 & -1 & -1 & 0\\-1 & -1 & -1 & 0\end{array}\right],$$

which, after performing two row operations, is

$$\left[\begin{array}{ccc|c}1 & 1 & 1 & 0\\0 & 0 & 0 & 0\\0 & 0 & 0 & 0\end{array}\right].$$

Only the first column is distinguished. We take u_2, u_3 arbitrary and $u_1 + u_2 + u_3 = 0$ gives $u_1 = -u_2 - u_3$. Thus,

$$\mathbf{u} = \begin{bmatrix}u_1\\u_2\\u_3\end{bmatrix} = \begin{bmatrix}-u_2 - u_3\\u_2\\u_3\end{bmatrix} = u_2\begin{bmatrix}-1\\1\\0\end{bmatrix} + u_3\begin{bmatrix}-1\\0\\1\end{bmatrix}.$$

Corresponding to the eigenvalue $\lambda_2 = -3$, we take the two linearly independent eigenvectors

$$\begin{bmatrix}-1\\1\\0\end{bmatrix}, \quad \begin{bmatrix}-1\\0\\1\end{bmatrix},$$

which give us the solutions

$$e^{-3t}\begin{bmatrix}-1\\1\\0\end{bmatrix}, \quad e^{-3t}\begin{bmatrix}-1\\0\\1\end{bmatrix}. \tag{15}$$

The matrix $\mathbf{A}$ is 3×3, and (13) and (15) provide us with three linearly independent solutions, so that

$$\mathbf{x} = \begin{bmatrix}x_1\\x_2\\x_3\end{bmatrix} = c_1\begin{bmatrix}1\\1\\1\end{bmatrix} + c_2e^{-3t}\begin{bmatrix}-1\\1\\0\end{bmatrix} + c_3e^{-3t}\begin{bmatrix}-1\\0\\1\end{bmatrix},$$

or, equivalently,

$$x_1 = c_1 - c_2 e^{-3t} - c_3 e^{-3t},$$
$$x_2 = c_1 + c_2 e^{-3t},$$
$$x_3 = c_1 + c_3 e^{-3t},$$

is the general solution of (10). ◄

Helpful Check If $\mathbf{A}$ is $n \times n$, then the sum of the diagonal entries is the **trace** of $\mathbf{A}$, denoted: trace($\mathbf{A}$). The trace is an easy-to-compute number that gives us two checks on our work.

$-$trace($\mathbf{A}$) is the coefficient of λ^{n-1} in the characteristic polynomial of $\mathbf{A}$; (16)

trace($\mathbf{A}$) is the sum of all the eigenvalues of $\mathbf{A}$ if they are repeated according to their multiplicities. (17)

EXAMPLE 5.10.7 **Trace**

If $\mathbf{A}$ is the matrix in Example 5.10.6,

$$\operatorname{trace}(\mathbf{A}) = -2 + -2 + -2 = \boxed{-6}.$$

The characteristic polynomial was

$$p(\lambda) = \lambda^3 + \boxed{6}\lambda^2 + 9\lambda.$$

The eigenvalues, repeated according to multiplicity, were $0, -3, -3$, and $0 + -3 + -3 = \boxed{-6}$. ◄

The discussion and examples of this section may be summarized as follows.

Summary of Method for Solving $\mathbf{x}' = \mathbf{A}\mathbf{x}$ for a Constant Matrix $\mathbf{A}$

1. Compute the characteristic polynomial of $\mathbf{A}$, $p(\lambda) = \det(\lambda\mathbf{I} - \mathbf{A})$. Use the trace as a check.
2. Find the roots of $p(\lambda)$; these are the eigenvalues $\lambda_1, \ldots, \lambda_m$ of $\mathbf{A}$. Check if trace($\mathbf{A}$) $= \lambda_1 + \cdots + \lambda_m$.

3. If an eigenvalue is complex, see Section 5.10.2.
4. For each eigenvalue λ_i, solve the system $(\lambda_i\mathbf{I} - \mathbf{A})\mathbf{u} = \mathbf{0}$ to get a linearly independent set of eigenvectors.
5. If the number of linear independent eigenvectors for any eigenvalue is less than the multiplicity of that eigenvalue, see Section 5.10.3.
6. If (3) and (5) have not occurred, we now have a fundamental set of solutions of the form $e^{\lambda t}\mathbf{u}$.

We close this section with an example that illustrates what can sometimes happen with eigenvalues whose multiplicities are greater than one. A complete solution to this problem will be given in Section 5.10.3.

EXAMPLE 5.10.8 Deficiency of Eigenvectors

Find all the solutions of

$$\begin{aligned} x_1' &= 3x_1 - x_2, \\ x_2' &= x_1 + x_2, \end{aligned} \tag{18}$$

that are possible, using the eigenvalue/eigenvector method of this section.

SOLUTION Let

$$\mathbf{x} = \begin{bmatrix} x_1 \\ x_2 \end{bmatrix}, \qquad \mathbf{A} = \begin{bmatrix} 3 & -1 \\ 1 & 1 \end{bmatrix},$$

so that (18) is $\mathbf{x}' = \mathbf{A}\mathbf{x}$. First we compute the characteristic polynomial of $\mathbf{A}$,

$$p(\lambda) = \det(\lambda\mathbf{I} - \mathbf{A}) = \det\begin{pmatrix} \lambda - 3 & 1 \\ -1 & \lambda - 1 \end{pmatrix} = \lambda^2 - 4\lambda + 4.$$

Since $p(\lambda) = (\lambda - 2)^2$, there is a single eigenvalue $\lambda_1 = 2$ of multiplicity 2. To find the eigenvectors associated with $\lambda_1 = 2$, we solve $(\lambda_1\mathbf{I} - \mathbf{A})\mathbf{u} = \mathbf{0}$ or

$$\begin{bmatrix} -1 & 1 \\ -1 & 1 \end{bmatrix}\mathbf{u} = \mathbf{0},$$

which is $-u_1 + u_2 = 0$. Thus we may take u_2 arbitrary, $u_1 = u_2$, and

$$\mathbf{u} = \begin{bmatrix} u_1 \\ u_2 \end{bmatrix} = \begin{bmatrix} u_2 \\ u_2 \end{bmatrix} = u_2\begin{bmatrix} 1 \\ 1 \end{bmatrix}.$$

In this case we have only one linearly independent eigenvector $\begin{bmatrix} 1 \\ 1 \end{bmatrix}$ corresponding to $\lambda = 2$. $\lambda = 2$ is an eigenvalue with multiplicity 2, but in this case there is only 1 eigenvector. We say the eigenvalue $\lambda = 2$ is **deficient**. We have then

$$e^{2t}\begin{bmatrix} 1 \\ 1 \end{bmatrix}$$

is one solution. However, this method does not provide a second linearly independent solution. The reason, as will be shown in Section 5.10.3, is that the other solutions are not of the form $e^{\lambda t}\mathbf{u}$. ◄

Symmetric Matrices

A square matrix is **symmetric** if the i,j entry equals the j,i entry. For example,

$$\begin{bmatrix} 1 & 2 \\ 2 & 3 \end{bmatrix}, \quad \begin{bmatrix} 0 & 1 & 0 \\ 1 & 2 & 3 \\ 0 & 3 & 4 \end{bmatrix}, \quad \begin{bmatrix} 1 & 2 & 3 \\ 2 & 4 & 5 \\ 3 & 5 & 6 \end{bmatrix}$$

are all symmetric matrices. The $\mathbf{A}$ of Example 5.10.5 is also symmetric. Symmetric matrices arise in many applications (see Exercises 5.5.15 through 5.5.17).

■ **THEOREM 5.10.3 Symmetric Matrix**

Suppose $\mathbf{A}$ is an $n \times n$ real symmetric matrix. Then

i. All eigenvalues of $\mathbf{A}$ are real.

ii. The number of linearly independent eigenvectors for each eigenvalue is the same as the multiplicity of the eigenvalue.

Thus all solutions of $\mathbf{x}' = \mathbf{A}\mathbf{x}$ can be found by the method of this section if $\mathbf{A}$ is symmetric. ■

Note that

$$\mathbf{A} = \begin{bmatrix} 3 & -1 \\ 1 & 1 \end{bmatrix}$$

of Example 5.10.8 is not symmetric, since the 1,2 entry is -1 whereas the 2,1 entry is 1.

Exercises

In Exercises 1 through 13, the method of Section 5.10.1 will provide a fundamental set of solutions and all eigenvalues have multiplicity one. Eigenvalues with multiplicity greater than one appear in Exercises 14 through 17.

In Exercises 1 through 17, solve the differential equation using the eigenvalue/eigenvector method. If no initial conditions are given, give the general solution. When checking your answers with those in the back of the book, keep in mind that, in the general solution, any nonzero multiple of the given eigenvectors may be used.

1. $x_1' = 3x_1 + x_2,$
$x_2' = -2x_1$

2. $x_1' = x_1,$
$x_2' = x_1 - x_2$

3. $x' = x - 2y, \quad x(0) = 1,$
$y' = -2x + 4y, \quad y(0) = 2$

4. $x' = 3x + y,$
$y' = -6x - 2y$

5. $x_1' = 6x_1 + 2x_2,$
$x_2' = 2x_1 + 3x_2$

6. $x_1' = 3x_1 + x_2, \quad x_1(0) = 1,$
$x_2' = -5x_1 - 3x_2, \quad x_2(0) = -2$

7. $v' = 2v - 2w,$
$w' = -2v + 5w$

8. $v' = -3v + 3w, \quad v(0) = 0,$
$w' = 3v + 5w, \quad w(0) = 2$

9. $x_1' = 2x_1 + x_3,$
$x_2' = 2x_2 + x_3,$
$x_3' = x_1 + x_2 + x_3$

10. $x' = x - 2y + z, \quad x(0) = 1,$
$y' = -2x + 4y + z, \quad y(0) = -1,$
$z' = 3z, \quad z(0) = 2$

11. $x_1' = x_1 + x_3,$
$x_2' = 3x_2,$
$x_3' = x_1 + x_3$

12. $u' = u + w, \quad u(0) = 0,$
$v' = v - w, \quad v(0) = 0,$
$w' = u - v, \quad w(0) = 2$

13. $x_1' = 2x_1 - 2x_2 - x_3,$
$x_2' = -2x_1 + 2x_2 + x_3,$
$x_3' = -x_1 + x_2 + 5x_3$

14. $x_1' = -2x_1 - x_2 + x_3,$
$x_2' = -x_1 - 2x_2 - x_3,$
$x_3' = x_1 - x_2 - 2x_3$

15. $x' = 2x + z,$
$y' = y,$
$z' = x + 2z$

16. $x_1' = x_1 + x_2 + 2x_3,$
$x_2' = x_1 + x_2 + 2x_3,$
$x_3' = 2x_1 + 2x_2 + 4x_3$

17. $x_1' = x_2 + x_4,$
$x_2' = x_1 - x_3,$
$x_3' = x_4,$
$x_4' = x_3$

18. Theorem 5.8.1 can sometimes be used on matrices with variable coefficients. In Example 5.10.6,

$$\lambda\mathbf{I} - \mathbf{A} = \begin{bmatrix} \lambda + 2 & -1 & -1 \\ -1 & \lambda + 2 & -1 \\ -1 & -1 & \lambda + 2 \end{bmatrix}.$$

Using Theorem 5.8.1, evaluate $\det(\lambda\mathbf{I} - \mathbf{A})$ by using the 1,3 entry to zero out the 2,3 and 3,3 entries. Then expand the determinant down the third column.

19. Evaluate

$$\det \begin{bmatrix} \lambda & 2 & 1 \\ 1 & \lambda - 1 & 1 \\ 1 & 2 & \lambda \end{bmatrix},$$

using Theorem 5.8.1 by using the 3,1 entry to zero out the 3,2 and 3,3 entries and then expanding the determinant along the bottom row.

20. Rewrite the differential equation $y'' + by' + cy = 0$, where b, c are constants, as a system $\mathbf{x}' = \mathbf{A}\mathbf{x}$ using

$$\mathbf{x} = \begin{bmatrix} y \\ y' \end{bmatrix}.$$

Show that the characteristic equation of $y'' + by' + cy = 0$, as defined in Chapter 2, is the same as the characteristic equation of $\mathbf{x}' = \mathbf{A}\mathbf{x}$ defined in Section 5.10.1.

In Exercises 21 through 30, you are given a matrix $\mathbf{A}$ and an eigenvalue λ. Find as many linearly independent solutions of the form $e^{\lambda t}\mathbf{u}$ as possible for $\mathbf{x}' = \mathbf{A}\mathbf{x}$.

21. $\mathbf{A} = \begin{bmatrix} 1 & 0 & 1 & 2 \\ 0 & 1 & 1 & 2 \\ 0 & 0 & 2 & 3 \\ 0 & 0 & 1 & 3 \end{bmatrix}, \quad \lambda = 1$

22. $\mathbf{A} = \begin{bmatrix} 1 & 0 & 1 & 2 \\ 0 & 1 & 1 & 2 \\ 0 & 0 & 2 & 3 \\ 0 & 0 & 1 & 4 \end{bmatrix}, \quad \lambda = 1$

23. $\mathbf{A} = \begin{bmatrix} 3 & 1 & 1 & 1 \\ 1 & 3 & 1 & 1 \\ 0 & 0 & 3 & 1 \\ 0 & 0 & 1 & 3 \end{bmatrix}, \quad \lambda = 2$

24. $\mathbf{A} = \begin{bmatrix} 3 & 2 & 0 & 0 \\ 2 & 3 & 0 & 0 \\ 0 & 0 & 3 & 2 \\ 0 & 0 & 2 & 3 \end{bmatrix}, \quad \lambda = 1$

25. $\mathbf{A} = \begin{bmatrix} 2 & 1 & 1 & 1 & 1 \\ 0 & 3 & 1 & 1 & 1 \\ 0 & 0 & 2 & 0 & 0 \\ 0 & 0 & 0 & 2 & 0 \\ 0 & 0 & 0 & 0 & 2 \end{bmatrix}, \quad \lambda = 2$

26. $\mathbf{A} = \begin{bmatrix} 3 & 1 & 1 & 1 & 1 \\ 0 & 4 & 1 & 1 & 1 \\ 0 & 0 & 3 & 1 & 1 \\ 0 & 0 & 2 & 4 & 1 \\ 0 & 0 & 0 & 0 & 3 \end{bmatrix}, \quad \lambda = 3$

27. $\mathbf{A} = \begin{bmatrix} 5 & 1 & 1 & 0 \\ 1 & 5 & 1 & 0 \\ 1 & 1 & 5 & 0 \\ 0 & 0 & 0 & 4 \end{bmatrix}, \quad \lambda = 4$

28. $\mathbf{A} = \begin{bmatrix} 1 & 1 & 2 & 3 \\ 1 & 1 & 2 & 3 \\ 0 & 0 & 1 & 1 \\ 0 & 0 & 1 & 1 \end{bmatrix}, \quad \lambda = 0$

29. $\mathbf{A} = \begin{bmatrix} 1 & 1 & 1 & 1 \\ 0 & 1 & 1 & 1 \\ 0 & 0 & 1 & 1 \\ 0 & 0 & 0 & 0 \end{bmatrix}, \quad \lambda = 0$

30. $\mathbf{A} = \begin{bmatrix} 6 & 0 & 1 & 1 \\ 0 & 6 & 2 & 3 \\ 0 & 0 & 6 & 0 \\ 0 & 0 & 0 & 6 \end{bmatrix}, \quad \lambda = 6$

5.10.2 Complex Eigenvalues

As we saw in the applications of linear, constant coefficient differential equations in Sections 2.11 and 2.12, solutions of the form $e^{\alpha t} \cos \beta t$, $e^{\alpha t} \sin \beta t$ are important physically; they occur when the characteristic equation has complex roots. The same is true for systems, as this section will show. (*Note*: Exercise 20 of Section 5.10.1 shows that both types of characteristic equations are actually the same.)

Assume, then, that $\mathbf{A}$ is an $n \times n$ **real matrix** (has real entries) and λ is a complex eigenvalue. Write λ as $\lambda = \alpha + i\beta$, where α, β are real numbers. Corresponding to λ there will be an eigenvector $\mathbf{u}$, so that $\mathbf{Au} = \lambda\mathbf{u}$. We may also break $\mathbf{u}$ into real and imaginary parts, $\mathbf{u} = \mathbf{a} + i\mathbf{b}$. For example,

$$\begin{bmatrix} 1 - 3i \\ 2 + i \end{bmatrix} = \begin{bmatrix} 1 \\ 2 \end{bmatrix} + i\begin{bmatrix} -3 \\ 1 \end{bmatrix}.$$

The key fact we need is the following:

> If $\mathbf{A}$ is an $n \times n$ real matrix with complex eigenvalue λ, and eigenvector $\mathbf{u}$,
>
> $$\lambda = \alpha + i\beta, \qquad \mathbf{u} = \mathbf{a} + i\mathbf{b}, \tag{19}$$
>
> then $\bar{\lambda}$ is another eigenvalue and $\bar{\mathbf{u}}$ its associated eigenvector, where
>
> $$\bar{\lambda} = \alpha - i\beta, \qquad \bar{\mathbf{u}} = \mathbf{a} - i\mathbf{b}.$$

(Recall that $\bar{\lambda}$ is called the **complex conjugate** of λ.) That $\bar{\lambda}$ is also an eigenvalue follows from the fact that $p(\lambda) = \det(\lambda\mathbf{I} - \mathbf{A})$ has real coefficients if $\mathbf{A}$ is real, and thus the complex roots of the characteristic polynomial will occur in conjugate pairs. On the other hand, if $\mathbf{Au} = \lambda\mathbf{u}$, then taking conjugates of both sides gives

$$\overline{\mathbf{Au}} = \overline{\lambda\mathbf{u}}$$

or

$$\mathbf{A}\bar{\mathbf{u}} = \bar{\lambda}\bar{\mathbf{u}},$$

since $\mathbf{A}$ is real, so that $\bar{\mathbf{u}}$ is an eigenvector for the eigenvalue $\bar{\lambda}$.

Fact (19) is very helpful since it means that we actually have to compute only the eigenvectors for *one* of the eigenvalues in a conjugate pair.

EXAMPLE 5.10.9 Complex Eigenvalues and Eigenvectors

Find the eigenvalues and associated eigenvectors for

$$\mathbf{A} = \begin{bmatrix} 1 & -1 \\ 1 & 1 \end{bmatrix}.$$

Solution The characteristic polynomial is

$$p(\lambda) = \det(\lambda\mathbf{I} - \mathbf{A}) = \det\begin{bmatrix} \lambda - 1 & 1 \\ -1 & \lambda - 1 \end{bmatrix} = \lambda^2 - 2\lambda + 2.$$

The roots of the characteristic polynomial can be found using the quadratic formula

$$\lambda = \frac{2 \pm \sqrt{4 - 8}}{2} = 1 \pm i.$$

The eigenvalues are thus $\lambda_1 = 1 + i$, $\lambda_2 = 1 - i$. First, we find the eigenvectors for λ_1. Solving $(\lambda_1\mathbf{I} - \mathbf{A})\mathbf{u} = \mathbf{0}$ gives

$$\begin{bmatrix} i & 1 \\ -1 & i \end{bmatrix}\mathbf{u} = \mathbf{0}.$$

The augmented matrix is

$$\left[\begin{array}{cc|c} i & 1 & 0 \\ -1 & i & 0 \end{array}\right].$$

There are several ways to proceed. Exchange rows and multiply the first row by -1,

$$\left[\begin{array}{cc|c} 1 & -i & 0 \\ i & 1 & 0 \end{array}\right].$$

Now add $-i$ times the first row to the second row:

$$\left[\begin{array}{cc|c} 1 & -i & 0 \\ 0 & 0 & 0 \end{array}\right].$$

Thus $u_1 - iu_2 = 0$ or $u_1 = iu_2$, u_2 arbitrary, and

$$\mathbf{u} = \begin{bmatrix} iu_2 \\ u_2 \end{bmatrix} = u_2\begin{bmatrix} i \\ 1 \end{bmatrix}.$$

We may take

$$\mathbf{u}_1 = \begin{bmatrix} i \\ 1 \end{bmatrix}$$

as the eigenvector associated with $\lambda_1 = 1 + i$. Then

$$\mathbf{u}_2 = \bar{\mathbf{u}}_1 = \overline{\begin{bmatrix} i \\ 1 \end{bmatrix}} = \begin{bmatrix} \bar{i} \\ \bar{1} \end{bmatrix} = \begin{bmatrix} -i \\ 1 \end{bmatrix}$$

will be the eigenvector associated with $\lambda_2 = 1 - i$. ◄

We still have $e^{\lambda t}\mathbf{u}$, $e^{\bar{\lambda} t}\bar{\mathbf{u}}$ as solutions but, as in Chapter 2, we wish to replace them by real solutions. Now, if $\lambda = \alpha + i\beta$, $\mathbf{u} = \mathbf{a} + i\mathbf{b}$, then

$$\begin{aligned} e^{\lambda t}\mathbf{u} &= e^{(\alpha+i\beta)t}(\mathbf{a} + i\,\mathbf{b}) = e^{\alpha t}(\cos\beta t + i\sin\beta t)(\mathbf{a} + i\,\mathbf{b}) \\ &= e^{\alpha t}(\cos\beta t\,\mathbf{a} - \sin\beta t\,\mathbf{b}) + e^{\alpha t}i(\cos\beta t\,\mathbf{b} + \sin\beta t\,\mathbf{a}), \end{aligned}$$

while

$$\begin{aligned} e^{\bar{\lambda} t}\bar{\mathbf{u}} &= e^{(\alpha-i\beta)t}(\mathbf{a} - i\,\mathbf{b}) = e^{\alpha t}(\cos\beta t - i\sin\beta t)(\mathbf{a} - i\,\mathbf{b}) \\ &= e^{\alpha t}(\cos\beta t\,\mathbf{a} - \sin\beta t\,\mathbf{b}) - e^{\alpha t}i(\cos\beta t\,\mathbf{b} + \sin\beta t\,\mathbf{a}). \end{aligned}$$

Since $e^{\lambda t}\mathbf{u}$, $e^{\bar{\lambda} t}\bar{\mathbf{u}}$ are solutions of $\mathbf{x}' = \mathbf{A}\mathbf{x}$, $c_1e^{\lambda t}\mathbf{u} + c_2e^{\bar{\lambda} t}\bar{\mathbf{u}}$ will also be a solution for any scalars c_1, c_2. Choosing

$$c_1 = c_2 = \frac{1}{2}$$

yields a real solution

$$e^{\alpha t}[\cos\beta t\,\mathbf{a} - \sin\beta t\,\mathbf{b}],$$

while

$$c_1 = \frac{1}{2i}, \qquad c_2 = -\frac{1}{2i}$$

yields another real solution

$$e^{\alpha t}[\cos\beta t\,\mathbf{b} + \sin\beta t\,\mathbf{a}].$$

In summary,

■ **THEOREM 5.10.4 Real Solutions for Complex Eigenvalues**

If $\mathbf{A}$ is an $n \times n$ real matrix with complex eigenvalue $\lambda = \alpha + i\beta$ and associated eigenvector $\mathbf{u} = \mathbf{a} + i\,\mathbf{b}$, then $\bar{\lambda} = \alpha - i\beta$ is also an eigenvalue with associated eigenvector $\mathbf{a} - i\,\mathbf{b}$. Furthermore,

$$\mathbf{x}_1(t) = e^{\alpha t}[\cos\beta t\mathbf{a} - \sin\beta t\,\mathbf{b}] \tag{20}$$

and

$$\mathbf{x}_2(t) = e^{\alpha t}[\cos \beta t \mathbf{b} + \sin \beta t \, \mathbf{a}] \tag{21}$$

are two linearly independent real solutions of $\mathbf{x}' = \mathbf{A}\mathbf{x}$. ■

Note that $\mathbf{x}_1$ and $\mathbf{x}_2$ are in the plane determined by $\mathbf{a}$, $\mathbf{b}$ for all t. For $\alpha = 0$, (20) and (21) trace out ovals as in Figure 5.10.1. For $\alpha < 0$, (20) and (21) are inward spirals as in Figure 5.10.2. For $\alpha > 0$, we have an outward spiral.

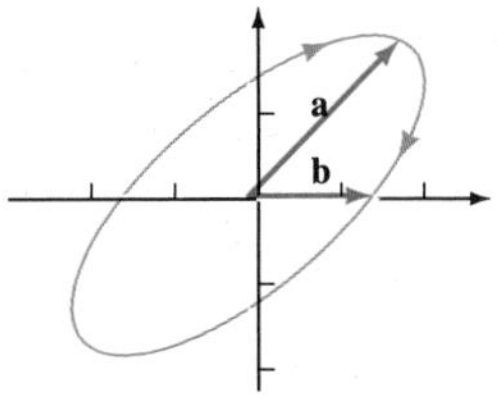

FIGURE 5.10.1

Graph of $\cos \beta t \, \mathbf{a} + \sin \beta t \, \mathbf{b}$.

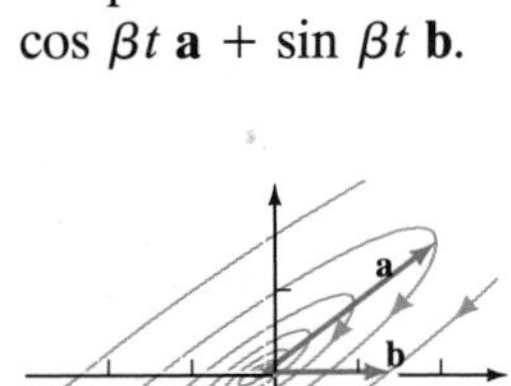

FIGURE 5.10.2

Graph of $e^{\alpha t}(\cos \beta t \, \mathbf{a} + \sin \beta t \, \mathbf{b})$ for $\alpha < 0$.

EXAMPLE 5.10.10 Solution of $\mathbf{x}' = \mathbf{A}\mathbf{x}$ with Complex Eigenvalues and Eigenvectors

Find a fundamental set of solutions of

$$\begin{aligned} x_1' &= x_1 - x_2, \\ x_2' &= x_1 + x_2. \end{aligned} \tag{22}$$

SOLUTION Let

$$\mathbf{x} = \begin{bmatrix} x_1 \\ x_2 \end{bmatrix}, \qquad \mathbf{A} = \begin{bmatrix} 1 & -1 \\ 1 & 1 \end{bmatrix},$$

so that (22) is $\mathbf{x}' = \mathbf{A}\mathbf{x}$. This $\mathbf{A}$ is the same as the $\mathbf{A}$ of Example 5.10.9, so that we have $\lambda = \alpha + \beta i = 1 + i$ is an eigenvalue and the associated eigenvector can be taken to be

$$\mathbf{u} = \mathbf{a} + i\,\mathbf{b} = \begin{bmatrix} i \\ 1 \end{bmatrix} = \begin{bmatrix} 0 \\ 1 \end{bmatrix} + i\begin{bmatrix} 1 \\ 0 \end{bmatrix}.$$

Thus,

$$\alpha = 1, \qquad \beta = 1, \qquad \mathbf{a} = \begin{bmatrix} 0 \\ 1 \end{bmatrix}, \qquad \mathbf{b} = \begin{bmatrix} 1 \\ 0 \end{bmatrix}.$$

The formulas (20) and (21) become

$$\mathbf{x}_1(t) = e^t\left[\cos t\begin{bmatrix} 0 \\ 1 \end{bmatrix} - \sin t\begin{bmatrix} 1 \\ 0 \end{bmatrix}\right] = \begin{bmatrix} -e^t \sin t \\ e^t \cos t \end{bmatrix},$$

$$\mathbf{x}_2(t) = e^t\left[\cos t\begin{bmatrix} 1 \\ 0 \end{bmatrix} + \sin t\begin{bmatrix} 0 \\ 1 \end{bmatrix}\right] = \begin{bmatrix} e^t \cos t \\ e^t \sin t \end{bmatrix},$$

and $\{\mathbf{x}_1(t), \mathbf{x}_2(t)\}$ is a fundamental set of solutions of (22). The general solution of (22) is $c_1\mathbf{x}_1 + c_2\mathbf{x}_2$ or

$$x_1(t) = -c_1 e^t \sin t + c_2 e^t \cos t,$$

$$x_2(t) = c_1 e^t \cos t + c_2 e^t \sin t.$$ ◄

Review Fact In calculating with complex numbers, the following algebraic fact is helpful in eliminating complex numbers in the denominator of a fraction. The key is to multiply the numerator and denominator by the conjugate of the denominator:

$$\frac{1}{\alpha + i\beta} = \frac{1}{\alpha + i\beta}\frac{\alpha - i\beta}{\alpha - i\beta} = \frac{\alpha - i\beta}{\alpha^2 + \beta^2} = \frac{\alpha}{\alpha^2 + \beta^2} - i\left(\frac{\beta}{\alpha^2 + \beta^2}\right). \tag{23}$$

EXAMPLE 5.10.11 Division of Complex Numbers

$$\begin{aligned}\frac{1 - i}{2 - i} &= \frac{1 - i}{2 - i}\frac{2 + i}{2 + i} = \frac{(1 - i)(2 + i)}{5} \\ &= \frac{2 - 2i + i - i^2}{5} = \frac{3 - i}{5} = \frac{3}{5} - \frac{i}{5}.\end{aligned}$$

In general, of course, we expect to see both real and complex eigenvalues.

EXAMPLE 5.10.12 Solution of $\mathbf{x}' = \mathbf{Ax}$ with Complex Eigenvalues and Eigenvectors

Find the general solution of

$$\begin{aligned} x_1' &= -x_2 + 2x_3, \\ x_2' &= -x_1, \\ x_3' &= -x_1 + x_2 - x_3. \end{aligned} \tag{24}$$

Solution Let

$$\mathbf{x} = \begin{bmatrix} x_1 \\ x_2 \\ x_3 \end{bmatrix}, \qquad \mathbf{A} = \begin{bmatrix} 0 & -1 & 2 \\ -1 & 0 & 0 \\ -1 & 1 & -1 \end{bmatrix},$$

so that (24) is $\mathbf{x}' = \mathbf{Ax}$. First we need the characteristic polynomial in order to find the eigenvalues:

$$p(\lambda) = \det(\lambda\mathbf{I} - \mathbf{A}) = \det\begin{bmatrix} \lambda & 1 & -2 \\ 1 & \lambda & 0 \\ 1 & -1 & \lambda + 1 \end{bmatrix},$$

(we continue by expanding down the third column)

$$\begin{aligned} &= -2(-1)^{3+1}\det\begin{bmatrix} 1 & \lambda \\ 1 & -1 \end{bmatrix} + (\lambda + 1)(-1)^{3+3}\det\begin{bmatrix} \lambda & 1 \\ 1 & \lambda \end{bmatrix} \\ &= -2(-1 - \lambda) + (\lambda + 1)(\lambda^2 - 1) \\ &= (\lambda + 1)(2 + \lambda^2 - 1) = (\lambda + 1)(\lambda^2 + 1). \end{aligned}$$

The roots are $\lambda_1 = -1$, $\lambda_2 = i$, $\lambda_3 = -i$. Now we need to find the eigenvectors. For $\lambda_1 = -1$, $(\lambda_1\mathbf{I} - \mathbf{A})\mathbf{u} = \mathbf{0}$ is

$$\begin{bmatrix} -1 & 1 & -2 \\ 1 & -1 & 0 \\ 1 & -1 & 0 \end{bmatrix}\mathbf{u} = \mathbf{0}$$

or, equivalently,

$$\left[\begin{array}{ccc|c} -1 & 1 & -2 & 0 \\ 1 & -1 & 0 & 0 \\ 1 & -1 & 0 & 0 \end{array}\right].$$

Exchange row 1 and row 2; then add the new row 1 to row 2 and -1 times row 1 to row 3, obtaining

$$\left[\begin{array}{ccc|c} 1 & -1 & 0 & 0 \\ 0 & 0 & -2 & 0 \\ 0 & 0 & 0 & 0 \end{array}\right].$$

That is, $u_1 - u_2 = 0$, $-2u_3 = 0$. Thus, $u_3 = 0$, u_2 is arbitrary, and $u_1 = u_2$, so that

$$\mathbf{u} = \begin{bmatrix} u_2 \\ u_2 \\ 0 \end{bmatrix} = u_2\begin{bmatrix} 1 \\ 1 \\ 0 \end{bmatrix},$$

and

$$\begin{bmatrix} 1 \\ 1 \\ 0 \end{bmatrix} \text{ is an eigenvector for } \lambda_1 = -1. \tag{25}$$

For $\lambda_2 = i$, $(\lambda_2\mathbf{I} - \mathbf{A})\mathbf{u} = \mathbf{0}$ is

$$\begin{bmatrix} i & 1 & -2 \\ 1 & i & 0 \\ 1 & -1 & i+1 \end{bmatrix}\mathbf{u} = \mathbf{0}.$$

We write down the augmented matrix, and, in order to reduce the complex arithmetic, exchange rows 1 and 2:

$$\left[\begin{array}{ccc|c} 1 & i & 0 & 0 \\ i & 1 & -2 & 0 \\ 1 & -1 & i+1 & 0 \end{array}\right].$$

Now add $-i$ times row 1 to row 2; then add -1 times row 1 to row 3:

$$\left[\begin{array}{ccc|c} 1 & i & 0 & 0 \\ 0 & 2 & -2 & 0 \\ 0 & -1-i & i+1 & 0 \end{array}\right].$$

Now multiply row 2 by $\frac{1}{2}$; then add $(1 + i)$ times row 2 to row 3:

$$\left[\begin{array}{ccc|c} 1 & i & 0 & 0 \\ 0 & 1 & -1 & 0 \\ 0 & 0 & 0 & 0 \end{array}\right].$$

Now add $-i$ times row 2 to row 1 to obtain the row echelon form:

$$\left[\begin{array}{ccc|c} 1 & 0 & i & 0 \\ 0 & 1 & -1 & 0 \\ 0 & 0 & 0 & 0 \end{array}\right]. \tag{26}$$

The solution of (26) is u_3 arbitrary, $u_2 = u_3$, $u_1 = -iu_3$. Thus,

$$\mathbf{u} = \begin{bmatrix} -iu_3 \\ u_3 \\ u_3 \end{bmatrix} = u_3 \begin{bmatrix} -i \\ 1 \\ 1 \end{bmatrix},$$

and we take

$$\begin{bmatrix} -i \\ 1 \\ 1 \end{bmatrix}$$

as the eigenvector associated with $\lambda_2 = i$. By Theorem 5.10.4, the eigenvector associated with $\lambda_3 = -i$ can be taken as

$$\begin{bmatrix} i \\ 1 \\ 1 \end{bmatrix}.$$

Thus, $\lambda = \alpha + i\beta = i$, so that $\alpha = 0$, $\beta = 1$, and

$$\begin{bmatrix} -i \\ 1 \\ 1 \end{bmatrix} = \begin{bmatrix} 0 \\ 1 \\ 1 \end{bmatrix} + i \begin{bmatrix} -1 \\ 0 \\ 0 \end{bmatrix} = \mathbf{a} + i\mathbf{b},$$

so that

$$\mathbf{a} = \begin{bmatrix} 0 \\ 1 \\ 1 \end{bmatrix}, \qquad \mathbf{b} = \begin{bmatrix} -1 \\ 0 \\ 0 \end{bmatrix}.$$

From (20), (21), and (25) we get the real solutions

$$e^{-t}\begin{bmatrix} 1 \\ 1 \\ 0 \end{bmatrix}, \quad \begin{bmatrix} \sin t \\ \cos t \\ \cos t \end{bmatrix}, \quad \begin{bmatrix} -\cos t \\ \sin t \\ \sin t \end{bmatrix},$$

which form a fundamental set of solutions for (24). The general solution is

$$\mathbf{x} = c_1e^{-t}\begin{bmatrix}1\\1\\0\end{bmatrix} + c_2\begin{bmatrix}\sin t\\\cos t\\\cos t\end{bmatrix} + c_3\begin{bmatrix}-\cos t\\\sin t\\\sin t\end{bmatrix}.$$

Several of these solutions for $c_1 = 10$, $0 \le c_2 \le 1$, $c_3 = 0$ are plotted in Figure 5.10.3. The component with c_1 goes to zero and the solution approaches an oval in the **a**, **b** plane. ◀

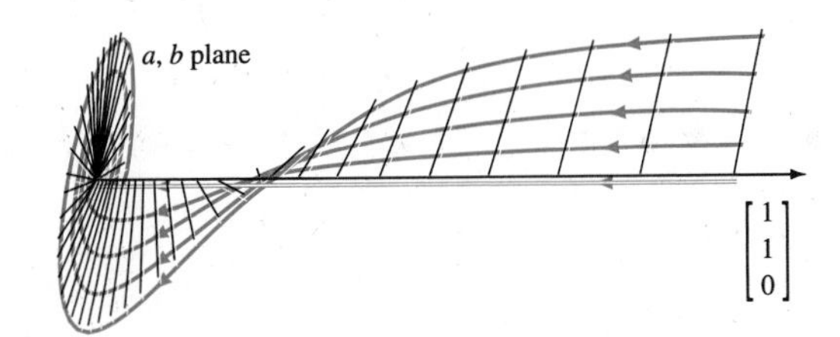

FIGURE 5.10.3

Several solutions of Example 5.10.12.

Exercises

In Exercises 31 through 38, solve the system of differential equations by the eigenvalue-eigenvector method. If no initial conditions are given, find the general solution.

31. $x_1' = -4x_2,$
$x_2' = x_1$

32. $x_1' = 3x_2,$
$x_2' = -3x_1$

33. $x_1' = -x_1 - x_2, \quad x_1(0) = 1,$
$x_2' = x_1 - x_2, \quad x_2(0) = -1$

34. $x_1' = 2x_1 - x_2,$
$x_2' = 2x_1$

35. $x_1' = 3x_1 + 5x_2,$
$x_2' = -x_1 + x_2$

36. $x_1' = 3x_1 - x_2, \quad x_1(0) = 0,$
$x_2' = 5x_1 - x_2, \quad x_2(0) = 3$

37. $x_1' = 5x_1 + 2x_2,$
$x_2' = -4x_1 + x_2$

38. $x_1' = x_1 - x_2,$
$x_2' = 5x_1 - x_2$

39. Recall that if $\mathbf{A}$ is 2×2, then trace($\mathbf{A}$) = $a_{11} + a_{22}$ is the *trace* of $\mathbf{A}$. Suppose $\mathbf{A}$ has a complex conjugate pair of eigenvalues and $\mathbf{A}$ is a real 2×2 matrix. Show that:

a) All solutions of $\mathbf{x}' = \mathbf{A}\mathbf{x}$ have $\lim_{t\to\infty} \mathbf{x}(t) = \mathbf{0}$ if and only if trace($\mathbf{A}$) < 0.

b) Show that all nonzero solutions of $\mathbf{x}' = \mathbf{A}\mathbf{x}$ are periodic with the same period if trace($\mathbf{A}$) $= 0$.

c) Show that, if trace($\mathbf{A}$) > 0, then there exist solutions whose entries are unbounded as $t \to \infty$.

40. Suppose that $\mathbf{A}$ is a real 2×2 matrix. Show that $\mathbf{A}$ has complex eigenvalues if and only if $[\text{trace}(\mathbf{A})]^2 < 4\det(\mathbf{A})$.

Problems involving complex numbers can often be reformulated as larger problems involving only real numbers. This is illustrated by Exercise 41.

41. Suppose that $\lambda = \alpha + i\beta$ is an eigenvalue of the real $n \times n$ matrix $\mathbf{A}$. Show that $\mathbf{u} = \mathbf{a} + i\mathbf{b}$ is an eigenvector for $\mathbf{A}$ corresponding to the eigenvalue $\lambda = \alpha + i\beta$ if and only if **a**, **b** are solutions of the system of equations:

$$\begin{aligned}[\mathbf{A} - \alpha\mathbf{I}]\mathbf{a} + \beta\mathbf{b} &= \mathbf{0},\\ \beta\mathbf{a} + [\alpha\mathbf{I} - \mathbf{A}]\mathbf{b} &= \mathbf{0}.\end{aligned} \tag{27}$$

or, equivalently,

$$\begin{bmatrix} \mathbf{A} - \alpha\mathbf{I} & \beta\mathbf{I} \\ \beta\mathbf{I} & \alpha\mathbf{I} - \mathbf{A} \end{bmatrix} \begin{bmatrix} \mathbf{a} \\ \mathbf{b} \end{bmatrix} = \begin{bmatrix} \mathbf{0} \\ \mathbf{0} \end{bmatrix}.$$

42. $\lambda_1 = 1 + i$ is an eigenvalue of $\mathbf{A} = \begin{bmatrix} 3 & -1 \\ 5 & -1 \end{bmatrix}$. Find the associated eigenvector $\mathbf{u} = \mathbf{a} + i\mathbf{b}$ by solving the system of equations (27). (*Note*: In solving (27) there will be two arbitrary constants. To find **a**, **b** you may give these constants any values that are not both zero.)

43. $\lambda_1 = 3 + 2i$ is an eigenvalue of $\mathbf{A} = \begin{bmatrix} 5 & 2 \\ -4 & 1 \end{bmatrix}$. Find the associated eigenvector $\mathbf{u} = \mathbf{a} + i\mathbf{b}$ by solving the system of equations (27). (See note for Exercise 42.)

44. Suppose that **A** is a 3×3 real matrix. Show **A** must have a real eigenvalue.

45. Show that if the $n \times n$ matrix **A** has more than one conjugate pair of eigenvalues, then $n \geq 4$.

For Exercises 46 and 47, find the general solution of the indicated system of differential equations.

46. $x_1' = 2x_1 - 2x_2 + 2x_3$
$x_2' = -x_2 + 5x_3$
$x_3' = -x_2 + x_3$

47. $x_1' = -x_2$
$x_2' = 2x_1 + 2x_2$
$x_3' = -2x_1 + 2x_3$

In Exercises 48 through 57, suppose that **A** is an $n \times n$ real matrix. You are given some of the eigenvalues λ_i and associated eigenvectors $\mathbf{u}_i$. Find a fundamental set of solutions for $\mathbf{x}' = \mathbf{A}\mathbf{x}$.

48. $n = 3$; $\lambda_1 = -2$, $\mathbf{u}_1 = \begin{bmatrix} 1 \\ 1 \\ 1 \end{bmatrix}$; $\lambda_2 = 2i$, $\mathbf{u}_2 = \begin{bmatrix} 3 \\ 1 - 2i \\ 2 \end{bmatrix}$

49. $n = 3$; $\lambda_1 = 2$, $\mathbf{u}_1 = \begin{bmatrix} 0 \\ 1 \\ 2 \end{bmatrix}$; $\lambda_2 = 1 + i$, $\mathbf{u}_2 = \begin{bmatrix} 3 + i \\ 1 \\ 0 \end{bmatrix}$

50. $n = 3$; $\lambda_1 = -2$, $\mathbf{u}_1 = \begin{bmatrix} 1 \\ 0 \\ -1 \end{bmatrix}$; $\lambda_2 = 2 - 4i$, $\mathbf{u}_2 = \begin{bmatrix} 6 \\ 1 - i \\ 2 + i \end{bmatrix}$

51. $n = 3$; $\lambda_1 = -1$, $\mathbf{u}_1 = \begin{bmatrix} 1 \\ 3 \\ -1 \end{bmatrix}$; $\lambda_2 = 4i$, $\mathbf{u}_2 = \begin{bmatrix} 2i \\ 1 \\ 2 + 3i \end{bmatrix}$

52. $n = 4$; $\lambda_1 = 1 + 3i$, $\mathbf{u}_1 = \begin{bmatrix} 1 \\ 0 \\ i \\ i + 1 \end{bmatrix}$; $\lambda_2 = 4i$, $\mathbf{u}_2 = \begin{bmatrix} 1 \\ i \\ 0 \\ -1 \end{bmatrix}$

53. $n = 4$; $\lambda_1 = 3i$, $\mathbf{u}_1 = \begin{bmatrix} 0 \\ 1 \\ i \\ 0 \end{bmatrix}$; $\lambda_2 = 1 - 2i$, $\mathbf{u}_2 = \begin{bmatrix} 1 \\ i \\ 1 - i \\ 1 \end{bmatrix}$

54. $n = 4$; $\lambda_1 = 2i$, $\mathbf{u}_1 = \begin{bmatrix} 1 \\ 1 \\ 1 + i \\ 1 + 2i \end{bmatrix}$; $\lambda_2 = 4i$, $\mathbf{u}_2 = \begin{bmatrix} 1 - i \\ 3 + i \\ 0 \\ 1 \end{bmatrix}$

55. $n = 4$; $\lambda_1 = 3i$, $\mathbf{u}_1 = \begin{bmatrix} 1 \\ 1 + i \\ 0 \\ 1 \end{bmatrix}$;

$\lambda_2 = 2i$, $\mathbf{u}_2 = \begin{bmatrix} 1+2i \\ 3i \\ 1 \\ 0 \end{bmatrix}$

56. $n = 4$; $\lambda_1 = 1$, $\mathbf{u}_1 = \begin{bmatrix} 1 \\ 0 \\ 0 \\ 1 \end{bmatrix}$; $\lambda_2 = -2$, $\mathbf{u}_2 = \begin{bmatrix} 1 \\ 3 \\ 0 \\ 0 \end{bmatrix}$;

$\lambda_3 = 1 + 2i$, $\mathbf{u}_3 = \begin{bmatrix} 1 \\ 0 \\ 1-i \\ 1 \end{bmatrix}$

57. $n = 4$; $\lambda_1 = -1$, $\mathbf{u}_1 = \begin{bmatrix} 0 \\ 1 \\ 0 \\ 0 \end{bmatrix}$; $\lambda_2 = 0$, $\mathbf{u}_2 = \begin{bmatrix} 2 \\ 1 \\ 3 \\ 0 \end{bmatrix}$;

$\lambda_3 = -1 + 3i$, $\mathbf{u}_3 = \begin{bmatrix} 1 \\ 1+i \\ 1-i \\ 1 \end{bmatrix}$

5.10.3 Eigenvalues of Higher Multiplicity

As was pointed out in Example 5.10.8, it is sometimes not possible to find all solutions of $\mathbf{x}' = \mathbf{A}\mathbf{x}$ by merely considering the eigenvalues and eigenvectors of $\mathbf{A}$. In this section we will discuss one method for obtaining the other solutions. An alternative method will be given in Section 5.12 on the matrix exponential. While all of the examples and exercises of this section deal only with real eigenvalues and eigenvector deficiencies of one or two, the approach may be generalized to cover the general case.

Suppose then that $\mathbf{A}$ is an $n \times n$ matrix and λ_0 is an eigenvalue of multiplicity m, but that, when we solve $(\lambda_0\mathbf{I} - \mathbf{A})\mathbf{u} = \mathbf{0}$, we get only r linearly independent eigenvectors $\{\mathbf{u}_1, \ldots, \mathbf{u}_r\}$. Then we shall call $m - r$ the **deficiency** of the eigenvalue λ_0. Since λ_0 has multiplicity m, we need to provide m solutions of $\mathbf{x}' = \mathbf{A}\mathbf{x}$. However, we are getting only the r solutions $\{e^{\lambda_0 t}\mathbf{u}_1, \ldots, e^{\lambda_0 t}\mathbf{u}_r\}$. The procedure we give will be applied only to eigenvalues of deficiency 1 and 2. Additional matrix theory is needed to efficiently handle problems with higher deficiency.

Deficiency One

In Chapter 2 we discussed how to solve the scalar differential equation

$$y'' - 2y' + y = 0. \tag{28}$$

The characteristic polynomial $r^2 - 2r + 1 = (r-1)^2$ has the root $r = 1$ of multiplicity 2. Thus a fundamental set of solutions for (28) is $\{e^t, te^t\}$. A similar result holds for systems.

If λ_0 is an eigenvalue of $\mathbf{A}$ of deficiency one, then there is a solution of $\mathbf{x}' = \mathbf{A}\mathbf{x}$ of the form

$$e^{\lambda_0 t}(\mathbf{u} + t\mathbf{v}) \tag{29}$$

with $\mathbf{v} \neq \mathbf{0}$.

Given the form (29) for a solution of $\mathbf{x}' = \mathbf{Ax}$, we may determine $\mathbf{u}$, $\mathbf{v}$ as follows: Assume that λ_0 is an eigenvalue of $\mathbf{x}' = \mathbf{Ax}$ of deficiency 1. Let

$$\mathbf{x} = e^{\lambda_0 t}(\mathbf{u} + t\mathbf{v}), \tag{30}$$

and substitute into $\mathbf{x}' = \mathbf{Ax}$:

$$[e^{\lambda_0 t}(\mathbf{u} + t\mathbf{v})]' = \mathbf{A}e^{\lambda_0 t}(\mathbf{u} + t\mathbf{v})$$

or

$$\lambda_0 e^{\lambda_0 t}(\mathbf{u} + t\mathbf{v}) + e^{\lambda_0 t}\mathbf{v} = e^{\lambda_0 t}(\mathbf{Au} + t\mathbf{Av}).$$

Divide by $e^{\lambda_0 t}$ and rearrange:

$$(\lambda_0 \mathbf{u} + \mathbf{v}) + t\lambda_0 \mathbf{v} = \mathbf{Au} + t\mathbf{Av}. \tag{31}$$

Since (31) holds for all t, we get the two equations

$$\lambda_0 \mathbf{u} + \mathbf{v} = \mathbf{Au}, \tag{32a}$$

$$\lambda_0 \mathbf{v} = \mathbf{Av}, \tag{32b}$$

or

$$(\lambda_0 \mathbf{I} - \mathbf{A})\mathbf{u} + \mathbf{v} = \mathbf{0}, \tag{33a}$$

$$(\lambda_0 \mathbf{I} - \mathbf{A})\mathbf{v} = \mathbf{0}. \tag{33b}$$

If $\mathbf{A}$ is $n \times n$, (32) or (33) gives $2n$ equations in the unknown components of $\mathbf{u}$, $\mathbf{v}$.

Comment Note that (32b) implies that $\mathbf{v}$ is an eigenvector. However, (32a) need not be consistent for every possible eigenvector $\mathbf{v}$. Thus, to be sure this method will *always* work, we must solve (32a), (32b) [or (33a), (33b)] at the same time. The exception is when λ_0 has multiplicity exactly 2 and deficiency one. Then we may solve (32b) first and then (32a). (See the special case after Algorithm 5.10.1.)

EXAMPLE 5.10.13 Solution of $\mathbf{x}' = \mathbf{Ax}$: Deficient Eigenvalues

Find a fundamental set of solutions of:

$$\begin{aligned} x_1' &= 3x_1 - x_2, \\ x_2' &= x_1 + x_2. \end{aligned} \tag{34}$$

Solution This is Example 5.10.8. System (34) is of the form $\mathbf{x}' = \mathbf{Ax}$, where

$$\mathbf{x} = \begin{bmatrix} x_1 \\ x_2 \end{bmatrix}, \quad \mathbf{A} = \begin{bmatrix} 3 & -1 \\ 1 & 1 \end{bmatrix}.$$

The characteristic polynomial is

$$p(\lambda) = \lambda^2 - 4\lambda + 4 = (\lambda - 2)^2,$$

so that $\lambda = 2$ is an eigenvalue of multiplicity 2. However, all eigenvectors for $\lambda = 2$ are of the form

$$u_2\begin{bmatrix}1\\1\end{bmatrix},$$

so that we have only one linearly independent eigenvector. The deficiency is $2 - 1 = 1$, and we have one solution:

$$e^{2t}\begin{bmatrix}1\\1\end{bmatrix}.$$

We shall illustrate the general procedure. To find additional solutions we solve (32). Note that (33) can be written

$$\begin{bmatrix}\lambda_0\mathbf{I} - \mathbf{A} & \mathbf{I}\\ \mathbf{0} & \lambda_0\mathbf{I} - \mathbf{A}\end{bmatrix}\begin{bmatrix}\mathbf{u}\\ \mathbf{v}\end{bmatrix} = \begin{bmatrix}\mathbf{0}\\ \mathbf{0}\end{bmatrix},$$

or, letting

$$\lambda_0 = 2, \qquad \mathbf{u} = \begin{bmatrix}u_1\\u_2\end{bmatrix}, \qquad \mathbf{v} = \begin{bmatrix}v_1\\v_2\end{bmatrix},$$

we have

$$\begin{bmatrix}-1 & 1 & 1 & 0\\ -1 & 1 & 0 & 1\\ 0 & 0 & -1 & 1\\ 0 & 0 & -1 & 1\end{bmatrix}\begin{bmatrix}u_1\\u_2\\v_1\\v_2\end{bmatrix} = \begin{bmatrix}0\\0\\0\\0\end{bmatrix}.$$

Adding (-1) times row 1 to row 2, and -1 times row 3 to row 4, we get a new augmented matrix,

$$\left[\begin{array}{cccc|c}-1 & 1 & 1 & 0 & 0\\ 0 & 0 & -1 & 1 & 0\\ 0 & 0 & -1 & 1 & 0\\ 0 & 0 & 0 & 0 & 0\end{array}\right].$$

Now add row 2 to row 1 and add -1 times row 2 to row 3 to give

$$\left[\begin{array}{cccc|c}-1 & 1 & 0 & 1 & 0\\ 0 & 0 & -1 & 1 & 0\\ 0 & 0 & 0 & 0 & 0\\ 0 & 0 & 0 & 0 & 0\end{array}\right], \tag{35}$$

or $-u_1 + u_2 + v_2 = 0$, $-v_1 + v_2 = 0$.

The first and third columns of (35) are distinguished, so we can take the variables corresponding to the second and fourth columns as arbitrary. That is, our solution is

$$u_2, v_2 \text{ arbitrary}, \qquad v_1 = v_2, \qquad u_1 = u_2 + v_2. \tag{36}$$

Thus all the solutions are of the form

$$\mathbf{x} = e^{\lambda t}(\mathbf{u} + t\mathbf{v}) = e^{2t}\left(\begin{bmatrix} u_1 \\ u_2 \end{bmatrix} + t\begin{bmatrix} v_1 \\ v_2 \end{bmatrix}\right),$$

which, from (36), is

$$= e^{2t}\left(\begin{bmatrix} u_2 + v_2 \\ u_2 \end{bmatrix} + t\begin{bmatrix} v_2 \\ v_2 \end{bmatrix}\right)$$

$$= u_2 e^{2t}\begin{bmatrix} 1 \\ 1 \end{bmatrix} + v_2 e^{2t}\begin{bmatrix} 1 + t \\ t \end{bmatrix}. \tag{37}$$

The solution

$$e^{2t}\begin{bmatrix} 1 \\ 1 \end{bmatrix}$$

had been found earlier. However,

$$e^{2t}\begin{bmatrix} 1 + t \\ t \end{bmatrix}$$

is a new solution in the form (29). Finally, (37) with u_2, v_2 arbitrary constants is the general solution of (34) and

$$\left\{e^{2t}\begin{bmatrix} 1 \\ 1 \end{bmatrix},\ e^{2t}\begin{bmatrix} 1 + t \\ t \end{bmatrix}\right\}$$

is a fundamental set of solutions. ◄

Comment If we are looking only for an additional (linearly independent) solution besides

$$e^{2t}\begin{bmatrix} 1 \\ 1 \end{bmatrix},$$

then in (36) we may take any values of u_2, v_2 for which $v_2 \neq 0$.

■ **ALGORITHM 5.10.1 Procedure for Finding Solutions of $\mathbf{x}' = \mathbf{Ax}$ for an Eigenvalue λ_0 of Deficiency One**

1. Set up the system of equations

$$\begin{aligned} (\lambda_0\mathbf{I} - \mathbf{A})\mathbf{u} + \mathbf{v} &= \mathbf{0}, \\ (\lambda_0\mathbf{I} - \mathbf{A})\mathbf{v} &= \mathbf{0}. \end{aligned} \tag{38}$$

2. Solve (38) for **u**, **v**.
3. $\mathbf{x} = e^{\lambda_0 t}(\mathbf{u} + t\mathbf{v})$ gives all the solutions of $\mathbf{x}' = \mathbf{Ax}$ that correspond to the eigenvalue λ_0. ■

Special Case If λ_0 is an eigenvalue of multiplicity 2 and deficiency 1, then (38) can be broken into the two steps:

1. Solve $(\lambda_0\mathbf{I} - \mathbf{A})\mathbf{v} = \mathbf{0}$ for $\mathbf{v}$.
2. Solve $(\lambda_0\mathbf{I} - \mathbf{A})\mathbf{u} = -\mathbf{v}$ to get $\mathbf{u}$.

Deficiency Two

Sometimes Algorithm 5.10.1 finds enough solutions even if the deficiency is greater than one. However, if λ_0 is an eigenvalue of deficiency two and Algorithm 5.10.1 does not find enough solutions, then we can get the needed number of solutions by considering solutions of the form

$$\mathbf{x} = e^{\lambda_0 t}(\mathbf{u} + t\mathbf{v} + t^2\mathbf{w}). \tag{39}$$

Substituting (39) into $\mathbf{x}' = \mathbf{A}\mathbf{x}$ gives

$$[e^{\lambda_0 t}(\mathbf{u} + t\mathbf{v} + t^2\mathbf{w})]' = \mathbf{A}e^{\lambda_0 t}(\mathbf{u} + t\mathbf{v} + t^2\mathbf{w})$$

or

$$\lambda_0 e^{\lambda_0 t}(\mathbf{u} + t\mathbf{v} + t^2\mathbf{w}) + e^{\lambda_0 t}(\mathbf{v} + 2t\mathbf{w}) = \mathbf{A}e^{\lambda_0 t}(\mathbf{u} + t\mathbf{v} + t^2\mathbf{w}).$$

Divide by $e^{\lambda_0 t}$ and equate like powers of t to get the following systems of algebraic equations:

$$\begin{aligned} \lambda_0\mathbf{u} + \mathbf{v} &= \mathbf{A}\mathbf{u}, \\ \lambda_0\mathbf{v} + 2\mathbf{w} &= \mathbf{A}\mathbf{v}, \\ \lambda_0\mathbf{w} &= \mathbf{A}\mathbf{w}; \end{aligned}$$

or equivalently,

$$(\lambda_0\mathbf{I} - \mathbf{A})\mathbf{u} + \mathbf{v} = \mathbf{0}, \tag{40a}$$

$$(\lambda_0\mathbf{I} - \mathbf{A})\mathbf{v} + 2\mathbf{w} = \mathbf{0}, \tag{40b}$$

$$(\lambda_0\mathbf{I} - \mathbf{A})\mathbf{w} = \mathbf{0}. \tag{40c}$$

In matrix notation, (40) is

$$\begin{bmatrix} \lambda_0\mathbf{I} - \mathbf{A} & \mathbf{I} & \mathbf{0} \\ \mathbf{0} & \lambda_0\mathbf{I} - \mathbf{A} & 2\mathbf{I} \\ \mathbf{0} & \mathbf{0} & \lambda_0\mathbf{I} - \mathbf{A} \end{bmatrix} \begin{bmatrix} \mathbf{u} \\ \mathbf{v} \\ \mathbf{w} \end{bmatrix} = \begin{bmatrix} \mathbf{0} \\ \mathbf{0} \\ \mathbf{0} \end{bmatrix}. \tag{41}$$

The system (40), or (41), is reasonably straightforward to solve on a computer. This approach can be used with any deficiency by using higher powers of t.

Exercises

In Exercises 58 through 65, the deficiency is one. Find the general solution of the differential equation.

58. $x_1' = x_1 - x_2,$
$x_2' = 4x_1 - 3x_2$

59. $x_1' = 3x_1 - 4x_2,$
$x_2' = x_1 - x_2$

60. $x_1' = 3x_1 - x_2,$
$x_2' = 9x_1 - 3x_2$

61. $x' = -x + y,$
$y' = -x + y$

62. $x' = y,$
$y' = -x + 2y$

63. $x_1' = -2x_1 + x_2,$
$x_2' = -16x_1 + 6x_2$

64. $x_1' = -2x_2 + x_3,$
$x_2' = x_1 - 3x_2 + x_3,$
$x_3' = x_1 - 2x_2$

65. $x_1' = 2x_1 - x_3,$
$x_2' = x_1 + x_2 - x_3,$
$x_3' = x_1$

In Exercises 66 through 69, the deficiency is two. Find the general solution of the differential equation.

66. $x_1' = 2x_1 - x_2,$
$x_2' = 2x_2 - 2x_3,$
$x_3' = 2x_3$

67. $x_1' = x_1,$
$x_2' = x_1 + x_2,$
$x_3' = x_1 + x_2 + x_3$

68. $x_1' = x_1 + x_2,$
$x_2' = 2x_2 + x_3,$
$x_3' = x_1 - x_2 + 3x_3$

69. $x_1' = -x_1 + 2x_2,$
$x_2' = x_2 + 2x_3,$
$x_3' = 2x_1 - 2x_2 + 3x_3$

In Exercises 70 through 77, suppose that $\mathbf{A}$ is an $n \times n$ real matrix. $\mathbf{A}$ has the indicated eigenvalue λ. There is a deficiency of one. The nonzero rows of the reduced row echelon form of (38) are given. Find a fundamental set of solutions for $\mathbf{x}' = \mathbf{A}\mathbf{x}$.

70. $n = 2;\ \lambda = -2,\ \begin{bmatrix} 1 & -1 & 0 & -1 \\ 0 & 0 & 1 & -1 \end{bmatrix}$

71. $n = 2;\ \lambda = -1,\ \begin{bmatrix} 1 & -1/2 & 0 & -1/4 \\ 0 & 0 & 1 & -1/2 \end{bmatrix}$

72. $n = 2;\ \lambda = 3,\ \begin{bmatrix} 1 & -1/2 & 0 & 1/8 \\ 0 & 0 & 1 & -1/2 \end{bmatrix}$

73. $n = 2;\ \lambda = 5,\ \begin{bmatrix} 1 & 0 & 0 & -1 \\ 0 & 0 & 1 & 0 \end{bmatrix}$

74. $n = 3;\ \lambda = 2,\ \begin{bmatrix} 0 & 1 & 0 & 0 & 0 & -1 \\ 0 & 0 & 0 & 1 & 0 & -1 \\ 0 & 0 & 0 & 0 & 1 & 0 \end{bmatrix}$

75. $n = 3;\ \lambda = 2,\ \begin{bmatrix} 1 & -1 & 0 & 0 & 0 & 1 \\ 0 & 0 & 0 & 1 & 0 & -1 \\ 0 & 0 & 0 & 0 & 1 & -1 \end{bmatrix}$

76. $n = 3;\ \lambda = -3,\ \begin{bmatrix} 1 & 1 & 0 & 0 & 0 & -1 \\ 0 & 0 & 0 & 1 & 0 & 0 \\ 0 & 0 & 0 & 0 & 1 & 0 \end{bmatrix}$

77. $n = 3;\ \lambda = 1,$
$$\begin{bmatrix} 1 & -1/2 & -1/2 & 0 & 0 & 1/2 \\ 0 & 0 & 0 & 1 & 0 & -1 \\ 0 & 0 & 0 & 0 & 1 & -1 \end{bmatrix}$$

5.11 Nonhomogeneous Systems (Undetermined Coefficients)

In the previous sections we have discussed how to solve the linear, homogeneous differential equation with constant coefficients,

$$\mathbf{x}'(t) = \mathbf{A}\mathbf{x}(t), \tag{1}$$

and find its general solution. In this section we will discuss how to solve the nonhomogeneous differential equation

$$\mathbf{x}'(t) = \mathbf{A}\mathbf{x}(t) + \mathbf{f}(t) \tag{2}$$

for a wide variety of forcing terms $\mathbf{f}(t)$ of interest. From Theorem 5.9.2, we know that the solution of (2) is of the form

$$\mathbf{x} = \mathbf{x}_p + \mathbf{x}_h,$$

where $\mathbf{x}_h$ is the general solution of the associated homogeneous equation (1) and $\mathbf{x}_p$ is a **particular solution** of (2). Since we know how to solve (1) we shall concentrate on how to find $\mathbf{x}_p$. The method presented in this section is an extension of the method of undetermined coefficients from Section 2.9.

■ ALGORITHM 5.11.1 Undetermined Coefficients for Systems

This method is useful for finding a particular solution $\mathbf{x}_p$ of $\mathbf{x}' = \mathbf{A}\mathbf{x} + \mathbf{f}$, that is

$$\begin{aligned} x_1' &= a_{11}x_1 + \cdots + a_{1n}x_n + f_1(t), \\ x_2' &= a_{21}x_1 + \cdots + a_{2n}x_n + f_2(t), \\ &\vdots \\ x_n' &= a_{n1}x_1 + \cdots + a_{nn}x_n + f_n(t). \end{aligned} \tag{3}$$

This method works if $\mathbf{A}$ is constant and each $f_i(t)$ is the type of function for which the method of undetermined coefficients of Section 2.9 works. That is, each $f_i(t)$ is a linear combination of functions $t^m e^{\alpha t}\cos\beta t$, $t^m e^{\alpha t}\sin\beta t$ for integers $m \geq 0$ and scalars α, β.

Step 1
Compute the characteristic polynomial $p(\lambda) = \det(\lambda\mathbf{I} - \mathbf{A})$ of $\mathbf{A}$, and determine its roots (eigenvalues).

Step 2
Construct a form for the particular solution

$$\mathbf{x}_p = \begin{bmatrix} x_{1p} \\ x_{2p} \\ \vdots \\ x_{np} \end{bmatrix},$$

as follows. For terms that appear in *any* of the $f_i(t)$, add terms into *all* the x_{ip} according to the rules of Section 2.9, with the exception that, if we multiply by t^k for a root of multiplicity $k+1$ of $p(\lambda)$, we also include the terms multiplied by the lower powers of t.

Step 3
Substitute the form for $\mathbf{x}_p$ into $\mathbf{x}'_p = \mathbf{A}\mathbf{x}_p + \mathbf{f}$ and solve for the constants in $\mathbf{x}_p$ by equating terms. The solution for the constants will be unique only if there are no terms in the form for $\mathbf{x}_p$ that also appear in the solution of the associated homogeneous equation (1). ■

We shall illustrate this algorithm with several examples.

EXAMPLE 5.11.1 Particular Solution

Find a particular solution of:

$$x'_1 = 2x_1 + x_2 + t, \tag{4}$$

$$x'_2 = x_1 + 2x_2 + 1 + e^{-t}. \tag{5}$$

SOLUTION

Step 1
The coefficient matrix

$$\mathbf{A} = \begin{bmatrix} 2 & 1 \\ 1 & 2 \end{bmatrix}$$

has characteristic polynomial

$$p(\lambda) = \det(\lambda\mathbf{I} - \mathbf{A}) = \det\begin{bmatrix} \lambda - 2 & -1 \\ -1 & \lambda - 2 \end{bmatrix} = \lambda^2 - 4\lambda + 3$$
$$= (\lambda - 3)(\lambda - 1),$$

so that the roots are $\lambda = 1, 3$.

Step 2
The forcing terms include 1, t, and e^{-t}. Since neither 0 nor -1 is a root of $p(\lambda)$, our particular solution will have the form

$$x_{1p} = A + Bt + Ce^{-t}, \tag{6}$$

$$x_{2p} = E + Ft + Ge^{-t}. \tag{7}$$

Step 3
We substitute the form (6), (7) for $\mathbf{x}_p$ into the differential equations (4) and (5), to get

$$B - Ce^{-t} = 2(A + Bt + Ce^{-t}) + (E + Ft + Ge^{-t}) + t, \tag{8}$$

$$F - Ge^{-t} = (A + Bt + Ce^{-t}) + 2(E + Ft + Ge^{-t}) + 1 + e^{-t}. \tag{9}$$

Equating coefficients of like terms in (8) and then (9) yields equations on the coefficients:

$$\begin{array}{rll} 1: & B = 2A + E, & F = A + 2E + 1, \\ t: & 0 = 2B + F + 1, & 0 = B + 2F, \\ e^{-t}: & -C = 2C + G, & -G = C + 2G + 1. \end{array}$$

This is six equations in six unknowns, which could be solved using augmented matrices. However, it is quicker to solve the bottom two equations for C and G, then the middle two for B and F, and finally the top two equations for A and E to get

$$A = -\tfrac{2}{9}, \quad B = -\tfrac{2}{3}, \quad C = \tfrac{1}{8}, \quad E = -\tfrac{2}{9}, \quad F = \tfrac{1}{3}, \quad G = -\tfrac{3}{8},$$

so that

$$\mathbf{x}_p = \begin{bmatrix} x_{1p} \\ x_{2p} \end{bmatrix} = \begin{bmatrix} -\frac{2}{9} - \frac{2}{3}t + \frac{1}{8}e^{-t} \\ -\frac{2}{9} + \frac{1}{3}t - \frac{3}{8}e^{-t} \end{bmatrix} \tag{10}$$

is a particular solution of (4) and (5). ◄

Note This problem can also be worked in vector notation. That is, we can take

$$\mathbf{x} = \mathbf{a} + t\mathbf{b} + e^{-t}\mathbf{c}, \tag{11}$$

where $\mathbf{a}$, $\mathbf{b}$, $\mathbf{c}$ are unknown 2×1 constant vectors. Substituting (11) into $\mathbf{x}' = \mathbf{A}\mathbf{x} + \mathbf{f}$ and rewriting $\mathbf{f}$ gives

$$\mathbf{b} - e^{-t}\mathbf{c} = \mathbf{A}\mathbf{a} + t\mathbf{A}\mathbf{b} + e^{-t}\mathbf{A}\mathbf{c} + \begin{bmatrix} 0 \\ 1 \end{bmatrix} + t\begin{bmatrix} 1 \\ 0 \end{bmatrix} + e^{-t}\begin{bmatrix} 0 \\ 1 \end{bmatrix}.$$

Equating corresponding coefficients gives the vector equations

$$1: \quad \mathbf{b} = \mathbf{A}\mathbf{a} + \begin{bmatrix} 0 \\ 1 \end{bmatrix} \quad \text{or} \quad \mathbf{A}\mathbf{a} = \mathbf{b} - \begin{bmatrix} 0 \\ 1 \end{bmatrix}. \tag{12}$$

$$t: \quad \mathbf{0} = \mathbf{A}\mathbf{b} + \begin{bmatrix} 1 \\ 0 \end{bmatrix} \quad \text{or} \quad \mathbf{A}\mathbf{b} = -\begin{bmatrix} 1 \\ 0 \end{bmatrix}, \tag{13}$$

$$e^{-t}: \quad -\mathbf{c} = \mathbf{A}\mathbf{c} + \begin{bmatrix} 0 \\ 1 \end{bmatrix} \quad \text{or} \quad (\mathbf{A} + \mathbf{I})\mathbf{c} = -\begin{bmatrix} 0 \\ 1 \end{bmatrix}. \tag{14}$$

This gives us three systems. First, solve (14) for $\mathbf{c}$, then (13) for $\mathbf{b}$, and finally (12) for $\mathbf{a}$. We leave the details to the interested reader (see Exercise 22 at the end of this section).

EXAMPLE 5.11.2 Particular Solution

Find a particular solution of

$$x_1' = -x_1 + 4x_2 + e^t, \tag{15}$$

$$x_2' = x_1 - x_2 + 1. \tag{16}$$

Solution

Step 1
The characteristic polynomial of

$$\mathbf{A} = \begin{bmatrix} -1 & 4 \\ 1 & -1 \end{bmatrix}$$

is

$$p(\lambda) = \det \begin{bmatrix} \lambda + 1 & -4 \\ -1 & \lambda + 1 \end{bmatrix} = \lambda^2 + 2\lambda - 3 = (\lambda + 3)(\lambda - 1),$$

with roots $\lambda = -3$, $\lambda = 1$.

Step 2
The functions 1 and e^t appear as forcing terms. 0 is not a root, while 1 is. Thus we take

$$x_{1p} = A + Be^t + Cte^t, \tag{17}$$

$$x_{2p} = E + Fe^t + Gte^t. \tag{18}$$

The inclusion of the Be^t, Fe^t terms in addition to the Cte^t, Gte^t terms is a major difference in the way the method of undetermined coefficients is applied to systems and the way it was applied in Section 2.9 to scalar equations. Substituting (17) and (18) into (15) and (16) gives us:

$$Be^t + C(e^t + te^t) = -(A + Be^t + Cte^t) + 4(E + Fe^t + Gte^t) + e^t,$$
$$Fe^t + G(e^t + te^t) = (A + Be^t + Cte^t) - (E + Fe^t + Gte^t) + 1.$$

Equating like coefficients we get the equations

$$\begin{array}{rlll} 1: & 0 = -A + 4E, & 0 = A - E + 1, & (19) \\ e^t: & B + C = -B + 4F + 1, & F + G = B - F, & (20) \\ te^t: & C = -C + 4G, & G = C - G. & (21) \end{array}$$

The first two equations (19) give $E = -\frac{1}{3}$, $A = -\frac{4}{3}$, while (21) implies only that $C = 2G$. Thus we can eliminate C in (20), and have left

$$\begin{aligned} 2B + 2G - 4F &= 1, \\ -B + G + 2F &= 0. \end{aligned}$$

This is only two equations in three unknowns. The augmented matrix is

$$\left[\begin{array}{rrr|r} 2 & 2 & -4 & 1 \\ -1 & 1 & 2 & 0 \end{array}\right] \quad \text{or} \quad \left[\begin{array}{rrr|c} 1 & 1 & -2 & 1/2 \\ -1 & 1 & 2 & 0 \end{array}\right].$$

Adding the first row to the second gives

$$\left[\begin{array}{rrr|c} 1 & 1 & -2 & \frac{1}{2} \\ 0 & 2 & 0 & \frac{1}{2} \end{array}\right].$$

Two more row operations yield the row echelon form

$$\left[\begin{array}{ccc|c} 1 & 0 & -2 & \frac{1}{4} \\ 0 & 1 & 0 & \frac{1}{4} \end{array}\right].$$

Thus $G = \frac{1}{4}$ and $B - 2F = \frac{1}{4}$ or $B = 2F + \frac{1}{4}$. Since we are looking for *any* particular solution of (15) and (16), we may set the arbitrary $F = 0$ and use $C = 2G$, to find

$$A = -\tfrac{4}{3}, \qquad B = \tfrac{1}{4}, \qquad C = \tfrac{1}{2}, \qquad E = -\tfrac{1}{3}, \qquad F = 0, \qquad G = \tfrac{1}{4}.$$

Thus

$$x_{1p} = -\frac{4}{3} + \frac{1}{4}e^t + \frac{1}{2}te^t, \tag{22}$$

$$x_{2p} = -\frac{1}{3} + \frac{1}{4}te^t \tag{23}$$

is a particular solution of (15) and (16). ◄

EXAMPLE 5.11.3 General Solution

Find the general solution of

$$x_1' = -x_1 + 4x_2 + e^t, \tag{24}$$

$$x_2' = x_1 - x_2 + 1. \tag{25}$$

SOLUTION The solution is $\mathbf{x} = \mathbf{x}_p + \mathbf{x}_h$, where $\mathbf{x}_p$ was computed in Example 5.11.2. We compute the homogeneous solution, using the eigenvector technique of Section 5.10. From Example 5.11.2, the eigenvalues of

$$\mathbf{A} = \begin{bmatrix} -1 & 4 \\ 1 & -1 \end{bmatrix}$$

are $\lambda = -3$, $\lambda = 1$. To find the eigenvectors, we must solve $(\lambda\mathbf{I} - \mathbf{A})\mathbf{u} = \mathbf{0}$. For $\lambda = 1$, we get

$$\left[\begin{array}{cc|c} 2 & -4 & 0 \\ -1 & 2 & 0 \end{array}\right] \quad \text{or} \quad \left[\begin{array}{cc|c} 1 & -2 & 0 \\ 0 & 0 & 0 \end{array}\right].$$

That is,

$$u_1 = 2u_2 \qquad \text{and} \qquad \mathbf{u} = \begin{bmatrix} 2u_2 \\ u_2 \end{bmatrix} = u_2\begin{bmatrix} 2 \\ 1 \end{bmatrix}.$$

For $\lambda = -3$, we get

$$\left[\begin{array}{cc|c} -2 & -4 & 0 \\ -1 & -2 & 0 \end{array}\right] \quad \text{or} \quad \left[\begin{array}{cc|c} 1 & 2 & 0 \\ 0 & 0 & 0 \end{array}\right].$$

Thus, $u_1 = -2u_2$ and

$$\mathbf{u} = \begin{bmatrix} -2u_2 \\ u_2 \end{bmatrix} = u_2\begin{bmatrix} -2 \\ 1 \end{bmatrix}.$$

Finally,

$$\mathbf{x}_h = c_1 e^t\begin{bmatrix} 2 \\ 1 \end{bmatrix} + c_2 e^{-3t}\begin{bmatrix} -2 \\ 1 \end{bmatrix} \tag{26}$$

is the general solution of the homogeneous equation. Combining (26), (22), and (23) yields the general solution of (24) and (25) as

$$\mathbf{x} = \mathbf{x}_p + \mathbf{x}_h$$

or

$$x_1 = -\frac{4}{3} + \frac{1}{4}e^t + \frac{1}{2}te^t + 2c_1e^t - 2c_2e^{-3t}, \tag{27}$$

$$x_2 = -\frac{1}{3} + \frac{1}{4}te^t + c_1e^t + c_2e^{-3t}, \tag{28}$$

with c_1, c_2 arbitrary constants. ◄

EXAMPLE 5.11.4 **Initial Conditions**

Find the solution of the initial value problem

$$x_1' = -x_1 + 4x_2 + e^t, \qquad x_1(0) = 0, \tag{29}$$

$$x_2' = x_1 - x_2 + 1, \qquad x_2(0) = 1. \tag{30}$$

SOLUTION First, we must find the general solution, which was done in the previous example. Then we apply the initial conditions in (29) and (30) to the general solution in (27) and (28), to obtain

$$0 = x_1(0) = -\frac{4}{3} + \frac{1}{4} + 2c_1 - 2c_2,$$

$$1 = x_2(0) = -\frac{1}{3} + c_1 + c_2.$$

That is,

$$\frac{13}{12} = 2c_1 - 2c_2,$$

$$\frac{4}{3} = c_1 + c_2,$$

which has the solution $c_1 = \frac{15}{16}$, $c_2 = \frac{19}{48}$, and the solution of (25) and (26) is

$$x_1 = -\frac{4}{3} + \frac{17}{8}e^t + \frac{1}{2}te^t - \frac{19}{24}e^{-3t},$$

$$x_2 = -\frac{1}{3} + \frac{1}{4}te^t + \frac{15}{16}e^t + \frac{19}{48}e^{-3t}. \quad ◄$$

We conclude with several additional examples of picking the form for $\mathbf{x}_p$.

EXAMPLE 5.11.5 Form of Particular Solution

Give the form for $\mathbf{x}_p$ if

$$x_1' = x_1 + 2x_2 + \sin t,$$
$$x_2' = 3x_1 + 4x_2 + \cos 2t$$

is to be solved by the method of undetermined coefficients.

Solution Since $\sin t$, $\cos 2t$ are forcing terms, we must include $\sin t$, $\cos t$, $\sin 2t$, $\cos 2t$ in the form for $\mathbf{x}_p$. The characteristic polynomial of the coefficient matrix is

$$p(\lambda) = \det \begin{bmatrix} \lambda - 1 & -2 \\ -3 & \lambda - 4 \end{bmatrix} = \lambda^2 - 5\lambda - 2.$$

Since neither i nor $2i$ is a root of $p(\lambda)$, we do not need any additional terms, and the form of $\mathbf{x}_p$ is

$$x_{1p} = A \cos t + B \sin t + C \cos 2t + D \sin 2t,$$

$$x_{2p} = E \cos t + F \sin t + G \cos 2t + H \sin 2t.$$

◄

EXAMPLE 5.11.6 Form of Particular Solution

Give the form for $\mathbf{x}_p$ if

$$x_1' = x_1 + 5x_2 + e^{2t},$$
$$x_2' = -x_1 - x_2 + \sin 2t$$

is to be solved by the method of undetermined coefficients.

Solution Since e^{2t}, $\sin 2t$, are the forcing terms, we must include e^{2t}, $\sin 2t$, $\cos 2t$ in the particular solution. The characteristic polynomial is

$$p(\lambda) = \det \begin{bmatrix} \lambda - 1 & -5 \\ 1 & \lambda + 1 \end{bmatrix} = \lambda^2 + 4,$$

which has roots $\pm 2i$. Thus we need to include not only a $\sin 2t$, $\cos 2t$, but also $t \sin 2t$, $t \cos 2t$, and

$$x_{1p} = Ae^{2t} + B \sin 2t + C \cos 2t + Dt \sin 2t + Et \cos 2t,$$

$$x_{2p} = Fe^{2t} + G \sin 2t + H \cos 2t + Jt \sin 2t + Kt \cos 2t$$

is the form for $\mathbf{x}_p$. ◄

EXAMPLE 5.11.7 Form of Particular Solution

Give the form for $\mathbf{x}_p$ if

$$x_1' = 3x_1 - 4x_2 + e^{3t},$$
$$x_2' = x_1 - x_2 + 6e^t$$

is to be solved by the method of undetermined coefficients.

SOLUTION The characteristic polynomial is

$$p(\lambda) = \det \begin{bmatrix} \lambda - 3 & 4 \\ -1 & \lambda + 1 \end{bmatrix} = \lambda^2 - 2\lambda + 1 = (\lambda - 1)^2,$$

so that $\lambda = 1$ is a root of multiplicity two. Since 3 is not a root, the e^{3t} forcing term means that we include e^{3t} in the form for $\mathbf{x}_p$. Since 1 is a root of multiplicity two, the e^t term means that we include e^t, te^t, t^2e^t. Thus the form for $\mathbf{x}_p$ is

$$x_{1p} = Ae^{3t} + Be^t + Cte^t + Dt^2e^t,$$

$$x_{2p} = Ee^{3t} + Fe^t + Gte^t + Ht^2e^t.$$

Exercises

For Exercises 1 through 8, find a particular solution using the method of undetermined coefficients. Then find the general solution.

1. $x_1' = x_1 + x_2 + e^t$
$x_2' = x_1 + x_2 - e^{-t}$

2. $x_1' = x_1 + 3x_2 + t,$
$x_2' = x_1 - x_2 - 1$

3. $x_1' = 2x_1 - x_2 + 2e^{-t},$
$x_2' = 2x_1 - x_2$

4. $x_1' = 3x_1 - 2x_2 + e^t,$
$x_2' = x_1$

5. $x_1' = 2x_1 + x_2 + \sin t,$
$x_2' = x_1 + 2x_2 + 3\cos t$

6. $x_1' = 2x_1 - 3x_2,$
$x_2' = x_1 - 2x_2 + 2e^{-t}$

7. $x_1' = 2x_2 + e^{2t},$
$x_2' = 2x_1$

8. $x_1' = x_2 + e^t,$
$x_2' = -x_1$

In Exercises 9 through 20, give the form for $\mathbf{x}_p$. You need not actually solve for the arbitrary constants.

9. $x_1' = -x_1 - 2x_2 + \sin t,$
$x_2' = x_1 + x_2 + \cos 2t$

10. $x_1' = x_1 - x_2 + t,$
$x_2' = x_1 - x_2 - 3$

11. $x_1' = 3x_1 - 2x_2 + e^t,$
$x_2' = 4x_1 - 3x_2$

12. $x_1' = x_1 + 2x_2 + e^{-t}\cos t,$
$x_2' = x_1 + 3x_2 + \cos t$

13. $x_1' = 4x_1 - 2x_2 + t^2e^{2t},$
$x_2' = 4x_1 - 2x_2 - e^t$

14. $x_1' = 2x_1 - x_2 + e^t,$
$x_2' = x_1$

15. $x_1' = 2x_1 - 4x_2 + t^2,$
$x_2' = x_1 - 2x_2 + t - 1$

16. $x_1' = x_1 - x_2 + \sin 2t,$
$x_2' = 5x_1 - x_2 - 3$

17. $x_1' = 2x_1 + 5x_2 + t\sin t,$
$x_2' = -x_1 - 2x_2 + \cos t$

18. $x_1' = 2x_1 - x_2 + t^2e^t,$
$x_2' = 2x_1 + 3x_2 + e^{-4t}$

19. $x_1' = x_1 + 2x_2 + e^{-t}\cos 3t,$
$x_2' = 3x_1 + 4x_2 + e^{-t}$

20. $x_1' = x_1 + 3x_2 + t\cos t,$
$x_2' = 9x_1 + 13x_2 + \cos t$

21. Suppose that λ is an eigenvalue of the $n \times n$ matrix **A** of multiplicity one. Let **v** be a fixed vector. Show that finding a particular solution of $\mathbf{x}' = \mathbf{A}\mathbf{x} + e^{\lambda t}\mathbf{v}$ of the form $\mathbf{x}_p = e^{\lambda t}\mathbf{a} + te^{\lambda t}\mathbf{b}$ is equivalent to solving the system of algebraic equations

$$\mathbf{A}\mathbf{b} = \lambda\mathbf{b},$$
$$(\mathbf{A} - \lambda\mathbf{I})\mathbf{a} = \mathbf{b} - \mathbf{v},$$

(Note that this is a nonhomogeneous version of system (6) of the previous section.)

or, equivalently,

$$\begin{bmatrix} \mathbf{A} - \lambda\mathbf{I} & \mathbf{0} \\ -\mathbf{I} & \mathbf{A} - \lambda\mathbf{I} \end{bmatrix} \begin{bmatrix} \mathbf{b} \\ \mathbf{a} \end{bmatrix} = \begin{bmatrix} \mathbf{0} \\ -\mathbf{v} \end{bmatrix}.$$

22. Solve the systems (14), (13), (12), and substitute the result into (11), to arrive at a particular solution to (4) and (5).

5.12 The Matrix Exponential

In Chapter 1, we found that the solution of $x' = ax$, a constant, was $x = e^{at}c$. The function e^{at} was then used in many of the calculations that followed. This section will show that similar notation is possible with the system $\mathbf{x}' = \mathbf{A}\mathbf{x}$.

The notation is very useful, both in working with constant coefficient differential equations and in motivating the theory for linear systems with time-varying coefficients. If α is a scalar and **A** is a matrix, then we shall use the notation $\mathbf{A}\alpha$ for $\alpha\mathbf{A}$. This is standard practice, and allows our formulas to appear almost exactly like the formulas for scalar differential equations.

Recall that

$$e^{at} = \sum_{n=0}^{\infty} \frac{a^n t^n}{n!}.$$

Suppose that **A** is an $n \times n$ matrix. Then the **matrix exponential** may be defined as

$$e^{\mathbf{A}t} = \sum_{n=0}^{\infty} t^n \frac{\mathbf{A}^n}{n!} = \mathbf{I} + t\mathbf{A} + \frac{t^2}{2}\mathbf{A}^2 + \frac{t^3}{6}\mathbf{A}^3 + \cdots, \tag{1}$$

where $\mathbf{A}^0 = \mathbf{I}$. This series converges to a matrix for all values of t and may be differentiated term by term, to show that

$$(e^{\mathbf{A}t})' = \mathbf{A}e^{\mathbf{A}t} = e^{\mathbf{A}t}\mathbf{A}.$$

This leads to the following key theorem.

■ **THEOREM 5.12.1 Matrix Exponential**

Suppose that $\mathbf{A}$ is an $n \times n$ matrix. Then $e^{\mathbf{A}t}$ is the unique solution of the matrix differential equation

$$\mathbf{X}' = \mathbf{A}\mathbf{X}, \qquad \mathbf{X}(0) = \mathbf{I}. \tag{2}$$

Furthermore, if $\mathbf{a}$ is a constant $n \times 1$ vector, then the unique solution of

$$\mathbf{x}' = \mathbf{A}\mathbf{x}, \qquad \mathbf{x}(0) = \mathbf{a} \tag{3}$$

is

$$\mathbf{x} = e^{\mathbf{A}t}\mathbf{a}. \tag{4}$$

■

Several other key facts about the matrix exponential are contained in the next theorem.

■ **THEOREM 5.12.2 Properties of Matrix Exponential**

Suppose that $\mathbf{A}$ is $n \times n$. Then

$$e^{\mathbf{A}t} \text{ is invertible} \quad \text{for all } t \quad \text{and} \quad (e^{\mathbf{A}t})^{-1} = e^{-\mathbf{A}t}, \tag{5}$$

$$e^{\mathbf{A}(t+s)} = e^{\mathbf{A}t}e^{\mathbf{A}s} \quad \text{for any scalars } s, t. \tag{6}$$

However, in general, if $\mathbf{B}$ is an $n \times n$ matrix that does not commute with $\mathbf{A}$ (that is, $\mathbf{AB} \neq \mathbf{BA}$), then (see Exercises 10 and 11 at the end of this section),

$$e^{(\mathbf{A}+\mathbf{B})t} \neq e^{\mathbf{A}t}e^{\mathbf{B}t}. \tag{7}$$

An outline of a proof of these theorems appears in Exercises 25 through 27.

■

As we shall see, computing the matrix exponential is, in general, as difficult as solving the differential equation, and is avoided in practice whenever possible. This is quite similar to the useful notation $\mathbf{A}^{-1}\mathbf{b}$ for the solution of $\mathbf{A}\mathbf{x} = \mathbf{b}$. One rarely computes $\mathbf{A}^{-1}$, and then $\mathbf{A}^{-1}\mathbf{b}$. Similarly, it is convenient to write $e^{\mathbf{A}t}\mathbf{a}$ for the solution of

$$\mathbf{x}' = \mathbf{A}\mathbf{x}, \qquad \mathbf{x}(0) = \mathbf{a}$$

even if the matrix exponential is not computed.

Before showing the usefulness of the matrix exponential notation, we shall compute a few examples. Occasionally it is possible to compute the series (1) directly.

EXAMPLE 5.12.1 **Matrix Exponential from Its Series Definition**

Let

$$\mathbf{A} = \begin{bmatrix} 0 & 1 \\ 1 & 0 \end{bmatrix}.$$

Compute $e^{\mathbf{A}t}$ from the series definition (1).

SOLUTION

$$e^{\mathbf{A}t} = \mathbf{I} + t\mathbf{A} + \frac{t^2}{2}\mathbf{A}^2 + \frac{t^3}{3!}\mathbf{A}^3 + \frac{t^4}{4!}\mathbf{A}^4 + \cdots$$

$$= \mathbf{I} + t\begin{bmatrix} 0 & 1 \\ 1 & 0 \end{bmatrix} + \frac{t^2}{2}\begin{bmatrix} 1 & 0 \\ 0 & 1 \end{bmatrix} + \frac{t^3}{3!}\begin{bmatrix} 0 & 1 \\ 1 & 0 \end{bmatrix} + \frac{t^4}{4!}\begin{bmatrix} 1 & 0 \\ 0 & 1 \end{bmatrix} + \cdots$$

$$= \mathbf{I} + \begin{bmatrix} 0 & t \\ t & 0 \end{bmatrix} + \begin{bmatrix} \frac{t^2}{2} & 0 \\ 0 & \frac{t^2}{2} \end{bmatrix} + \begin{bmatrix} 0 & \frac{t^3}{3!} \\ \frac{t^3}{3!} & 0 \end{bmatrix} + \begin{bmatrix} \frac{t^4}{4!} & 0 \\ 0 & \frac{t^4}{4!} \end{bmatrix} + \cdots$$

$$= \begin{bmatrix} 1 + \frac{t^2}{2!} + \frac{t^4}{4!} + \cdots, & t + \frac{t^3}{3!} + \frac{t^5}{5!} + \cdots \\ t + \frac{t^3}{3!} + \frac{t^5}{5!} + \ldots, & 1 + \frac{t^2}{2} + \frac{t^4}{4!} + \cdots \end{bmatrix}$$

$$= \begin{bmatrix} \frac{1}{2}e^t + \frac{1}{2}e^{-t} & \frac{1}{2}e^t - \frac{1}{2}e^{-t} \\ \frac{1}{2}e^t - \frac{1}{2}e^{-t} & \frac{1}{2}e^t + \frac{1}{2}e^{-t} \end{bmatrix} = \begin{bmatrix} \cosh t \sinh t \\ \sinh t \cosh t \end{bmatrix}.$$

However, it is usually easier to proceed in one of three ways.

1. Utilize some version of the Cayley–Hamilton Theorem (Exercises 23 and 24).
2. Compute eigenvectors and eigenvalues for $\mathbf{A}$ and use a change of coordinates (Exercises 13 through 21).
3. Solve the system of differential equations, $\mathbf{x}' = \mathbf{A}\mathbf{x}$ for several initial conditions.

We shall illustrate the third method of finding $e^{\mathbf{A}t}$, and leave the other two methods to the exercises.

■ **THEOREM 5.12.3 Matrix Exponential is a Solution of a Differential Equation**

Suppose $\mathbf{A}$ is an $n \times n$ matrix. Let $\mathbf{X} = e^{\mathbf{A}t}$. Let $\mathbf{x}_i$ be the ith column of $e^{\mathbf{A}t}$. Then $\mathbf{x}_i$ is the solution of

$$\mathbf{x}' = \mathbf{A}\mathbf{x}, \qquad \mathbf{x}(0) = \mathbf{e}_i, \tag{8}$$

where $\mathbf{e}_i$ is the ith column of the $n \times n$ identity matrix (all entries are zero except for the ith which is one). ■

Verification If $\mathbf{X} = e^{\mathbf{A}t}$, then

$$\mathbf{X}' = \mathbf{A}\mathbf{X}, \qquad \mathbf{X}(0) = \mathbf{I}.$$

Let $\mathbf{X} = [\mathbf{x}_1, \ldots, \mathbf{x}_n]$, where the x_i are the columns of $\mathbf{X}$. Then $\mathbf{X}' = \mathbf{A}\mathbf{X}$ becomes

$$[\mathbf{x}_1', \ldots, \mathbf{x}_n'] = \mathbf{A}[\mathbf{x}_1, \ldots, \mathbf{x}_n] = [\mathbf{A}\mathbf{x}_1, \ldots, \mathbf{A}\mathbf{x}_n],$$

while

$$\mathbf{X}(0) = [\mathbf{x}_1(0), \ldots, \mathbf{x}_n(0)] = \mathbf{I} = [\mathbf{e}_1, \mathbf{e}_2, \ldots, \mathbf{e}_n].$$

Thus, we have verified (8) $\mathbf{x}_i' = \mathbf{A}\mathbf{x}_i$, and $\mathbf{x}_i(0) = \mathbf{e}_i$.

EXAMPLE 5.12.2 Matrix Exponential from Fundamental Set

Let

$$\mathbf{A} = \begin{bmatrix} 0 & 1 \\ 1 & 0 \end{bmatrix}.$$

Find $e^{\mathbf{A}t}$, using Theorem 5.12.3.

SOLUTION First, we need to find the general solution of $\mathbf{x}' = \mathbf{A}\mathbf{x}$. The characteristic polynomial of $\mathbf{A}$ is

$$p(\lambda) = \det\begin{bmatrix} \lambda & -1 \\ -1 & \lambda \end{bmatrix} = \lambda^2 - 1,$$

so that the eigenvalues are $\lambda = \pm 1$. Computing the eigenvectors, we have an eigenvector

$$\begin{bmatrix} 1 \\ 1 \end{bmatrix}$$

associated with $\lambda = 1$, and an eigenvector

$$\begin{bmatrix} -1 \\ 1 \end{bmatrix}$$

associated with $\lambda = -1$. Thus,

$$\mathbf{x} = c_1 e^{t}\begin{bmatrix} 1 \\ 1 \end{bmatrix} + c_2 e^{-t}\begin{bmatrix} -1 \\ 1 \end{bmatrix} \tag{9}$$

is the general solution of $\mathbf{x}' = \mathbf{A}\mathbf{x}$. The columns of the matrix exponential are the solutions of

$$\mathbf{x}_1' = \mathbf{A}\mathbf{x}_1, \qquad \mathbf{x}_1(0) = \begin{bmatrix} 1 \\ 0 \end{bmatrix}, \tag{10}$$

$$\mathbf{x}_2' = \mathbf{A}\mathbf{x}_2, \qquad \mathbf{x}_2(0) = \begin{bmatrix} 0 \\ 1 \end{bmatrix}. \tag{11}$$

To find $\mathbf{x}_1$ we apply the initial condition (10) to the general solution (9), to get

$$\begin{aligned} c_1 - c_2 &= 1, \\ c_1 + c_2 &= 0, \end{aligned} \tag{12}$$

or $c_1 = \frac{1}{2}$, $c_2 = -\frac{1}{2}$. To find $\mathbf{x}_2$ we apply the initial condition (11) to the solution (9), to get

$$\begin{aligned} c_1 - c_2 &= 0, \\ c_1 + c_2 &= 1, \end{aligned} \tag{13}$$

or $c_1 = \frac{1}{2}$, $c_2 = \frac{1}{2}$. Thus,

$$e^{\mathbf{A}t} = [\mathbf{x}_1, \mathbf{x}_2] = \left[\begin{bmatrix} \frac{1}{2}e^t + \frac{1}{2}e^{-t} \\ \frac{1}{2}e^t - \frac{1}{2}e^{-t} \end{bmatrix}, \begin{bmatrix} \frac{1}{2}e^t - \frac{1}{2}e^{-t} \\ \frac{1}{2}e^t + \frac{1}{2}e^{-t} \end{bmatrix} \right]$$

as in Example 5.12.1. ◄

Note that in Example 5.12.2, the systems (12) and (13) had the same coefficients. This observation is closely related to the following fact.

■ **THEOREM 5.12.4 Fundamental Set of Solutions and Matrix Exponential**

Suppose $\mathbf{A}$ is $n \times n$ and $\{\mathbf{x}_1, \ldots, \mathbf{x}_n\}$ is a fundamental set of solutions of $\mathbf{x}' = \mathbf{A}\mathbf{x}$. Then,

$$e^{\mathbf{A}t} = [\mathbf{x}_1, \ldots, \mathbf{x}_n][\mathbf{x}_1(0), \ldots, \mathbf{x}_n(0)]^{-1}. \tag{14}$$

■

Verification Let

$$\mathbf{Z}(t) = [\mathbf{x}_1, \ldots, \mathbf{x}_n][\mathbf{x}_1(0), \ldots, \mathbf{x}_n(0)]^{-1},$$

where $\{\mathbf{x}_1, \ldots, \mathbf{x}_n\}$ is a fundamental set of solutions of $\mathbf{x}' = \mathbf{A}\mathbf{x}$. Then

$$\begin{aligned} \mathbf{Z}' &= [\mathbf{x}_1', \ldots, \mathbf{x}_n'][\mathbf{x}_1(0), \ldots, \mathbf{x}_n(0)]^{-1} \\ &= [\mathbf{A}\mathbf{x}_1, \ldots, \mathbf{A}\mathbf{x}_n][\mathbf{x}_1(0), \ldots, \mathbf{x}_n(0)]^{-1} \\ &= \mathbf{A}[\mathbf{x}_1, \ldots, \mathbf{x}_n][\mathbf{x}_1(0), \ldots, \mathbf{x}_n(0)]^{-1} = \mathbf{A}\mathbf{Z}, \end{aligned}$$

and

$$\mathbf{Z}(0) = [\mathbf{x}_1(0), \ldots, \mathbf{x}_n(0)][\mathbf{x}_1(0), \ldots, \mathbf{x}_n(0)]^{-1} = \mathbf{I}.$$

Thus, we have verified (14), $\mathbf{Z}(t) = e^{\mathbf{A}t}$, by Theorem 5.12.1.

The invertibility of the matrix of initial values $[\mathbf{x}_1(0), \ldots, \mathbf{x}_n(0)]$ occurs because $\mathbf{x}_1(t), \ldots, \mathbf{x}_n(t)$ are a fundamental set of solutions.

EXAMPLE 5.12.3 Matrix Exponential from Fundamental Set

Given that we know, from (9), that

$$\mathbf{x}_1(t) = e^t\begin{bmatrix} 1 \\ 1 \end{bmatrix} = \begin{bmatrix} e^t \\ e^t \end{bmatrix}, \qquad \mathbf{x}_2(t) = e^{-t}\begin{bmatrix} -1 \\ 1 \end{bmatrix} = \begin{bmatrix} -e^{-t} \\ e^{-t} \end{bmatrix} \tag{15}$$

are a fundamental set of solutions of $\mathbf{x}' = \mathbf{A}\mathbf{x}$, where

$$\mathbf{A} = \begin{bmatrix} 0 & 1 \\ 1 & 0 \end{bmatrix},$$

compute $e^{\mathbf{A}t}$ using Theorem 5.12.4.

SOLUTION By (14),

$$\begin{aligned} e^{\mathbf{A}t} &= [\mathbf{x}_1, \mathbf{x}_2][\mathbf{x}_1(0), \mathbf{x}_2(0)]^{-1} \\ &= \begin{bmatrix} e^t & -e^{-t} \\ e^t & e^{-t} \end{bmatrix}\begin{bmatrix} 1 & -1 \\ 1 & 1 \end{bmatrix}^{-1}. \end{aligned} \tag{16}$$

Using the technique of Section 5.7 (or a good calculator), we compute that

$$\begin{bmatrix} 1 & -1 \\ 1 & 1 \end{bmatrix}^{-1} = \begin{bmatrix} \frac{1}{2} & \frac{1}{2} \\ -\frac{1}{2} & \frac{1}{2} \end{bmatrix},$$

so that by (16)

$$e^{\mathbf{A}t} = \begin{bmatrix} e^t & -e^{-t} \\ e^t & e^{-t} \end{bmatrix}\begin{bmatrix} \frac{1}{2} & \frac{1}{2} \\ -\frac{1}{2} & \frac{1}{2} \end{bmatrix} = \begin{bmatrix} \dfrac{e^t + e^{-t}}{2} & \dfrac{e^t - e^{-t}}{2} \\ \dfrac{e^t - e^{-t}}{2} & \dfrac{e^t + e^{-t}}{2} \end{bmatrix}.$$ ◄

Nonhomogeneous Systems with Constant Coefficients

With these preliminaries out of the way, we may use the matrix exponential to solve linear, constant coefficient systems of differential equations. Consider, then, the nonhomogeneous problem

$$\mathbf{x}'(t) = \mathbf{A}\mathbf{x}(t) + \mathbf{f}(t)$$

or

$$\mathbf{x}' - \mathbf{A}\mathbf{x} = \mathbf{f}. \tag{17}$$

Proceeding as in Section 1.6.3 where an integrating factor was used to solve first-order linear equations, multiply (17) by $e^{-\mathbf{A}t}$ on the left, to obtain

$$e^{-\mathbf{A}t}(\mathbf{x}' - \mathbf{A}\mathbf{x}) = e^{-\mathbf{A}t}\mathbf{f}, \tag{18}$$

which is

$$(e^{-\mathbf{A}t}\mathbf{x})' = e^{-\mathbf{A}t}\mathbf{f}. \tag{19}$$

Antidifferentiate both sides of (19), to yield

$$e^{-\mathbf{A}t}\mathbf{x} = \int_a^t e^{-\mathbf{A}s}\mathbf{f}(s)\,ds + \mathbf{c}, \tag{20}$$

where $\mathbf{c}$ is an arbitrary constant vector. Now multiply both sides of (20) by the inverse of $e^{-\mathbf{A}t}$, which is $e^{\mathbf{A}t}$, to get the following key result, which should be compared to that of Section 1.6.3.

■ **THEOREM 5.12.5 General Solution of $\mathbf{x}' = \mathbf{A}\mathbf{x} + \mathbf{f}$**

Suppose that $\mathbf{A}$ is an $n \times n$ constant matrix and $\mathbf{f}(t)$ is a continuous $n \times 1$ vector-valued function on the interval I containing t_0. (Equivalently, $\mathbf{f}$ is a vector of continuous functions.) Then the general solution of

$$\mathbf{x}' = \mathbf{A}\mathbf{x} + \mathbf{f}$$

is

$$\begin{aligned} x(t) &= e^{\mathbf{A}t}\int_a^t e^{-\mathbf{A}s}\mathbf{f}(s)\,ds + e^{\mathbf{A}t}\mathbf{c} \\ &= \int_a^t e^{\mathbf{A}(t-s)}\mathbf{f}(s)\,ds + e^{\mathbf{A}t}\mathbf{c}, \end{aligned} \tag{21}$$

where $\mathbf{c}$ is an arbitrary constant vector.

In particular, the solution of

$$\mathbf{x}' = \mathbf{A}\mathbf{x} + \mathbf{f}, \qquad \mathbf{x}(t_0) = \mathbf{a}$$

is

$$\mathbf{x}(t) = \int_{t_0}^t e^{\mathbf{A}(t-s)}\mathbf{f}(s)\,ds + e^{\mathbf{A}(t-t_0)}\mathbf{a}. \tag{22}$$

■

EXAMPLE 5.12.4 General Solution and Initial Conditions

Use Theorem 5.12.5 to solve the nonhomogeneous differential equation

$$\begin{aligned} x_1' &= x_2 + 1, \qquad x_1(0) = a_1, \\ x_2' &= x_1 + e^t, \qquad x_2(0) = a_2 \end{aligned} \tag{23}$$

on the interval $[0, \infty)$.

Solution We already have computed in Example 5.12.3 that

$$e^{\mathbf{A}t} = \begin{bmatrix} \frac{1}{2}e^t + \frac{1}{2}e^{-t} & \frac{1}{2}e^t - \frac{1}{2}e^{-t} \\ \frac{1}{2}e^t - \frac{1}{2}e^{-t} & \frac{1}{2}e^t + \frac{1}{2}e^{-t} \end{bmatrix}$$

so that

$$e^{-\mathbf{A}t} = e^{\mathbf{A}(-t)} = \begin{bmatrix} \frac{1}{2}e^t + \frac{1}{2}e^{-t} & \frac{1}{2}e^{-t} - \frac{1}{2}e^t \\ \frac{1}{2}e^{-t} - \frac{1}{2}e^t & \frac{1}{2}e^t + \frac{1}{2}e^{-t} \end{bmatrix}$$

By (22), the solution of (23) is

$$\begin{aligned}
\mathbf{x} &= e^{\mathbf{A}t}\int_0^t e^{-\mathbf{A}s}\mathbf{f}(s)\,ds + e^{\mathbf{A}t}\mathbf{a} \\
&= e^{\mathbf{A}t}\int_0^t \begin{bmatrix} \frac{1}{2}e^s + \frac{1}{2}e^{-s} & \frac{1}{2}e^{-s} - \frac{1}{2}e^s \\ \frac{1}{2}e^{-s} - \frac{1}{2}e^s & \frac{1}{2}e^s + \frac{1}{2}e^{-s} \end{bmatrix} \begin{bmatrix} 1 \\ e^s \end{bmatrix} ds + e^{\mathbf{A}t}\mathbf{a} \\
&= \underbrace{e^{\mathbf{A}t}}\ \underbrace{\int_0^t \frac{1}{2}\begin{bmatrix} e^s + e^{-s} - e^{2s} + 1 \\ -e^s + e^{-s} + e^{2s} + 1 \end{bmatrix} ds} + \underbrace{e^{\mathbf{A}t}\mathbf{a}} \\
&= \underbrace{\begin{bmatrix} \frac{1}{2}e^t + \frac{1}{2}e^{-t} & \frac{1}{2}e^t - \frac{1}{2}e^{-t} \\ \frac{1}{2}e^t - \frac{1}{2}e^{-t} & \frac{1}{2}e^t + \frac{1}{2}e^{-t} \end{bmatrix}} \cdot \underbrace{\frac{1}{2}\begin{bmatrix} e^t - e^{-t} - \frac{e^{2t}}{2} + t + \frac{1}{2} \\ -e^t - e^{-t} + \frac{e^{2t}}{2} + t + \frac{3}{2} \end{bmatrix}} \\
&\quad + \underbrace{\begin{bmatrix} \frac{1}{2}e^t + \frac{1}{2}e^{-t} & \frac{1}{2}e^t - \frac{1}{2}e^{-t} \\ \frac{1}{2}e^t - \frac{1}{2}e^{-t} & \frac{1}{2}e^t + \frac{1}{2}e^{-t} \end{bmatrix}} \begin{bmatrix} a_1 \\ a_2 \end{bmatrix} \\
&= \frac{1}{4}\begin{bmatrix} 2te^t + e^t - e^{-t} \\ -4 + 2te^t + 3e^t + e^{-t} \end{bmatrix} + a_1\begin{bmatrix} \frac{1}{2}e^t + \frac{1}{2}e^{-t} \\ \frac{1}{2}e^t - \frac{1}{2}e^{-t} \end{bmatrix} + a_2\begin{bmatrix} \frac{1}{2}e^t - \frac{1}{2}e^{-t} \\ \frac{1}{2}e^t + \frac{1}{2}e^{-t} \end{bmatrix}.
\end{aligned} \tag{24}$$

◄

Note that this technique works for forcing terms not included by the method of Section 5.11. If we cannot evaluate the integrals, the definite integral is included in the solution.

Exercises

In Exercises 1 through 6, write the system as $\mathbf{x}' = \mathbf{Ax} + \mathbf{f}$. Find $e^{\mathbf{A}t}$ using Theorem 5.12.3 or 5.12.4. Then solve the differential equation, using Theorem 5.12.5. If no initial conditions are given, find the general solution.

1. $x_1' = 3x_1 + x_2 + 1, \quad x_1(0) = 1,$
$x_2' = -2x_1 - 2, \quad x_2(0) = 2$

2. $x_1' = x_1 + 2,$
$x_2' = x_1 - x_2 + 3$

3. $x' = x - 2y + 6e^{-t},$
$y' = -2x + 4y + e^t$

4. $x' = 3x + y + 1, \quad x(0) = 0,$
$y' = -6x - 2y + 1, \quad y(0) = 0$

5. $x_1' = 6x_1 + 2x_2 + e^{2t},$
$x_2' = 2x_1 + 3x_2$

6. $x_1' = 3x_1 + x_2,$
$x_2' = -5x_1 - 3x_2 + e^t$

7. Let $\mathbf{A} = \begin{bmatrix} 0 & 1 \\ 0 & 0 \end{bmatrix}$.

i) Compute $e^{\mathbf{A}t}$, using the series (1).

ii) Compute $(e^{\mathbf{A}t})^{-1}$ and verify $(e^{\mathbf{A}t})^{-1} = e^{-\mathbf{A}t}$.

8. Let $\mathbf{A} = \begin{bmatrix} 0 & 1 & 0 \\ 0 & 0 & 1 \\ 0 & 0 & 0 \end{bmatrix}$. Compute $e^{\mathbf{A}t}$, using the series (1).

9. Let $\mathbf{A} = \begin{bmatrix} 0 & 1 \\ -1 & 0 \end{bmatrix}$. Compute $e^{\mathbf{A}t}$, using the series (1).

10. Let $\mathbf{A} = \begin{bmatrix} 0 & 1 \\ 0 & 0 \end{bmatrix}$, $\mathbf{B} = \begin{bmatrix} 0 & 0 \\ 1 & 0 \end{bmatrix}$,

and $\mathbf{C} = \mathbf{A} + \mathbf{B} = \begin{bmatrix} 0 & 1 \\ 1 & 0 \end{bmatrix}$. Compute $e^{\mathbf{A}t}$, $e^{\mathbf{B}t}$, $e^{\mathbf{C}t}$, and verify $e^{\mathbf{A}t}e^{\mathbf{B}t} \neq e^{(\mathbf{A}+\mathbf{B})t}$.

11. Suppose that $e^{\mathbf{A}t}e^{\mathbf{B}t} = e^{(\mathbf{A}+\mathbf{B})t}$ for two $n \times n$ matrices $\mathbf{A}$, $\mathbf{B}$. Show that $\mathbf{AB} = \mathbf{BA}$. (*Hint*: Differentiate twice and evaluate at zero.)

12. Suppose $\mathbf{A}$ is an invertible $n \times n$ matrix. Verify that

$$\int e^{\mathbf{A}t}\, dt = \mathbf{A}^{-1}e^{\mathbf{A}t} + \mathbf{C},$$

where $\mathbf{C}$ is an arbitrary $n \times n$ constant matrix.

Exercises 13 through 21 provide an alternative way to compute $e^{\mathbf{A}t}$ and introduce the important concept of a **similarity transformation.**

13. Suppose $\mathbf{A}$ is a 2×2 matrix with distinct eigenvalues λ_1, λ_2 and nonzero eigenvectors $\mathbf{u}_1$, $\mathbf{u}_2$. Let $\mathbf{U} = [\mathbf{u}_1, \mathbf{u}_2]$ be the 2×2 matrix with $\mathbf{u}_1$, $\mathbf{u}_2$ as columns. Verify that

$$\mathbf{A}[\mathbf{u}_1, \mathbf{u}_2] = [\mathbf{u}_1, \mathbf{u}_2]\begin{bmatrix} \lambda_1 & 0 \\ 0 & \lambda_2 \end{bmatrix}, \tag{25}$$

or, equivalently,

$$\mathbf{AU} = \mathbf{U\Lambda},$$

where

$$\mathbf{\Lambda} = \begin{bmatrix} \lambda_1 & 0 \\ 0 & \lambda_2 \end{bmatrix}$$

is the diagonal matrix with the eigenvalues as entries.

14. Suppose that $\mathbf{U}$ is an invertible $m \times m$ matrix and $\mathbf{B}$ is an $m \times m$ matrix. Verify that, for any integer $n \geq 0$,

$$(\mathbf{UBU}^{-1})^n = \mathbf{UB}^n\mathbf{U}^{-1}.$$

Then conclude from (1) that

$$e^{\mathbf{UBU}^{-1}t} = \mathbf{U}e^{\mathbf{B}t}\mathbf{U}^{-1}. \tag{26}$$

15. Show that, if

$$\Lambda = \begin{bmatrix} \lambda_1 & & 0 \\ & \ddots & \\ 0 & & \lambda_n \end{bmatrix}$$

is a diagonal matrix, then

$$e^{\Lambda t} = \begin{bmatrix} e^{\lambda_1 t} & & 0 \\ & \ddots & \\ 0 & & e^{\lambda_n t} \end{bmatrix}. \tag{27}$$

Exercises 13, 14, and 15 show that, if $\mathbf{A}$ is a 2×2 matrix and has two distinct eigenvalues λ_1, λ_2 with distinct eigenvectors $\mathbf{u}_1$, $\mathbf{u}_2$, then

$$e^{\mathbf{A}t} = \mathbf{U}\begin{bmatrix} e^{\lambda_1 t} & 0 \\ 0 & e^{\lambda_2 t} \end{bmatrix}\mathbf{U}^{-1}, \tag{28}$$

where $\mathbf{U}$ has for columns the eigenvectors $\mathbf{u}_1$, $\mathbf{u}_2$.

16. Let $\mathbf{A} = \begin{bmatrix} 3 & 1 \\ -2 & 0 \end{bmatrix}$.
Compute $e^{\mathbf{A}t}$, using (28).

17. Let $\mathbf{A} = \begin{bmatrix} 1 & -2 \\ -2 & 4 \end{bmatrix}$.
Compute $e^{\mathbf{A}t}$, using (28).

18. Let $\mathbf{A} = \begin{bmatrix} 3 & 1 \\ -6 & -2 \end{bmatrix}$.
Compute $e^{\mathbf{A}t}$, using (28).

19. Let $\mathbf{A} = \begin{bmatrix} 6 & 2 \\ 2 & 3 \end{bmatrix}$.
Compute $e^{\mathbf{A}t}$, using (28).

20. Let $\mathbf{A} = \begin{bmatrix} 3 & 1 \\ -5 & -3 \end{bmatrix}$.
Compute $e^{\mathbf{A}t}$, using (28).

21. Let $\mathbf{A} = \begin{bmatrix} -3 & 3 \\ 3 & 5 \end{bmatrix}$.
Compute $e^{\mathbf{A}t}$, using (28).

22. Show that if $\mathbf{A}$ is an $n \times n$ matrix with n distinct eigenvalues $\lambda_1, \ldots, \lambda_n$ with corresponding eigenvectors $\mathbf{u}_1, \ldots, \mathbf{u}_n$, then

$$e^{\mathbf{A}t} = \mathbf{U}\begin{bmatrix} e^{\lambda_1 t} & & 0 \\ & \ddots & \\ 0 & & e^{\lambda_n t} \end{bmatrix}\mathbf{U}^{-1}, \tag{29}$$

where $\mathbf{U} = [\mathbf{u}_1, \ldots, \mathbf{u}_n]$ has the eigenvectors as columns.

The **Cayley–Hamilton Theorem** has a consequence that if $\mathbf{A}$ is an $n \times n$ matrix, then

$$e^{\mathbf{A}t} = \alpha_0(t)\mathbf{I} + \alpha_1(t)\mathbf{A} + \cdots + \alpha_{n-1}(t)\mathbf{A}^{n-1} \tag{30}$$

for some scalar functions $\alpha_0(t), \ldots, \alpha_{n-1}(t)$. If $\mathbf{u}$ is an eigenvector for the eigenvalue λ, then

$$e^{\mathbf{A}t}\mathbf{u} = \alpha_0(t)\mathbf{I}\mathbf{u} + \alpha_1(t)\mathbf{A}\mathbf{u} + \cdots + \alpha_{n-1}(t)\mathbf{A}^n\mathbf{u}$$

or

$$e^{\lambda t}\mathbf{u} = [\alpha_0(t) + \alpha_1(t)\lambda + \cdots + \alpha_{n-1}(t)\lambda^n]\mathbf{u}. \tag{31}$$

23. If $\mathbf{A}$ is a 2×2 matrix with distinct eigenvalues λ_1, λ_2, then (30) becomes

$$e^{\mathbf{A}t} = \alpha_0(t)\mathbf{I} + \alpha_1(t)\mathbf{A}, \tag{32}$$

and (31) yields

$$\alpha_0(t) + \lambda_1\alpha_1(t) = e^{\lambda_1 t},$$
$$\alpha_0(t) + \lambda_2\alpha_1(t) = e^{\lambda_2 t},$$

or

$$\begin{bmatrix} 1 & \lambda_1 \\ 1 & \lambda_2 \end{bmatrix}\begin{bmatrix} \alpha_0(t) \\ \alpha_1(t) \end{bmatrix} = \begin{bmatrix} e^{\lambda_1 t} \\ e^{\lambda_2 t} \end{bmatrix}. \tag{33}$$

Thus, to find $e^{\mathbf{A}t}$ for a 2×2 matrix $\mathbf{A}$, one can find the eigenvalues λ_1, λ_2, solve (33) for α_0, α_1, and then use (32).

a) Let $\mathbf{A} = \begin{bmatrix} 3 & 1 \\ -2 & 0 \end{bmatrix}$.
Find $e^{\mathbf{A}t}$ by this method.

b) Let $\mathbf{A} = \begin{bmatrix} 1 & -2 \\ -2 & 4 \end{bmatrix}$.
Find $e^{\mathbf{A}t}$ by this method.

c) Let $\mathbf{A} = \begin{bmatrix} 3 & 1 \\ -6 & -2 \end{bmatrix}$.
Find $e^{\mathbf{A}t}$ by this method.

24. If $\mathbf{A}$ is a 3×3 matrix with distinct eigenvalues λ_1, λ_2, λ_3, then (30) is

$$e^{\mathbf{A}t} = \alpha_0(t)\mathbf{I} + \alpha_1(t)\mathbf{A} + \alpha_2(t)\mathbf{A}^2, \tag{34}$$

where α_0, α_1, α_2 are the solutions of

$$\begin{bmatrix} 1 & \lambda_1 & \lambda_1^2 \\ 1 & \lambda_2 & \lambda_2^2 \\ 1 & \lambda_3 & \lambda_3^2 \end{bmatrix} \begin{bmatrix} \alpha_0 \\ \alpha_1 \\ \alpha_2 \end{bmatrix} = \begin{bmatrix} e^{\lambda_1 t} \\ e^{\lambda_2 t} \\ e^{\lambda_3 t} \end{bmatrix}. \quad (35)$$

Find $e^{\mathbf{A}t}$ for

$$\mathbf{A} = \begin{bmatrix} 1 & 1 & 0 \\ 0 & 0 & 1 \\ 0 & 0 & 2 \end{bmatrix}$$

by this method.

25. Verify $(e^{\mathbf{A}t})' = \mathbf{A}e^{\mathbf{A}t}$ by differentiating the series (1) with respect to t.

26. Verify that $(e^{\mathbf{A}t})^{-1} = e^{-\mathbf{A}t}$ by multiplying the series for $e^{\mathbf{A}t}$ and $e^{-\mathbf{A}t}$ to get $\mathbf{I}$.

27. Let $\mathbf{F}(t) = e^{\mathbf{A}(t+s)}$, $\mathbf{G}(t) = e^{\mathbf{A}t}e^{\mathbf{A}s}$ for a constant $n \times n$ matrix $\mathbf{A}$ and fixed but unknown scalar s. Verify that $\mathbf{F}^{(n)}(0) = \mathbf{G}^{(n)}(0) = \mathbf{A}^n e^{\mathbf{A}s}$, and conclude, by the uniqueness of power series, that $e^{\mathbf{A}(t+s)} = e^{\mathbf{A}t}e^{\mathbf{A}s}$.

Comment 1 The method of Exercises 23 and 24 seems so simple that we should probably explain why it is relegated to the exercises. It is not very practical for the larger-sized matrices that occur in real applications. There are two reasons for this. First, computing the powers of $\mathbf{A}$ becomes a lot of work. Surprisingly enough, computing $\mathbf{A}^2$ involves about as much "work" as inverting $\mathbf{A}$. Second, while the matrix

$$\begin{bmatrix} 1 & \lambda_1 & \cdots & \lambda_1^{n-1} \\ 1 & \lambda_2 & \cdots & \lambda_2^{n-1} \\ \vdots & & & \vdots \\ 1 & \lambda_n & \cdots & \lambda_n^{n-1} \end{bmatrix}$$

(known as the Vandermond matrix) is known to be invertible if the λ_i are distinct, it is also known to become very *ill-conditioned* as n increases. In other words, the equation for the α_i can be difficult to work with and are very susceptible to round-off error when they are solved for numerically.

Comment 2 The eigenvalue/eigenvector approach of Exercises 13 through 21 is very important even though the matrix exponential is usually not computed.

CHAPTER

6

Numerical Methods

6.1 Introduction

Given a differential equation, there are several different ways we can begin to analyze it. In Chapters 1, 2, 3, and 5 we obtained explicit formulas for the solutions in terms of known functions. In Chapter 4 we obtained series solutions. These methods are often collectively referred to as the **analytic** solution of a differential equation. In Chapters 7, 8, and 9 we try to determine **qualitative** properties of the solutions, such as boundedness and stability of equilibria.

This chapter will introduce the **numerical** approach. Here we are interested in obtaining estimates of the values of solutions at certain discrete points. With the advent of modern, widely accessible digital computers, many very complicated differential equations can now be solved numerically, and a good understanding of numerical methods is increasingly important. A large variety of numerical methods have become available. Some are included in software packages such as MAPLE, MATHEMATICA, MATLAB, or one of the various simulation packages such as ACSL, PSILAB, or SIMULINK. Other software which is designed to deal with more complex differential equations can be found in software libraries, such as NETLIB, on the Internet.

These numerical procedures, however, have not changed the need for the other approaches. A qualitative analysis can show what types of numerical methods should be used, and help to determine whether the numerical results are reasonable.

On the other hand, plunging immediately into numerical methods can lead to unnecessary expense. We are aware of a computer program that was being used on a daily basis at a large corporation. Every time the program was used it numerically solved a system of differential equations several times. A mathematics major replaced the subroutine that solved the differential equation with a formula for the exact solution, obtained using the methods of Chapter 5. The result was a savings of several thousand dollars a year in computer time, and a pay raise for the programmer.

Comment on Computations It is not necessary to have access to a computer in order to utilize this chapter. All of the work may, in principle, be done on a pocket calculator, and some may be done by hand (but that is not advised except for the simplest problems). However, most of the homework is best done on at least a programmable pocket calculator. Such homework problems are denoted by an asterisk (*). ◄

When comparing your answers with those given in the text, keep in mind that any of the following may affect the number of significant digits in your answers:

- Precision of the arithmetic on your machine.
- How round-off and floating-point operations are carried out.
- Order in which computation has been done (adding large numbers to small numbers and subtracting almost equal numbers will often give a loss of significance).
- Built-in functions, such as sin, cos, exp, log, a^x, may have less precision than arithmetic operations.
- How underflow and overflow are handled.

Unless noted otherwise, all of the numbers in this chapter were computed in double precision, even if only the first few (rounded) digits are given.

The basic problem of the first six sections of this chapter is:

Given the initial value problem

$$y' = f(x, y), \qquad y(a) = y_0, \tag{1}$$

estimate the value of y at a later time $x = b$.

In some cases we are interested only in $y(b)$. In others, we also want some intermediate values of y. All of our estimates of the values of the solution of (1) will be arrived at using algorithms that follow the following pattern:

1. $N + 1$ values of the independent variable x will be chosen from a to b:

$$a = x_0 < x_1 < x_2 < \cdots < x_N = b.$$

2. Let $y_0 = y(x_0)$ be the initial condition.
3. Given $\{y_0, \ldots, y_n\}$, let the estimate y_{n+1} for $y(x_{n+1})$, the value of the solution at time x_{n+1}, be computed by some method. We now have $\{y_0, \ldots, y_n, y_{n+1}\}$. Repeat this step until we have computed $\{y_0, \ldots, y_N\}$.

4. Then y_N is an estimate of $y(x_N) = y(b)$. Because the estimates $y_1, \ldots, y_N$ are computed one after another, we shall refer to each time (3) is done as one **step** of the method.

The amount the independent variable changes each time is $x_n - x_{n-1}$ and is called the **step size**. We shall use h to denote the step size, so that

$$h_n = x_n - x_{n-1}.$$

Most numerical methods for solving Eq. (1), when actually used in practice, employ changes in step size. However, except in a couple of the homework exercises, we assume a **constant step size**.

Notation
The following notation will be used throughout this chapter. The differential equation is

$$y'(x) = f(x, y(x)), \qquad y(a) = y_0. \tag{2}$$

The points $x_0, \ldots, x_N$ are a **partition** (also a grid, or mesh) of the interval $[a, b]$:

$$a = x_0 < x_1 < \cdots < x_N = b.$$

The step size is h and

$$x_{n+1} = x_n + h.$$

The function $y(x)$ is the solution of the differential equation (2). The estimate of $y(x_n)$ provided by the numerical method is y_n. The **error** e_n at the nth step is the difference between the true solution $y(x_n)$ and our estimate y_n so that

$$e_n = y(x_n) - y_n.$$

How then do we compute the estimate y_{n+1} for the value of the solution y at time x_{n+1} if we already have the estimates $\{y_0, \ldots, y_n\}$? All the techniques described in subsequent sections are based on combinations of two ideas:

Taylor Series
Estimate $y(x_{n+1}) = y(x_n + h)$, using the Taylor series of the solution y at time x_n.

Integral Formula
Integrating Eq. (2) from x_n to x_{n+1} gives

$$y(x_{n+1}) - y(x_n) = \int_{x_n}^{x_{n+1}} \frac{dy}{dx}\, dx = \int_{x_n}^{x_n+h} f(x, y)\, dx,$$

so that

$$y(x_{n+1}) = y(x_n) + \int_{x_n}^{x_n+h} f(x, y)\, dx. \tag{3}$$

Estimating the integral in Eq. (3) then gives an estimate y_{n+1} of $y(x_{n+1})$.

6.2 Euler's Method

We now develop our first method. Suppose that we wish to numerically solve

$$y' = f(x, y), \qquad y(a) = y_0, \tag{1}$$

and obtain an estimate for $y(b)$. We assume that f, f_y are continuous, so that the initial value problem (1) has a unique solution. (See Section 1.5.) Subdivide the interval $[a, b]$ with $N + 1$ mesh points $x_0, \ldots, x_N$, with $x_0 = a$, $x_N = b$. Let $h = (b - a)/N$ be the step size, so that $x_n = a + hn$. In some problems h is given and $N = (b - a)/h$.

Let

$$y_0 = y(a) = y(x_0).$$

The first-order Taylor polynomial (Section 4.2) for the solution y of (1) at a point $\hat{x}$ is

$$y(x) \approx y(\hat{x}) + y'(\hat{x})(x - \hat{x}). \tag{2}$$

[This is the same as using the tangent line at $(\hat{x}, y(\hat{x}))$ as an approximation for $y(x)$ near $(\hat{x}, y(\hat{x}))$.] Thus, letting $x = \hat{x} + h$ in (2) gives

$$y(\hat{x} + h) \approx y(\hat{x}) + y'(\hat{x})h. \tag{3}$$

Pictorially, we have Figure 6.2.1. But $y'(\hat{x}) = f(\hat{x}, y(\hat{x}))$, since y is assumed to be a solution of the differential equation (1). Thus (3) becomes

$$y(\hat{x} + h) \approx y(\hat{x}) + f(\hat{x}, y(\hat{x}))h. \tag{4}$$

This formula gives a way of estimating y at time x_{n+1} given y at time x_n, by letting $\hat{x} = x_n$ in Eq. (4) and using our previous estimate y_n for $y(x_n)$. The resulting numerical method is called **Euler's method.**

FIGURE 6.2.1

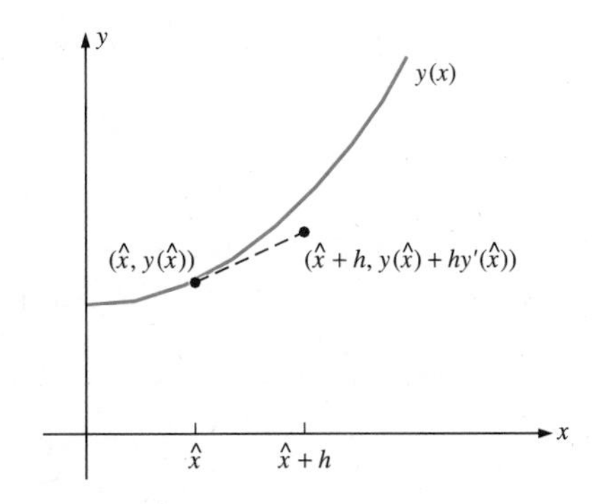

Euler's Method

For the solution of $y' = f(x, y)$, $y(x_0) = y_0$,

1. Let $y_0 = y(a)$.
2. Recursively compute $y_1, \ldots, y_N$ by

$$y_{n+1} = y_n + hf(x_n, y_n) \tag{5}$$

where

$$x_n = a + hn, \qquad h = \frac{b - a}{N}.$$

3. y_N is an estimate of $y(b)$.

Note that (5) is a difference equation as discussed in Chapter 8.

EXAMPLE 6.2.1 Euler's Method

Let y be the solution of the differential equation

$$y' - xy = x, \qquad y(0) = 1. \tag{6}$$

Estimate $y(1)$ by Euler's method using a step size of $h = 0.25$.

SOLUTION The differential equation is

$$y' = x + xy, \qquad y(0) = 1,$$

so that $f(x, y) = x + xy$. We have [from Eq. (6)] that

$$x_0 = 0, \qquad y_0 = y(0) = 1.$$

The recursive relationship (5) is then

$$y_{n+1} = y_n + hf(x_n, y_n) = y_n + h(x_n + x_n y_n). \tag{7}$$

Thus we get

$$y_1 = y_0 + h(x_0 + x_0 y_0) = 1 + 0.25(0 + 0 \cdot 1) = 1;$$

$$x_1 = x_0 + h = 0.25; \qquad y_2 = y_1 + h(x_1 + x_1 y_1) = 1 + 0.25(0.25 + 0.25 \cdot 1) = 1.125;$$

$$x_2 = x_1 + h = 0.5; \qquad y_3 = y_2 + h(x_2 + x_2 y_2) = 1.125 + 0.25(0.5 + 0.5 \cdot 1.125) = 1.391;$$

$$x_3 = x_2 + h = 0.75; \qquad y_4 = y_3 + h(x_3 + x_3 y_3) = 1.391 + 0.25(0.75 + 0.75 \cdot 1.391) = 1.839.$$

The estimate for $y(1)$ is $y_4 = 1.839$ (to three places). ◄

FIGURE 6.2.2

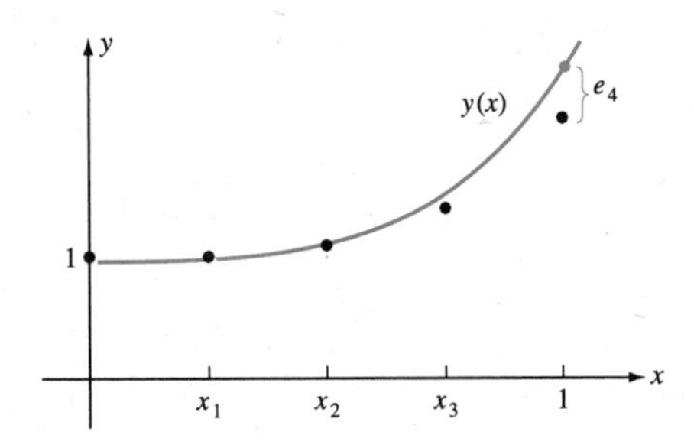

Equation (6) can be solved using separation of variables or an integrating factor to get the actual solution:

$$y = 2e^{x^2/2} - 1, \tag{8}$$

so that

$$y(1) = 2e^{0.5} - 1 \approx 2.2974425. \tag{9}$$

The error in our estimate of $y(1)$ is then

$$e_4 = y(1) - y(x_4) = 0.458. \tag{10}$$

These calculations are pictorially represented in Figure 6.2.2.

It seems natural to try to get a smaller error by taking smaller steps.

EXAMPLE 6.2.2 Euler's Method

Again estimate $y(1)$ using Euler's method, where y is the solution of (6), but this time use a step size of $h = 0.125$.

Solution We have $a = 0$, $b = 1$, $h = 0.125$, and $N = (b - a)/n = 8$. Thus

$$x_0 = 0, \quad x_n = nh, \quad x_8 = 1.$$

Again use the difference relationship (7) and compute (x_n, y_n) for $n = 1, \ldots, 8$. The computed y_n and the actual solution are graphed in Figure 6.2.3. The estimate for $y(1)$ is

$$y(1) = y(x_8) \approx y_8 = 2.048. \tag{11}$$

FIGURE 6.2.3

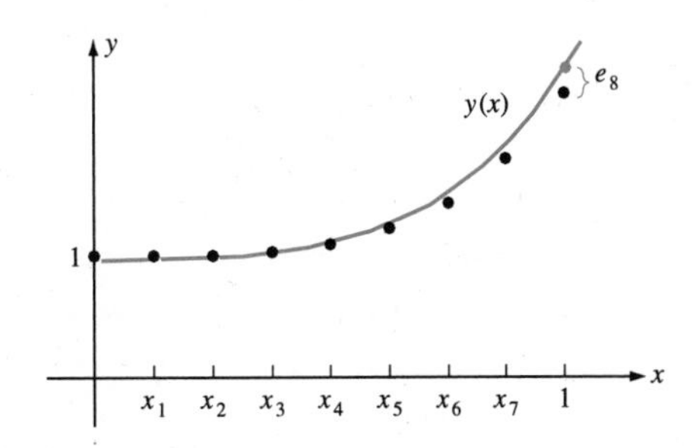

Note that the error in the estimate (11) is

$$e_8 = y(1) - y_8 = 0.249. \tag{12}$$

Thus reducing the step size by a factor of one-half has reduced the error (10) by about one-half also. This in fact turns out to be a general property of Euler's method.

A method for solving (1) is said to be of **order** r [written $O(h^r)$] on the interval $[a, b]$ if there is a constant M depending on a, b, f, but not the step size h or the point $x_n \in [a, b]$, such that

$$|e_n| = |y(x_n) - y_n| \leq Mh^r. \tag{13}$$

THEOREM 6.2.1 Order of Euler's Method

Euler's method is a first-order method. That is, if $x_n \in [a, b]$ for a given step size h, then

$$|e_n| = |y(x_n) - y_n| \leq Mh \tag{14}$$

for some constant M independent of n, h.

This result, which of course requires some technical assumptions on $f(x, y)$ and a careful consideration of the different types of error, will be proved in Section 6.3. Note that Eq. (14) says that, by taking h to be half as large, we should expect about half as much error, and that is what was observed in (10), (12). Theorem 6.2.1 also seems to suggest that, in order to get answers accurate to several significant figures using Euler's method, we would have to take very small steps, and hence a large number of steps. Table 6.2.1 gives the result of solving Eq. (6) by Euler's method for several step sizes.

We can give a *heuristic* argument for Theorem 6.2.1 as follows. At any value of $\hat{x}$ we have, by Taylor's Theorem from calculus, that the error in using

$$y(\hat{x}) + y'(\hat{x}) \qquad \text{for } y(\hat{x} + h) \tag{15}$$

is

$$R_1(x) = \frac{1}{2}y''(\xi)(x - \hat{x})^2, \tag{16}$$

TABLE 6.2.1 EULER'S METHOD FOR (6)

Step Size h	Steps N	Estimate y_N for $y(1)$	Error at $b = 1$ $e_N = y(1) - y_N$
0.1	10	2.09422	0.20322
0.01	100	2.27564	0.02180
0.001	1000	2.29525	0.00220
0.0001	10000	2.29722	0.00022

where ξ is some number between x and $\hat{x}$. Since $x - \hat{x} = h$, the error induced at step $n + 1$ (*if* y_n *is exact*) can be estimated by

$$|R_1(x)| \leq \frac{1}{2}|y''(\xi)|h^2.$$

Let $\tilde{M}$ be the maximum of $\frac{1}{2}|y''(x)|$ on $[a, b]$. Then the error at each step is estimated to be $\tilde{M}h^2$. The number of steps is $N = (b - a)/h$. Thus the error could be estimated as the amount of error at each step times the number of steps, or

$$\tilde{M}h^2N = \tilde{M}h^2\frac{(b-a)}{h} = \tilde{M}(b-a)h.$$

Taking $M = \tilde{M}(b - a)$ would give Eq. (14).

This way of estimating the error is incorrect, however, since it overlooks the fact that we have to use an estimate y_n for $y(x_n)$ in Eq. (15) when $x = x_n$ and thus the error is **compounded**. It also overlooks roundoff error. However, it still remains a reasonable rule of thumb that, if the error on a given step is $O(h^{r+1})$ (sometimes called the **truncation** or **local error**), then one should expect the method to be $O(h^r)$ for h not so small that roundoff error becomes important.

The exercises that follow illustrate several aspects of Euler's method that are typical of numerical methods. Additional properties are developed in the (optional) Section 6.3.

Exercises

[Exercises marked with an asterisk are suggested only for programmable calculators and computers.] For Exercises 1 through 11, compute an estimate for the value of $y(b)$, using Euler's method, given the indicated step size.

1. $y' = y - x$, $y(0) = 1$, $b = 2$, $h = 0.5$

2. $y' = xy$, $y(0) = 1$, $b = 1$, $h = 0.2$

3. $y' = -y^2x$, $y(0) = 1$, $b = 2$, $h = 1$

4. $y' = 3y - 2x$, $y(0) = 0$, $b = 2$, $h = 0.5$

5. $y' = 2y - 4x$, $y(0) = 1$, $b = 2$, $h = 0.5$

6. $y' = xy - x$, $y(0) = 1$, $b = 4$, $h = 0.2$

7. $y' = \sin y$, $y(0) = 0$, $b = 4$, $h = 0.5$

8. $y' = \sin y$, $y(0) = 1$, $b = 4$, $h = 0.2$

9. **a)** $y' = -20y$, $y(0) = 1$, $b = 2$, $h = 0.2$

b) On the same (x, y) axis, plot the points (x_n, y_n) from part (a) and the true solution.

10. (Continuation of Exercise 9)

a) $y' = -20y$, $y(0) = 1$, $h = 0.1$, $b = 2$

b)* On the same (x, y) axis, plot the points (x_n, y_n) from part (a) and the true solution.

11.* (Continuation of Exercises 9, 10)

a) $y' = -20y$, $y(0) = 1$, $h = 0.01$, $b = 2$

b) On the same (x, y) axis, plot the points (x_n, y_n) from part (a) and the true solution.

Exercises 9, 10, 11 show that for some equations, Euler's method produces numerical solutions that resemble the actual solution only for small step sizes. In this example, this behavior is related to a property called **stiffness**, which is discussed in Section 6.3.

12. Estimate $y(1)$ for the solution of $y' = y^2$, $y(0) = 1$, using Euler's method with a step size of $h = 0.5$.

13. (Continuation of Exercise 12). Estimate $y(1)$ for the solution of $y' = y^2$, $y(0) = 1$, using Euler's method with step sizes of $h = 0.2$, $h = 0.1$, $h = 0.01$,* $h = 0.001$.*

14. (Continuation of Exercises 12, 13). Explain what you observed about the estimates for $y(1)$ by solving the differential equation $y' = y^2$, $y(0) = 1$, by separation of variables.

Exercises 15 through 17 illustrate the point that, while analytically obtained solutions cannot cross equilibria if $f(x, y)$, $f_y(x, y)$ are continuous, the numerical solution can "jump" over an equilibrium.

15. Using Euler's method, calculate estimates for the solution of the differential equation

$$\begin{aligned} y' &= 1 - 2y + y^2, \\ y(a) &= -5, \qquad a \le x \le b, \end{aligned} \tag{17}$$

at points (x_n, y_n) where $a = 0$, $b = 2$, $h = 0.2$. Graph these points (x_n, y_n) and sketch what you think the solution would look like.

16. (Continuation of Exercise 15). Sketch the solutions of $y' = 1 - 2y + y^2$ using the techniques of Section 1.4. Observe that $y = 1$ is an equilibrium solution. Compare this picture with that of Exercise 15.

17. (Continuation of Exercises 15, 16). Solve (17) with $a = 0$, $b = 2$, using Euler's method with step of $h = 0.1$, plot* all values of (x_n, y_n), and compare to the analysis of Exercise 16.

Euler's method can also be derived using integration.

18. **i)** By antidifferentiating both sides of $y' = f(x, y)$ with respect to x, show that

$$y(\hat{x} + h) - y(\hat{x}) = \int_{\hat{x}}^{\hat{x}+h} f(x, y(x))\, dx. \tag{18}$$

ii) One estimate for an integral $\int_c^d g(x)\, dx$ (**constant-slope estimate**) is $g(\xi)(d - c)$, where ξ is a point between c and d. Using $c = \hat{x}$, $d = \hat{x} + h$, and $\xi = \hat{x}$, obtain

$$y(\hat{x} + h) - y(\hat{x}) \approx f(\hat{x}, y(\hat{x}))h \tag{19}$$

from (18) using the constant-slope estimate.

iii) Let $\hat{x} = x_n$ and use the estimate y_n for $y(x_n)$ to derive Euler's formula (5) from (19).

Suppose that we are using an rth-order method with N_1 steps, so that

$$y_{N_1} - y(b) \approx Mh^r. \tag{20}$$

If we also use a step size that is one-half as big, we need to take $N_2 = 2N_1$ steps to reach b from a. Then we have

$$y_{N_2} - y(b) \approx M\left(\frac{h}{2}\right)^r. \tag{21}$$

From (20) and (21) we find, after a bit of algebra, that we might expect that

$$\frac{y_{N_1} - y_{N_2}}{y_{N_2} - y_{N_3}} \approx 2^r, \tag{22}$$

where $N_3 = 2N_2$. This heuristic argument suggests that one can examine the order of the method by examining the left-hand side of (22).

19. Assume equality holds in (20) and (21), and derive (22) with $\approx$ replaced by $=$, to heuristically justify (22).

20. Solve the differential equation $y' = y + x^2$, $y(0) = 0$, to obtain an estimate y_N for $y(2)$.

Using each of the steps $h = 0.2$, $h = 0.1$, $h = 0.05$, $h = 0.025$, and $h = 0.0125$, compute the respective estimates y_{N_1}, y_{N_2}, y_{N_3}, y_{N_4}, and y_{N_5} for $y(2)$, where $N_1 = 10$, $N_2 = 20$, $N_3 = 40$, $N_4 = 80$, and $N_5 = 160$. Then compute

$$\frac{y_{N_1} - y_{N_2}}{y_{N_2} - y_{N_3}}, \quad \frac{y_{N_2} - y_{N_3}}{y_{N_3} - y_{N_4}}, \quad \frac{y_{N_3} - y_{N_4}}{y_{N_4} - y_{N_5}}.$$

If Euler's is a first-order method, from (22) with $r = 1$ we would expect these ratios to be approaching 2. Do they appear to be doing so?

6.3 An Analysis of Euler's Method (Optional)

The other sections in this chapter present various numerical methods for solving ordinary differential equations. This section, which may be covered any time after Section 6.2, analyzes the Euler method of Section 6.2, and introduces several important concepts. Recall from Section 6.2 that $y(x)$ is the solution of

$$y' = f(x, y), \qquad y(a) = y_0, \tag{1}$$

where $x_0 = a$, $x_n = a + nh$, $x_N = b$, $h = (b - a)/N$, and Euler's method is

$$y_0 = y(a), \qquad y_{n+1} = y_n + hf(x_n, y_n), \qquad n \geq 0. \tag{2}$$

The error in this estimate of $y(x_n)$ (ignoring round-off) is

$$e_n = y(x_n) - y_n. \tag{3}$$

6.3.1 Discretization Error

In an actual computation we have two sources of error. One is due to the fact that the difference in Eq. (2) is only approximating the differential equation (1). This is called **discretization error**. The other major source is **rounding error** when doing the computations.

We consider the discretization error first. Let

$$E(h) = \max_{1 \leq n \leq N} |y_n - y(x_n)| = \max_{1 \leq n \leq N} |e_n|. \tag{4}$$

Then $E(h)$ is the **global discretization** (or **truncation error**). For a given h, $E(h)$ is the largest error that occurs in our estimate at any mesh point. In order to estimate $E(h)$ for Euler's method, we need the following facts:

Fact 1
For constants α, β,

$$\lim_{h \to 0} (1 + h\alpha)^{\beta/h} = e^{\alpha\beta}. \tag{5}$$

(This limit may be verified using l'Hôpital's rule; see Exercise 1 at the end of this section.)

Fact 2
Taylor's Theorem applied to the solution $y(x)$ gives (requires y being twice differentiable):

$$y(x+h) = y(x) + hy'(x) + R_1(x,h), \tag{6}$$

where

$$R_1(x,h) = \frac{1}{2}y''(\xi)h^2 \tag{7}$$

with ξ between x and $x+h$.

Fact 3
The Mean-Value Theorem (which is, of course, related to Taylor's Theorem) says

$$G(z) - G(w) = G'(\xi)(z-w), \tag{8}$$

with ξ between z and w if $G(z)$ is continuously differentiable.

We are now ready to prove Theorem 6.2.1, which we restate here as Theorem 6.3.1. The proof is complete except that we do not carefully specify some of the sets over which we take maxima.

■ THEOREM 6.3.1 Order of Euler's Method

Suppose that $f(x, y)$ is continuously differentiable with respect to both x and y. Suppose also that the solution y of

$$y' = f(x,y), \qquad y(a) = y_0, \tag{9}$$

exists on the interval $a \le x \le b$. Then there is a constant M, so that the global truncation error [Eqs. (3), (4)] for Euler's method satisfies

$$E(h) \le Mh. \tag{10}$$

Thus Euler's method is a first-order method. ■

Proof
Since $f(x, y)$ is continuously differentiable and $y(x)$ is continuously differentiable, we see that y is *twice* differentiable, since differentiating (9) with respect to x gives

$$y'' = f_x(x,y) + f_y(x,y)y'. \tag{11}$$

Thus Facts 2 and 3 can be used. We now begin to estimate the error. Our immediate goal is to construct a difference equation to bound the error. From (3) we have

$$e_{n+1} = \underbrace{y(x_{n+1})} - \underbrace{y_{n+1}}. \tag{12}$$

Using (6) on $y(x_{n+1}) = y(x_n + h)$ and the difference equation (2) from Euler's method for y_{n+1}, Eq. (12) can be rewritten as

$$\begin{aligned} e_{n+1} &= \underbrace{y(x_n) + hy'(x_n) + R_1(x_n, h)} - \underbrace{[y_n + hf(x_n, y_n)]} \\ &= y(x_n) - y_n + h[y'(x_n) - f(x_n, y_n)] + R_1(x_n, h) \\ &= e_n + h[y'(x_n) - f(x_n, y_n)] + R_1(x_n, h). \end{aligned} \tag{13}$$

We need to estimate the second and third terms in (13). Using the fact that $y(x)$ is a solution of the differential equation (9) we get, for the second term,

$$|y'(x_n) - f(x_n, y_n)| = |f(x_n, y(x_n)) - f(x_n, y_n)|. \tag{14}$$

But $f(x_n, z)$ is a function of z. Applying the Mean-Value Theorem (8) to the right side of (14) gives

$$|y'(x_n) - f(x_n, y_n)| \leq M_1|y(x_n) - y_n| = M_1|e_n|, \tag{15}$$

where M_1 is an upper bound for $|f_y|$, taken over both x and y. We estimate $R_1(x_n, h)$ in (13) by using (7), to get

$$|R_1(x_n, h)| \leq M_2 h^2, \tag{16}$$

where $2M_2$ is the maximum of $|y''|$ on $[a, b]$.

Now, taking absolute values in (13),

$$|e_{n+1}| \leq |e_n| + h|y'(x_n) - f(x_n, y_n)| + |R_1(x_n, h)|,$$

which by (15), (16) implies

$$|e_{n+1}| \leq |e_n| + hM_1|e_n| + h^2 M_2. \tag{17}$$

Let ε_n be the solution of the difference equation

$$\varepsilon_{n+1} = \varepsilon_n + hM_1\varepsilon_n + h^2 M_2, \qquad \varepsilon_0 = |e_0|. \tag{18}$$

It is easy to show by induction using (17), (18) that

$$\varepsilon_n \geq |e_n| \qquad \text{for all } n. \tag{19}$$

Thus, to estimate $|e_n|$ it suffices to solve the first-order difference equation (18). From Exercise 23 of Section 8.2 (or by direct verification), the solution of (18) is

$$\varepsilon_n = (1 + hM_1)^n \varepsilon_0 + \left[\sum_{i=0}^{n-1} (1 + hM_1)^i\right] h^2 M_2. \tag{20}$$

Since from (18) we have $\varepsilon_n \leq \varepsilon_{n+1}$, it follows that the largest ε_n is the last one, ε_N, so that (4) and (19) give

$$E(h) \leq \varepsilon_N.$$

(It is not true, in general, that $E(h) \le |e_N|$.) Thus to estimate $E(h)$ it suffices to estimate

$$\varepsilon_N = (1 + hM_1)^N \varepsilon_0 + \left[\sum_{i=0}^{N-1} (1 + hM_1)^i \right] h^2 M_2. \tag{21}$$

The first term in (21) is easily handled by the fact that

$$(1 + hM_1)^N = (1 + hM_1)^{(b-a)/h},$$

which, by Eq. (5), converges to $e^{M_1(b-a)}$. In fact, it is monotonically increasing (see Exercise 2 at the end of this section) as $h \to 0^+$, so that

$$(1 + hM_1)^N \leqslant e^{M_1(b-a)}. \tag{22}$$

The second term in Eq. (21) is estimated by using the fact that, if $u \neq 1$, then

$$\sum_{i=0}^{n-1} u^i = \frac{u^n - 1}{u - 1},$$

so that

$$\begin{aligned}
\left[\sum_{i=0}^{N-1} (1 + hM_1)^i \right] h^2 M_2 &= \left[\frac{(1 + hM_1)^N - 1}{(1 + hM_1) - 1} \right] h^2 M_2 \\
&= \frac{(1 + hM_1)^{(b-a)/h} - 1}{hM_1} h^2 M_2 \\
&\le [e^{M_1(b-a)} - 1] h \left(\frac{M_2}{M_1} \right) \le e^{M_1(b-a)} h \left(\frac{M_2}{M_1} \right).
\end{aligned}$$

This yields the final result

$$\varepsilon_N \le e^{M_1(b-a)} e_0 + e^{M_1(b-a)} \left(\frac{M_2}{M_1} \right) h. \tag{23}$$

If $e_0 = 0$, then

$$\varepsilon_N \le Mh,$$

where $M = e^{M_1(b-a)} M_2 / M_1$, which completes the proof of Theorem 6.3.1. ◄

As a bonus we get, from Eq. (23), an estimate of how much an error or change of the starting value y_0 affects the numerical solution (take $\varepsilon_0 = |e_0| \neq 0$).

6.3.2 Round-Off Error

Suppose that at each step there is an additional error ρ_i due to round-off and function evaluation. For convenience, assume that there is a bound ρ, so that

$$|\rho_i| \le \rho. \tag{24}$$

Note that ρ is independent of the step size and is at least as large as the machine round-off error. Let Y_n be the actual value calculated for y_n using Euler's method. Thus the values Y_n satisfy

$$Y_{n+1} = Y_n + hf(x_n, Y_n) + \rho_{n+1}. \tag{25}$$

The theoretical values of y_n satisfy

$$y_{n+1} = y_n + hf(x_n, y_n). \tag{26}$$

Let $E_n = Y_n - y_n$ be the difference between the computed and theoretical value of y_n. Subtracting (26) from (25) gives

$$E_{n+1} = E_n + h[f(x_n, Y_n) - f(x_n, y_n)] + \rho_{n+1}. \tag{27}$$

But $f(x_n, Y_n) - f(x_n, y_n) = f_y(x_n, \xi)(Y_n - y_n)$, with ξ between Y_n and y_n by the Mean-Value Theorem (8). Thus, taking absolute values of (27) and using (24), we get

$$|E_{n+1}| \le |E_n| + hM_1|E_n| + \rho, \tag{28}$$

where M_1 is again an upper bound for $f_y(x, y)$. Again we solve the difference equation (see Exercise 5 at the end of this section) with equality in (28) to get

$$|E_N| \le e^{M_1(b-a)}|E_0| + e^{M_1(b-a)}\frac{\rho}{M_1 h}. \tag{29}$$

Note a significant difference in this new estimate from that of (23). As h decreases, the estimate for the round-off error *increases*, due to the fact that smaller steps require more calculations. The number ρ is usually quite small, between 10^{-8} and 10^{-17} on most machines. However, at some point the decreasing discretization error due to smaller steps is offset by increasing round-off error, and still smaller steps will only make things worse. Combining (23) and (29) with $e_0 = 0$, we see that if round-off error is included, then the global error for Euler's method is

$$\tilde{E}(h) = \max_{1 \le n \le N} |y(x_n) - Y_n| \approx \alpha h + \frac{\beta}{h},$$

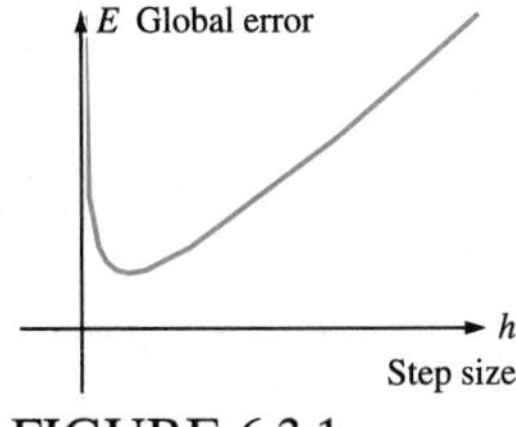

FIGURE 6.3.1

Graph of $\alpha h + \beta/h$ with $\alpha = 1$, $\beta = 0.01$.

which is graphed in Figure 6.3.1.

In fact, Figure 6.3.1 is appropriate for many numerical algorithms except that one gets $E(h) \approx \alpha h^r + \beta/h^s$, where r, s are constants depending on the method.

6.3.3 Stiffness

We conclude this section with a brief discussion of an important property called **stiffness**. This discussion will also illustrate that a theorem like Theorem 6.3.1 may require small h before it is applicable.

Consider the differential equation

$$y' = -40y + 40x + 1, \qquad y(0) = 4. \tag{30}$$

The solution of this linear differential equation is

$$y = x + 4e^{-40x}. \tag{31}$$

The solution (31) consists of two parts. A slowly varying portion, x, and a faster (transient) part, $4e^{-40x}$. The graph of Eq. (31) is given in Figure 6.3.2.

These types of solution arise frequently in electric circuits, in chemical reactions, and in "boundary layers" in fluid problems. Suppose that we attempt to use Euler's method on Eq. (30) for $0 \leq x \leq 2$, and take a step size of $h = 0.1$. Let

$$E_n(h) = \max_{0 \leq i \leq n} |e_i|.$$

The result is Table 6.3.1.

This is not even close. Trying again with $h = 0.05$, we get Table 6.3.2. The error is no longer increasing unboundedly, but it still gives a poor estimate. Letting $h = 0.01$, we get Table 6.3.3. This gives excellent estimates for $x \geq 0.5$ but less accuracy on the first few steps, since $E(0.01) = 0.357$.

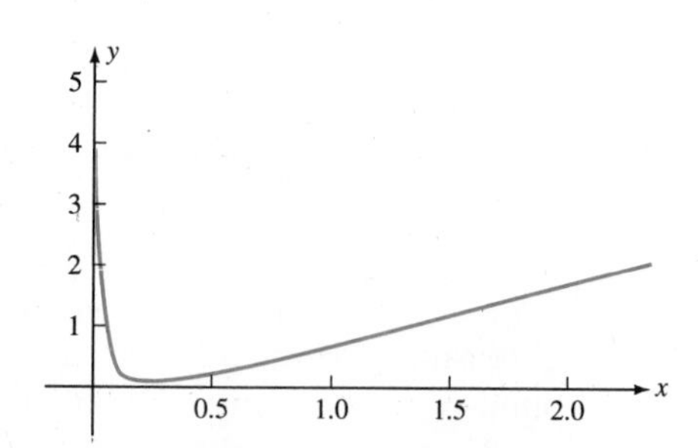

FIGURE 6.3.2

Graph of $y = x + 4e^{-40x}$.

TABLE 6.3.1 SOLUTION OF (30) WITH EULER'S METHOD, AND $h = 0.1$

x	n	Y_n	e_n	$E_n(0.1)$
0.0	0	4	0	0
0.5	5	-971.5	972	972
1.0	10	236,197	236,196	236,196
1.5	15	-5.7×10^7	5.7×10^7	5.7×10^7
2.0	20	1.4×10^{10}	1.4×10^{10}	1.4×10^{10}

TABLE 6.3.2 SOLUTION OF (30) WITH EULER'S METHOD, AND $h = 0.05$

x	n	Y_n	e_n	$E_n(0.05)$
0.0	0	4.0	0	0
0.5	10	4.5	4	4.54
1.0	20	5.0	4	4.54
1.5	30	5.5	4	4.54
2.0	40	6.0	4	4.54

TABLE 6.3.3 SOLUTION OF (30) WITH EULER'S METHOD, AND $h = 0.01$

x	n	Y_n	e_n	$E_n(0.01)$
0.0	0	4.0	0	0
0.5	50	0.5	8.2×10^{-1}	0.357
1.0	100	1.0	3.4×10^{-17}	0.357
1.5	150	1.5	0	0.357
2.0	200	2.0	5.6×10^{-17}	0.357

This is somewhat frustrating. We are forced to use small steps just to approximate a function that is quickly almost the straight line $y = x$.

A natural thing to attempt is to use smaller steps when the solution is changing rapidly and large steps later. If we take $h = 0.01$ for $0 \leq x \leq 0.2$ and $h = 0.1$ for $0.2 \leq x \leq 2$, we get the result in Table 6.3.4. Again we have a poor numerical estimate.

To understand what is happening, we need to examine Euler's method more carefully. Euler's method applied to Eq. (30) gives

$$y_{n+1} = y_n + h(-40y_n + 40nh + 1), \qquad y_0 = 4, \tag{32}$$

or

$$y_{n+1} = (1 - h40)y_n + 40nh^2 + h, \qquad y_0 = 4; \tag{33}$$

the solution of (33), as can be directly verified, is

$$y_n = (1 - h40)^n y_0 + hn. \tag{34}$$

TABLE 6.3.4

x	Y_n	e_n
0.0	4	0
0.5	0.496	0.004
1.0	1.960	0.960
1.5	−231.664	233.164
2.0	56,660	56,658

It is true that $\lim_{h \to 0^+} (1 - h40)^n = 1$ for fixed n, but if h is not very small, then $(1 - h40)^n$ can be quite large for large n. In fact, if $h = 0.1$, then $(1 - h40)^n y_0 = (-3)^n y_0$ oscillates with growing amplitude as n increases. Only if $0 < h < 1/20$ is $|1 - h40| < 1$, so that the numerical solution resembles the real solution.

Backward Euler

Seemingly small differences in a numerical method can greatly affect its behavior. Euler's method for $y' = f(x, y)$ may be written as

$$\frac{y_{n+1} - y_n}{h} = f(x_n, y_n), \tag{35}$$

where $(y_{n+1} - y_n)/h$ is an approximation to y'. The **backward** or **implicit Euler** uses a backward difference to approximate y', and is given by

$$\frac{y_n - y_{n-1}}{h} = f(x_n, y_n),$$

or, equivalently,

$$\frac{y_{n+1} - y_n}{h} = f(x_{n+1}, y_{n+1}). \tag{36}$$

It is more effort to solve for y_{n+1} in (36) than in (35), since y_{n+1} appears inside f. Frequently a numerical method like Newton's must be used to solve for y_{n+1}. To illustrate (36), we solve Eq. (30) by the backward Euler's (36). This time the difference equation is

$$\frac{y_{n+1} - y_n}{h} = -40y_{n+1} + 40(n + 1)h + 1,$$

or

$$(1 + 40h)y_{n+1} = y_n + 40(n + 1)h^2 + h,$$

and finally

$$y_{n+1} = \frac{1}{1 + 40h} y_n + h + \frac{40nh^2}{1 + 40h}. \tag{37}$$

When this difference equation is solved, one gets a $(1 + 40h)^{-n}$ term. But

$$\left| \frac{1}{1 + 40h} \right| < 1 \qquad \text{for any } h > 0,$$

so we expect better convergence. The result of using the backward Euler method (37) on (30) with a step size of $h = 0.1$ is given in Table 6.3.5.

Comparing Table 6.3.5 to Table 6.3.1, we see that changing from a forward difference to a backward difference has greatly improved our estimate for $y(2)$ in this example.

TABLE 6.3.5 BACKWARD EULER ON (30), WITH $h = 0.1$

x	n	Y_n	e_n	$E_n(0.1)$
0.0	0	4	0	0.727
0.5	5	0.501	1.2×10^{-3}	0.727
1.0	10	1.000	4.1×10^{-7}	0.727
1.5	15	1.500	1.3×10^{-10}	0.727
2.0	20	2.000	4.2×10^{-14}	0.727

Exercises

1. Using l'Hôpital's rule, verify that $\lim_{h \to 0} (1 + h\alpha)^{\beta/h} = e^{\alpha\beta}$, as stated in eq. (5).
2. Show that if α, $\beta > 0$, then, as $h \to 0$, $(1 + h\alpha)^{\beta/h}$ is monotonically increasing. [*Hint*: Show that, if $g(h) = (1 + h\alpha)^{\beta/h}$, then $g'(h) < 0$ for all $h > 0$.]
3. Let $|e_n|$, ε_n be defined as in Section 6.3. Using Eqs. (17) and (18), verify by induction that $\varepsilon_n \geq |e_n|$ for all $n \geq 0$ and thus (19) holds.
4. Verify that the sequence (20) is a solution of the difference equation (18).
5. Let $T_{n+1} = T_n + hM_1T_n + \rho$, with $M_1 > 0$, $T_0 \geq 0$. Show that
$$T_N \leq e^{M_1(b-a)}T_0 + e^{M_1(b-a)}\frac{\rho}{M_1 h},$$
where $N = (b - a)/h$. This verifies (29).
6. Note that equation (33) can be written as $y_{n+1} = \alpha y_n + \beta n + \gamma$. Verify that (34) is the solution to (33).

The function $g(h) = \alpha h + \beta/h$, with $\alpha > 0$, $\beta > 0$, and $h > 0$, attains its minimum at $\hat{h} = \sqrt{\beta/\alpha}$. In the error estimate (23) we usually have $\alpha > 1$ so that $\hat{h} \geq \sqrt{\beta}$. Since β is at least as big as machine precision, we have that round-off error can become a major factor in Euler's method by the time the step size nears the square root of machine precision.

7.* For a given step size h, solve
$$y' = 2y + x, \quad y(0) = 1, \quad 0 \leq x \leq 1, \quad (38)$$
by Euler's method. At each step compute the absolute value of the difference between the computed Y_n and the true $y(x_n)$. Let $\hat{E}(h)$ be the largest of these errors. Compute $\hat{E}(h)$ for progressively smaller h's, starting with $h = 0.1$, $h = 0.01$, etc. You should observe that $\hat{E}(h)$ decreases for a while and then increases. For what value of h was $\hat{E}(h)$ smallest? How does that compare to the square root of your machine precision? (It is recommended that you print out only the $\hat{E}(h)$ and not all the y_n values.)

8. Verify (15).

9.* a) Solve the differential equation
$$y' = -30y + 30x + 31, \quad y(0) = 4, \quad (39)$$
on the interval $0 \leq x \leq 2$ using both Euler's method and the backward Euler's method with $h = 0.1$.

b) Find the formula for the solution of (39).

c) Graph the values of (x_n, y_n) from both methods in part (a) and the solution from part (b) on the same axis.

10.* **Newton's method** for solving $g(z) = 0$, given an initial guess, $\hat{z}$, is to let $z_0 = \hat{z}$ and consider the iteration
$$z_{m+1} = z_m - g'(z_m)^{-1}g(z_m). \quad (40)$$
The iteration (40) is repeated for M steps, and z_M is taken as the estimate of a solution of $g(z) = 0$. Depending on the application, M can range from 1 to a large number. This

exercise is to solve

$$y' = 30\cos y, \qquad y(0) = 1,$$
$$0 \le x \le 2 \qquad (41)$$

using the backward Euler method (36) with $h = 0.1$. This will require solving

$$y_{n+1} - y_n - h30\cos y_{n+1} = 0 \qquad (42)$$

for y_{n+1}, given y_n. Do this using Newton's method. This entails, at each n, having an initial guess for y_{n+1}, denoted $\hat{y}_{n+1}$, and a rule for when to stop the Newton iteration. Experiment with these choices for $\hat{y}_{n+1}$;

i) $\hat{y}_{n+1} = y_n$
ii) $\hat{y}_{n+1} = y_n + h30\cos y_n$ (Euler predictor), and the following rules for M;
i) Stop after one iteration ($M = 1$).
ii) Stop after two iterations ($M = 2$).
iii) Stop when $|z_{m+1} - z_m| \le 0.01$.

11. (Continuation of Exercise 10)

a) Solve (41) using Euler's method with $h = 0.1$. Compare to what you observed in Exercise 10.

b) Sketch the solutions of (41) and compare to the values of the two computed approximations.

6.4 Second-Order Methods

As noted in Sections 6.2 and 6.3, in order to get sufficient accuracy with Euler's method we may have to take a large number of steps with a small step size. Such calculations can be time-consuming, even on a computer, and present difficulties caused by the accumulation of round-off error. One way to circumvent this difficulty is to use higher-order methods. This section will present two second-order methods.

Perhaps the most natural way to improve on Euler's method is to replace the first-order Taylor polynomial used in Euler's method with a second-order Taylor polynomial.

If y is the solution of

$$y' = f(x, y), \qquad y(x_0) = y_0, \qquad (1)$$

then the second-order Taylor approximation (Section 4.2) for y at x is

$$y(x + h) = y(x) + hy'(x) + \frac{h^2}{2}y''(x). \qquad (2)$$

From (1) we have $y'(x) = f(x, y)$. Differentiating both sides of (1) with respect to x yields

$$y'' = f_x(x, y) + f_y(x, y)y' = f_x(x, y) + f_y(x, y)f(x, y). \qquad (3)$$

6.4.1 Second-Order Taylor Method

Substituting (1) and (3) into (2) gives the **second-order Taylor method** for solving $y' = f(x, y)$, $y(a) = y_0$ with a fixed step size of h.

1. Let y_0 be as in (1), $x_0 = a$.
2. Compute x_n, y_n for $n \geq 1$ by $x_{n+1} = x_n + h$:

$$y_{n+1} = y_n + hf(x_n, y_n) + \frac{h^2}{2}\left[f_x(x_n, y_n) + f_y(x_n, y_n)f(x_n, y_n)\right]. \quad (4)$$

EXAMPLE 6.4.1 Second-Order Taylor Method

If we are going to solve

$$y' = xy + x, \quad (5)$$

using this second-order Taylor method, we have

$$f(x, y) = xy + x, \qquad f_x(x, y) = y + 1, \qquad f_y(x, y) = x.$$

Thus (4) becomes

$$y_{n+1} = y_n + h(x_n y_n + x_n) + \frac{h^2}{2}[(y_n + 1) + x_n(x_n y_n + x_n)]. \quad (6)$$

These calculations will be carried out in Example 6.4.3. ◄

Unfortunately, the Taylor method requires the calculation of two partials and three function evaluations. In many applications, these partials are not easily available, and the function evaluations are computationally expensive. The next approach is closer in spirit to those usually used in practice.

To motivate this next method, integrate (1) from x_n to x_{n+1} to get

$$\int_{x_n}^{x_{n+1}} y'\,dx = \int_{x_n}^{x_{n+1}} f(x, y)\,dx \quad (7)$$

or

$$y(x_{n+1}) - y(x_n) = \int_{x_n}^{x_n+1} f(x, y(x))\,dx. \quad (8)$$

Estimate this integral by using the trapezoid rule with just two mesh points, x_n, x_{n+1}. Equation (8) becomes

$$y(x_{n+1}) - y(x_n) \approx \frac{h}{2}[f(x_n, y(x_n)) + f(x_{n+1}, y(x_{n+1}))]$$

or

$$y(x_{n+1}) \approx y(x_n) + \frac{h}{2}[f(x_n, y(x_n)) + f(x_{n+1}, y(x_{n+1}))]. \quad (9)$$

The only difficulty with (9) is that we do not know $y(x_{n+1})$ when evaluating the right-hand side, since $y(x_{n+1})$ is what y_{n+1} is supposed to estimate. This difficulty can be circumvented by estimating $y(x_{n+1})$ in the right-hand side of (9), using Euler's method. The result is the **modified Euler method.**

6.4.2 Modified Euler Method

To solve $y' = f(x, y)$, $y(a) = y_0$ with step size h, by the modified Euler method:

1. Let y_0 be as in (1), $x_0 = a$.
2. For $n \geq 0$, define x_n, y_n by the recursion

$$x_{n+1} = x_n + h,$$

$$z_{n+1} = y_h + hf(x_n, y_n), \tag{10}$$

$$y_{n+1} = y_n + \frac{h}{2}[f(x_n, y_n) + f(x_{n+1}, z_{n+1})]. \tag{11}$$

The modified Euler method is also called the **Heun** (or **improved Euler**) **method**. It is an example of a **second-order Runge–Kutta method** (the reasons for this will be mentioned in Section 6.5). The modified Euler method illustrates the idea of a **predictor-corrector** method. Given the values x_n, y_n, we **predict** the value of $y(x_{n+1})$. This prediction is denoted by z_{n+1} in (10). Then in (11) the prediction z_{n+1} is **corrected** to give the final estimate y_{n+1} for $y(x_{n+1})$. In the modified Euler method, the "predictor" is Euler's method. The "corrector" is integrating (8) by the trapezoid rule.

EXAMPLE 6.4.2 **Modified Euler**

If we are to solve $y' = xy + x$, $y(0) = 1$, by the modified Euler method with step size h, we have

$$x_n = nh,$$

$$z_{n+1} = y_n + h(x_n y_n + x_n), \tag{12}$$

$$y_{n+1} = y_n + \frac{h}{2}(x_n y_n + x_n + x_{n+1} z_{n+1} + x_{n+1}). \tag{13}$$

In this case $x_0 = 0$, $y_0 = 1$. Thus $x_1 = h$. We shall calculate y_1 using $h = 0.1$. (Additional calculations appear in Example 6.4.3.) From (12),

$$z_1 = y_0 + h(x_0 y_0 + x_0) = 1 + (0.1)0 = 1.$$

Then, from (13),

$$y_1 = y_0 + \frac{h}{2}[x_0 y_0 + x_0 + x_1 z_1 + x_1] = 1 + \frac{0.1}{2}[0 + 0 + (0.1)1 + 0.1]$$

$$= 1.01.$$

◄

The Taylor, Euler, and modified Euler methods are examples of **one-step methods**. That is, to compute the estimate y_{n+1} for $y(x_{n+1})$, we need only know the values x_n, y_n from *one* step earlier. For example, in Example 6.4.2, (12) and (13) may be combined to express y_{n+1} directly in terms of y_n, as

$$y_{n+1} = y_n + \frac{h}{2}\{x_n y_n + x_n + x_{n+1}[y_n + h(x_n y_n + x_n)] + x_{n+1}\}. \quad (14)$$

We note without proof that, under the appropriate smoothness assumptions on $f(x, y)$:

THEOREM 6.4.1 Order of Taylor (4) and Modified Euler

Both the Taylor method (4) and the modified Euler method are second-order methods on a finite interval $[a, b]$. That is, there exists a constant M depending on the method, $f(x, y)$, the interval $[a, b]$, and the initial condition, but not on the step size h, such that the global error,

$$E(h) = \max_{1 \le n \le (b-a)/h} |y(x_n) - y_n| = \max_{1 \le n \le (b-a)/h} |e_n| \quad (15)$$

satisfies

$$E(h) \le Mh^2. \quad (16)$$

■

EXAMPLE 6.4.3 Comparing Methods

Table 6.4.1 gives the errors that result in solving

$$y' = xy + x, \qquad y(0) = 1, \qquad 0 \le x \le 1, \quad (17)$$

for $y(1)$, using Euler's, second-order Taylor, and modified Euler method with a step size of $h = 0.1$ and $h = 0.01$.

The dramatic reduction in error for only slightly more computational effort shows why higher-order methods are almost always chosen over Euler's method. ◄

TABLE 6.4.1 ERROR $|e_N|$ IN ESTIMATE OF $y(1)$ FOR (17)

h	Euler	Second-Order Taylor	Modified Euler
0.1	0.203	0.008602	0.001680
0.01	0.022	0.000095	0.000015

Exercises

In Exercises 1 through 7, compute the solution on the interval $[0, 1]$ with the given step size using Euler's, second-order Taylor, and modified Euler's methods. For Exercises 1 through 5, compare your result to the true solution. Values for y_N and e_N are given in the answer section at the end of this book.

1. $y' = 2y, \quad y(0) = 1, \quad h = 0.1$

2.* $y' = 2y, \quad y(0) = 1, \quad h = 0.01$

3. $y' = -xy, \quad y(0) = 1, \quad h = 0.2$

4. $y' = 1 - y^2, \quad y(0) = 0, \quad h = 0.1$

5.* $y' = 1 - y^2, \quad y(0) = 0, \quad h = 0.01$

6. $y' = \cos y, \quad y(0) = 0, \quad h = 0.1$

7.* $y' = \cos y, \quad y(0) = 0, \quad h = 0.01$

8. Let λ be a constant.

i) Show that the second-order Taylor and modified Euler methods give the same result when applied to the linear constant-coefficient problem

$$y' = \lambda y, \qquad y(a) = y_0. \tag{18}$$

ii) Show that both these methods applied to (18) give the recursion relationship

$$y_{n+1} = \left[1 + h\lambda + \frac{h^2\lambda^2}{2}\right] y_n. \tag{19}$$

iii) Show that $1 + h\lambda + h^2\lambda^2/2$ are the first three terms in a series expansion for $e^{h\lambda}$.

9. Take x_n, y_n as given numbers.

i) Show that the difference relations in solving (5) of Example 6.4.1 by the second-order Taylor (6) and the modified Euler (14) methods give different values of y_{n+1}.

ii) Show that these two estimates for y_{n+1} differ by terms all of which have an h^3 in them.

10. Derive a third-order Taylor method for solving $y' = f(x, y)$, $y(a) = y_0$ similar to the way (4) was derived.

Using the **third-order Taylor method** defined in Exercise 10, estimate $y(1)$ for the following differential equations with the indicated step sizes.

11. $y' = 2y, \quad y(0) = 1, \quad h = 0.2, \ 0 \le x \le 1$

12. $y' = xy, \quad y(0) = 1, \quad h = 0.1, \ 0 \le x \le 1$

13. $y' = xy + x^2, \quad y(0) = 1,$
$h = 0.1, \ 0 \le x \le 1$

14. It can be shown that if round-off error is considered, then the global error from either the modified Euler or second-order Taylor method has an estimate of the form

$$\hat{E}(h) = Mh^2 + \frac{\beta}{h}. \tag{20}$$

Assume that $M = 1$, and β is a small number approximately equal to machine precision.

a) Show that $\hat{E}(h)$ attains its minimum at about $h = \sqrt[3]{\beta}$.

b) Let $\beta = 10^{-16}$. Compute the value of $\hat{E}(h)$ at the minimum from part (a).

c) Accepting that the minimum error from Euler's method is approximately $\sqrt{\beta}$, can you (heuristically) conclude that the second-order methods not only can be expected to give greater accuracy at less computational cost but are capable of attaining greater accuracy before the effect of round-off error dominates the calculation?

6.5 Fourth-Order Runge–Kutta

In some problems the numerical (or actual) cost of evaluating $f(x, y)$ can be high. In other problems we must worry about rapid transients (see Section 6.3). Finally, many modern numerical packages for solving differential equations not only automatically vary the step size but also change the order, and even the type of method being used. However, among the more popular general-purpose methods used are the Runge–Kutta methods. These are one-step methods and can thus easily vary the step size from one step to another to better control errors. One feature of Runge–Kutta methods is that they often compute intermediate values of the independent variable.

There are actually several fourth-order Runge–Kutta methods. One of the more popular is given below.

Fourth-Order Runge–Kutta

Fourth-order Runge–Kutta for solving

$$y' = f(x, y), \qquad y(a) = y_0 \tag{1}$$

with step size h proceeds as follows.

1. Let $x_0 = a$, $x_{n+1} = x_n + h$, $y_0 = y(a)$.
2. For $n \geq 0$ define y_{n+1} in terms of y_n by:
   a. Compute

$$F_1(n) = hf(x_n, y_n), \tag{2}$$

$$F_2(n) = hf\left(x_n + \frac{h}{2}, y_n + \frac{F_1}{2}\right), \tag{3}$$

$$F_3(n) = hf\left(x_n + \frac{h}{2}, y_n + \frac{F_2}{2}\right), \tag{4}$$

$$F_4(n) = hf(x_n + h, y_n + F_3). \tag{5}$$

   b. Let

$$y_{n+1} = y_n + \frac{1}{6}[F_1(n) + 2F_2(n) + 2F_3(n) + F_4(n)]. \tag{6}$$

EXAMPLE 6.5.1 Fourth-Order Runge–Kutta

To illustrate, again consider

$$y' = xy + x, \qquad y(0) = 1, \tag{7}$$

so that $f(x, y) = xy + x$ and $x_0 = 0$, $y_0 = 1$. We shall compute y_1 using the fourth-order Runge–Kutta method [(2) through (6)] with $h = 0.1$. From (2),

$$F_1 = 0.1(x_0 y_0 + x_0) = 0.1(0 \cdot 1 + 0) = 0,$$

so that

$$F_2 = 0.1f\left(x_0 + \frac{h}{2}, y_0 + \frac{F_1}{2}\right) = 0.1f(0.05, 1)$$

$$= 0.1[0.05(1) + 0.05] = 0.01,$$

$$F_3 = 0.1f\left(x_0 + \frac{h}{2}, y_0 + \frac{F_2}{2}\right) = 0.1f(0.05, 1.005)$$

$$= 0.1[0.05(1.005) + 0.05] = 0.010025,$$

$$F_4 = 0.1f(x_0 + h, y_0 + F_3) = 0.1f(0.1, 1.010025)$$

$$= 0.1[0.1(1.010025) + 0.1] = 0.02010025,$$

and then from (6)

$$y_1 = 1 + \frac{1}{6}(0 + 0.02 + 0.02005 + 0.02010025) = 1.010025.$$ ◀

It is interesting to compare this result with the solution of (7) by the methods of Sections 6.2 and 6.4. For (7) we have

True rounded value of $y(0.1) = 1.010025042$;
y_1 from Euler's method $= 1$,
y_1 from modified Euler $= 1.01$,
y_1 from second-order Taylor $= 1.01$,
y_1 from fourth-order Runge–Kutta $= 1.010025$.

The fourth-order Runge–Kutta method can be understood much as the modified Euler was. Again, antidifferentiate both sides of (1) from x_n to $x_n + 1$, to give

$$y(x_{n+1}) = y(x_n) + \int_{x_n}^{x_n+1} f(x, y)\,dx. \tag{8}$$

In the modified Euler we estimate the integral using the trapezoid rule. In the fourth-order Runge–Kutta a variation of Simpson's rule is used to estimate the integral in (8).

The Runge–Kutta methods may also be derived using a Taylor-series argument.

Exercises

For Exercises 1 through 8, estimate $y(1)$ using a fourth-order Runge–Kutta with the indicated step size. Values of y_1 and y_N are given in the solution appendix.

1. $y' = xy + x, \quad y(0) = 1, \quad h = 0.1$
2.* $y' = xy + x, \quad y(0) = 1, \quad h = 0.01$
3. $y' = 1 - y^2, \quad y(0) = 0.5, \quad h = 0.1$
4. $y' = xy + x^2, \quad y(0) = 0, \quad h = 0.1$
5. $y' = \cos y, \quad y(0) = 1, \quad h = 0.1$
6. $y' = y - xy^2, \quad y(0) = 1, \quad h = 0.1$
7. $y' = -40y + 40x + 1, \quad y(0) = 4,$ $h = 0.1$ (See the discussion of stiffness in Section 6.3.)
8.* $y' = -40y + 40x + 1, \quad y(0) = 4, \quad h = 0.01$
9. **a)** Show that if the fourth-order Runge–Kutta method of this section is applied to $y' = \lambda y$, $y(0) = 1$, λ a constant, then

$$y_{n+1} = \left[1 + h\lambda + \frac{(h\lambda)^2}{2} + \frac{(h\lambda)^3}{3!} + \frac{(h\lambda)^4}{4!}\right] y_n.$$

   b) Observe that

$$1 + h\lambda + \frac{(h\lambda)^2}{2} + \frac{(h\lambda)^3}{3!} + \frac{(h\lambda)^4}{4!}$$

   is the first five terms in an expansion of $e^{\lambda h}$ and the Runge–Kutta method is *fourth-order*. Compare to Exercise 8 at the end of Section 6.4.

(Runge–Kutta methods are also used to compute starting values in some of the Exercises of Section 6.6.)

6.6 Multistep Methods

In all of the methods examined so far for solving

$$y' = f(x, y), \qquad y(a) = y_0, \tag{1}$$

higher accuracy was obtained by performing additional evaluations of the function $f(x, y)$. The Euler method required one evaluation, the modified Euler required two, and the fourth-order Runge–Kutta required four evaluations per step.

In many problems, computing these values of f requires a great deal of computational effort. One way to circumvent this difficulty is by using earlier values of f rather than more function evaluations of f to get higher-order methods. One such family of methods is "The Adams Family" for solving (1). Three of these methods are the:

Second-Order Adams–Bashforth

Given y_0, y_1, then for $n \geq 1$,

$$y_{n+1} = y_n + \frac{h}{2}[3f(x_n, y_n) - f(x_{n-1}, y_{n-1})]. \tag{2}$$

Third-Order Adams–Bashforth

Given y_0, y_1, y_2, then for $n \geq 2$,

$$y_{n+1} = y_n + \frac{h}{12}[23f(x_n, y_n) - 16f(x_{n-1}, y_{n-1}) + 5f(x_{n-2}, y_{n-2})]. \quad (3)$$

Fourth-Order Adams–Bashforth

Given y_0, y_1, y_2, y_3, then for $n \geq 3$,

$$y_{n+1} = y_n + \frac{h}{24}(55w_n - 59w_{n-1} + 37w_{n-2} - 9w_{n-3}), \quad (4)$$

where

$$w_n = f(x_n, y_n).$$

Note that while (4), for example, involves four evaluations, only one of them, $f(x_n, y_n) = w_n$, has to be computed at each step. The other three, w_{n-1}, w_{n-2}, w_{n-3}, have been computed on previous steps.

It is important that the starting values be as accurate as the method is expected to be. Thus, in using the fourth-order method (4), we would probably use a fourth-order one-step method such as a fourth-order Runge–Kutta, to compute y_1, y_2, y_3 from y_0. Other options are to use an Euler method with step size h^4 or a modified Euler with step size h^2.

EXAMPLE 6.6.1 Second-Order Adams–Bashforth

Estimate $y(1)$ for the solution of

$$y' = xy + x, \qquad y(0) = 1, \quad (5)$$

using the second-order Adams–Bashforth method (2) with step size $h = 0.1$. Estimate y_1 using a modified Euler method.

Solution Let

$$w_n = f(x_n, y_n) = x_n y_n + x_n.$$

Then (2) becomes

$$y_{n+1} = y_n + \frac{h}{2}(3w_n - w_{n-1}). \quad (6)$$

The initial conditions

$$x_0 = 0, \qquad y_0 = 1$$

imply that $w_0 = 0$. We compute y_1 using a modified Euler (see Example 6.4.2), so that

$$x_1 = 0.1, \qquad y_1 = 1.01.$$

Thus $w_1 = 0.201$.

We can now use (6) to compute the other y_n. First

$$y_2 = y_1 + \frac{0.1}{2}(3w_1 - w_0) = 1.04015.$$

Then

$$x_2 = 0.2, \qquad w_2 = x_2(y_2 + 1) = 0.40803.$$

Continuing,

$$y_3 = y_2 + \frac{0.1}{2}(3w_2 - w_1) = 1.0913045,$$

and finally after seven more iterations of this process,

$$y_{10} = 2.27768.$$

The actual value of $y(1)$ is approximately $y(1) = 2.2974$. ◄

Second-Order Adams–Moulton

The starting value y_1 is computed by a second-order one-step method. Then given y_n, y_{n-1}, the predicted value of y_{n+1} is given by the second-order Adams–Bashforth,

$$z_{n+1} = y_n + \frac{h}{2}[3f(x_n, y_n) - f(x_{n-1}, y_{n-1})]. \tag{7}$$

The prediction z_{n+1} is corrected using the trapezoid rule,

$$y_{n+1} = y_n + \frac{h}{2}[f(x_n, y_n) + f(x_{n+1}, z_{n+1})]. \tag{8}$$

The Adams–Moulton method (7), (8) requires two function evaluations at each step. However, the method can sometimes have numerical advantages. (Note Exercise 5.)

Exercises

In Exercises 1 through 4, estimate $y(1)$ using a second-order Adams–Bashforth with a step size of 0.1. Calculate the second starting value y_1 using a modified Euler method.

1. $y' = 3y + 1, \quad y(0) = 1$

2. $y' = 1 - y^2, \quad y(0) = 2$

3. $y' = \cos y, \quad y(0) = 0$

4. $y' = xy - x^2, \qquad y(0) = 1$

5.* Find the actual solution of

$$y' = -40y + 40x + 1, \qquad y(0) = 4.$$

Using step sizes of $h = 0.1, 0.05, 0.04, 0.01$, estimate $y(1)$, using both a second-order Adams–Bashforth and a second-order Adams–Moulton. Compare the result to the actual answer. [This is the stiffness example of Section 6.3.]

6. Let y be the solution of $y' = f(x, y)$, $y(x_0) = y_0$. Using the Taylor expansion of $y(x)$ centered at x_n, show that

$$y(x_n + h) - y(x_n - h) = 2y'(x_n)h + O(h^3). \tag{9}$$

The resulting two-step method

$$y_{n+1} = y_{n-1} + 2hf(x_n, y_n) \tag{10}$$

is called the **centered-difference** method.

7. Estimate $y(1)$ for $y' = -xy^2$, $y(0) = 1$, by the centered-difference method with a step size of $h = 0.1$. (Find y_1 using a modified Euler's.)

8.* For the differential equation

$$y' = \sin(xy) + \cos(y^2) + e^{-x^2}, \qquad y(0) = 1,$$

using a step size of $h = 0.01$, estimate $y(10)$ using both:

i) A fourth-order Runge–Kutta.

ii) A fourth-order Adams–Bashforth with starting values, y_1, y_2, y_3, determined by a fourth-order Runge–Kutta.

Compare the amount of computer time each method took.

9.* Verify that $y_{10} = 2.27768$ for Example 6.6.1.

10.* Estimate $y(1)$ for (5) using the fourth-order Adams–Bashforth method (4) with $h = 0.01$. Find starting values with a fourth-order Runge–Kutta method. Find the actual solution to (5) and determine $e_{100} = y(1) - y_{100}$.

6.7 Systems

There are at least three ways a computer can be used to solve systems of differential equations. One is by use of **symbolic** languages or programs such as MATHEMATICA, MACSYMA, or MAPLE. The second, if considering linear constant coefficient differential equations, is to compute eigenvalues and eigenvectors as in Chapter 5. Numerical routines for doing this may be found in such packages as MATLAB and EISPACK, and they even appear as primitives in such languages as APL. The third method, and the one we shall briefly discuss, is by extending the numerical methods of Sections 6.1 through 6.6 to systems. This turns out to be notational. In fact, most software for numerically solving differential equations is designed to solve systems of differential equations.

Given a system of m equations,

$$\begin{aligned}
y_1'(t) &= f_1(t, y_1, \ldots, y_m), & y_1(a) &= y_{10}, \\
y_2'(t) &= f_2(t, y_1, \ldots, y_m), & y_2(a) &= y_{20}, \\
&\ \vdots & &\ \vdots \\
y_m'(t) &= f_m(t, y_1, \ldots, y_m), & y_m(a) &= y_{m0},
\end{aligned} \tag{1}$$

we adopt the vector notation

$$\mathbf{y} = \begin{bmatrix} y_1 \\ \vdots \\ y_m \end{bmatrix}, \qquad \mathbf{f} = \begin{bmatrix} f_1 \\ \vdots \\ f_m \end{bmatrix}, \qquad \mathbf{y}_0 = \begin{bmatrix} y_{10} \\ \vdots \\ y_{m0} \end{bmatrix}. \tag{2}$$

Then (1) becomes

$$\mathbf{y}' = \mathbf{f}(t, \mathbf{y}), \qquad \mathbf{y}(a) = \mathbf{y}_0. \tag{3}$$

All of the previous techniques may be applied to (3) by using boldface* type for the y and f wherever they appear. For example, the Euler method of Section 6.2 is

$$\mathbf{y}_{n+1} = \mathbf{y}_n + h\mathbf{f}(t_n, \mathbf{y}_n) \qquad \text{for } n \geq 0. \tag{4}$$

[Note that the n in $\mathbf{y}_n$ denotes the nth step, whereas y_n in (1) is the nth function in the system. To avoid this notational confusion, we shall consider only small-order systems and give each dependent function a different letter.]

EXAMPLE 6.7.1 Modified Euler's Method on a System

Estimate $x(1)$, $y(1)$ for the solution of the system of differential equations

$$\begin{aligned} x'(t) &= x(t) + 3y(t) - 1, \qquad & x(0) &= 1, \\ y'(t) &= x(t) - 2y(t) + t, \qquad & y(0) &= 3, \end{aligned} \tag{5}$$

using a modified Euler method with step size $h = 0.1$.

SOLUTION We have $t_0 = 0$ and $t_n = nh = 0.1n$. Also,

$$N = \frac{b-a}{h} = \frac{1}{0.1} = 10.$$

Let

$$\mathbf{y} = \begin{bmatrix} x \\ y \end{bmatrix}, \qquad \mathbf{f}(t, \mathbf{y}) = \begin{bmatrix} x + 3y - 1 \\ x - 2y + t \end{bmatrix}. \tag{6}$$

We also need the predictor variable $\mathbf{z}$, which we denote

$$\mathbf{z} = \begin{bmatrix} u \\ v \end{bmatrix}. \tag{7}$$

**In handwritten calculations, other conventions, such as underlining, are used to denote vectors.*

The modified Euler method in vector form is then

$$\mathbf{z}_{n+1} = \mathbf{y}_n + h\mathbf{f}(t_n, \mathbf{y}_n), \tag{8}$$

$$\mathbf{y}_{n+1} = \mathbf{y}_n + \frac{h}{2}\Big[\mathbf{f}(t_n, \mathbf{y}_n) + \mathbf{f}(t_{n+1}, \mathbf{z}_{n+1})\Big]. \tag{9}$$

In terms of (6) and (7), the equations (8) and (9) become

$$\begin{aligned} u_{n+1} &= x_n + h(x_n + 3y_n - 1), \\ v_{n+1} &= y_n + h(x_n - 2y_n + t_n), \end{aligned} \tag{10}$$

and

$$\begin{aligned} x_{n+1} &= x_n + \frac{h}{2}[(x_n + 3y_n - 1) + (u_{n+1} + 3v_{n+1} - 1)], \\ y_{n+1} &= y_n + \frac{h}{2}[(x_n - 2y_n + t_n) + (u_{n+1} - 2v_{n+1} + t_{n+1})], \end{aligned} \tag{11}$$

respectively.

For example, given $x_0 = 1$, $y_0 = 3$, $h = 0.1$, we get from (10)

$$u_1 = x_0 + (0.1)(x_0 + 3y_0 - 1) = 1 + 0.1(1 + 9 - 1) = 1.9,$$

$$v_1 = y_0 + (0.1)(x_0 - 2y_0 + 0) = 3 + 0.1(1 - 6) = 2.5.$$

Then from (11),

$$x_1 = 1 + \frac{0.1}{2}\Big\{[1 + 3(3) - 1] + [1.9 + 3(2.5) - 1]\Big\} = 1.87,$$

$$y_1 = 3 + \frac{0.1}{2}\Big\{[1 - 2(3) + 0] + [1.9 - 2(2.5) + 0.1]\Big\} = 2.6.$$

Given (x_1, y_1) we may now compute (x_2, y_2), (x_3, y_3), etc., recursively, from (10) and (11). The estimates for $x(1)$, $y(1)$ are $x_{10} = 14.630$, $y_{10} = 4.361$. ◄

If a programming language with vector variables and vector operators is used, then the previous methods on systems are programmed almost exactly as they were for a single equation.

Higher-Order Equations

Our numerical methods may be applied to higher-order equations by rewriting them as systems. (See Chapter 5.)

EXAMPLE 6.7.2 **Modified Euler's Method on a Second-Order Equation**

Find $y(1)$ for the solution of

$$y''(t) + 4y(t) = 0, \qquad y(0) = 1, \qquad y'(0) = 0, \tag{12}$$

using a modified Euler with step size of 0.1.

SOLUTION To convert (12) to a system, we let $x(t) = y'(t)$. Then (12) becomes $x' + 4y = 0$, so that we have the first-order linear system

$$\begin{aligned} x' &= -4y, \qquad & x(0) &= 0, \\ y' &= x, & y(0) &= 1. \end{aligned} \tag{13}$$

Let $t_0 = 0$, $t_n = n(0.1)$. If we let

$$\mathbf{z}_n = \begin{bmatrix} u_n \\ v_n \end{bmatrix}$$

be the predictor variable, the predictor equations are

$$\begin{aligned} u_{n+1} &= x_n + h(-4y_n), \\ v_{n+1} &= y_n + hx_n, \end{aligned} \tag{14}$$

and the corrector equations are

$$\begin{aligned} x_{n+1} &= x_n + \frac{h}{2}[-4y_n + (-4v_{n+1})], \\ y_{n+1} &= y_n + \frac{h}{2}[x_n + u_{n+1}]. \end{aligned} \tag{15}$$

The actual solution is $x = -2\sin 2t$, $y = \cos 2t$. Table 6.7.1 gives the computed values and their error at several steps. The actual solution is periodic with period π and traces out an ellipse. The numerical solution, however, is not periodic. ◄

TABLE 6.7.1 SOLUTION OF EQ. (13) BY MODIFIED EULER

n	t_n	x_n	y_n	Error in x $x(t_n) - x_n$	Error in y $y(t_n) - y_n$
0	0	0	1	0	0
1	0.1	−0.40000	0.98000	0.00267	0.00007
5	0.5	−1.69171	0.53528	0.00877	0.00502
10	1	−1.81109	−0.42894	−0.00750	0.01280

Exercises

For the systems in Exercises 1 through 6, estimate $x(1)$, $y(1)$, using both an Euler and a modified Euler method.

1. $x' = x + y + 1, \quad h = 0.2,$
   $y' = x + 3y + t^2, \quad x(0) = 0, \quad y(0) = 1$
2. $x' = xy, \quad h = 0.2,$
   $y' = -xy, \quad x(0) = 1, \quad y(0) = 1$
3. $x' = tx - ty, \quad h = 0.1,$
   $y' = x - y + t, \quad x(0) = 1, \quad y(0) = 0$
4. $x' = ty, \quad h = 0.1,$
   $y' = -tx, \quad x(0) = 1, \quad y(0) = 1$
5. $x' = -y^2, \quad h = 0.1,$
   $y' = x, \quad x(0) = y(0) = 1$
6. $x' = x - y - t, \quad h = 0.1,$
   $y' = y + 3x + t, \quad x(0) = y(0) = 0$

In Exercises 7 through 10, find the exact solution of the differential equation. Then compute a numerical estimate for $y(1)$, using an Euler and a modified Euler. Find the error in your estimate by comparing to the true solution.

7. $y'' + y = 0, \quad y(0) = 0, \quad y'(0) = 1, \quad h = 0.1$
8. $y'' - y = t, \quad y(0) = y'(0) = 1, \quad h = 0.1$
9. $y'' - 5y' + 6y = 0, \quad y(0) = 1,$
   $y'(0) = 2, \quad h = 0.1$
10. $y'' + y' = 1, \quad y(0) = y'(0) = 1, \quad h = 0.1$
11. Estimate $y(1)$ for Exercise 9, using a fourth-order Runge–Kutta, and compare to the true solution.

CHAPTER

7

Mostly Nonlinear First-Order Differential Equations

7.1 First-Order Differential Equations

In this chapter, we will solve and analyze first-order differential equations in which the solution x will be a function of time t. Earlier, in Sections 1.3 and 1.7, we studied first-order differential equations

$$\frac{dx}{dt} = f(x, t)$$

when the differential equation was separable,

$$\frac{dx}{dt} = h(t)g(x),$$

or the differential equation was linear,

$$\frac{dx}{dt} + p(t)x = q(t).$$

For separable differential equations, explicit solutions can be obtained by direct integration. For linear differential equations, the general solution, $x(t) = x_p(t) + cx_1(t)$, is always in the form of a particular solution $x_p(t)$ plus an arbitrary multiple of a solution of the associated homogeneous equation $x_1(t)$. Explicit solutions of first-order linear differential equations can always be obtained using an integrating factor, $\exp(\int p(t)\,dt)$. If $p(t)$ is a constant (in which case we said that the linear differential equation had constant coefficients) and $q(t)$ is a polynomial, exponential, or sinusoidal function, or a product or sum of these, it is usually easier to obtain a particular solution by the method of undetermined coefficients (Section 2.9).

In this chapter we will discuss other properties of first-order differential equations, mostly for differential equations that are nonlinear. In Sections 7.1 through 7.4 we will discuss qualitative properties of solutions of first-order differential equations that are called **autonomous**:

$$\frac{dx}{dt} = f(x). \tag{1}$$

First, we will study equilibrium solutions and their stability, introducing ideas that will be discussed further in Chapter 9 for more complicated systems of autonomous differential equations. Although (1) is separable, we will introduce elementary geometric methods to describe the qualitative behavior of the solution, obtaining different information from that obtained by integration. Applications of these ideas to nonlinear models of population dynamics, electronic circuits, and chemical reactions will be given in Sections 7.5 through 7.7. Some techniques to obtain explicit solutions of nonlinear differential equations in some special cases were discussed in Sections 1.13 and 1.14.

7.2 Autonomous Equations and Their Equilibria

The first-order differential equation

$$\frac{dx}{dt} = f(x) \tag{1}$$

is called **autonomous** (meaning time-independent). Autonomous equations have the property that the rate of change dx/dt depends only on the value of the unknown solution x and not on time t. Many physical systems have this property. If the nine o'clock laboratory class is expected to get the same experimental results as the ten o'clock class, then the physical experiment can be quite complicated, but the results do not depend on the starting time. We have assumed, of course, that both classes have the same initial experimental configuration, which means that the initial conditions for the differential equation must be the same.

7.2.1 Equilibrium

A specific solution of an autonomous differential equation

$$\frac{dx}{dt} = f(x) \tag{2}$$

is called an **equilibrium** or **steady-state** solution if the solution is constant in time (it does not depend on time t):

$$x(t) = x_e = \text{constant}.$$

In order for a constant to be an equilibrium solution, it must satisfy the differential equation. Since x_e does not depend on time, $dx_e/dt = 0$. Letting $x = x_e$ in (2) gives $0 = f(x_e)$. Thus

$$x_e \text{ is an } \textbf{equilibrium} \text{ of } dx/dt = f(x) \text{ precisely when } 0 = f(x_e). \tag{3}$$

Thus, all zeros of the function $f(x)$ are equilibrium solutions.

Think of (2) as modeling a physical process and assume that we are at an equilibrium x_e. That is, $x(t_0) = x_e$. Then $dx/dt = 0$ says that everything is in perfect balance (the forces, pressures, etc., balance), and the process $x(t)$ stays at x_e. We say that x_e is an equilibrium solution.

EXAMPLE 7.2.1 **Equilibria**

Find the equilibria of the first-order nonlinear autonomous differential equation

$$\frac{dx}{dt} = x - x^3. \tag{4}$$

SOLUTION In order for a constant x_e to be an equilibrium solution, it must satisfy the differential equation (4). Letting $x = x_e$ in (4) gives

$$0 = x_e - x_e^3 = x_e(1 - x_e^2).$$

In this case, there are three equilibrium solutions:

$$x_e = 0 \qquad \text{and} \qquad x_e = \pm 1.$$

◄

Exercises

In Exercises 1 through 7, determine all equilibria.

1. $\dfrac{dx}{dt} = x(x - 1)$
2. $\dfrac{dx}{dt} = x + x^3$
3. $\dfrac{dx}{dt} = (x - 1)(x - 2)^2$
4. $\dfrac{dx}{dt} = (x^2 - 1)(x^2 - 3)$
5. $\dfrac{dx}{dt} = \sin x$
6. $\dfrac{dx}{dt} = \cos x$
7. $\dfrac{dx}{dt} = x(a - bx)$, where a and b are positive constants

7.3 Stability and Linear Stability Analysis

A physical system in perfect balance may be very difficult to achieve (a pencil standing on its point is unstable) or easy to maintain (a marble at the bottom of a bowl is stable). This section introduces the fundamental idea of stability, which plays an important role in such areas as control systems and fluid mechanics. A solution in perfect balance—that is, an equilibrium solution—may be stable or unstable. What we mean by this is that if the initial conditions are nudged a little bit, changed a small amount from the equilibrium, then the solution of the differential equation may remain near the equilibrium or the differential equation may take the solution away from the equilibrium.

There are more precise mathematical definitions of these ideas, but we will be satisfied with the following definition of stability of an equilibrium:

An equilibrium x_e is **stable**:

- If all solutions starting near x_e stay nearby and
- The closer the solution starts to x_e, the closer nearby it stays.

Otherwise the equilibrium is said to be **unstable**. To be stable, the solution should stay near the equilibrium for all nearby initial conditions. For example, if the solution does not stay near the equilibrium for only one nearby initial condition, we still say that the equilibrium is unstable. Although these ideas are somewhat vague, we will develop more mathematical details shortly.

Sometimes it is helpful to also define a stronger sense of stability. An equilibrium x_e is **asymptotically stable** if the equilibrium is stable and in addition the solution approaches the equilibrium for all nearby initial conditions, $x(t) \to x_e$ as $t \to \infty$. There will be simple examples in which the equilibrium is stable, but not asymptotically stable.

7.3.1 Review of Taylor Polynomial Approximation (Taylor Series)

To analyze the stability of an equilibrium, we must consider initial conditions that are near to the equilibrium x_e. One way to do this analysis is to approximate the differential equation in the neighborhood of the equilibrium. In this way we hope to simplify the differential equation. Mathematically, if we wish to restrict our attention to the neighborhood of some point, then the function $f(x)$ that appears in the differential equation may be approximated by a polynomial, the Taylor polynomial used in Taylor series analysis. Recall from calculus that the Maclaurin polynomial for approximating a function $f(x)$ around $x = 0$ is

$$f(x) \approx f(0) + xf'(0) + \frac{1}{2}x^2 f''(0) + \cdots.$$

This can be generalized to approximating a function around $x = x_e$ and is called the **Taylor polynomial**:

$$f(x) \approx f(x_e) + (x - x_e)f'(x_e) + \frac{1}{2}(x - x_e)^2 f''(x_e) + \cdots.$$

The linear term, $f(x_e) + (x - x_e)f'(x_e)$, including the constant $f(x_e)$, is the **linearization** or **tangent-line approximation** to the function $f(x)$ in the neighborhood of the equilibrium x_e. Most of the time, when we are near the equilibrium, we will be satisfied with approximating our function by its linearization:

$$f(x) \approx f(x_e) + (x - x_e)f'(x_e). \tag{1}$$

7.3.2 Linear Stability Analysis

An equilibrium x_e is stable or unstable depending on the behavior of solutions of the differential equation

$$\frac{dx}{dt} = f(x) \tag{2}$$

near the equilibrium. Recall that the equilibrium x_e satisfies $f(x_e) = 0$. We first assume that x is near x_e and, as an approximation, replace $f(x)$ by its linearization:

$$\frac{dx}{dt} = f(x_e) + (x - x_e)f'(x_e). \tag{3}$$

The differential equation (3) simplifies because x_e is an equilibrium and $f(x_e) = 0$, so that

$$\frac{dx}{dt} = (x - x_e)f'(x_e). \tag{4}$$

Differential equation (4) is actually quite simple. It is linear with constant coefficients. However, that may not be particularly apparent. It is usual to introduce a new dependent variable

$$y \equiv x - x_e,$$

which is the **displacement from the equilibrium** of x. Letting $x = y + x_e$, the differential equation (4) becomes

$$\frac{dy}{dt} = f'(x_e)y, \tag{5}$$

since x_e is a constant and hence $dy/dt = dx/dt$.

Equation (5) is a linear differential equation with constant coefficient $f'(x_e)$. We easily obtain its general solution to be

$$y = x - x_e = ce^{f'(x_e)t}. \tag{6}$$

From (6), we see that the behavior of the solution of (5) near the equilibrium depends in a very simple way on the sign of $f'(x_e)$:

If $\mathbf{f'(x_e) > 0}$,

the solutions exponentially grow in time away from the equilibrium and we say that the equilibrium $x = x_e$ is **unstable** (sometimes we say that the equilibrium $y = 0$ is unstable).

If $\mathbf{f'(x_e) < 0}$,

the solutions exponentially decay in time toward the equilibrium, in which case we say that the equilibrium $x = x_e$ is **stable** (sometimes we say that the equilibrium $y = 0$ is stable). In fact, if $f'(x_e) < 0$, the equilibrium $x = x_e$ is asymptotically stable, since from (6), $x(t) \to x_e$ as $t \to \infty$.

EXAMPLE 7.3.1 Linear Stability

We reconsider the first-order nonlinear differential equation from Example 7.2.1,

$$\frac{dx}{dt} = f(x) = x - x^3, \tag{7}$$

which has three equilibria, $x_e = 0$ and $x_e = \pm 1$. From the linear stability analysis, the stability of the equilibria is determined from the sign of $f'(x_e)$. In general, $f'(x) = 1 - 3x^2$. Since $f'(0) = 1 > 0$, we conclude that the equilibrium $x_e = 0$ is unstable; the linearized equation $y' = y$ has solutions $y = ce^t$ that grow with t. The other two equilibria are stable, since $f'(\pm 1) = -2 < 0$. They each have the linearized equation $y' = -2y$, which has the solution $y = ce^{-2t}$. Here $\lim_{t\to\infty} y = 0$. The solutions to (7) are graphed later in Figure 7.4.4 on page 512. Note in that figure that although $x_e = 1$ is stable, if we start far enough from that equilibrium, say $x_0 = -1.5$, then $x(t)$ need not go to the equilibrium $x_e = 1$. We still refer to $x_e = 1$ as being stable, since stability (as we describe it) depends only on nearby initial conditions. ◄

Validity of Linearization and the Borderline Cases (Neutral Stability)

If $f'(x_e) = 0$, the equilibrium x_e is on the borderline between the clearly stable and unstable cases. This case is of less importance in a first exposure to the concepts of stability, and it is somewhat subtle and may lead to some confusion. Technically, for the *linear* differential equation (5), since $f'(x_e) = 0$, the linearized differential equation is $y' = 0$. The equilibrium $y = 0$ is stable (but not asymptotically stable), since from the solution of the linear differential equation, (6), $y(t) = c$ or $x(t) = x_e + c$. In other words, if the solution starts near the equilibrium, it doesn't move at all according to the *linear* differential equation, since $y' = 0$. (It thus stays near the equilibrium according to the *linear* equation, since it is assumed to start near the equilibrium.)

It will be shown in the next section that our conclusions concerning the stability of an equilibrium x_e based on the linearization are always valid for the real nonlinear problem in the clear cases, $f'(x_e) > 0$ and $f'(x_e) < 0$. To be more explicit, the equilibrium x_e is unstable if $f'(x_e) > 0$ and stable if $f'(x_e) < 0$. However, in the borderline case between stable and unstable in which $f'(x_e) = 0$, the stability of the real nonlinear problem may differ from the simplistic conclusion reached from its linearization. When $f'(x_e) = 0$, we will show that the equilibrium solution of the real problem may be stable or unstable. Its stability is not determined by the linearization, but is determined instead by keeping the first nonzero higher-order nonlinear terms in the Taylor polynomial approximation of $f(x)$ around its equilibrium. This case, which is linearly stable (but not asymptotically stable), $f'(x_e) = 0$, is sometimes called **neutrally stable**. [Other scientists refer to the case $f'(x_e) \neq 0$ as being **hyperbolic**, and hence to the case $f'(x_e) = 0$ as being **nonhyperbolic**.)

Summary of Linear Stability Analysis

We summarize these results, which we refer to as a **linearized stability analysis**.

Suppose x_e is an equilibrium of $dx/dt = f(x)$.

If $f'(x_e) < 0$, the equilibrium $x = x_e$ is stable.

If $f'(x_e) > 0$, the equilibrium $x = x_e$ is unstable. (8)

If $f'(x_e) = 0$, the equilibrium $x = x_e$ may be stable or unstable, depending on the nonlinear terms.

Exercises

In Exercises 1 through 8, determine all equilibria and their linearized stability. If the linearized analysis is inconclusive, say so.

1. $\dfrac{dx}{dt} = x(x - 1)$

2. $\dfrac{dx}{dt} = x(x - 1)^2$

3. $\dfrac{dx}{dt} = (x - 1)(x - 2)^2$

4. $\dfrac{dx}{dt} = (x^2 - 1)(x^2 - 3)$

5. $\dfrac{dx}{dt} = \sin x$

6. $\dfrac{dx}{dt} = \tan x$

7. $\dfrac{dx}{dt} = e^{3x} - 7$

8. $\dfrac{dx}{dt} = e^{-3x} - 5$

7.4 One-Dimensional Phase Portraits

In the previous section, the nonlinear autonomous first-order differential equation

$$\frac{dx}{dt} = f(x) \tag{1}$$

was discussed by analyzing the linear stability of equilibrium solutions. It is usually easy to determine whether each equilibrium is stable or unstable. However, the linear analysis describes only the behavior of the solution near the equilibrium. In the present section, we will introduce a geometric method to understand the behavior of solutions of (1) that is valid not only near an equilibrium but also far away from all equilibria.

The differential equation (1) defines the relationship between x and dx/dt. That relationship can often be graphed easily, since $f(x)$ is given. We simply graph dx/dt as a function of x. A typical functional relationship is graphed in Figure 7.4.1 (see page 511). At time t, the solution $x(t)$ is a point on the x-axis and the point $(x(t), x'(t))$ lies on the graph. As time increases, $x(t)$ moves along the x-axis and $(x(t), x'(t))$ moves along the curve. The arrows in the figure show the manner in which $(x(t), x'(t))$ moves along the curve as time advances. The differential equation easily determines the direction of these arrows. In the upper half plane, where $dx/dt > 0$, it follows that x is an increasing function of time. So the arrows in the graph point to the right, indicating that $(x(t), x'(t))$ moves to the right as time increases. Similarly, in the lower half plane, where $dx/dt < 0$, the arrows show motion to the left as time increases, since x is a decreasing function of time there.

The places where the curve intersects the x-axis are the equilibrium points, since $dx/dt = 0$ there. Not only do we determine graphically where the equilibrium is located, but it is easy to determine in all cases whether the equilibrium is stable or unstable. If the arrows indicate that all solutions near an equilibrium move toward the equilibrium as time increases, then the equilibrium is stable, as shown in Figure 7.4.1. If some solutions near an equilibrium move away from that equilibrium as time increases, then that equilibrium is unstable.

The stability of an equilibrium can be easily determined from the graph whether or not that stability can be determined by linearization. In the situation assumed in Figure 7.4.1, two of the equilibria are clearly stable and one is unstable, as indicated on the figure. If the slope of the curve is positive when it crosses the x-axis at the equilibrium $[f'(x_e) > 0]$, then the equilibrium is unstable. If the slope of the curve is negative when it crosses the x-axis $[f'(x_e) < 0]$, then the equilibrium is stable. Thus, the geometric analysis has verified the conclusions of the linear stability analysis in the usual simple cases in which $f'(x_e) \neq 0$. In the example graphed in Figure 7.4.1, we assumed that the three equilibria all satisfied the simple criterion. However, as is developed further in the exercises, this geometric method can also be used to determine the stability of an equilibrium in the more difficult case in which the linearization does not determine the stability $[f'(x_e) = 0]$.

FIGURE 7.4.1

Phase portrait.

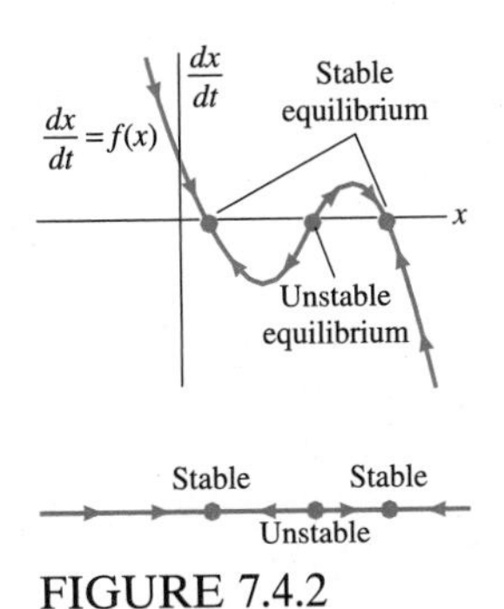

FIGURE 7.4.2

Phase portrait.

Furthermore, we have easily determined the qualitative behavior of the solution to the differential equation corresponding to all initial conditions. We know qualitatively how the solution approaches an equilibrium if the solution starts near an equilibrium, and we even know how the solution behaves when it starts far away from an equilibrium. If there is more than one stable equilibrium, we can determine which initial conditions approach which equilibrium. These results can be summarized in a **one-dimensional phase portrait**, sketched in Figure 7.4.2. (Figure 7.4.1 can also be called a one-dimensional phase portrait.) In Figure 7.4.2 we show just the movement of x on the x-axis instead of (x, x') as in Figure 7.4.1.

We discuss two illustrative examples.

EXAMPLE 7.4.1 Two Stable Equilibria

As a specific example to illustrate the geometric method of a one-dimensional phase portrait, reconsider the first-order nonlinear differential equation

$$\frac{dx}{dt} = f(x) = x - x^3,$$

which has three equilibria, $x_e = 0$ and $x_e = \pm 1$. The phase portrait requires the graph of the cubic $f(x) = x - x^3$, which is given in Figure 7.4.3. The graph sketched in Figure 7.4.3 has two critical points at $x = \pm 1/\sqrt{3}$, one a local maximum and the other a local minimum. The one-dimensional phase portrait shows that the equilibrium $x = 0$ is clearly unstable, whereas the other two equilibria $x = \pm 1$ are clearly stable. Since $f'(0) = 1 > 0$, we conclude from a linear stability analysis that the equilibrium $x_e = 0$ is unstable. The other two equilibria are stable, since $f'(\pm 1) = -2 < 0$. This shows that the stability of the equilibrium determined from the phase portrait

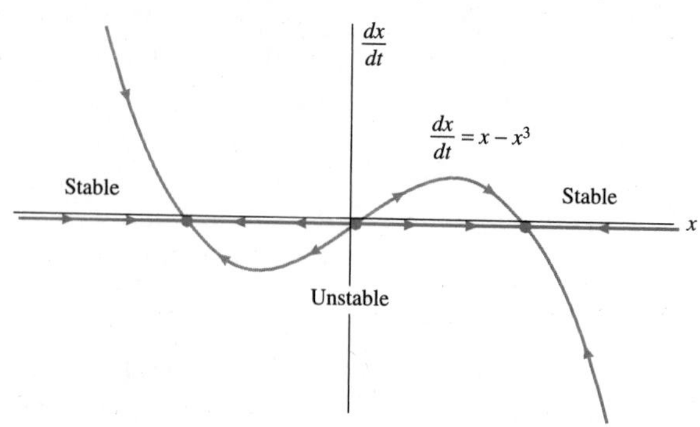

FIGURE 7.4.3

Phase portrait.

agrees with the linear stability analysis. The other form of the one-dimensional phase portrait, showing how x changes as a function of time, can be drawn on the same figure.

In Figure 7.4.4, we have sketched the qualitative behavior of x as a function of time for various initial conditions. (This was called the direction field in Chapter 1.) For example, we see that the solution approaches 1 as $t \to +\infty$ for all initial conditions that satisfy $x(0) > 0$. For initial conditions such that $x(0) < 0$, $x(t) \to -1$ as $t \to +\infty$. For the initial condition $x(0) = 0$, the unstable equilibrium, the solution stays at the unstable equilibrium for all time, $x(t) = x(0) = 0$. This is the only solution that approaches or is at the unstable equilibrium as $t \to \infty$. From this method we can determine only the qualitative behavior of these curves. The precise dependence of x on t can be determined by integration of this separable equation (or accurately approximated using contemporary numerical methods). ◄

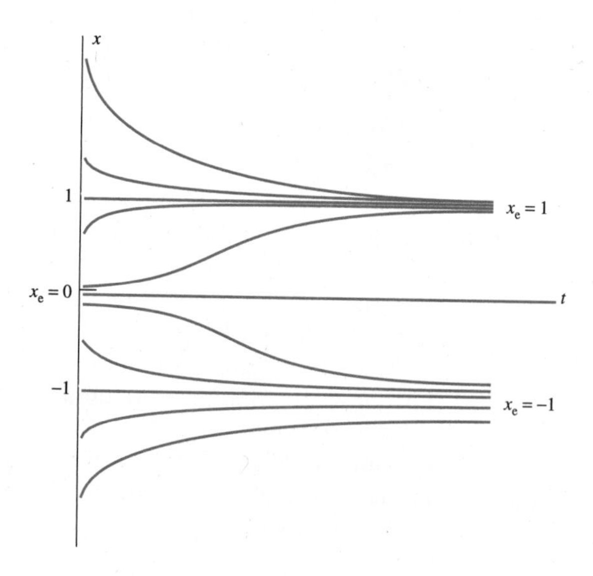

FIGURE 7.4.4

The advantage of the present geometric method is the simplicity by which we obtain an understanding of fundamental relationships of equilibrium, stability, and the manner in which the solution depends on the initial conditions. If the differential equation is more complicated—for example, if it depends on some parameter—then the qualitative method based on the one-dimensional phase portrait is particularly effective. If more quantitative details are needed, the one-dimensional phase portrait can always be supplemented by further analytic work or numerical computations.

Two graphs of solutions cannot intersect. If two different solutions intersected in a finite time, it would violate the uniqueness theorem of the initial value problem. There would be two solutions with the same initial conditions. This proves, for example, that the solution takes an infinite amount of time to reach $x = 1$, since $x = 1$ is a solution of the differential equation.

In general, for first-order autonomous equations, only very simple qualitative behavior is possible. If a solution approaches an equilibrium, it does so directly without any oscillations.

EXAMPLE 7.4.2 An Explosion

Let us consider the simple nonlinear autonomous first-order differential equation

$$\frac{dx}{dt} = x^2. \tag{2}$$

Although this problem can be explicitly solved relatively easily by separation (see the exercises), it is interesting to note that the one-dimensional phase portrait is a quick and easy method that probably gives a better *understanding* of some of the behavior of the solutions to this differential equation. In Figure 7.4.5, dx/dt is graphed as a function of x. From this figure, we see immediately that $x = 0$ is an unstable equilibrium. If the initial conditions are less than zero, $x(0) < 0$, then the solution approaches 0 as $t \to \infty$. (It cannot reach the equilibrium, $x = 0$, in a finite time because of the uniqueness theorem.) For these initial conditions, the equilibrium appears stable. However, for the equilibrium to be stable, the solutions for all nearby initial conditions, not just for those initial conditions less than the equilibrium, must approach the equilibrium. From Figure 7.4.5, we see that $x(t)$ goes away from zero for initial conditions that are positive, $x(0) > 0$. Thus, $x = 0$ is an unstable equilibrium. Furthermore, for initial conditions $x(0) > 0$, it is clear that the solution approaches infinity, $x(t) \to \infty$. However, from the phase portrait one cannot know that the solution explodes in a finite time (has a vertical asymptote). This can most easily be seen from the explicit solution obtained by separation in Exercise 17.

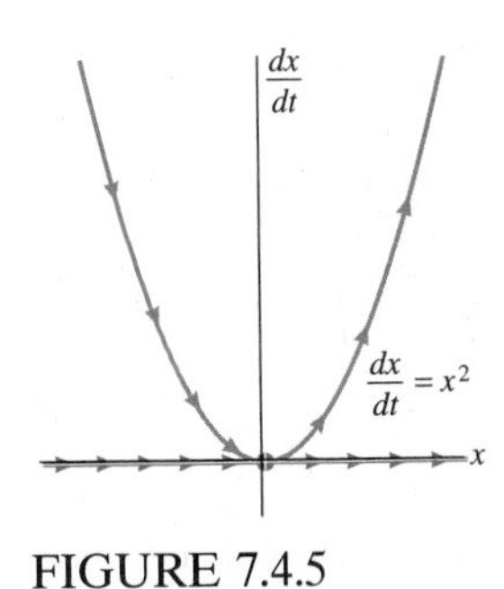

FIGURE 7.4.5

Phase portrait.

The stability of equilibria is usually determined by the linearization. However, in this example, the linearization approximation in the neighborhood of the equilibrium $x = 0$ is

$$\frac{dx}{dt} = f'(0)x = 0x = 0.$$

In this case, the linearization is on the border between being stable and unstable. The linearization is neutrally stable, and the stability of $x = 0$ cannot be determined from the linearization. We have shown above that the equilibrium $x = 0$ is unstable based on the geometric analysis of the nonlinear differential equation. ◀

Exercises

In Exercises 1 through 11, graph the one-dimensional phase portraits. In doing so, determine all equilibria and their stability. If the linearized stability analysis is inconclusive, say so.

1. $\dfrac{dx}{dt} = x(x-1)$

2. $\dfrac{dx}{dt} = x(x-1)^2$

3. $\dfrac{dx}{dt} = (x-1)(x-2)^2$

4. $\dfrac{dx}{dt} = (x^2-1)(x^2-3)$

5. $\dfrac{dx}{dt} = \sin x$

6. $\dfrac{dx}{dt} = \tan x$

7. $\dfrac{dx}{dt} = x^2 + 1$

8. $\dfrac{dx}{dt} = -x^2$

9. $\dfrac{dx}{dt} = x^3$

10. $\dfrac{dx}{dt} = -x^3$

11. $\dfrac{dx}{dt} = (x-3)^4$

In Exercises 12 through 15, sketch time-dependent solutions based on one-dimensional phase portraits.

12. $\dfrac{dx}{dt} = x(x-1)$

13. $\dfrac{dx}{dt} = x(x-1)^2$

14. $\dfrac{dx}{dt} = (x-1)(x-2)^2$

15. $\dfrac{dx}{dt} = (x^2-1)(x^2-3)$

16. In this problem, we will consider equilibria x_e for the differential equation $dx/dt = f(x)$ in the case in which $f'(x_e) = 0$.

a) Determine the stability of x_e if $f'(x_e) = 0$, but $f''(x_e) \neq 0$.

b) Determine the stability of x_e if $f'(x_e) = 0$, $f''(x_e) = 0$, but $f'''(x_e) \neq 0$.

c) Generalize the results of part (b) to even and odd n.

d) Generalize the results of part (b) based on sign changes of $f(x)$.

e) Simplify the criteria in (d) by applying the first derivative test for a maximum or minimum to $-\int f(x)\,dx$.

17. Use separation to determine the stability of $x = 0$ for $dx/dt = x^2$. Compare your results to those obtained in this section.

18. Use separation to show that solutions of $dx/dt = x^2$ explode in a finite time if the initial condition is positive.

19. Use separation to show that solutions of $dx/dt = x^2$ do not have a vertical asymptote but are well behaved for all time if the initial condition is negative. Show that the solution approaches $x = 0$, but never reaches $x = 0$ in a finite time.

7.5 Application to Population Dynamics: The Logistic Equation

The simplest model for the growth of a population x results from assuming that the growth rate k is a positive constant:

$$\frac{dx}{dt} = kx.$$

In this case, the population grows exponentially, $x(t) = x(0)e^{kt}$, and eventually becomes indefinitely large. A more realistic model of population growth is needed when the population becomes sufficiently large. Experiments have shown that there is a crowding effect. The growth rate $(1/x)(dx/dt)$ is not a constant, but depends on the population itself, diminishing as the population increases. For simplicity, we choose to model the growth rate itself as a linear function of the population:

$$\frac{1}{x}\frac{dx}{dt} = a - bx.$$

Equivalently, we have

The logistic equation:

$$\frac{dx}{dt} = x(a - bx) = ax - bx^2. \tag{1}$$

Equation (1) is a first-order nonlinear differential equation that is autonomous. The constants a and b are assumed to be positive. The growth rate, $a - bx$, depends on the population x. When the population is zero, the growth rate is a. This represents the growth rate without the crowding effect of environmental influences (and corresponds to the growth rate we called k previously). The constant b is the decrease in the growth rate per individual. According to this model, if the population becomes sufficiently large, the growth rate becomes negative. This model was first investigated in the late 1830s by Verhulst and is known as the **logistic equation.**

Although the logistic equation (1) is separable, the qualitative behavior of the solutions can be determined more readily using the one-dimensional phase portrait and other ideas of this chapter. To analyze a nonlinear equation such as (1), we first look for equilibrium populations (equilibrium solutions of the differential equation). The rate of change of the population will be zero if $0 = x(a - bx)$. The population can be in equilibrium in two different ways:

$$x = 0 \quad \text{and} \quad x = \frac{a}{b}.$$

Zero population is certainly an equilibrium population. If the population is initially zero, it will stay zero. The other equilibrium, $x = a/b$, is the largest population that the environment can sustain without loss, since the growth rate will be negative if $x > a/b$. This equilibrium is called the **carrying capacity** of the environment.

The stability of the equilibrium populations can be determined by a linearized stability analysis. The nonlinear differential equation $dx/dt = f(x) = ax - bx^2$ can be approximated by its linearization in the neighborhood of each equilibrium,

$$\frac{dy}{dt} = f'(x_e)y,$$

where $y = x - x_e$ is the displacement from an equilibrium and x_e denotes an equilibrium. In this case, $f'(x_e) = a - 2bx_e$. The zero population is unstable, since $f'(0) = a > 0$. The small populations we consider here grow exponentially. However, the environment's carrying capacity is stable, since $f'(a/b) = a - 2b \cdot a/b = -a < 0$.

To obtain the qualitative behavior of the populations away from the equilibria, we first sketch the one-dimensional phase portrait in Figure 7.5.1. The derivative dx/dt is a function of x, and the graph of this dependence is a parabola with intercepts at the equilibria $x = 0$ and $x = a/b$. Since x is a population, we are concerned only with $x \geq 0$. Arrows are introduced, indicating how the solution changes in time. In the upper half plane, the

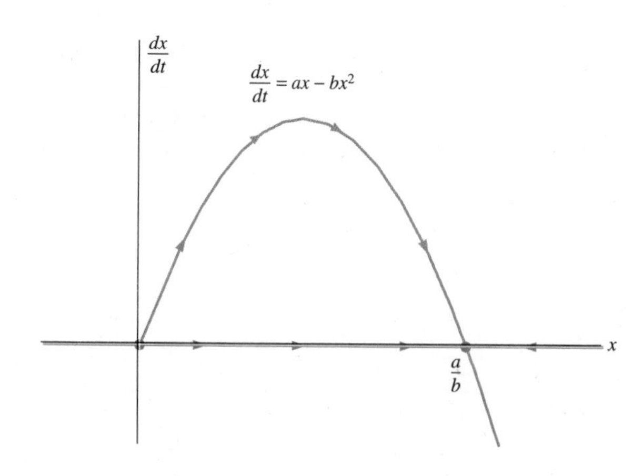

FIGURE 7.5.1

Phase portrait.

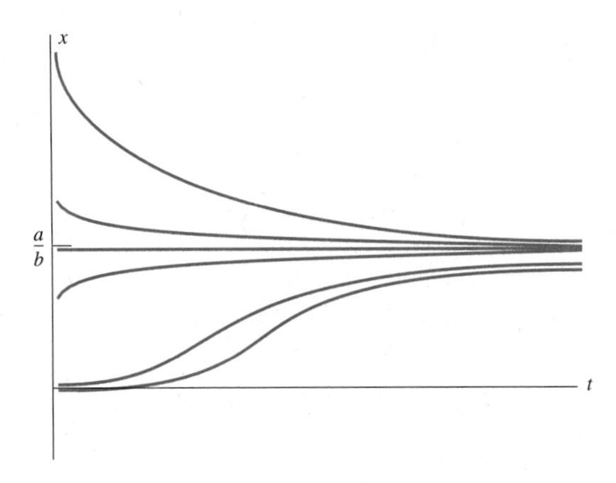

FIGURE 7.5.2

Solutions of the logistic equation.

arrows point to the right, since x is increasing there. In the lower half plane, the arrows point to the left. The instability of $x = 0$ and the stability of $x = a/b$ are seen. For populations less than the carrying capacity, the population increases toward the carrying capacity as a function of time (but the population never actually reaches the carrying capacity). For initial populations greater than the carrying capacity, the population decreases toward the carrying capacity without reaching it. The qualitative sketch of the population as a function of time is shown in Figure 7.5.2 for various initial populations. The analysis of possible inflection points is left for an exercise.

Explicit Solution of the Logistic Equation by Separation

An explicit solution to the logistic equation can be obtained, since the equation is separable. After using partial fractions and a considerable amount of algebra, we find that if $x(0) \neq 0$, then

$$x(t) = \frac{\dfrac{a}{b}}{1 + \left[\dfrac{a - bx(0)}{bx(0)}\right] e^{-at}}, \tag{2}$$

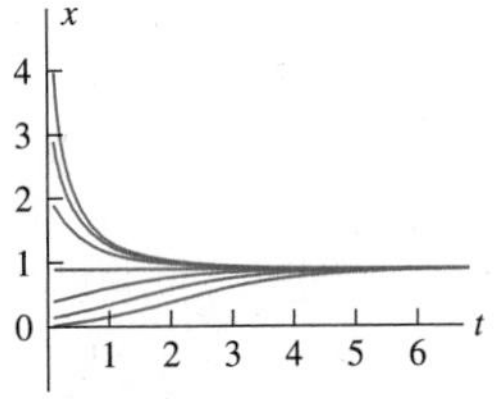

FIGURE 7.5.3

Graph of (2) with $a = b = 1$.

in terms of the initial population $x(0)$. In the exercises, you are asked to show how this explicit solution verifies the qualitative results obtained from the one-dimensional phase portrait. Specific logistic curves depend on three parameters a, b, and $x(0)$. A few examples are sketched in Figure 7.5.3 and are seen to be consistent with the qualitative behavior shown in Figure 7.5.2.

Exercises

1. Consider $dx/dt = ax - bx^3$ (with $a > 0$, $b > 0$, and $x \geq 0$).

a) How does the growth rate depend on the population?

b) Sketch the one-dimensional phase portrait.

c) Sketch the time-dependent solution qualitatively.

2. Suppose that we have a model of population growth that has a carrying capacity, in the sense that the growth rate is negative for populations greater than the carrying capacity and positive for populations less than the carrying capacity. Show that the qualitative behavior of this model will be the same as the qualitative behavior of the quadratic logistic model analyzed in this section.

3. Derive the general solution (2) from the differential equation (1) by separation.

4. Show from the exact solution (2) that by rescaling the population, there are only two parameters.

5. Show from the differential equation (1) that by rescaling the population, there are only two parameters.

6. Consider $dx/dt = ax^2 - bx$ (with $a > 0$, $b > 0$, and $x \geq 0$).

a) How does the growth rate depend on the population?

b) Sketch the one-dimensional phase portrait.

c) Sketch the time-dependent solution qualitatively.

d) Obtain the exact solution.

e) Show that the exact solution has the qualitative behavior obtained in (c).

7. Consider $dx/dt = ax + bx^2$ (with $a > 0$, $b > 0$, and $x \geq 0$).

a) How does the growth rate depend on the population?

b) Sketch the one-dimensional phase portrait.

c) Sketch the time-dependent solution qualitatively.

d) From the exact solution, show that the population reaches infinity in a finite time, what might be called a population explosion.

8. From the differential equation (1), investigate the concavity of $x(t)$. Where will an inflection point occur? Compare to the sketches in Figure 7.5.2.

9. The logistic curve, (2), is sometimes referred to as an S-curve for reasons to be described.

a) Show that

$$x(t) = \alpha + \beta \frac{e^{(a/2)(t-t_0)} - e^{-(a/2)(t-t_0)}}{e^{(a/2)(t-t_0)} + e^{-(a/2)(t-t_0)}}. \quad (3)$$

What are α, β, and t_0? [*Hint*: Put (3) over a common denominator. Multiply numerator and denominator by $e^{-(a/2)(t-t_0)}$.]

b) Recall that the hyperbolic functions are defined as follows:

$$\sinh t = \frac{e^t - e^{-t}}{2}, \qquad \cosh t = \frac{e^t + e^{-t}}{2},$$

$$\tanh t = \frac{\sinh t}{\cosh t}.$$

Show that

$$x(t) = \alpha + \beta \tanh \frac{a}{2}(t - t_0).$$

c) Sketch $\tanh t$ as a function of t. Show that it might be called S-shaped. Consider the asymptotic behavior as $t \to \pm\infty$.

d) Now sketch the logistic curve.

e) Show that $\alpha + \beta = a/b$ and $\alpha - \beta = 0$.

10. Let $Q(t)$ be the population of a particular species at time t. Suppose that the rate of population growth depends only on the population size, so that

$$\frac{dQ}{dt} = f(Q). \quad (4)$$

Explain, in biological terms, why each of the following three statements could be reasonable assumptions.

a) $f(0) = 0$

b) $f'(Q) > 0$ for small positive Q

c) $f'(Q) < 0$ for large Q

11. Suppose that $f(Q) = a + bQ + cQ^2$ is a second-degree polynomial in Q. Show that if f satisfies (a), (b), and (c) of Exercise 10, then $a = 0$, $b > 0$, and $c < 0$, so that (4) becomes (1).

7.6 Application to Circuit Theory

The relationship between the voltage v and the current i for a resistor is usually $v = iR$ if the current is not too large. This is a linear relationship if R does not depend on i. A nonlinear resistor is one in which the voltage depends on the current in a nonlinear way. For example, if the current gets large enough, heating effects could cause the resistance to increase. Then R would depend on i. Nonlinear v–i characteristics also occur in the mathematical models for such devices as transistors and diodes. We consider two examples here and leave others for the exercises. Electric circuits were discussed earlier in Sections 1.10 and 2.12.

EXAMPLE 7.6.1 **Stability of Equilibria for Nonlinear Circuit**

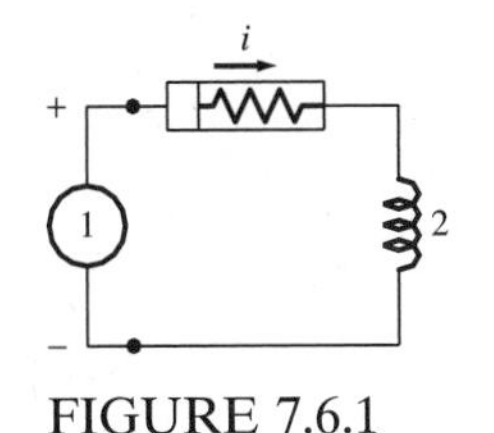

FIGURE 7.6.1

In the circuit shown in Figure 7.6.1, the current is initially zero, and there is a voltage source of 1 V, an inductor with inductance 2 H, and a nonlinear resistor whose v–i characteristic is $v = i^2$. Find the current as a function of time t. Also determine any equilibria and their stability.

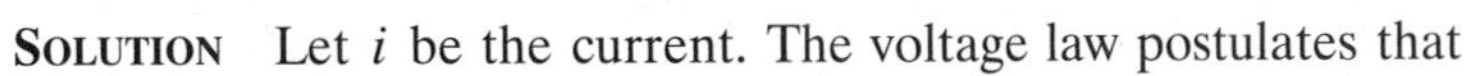

Solution Let i be the current. The voltage law postulates that

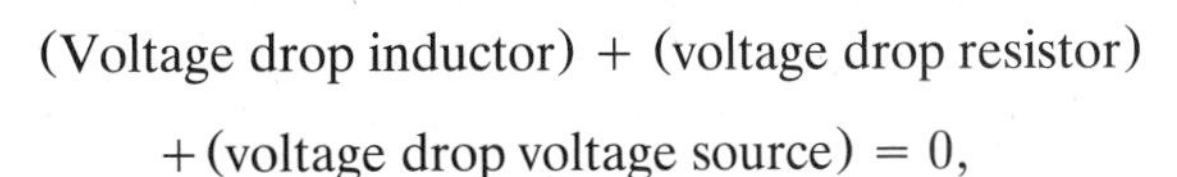

$$\text{(Voltage drop inductor)} + \text{(voltage drop resistor)}$$
$$+\,\text{(voltage drop voltage source)} = 0,$$

or

$$2\frac{di}{dt} + i^2 - 1 = 0.$$

This is a nonlinear equation that can be solved by separation of variables. The solution for $i(0) = 0$ is

$$i(t) = \frac{e^t - 1}{e^t + 1}. \tag{1}$$

The differential equation is

$$\frac{di}{dt} = \frac{1}{2}(1 - i^2) = f(i).$$

The equilibria are given by $f(i) = 0$ or

$$i = 1, \qquad i = -1.$$

Also, $f'(i) = -i$. Thus the linear stability analysis of Section 7.3 gives

$$f'(1) = -1 < 0 \qquad i = 1 \text{ is stable},$$
$$f'(-1) = 1 > 0 \qquad i = -1 \text{ is unstable}.$$

Note that as $t \to \infty$, the solution (1) approaches the stable equilibrium $i = 1$. This conclusion can also be reached using the one dimensional phase portrait of Section 7.4. ◄

EXAMPLE 7.6.2 Solutions and Equilibria for Nonlinear Circuit

As our last example, let us consider the circuit shown in Figure 7.6.2. The figure shows a current source of 3 A, an inductor of 1 H, and a nonlinear resistor with v–i characteristic $v = i^3 - 4i$. Set up a differential equation for the current in the resistor. Sketch the solutions and find all equilibria. Then find all solutions.

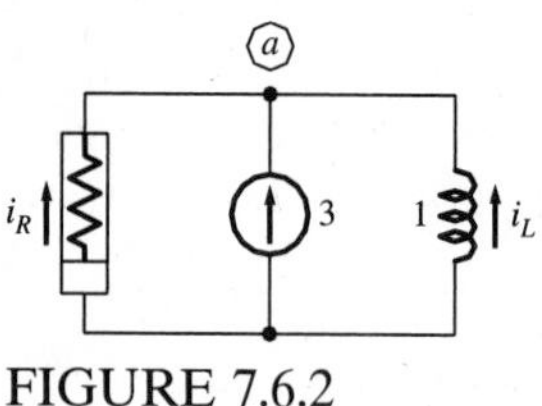

FIGURE 7.6.2

Solution Let i_R, v_R, i_L, v_L be the currents and voltages in the resistor and the inductor. Applying the current law at node ⓐ, we get

$$3 + i_R + i_L = 0, \tag{2}$$

and, applying the voltage law to the outside loop consisting of just the inductor and the resistor, we have

$$v_L - v_R = 0. \tag{3}$$

Since

$$v_L = L\frac{di_L}{dt} \qquad \text{and} \qquad v_R = i_R^3 - 4i_R,$$

(3) becomes

$$\frac{di_L}{dt} - (i_R^3 - 4i_R) = 0. \tag{4}$$

This equation has two dependent variables, i_R and i_L. We use (2) to eliminate i_L by letting $i_L = -3 - i_R$, so that finally (4) is

$$-\frac{di_R}{dt} - (i_R^3 - 4i_R) = 0$$

or

$$\frac{di_R}{dt} = 4i_R - i_R^3. \tag{5}$$

This is a nonlinear equation in i_R. To simplify what follows, let $i = i_R$. To find the equilibrium points (also known as steady-state operating points), set

$di/dt = 0$, that is

$$0 = \frac{di}{dt} = 4i - i^3 = i(4 - i^2).$$

The equilibrium points are $i = 0$ and $i = \pm 2$. If we analyze Eq. (5) as in Section 7.4, we see that

$$\frac{di}{dt} > 0 \quad \text{if} \quad 0 < i < 2, \quad \text{or} \quad i < -2,$$
$$\frac{di}{dt} < 0 \quad \text{if} \quad i > 2, \quad \text{or} \quad -2 < i < 0.$$

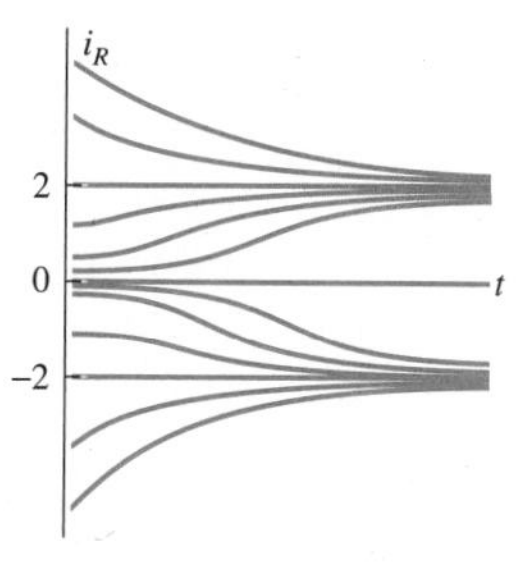

FIGURE 7.6.3

Solutions of (5).

The solutions are graphically represented in Figure 7.6.3.

Looking at Figure 7.6.3, we see that the circuit shown in Figure 7.6.2 always approaches an equilibrium—an equilibrium, however, that depends on the initial currents. Since the zero equilibrium is unstable, it would be difficult to physically observe it, and so there would appear to be only the two equilibria, ± 2.

To actually solve (5), use separation of variables. Note that separation of variables works only if i is not an equilibrium solution. The solution can be found to be

$$i = \pm 2\left[\frac{e^{8t}k}{e^{8t}k - 1}\right]^{1/2} \quad \text{and} \quad i = \pm 2.$$

The v–i characteristic of the resistor in Figure 7.6.2 was chosen to simplify the calculations. However, similarly shaped v–i characteristics occur in certain diodes and transistors.

Exercises

Exercises 1 through 4 refer to the circuit given in Figure 7.6.4. The inductor is linear with inductance of L. The voltage source is $e(t)$.

1. The voltage is constant, $e = 0$. The resistor is nonlinear with v–i characteristic $v = -i^2$, and $L = 1$.

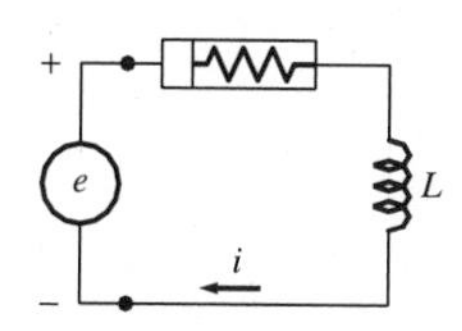

FIGURE 7.6.4

a) Set up the differential equation for the current.

b) Sketch the solution.

c) Solve the differential equation. (That the sign of the voltage in the resistor is opposite to that of the current merely means that we are dealing with the model of a device that puts power into the circuit rather than one that dissipates power.)

2. The voltage is constant, $e = 1$. The v–i characteristic of the resistor is $v = -i^2$, and $L = 1$.

a) Set up the differential equation for the current.

b) Sketch the solutions.

c) Solve the differential equation.

3. The voltage is constant, $e = 0$. The v–i characteristic of the resistor is $v = i^2$ and $L = 1$.

a) Set up the differential equation for the current.

b) Sketch the solutions.

c) Suppose that the initial current is 4. Solve the differential equation for the current.

4. The resistor has v–i characteristic $v = i^3$, $e = 0$, and $L = 1$.

a) Set up the differential equation for the current.

b) Sketch the solution.

c) Solve the differential equation if the initial current is 4.

Exercises 5 through 7 refer to the circuit in Figure 7.6.5. The capacitor is linear and has capacitance C. The voltage source is $e(t)$.

5. For the circuit in Figure 7.6.5, suppose that C is constant and the v–i characteristic of the resistor is $f(i)$. Show that the current satisfies the differential equation $f'(i)\frac{di}{dt} + \frac{1}{C}i = e'(t)$.

6. The capacitance is 1, the nonlinear resistor has v–i characteristic $v = i^2$, and the voltage source is $e(t) = t$. Suppose that the current in the resistor is initially 2. Apply the voltage law, differentiate the resulting equation with respect to t, and use $dq/dt = i$ to get a first-order equation in the current. Solve this equation for the relationship (which may be implicit) between current and time.

7. The capacitance is 1, and the resistor is nonlinear with v–i characteristic $v = \frac{1}{3}i^3$. The voltage source is $e(t) = t$. The current in the resistor at $t = 1$ is $i(1) = 2$. Derive a differential equation for the current in the resistor and solve it (note Exercise 6).

8. Consider the circuit in Figure 7.6.6. The capacitor has capacitance $C = 1$. The current source is 1, and the resistor's v–i characteristic is $v = \ln i$ for $i > 0$. Proceeding as in Example 7.6.2, set up a differential equation for the current in the resistor, and solve the resulting differential equation, given that the current in the resistor is initially 2.

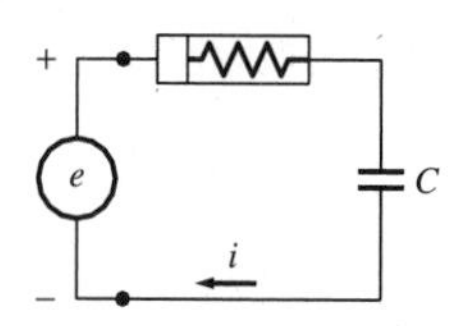

FIGURE 7.6.5

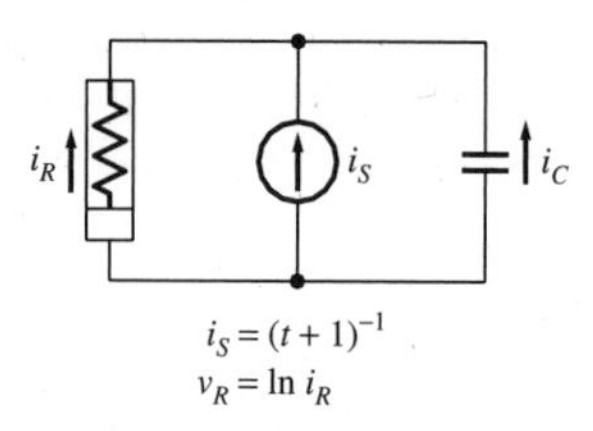

FIGURE 7.6.6

7.7 Application to Chemical Reactions

Consider a chemical reaction occurring in a well-mixed solution. Such reactions can be quite complicated. We shall present a model that is sometimes sufficiently accurate.

It will be assumed that the reaction is irreversible and that no other processes take place to affect the amount of each reactant. Temperature and volume are assumed constant. It will be convenient to measure amounts in moles and concentrations in moles/liter, since a mole of a given chemical always contains the same number of molecules.

Suppose that at the start, our reaction involves two reactants, A and B, and a product E. Furthermore, one molecule of A and one molecule of B produce one molecule of E. This could be written as

$$\mathrm{A} + \mathrm{B} \to \mathrm{E}. \tag{1}$$

Let a and b be the concentrations at time t of reactants A and B. The probability of a collision between a molecule of A and a molecule of B should be proportional to the product of the concentrations. Similarly, a certain portion of the collisions between molecules of A and B would be expected to react to form a molecule of E. This leads us to the **mass action law**.

> The instantaneous rate of production of product per unit volume is proportional to the product of the concentrations of the reactants.

Let $x(t)$ be the amount of product E per unit volume produced by time t. Then the mass action law for (1) would be

$$\frac{dx}{dt} = kab; \tag{2}$$

k is always positive and is called the **rate constant**. Since, according to (1), one mole of both A and B is needed to produce one mole of E, we have $a = a(0) - x$, $b = b(0) - x$, and (1) may be rewritten in terms of one dependent variable as

$$\frac{dx}{dt} = k[a(0) - x][b(0) - x], \qquad x(0) = 0, \tag{3}$$

which is a first-order nonlinear equation.

EXAMPLE 7.7.1 **Solution for Chemical Reactants**

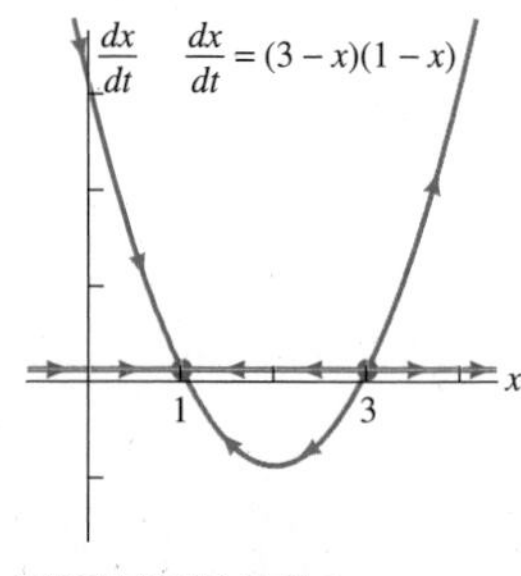

FIGURE 7.7.1

Phase portrait for Example 7.7.1.

Two reactants, A and B, are in solution. The initial concentrations are $a(0) = 3$ mol/L, $b(0) = 1$ mol/L, and the rate constant $k = 1$ L/mol · s. Assume that the mass action law is applicable.

a. Find all equilibria and analyze their stability.

b. Determine a one-dimensional phase portrait (see Section 7.4).

c. From (b), sketch the time dependence of solutions to the differential equation.

d. What are the limiting concentrations of the product E and the reactant B?

Solution Our model (3) gives the differential equation

$$\frac{dx}{dt} = (3 - x)(1 - x), \qquad x(0) = 0. \tag{4}$$

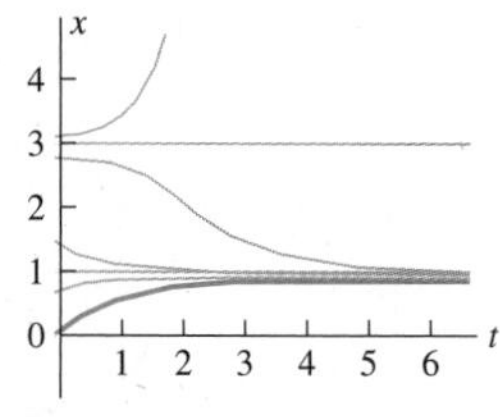

FIGURE 7.7.2

Solutions for Example 7.7.1.

a. The differential equation is $dx/dt = f(x) = (3 - x)(1 - x)$. Setting $f(x) = 0$ and solving for x, we get that $x = 1$ and $x = 3$ are the two equilibrium concentrations. Computing $f'(x) = -4 + 2x$, we have

$$f'(1) = -2 < 0, \quad \text{which implies that } x = 1 \text{ is stable;}$$
$$f'(3) = 2 > 0, \quad \text{which implies that } x = 3 \text{ is unstable.}$$

b. The one-dimensional phase portrait is sketched in Figure 7.7.1.

c. The time dependence of solutions of (4) is given in Figure 7.7.2. The solution corresponding to the initial condition $x(0) = 0$ is the heavier line in Figure 7.7.2.

d. From Figure 7.7.1 or 7.7.2, it can be seen that the limiting concentration of the product is 1. That is, $\lim_{t\to\infty} x(t) = 1$. Since $b(t) = b(0) - x(t)$, the limiting concentration of b is $b(0) - 1 = 1 - 1 = 0$. ◀

Exercises

1. The chemical reaction A + B → E can be modeled by (3). Suppose that the initial concentrations are $a(0) = 3$ mol/L and $b(0) = 4$ mol/L.

a) Find all equilibria and analyze their stability.

b) Determine the one-dimensional phase portrait.

c) Sketch the time dependence of solutions to the differential equation using the phase portrait.

d) What are the limiting concentrations of the product E and the reactant B?

2. The chemical reaction A + B → E can be modeled by (3). Suppose that the initial concentrations are $a(0) = 5$ mol/L and $b(0) = 2$ mol/L.

a) Find all equilibria and analyze their stability.

b) Determine the one-dimensional phase portrait.

c) Sketch the time dependence of solutions to the differential equation using the phase portrait.

d) What are the limiting concentrations of the product E and the reactant B?

3. The chemical reaction A + B → E can be modeled by (3). Suppose that the initial concentrations are $a(0) = \alpha$ mol/L and $b(0) = \beta$ mol/L with $\alpha < \beta$.

a) Find all equilibria and analyze their stability.

b) Determine the one-dimensional phase portrait.

c) Sketch the time dependence of solutions to the differential equation using the phase portrait.

d) What are the limiting concentrations of the product E and the reactant B?

4. The chemical reaction A + B → E can be modeled by (3). Suppose that the initial concentrations are $a(0) = b(0) = \alpha$ mol/L.

a) Find all equilibria and analyze their stability.

b) Determine the one-dimensional phase portrait.

c) Sketch the time dependence of solutions to

the differential equation using the phase portrait.

d) What are the limiting concentrations of the product E and the reactant B?

In Exercises 5 through 10, let x be the concentration of product E produced by the chemical reaction A + B + C → E. suppose that the reaction can be modeled by the differential equation

$$\frac{dx}{dt} = k[a(0) - x][b(0) - x][c(0) - x], \quad x(0) = 0, \tag{5}$$

where $a(0)$, $b(0)$, $c(0)$ are the initial concentrations of the reactants A, B, C and $k > 0$ is the rate constant.

5. Suppose the initial concentrations are $a(0) = 1$ mol/L, $b(0) = 2$ mol/L, $c(0) = 3$ mol/L.

a) Find all equilibria and analyze their stability.

b) Determine the one-dimensional phase portrait.

c) Sketch the time dependence of solutions to the differential equation using the phase portrait.

d) What are the limiting concentrations of the product E and the reactants B and C?

6. Suppose the initial concentrations are $a(0) = 1$ mol/L, $b(0) = 2$ mol/L, $c(0) = 2$ mol/L.

a) Find all equilibria and analyze their stability.

b) Determine the one-dimensional phase portrait.

c) Sketch the time dependence of solutions to the differential equation using the phase portrait.

d) What are the limiting concentrations of the product E and the reactants B and C?

7. Suppose the initial concentrations are $a(0) = \alpha$ mol/L, $b(0) = \beta$ mol/L, $c(0) = \gamma$ mol/L with $\alpha < \beta < \gamma$.

a) Find all equilibria and analyze their stability.

b) Determine the one-dimensional phase portrait.

c) Sketch the time dependence of solutions to the differential equation using the phase portrait.

d) What are the limiting concentrations of the product E and the reactants B and C?

8. Suppose the initial concentrations are $a(0) = b(0) = \alpha$ mol/L, $c(0) = \gamma$ mol/L with $\alpha < \gamma$.

a) Find all equilibria and analyze their stability.

b) Determine the one-dimensional phase portrait.

c) Sketch the time dependence of solutions to the differential equation using the phase portrait.

d) What are the limiting concentrations of the product E and the reactants B and C?

9. Suppose the initial concentrations are $a(0) = \alpha$ mol/L, $b(0) = c(0) = \beta$ mol/L with $\alpha < \beta$.

a) Find all equilibria and analyze their stability.

b) Determine the one-dimensional phase portrait.

c) Sketch the time dependence of solutions to the differential equation using the phase portrait.

d) What are the limiting concentrations of the product E and the reactants A and B?

10. Suppose that the initial concentrations are $a(0) = b(0) = c(0) = \alpha$.

a) Find all equilibria and analyze their stability.

b) Determine the one-dimensional phase portrait.

c) Sketch the time dependence of solutions to the differential equation using the phase portrait.

d) What are the limiting concentrations of the product E and the reactant A?

CHAPTER

8

Discrete Dynamical Systems

8.1 Introduction

As we have seen in previous chapters, differential equations play an important role in problems where there is a quantity y that depends on a continuous independent variable t. In many applications, however, the independent variable, denoted in this section by k, takes on only isolated or discrete values, and we are led to consider what are known as difference equations. The increased use of digital circuits, computer simulations, and numerical methods (see Chapter 6) has made this topic even more important, since they all use difference equations as a fundamental concept.

There is a discrete (or difference) version of most differential-equation techniques. For example, there are difference-equation versions of undetermined coefficients, variation of parameters, resonance, the Laplace transform, systems, equilibrium, and stability. We shall touch on only a few of these here.

A **discrete dynamical system** or **difference equation** relates different values of an unknown sequence $\{y_k\}$. Here k is an integer variable, and y is the quantity of interest. We usually assume $k \geq M$ for some integer M. Initial values of the sequence are also sometimes specified.

Four fundamental ways in which difference equations arise are:

1. Some physical problems are directly modeled in terms of difference equations.
2. Difference equations may be introduced by numerical methods.

3. A continuous process is sampled (measured) at discrete times.
4. In analyzing periodic solutions of differential equations, the mapping over the periodic interval of time is a difference equation called the Poincare map.

We present as motivation a brief example of each type.

1. Discrete Model of a Physical Problem

In population dynamics, modeling the rate of change of the population or modeling the growth rate yields a differential equation for the population y as a function of time t:

$$\frac{dy}{dt} = f(y, t).$$

In differential-equation models, usually the population is assumed to vary continuously in time. Difference-equation models arise when the population is modeled only at certain discrete times. For example, in some situations only the population of a species each year is of biological relevance. The concern might be with how the number of salmon in some region varies from year to year. We let y_k be the population of salmon in the kth year (now). k is the discrete time variable. The simplest mathematical model arises if we assume that the change in the population during the coming year $(y_{k+1} - y_k)$ is proportional to the present year's population:

$$y_{k+1} - y_k = ay_k. \tag{1}$$

Here a is the known growth rate, which we have assumed to be constant from year to year. More accurate (and usually more complicated) models are possible.

Equation (1) is called a first-order difference equation because the equation involves y at years (k and $k + 1$) that differ by one year. Instead of modeling the growth rate as a constant, we might model the growth rate as a function of the present year's population. The simplest model of this would be to assume that the growth rate is $a - by_k$. In this case, the following difference equation would be used:

$$y_{k+1} - y_k = (a - by_k)y_k. \tag{2}$$

This discrete logistic equation (2), as simple as it is, turns out to be a mathematical equation with extraordinary complex and interesting properties. We examine (2) in Section 8.4.

If the growth rate over the next year depends on the previous year's population instead of this year's, then one simple model of this would be the following difference equation:

$$y_{k+1} - y_k = (a - by_{k-1})y_k. \tag{3}$$

This is a second-order difference equation because the difference between the largest $(k + 1)$ and smallest $(k - 1)$ subscript of the population is two. Equivalently, (3) involves y at times two years apart.

2. Difference Equations as the Numerical Approximation of Differential Equations

Newton's law of cooling implies that the rate of change of the temperature of an object is proportional to the temperature difference between the object $T(t)$ and its outside environment. For definiteness we assume the temperature of the outside environment to be 30°C, the proportionality constant to be given as 0.2, and the initial temperature to be T_0 at time $t = 0$. The differential equation would then be

$$\frac{dT}{dt} = -0.2(T - 30), \qquad T(0) = T_0. \tag{4}$$

Although it is easy to solve this differential equation, we wish to understand what occurs if Euler's numerical method (not a particularly good method) is used. It is shown in Chapter 6 that in Euler's numerical method, the derivative is approximated by the forward difference:

$$\frac{dT}{dt} \approx \frac{T_{k+1} - T_k}{\Delta t}.$$

Here the discrete time interval is Δt and T_k is the estimate for the temperature $T(k\,\Delta t)$ at the end of the kth time interval. Thus, with this approximation, Newton's law of cooling (4) becomes

$$T_{k+1} - T_k = -0.2\,\Delta t(T_k - 30), \tag{5}$$

which is a first-order difference equation instead of the first-order differential equation.

3. Difference Equations by Sampling of Differential Equations

Consider again (4). Suppose, however, that we are only interested in the temperature every hour. Let T_k be the temperature at the end of the kth hour. To find T_{k+1} we solve the differential equation

$$\frac{dT}{dt} = -0.2(T - 30), \qquad T(k) = T_k$$

to get $T(k + 1)$. The result is

$$T_{k+1} = e^{-0.02}(T_k - 30) + 30, \tag{6}$$

which is a first-order difference equation.

In the difference equation (5) derived from a differential equation (4), the interval Δt has to be small in order to ensure convergence of the numerical method. In (6), Δt does not have to be small. Sampling is often used in the control of industrial processes where the process itself varies continuously but the controller is held constant between the times the process is sampled (measured).

4. Poincare Map of a Periodic Differential Equation

The solution of a first-order periodic differential equation satisfies a Poincare map, which is a first-order difference equation. We briefly summarize the ideas. We consider a first-order differential equation

$$\frac{dx}{dt} = f(x, t), \tag{7}$$

where the differential equation itself is assumed to be periodic with period 1:

$$f(x, t + 1) = f(x, t).$$

The solution $x(1) = x_1$ of the differential equation at $t = 1$ is determined from the given initial condition $x(0) = x_0$. Thus, x_1 is a function of x_0,

$$x_1 = \Pi(x_0).$$

Here $\Pi(x_0)$ is a function of the initial condition known as the Poincare map. The differential equation repeats its effect every unit of time (since the period of f is 1). Thus, the solution at the $k + 1$st time is the same function of the solution at the kth time:

$$x_{k+1} = \Pi(x_k), \tag{8}$$

a first-order difference equation. However, the Poincare map (8) can only be determined by solving the differential equation (7) over one period of time (which must usually be done numerically).

Difference equations may also be used to analyze differential equations when it is known that one solution of the differential equation is periodic. This situation is more complicated, and we do not discuss it here.

8.2 First-Order Linear Difference Equations

A first-order difference equation

$$x_{k+1} = f(x_k, k)$$

is linear if the function f depends on x_k in a linear way. Thus, a first-order **linear** difference equation can be written in the form

$$x_{k+1} = b_k x_k + q_k, \tag{1}$$

where b_k, q_k depend only on k. The initial condition x_0 may be taken arbitrarily and succeeding terms found recursively from (1). This shows that there is a unique solution for the initial value problem for the linear difference equation (1).

It can be shown, as for first-order linear differential equations, that the general solution of (1) is in the form

$$x_k = p_k + ch_k, \tag{2}$$

a particular solution (sequence) p_k of (1) plus an arbitrary constant c times a solution (sequence) h_k that solves the **associated homogeneous equation**

$$x_{k+1} - b_k x_k = 0. \tag{3}$$

We may take h_k as any solution of (3) that is not identically zero. In general, the coefficients b_k in the linear difference equation are not constant. If the coefficient b_k is a constant ($b_k = b$), the linear first-order difference equation (1) is said to have a constant coefficient:

$$x_{k+1} - bx_k = q_k. \tag{4}$$

8.2.1 Solutions of Homogeneous Difference Equations with Constant Coefficients

When the difference equation has constant coefficients, solutions of the homogeneous equation satisfy

$$x_{k+1} - bx_k = 0, \qquad \text{or equivalently,} \qquad x_{k+1} = bx_k. \tag{5}$$

Solutions can be obtained by directly iterating (5).

EXAMPLE 8.2.1 Direct Iteration

Find the general solution of

$$x_{k+1} = -5x_k. \tag{6}$$

Solution By iterating (6),

$$\begin{aligned} x_1 &= -5x_0, \\ x_2 &= -5x_1 = (-5)^2 x_0, \\ x_3 &= -5x_2 = (-5)^3 x_0, \\ &\cdots \end{aligned}$$

Thus, the solution of (6) is

$$x_k = (-5)^k x_0. \tag{7}$$

◄

For homogeneous linear differential equations with constant coefficients, the simplest method of obtaining solutions is to substitute e^{rt} into the differential equation and determine r so that e^{rt} is a solution of the homogeneous differential equation. As can be seen from (7), solutions of homogeneous linear difference equations are in the form of powers. The simplest method of obtaining solutions of homogeneous linear difference equations is to substitute

$$x_k = \lambda^k \tag{8}$$

with $\lambda \neq 0$ into the difference equation (5) in order to determine λ.

EXAMPLE 8.2.2 Power Solutions

Find the general solution of

$$x_{k+1} = 7x_k. \tag{9}$$

SOLUTION Substituting $x_k = \lambda^k$ into the difference equation (9) yields

$$\lambda^{k+1} = 7\lambda^k.$$

Canceling λ^k shows that

$$\lambda = 7,$$

so that solutions exist in the form $x_k = 7^k$. Since (9) is linear, general solutions of the homogeneous equation are an arbitrary multiple times this solution:

$$x_k = c7^k. \tag{10}$$

(10) satisfies the initial condition x_0 if $c = x_0$. For this value of c, (10) becomes

$$x_k = x_0 7^k.$$

8.2.2 Particular Solution of Linear Difference Equations with a Constant Coefficient and Constant Input

A particular solution of a linear first-order difference equation (with a constant coefficient),

$$x_{k+1} - bx_k = q_k, \tag{11}$$

can always be obtained by a method analogous to an integrating factor for a first-order linear differential equation. The solution obtained this way is rather complicated, and we do not discuss it here. Instead, we consider here only the elementary but important case in which the input q_k is a constant, $q_k = q$.

Here, we will obtain the general solution of a linear first-order difference equation with a constant coefficient (11) when the input q_k is a constant:

$$x_{k+1} - bx_k = q. \tag{12}$$

When the input is a constant, the **method of undetermined coefficients** may be used to obtain a particular solution. Since the input is a constant, there is always a particular solution of the constant coefficient linear difference equation (12) in the form of a constant:

$$x_k = A, \tag{13}$$

provided $b \neq 1$.

EXAMPLE 8.2.3 Constant Input

Find the solution of the initial value problem for

$$x_{k+1} + 2x_k = 4, \qquad x_0 = 3. \tag{14}$$

SOLUTION Since this is a linear difference equation with a constant coefficient, a solution of the associated homogeneous equation is $(-2)^k$. Since the input is a constant, we look for a particular solution of (14) that is just a constant, $x_k = A$. Substitution for x_k in (14) gives $A + 2A = 4$ or $A = \frac{4}{3}$. Thus, the general solution of (14) is

$$x_k = \frac{4}{3} + c(-2)^k. \tag{15}$$

The initial condition ($k = 0$) implies that c satisfies $x_0 = \frac{4}{3} + c = 3$. Thus $c = \frac{5}{3}$ and the solution of (14) is $x_k = \frac{4}{3} + \frac{5}{3}(-2)^k$. ◄

EXAMPLE 8.2.4 Thermal Cooling

The initial temperature of an object is measured as 100°C. The temperature is measured each hour afterwards. It is observed that the amount the temperature changes each hour is 20% of the difference between the previous hour's temperature and the room temperature, which is 60°C. Model this problem as a difference equation, and determine the solution.

Solution Let T_k be the temperature at the end of k hours. The following equation describes the given observation on the change in temperature:

$$T_{k+1} - T_k = -0.2(T_k - 60). \tag{16}$$

The minus sign in the right hand side of (16) is introduced because we know that if an object's temperature is hotter than the room, then the object loses temperature to the room. We rewrite the linear difference equation (16) in the form we prefer,

$$T_{k+1} - 0.8T_k = 12. \tag{17}$$

To solve (17), we note that the solution of the associated homogeneous equation is $c(0.8)^k$. Since the right-hand side is a constant, we attempt to obtain an elementary particular solution that is a constant, $T_k = A$. We can determine A from (17), but it is easier to obtain A directly from the equation with more physical meaning, (16). Substituting $T_k = A$ into (16) gives $A - A = -0.2(A - 60)$. Clearly, $A = 60$. The room temperature is a particular solution. The general solution of (17) is

$$T_k = 60 + c(0.8)^k. \tag{18}$$

Formula (18) is particularly convenient. It shows that as $k \to \infty$, then $T_k \to 60$ no matter what the initial condition is. As time increases, the temperature of the object approaches the room temperature, as should be expected. In other sections, constant solutions of difference or differential equations are called **equilibrium** solutions. In this case, the temperature of the object approaches its equilibrium temperature.

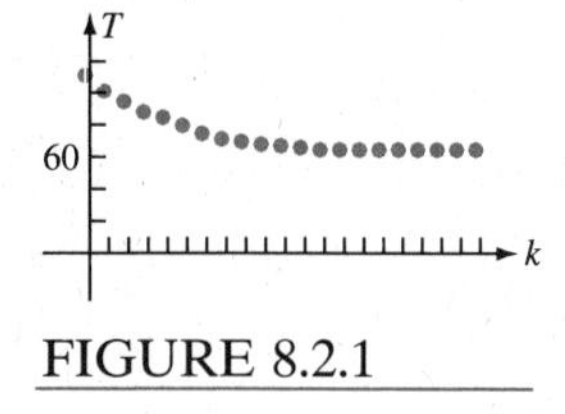

FIGURE 8.2.1

First 20 values of $\{T_k\}$ in (19).

In this thermal problem, the initial temperature was given, $T_0 = 100$. We can determine the constant c in the general solution (18) by letting $k = 0$ to get $100 = 60 + c$ or $c = 40$. Thus, the solution of the initial value problem is

$$T_k = 60 + 40(0.8)^k. \tag{19}$$

This solution is graphed in Figure 8.2.1. Note that the temperature of the object approaches the room temperature. ◄

Exercises

In Exercises 1 through 15, find the solution of the initial value problem if an initial condition is given. Otherwise, find the general solution.

1. $x_{k+1} = \frac{1}{2}x_k$

2. $x_{k+1} = -\frac{1}{3}x_k$

3. $x_{k+1} = -3x_k$

4. $x_{k+1} = 6x_k$

5. $x_{k+1} = 2x_k + 1, \qquad x_0 = 1$

6. $x_{k+1} = -x_k + 3, \quad x_0 = 0$

7. $2x_{k+1} = 3x_k$

8. $4x_{k+1} - 2x_k = 7, \quad x_0 = 2$

9. $x_{k+1} - 2x_k = 3$

10. $3x_{k+1} = 7x_k, \quad x_0 = 2$

11. $2x_{k+1} + 2x_k = 6$

12. $3x_{k+1} - 2x_k = 9$

13. $x_{k+1} - 4x_k = 2, \quad x_0 = 6$

14. $x_{k+1} + 7x_k = 3$

15. $5x_{k+1} + 4x_k = 14$

16. Show that the method of this section for finding a particular solution does not work for

$$x_{k+1} - x_k = 3.$$

Find a particular solution by iterating beginning with $x_0 = 0$. Find the general solution.

17. The initial temperature of an object is measured as 60°C. The temperature is measured each hour afterwards. It is observed that the amount the temperature changes each hour is 40% of the difference between the previous hour's temperature and the room temperature, which is 100°C. Model this problem as a difference equation, and determine the solution.

18. An object is removed from an ice bath at $t = 0$ (after spending a very long time there) and hence has temperature 0°C. It is then placed in a very large vat of boiling water whose temperature is 100°C and is unaffected by the frozen object. The temperature of the object is measured each hour afterwards. It is observed that the amount the temperature changes each hour is 10% of the difference between the previous hour's temperature and the temperature of the boiling water. Model this problem as a difference equation, and determine the solution.

19. Show that $x_{k+1} - bx_k = q$ has a unique equilibrium if and only if $b \neq 1$, and in this case the equilibrium is $x_k = q/(1-b)$. Show that if $b = 1$ and $q \neq 0$, then there is no equilibrium, and if $b = 1$ and $q = 0$, then every solution is an equilibrium solution.

20. Let p_k be a particular solution of $x_{k+1} - b_k x_k = q_k$. Show that, if z_k is any other solution, then $z_k = p_k + w_k$, where w_k is a solution of $x_{k+1} - b_k x_k = 0$. Conversely, show that, if p_k and w_k are as described, then $z_k = p_k + w_k$ is a solution of $x_{k+1} - b_k x_k = q_k$.

21. Let h_k be a not identically zero solution of $x_{k+1} - b_k x_k = 0$. Show that every other solution of $x_{k+1} - b_k x_k = 0$ is of the form ch_k, where c is a constant.

22. If $b \neq 1$, show that the general solution of (12) is

$$x_k = \frac{q}{1-b} + cb^k. \tag{20}$$

23. Using (20), show that if $b \neq 1$,

$$x_k = b^k x_0 + q\frac{b^k - 1}{b - 1}.$$

8.3 First-Order Autonomous Mostly Nonlinear Difference Equations

The most general first-order difference equation would be $x_{k+1} = f(x_k, k)$. That is, x_{k+1} can depend on both the value of x_k and the discrete timelike variable k. Here, we will only consider difference equations that are **autonomous**. That is, the function in the difference equation is the same for all k:

$$x_{k+1} = f(x_k). \tag{1}$$

This difference equation is usually nonlinear. Unlike the case of a differential equation, an explicit solution of (1) can be obtained by recursion:

$$\begin{aligned} x_1 &= f(x_0), \\ x_2 &= f(x_1) = f(f(x_0)), \\ x_3 &= f(x_2) = f(f(f(x_0))), \\ &\vdots \end{aligned}$$

$f(f(x_0))$ is called the **second iterate**. We introduce the notation

$$f^2(x_0) \equiv f(f(x_0)).$$

This must not be confused with squaring the function. Similarly, $f(f(f(x_0)))$ is called the third iterate, and we introduce the notation

$$f^3(x_0) \equiv f(f(f(x_0))).$$

The general solution of the first-order autonomous difference equation (1) is that x_k is the kth iterate of the initial condition:

$$x_k = f^k(x_0). \tag{2}$$

EXAMPLE 8.3.1 Third Iterate

Let $f(x) = 2x + 1$. Compute $f^3(1)$.

Solution We have that

$$\begin{aligned} f(1) &= 3, \\ f^2(1) &= f(f(1)) = f(3) = 2(3) + 1 = 7, \\ f^3(1) &= f(f(f(1))) = f(7) = 2(7) + 1 = 15. \end{aligned}$$

Note that $[f(1)]^3 = 3^3 = 27$, which is different from $f^3(1)$. ◄

Equilibrium or Fixed Points

The difference equation,

$$x_{k+1} = f(x_k), \tag{3}$$

has a solution that is a sequence, $x_0, x_1, x_2, x_3, \ldots$. The simplest kind of solution of a difference equation would be a sequence in which all the values are the same: $x_0, x_0, x_0, \ldots$. Such a sequence is called an **equilibrium** solution of the difference equation.

For a given difference equation (3), we want to determine all equilibrium solutions $\bar{x}$. We want a constant sequence $\bar{x}, \bar{x}, \bar{x}, \bar{x}, \ldots$ to solve the difference equation (3). Thus, in (3), $x_k = \bar{x}$ and $x_{k+1} = \bar{x}$. For this sequence to be an equilibrium solution, the next value must be the same as the first:

$$\bar{x} = f(\bar{x}). \tag{4}$$

Under this condition, all subsequent values will be the same. All solutions $\bar{x}$ of (4) are also called **fixed points** of the function $f(x)$ or fixed points of the difference equation (3).

An equilibrium or fixed-point solution of the difference equation is a sequence whose values are the same at each $k = 0, 1, 2, 3, 4, 5, \ldots$. Thus, an equilibrium [fixed point of $f(x)$] solution is also called a **period 1 solution** of the difference equation (3). A solution of the difference equation that is not an equilibrium but has the same value when $k = 0, 2, 4, 6, \ldots$ is periodic with period 2.

An equilibrium (fixed-point) or period 1 solution is a special and a very simple solution of the difference equation. It may be stable or unstable. If all solutions that are initially near the equilibrium stay near the equilibrium, then we say that the equilibrium is **stable**. Otherwise the equilibrium solution is said to be **unstable**.

Linear Stability of an Equilibrium Solution (Fixed Point)

A fixed-point $\bar{x}$ is a very special solution of the difference equation

$$x_{k+1} = f(x_k). \tag{5}$$

A fixed-point $\bar{x}$ satisfies

$$\bar{x} = f(\bar{x}). \tag{6}$$

To determine the stability of a fixed point, we analyze the difference equation (5) assuming that the solution x_k is near the fixed point $\bar{x}$. As with the stability analysis in Section 7.3, we approximate the nonlinear function $f(x_k)$ by its linearization in the neighborhood of the fixed point $\bar{x}$. The difference equation (5) is approximated by

$$x_{k+1} = f(\bar{x}) + (x_k - \bar{x})f'(\bar{x}). \tag{7}$$

It is convenient to consider the displacement from the fixed point

$$y_k = x_k - \bar{x}, \tag{8}$$

in which case (7) becomes

$$y_{k+1} = f'(\bar{x})y_k, \tag{9}$$

since the fixed point satisfies $\bar{x} = f(\bar{x})$. This difference equation (9) is linear and homogeneous with constant coefficients.

The general solution of the linearized equation (9) is

$$y_k = [f'(\bar{x})]^k y_0. \tag{10}$$

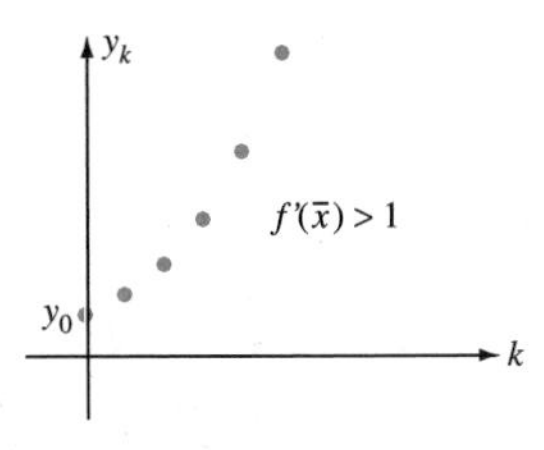

FIGURE 8.3.1

Exponential growth.

If $f'(\bar{x}) > 1$, the solution (10) diverges as k increases, and we say that the fixed point ($x = \bar{x}$ or $y = 0$) is unstable. The solution can be shown to lie on a simple exponentially growing curve, as sketched in Figure 8.3.1.

If $0 < f'(\bar{x}) < 1$, the solution converges to zero, lying on an exponentially decaying curve, and we say that the fixed point ($x = \bar{x}$ or $y = 0$) is stable, as graphed in Figure 8.3.2.

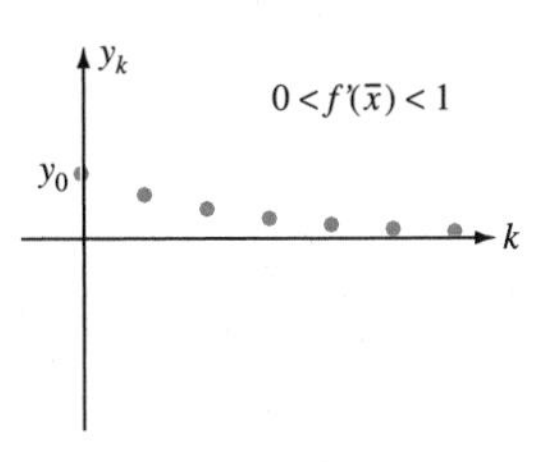

FIGURE 8.3.2

Exponential decay.

In these cases, the behavior of the difference equation is quite similar to the behavior of corresponding differential equations. However, if $f'(\bar{x}) < 0$, behavior occurs that does not occur for first-order differential equations.

If $-1 < f'(\bar{x}) < 0$, then the solution (10) converges to the fixed point ($x = \bar{x}$ or $y = 0$), as graphed in Figure 8.3.3, with an oscillation corresponding to flipping back and forth across $y = 0$ for every increase in k. We again say that the solution is stable.

If $f'(\bar{x}) < -1$, then the solution diverges with an oscillatory flip, as graphed in Figure 8.3.4. The solution is unstable. [The simplest way to graph these last two cases is to first graph the corresponding exponential $|f'(\bar{x})|^k y_0$ and its negative. The solution simply alternates between the two elementary curves.]

These four cases can be summarized into two cases:

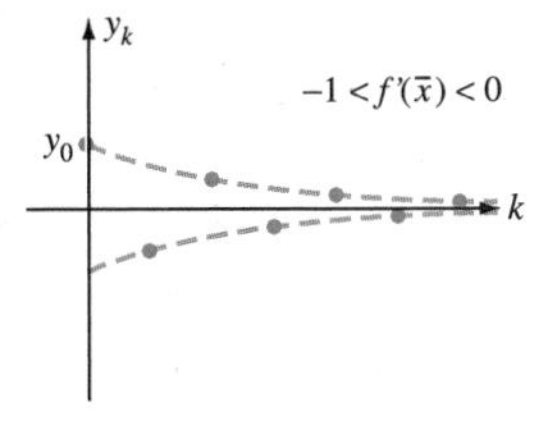

FIGURE 8.3.3

Exponential decay with flip.

Summary of Linear Stability for Fixed Points of
$$x_{k+1} = f(x_k)$$

The fixed point $\bar{x}$ is:

Stable if $|f'(\bar{x})| < 1$

Unstable if $|f'(\bar{x})| > 1$.

In these cases, the stability of the fixed point of the nonlinear problem is the same as the stability of its linearization.

There are two cases that are on the border between being stable and unstable. If $f'(\bar{x}) = 1$, then, as graphed in Figure 8.3.5, the solution of the linearization (9) is a constant, staying near the fixed point if the initial condition is near the fixed point. If $f'(\bar{x}) = -1$, then the solution of the linearization (9) changes sign (flips) each time k is increased but maintains the same distance from the fixed point, as graphed in Figure 8.3.6. In these borderline cases, the stability of the fixed point for the nonlinear problem is determined from neglected nonlinear terms. These cases are stable by the reasoning provided by the linearization, but the actual stability of the fixed point may be different.

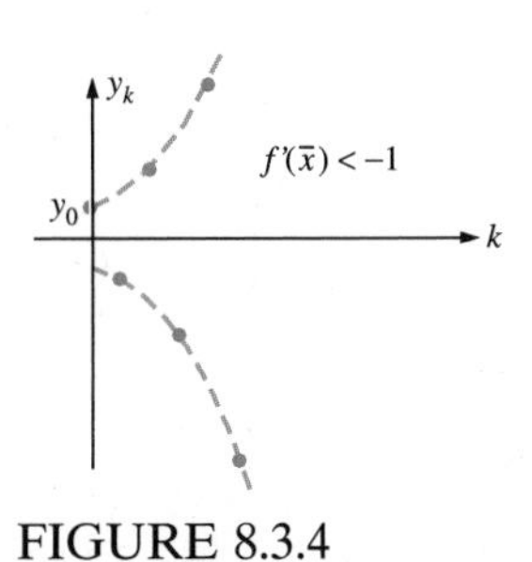

FIGURE 8.3.4

Exponential growth with flip.

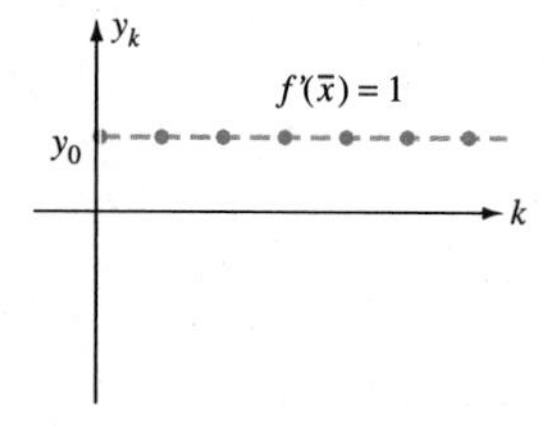

FIGURE 8.3.5

Constant.

EXAMPLE 8.3.2 **Fixed Points and Stability**

Find all fixed points and determine their stability for

$$x_{k+1} = -3x_k + 2. \tag{11}$$

SOLUTION Equation (11) is $x_{k+1} = f(x_k)$ with $f(x) = -3x + 2$. The fixed points satisfy $\bar{x} = f(\bar{x})$ or $\bar{x} = \frac{1}{2}$. Since (11) is linear, its general solution is

$$x_k = \frac{1}{2} + c(-3)^k.$$

Here the fixed point is unstable. Alternatively, $f'(x) = -3$ so that $f'(\frac{1}{2}) = -3$, in which case we conclude that $x = \frac{1}{2}$ is unstable. ◄

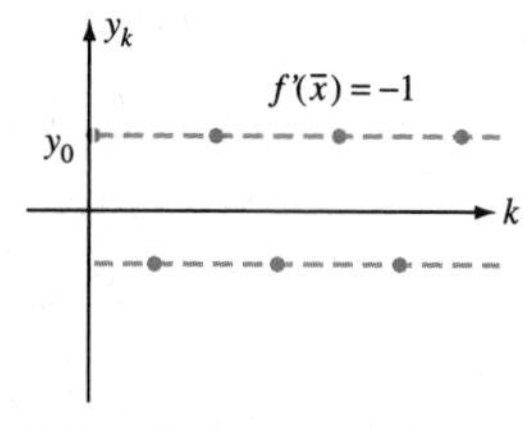

FIGURE 8.3.6

Constant with flip.

EXAMPLE 8.3.3 **Fixed Points and Stability**

Find all fixed points and determine their stability for

$$x_{k+1} = x_k^2 + x_k - \frac{1}{9}. \tag{12}$$

SOLUTION Equation (12) is $x_{k+1} = f(x_k)$ with $f(x) = x^2 + x - \frac{1}{9}$. We first find the fixed points. These are the solutions of $x = f(x)$ or, after simplification, $x^2 = \frac{1}{9}$. Thus there are two fixed points, $\bar{x} = \pm\frac{1}{3}$.

To determine the stability of each fixed point, we compute that $f'(x) = 2x + 1$. Then $|f'(\frac{1}{3})| = \frac{5}{3} > 1$, so that $\bar{x} = \frac{1}{3}$ is an unstable fixed point. On the other hand, $|f'(-\frac{1}{3})| = \frac{1}{3} < 1$, so that $\bar{x} = \frac{1}{3}$ is a stable fixed point. ◄

8.3.1 Geometric Method: Stair-Stepping / Cobwebbing

There is a geometric method for graphing the solution of a difference equation,

$$x_{k+1} = f(x_k), \tag{13}$$

that gives geometric insight. This geometric method can be done on a graphing calculator. We graph the successive iterations implied by the difference equation (13), utilizing the graph of $f(x)$.

The procedure consists of a sequence of vertical and horizontal lines that sometimes resembles the steps of a flight of stairs. The procedure is sometimes called **stair-stepping**. Other times the graph looks more like the web of a spider, so that the geometrical method is sometimes called **cobwebbing**. The procedure is straightforward:

Stair-Step / Cobweb Procedure:

1. Plot $y = f(x)$ and the straight 45° line $y = x$.
2. Start at a given initial point x_0.
3. Draw a vertical line up or down to the graph of $f(x)$.
4. Then (without picking the pencil up) draw a horizontal line across (left or right) to the 45° line $y = x$.
5. Repeat steps 3 and 4 alternately.

For clarity, we add arrows in the direction in which the lines are drawn. The first vertical line represents the initial point x_0. We claim that the next vertical line represents the first value of the solution of the difference equation $x_1 = f(x_0)$. To see this, recall that from x_0 we move vertically to the point $(x_0, f(x_0))$. We then move horizontally along $y = f(x_0)$ until $y = x$. We are now at $(f(x_0), f(x_0))$ when we switch to a vertical line. The equation for that vertical line will be $x = f(x_0)$. What happens next? Since $x = f(x_0)$, the y value will be $y = f(f(x_0))$, the second iterate. In this way the vertical lines that are generated represent the higher iterates, the sequence that is the solution of the initial value problem for the difference equation

$$x_0, f(x_0), f(f(x_0)), f(f(f(x_0))), f(f(f(f(x_0)))), \ldots.$$

The horizontal lines represent the same sequence starting from $y = f(x_0)$. ◄

The method begins by plotting $y = f(x)$ and the line $y = x$. Any intersections of $y = f(x)$ and $y = x$ are fixed points of the difference equation, since they satisfy $x = f(x)$.

EXAMPLE 8.3.4 **Stair-Step**

We will look at the difference equation $x_{k+1} = -(x_k - 1)^2$ or, equivalently, $x_{k+1} = f(x_k)$ with $f(x) = -(x - 1)^2$. See Figure 8.3.7. This function $f(x)$ has no fixed points, since there are no intersections of $y = f(x)$ with $y = x$. The stair-step diagram for one particular initial condition is given in Figure 8.3.7. Note that the solution marches off toward $-\infty$. Other initial conditions also go to $+\infty$ or $-\infty$. ◄

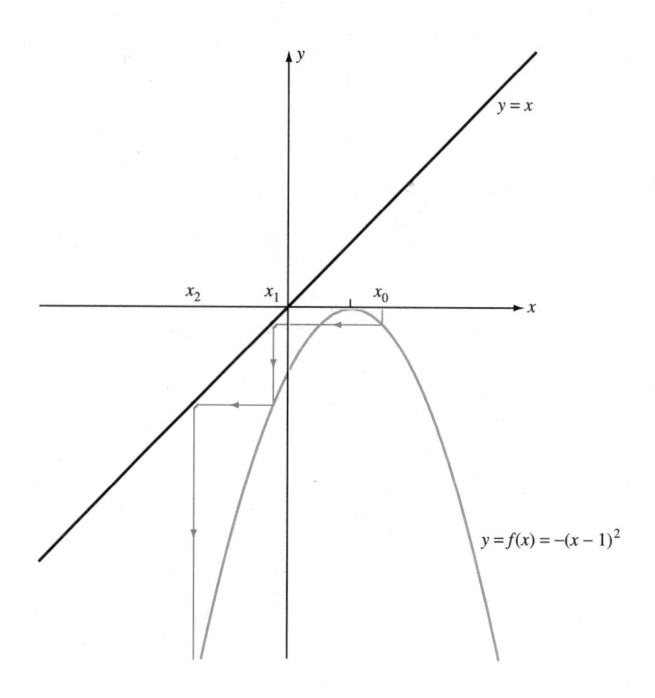

FIGURE 8.3.7

Stair-step (cobweb) for $x_{k+1} = -(x_k - 1)^2$.

EXAMPLE 8.3.5 **Finding Equilibria Graphically**

In the previous example, the graph of $y = f(x)$ did not intersect the line $y = x$. In Figure 8.3.8 we graph a function $y = f(x)$ whose graph intersects the line $y = x$ twice. In this case, the difference equation $x_{k+1} = f(x_k)$ has two equilibria (fixed points). ◄

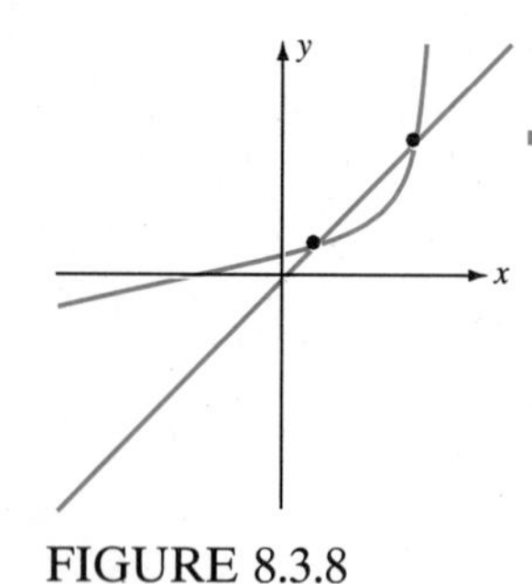

FIGURE 8.3.8

Two fixed points.

Cobwebbing for Linear Difference Equations

In the neighborhood of a fixed point ($\bar{x} = f(\bar{x})$), the difference equation $x_{k+1} = f(x_k)$ can be approximated by its linearization

$$y_{k+1} = f'(\bar{x})y_k, \tag{14}$$

where the displacement from the fixed point is $y_k = x_k - \bar{x}$. To understand what the stair-step or cobweb diagram looks like near a fixed point, we analyze the linear difference equation (linear map),

$$x_{k+1} = ax_k, \tag{15}$$

for various constant a's, corresponding to different possibilities for $f'(\bar{x})$. Here, in (14) and (15), the origin is the fixed point. Thus, our cobweb diagrams for the linear equations appear only near the fixed point of the original equation (translated to the fixed point). The four main cases are $a > 1, 0 < a < 1, -1 < a < 0, a < -1$. In all cases, the function $y = f(x)$ is a straight line $y = ax$, with slope a.

If $a > 1$, the case in which the analytic solution exponentially grows, the geometric solution marches away from the fixed point, looking (Figure 8.3.9) like the steps of a flight of stairs.

When the solution of the difference equation exponentially decays, $0 < a < 1$, the geometric figure (Figure 8.3.10) is again steplike, but here the arrows flow toward the fixed point.

The stair-step diagrams are more interesting for linear difference equations when $a < 0$.

If $-1 < a < 0$, then the solution of the difference equation exhibits an exponential decay with a flip (oscillation). In this case, Figure 8.3.11 looks like a cobweb. This stable solution looks like a cobweb that is spiraling in clockwise toward and around the fixed point (with a rectangular spiral).

If $a < -1$, then the cobweb spirals out clockwise away from but around the fixed point, as shown in Figure 8.3.12. This corresponds to the exponential growth with a flip of the solution of the difference equation in this unstable case.

The case in which $a = -1$ deserves special attention, not only because it is on the border between being stable and unstable, but because of its geometric appearance and its significance in the next section. In this case, the solution of the initial value problem for (15) is

$$x_0, \quad -x_0, \quad x_0, \quad -x_0, \quad x_0, \ldots \tag{16}$$

The solution (16) repeats with period 2. It is called a **period 2 solution.** In Figure 8.3.13, we show the cobweb diagram for this case. The pattern of a box

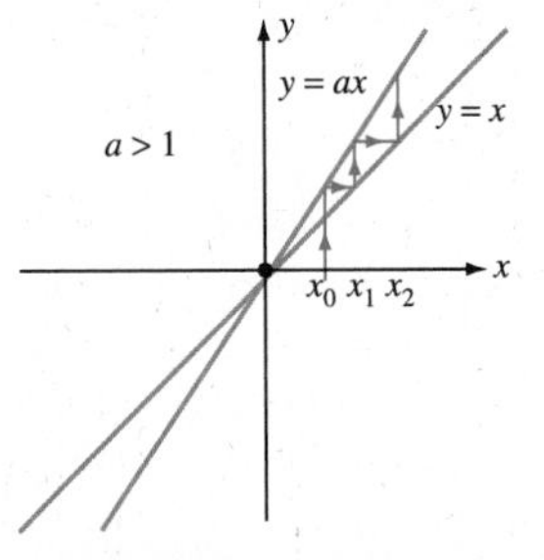

FIGURE 8.3.9

Cobweb for exponential growth.

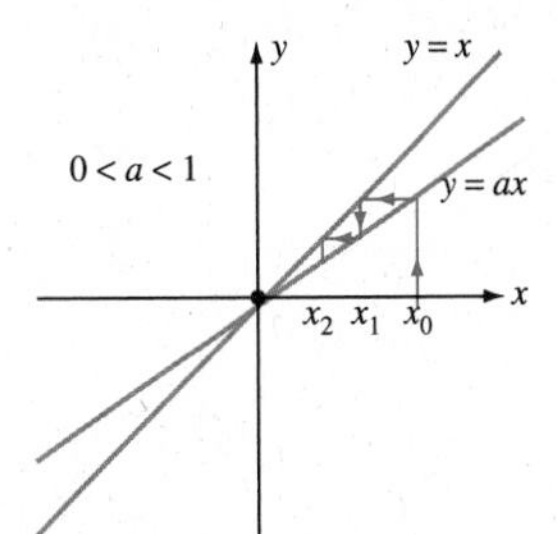

FIGURE 8.3.10

Cobweb for exponential decay.

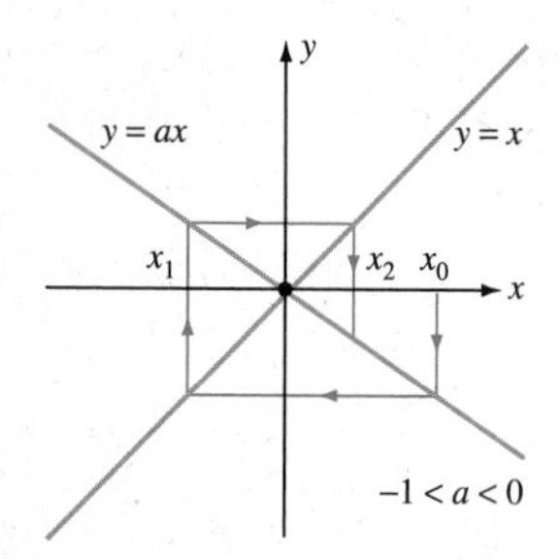

FIGURE 8.3.11

Cobweb for exponential decay with flip.

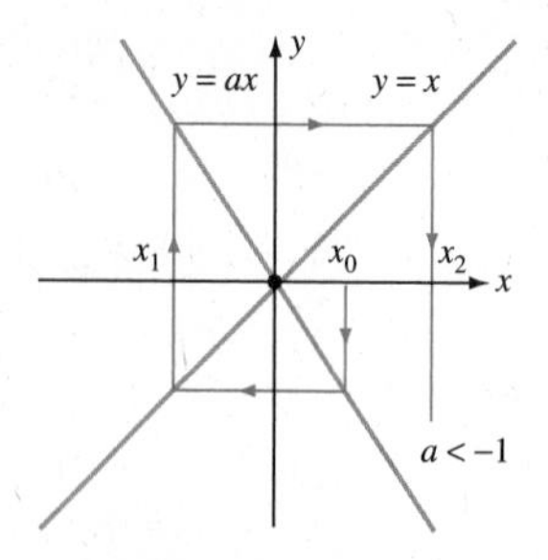

FIGURE 8.3.12

Cobweb for exponential growth with flip.

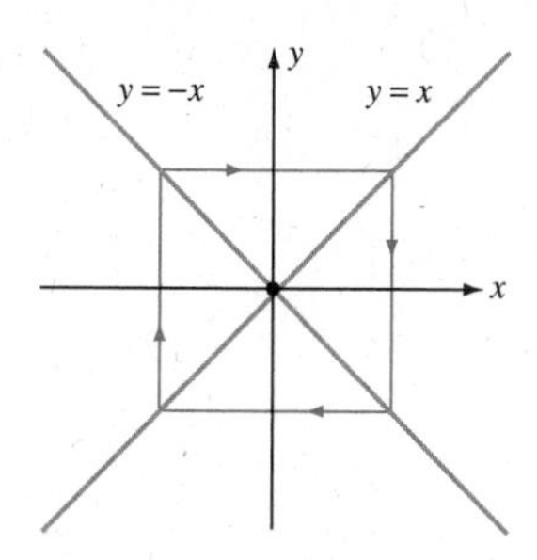

FIGURE 8.3.13

Cobweb for flip (period 2 solution).

repeats. The solution flips back and forth. The cobweb pattern for a period 2 solution will always appear boxlike.

Exercises

In Exercises 1 through 9, find all fixed points and their linear stability.

1. $x_{k+1} = 2x_k - 3$

2. $x_{k+1} = -3x_k + 6$

3. $2x_{k+1} - x_k = 8$

4. $3x_{k+1} + x_k = 9$

5. $x_{k+1} = \frac{4}{3}x_k(1 - x_k)$

6. $x_{k+1} = 2x_k - 1 + \frac{1}{2}(1 - x_k)^2$

7. $x_{k+1} = \frac{10}{3}x_k(1 - x_k)$

8. $x_{k+1} = x_k - \frac{1}{2}(e^{x_k} - 1)$

9. $x_{k+1} = x_k + (e^{x_k} + \frac{1}{2})$

In Exercises 10 through 13, find the analytic solution and then fit the solution to an exponential $|y_k| = |y_0|e^{rt}$, where $t = k$. In other words, solve the difference equation and determine r.

10. $y_{k+1} = 3y_k$

11. $y_{k+1} = \frac{1}{7}y_k$

12. $y_{k+1} = -5y_k$

13. $y_{k+1} = -\frac{1}{2}y_k$

14. The stability of the fixed point $\bar{x}$ is often governed by $y_{k+1} = f'(\bar{x})y_k$. Fit this solution to an exponential $|y_k| = |y_0|e^{rt}$, where $t = k$. Show that $r = \ln|f'(\bar{x})|$. r is called the **Liapunov exponent**. Briefly discuss the stability of the fixed point in terms of the value of the Liapunov exponent.

In Exercises 15 through 23, draw the stair-step diagram of the linear difference equation.

15. $x_{k+1} = \frac{3}{2}x_k$

16. $x_{k+1} = -\frac{1}{3}x_k$

17. $x_{k+1} = -x_k$

18. $x_{k+1} = -\frac{3}{2}x_k$

19. $x_{k+1} = \frac{2}{3}x_k$

20. $x_{k+1} + x_k = 2$

21. $x_{k+1} - 4x_k = 6$

22. $5x_{k+1} + x_k = 12$

23. $7x_{k+1} + 2x_k = 9$

In Exercises 24 through 27, the stair-step (cobwebbing) analysis is applied to determine the stability of fixed points in one of the cases in which the linearized analysis fails. In these exercises, first show that $\bar{x} = 0$ is a fixed point with $f'(\bar{x}) = 1$. Then sketch the geometric method (stair-step) for the iterations and determine the stability of the fixed point $\bar{x} = 0$.

24. $x_{k+1} = x_k + x_k^2$

25. $x_{k+1} = x_k - x_k^2$

26. $x_{k+1} = x_k + x_k^3$

27. $x_{k+1} = x_k - x_k^3$

8.4 Quadratic Map and Period Doubling Bifurcation Sequence to Chaos

In this section, we wish to consider the first-order nonlinear autonomous difference equation

$$x_{k+1} = f(x_k) = x_k + \gamma x_k(1 - x_k), \tag{1}$$

known as the **quadratic map**. The term **map** is used in mathematics when thinking of a function geometrically. Here we use map as another phrase for a difference equation $x_{k+1} = f(x_k)$, where we think of the difference equation as sending the point x_k to the point x_{k+1}.* There is one parameter γ, and we wish to show how the solution depends on this parameter.

There are many motivations for studying this equation. For one thing, it introduces the important idea of bifurcation, which plays a role in areas as diverse as mechanics and chemical reactions. It also introduces the idea of "chaos." In Section 7.5 we considered the logistic differential equation from population dynamics,

$$\frac{dx}{dt} = ax(1 - x).$$

Similar equations occur in chemical reactions. The instantaneous growth rate, $(1/x)(dx/dt) = a(1 - x)$, diminishes from the intrinsic growth rate a as the population increases. Here, the environmental carrying capacity has been normalized to 1. Instead of this differential equation and the function $x(t)$, we will consider population growth over a finite discrete time interval Δt. The growth rate measured over this time interval is

$$\frac{1}{x}\frac{\Delta x}{\Delta t} = \frac{1}{x_k}\frac{x_{k+1} - x_k}{\Delta t},$$

where x_k is the population at one time and x_{k+1} is the population at the next time. We assume that the growth that occurs over this discrete time interval

**As a word of warning, our quadratic map is equivalent to what other authors call the quadratic map, but the form that we have chosen differs from most. One difference is that our parameter γ is one less than other authors' parameter.*

depends on the population at the earlier time. We model the growth rate in the same way as in the modeling of the logistic equation, so that we have

$$\frac{1}{x_k}\frac{x_{k+1}-x_k}{\Delta t} = a(1-x_k).$$

In this way, we derive a nonlinear difference equation that is a model for environmentally limited population growth over a discrete time interval Δt:

$$x_{k+1} - x_k = a\,\Delta t x_k(1-x_k). \tag{2}$$

This is (1) with $\gamma = a\,\Delta t$. With this example in mind, we restrict our attention in this section to the very interesting case in which $\gamma > 0$.

There are two fixed points of (1), $x = 0$ and $x = 1$. They correspond to zero population and the environment's carrying capacity. For difference equations, these fixed points are referred to as period 1 solutions, since they repeat every $1 \cdot \Delta t$ period of time. In preparation for the geometric iteration of the nonlinear difference equation, we graph $y = f(x)$ and $y = x$ in Figure 8.4.1. This model can predict negative populations. We wish to avoid this by choosing our initial condition x_0 such that $x_1 > 0$. From (1), the quadratic $f(x) = x + \gamma x(1-x)$ has intercepts at $x = 0$ and at $x = 1 + 1/\gamma$. We will restrict our initial condition so that $x_0 < 1 + 1/\gamma$. We note that x_1 can be as large as $\frac{1}{4}(1 + 1/\gamma)(1+\gamma)$. We can guarantee that x_2 and all the other subsequent populations are positive if $\frac{1}{4}(1+1/\gamma)(1+\gamma) < (1 + 1/\gamma)$. This can be accomplished only if we assume that

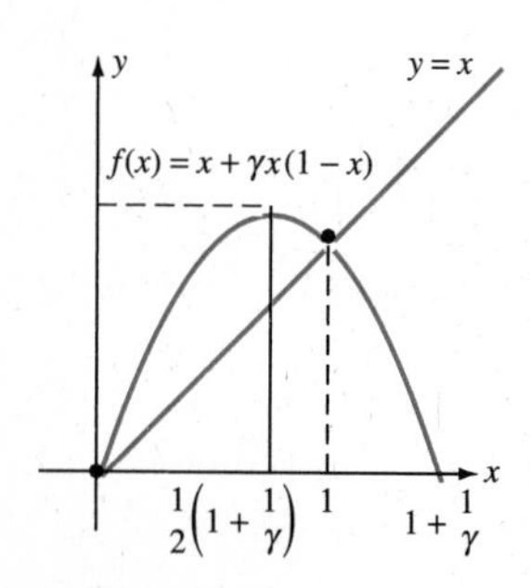

FIGURE 8.4.1

Fixed points of quadratic map.

$$\gamma < 3.$$

All the interesting dynamics we describe in the remaining part of this section correspond to

$$0 < \gamma < 3.$$

We begin by analyzing the stability of the fixed point $\bar{x} = 0$ (period 1 solution). From Section 8.3 the linearized difference equation valid in the neighborhood of a fixed point is

$$y_{k+1} = f'(\bar{x})y_k, \tag{3}$$

where $y_k = x_k - \bar{x}$ is the displacement from the equilibrium. In this case, from (1), $f(x_k) = x_k + \gamma x_k(1-x_k)$. Hence it follows that

$$f'(\bar{x}) = 1 + \gamma(1 - 2\bar{x}).$$

The zero population ($\bar{x} = 0$) is always unstable because $f'(0) = 1 + \gamma > 1$. Small populations grow exponentially, since $\gamma = a\,\Delta t > 0$. Small populations for the difference equation behave the same as small populations for the differential equation.

The stability of $\bar{x} = 1$, the environment's carrying capacity, is more interesting for this difference equation model. The stability is determined from (3). Since $f'(1) = 1 - \gamma$ and $\gamma > 0$, it follows that $f'(1) < 1$. Nearby solutions exponentially decay (just like solutions to the differential equation) to the fixed point $\bar{x} = 1$ if $\gamma < 1$ and exponentially decay with a flip to the fixed point if $1 < \gamma < 2$. Thus, the fixed point $\bar{x} = 1$ is stable if $0 < \gamma < 2$. However, at $\gamma = 2$, $f'(1) = -1$. At that parameter value, the fixed point is on the border between being stable and being unstable. The solution to the linearized difference equation (3) flips back and forth. In fact, the fixed point $\bar{x} = 1$ is unstable if $2 < \gamma < 3$, since there $f'(-1) < -1$. In this case the population oscillates wildly around the environment's carrying capacity (if the population is near that level).

For the logistic differential equation, (2), the equilibrium $\bar{x} = 1$ is stable. Thus, the difference equation model (1) is a poor one when $\gamma = a\,\Delta t > 2$ if the difference equation is derived by a difference approximation of the differential equation. The discretization time Δt is too large. However, we are assuming here that the difference equation (1) correctly models some discrete growth process. The difference equation (1) could arise either as a direct model or by sampling over a longer time interval.

Quadratic Map When One Fixed Point Is Stable ($0 < \gamma < 2$)

The environment's carrying capacity $\bar{x} = 1$ is stable if $0 < \gamma < 2$. We illustrate the iterations of the nonlinear difference equation with the geometric method as well as graphing the solution as a function of time. In Figure 8.4.2,

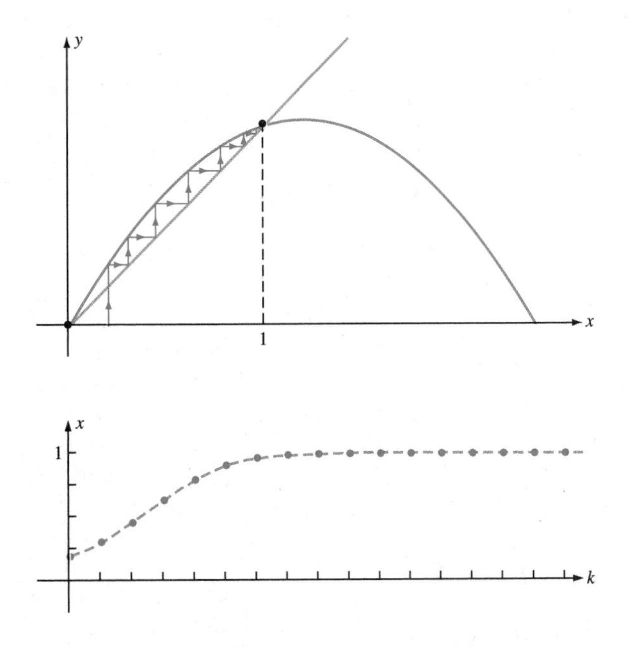

FIGURE 8.4.2

$0 < \gamma < 1$
Quadratic map.

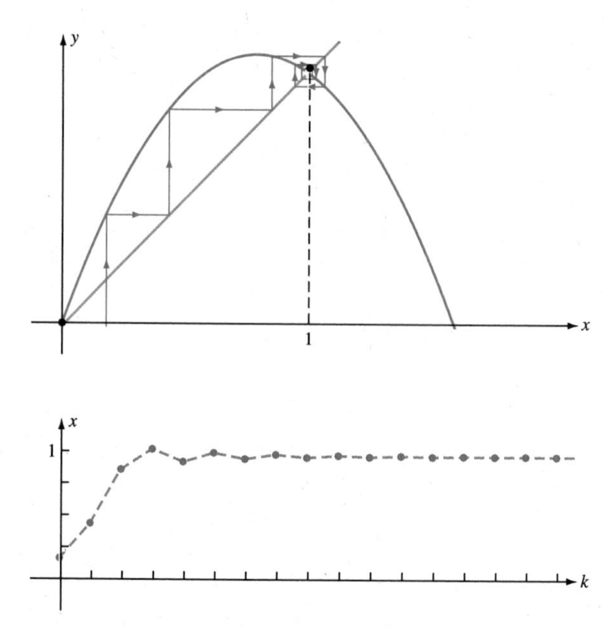

FIGURE 8.4.3

$1 < \gamma < 2$
Quadratic map.

we chose an example with $0 < \gamma < 1$ because in this case the solution approaches the stable fixed point $\bar{x} = 1$ in a manner similar to the differential equation. Note that the slope of $f(x)$ when it crosses the 45° line satisfies $0 < f'(\bar{x} = 1) < 1$, so that the solution exponentially decays to the fixed point $\bar{x} = 1$. Other initial conditions are discussed in the exercises. However, if $1 < \gamma < 2$, then there is an oscillatory (flip) decay in the neighborhood of the fixed point, as illustrated by the example in Figure 8.4.3. Note in the figure that $-1 < f'(\bar{x} = 1) < 0$. For a first-order autonomous difference equation, it is possible for the solution to oscillate around the fixed point, a phenomenon that is impossible for first-order autonomous differential equations.

8.4.1 Period 2 Solutions

For the quadratic map, $x_{k+1} = f(x_k) = x_k + \gamma x_k(1 - x_k)$, there are two fixed points. If $2 < \gamma < 3$, both are unstable. An interesting question is what happens if $2 < \gamma < 3$, since both fixed points are unstable. We claim that very complex behavior is possible. However, some of the interesting behavior is not particularly difficult to analyze. When $\gamma = 2$, the behavior near the environment's carrying capacity $\bar{x} = 1$ is oscillatory with a simple flip back and forth. This suggests that we might analyze what happens every second iteration. We will show that period 2 solutions are possible for the quadratic map, and in addition we will analyze their stability.

We begin by analyzing any first-order nonlinear difference equation,

$$x_{k+1} = f(x_k). \tag{4}$$

The solution every other time is easily determined:

$$x_{k+2} = f(x_{k+1}) = f(f(x_k)), \tag{5}$$

which we call the **second iterate**. Equation (5) is just a different nonlinear difference equation.

We investigate whether there are any very special simple solutions of the second iterate. We look for fixed points of the second iterate, that is, where $x_{k+2} = x_k$. A fixed point x^* of the second iterate satisfies

$$x^* = f(f(x^*)). \tag{6}$$

Suppose we return to the usual first iterate (4). If we start with any fixed point x^* of the second iterate, then the first three terms of the sequence generated are

$$x^*, f(x^*), f(f(x^*)), \ldots.$$

However, x^* is a fixed point of the second iterate and hence satisfies (6). Thus, the third term in the sequence is automatically the same as the first term:

$$x^*, f(x^*), x^*, \ldots.$$

Continuing with the iterations, we see that the sequence is

$$x^*, f(x^*), x^*, f(x^*), x^*, f(x^*), \ldots; \tag{7}$$

the solution generated is a **period 2 solution**. We learn the important fact that **fixed points of the second iterate correspond to period 2 solutions.** Technically speaking, we call this solution a period 2 solution only if $f(x^*) \neq x^*$. If $f(x^*) = x^*$ (which will happen in practice), then this is just the same period 1 solution that we have already discussed. When there is a period 2 solution x^*, then there is at least a second period 2 solution $f(x^*)$, since (somewhat obviously) the following sequence also solves the difference equation (4):

$$f(x^*), x^*, f(x^*), x^*, f(x^*), \ldots.$$

That is, real period 2 solutions come in pairs.

Stability of a Period 2 Solution

Period 2 solutions $(x^*, f(x^*))$ are fixed points of the second iterate

$$x_{k+2} = f(f(x_k)). \tag{8}$$

Thus, the stability of period 2 solutions can be determined from the linearization of the second iterate in the neighborhood of the period 2 solution:

$$y_{k+2} = \frac{d}{dx} f(f(x))\bigg|_{x=x^*} y_k, \tag{9}$$

where $y_k = x_k - x^*$ is the displacement from x^*. We proceed in the same way as we have done before. The stability criterion depends on the first derivative (here the first derivative of the second iterate):

$$\begin{aligned} &\text{If } \left|\frac{d}{dx}f(f(x))\right|_{x=x^*} < 1, \qquad \text{the period 2 solution is stable.} \\ &\text{If } \left|\frac{d}{dx}f(f(x))\right|_{x=x^*} > 1, \qquad \text{the period 2 solution is unstable.} \end{aligned} \tag{10}$$

We must remember to evaluate the derivative at x^*.

There is an interesting result from calculus concerning the derivative of the second iterate that can be applied to the stability criterion (10) for period 2 solutions. From the usual chain rule of calculus,

$$\frac{d}{dx}f(f(x)) = \frac{df(f(x))}{df(x)} \cdot \frac{df(x)}{dx} = f'(f(x)) \cdot f'(x).$$

However, we wish to evaluate this at $x = x^*$, in which case $f(x) = f(x^*)$:

$$\frac{d}{dx}f(f(x))\bigg|_{x=x^*} = f'(f(x^*)) \cdot f'(x^*) = \prod_{i=1}^{2} f'(x_i). \tag{11}$$

Thus, the derivative of the second iterate equals the product of the first derivative evaluated at all values of the periodic sequence $x_1 = x^*$, $x_2 = f(x^*)$. These results can be generalized to higher-period solutions and their stability, but we discuss this only in the exercises.

Period 2 Solutions for the Quadratic Map

Computer software can easily determine and graph the second (and higher) iterate. As an example (without a computer) of the second iterate, we determine the second iterate for the quadratic map:

$$x_{k+1} = f(x_k) = x_k + \gamma x_k(1 - x_k) = x_k[1 + \gamma(1 - x_k)]. \tag{12}$$

The second iterate will be a fourth degree polynomial:

$$\begin{aligned} x_{k+2} &= f(f(x_k)) = f(x_k)[1 + \gamma(1 - f(x_k))] \\ &= x_k[1 + \gamma(1 - x_k)]\left[1 + \gamma(1 - x_k) - \gamma^2 x_k(1 - x_k)\right]. \end{aligned} \tag{13}$$

Fixed points of the second iterate (13) will be solutions of a fourth degree equation:

$$x = x[1 + \gamma(1 - x)][1 + \gamma(1 - x) - \gamma^2 x(1 - x)]. \tag{14}$$

However, it will be easy to solve (14) because period 1 solutions automatically are also periodic with period 2. We already know two period 1 solutions,

$\bar{x} = 0$ and $\bar{x} = 1$. Canceling the x in (14) yields a "messy" cubic equation:

$$\begin{aligned} 1 &= [1 + \gamma(1 - x)][1 + \gamma(1 - x) - \gamma^2 x(1 - x)] \\ &= 1 + 2\gamma(1 - x) + \gamma^2(1 - x)^2 - \gamma^2 x(1 - x) - \gamma^3 x(1 - x)^2. \quad (15) \end{aligned}$$

Canceling the 1 and then canceling a factor of $\gamma(1 - x)$ in (15) yields a quadratic equation for the period 2 solutions of the quadratic map:

$$0 = 2 + \gamma(1 - x) - \gamma x - \gamma^2 x(1 - x) = \gamma^2 x^2 - \gamma(2 + \gamma)x + \gamma + 2. \quad (16)$$

Using the quadratic formula, we obtain both period 2 solutions:

$$x = \frac{\gamma(2 + \gamma) \pm \gamma\sqrt{(\gamma + 2)(\gamma - 2)}}{2\gamma^2} = \frac{(2 + \gamma) \pm \sqrt{(\gamma + 2)(\gamma - 2)}}{2\gamma}. \quad (17)$$

We first observe from (17) that these period 2 solutions do not exist for $\gamma < 2$, since the term inside the square root is negative. Period 2 solutions exist only for $\gamma > 2$. This is very convenient because the period 1 solution corresponding to the environment's carrying capacity loses its stability at $\gamma = 2$. In deeper mathematics, it is shown that when period 1 solutions lose their stability in this way, period 2 solutions are usually created. This is called a **period doubling bifurcation**.

The cobweb of the quadratic map is graphed in Figure 8.4.4 when $\gamma = 2.05$. Note that the period 2 solution appears as a rectangular box. The time-dependent iterations are more complicated. What is observed in Figure 8.4.4 is the convergence of the solution to the period 2 solution.

FIGURE 8.4.4

$\gamma = 2.05$ (Stable period 2).

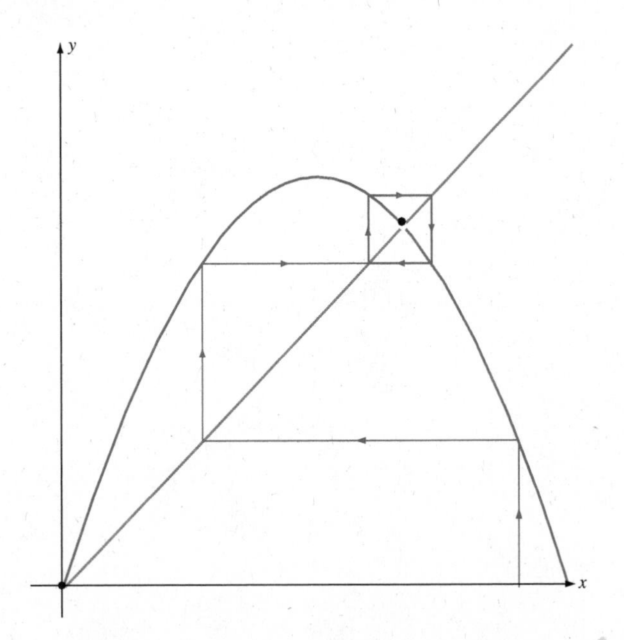

Stability of the Period 2 Solution for the Quadratic Map

The stability of the period 2 solution is determined from the linearization of the second iterate, (9), near the fixed point of the second iterate x^*:

$$y_{k+2} = \frac{d}{dx} f(f(x))\bigg|_{x=x^*} y_k. \tag{18}$$

We have shown in (11) that the derivative of the second iterate is the product of the first derivatives:

$$\frac{d}{dx} f(f(x))\bigg|_{x=x^*} = f'(f(x^*)) \cdot f'(x^*) = \prod_{i=1}^{2} f'(x_i). \tag{19}$$

For the quadratic map,

$$f(x_k) = x_k + \gamma x_k(1 - x_k),$$

and thus the first derivative is easy:

$$f'(x) = 1 + \gamma(1 - 2x) = 1 + \gamma - 2\gamma x. \tag{20}$$

The first derivative must be evaluated at both points of the periodic sequence (orbit) given by (17):

$$2\gamma x^* = (2 + \gamma) \pm \sqrt{(\gamma + 2)(\gamma - 2)},$$

so that the first derivatives at the two points in the period 2 orbit are

$$f'(x^*) = -1 \pm \sqrt{(\gamma + 2)(\gamma - 2)}. \tag{21}$$

Because of the $\pm$ in (21), the product of the two first derivatives given in (21) is easy:

$$\prod_{i=1}^{2} f'(x_i^*) = 1 - (\gamma + 2)(\gamma - 2). \tag{22}$$

The period 2 solution is stable when $|1 - (\gamma + 2)(\gamma - 2)| < 1$. This stability criterion is satisfied when γ is near 2 with $\gamma > 2$. When the period 2 solution is created, it is created stable. However, as γ increases from 2, it will reach a point where $1 - (\gamma + 2)(\gamma - 2) = -1$, and the period 2 solution will become unstable. The period 2 solution becomes unstable when

$$1 - (\gamma + 2)(\gamma - 2) = -1. \tag{23}$$

This equation (23) can be rewritten as

$$\gamma^2 = 6.$$

Thus, the period 2 solution becomes unstable at

$$\gamma = \sqrt{6} \approx 2.44949, \tag{24}$$

and the period 2 solution will be unstable for $\gamma > \sqrt{6}$.

8.4.2 Quadratic Map for $2 < \gamma < 3$

Let us summarize the behavior that we have obtained for the quadratic map

$$x_{k+1} = x_k + \gamma x_k(1 - x_k). \tag{25}$$

The zero solution is unstable for $\gamma > 0$. The environment's carrying capacity ($\bar{x} = 1$), what we call a period 1 solution, is stable for $\gamma < 2$, but is unstable for $\gamma > 2$. We have analyzed the existence of period 2 solutions and found that they exist only for $\gamma > 2$. We refer to $\gamma = 2$ as the value of the first bifurcation, at which the stable solution changes from a period 1 solution to a period 2 solution. By analyzing the stability of the period 2 solutions, we learned that they were stable for $\gamma < \sqrt{6}$. Thus, we expect and will verify here that when $2 < \gamma < \sqrt{6}$, periodic solutions of period 2 will be observed. However, when $\gamma > \sqrt{6}$, the period 2 solutions are unstable. We claim (but have not shown) that a stable period 4 solution is created at $\gamma = \sqrt{6}$, and so we refer to $\gamma = \sqrt{6} \approx 2.44949$ as the **second bifurcation point**. We claim that this process continues as we increase γ. The stable period 4 solution in turn becomes unstable, and a stable period 8 solution is created. This is observed numerically to occur at approximately $\gamma = 2.544$. The period 8 solution becomes unstable and changes to a stable period 16 solution, which becomes unstable and changes to a stable period 32 solution. This **period-doubling bifurcation sequence** continues forever. There is an infinite sequence of bifurcation points ($\gamma_1 = 2$, $\gamma_2 = \sqrt{6} \approx 2.44949$, $\gamma_3 = 2.54409, \ldots$). Interestingly enough, all of these bifurcations occur before $\gamma = 3$. The intervals in γ over which a particular period solution is stable get smaller and smaller as γ increases. In fact, the infinite sequence of values of γ at which period-doubling bifurcations continue to occur are observed numerically to converge to $\gamma \approx 2.57$. Feigenbaum showed that there is a universal shrinking factor for the region of stability valid not only for the quadratic map but in many other period-doubling bifurcations:

$$\lim_{n \to \infty} \frac{\gamma_n - \gamma_{n-1}}{\gamma_{n+1} - \gamma_n} \approx 4.6692. \tag{26}$$

We can observe the period-doubling bifurcation sequence with the geometric method. The stable period 2 orbit is shown in Figure 8.4.4 at $\gamma = 2.05$. To observe the period 4 orbit, we have to be more careful. We claim that the period 4 orbit exists for $\gamma > \sqrt{6} = 2.44949\ldots$, but it is stable (and observable) only in a short interval ($2.44949\ldots < \gamma < 2.54409\ldots$). It

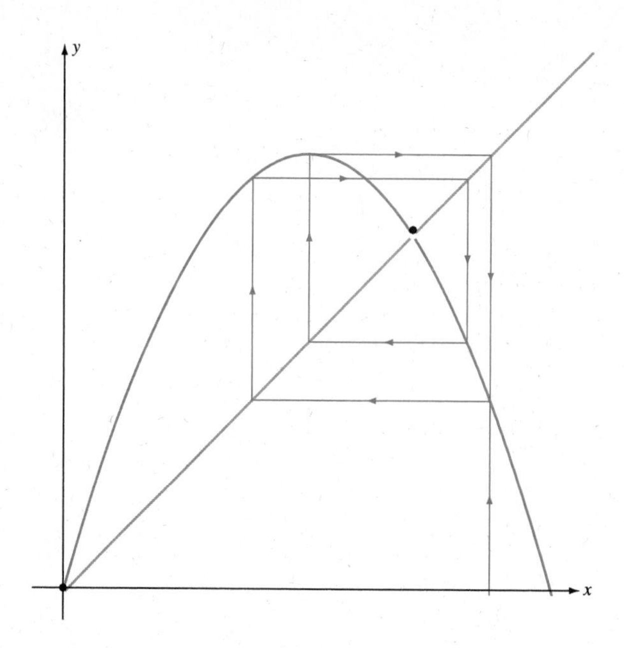

FIGURE 8.4.5

$\gamma = 2.5$ ($t \geq 30$),
Stable period 4.

may take a while for the orbit to converge to the stable period 4 orbit. In Figure 8.4.5, the stair-step diagram for $\gamma = 2.5$ is computed, but the first 30 iterations are not sketched. Thus, Figure 8.4.5 illustrates the period 4 stable solution when $\gamma = 2.5$. As γ is increased by a very small amount, we can observe the period 4 solution becoming unstable and period-doubling to a stable period 8 orbit. We illustrate a stable period 8 orbit at $\gamma = 2.56$ in Figure 8.4.6.

But even stranger phenomena are observed for $\gamma > 2.57$. Figure 8.4.7 shows the cobweb pattern for $\gamma = 2.60$ (with 1000 iterations). The pattern is quite complicated. It is difficult to detect from the figure whether a stable periodic orbit with large period exists. The solution is clearly not periodic with a small period. Most of the thin gaps fill up as more iterations are computed. The solution after a long period of time covers all values of x in some region. However, other regions of x are not part of the long-term solution when $\gamma = 2.60$. This type of behavior is called **chaotic**. Solutions are bounded but not periodic and do not have a limit. It is very difficult to give a precise meaning of chaos (especially in an introductory course). Chaotic solutions, for example, are unstable in the sense we have discussed. If the initial conditions are changed slightly, the time-dependent solution will eventually depart significantly. This gives rise to the so-called **butterfly effect**, described by the meteorologist Lorenz in his 1972 talk "Predictability: does the flap of a butterfly's wings in Brazil set off a tornado in Texas." In the 1960s Lorenz studied mathematical models of the physics of weather that had these types of chaotic solutions (way before any other scientists believed him).

FIGURE 8.4.6

$\gamma = 2.56$,
Stable period 8.

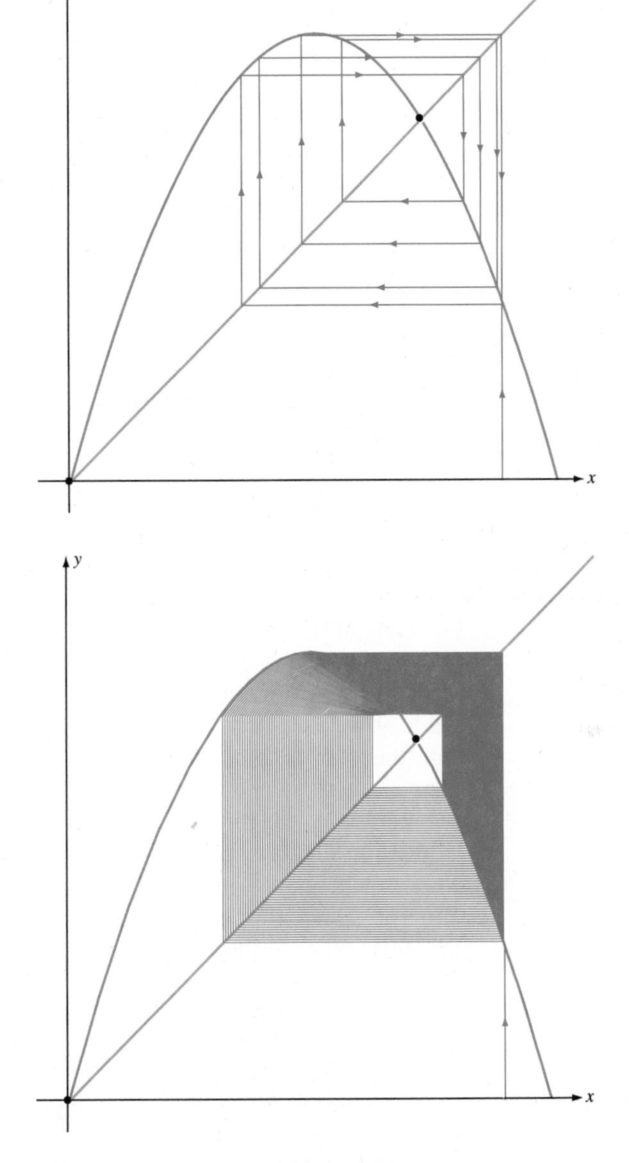

FIGURE 8.4.7

$\gamma = 2.60$
(1000 iterations).

Perhaps the ability to accurately predict weather for only 3 to 5 days is a result of the physics of weather being chaotic.

Not all solutions for $\gamma > 2.57$ are chaotic. For some values of γ in this range, stable nonchaotic solutions exist with periods $3, 5, 7, \ldots$. For example, when $\gamma = 2.83$, as shown in Figure 8.4.8, there is a stable period 3 solution.

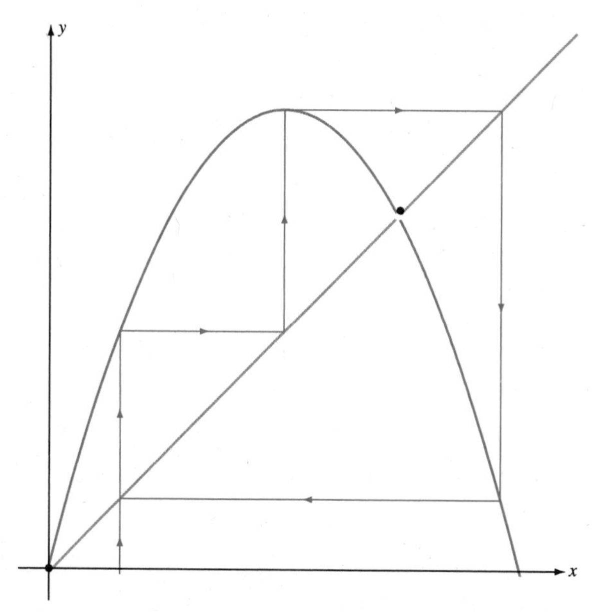

FIGURE 8.4.8

$\gamma = 2.83$,
Stable period 3.

However, in this example, the period 3 solution loses its stability through the same period-doubling bifurcation sequence; the sequence of periods is 3, $3 \cdot 2$, $3 \cdot 2^2$, $3 \cdot 2^3, \ldots$. The intervals of stability shrink and satisfy the same Feigenbaum number (26).

Sometimes a **bifurcation diagram** is drawn in which stable solutions are graphed as a function of the parameter γ. In the case discussed here, the stable solution is just $x = 1$ for $\gamma < 2$. However, for $\gamma > 2$, we have shown that the stable solutions are much more complicated. A period-doubling bifurcation sequence occurs, as do chaotic solutions for $\gamma > 2.57\ldots$. For an elementary nonlinear difference equation such as the quadratic map, the bifurcation diagram may be obtained easily in a numerical way. Points (pixels on a screen) may be plotted in the following way. First, one freely chooses a value of γ and simply computes the solution from the difference equation. The solution may move around in a complicated way, but eventually it may converge to a stable solution, for example, a period 7 solution. We observe numerically a periodic solution when the pattern of numbers seems to be repeating (or approximately repeating). The solution would be approaching the desired seven points of the period 7 solution, all corresponding to the same value of γ. The solution would be an infinite sequence of numbers that eventually approached seven points. To obtain only the desired seven points (or an approximation of them), the trick is to not print the first 50 or 100 values. Then, only the stable solutions would be observed. That is the way the bifurcation diagram for the quadratic map has been produced in Figure 8.4.9. Once you have the solutions for one value of γ, you just change γ as you wish to obtain the next set of points.

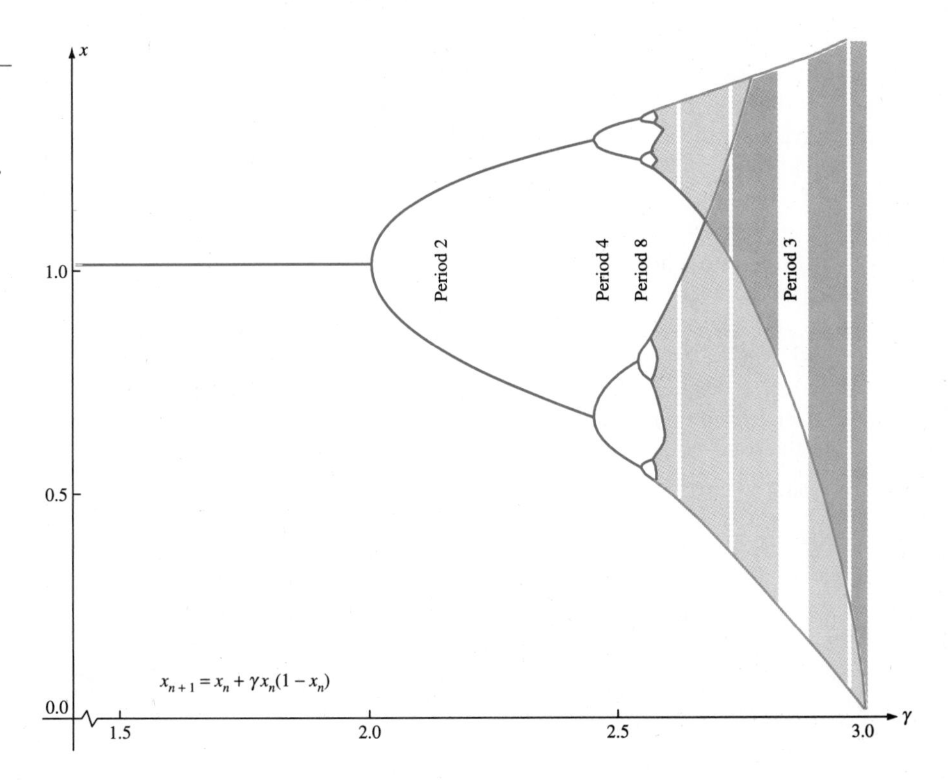

FIGURE 8.4.9

Bifurcation diagram for quadratic map (thanks to E. K. Ho).

Exercises

In Exercises 1 and 2, consider the quadratic map $x_{k+1} = f(x_k) = x_k + \gamma x_k(1 - x_k)$.

1. Consider $0 < \gamma < 1$, as shown in Figure 8.4.2. Choose other kinds of initial conditions, in particular much greater than the environment's carrying capacity. Sketch cobweb solutions and time-dependent solutions.

2. Consider $1 < \gamma < 2$, as shown in Figure 8.4.3. Choose other kinds of initial conditions, in particular much greater than the environment's carrying capacity. Sketch cobweb solutions and time-dependent solutions.

3. Calculate the shrinking factor from the first three bifurcation values for the quadratic map and compare it to Feigenbaum's limiting value.

4. Using some available software, reproduce Figures 8.4.4, 8.4.5, 8.4.6, and 8.4.7.

5. Using a graphing calculator, program the cobweb, and reproduce Figures 8.4.4, 8.4.5, 8.4.6, and 8.4.7.

Exercises 6 through 9 investigate period 3 solutions by graphing the third iterate for the quadratic map using some form of software or graphing calculator.

6. Show that there are only two fixed points for the third iterate when $\gamma = 2.5$. What are those two fixed points?

7. Show that there are eight fixed points for the third iterate when $\gamma = 2.9$. What is the significance of the eight fixed points? Determine the stability from the fixed-point diagram only.

8. By using a bisection search between $\gamma = 2.5$ and $\gamma = 2.9$, determine an approximation to the value where the period 3 solution first appears.

9. Show the beginning of a period-doubling sequence starting from $\gamma = 2.83$, where a period

3 solution is known to exist. Use the third iterate.

Exercises 10 through 12 investigate period 5 solutions by graphing the fifth iterate for the quadratic map using some form of software or graphing calculator.

10. Show that there are only two fixed points for the fifth iterate when $\gamma = 2.5$. What are those two fixed points?

11. Show that (or at least try to understand why) there are twelve fixed points for the fifth iterate when $\gamma = 2.9$. What is the significance of the twelve fixed points? Determine the stability from the fixed-point diagram only.

12. By using a bisection search between $\gamma = 2.5$ and $\gamma = 2.9$, determine an approximation to the value where the period 5 solution first appears.

13. Consider a general first-order autonomous difference equation $x_{k+1} = f(x_k)$. By considering the third iterate, find a formula for the stability of a period 3 solution.

14. Consider a general first-order autonomous difference equation $x_{k+1} = f(x_k)$. By considering the nth iterate, find a formula for the stability of a period n solution.

In Exercises 15 through 18, we will analyze the stability of solutions to the first-order autonomous difference equation $x_{k+1} = f(x_k)$ by fitting an exponential to the linearization. The resulting exponential growth coefficient is called the Liapunov exponent.

15. The stability of the fixed point $\bar{x}$ is governed by $y_{k+1} = f'(\bar{x})y_k$. Fit this solution to an exponential $|y_k| = |y_0|e^{rt}$, where $t = k$. Show that $r = \ln|f'(\bar{x})|$. r is called the **Liapunov exponent**. Briefly discuss the stability of the fixed point in terms of the value of the Liapunov exponent.

16. To investigate period 2 solutions, we introduce the second iterate $x_{k+2} = f(f(x_k))$. Assume that a period 2 solution exists, with the convenient notation $\bar{x}_1 = \bar{x}$ and $\bar{x}_2 = f(\bar{x})$. The stability of the period 2 solution is governed by $y_{k+2} = (d/dx)f(f(\bar{x}))y_k$. Solve for y_{2k} and fit it to an exponential

$$|y_{2k}| = |y_0|e^{rt}.$$

Show that $r = \frac{1}{2}\Sigma_{i=1}^{2} \ln|f'(\bar{x}_i)|$, since $t = 2k$. r is again called the Liapunov exponent. Briefly discuss the stability of the period 2 solution in terms of the value of the Liapunov exponent.

17. Generalize the previous problem to period m solutions. Show that the Liapunov exponent is $r = (1/m)\Sigma_{i=1}^{m} \ln|f'(\bar{x}_i)|$, where $t = mk$.

18. Consider the nonlinear difference equation $x_{k+1} = f(x_k)$. Consider a solution $\bar{x}_0, \bar{x}_1, \bar{x}_2, \ldots$ that is bounded but not periodic.

a) Show that the stability of this nonperiodic solution is governed by $y_{k+1} = f'(\bar{x}_k)y_k$. How is y_k defined?

b) Show that the solution of the linearization is $y_k = \Pi_{i=0}^{k-1} |f'(\bar{x}_i)| y_0$.

c) Here we wish to fit this solution to an exponential $|y_k| = |y_0|e^{rt}$. Find r. In the limit as $k \to \infty$, this is called the Liapunov exponent. If $r > 0$, the bounded nonperiodic solution is unstable, and the solution is called chaotic. [Hint: $t = k$]

CHAPTER

9

Qualitative Analysis of Nonlinear Equations in the Plane

9.1 Introduction

In Chapter 7 we introduced the qualitative analysis of a single differential equation. This chapter will discuss the qualitative analysis of systems of **autonomous differential equations** of the form

$$\begin{aligned} \frac{dx}{dt} &= f(x, y), \\ \frac{dy}{dt} &= g(x, y). \end{aligned} \tag{1}$$

The system (1) is called autonomous (time-independent) since the equations (1) for the velocities

$$\frac{dx}{dt} \quad \text{and} \quad \frac{dy}{dt}$$

depend explicitly only on the values of x and y and not on the time t. In a qualitative analysis, one is interested in questions such as:

- Are there any equilibria (constant solutions)?
- Are these equilibria stable or unstable?
- Are solutions bounded or unbounded?
- Are there oscillations (periodic solutions)?

The answers to these questions can be important not only to verify numerical results, but also to explain physical behavior when (1) cannot actually be solved or when only forms of f and g are known. Such an analysis underlies many physical theories.

Before beginning this analysis in the next section, we shall give several examples of nonlinear systems, some of which will be examined more carefully later in this chapter.

As noted before, a single second-order differential equation,

$$\frac{d^2x}{dt^2} + f\left(x, \frac{dx}{dt}\right) = 0, \qquad x(0) = x_0, \qquad x'(0) = y_0, \tag{2}$$

may be rewritten as a first-order system:

$$\begin{aligned} \frac{dx}{dt} &= y, & x(0) &= x_0, \\ \frac{dy}{dt} &= -f(x, y), & y(0) &= y_0. \end{aligned} \tag{3}$$

Electric Circuits

(Note Sections 1.10 and 2.12.) In many devices, such as diodes, the voltage v and current i satisfy a nonlinear relationship. That is, the device has a nonlinear v–i characteristic. Assuming that the device is current-controlled, that is, the voltage drop is a function of current, we get circuit models of the form (4)

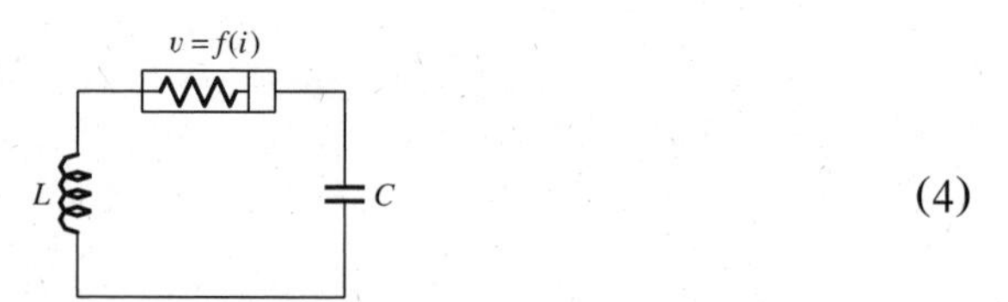

(4)

which, by Kirchhoff's voltage law, leads to the differential equation

$$L\frac{di}{dt} + f(i) + \frac{1}{C}q = 0, \tag{5}$$

or, equivalently, the system

$$\begin{aligned} \frac{di}{dt} &= -\frac{1}{L}f(i) - \frac{1}{LC}q, \\ \frac{dq}{dt} &= i, \end{aligned} \tag{6}$$

where i is the current in the loop and q is the charge on the capacitor.

Additional electrical examples will be given later.

Mechanical Systems

(Note Sections 2.8 and 2.11.) If we no longer assume small velocities and small displacements, we expect the resistance and spring force to vary in a nonlinear manner. ("Linear behavior is only for small variations" is a heuristic physical version of Taylor's theorem.) This would lead to a spring-mass system with differential equation

$$m\frac{d^2x}{dt^2} + f\left(\frac{dx}{dt}\right) + g(x) = 0,$$

or the equivalent system

$$\begin{aligned}\frac{dx}{dt} &= y,\\ \frac{dy}{dt} &= -\frac{1}{m}f(y) - \frac{1}{m}g(x).\end{aligned}$$

Chemical Reactions

Chemical reactions involving two substances with concentrations x and y, which can not only combine but also disassociate, lead to differential equations of the form

$$\begin{aligned}\frac{dx}{dt} &= ax + bxy + cy + d,\\ \frac{dy}{dt} &= ex + fxy + gy + h,\end{aligned}$$

where a, b, c, d, e, f, g, h are constants. [A special case is Eq. (3) of Section 7.7.]

A closely related idea is that of populations (which could be nonbiological) that are either competing or feeding on each other. Under the appropriate modeling assumptions, these often lead to models of the general form

$$\begin{aligned}\frac{dx}{dt} &= ax + bxy + cx^2,\\ \frac{dy}{dt} &= dy + exy + fy^2,\end{aligned} \tag{7}$$

with a, b, c, d, e, f constants.

9.2 The Phase Plane

This section will present some basic results for systems of the form

$$\begin{aligned}x' &= f(x, y)\\ y' &= g(x, y)\end{aligned} \tag{1}$$

and introduce the important concept of the **phase plane** of (1). The basic existence and uniqueness theorem for (1) is:

■ **THEOREM 9.2.1 Existence and Uniqueness**

Suppose that f, g, f_x, f_y, g_x, g_y are continuous in a region R (a connected open set) of the xy plane containing the point (x_0, y_0) in its interior. Let t_0 be a fixed value of t. Then there is a unique solution to (1) in the region R such that $x(t_0) = x_0$, $y(t_0) = y_0$. ■

A solution $(x(t), y(t))$ of (1) traces out a curve in the xy plane. This parameterized curve, along with an indication of the direction the solution moves along the curve, is called a **trajectory** or **orbit**. The term **solution curve** is also sometimes used. (The curve without a sense of direction is also referred to as an **integral curve**.) The set of trajectories in the xy plane, together with an indication of the solutions' directions along them, is the **phase plane**. Usually only a few representative solutions are drawn. This sketch is sometimes called a **phase portrait**.

The fact that (1) is autonomous has several implications for the phase portrait if Theorem 9.2.1 holds. First, observe that if (x, y) is a solution of (1) defined for $t \in (\alpha, \beta)$, then $\hat{x}(t) = x(t - c)$, $\hat{y}(t) = y(t - c)$ is a solution for $t \in (\alpha + c, \beta + c)$ that gives rise to the same trajectory. Conversely, every solution giving rise to the same trajectory is of the form $(x(t - c), y(t - c))$ for some constant c.

A curve $\phi(x, y) = C$ is an **invariant curve** for the autonomous system (1) if for any point (x_0, y_0) on the curve, and for any t_0, the solution to (1) satisfying $x(t_0) = x_0$, $y(t_0) = y_0$, stays on the curve. That is, solutions starting on the curve stay on the curve. The points on a trajectory are an invariant curve. In general, an invariant curve is the union of the points making up one or more trajectories.

To find the invariant curves of (1), we can use the fact that if $x(t)$, $y(t)$ is a solution of (1), then

$$\frac{dy}{dx} = \frac{dy/dt}{dx/dt} = \frac{g(x, y)}{f(x, y)} \tag{2}$$

or

$$f(x, y)\, dy - g(x, y)\, dx = 0.$$

■ **THEOREM 9.2.2 Invariant Curves**

A curve $\phi(x, y) = c$ is an invariant curve for the autonomous system (1) if and only if $\phi(x, y) = c$ is a solution of $f(x, y)\, dy - g(x, y)\, dx = 0$. ■

EXAMPLE 9.2.1 **Invariant Curves and Phase Portrait**

Find the invariant curves and draw the phase portrait for the system

$$\frac{dx}{dt} = 4y, \qquad \frac{dy}{dt} = -x. \tag{3}$$

SOLUTION The invariant curves satisfy

$$\frac{dy}{dx} = \frac{dy/dt}{dx/dt} = \frac{-x}{4y}.$$

Solving this differential equation by separation of variables, we get that the invariant curves are

$$x^2 + 4y^2 = c.$$

If $c < 0$, there is no curve. If $c = 0$, the curve is a point representing the constant solution $x = 0$, $y = 0$ of (3). If $c > 0$, the invariant curve is an ellipse. Suppose $c > 0$. Then

$$\mathbf{v} = \left[\frac{dx}{dt}, \frac{dy}{dt}\right] = [4y, -x]$$

is the **velocity vector** and is tangent to the curve. The length of the velocity vector is the **speed** $s(t)$. Since

$$s(t) = \sqrt{\left(\frac{dx}{dt}\right)^2 + \left(\frac{dy}{dt}\right)^2} = \sqrt{16y^2 + x^2} > \sqrt{4y^2 + x^2} = \sqrt{c},$$

we see that the solutions keep moving along the ellipse without ever slowing down below $\sqrt{c}$. Thus they repeatedly traverse the ellipse. (A more sophisticated technique to reach this same conclusion is given in Section 9.5.) The direction of movement can be determined by noting that, if $x > 0$, $y > 0$, then the velocity vector points down and to the right. Thus the ellipses are traversed clockwise. The phase portrait is then Figure 9.2.1. ◄

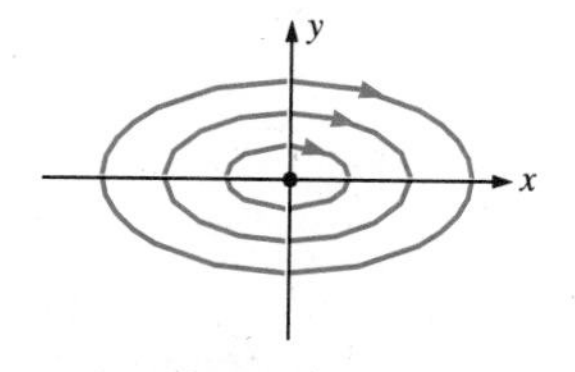

FIGURE 9.2.1

Phase portrait for system (3).

While Figure 9.2.1 is informative, some information is lost because of the suppression of the t variable. For example, Figure 9.2.1 does not indicate how fast solutions go around the ellipses.

In many applications, an invariant curve $\phi(x, y) = c$ is referred to as a **conservation law**, since the quantity $\phi(x, y)$ remains constant. Conservation of energy is one such law.

In drawing these phase planes, it is helpful to realize that, if f, g, f_x, f_y, g_x, g_y are continuous for all x, y, as they are for most of our problems, then the only way a solution can fail to exist for all t is if (x, y) goes to infinity. Also, under these same circumstances, two distinct solution curves can never cross, since to do so would violate the basic existence and uniqueness Theorem 9.2.1. On the other hand, suppose that a trajectory given by the solution $(x(t), y(t))$ takes on the same value at time t_1 that it took on at an earlier time. Let t_0 be the previous time the solution was at this same point, so that $(x(t_0), y(t_0)) = (x(t_1), y(t_1))$. But then $(x(t), y(t))$ and $(x(t + t_0 - t_1), y(t + t_0 - t_1))$ are both solutions of (1) that satisfy the same initial condition at t_1. By Theorem 9.2.1, we have $x(t) = x(t + t_0 - t_1)$, $y(t) = y(t + t_0 - t_1)$, and the solution $(x(t), y(t))$ is periodic with period $t_0 - t_1$.

This means that in our problems there are three possibilities for a solution curve:

- A point (constant solution); (Figure 9.2.1, $x = 0$, $y = 0$)
- A simple closed curve (like an ellipse but not like a figure eight); (Figure 9.2.1)
- A curve that at each end either goes to infinity or does not include its endpoint (examples of this appear in the next section); (Figure 9.2.2b)

If an equilibrium lies on an invariant curve, then we get several trajectories on the invariant curve, as shown in the next example.

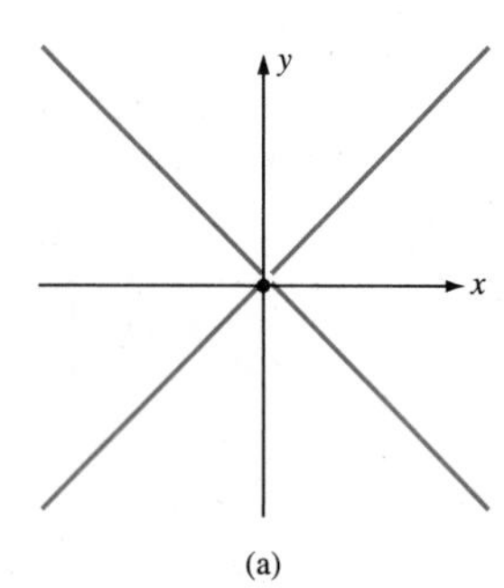

FIGURE 9.2.2

EXAMPLE 9.2.2 Equilibrium Point on Invariant Curve

Note that

$$x^4 - y^4 = 0 \tag{4}$$

is one of the invariant curves of

$$\frac{dx}{dt} = y^3, \qquad \frac{dy}{dt} = x^3. \tag{5}$$

The graph of (4) is shown in Figure 9.2.2a. The origin (0, 0) is a constant solution of (5). Each of the line segments that are left is a separate trajectory. Checking the direction on each we get figure 9.2.2b. For example, if $x > 0$, $y > 0$, then (5) says that $x' > 0$, $y' > 0$. Thus the trajectory in the first quadrant moves up ($y' > 0$) and to the right ($x' > 0$) away from the origin. Similarly, if $x > 0$, $y < 0$, then from (5), $x' < 0$, $y' > 0$, and the trajectory in the fourth quadrant moves up ($y' > 0$) and to the left ($x' < 0$). ◄

Note The precise definitions of solution curve and trajectory vary slightly in some texts. However, all treatments lead to the same phase portrait [Figure 9.2.2 for (4)].

Since $[f(x_0, y_0), g(x_0, y_0)]$ is the tangent vector to the solution at (x_0, y_0), we can also draw vector fields as in Section 1.5. Care must be taken if tangent lines are plotted. If, for example, the tangent vectors are all drawn with the same length, we are really sketching tangents to invariant curves, rather than trajectories. The resulting sketch, while helpful, may not indicate all equilibrium.

Exercises

For Exercises 1 through 11, (a) find and sketch the invariant curves, (b) determine the equilibrium point, and (c) find the direction traveled by the trajectories.

1. $\frac{dx}{dt} = x, \quad \frac{dy}{dt} = -y$

2. $\frac{dx}{dt} = y, \quad \frac{dy}{dt} = x$

3. $\frac{dx}{dt} = xy, \quad \frac{dy}{dt} = x$

4. $\frac{dx}{dt} = xy, \quad \frac{dy}{dt} = xy^3 + xy$

5. $\frac{dx}{dt} = 3, \quad \frac{dy}{dt} = 1$

6. $\frac{dx}{dt} = y^3, \quad \frac{dy}{dt} = -x^3$

7. $\frac{dx}{dt} = 2 - 2y, \quad \frac{dy}{dt} = 2x - 2$

8. $\frac{dx}{dt} = y, \quad \frac{dy}{dt} = xy + y$

9. $\frac{dx}{dt} = x^2, \quad \frac{dy}{dt} = 3yx$

10. $\frac{dx}{dt} = -4x, \quad \frac{dy}{dt} = x$

11. $\frac{dx}{dt} = -x^2, \quad \frac{dy}{dt} = yx$

12. Rewrite the second-order differential equation

$$x'' + f(x) = 0$$

as a first-order system using $y = x'$ and show that the integral curves can always, in principle, be found by separation of variables.

For Exercises 13 through 20, use Exercise 12 to rewrite the second-order nonlinear equation as a first-order system, and then find its invariant curves.

13. $x'' + \sin x = 0$

14. $x'' + e^x = 0$

15. $x'' + x^2 = 0$

16. $x'' + \frac{1}{1 + x^2} = 0$

17. $x'' + \frac{x}{1 + x^2} = 0$

18. $x'' + x^4 + 1 = 0$

19. $x'' - x^2 = 0$

20. $x'' + x^3 = 0$

21 a) Show that $x^2 + y^2 = 1$ is an invariant curve of

$$\frac{dx}{dt} = -y^2, \quad \frac{dy}{dt} = xy. \tag{6}$$

b) Find all equilibria of (6) on $x^2 + y^2 = 1$.

c) Sketch the trajectories of (6) on $x^2 + y^2 = 1$.

22. a) Show that $x^2 + y^2 = 1$ is an invariant curve of

$$\frac{dx}{dt} = -xy^2, \quad \frac{dy}{dt} = x^2y. \tag{7}$$

b) Find all equilibria of (7) on $x^2 + y^2 = 1$.

c) Sketch the trajectories of (7) on $x^2 + y^2 = 1$.

23. a) Show that $x^2 + y^2 = 1$ is an invariant curve of

$$\frac{dx}{dt} = -y(y - 1), \quad \frac{dy}{dt} = x(y - 1). \tag{8}$$

b) Find all equilibria of (8) on $x^2 + y^2 = 1$.

c) Sketch the trajectories of (8) on $x^2 + y^2 = 1$.

24. Use Theorem 9.2.1 to show that if a solution stays in the open region R, then the solution exists for all time.

25. Show that if $(x(t), y(t))$ is a solution of (1) and the speed is zero at any time t_0, then (x, y) is a constant solution.

26. Prove Theorem 9.2.2.

In Exercises 27 through 30, find equilibria for the given system. Then sketch the vector field, indicating the direction of each vector. (You may wish to experiment with variable length vectors also.)

27. $\dfrac{dx}{dt} = -y, \quad \dfrac{dy}{dt} = x$

28. $\dfrac{dx}{dt} = y, \quad \dfrac{dy}{dt} = x$

29. $\dfrac{dx}{dt} = y^2, \quad \dfrac{dy}{dt} = x^2$

30. $\dfrac{dx}{dt} = 2 - 2y, \quad \dfrac{dy}{dt} = 2x - 2$

9.3 Linear Systems

One of the keys to drawing phase portraits for nonlinear systems turns out to be understanding linear systems. In this section we shall derive almost all possible phase portraits for linear homogeneous systems in the form:

$$\begin{aligned} \frac{dx}{dt} &= ax + by, \\ \frac{dy}{dt} &= cx + dy, \end{aligned} \tag{1}$$

where a, b, c, d are constants. Recall from Chapter 5 that if we introduce the vector $\mathbf{x} = [\begin{smallmatrix} x \\ y \end{smallmatrix}]$ and the matrix $\mathbf{A}$,

$$\mathbf{A} = \begin{bmatrix} a & b \\ c & d \end{bmatrix},$$

then (1) may be written

$$\frac{d\mathbf{x}}{dt} = \mathbf{A}\mathbf{x}.$$

Also recall that solutions of (1) of the form

$$\mathbf{x} = \mathbf{u}e^{\lambda t}$$

exist if

$$\lambda\mathbf{u} = \mathbf{A}\mathbf{u}.$$

Nonzero solutions exist only if λ_1, λ_2 are **eigenvalues** of the matrix $\mathbf{A}$ satisfying the **characteristic equation**

$$p(\lambda) = \det \begin{bmatrix} \lambda - a & -b \\ -c & \lambda - d \end{bmatrix} = \lambda^2 - (a + d)\lambda + (ad - bc) = 0.$$

For some questions, just knowing λ_1, λ_2 is sufficient. However, we shall point out how extra information, such as the eigenvectors **u** (Section 5.10) of **A** can be useful.

It turns out, for our later purposes, that if either or both eigenvalues are zero, our analysis cannot proceed. (Note Exercises 17 and 18.) Thus we assume

$$\lambda_1 \neq 0, \qquad \lambda_2 \neq 0. \tag{2}$$

An **equilibrium** is a constant solution $x(t) \equiv r$, $y(t) \equiv s$. Substituting into (1), we find that r, s must satisfy:

$$\begin{aligned} 0 &= ar + bs, \\ 0 &= cr + ds. \end{aligned} \tag{3}$$

By solving (3) we can show that (2) implies that $r = s = 0$ (see Exercise 19 at the end of this section, or Theorem 5.8.2). Thus the only equilibrium is the origin. Note that the curve $e^{\lambda t}\mathbf{u}$ for a scalar λ and vector **u** is a ray from the origin parallel to **u** in the (x, y) plane.

Case 1 Positive Eigenvalues

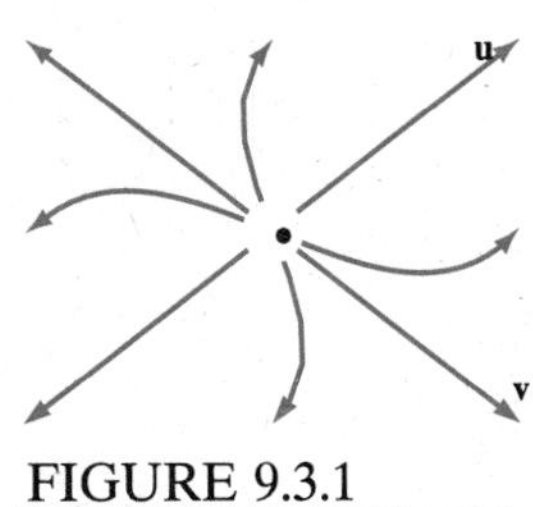

FIGURE 9.3.1

If $\lambda_1 > \lambda_2 > 0$, then the nonzero solutions involve $e^{\lambda_1 t}$, $e^{\lambda_2 t}$, and all solutions except the equilibrium tend to infinity as $t \to \infty$. The resulting phase portrait is Figure 9.3.1. The origin is an **unstable** or **repelling** equilibrium, since all the other trajectories move away from it. The two straight trajectories are in the directions of the eigenvectors of

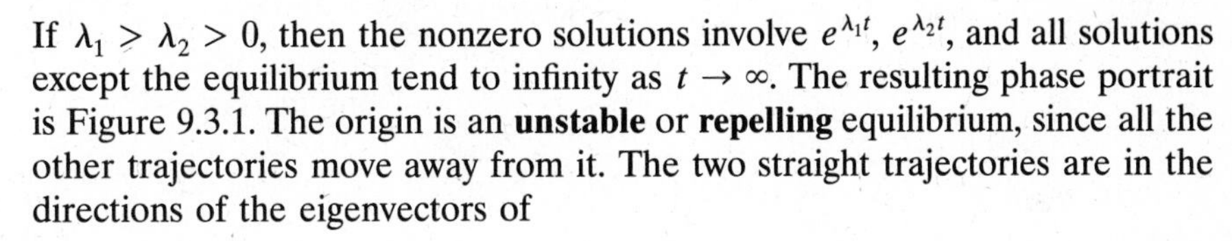

$$\begin{bmatrix} a & b \\ c & d \end{bmatrix}$$

with **u** being the eigenvector for λ_1 and **v** the eigenvector for λ_2. (See Exercises 20 and 21 at the end of this section.) If $\lambda_1 = \lambda_2$ and $\lambda_1 > 0$, then all nonzero trajectories still go to infinity but they may bend differently.

Case 2 Negative Eigenvalues

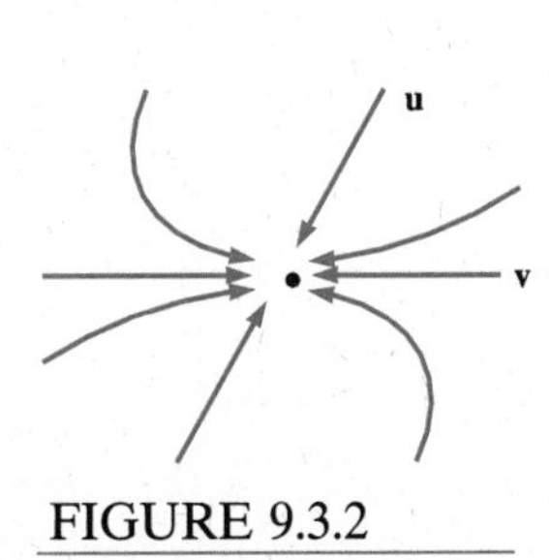

FIGURE 9.3.2

If $\lambda_1 < \lambda_2 < 0$, then the solutions involve $e^{\lambda_1 t}$, $e^{\lambda_2 t}$ which now go to zero as $t \to \infty$. Thus, all solutions approach the equilibrium and it is called an **attractor**, or an **asymptotically stable equilibrium**. The phase portrait is given in Figure 9.3.2. The trajectories that are approaching the equilibrium solution (origin) get arbitrarily close to the origin but never reach it. Again **u** is in the direction of an eigenvector corresponding to λ_1 and **v** an eigenvector corresponding to λ_2. If $\lambda_1 = \lambda_2$ but both are negative, the curves may bend differently but they still approach the origin.

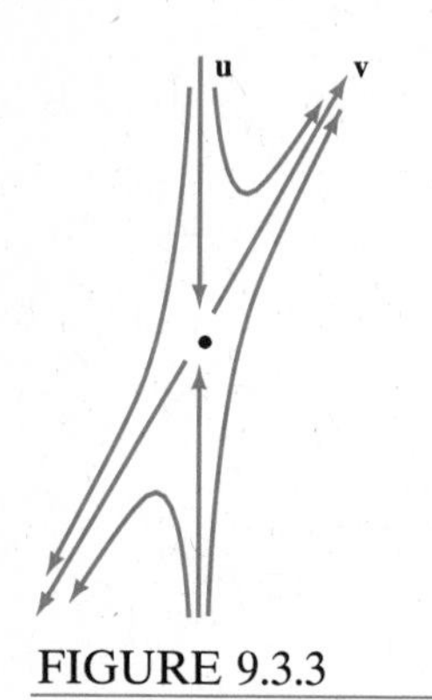

FIGURE 9.3.3

Case 3 One Positive and One Negative Eigenvalue

If $\lambda_1 < 0 < \lambda_2$, the equilibrium is a **saddle point**. It has solutions with both decaying and increasing terms. Case 3 is illustrated in Figure 9.3.3. Again **u**, **v** are directions of eigenvectors corresponding to λ_1 and λ_2, respectively.

Before considering what happens with complex eigenvalues, we shall give an example with real eigenvalues.

EXAMPLE 9.3.1 Phase Portrait of a Saddle Point

Draw the phase portrait for the linear system

$$x' = -7x + 6y,$$
$$y' = 6x + 2y.$$

SOLUTION This is $\mathbf{x}' = \mathbf{A}\mathbf{x}$ with

$$\mathbf{x} = \begin{bmatrix} x \\ y \end{bmatrix} \quad \text{and} \quad \mathbf{A} = \begin{bmatrix} -7 & 6 \\ 6 & 2 \end{bmatrix}.$$

The characteristic polynomial of **A** is

$$p(\lambda) = \det \begin{bmatrix} \lambda + 7 & -6 \\ -6 & \lambda - 2 \end{bmatrix} = \lambda^2 + 5\lambda - 50 = (\lambda - 5)(\lambda + 10).$$

Thus the eigenvalues are $\lambda_1 = -10$, $\lambda_2 = 5$, which have opposite sign so that we are in Case 3, and will have a saddle point at the origin. Using the techniques of Chapter 5, we compute eigenvectors of

$$\mathbf{u} = \begin{bmatrix} -2 \\ 1 \end{bmatrix} \quad \text{for } \lambda_1 = -10 \quad \text{and} \quad \mathbf{v} = \begin{bmatrix} 1 \\ 2 \end{bmatrix} \quad \text{for } \lambda_2 = 5.$$

We know that $(0, 0)$ is an equilibrium. The trajectories labeled 1 through 4 in Figure 9.3.4 are the solutions

$$e^{5t}\mathbf{v}, \quad e^{-10t}\mathbf{u}, \quad -e^{5t}\mathbf{v}, \quad -e^{-10t}\mathbf{u},$$

respectively. Any other solution is of the form

$$c_1 e^{5t}\mathbf{v} + c_2 e^{-10t}\mathbf{u}$$

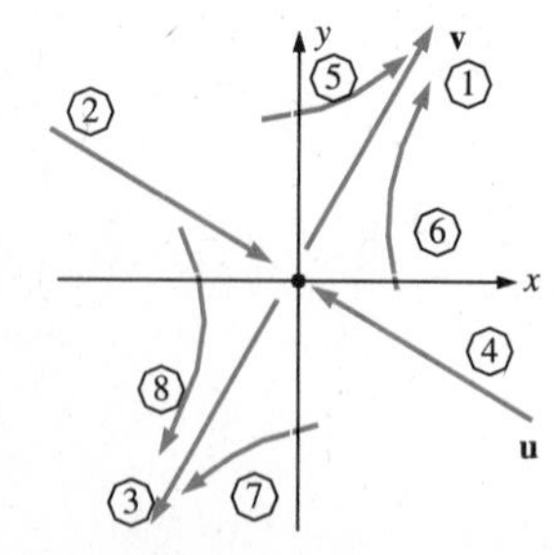

FIGURE 9.3.4

for constants c_1, c_2. As t increases, we have

$$c_2 e^{-10t}\mathbf{u} \to \mathbf{0} \quad \text{and} \quad c_1 e^{5t}\mathbf{v} + c_2 e^{-10t}\mathbf{u} \approx c_1 e^{5t}\mathbf{v}.$$

This is illustrated by trajectories 5, 6, 7, and 8 in Figure 9.3.4. ◄

Case 4 Complex Eigenvalues (Zero Real Part)

$\lambda_1 = \beta i$, $\lambda_2 = -\beta i$ with β a real nonzero number. The solutions involve $\sin \beta t$, $\cos \beta t$, and are "skewed ellipses." Motion is periodic. The equilibrium

FIGURE 9.3.5

is **stable**, since nearby solutions do not move very far away. However, since the solutions do not approach the equilibrium, the equilibrium is not asymptotically stable. See Figure 9.3.5.

Case 5 Complex Eigenvalues (Nonzero Real Part)

$\lambda_1 = \alpha + \beta i$, $\lambda_2 = \alpha - \beta i$, $\beta \neq 0$, $\alpha \neq 0$. The solutions of (1) now involve $e^{\alpha t} \cos \beta t$, $e^{\alpha t} \sin \beta t$. The result is a **spiral** centered at the origin. The solution moves out from the origin if $\alpha > 0$ (since $e^{\alpha t}$ increases if $\alpha > 0$) and inward if $\alpha < 0$. The solutions may also spiral either clockwise or counterclockwise but that is not usually an important consideration. See Figure 9.3.6.

In Figure 9.3.6 one should visualize an infinite number of spirals both inside and outside of those drawn. If $\alpha > 0$, then the equilibrium is unstable, or a **repeller**. If $\alpha < 0$, the equilibrium is asymptotically stable, or an **attractor**.

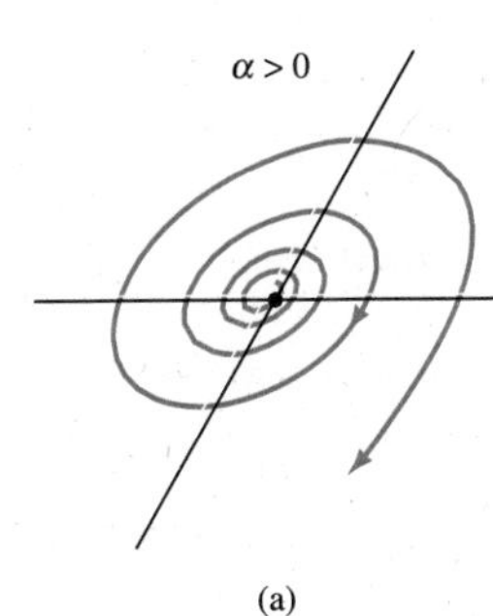

(a)

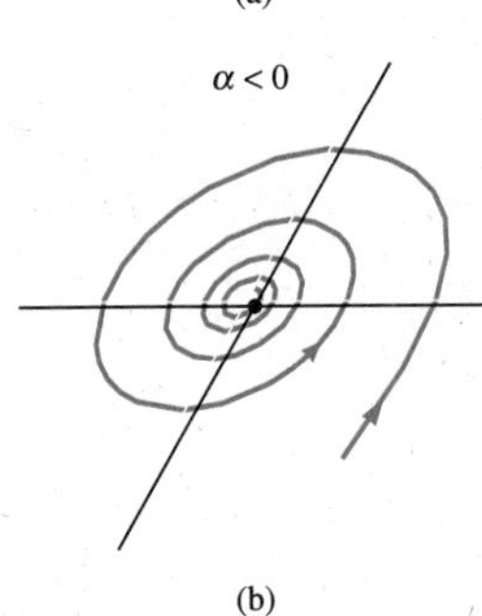

(b)

FIGURE 9.3.6

EXAMPLE 9.3.2 **Phase Portrait of a Spiral**

Determine the phase portrait for

$$\begin{aligned} x' &= 2x + y, \\ y' &= -x + 2y. \end{aligned} \tag{4}$$

Solution The characteristic polynomial is

$$p(\lambda) = \det \begin{bmatrix} \lambda - 2 & -1 \\ 1 & \lambda - 2 \end{bmatrix} = \lambda^2 - 4\lambda + 5.$$

The roots are $\lambda_1 = 2 + i$, $\lambda_2 = 2 - i$. Since $\alpha = 2 > 0$, we have an unstable outward spiral. Taking a nonzero point, say $x_0 = 1$, $y_0 = 0$, we get the tangent vector

$$[x', y'] = [2, -1],$$

which points down and to the right from $(1, 0)$. Thus the spiral is clockwise and we have precisely Figure 9.3.6a. ◄

In concluding this section, note that the invariant curves for these linear homogeneous differential equations always satisfy

$$\frac{dy}{dx} = \frac{cx + dy}{ax + by}. \tag{5}$$

This type of differential equation is called a **nonlinear homogeneous equation**. Its solution is discussed in Exercises 22–24.

Exercises

In Exercises 1 through 16, determine the behavior of the solutions near the equilibrium $x = 0$, $y = 0$, and sketch the phase portrait.

1. $x' = x, \quad y' = x + 2y$

2. $x' = 2x - y, \quad y' = 3x - 2y$

3. $x' = -x - 5y, \quad y' = x + y$

4. $x' = 2x - y, \quad y' = 2x + 5y$

5. $x' = x - y, \quad y' = x + y$

6. $x' = -2x + 2y, \quad y' = -x$

7. $x' = -5x - 4y, \quad y' = 2x + y$

8. $x' = x + 5y, \quad y' = -2x - y$

9. $x' = y, \quad y' = 2x + y$

10. $x' = -x - 2y, \quad y' = 2x - y$

11. $x' = -5x - y, \quad y' = 3x - y$

12. $x' = x + 2y, \quad y' = -4x - 3y$

13. $x' = -x + 4y \quad y' = -4x - y$

14. $x' = 3x + 2y, \quad y' = -2x + 3y$

15. $x' = 4x + 3y, \quad y' = 3x + 4y$

16. $x' = 2x + 3y, \quad y' = 3x + 2y$

In Exercises 17 through 19, λ_1, λ_2 are the roots of the characteristic equation for system (1).

17. Show that, if $\lambda_1 = \lambda_2 = 0$, then either all solutions of (3) are constant or constants and straight lines.

18. Show that, if $\lambda_1 = 0$ but $\lambda_2 \neq 0$, then there is a line of equilibrium points and

a) all other solutions tend toward this line if $\lambda_2 < 0$;

b) all other solutions go away from this line if $\lambda_2 > 0$.

19. Show that if $\lambda_1 \neq 0$, $\lambda_2 \neq 0$, then the only equilibrium of (1) is $x = 0$, $y = 0$.

20. Show that if λ is a nonzero scalar and $\mathbf{u} \neq \mathbf{0}$ is a two-vector, then the curve

$$\begin{bmatrix} x \\ y \end{bmatrix} = e^{\lambda t}\mathbf{u}, \quad -\infty < t < \infty$$

in the x, y plane is a ray extending from (but not including) the origin.

21. (Continuation of Exercise 20.) Show that if $\lambda > 0$, then (x, y) moves away from the origin along the ray, while if $\lambda < 0$, then (x, y) moves toward the origin along the ray.

In Exercises 22 through 24 find the invariant curves as follows. First form the differential equation (5) for the invariant curves. Let $y = vx$, where v is an unknown function of x. The resulting differential equation in v and x can be solved by separation. Then let $v = y/x$ to get the solution in terms of x and y.

22. $x' = -x - 5y$ (Exercise 3)
$y' = x + y$

23. $x' = -x - 2y$ (Exercise 10)
$y' = 2x - y$

24. $x' = x - y$ (Exercise 5)
$y' = x + y$

9.4 Equilibria of Nonlinear Systems

In this section we return to studying the nonlinear autonomous systems

$$\begin{aligned} x' &= f(x, y), \\ y' &= g(x, y). \end{aligned} \tag{1}$$

If $x(t)$, $y(t)$ is an **equilibrium**, or constant solution, then $x(t) = r$, $y(t) = s$ for constants r, s. Substituting into (1) we find that r, s must satisfy

$$\begin{aligned} 0 &= f(r, s), \\ 0 &= g(r, s). \end{aligned} \tag{2}$$

Conversely, if r, s are constants satisfying (2), then $x(t) = r$, $y(t) = s$ is a solution of (1). That is,

The equilibrium solutions $x = r$, $y = s$ of (1) are exactly the solutions (r, s) of the system of nonlinear algebraic equations (2).

EXAMPLE 9.4.1 Equilibria

Find all equilibria of

$$\begin{aligned} x' &= -x + xy, \\ y' &= -y + 2xy. \end{aligned} \tag{3}$$

SOLUTION Let $x(t) = r$, $y(t) = s$, where r, s are constants. Then (3) becomes

$$0 = -r + rs = r(-1 + s), \tag{4a}$$

$$0 = -s + 2rs = s(-1 + 2r). \tag{4b}$$

Equation (4a) implies that $r = 0$ or $s = 1$. Then,

If $r = 0$, (4b) implies that $s = 0$, while

If $s = 1$, (4b) implies that $-1 + 2r = 0$, so that $r = \frac{1}{2}$.

Thus

$$x = 0, y = 0 \quad \text{and} \quad x = \frac{1}{2}, y = 1 \tag{5}$$

are the only two equilibria of (3). ◀

We now wish to determine the behavior of solutions of (1) near an equilibrium. Suppose that $x(t) = r$, $y(t) = s$ is an equilibrium. If we use the two-dimensional version of Taylor's approximations learned in calculus, we have

$$\begin{aligned} f(x, y) &= f(r, s) + f_x(r, s)(x - r) + f_y(r, s)(y - s) + \theta_1, \\ g(x, y) &= g(r, s) + g_x(r, s)(x - r) + g_y(r, s)(y - s) + \theta_2, \end{aligned} \tag{6}$$

where θ_1, θ_2 involve quadratic terms in $x - r$, $y - s$ and are thus "small" if x is close to r and y is close to s. But $f(r, s) = 0$, $g(r, s) = 0$ by (2). This suggests that, near the equilibrium, the solutions of (1) should resemble those of the **linearized differential equation,**

$$x' = f_x(r,s)(x-r) + f_y(r,s)(y-s),$$
$$y' = g_x(r,s)(x-r) + g_y(r,s)(y-s). \tag{7}$$

Letting $z = x - r$, $w = y - s$ (which translates the equilibrium point r, s to the origin $z = 0$, $w = 0$), (7) becomes

$$z' = az + bw,$$
$$w' = cz + dw, \tag{8a}$$

where a, b, c, and d are constants given by

$$a = f_x(r,s), \qquad b = f_y(r,s), \qquad c = g_x(r,s), \qquad d = g_y(r,s). \tag{8b}$$

But this is a linear homogeneous system like that discussed in the last section. This motivates, but does not prove, the following.

■ THEOREM 9.4.1 Stability of Equilibria

Suppose that (r, s) is an equilibrium point of (1). Define a, b, c, d by (8b). Let λ_1, λ_2 be the roots of the characteristic polynomial of (8a),

$$\lambda^2 - (a+d)\lambda + ad - bc.$$

Then

1. If $\lambda_1, \lambda_2 > 0$, the equilibrium is unstable and the local phase portrait resembles Figure 9.3.1.
2. If $\lambda_1, \lambda_2 < 0$, the equilibrium is asymptotically stable and the local phase portrait resembles Figure 9.3.2.
3. If λ_1, λ_2 are nonzero and of opposite signs, the equilibrium is unstable and the local phase portrait resembles Figure 9.3.3 (is a saddle point).
4. If $\lambda_1 = \alpha + \beta i$, $\lambda_2 = \alpha - \beta i$ with $\beta \neq 0$, then $\alpha > 0$ means the equilibrium is unstable and a spiral repeller (Figure 9.3.6a), while $\alpha < 0$ means that the equilibrium is asymptotically stable and a spiral attractor (Fig. 9.3.6b).
5. If $\alpha = 0$, this procedure gives no immediate information. If $\lambda_1 = \beta i$, β real, then trajectories may spiral or be "ellipselike." The equilibrium will be stable or unstable depending on the neglected nonlinear terms. ■

Local means that the picture is valid only near the equilibrium point. **Resembles** means that the axis can be bent and the picture somewhat distorted.

EXAMPLE 9.4.2 **Phase Portrait Near Equilibria**

Determine the phase portrait of (3) near the equilibria (5) found in Example 9.4.1.

Solution We have that

$$f_x = -1 + y, \qquad f_y = x, \qquad g_x = 2y, \qquad g_y = -1 + 2x. \tag{9}$$

Consider first the equilibrium $x = 0$, $y = 0$. By Theorem 9.4.1, the behavior of (3) near $(0, 0)$ is that of (8a) which is

$$\begin{aligned} z' &= (-1)z + 0w, \\ w' &= 0z + (-1)w. \end{aligned} \tag{10}$$

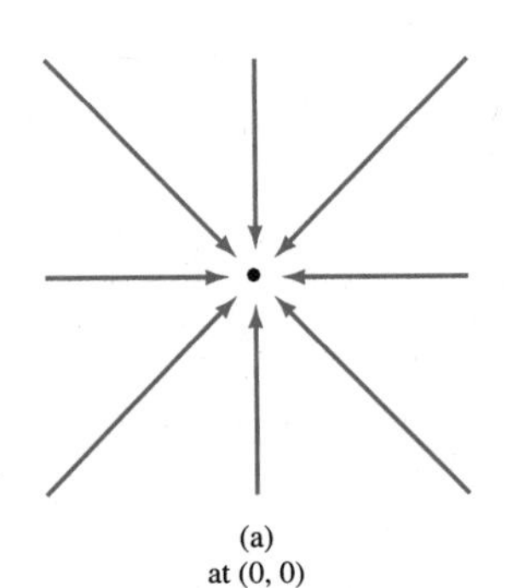

(a)
at (0, 0)

The characteristic polynomial is $(\lambda + 1)^2 = \lambda^2 + 2\lambda + 1$, with roots $-1, -1$. Thus, $(0, 0)$ is an attractor. Near $(0, 0)$ the phase portrait of (3) will resemble that of (10), which is shown in Figure 9.4.1a.

Near the other equilibrium point $x = \frac{1}{2}$, $y = 1$, the behavior of (3) is that of (8a) with $x = \frac{1}{2}$, $y = 1$, or

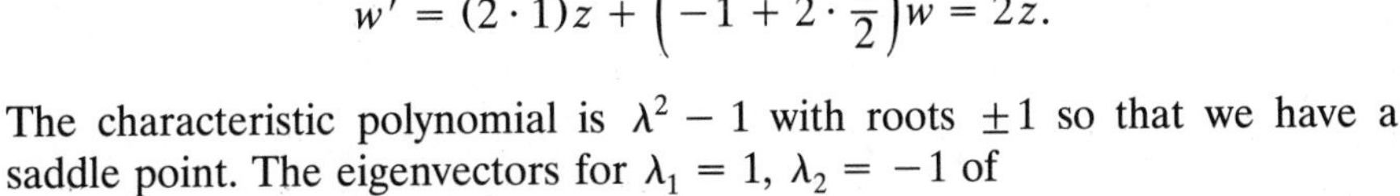

$$\begin{aligned} z' &= (-1 + 1)z + \frac{1}{2}w = \frac{1}{2}w, \\ w' &= (2 \cdot 1)z + \left(-1 + 2 \cdot \frac{1}{2}\right)w = 2z. \end{aligned}$$

The characteristic polynomial is $\lambda^2 - 1$ with roots ± 1 so that we have a saddle point. The eigenvectors for $\lambda_1 = 1$, $\lambda_2 = -1$ of

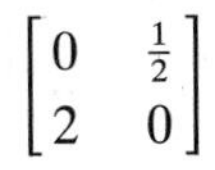

$$\begin{bmatrix} 0 & \frac{1}{2} \\ 2 & 0 \end{bmatrix}$$

are

$$\mathbf{u} = \begin{bmatrix} 1 \\ 2 \end{bmatrix} \text{ for } \lambda = 1 \qquad \text{and} \qquad \mathbf{v} = \begin{bmatrix} -1 \\ 2 \end{bmatrix} \text{ for } \lambda = -1,$$

leading to the local picture in Figure 9.4.1b. ◀

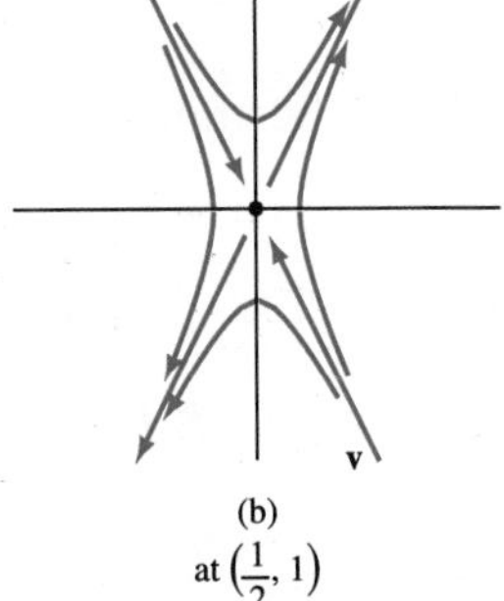

(b)
at $\left(\frac{1}{2}, 1\right)$

FIGURE 9.4.1

The actual full phase portrait for (3) is given in Figure 9.4.2.

It is important to keep in mind that Theorem 9.4.1 describes only the behavior near the equilibria as indicated in Figure 9.4.2 by the dashed rectangles. The extra detail in Figure 9.4.2, such as the trajectories first approaching $(\frac{1}{2}, 1)$ and then veering off toward $(0, 0)$, come from a more sophisticated analysis and the use of more advanced theorems. Some of these theorems and facts will be presented in the next section. Alternatively, the

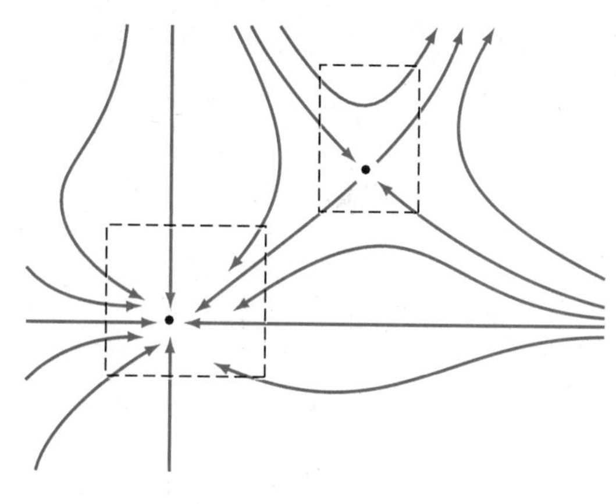

FIGURE 9.4.2

Phase portrait for (3).

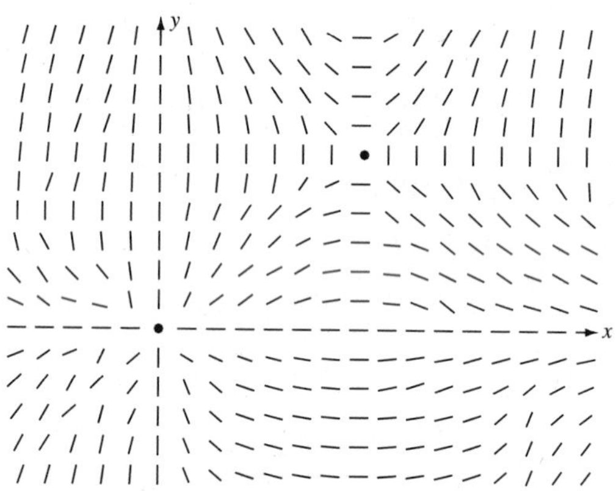

FIGURE 9.4.3

techniques of Section 1.4 can be used to help sketch the phase portrait of (3). For example, the invariant curves of (3) satisfy

$$\frac{dy}{dx} = \frac{-y + 2xy}{-x + xy}. \tag{11}$$

The slope portrait (direction field) for (11) is shown in Figure 9.4.3. Figure 9.4.3 should be compared to Figure 9.4.2.

Exercises

In Exercises 1 through 12, determine all equilibria and the behavior of solutions near them.

1. $x' = x + xy, \quad y' = y - 2xy$

2. $x' = -x + xy, \quad y' = -y + 4xy$

3. $x' = x - xy, \quad y' = y - xy$

4. $x' = 1 - x^2, \quad y' = y + 1$

5. $x' = x^2 - y^2, \quad y' = x - xy$

6. $x' = y^3 + 1, \quad y' = x^2 + y$

7. $x' = 1 - y^2, \quad y' = 1 - x^2$

8. $x' = x(1 - y^2), \quad y' = x + y$

9. $x' = x - y + x^2, \quad y' = x + y$

10. $x' = 2x - y - xy, \quad y' = x + 2y$

11. $x' = -x - 2y, \quad y' = 2x - y + xy^2$

12. $x' = -2x + y, \quad y' = -x - 2y + y^3$

Exercises 13 through 17 refer to

$$x' = x - xy + \gamma x^2,$$
$$y' = -y + xy.$$

In each case find all equilibrium points and determine the behavior near each equilibrium using Theorem 9.4.1.

13. $\gamma = -8$

14. $\gamma = -\frac{1}{3}$

15. $\gamma = \frac{1}{3}$

16. $\gamma = 1$

17. $\gamma = 8$

Exercises 18 through 22 show that, if the roots of the characteristic polynomial are pure imaginary, then a nonlinear system is different from a linear system in that there need not be a periodic orbit which is not an equilibrium. First consider the system

$$x' = -x^3 - y, \qquad y' = x - y^3. \tag{12}$$

18. Show that $x = 0$, $y = 0$ is the only equilibrium of system (12), and that the characteristic polynomial at this equilibrium is $\lambda^2 + 1$ with roots $\pm i$.

19. Let $x(t)$, $y(t)$ be a nonzero trajectory of (12). Define

$$d(t) = \left[x(t)^2 + y(t)^2\right]^{1/2},$$

which is the distance from $(x(t), y(t))$ to $(0, 0)$. Differentiate $d(t)$ with respect to t and use (12) for x', y', to conclude that $d'(t) < 0$ for all t. Thus the trajectory cannot be periodic since it always moves toward the origin.

Now consider

$$x' = x^3 - y, \qquad y' = x + y^3. \tag{13}$$

20. Show that $x = 0$, $y = 0$ is the only equilibrium of (13) and that the characteristic polynomial at this equilibrium is $\lambda^2 + 1$ with roots $\pm i$.

21. Let $(x(t), y(t))$ be a nonzero trajectory of (13). Define

$$d(t) = \left[x(t)^2 + y(t)^2\right]^{1/2},$$

which is the distance from $(x(t), y(t))$ to the origin at time t. Differentiate $d(t)$ with respect to t and use (13) for x', y' to conclude that $d'(t) > 0$ for all t. Thus the trajectory cannot be periodic since it always moves away from the origin.

22. Let d be as in Exercise 21. Show that $\lim_{t\to\infty} d(t) = +\infty$.

The function $d(t)$ in Exercises 19 through 22 is an example of a **Liapunov function**. Liapunov functions are very important in studying the stability of nonlinear systems, but we will not go into the topic in much greater detail since the construction of Liapunov (also sometimes called Lyapunov) function takes some experience. (See Section 9.5.)

Let $x' = y$ and rewrite the second-order equation $x'' = g(x, x')$ as the system

$$x' = y, \qquad y' = g(x, y). \tag{14}$$

23. a) Show that the equilibria of $x'' = g(x, x')$ are those r such that $g(r, 0) = 0$ and that in this case $(r, 0)$ is an equilibrium also of (14). Conversely, show that if (r, s) is an equilibrium of (13), then $s = 0$ and r is an equilibrium of $x'' = g(x, x')$.

b) Let $(r, 0)$ be an equilibrium of (14). Show that the characteristic polynomial at this equilibrium point is

$$\lambda^2 - g_y(r, 0)\lambda - g_x(r, 0).$$

24. Using Exercise 23, show that if $x'' = g(x)$, then Theorem 9.4.1 either guarantees the equilibrium $(r, 0)$ is a saddle point of (14) or fails to determine the behavior of trajectories near the equilibrium.

If a point (r, s) is not an equilibrium of (1), one may still use (6) to get a linear differential equation that approximates (1) as long as trajectories stay near (r, s). For each of the following systems, determine the linear system that approximates it near the given point.

25. $x' = x^2 + xy^4, \quad r = 1, \quad s = 1$
$y' = x - y^3$

26. $x' = x^2 + xy^4, \quad r = 0, \quad s = 0$
$y' = x - y^3$

27. $x' = \sin(x + y), \quad r = 0, \quad s = 0$
$y' = \cos\left(x + y + \frac{\pi}{2}\right)$

28. $x' = -xy, \quad r = 1, \quad s = -1$
$y' = x^2 + y^2$

9.5 Periodic Solutions

Oscillations are of interest in many applications. In the case of linear autonomous systems (Section 9.3), the existence of pure imaginary roots guaranteed the existence of periodic trajectories. Unfortunately, as noted in Theorem 9.4.1, and Exercises 9.4.18 through 9.4.21, this technique breaks down for nonlinear systems. Several theorems have been developed to overcome this difficulty. We shall discuss one of these.

Assumptions
Suppose that $x(t)$, $y(t)$ is a solution of

$$\begin{aligned} x' &= f(x, y), \\ y' &= g(x, y), \end{aligned} \tag{1}$$

and that $(x(t), y(t))$ is always in the interior of a *bounded* region R of the x,y phase plane. Assume also that f, g, f_x, f_y, g_x, g_y are continuous on an open set containing R.

With these assumptions what can the trajectory $x(t)$, $y(t)$ do? It will exist for all time. Intuitively, there are two possibilities. It will approach an equilibrium or wander about the region R. It turns out that in the second case it must eventually become almost periodic. The formal statement is:

■ **THEOREM 9.5.1 Poincaré-Bendixson**

Under the assumptions just described there are three possibilities for a trajectory $(x(t), y(t))$:

i. There exists an equilibrium point toward which the trajectory approaches arbitrarily close ($(0, 0)$ in Fig. 9.4.2 or Fig. 9.3.6b)
ii. The trajectory is periodic (Fig. 9.5.1)
iii. The trajectory approaches a periodic orbit asymptotically (Fig. 9.5.2). ■

There are two ways to use this theorem. The simplest is the following.

■ **THEOREM 9.5.2 Invariant Curve Without Equilibria**

Suppose that $F(x, y) = c$ gives an invariant curve for (1) that does not contain any equilibrium points. Then, for each connected component of $F(x, y) = c$, there is a solution of (1) with that trajectory. In particular, if a connected piece of $F(x, y) = c$ is a closed curve, then that trajectory will be periodic. ■

By **connected**, we mean that the set can be continuously parameterized by a variable t on a connected interval. This is sometimes referred to as being **arcwise connected**.

EXAMPLE 9.5.1 Invariant Curve Without Equilibria

Consider the nonlinear system

$$\begin{aligned} x' &= y^3, \\ y' &= -x^3. \end{aligned} \tag{2}$$

The only equilibrium of (2) is $x = 0,\ y = 0$ and Theorem 9.4.1 gives no information about the behavior near these points since the roots of the characteristic polynomial λ^2 are 0, 0.

The invariant curves satisfy

$$\frac{dy}{dx} = \frac{dy/dt}{dx/dt} = \frac{-x^3}{y^3}, \tag{3}$$

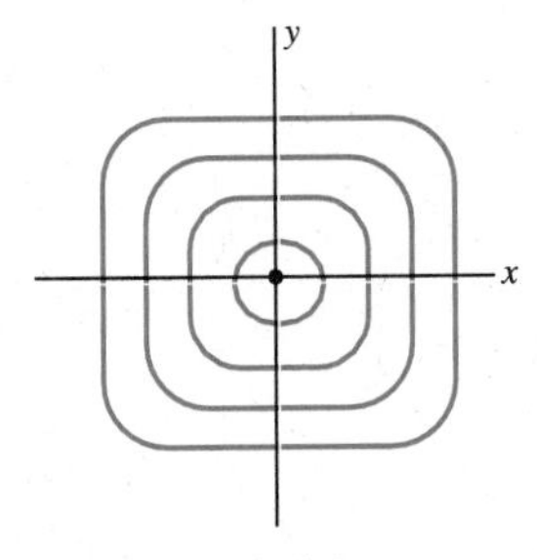

FIGURE 9.5.1

Phase portrait for system (2).

which can be solved by separation of variables as

$$\frac{y^4}{4} + \frac{x^4}{4} = c. \tag{4}$$

For any $c > 0$, the equation (4) defines a closed curve about the origin. (Note that $|x| \leq (4c)^{1/4}$, $|y| \leq (4c)^{1/4}$.) The only equilibrium $(0, 0)$ is not on the curve. Thus by Theorem 9.5.2 we have that (4) with $c > 0$ defines a periodic trajectory. Figure 9.5.1 gives the phase portrait. ◄

9.5.1 Liapunov Functions (Optional)

It is not always possible to find invariant curves as in Example 9.5.1. In that case, one can use a different approach in order to apply Theorem 9.5.1.

A set S in the phase plane is an **invariant set** for the autonomous system (1) if for any solution of (1) such that $x(t_0) = x_0$, $y(t_0) = y_0$, and $(x_0, y_0) \in S$, then $(x(t), y(t)) \in S$ for all $t \geq t_0$. That is, solutions that start in S, or enter S, stay in S. An invariant curve is a special example of an invariant set.

THEOREM 9.5.3 Invariant Set Without Equilibrium

Suppose that the assumptions of Theorem 9.5.1 hold and that (1) has a *closed* invariant set $S \subseteq R$. If S does not contain any equilibrium points, then (1) has a periodic orbit in S.

EXAMPLE 9.5.2 Invariant Set Without Equilibrium

Consider the system

$$x' = y + x^3 \cos(x^2 + y^2), \tag{5a}$$

$$y' = -x + y^3 \cos(x^2 + y^2). \tag{5b}$$

In order to concentrate on the application of Theorem 9.5.1, we shall leave some of the details to the exercises.

First, note that $x = 0$, $y = 0$ is the only equilibrium (see Exercise 16 at the end of this section) and that the roots there are $\pm i$.

Let $d(t) = x^2 + y^2$ be the bounded square of the distance of (x, y) from the origin at time t. From (5), then

$$d' = 2xx' + 2yy' = 2(x^4 + y^4)\cos(d) \tag{6}$$

Let R be the bounded region in the x, y plane:

$$R = \{(x, y): d = x^2 + y^2 < \pi\}. \tag{7}$$

This is a disk around the origin of radius $\sqrt{\pi}$. If (x, y) is close to the origin, say $d < \pi/2$, then (6) shows that $d' > 0$, and the trajectory moves farther away from the equilibrium point. On the other hand, if $\pi/2 < d < \pi$, then $d' < 0$ so that (x, y) moves closer to the origin.

Let $S = \{(x, y) \mid \pi/2 \leq d = x^2 + y^2 \leq 3/4\pi\}$. Then S is invariant and closed. Thus S contains a periodic orbit for (5) by Theorem 9.5.3. Alternatively, we can argue directly that the trajectory can neither leave R nor approach the equilibrium. Thus by Theorem 9.5.1, there is a periodic orbit. Pictorially we have Figure 9.5.2.

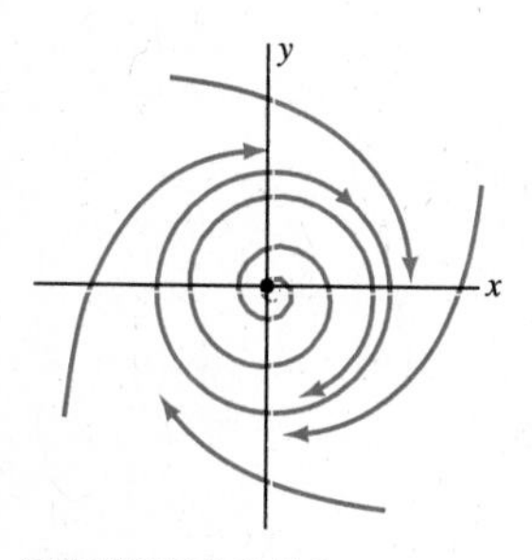

FIGURE 9.5.2

The function $d(t)$ in this example is a special case of a **Liapunov function**. In general, if $d(t)$ is a function of x, y, and $d(t) = c$ is a level curve of this function, and $d'(t) < 0$ on this level curve, then the trajectories can never cross $d(t) = c$ in the direction that would increase $d(t)$. Those level curves can then often be used to define either R or an invariant set S.

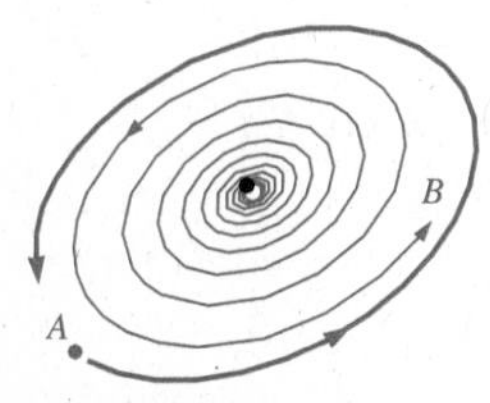

FIGURE 9.5.3

Comment Possibility i of Theorem 9.5.1 does not say the trajectory has the equilibrium point as its limit. The trajectory may just come closer and closer to the equilibrium point (trajectory B and equilibrium A in Figure 9.5.3).

Exercises

In Exercises 1 through 13, find all equilibria and determine their local behavior. Then use Theorem 9.5.2 to show the existence of periodic trajectories. Note that some of these are linear and could be done by Section 9.3, but these exercises provide practice in using Theorem 9.5.2.

1. $x' = 8y, \quad y' = -2x$
2. $x' = -2y, \quad y' = 8x$
3. $x' = 4y - 8, \quad y' = -2x + 2$
4. $x' = -2y + 6, \quad y' = 2x - 4$
5. $x' = 4y^3 - 12y^2 + 12y - 4,$

   $y' = -4x^3 + 12x^2 - 12x + 4$
6. $x' = -6y^5, \quad y' = 6x^5$
7. $x' = -4x + 10y, \quad y' = -16x + 4y$
8. $x' = 4y^3, \quad y' = -2x$
9. $x' = -4y, \quad y' = 4x^3$
10. $x' = -8y^3, \quad y' = 2x + 4x^3$
11. $x' = 8y + 32y^3, \quad y' = -5x^3$
12. $x' = 7y + 13y^3, \quad y' = -9x$
13. $x' = 6y, \quad y' = -3x - 5x^3$

To illustrate it is important to know not only the integral curves but also the equilibria, consider

$$
\begin{aligned}
x' &= -xy, \\
y' &= x^2.
\end{aligned} \tag{8}
$$

14. Show that $(0, y_0)$ is an equilibrium of (8) for any $y_0 \neq 0$.
15. Show that (8) has integral curves that are circles but that on every circle there are two equilibrium points. Conclude that there are no nonconstant periodic trajectories. Sketch the phase portrait of (8).
16. Verify that $(0, 0)$ is the only equilibrium of (5).
17. Verify that $x = a \sin t$, $y = a \cos t$ is a solution of (5) for any a such that $\cos a^2 = 0$. With this added information, sketch the phase portrait for $|x|, |y| \leq 3\sqrt{\pi}$.

Autonomous systems in other coordinate systems are also important. The remaining exercises concern systems

$$
\begin{aligned}
\frac{dr}{dt} &= f(r, \theta), \\
\frac{d\theta}{dt} &= g(r, \theta)
\end{aligned} \tag{9}
$$

with the trajectory $(r(t)\ \theta(t))$ being in polar coordinates. The existence and uniqueness theorems still hold if f, g satisfy their assumptions. However, it is now possible for trajectories to cross, since different values of (r, θ) can be coordinates of the same point.

18. Find the invariant curves for
$$
\begin{aligned}
r' &= -\sin \theta, \\
\theta' &= 1
\end{aligned} \tag{10}
$$
Sketch the curves through $(1.5, 0)$, $(2.3, 0)$, $(3.1, 0)$. (Graph with solutions in the solutions section at the end of this book.)
19. Show that if $r(0) > 2$, then the trajectory for (10) starting at $(r(0), 0)$ is periodic (in polar coordinates).
20. Show that if $0 < r(0) < 2$, then the trajectory for (10) starting at $(r(0), 0)$ crosses itself (in polar coordinates).

9.6 Population Models

One of the many interesting applications of systems has been in developing population models for interacting species. Actually the species could be chemical, economic, or other nonbiological quantities. However, we shall consider two species and speak of them as **biological species**.

Assume that there are two species whose populations at time t are given by $x(t)$, $y(t)$. In order to consider x, y as continuous functions, we must assume that the populations are fairly large. It is also often more appropriate to measure x, y in **mass units** rather than to count individuals. There are two reasons for this. First, the **biomass** of a species sometimes varies more continuously. Second, what is important for a predator is not how *many* things it eats but how *much* it eats. Nonetheless, we shall talk of the numbers of a given species. We also assume that:

> At time t, the rates of changes of the populations of x and y depend only on the number of x and y at time t and on no other factors.

Thus we get the first-order system:

$$\frac{dx}{dt} = f(x, y), \tag{1a}$$

$$\frac{dy}{dt} = g(x, y). \tag{1b}$$

f and g are the **rates of change** of x and y with respect to time. It should be stressed that merely writing down (1) has ruled out any effects due to time (seasonal variations), delay effects, and external factors such as a fluctuating food supply. However, models like (1) do provide some insight into the dynamics of populations. Two special cases will illustrate how a model might be developed.

Case 1: Two-Prey Species

In this example, consider two populations that are competing for the same limited food and shelter. In this situation, it is often reasonable to make the following assumptions (we assume $x \geq 0$, $y \geq 0$):

Assumption 1 If one species is absent, there can be no change in that population. Thus,

$$f(0, y) = 0, \qquad g(x, 0) = 0 \qquad \text{for all } x, y. \tag{2}$$

Assumption 2 Increasing the numbers of either species decreases the rate of change with respect to time of the other species. Since f_y is the rate of change of f with respect to a change in y (and similarly for g_x), this says

$$f_y(x, y) < 0, \qquad g_x(x, y) < 0 \qquad \text{if } x > 0, y > 0. \tag{3}$$

Assumption 3 If the other species is absent, and the population is small, the first species will grow. Thus

$$f(x, 0) > 0, \qquad g(0, y) > 0 \qquad \text{for small } x, y. \tag{4}$$

Assumption 4 If the population of either species is too large, the rate of change of that species is negative. That is, there is a maximum number of each species that can be supported by the available food and shelter. Thus,

$$\begin{aligned} f(x, y) &< 0 \quad \text{for large } x, \\ g(x, y) &< 0 \quad \text{for large } y. \end{aligned} \tag{5}$$

Ideally, we should now consider (1) using only these assumptions and not relying on any special choice of f, g. However, to simplify the analysis, we shall look for functions f, g that are quadratics in x, y and satisfy Eqs. (2) through (5). Intuitively, we are taking second-order approximations to $f(x, y)$, $g(x, y)$. Thus suppose

$$\begin{aligned} f(x, y) &= a_1 + a_2 x + a_3 y + a_4 x^2 + a_5 xy + a_6 y^2, \\ g(x, y) &= b_1 + b_2 x + b_3 y + b_4 x^2 + b_5 xy + b_6 y^2. \end{aligned}$$

Assumption (2) implies that

$$a_1 + a_3 y + a_6 y^2 = 0, \qquad b_1 + b_2 x + b_4 x^2 = 0 \qquad \text{for all } y, x.$$

Thus $a_1 = a_3 = a_6 = b_1 = b_2 = b_4 = 0$. Now

$$f_y = a_5 x, \qquad g_x = b_5 y.$$

Thus Assumption (3) implies that $a_5 < 0$, $b_5 < 0$. Similarly (4) implies that $a_2 > 0$, $b_3 > 0$, while (5) implies $a_4 < 0$, $b_6 < 0$. This leads to the nonlinear system:

$$x' = ax - bxy - cx^2, \tag{6a}$$
$$y' = qy - rxy - sy^2, \tag{6b}$$

where a, b, c, q, r, s are positive constants. The equilibria of (6) are given by

$$x = 0 \quad \text{or} \quad a - by - cx = 0$$

and

$$y = 0 \quad \text{or} \quad q - rx - sy = 0,$$

so that the equilibria are

$$(0, 0), \qquad \left(0, \frac{q}{s}\right), \qquad \left(\frac{a}{c}, 0\right), \qquad (\alpha, \beta),$$

where α, β is the solution of

$$\begin{aligned} cx + by &= a, \\ rx + sy &= q. \end{aligned}$$

[We assume that $cs - br \neq 0$, so that the point (α, β) exists and is unique.] By Cramer's rule (or some algebra),

$$\alpha = \frac{as - bq}{cs - br}, \qquad \beta = \frac{cq - ar}{cs - br}. \tag{7}$$

Note that if $x(t)$ satisfies

$$x' = ax - cx^2, \tag{8}$$

then $(x, 0)$ is a solution of (6), and if $y(t)$ satisfies

$$y' = qy - sy^2, \tag{9}$$

then $(0, y)$ is a solution of (6). Thus the x,y axis contains trajectories. The equations (8) and (9) are sometimes called **logistics equations**. Since the x- and y-axes contain trajectories, trajectories beginning with $x(0) > 0$ and $y(0) > 0$ must stay in the first quadrant. The only portion of the phase plane of physical interest is $x \geq 0$, $y \geq 0$. Depending on the values of the constants, the equilibrium (α, β) may not be of physical interest.

EXAMPLE 9.6.1 Two-Prey Species

Suppose that $b = q = r = s = 1$, and $c = 3$, $a = 2$. Then (6) is

$$\begin{aligned} x' &= 2x - xy - 3x^2, \\ y' &= y - xy - y^2. \end{aligned} \tag{10}$$

The equilibria are

$$(0, 0), \qquad (0, 1), \qquad (\tfrac{2}{3}, 0), \qquad \text{and} \qquad (\tfrac{1}{2}, \tfrac{1}{2}).$$

From Section 9.4 the characteristic polynomial is

$$\begin{aligned} p(\lambda) &= (\lambda - f_x)(\lambda - g_y) - f_y g_x \\ &= \lambda^2 - (2 - y - 6x + 1 - x - 2y)\lambda \\ &\quad + [(2 - y - 6x)(1 - x - 2y) - xy]. \end{aligned}$$

We determine the behavior near each equilibrium using Theorem 9.4.1.

1. $(0, 0)$ gives $p(\lambda) = \lambda^2 - 3\lambda + 2 = (\lambda - 1)(\lambda - 2)$, so that

   $$\lambda_1 = 1 > 0, \qquad \lambda_2 = 2 > 0,$$

   and the equilibrium is a repellor (unstable).
2. $(0, 1)$ gives $p(\lambda) = \lambda^2 - 1$, so that $\lambda = \pm 1$ and the equilibrium is a saddle.
3. $(\frac{2}{3}, 0)$ gives $p(\lambda) = (\lambda + 2)(\lambda - \frac{1}{3})$ with roots -2, $\frac{1}{3}$, and the equilibrium is a saddle.
4. $(\frac{1}{2}, \frac{1}{2})$ gives $p(\lambda) = \lambda^2 + 2\lambda + \frac{1}{2}$, so that $\lambda = -1 \pm 1/\sqrt{2}$ and the equilibrium is an attractor.

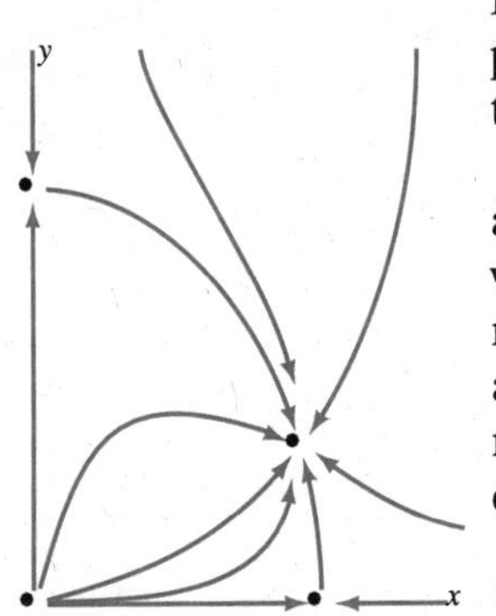

FIGURE 9.6.1

Phase portrait for (10).

By checking the directions of the different trajectories, we get the phase portrait in Figure 9.6.1. (Again, techniques besides Theorem 9.4.1 are needed to determine trajectories away from the equilibria.)

This model suggests that, for these species, the populations should approach a nonzero equilibrium. This model runs counter to the **niche theory**, which says that, given two essentially similar species, one will eventually replace the other. The reason for this difference is that the terms $-cx^2$, $-sy^2$ act as limitations on the population sizes and prevent either species from running the other one off. What happens if $c = s = 0$ is covered in the exercises. ◀

Case 2: Predator–Prey Models

Now suppose that we have a **prey** species x that is eaten by a **predator** species y. We again assume (1), (2), and $x \geq 0$, $y \geq 0$ so that

$$f(0, y) = 0, \qquad g(x, 0) = 0 \qquad \text{for all } x, y.$$

However, we now assume the following:

Assumption 5 Increasing the prey population increases the rate of change of the predator (more food yields faster growth):

$$g_x(x, y) > 0 \qquad \text{for all } x, y. \tag{11}$$

Assumption 6 Increasing the predator population decreases the rate of change of the prey:

$$f_y(x, y) < 0 \qquad \text{for all } x, y. \tag{12}$$

Assumption 7 If either population is large enough, its rate of change will be negative (there's a finite amount of food for prey and a finite amount of territory/shelter for predators):

$$f(x, y) < 0 \quad \text{for large } x \qquad g(x, y) < 0 \quad \text{for large } y. \tag{13}$$

Assumption 8 If there are no predators and a few prey, the prey increases:

$$f(x, 0) > 0 \qquad \text{for small } x. \tag{14}$$

Assumption 9 If there are no prey, the predator decreases:

$$g(0, y) < 0 \qquad \text{for all } y. \tag{15}$$

If we now seek a quadratic equation for f, g satisfying these assumptions, we get, after some calculation (see Exercise 17 at the end of this section), the model

$$\begin{aligned} x' &= ax - bxy - cx^2, \\ y' &= -qy + rxy - sy^2, \end{aligned} \tag{16}$$

with a, b, c, q, r, s positive constants. Again the equilibria are given by

$$x = 0 \quad \text{or} \quad a - by - cx = 0,$$

and

$$y = 0 \quad \text{or} \quad -q + rx - sy = 0,$$

which yields

$$(0,0), \quad \left(0, -\frac{q}{s}\right), \quad \left(\frac{a}{c}, 0\right), \quad \text{and} \quad (\alpha, \beta),$$

where α, β satisfy

$$\begin{aligned} c\alpha + b\beta &= a, \\ r\alpha - s\beta &= q, \end{aligned}$$

so that

$$\alpha = \frac{as + bq}{cs + br}, \qquad \beta = \frac{ra - cq}{cs + br},$$

assuming that $cs + br \neq 0$.

EXAMPLE 9.6.2 Predator–Prey

Suppose that $b = q = r = s = 1$, $c = 2$, $a = 3$, so that (16) is

$$\begin{aligned} x' &= 3x - xy - 2x^2, \\ y' &= -y + xy - y^2. \end{aligned}$$

The equilibria are

$$(0,0), \quad (0, -1), \quad \left(\frac{3}{2}, 0\right), \quad \left(\frac{4}{3}, \frac{1}{3}\right).$$

The $(0, -1)$ equilibrium is nonphysical because of the negative y value. The characteristic polynomial for the linearized differential equation at each point (x, y) is

$$\begin{aligned} &\lambda^2 - [(3 - y - 4x) + (-1 + x - 2y)]\lambda \\ &+(3 - y - 4x)(-1 + x - 2y) - (-x)(y). \end{aligned}$$

Thus, by Theorem 9.4.1, at each equilibrium we have:

1. $(0,0)$ gives $p(\lambda) = \lambda^2 - 2\lambda - 3 = (\lambda - 3)(\lambda + 1)$ and $\lambda_1 = -1$, $\lambda_2 = 3$, so that $(0,0)$ is a saddle.
2. $(\frac{3}{2}, 0)$ gives $p(\lambda) = \lambda^2 + \frac{5}{2}\lambda - \frac{3}{2}$, so that $\lambda_1 = \frac{1}{2}$, $\lambda_2 = -3$, which also gives a saddle.

3. $(\frac{4}{3}, \frac{1}{3})$ gives $p(\lambda) = \lambda^2 + 3\lambda + \frac{4}{3}$ so that $\lambda_1 \approx -0.54$, $\lambda_2 \approx -2.45$, and the equilibrium is an attractor.

A complete phase portrait is Figure 9.6.2. ◄

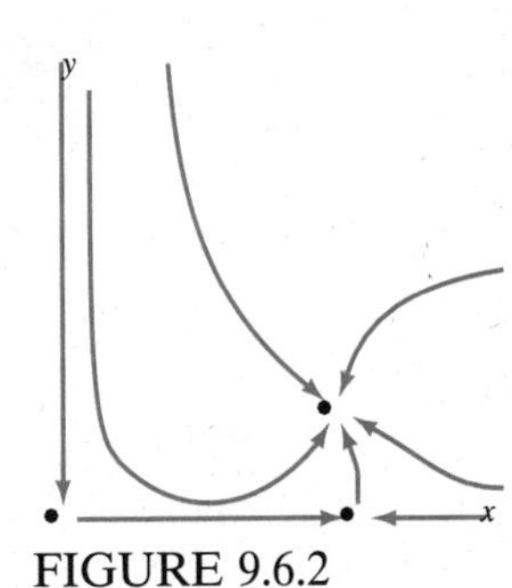

FIGURE 9.6.2

Phase portrait of Example 9.6.2.

As will be shown in the exercises, if all resource limitations are removed from (16) (take $c = 0$, $s = 0$), then periodic oscillations exist.

In the predator–prey model (16) there is a tendency by some to refer to the xy term as the **predation term**. However, it is really not exactly a predation term. Suppose that $-bxy$ was the loss rate of (say) rabbits and $+sxy$ was a gain (in weight) for foxes. This would imply that, given a *fixed* number (or weight) of foxes and any number of rabbits, doubling the number of rabbits would double the amount of predation. This is clearly nonsense. A given number of foxes can eat only so many rabbits per day. Thus, as x increases, the loss of rabbits due to just predation would approach a multiple of y, the number of foxes. Typical functions used in the biological literature are

$$-b\frac{x}{1+x}y, \qquad -b\frac{x^2}{1+x^2}y.$$

These functions are called **uptake functions**. The effect of trying to actually model predation is examined in Exercises 12 through 16 at the end of this section.

The key thing to remember is that one is often better off thinking of $f(x, y)$, $g(x, y)$ as **representative** functions satisfying the right sort of change conditions [such as Assumptions 1 through 4 under Case 1, Eqs. (2) through (5)] rather than trying to impart special biological significance directly to each term.

Exercises

In Exercises 1 through 17, the x- and y-axes contain trajectories and you are only to consider the trajectories for $x \geq 0$, $y \geq 0$.

Two-Prey Models

In Exercises 1 through 3, determine the behavior near all equilibria.

1. In Example 9.6.1, decrease the limitation term on x by making $c = 2$, so that

$$a = c = 2, \qquad q = r = s = b = 1.$$

2. In Example 9.6.1, increase the effect of competition on x by changing b to 2. Thus

$$a = b = 2, \qquad c = 3, \qquad q = r = s = 1.$$

This makes the competition between x and y hurt x more than y.

3. In Example 9.6.1, reduce the inherent growth rate of x by changing a to 1, increase the

effect of competition on x by changing b to 2, and increase the resource limitations on x by changing c to 4, so that $b = 2$, $c = 4$, and $a = q = r = s = 1$.

4. (Continuation of Exercise 3.) Assuming that, in fact, $(0, 1)$ is an attractor for the entire open first quadrant ($x > 0$, $y > 0$), sketch the phase portrait. Explain in biological terms what this represents.

5. In the system (6), if there is ample food and shelter, then one might set $c = 0$, $s = 0$. Show that there are only two equilibria, $(0, 0)$ and $(q/r, a/b)$, which are a repellor and a saddle, respectively.

6. (Continuation of Exercise 5.) Sketch the phase portrait for Exercise 5, and explain why this is compatible with the niche theory. Considering Exercises 1, 4, 5, and Example 9.6.1, what do you think are the limitations and possible truth in the niche theory?

Predator–Prey Problems

In Exercises 7 through 10, determine the behavior near all physically meaningful equilibria.

7. In Example 9.6.2, effectively decrease the available prey by increasing c to 6, so that $q = r = s = b = 1$, $a = 3$, $c = 6$.

8. In Example 9.6.2, remove the limitation on prey population by setting $c = 0$, so that $q = r = s = b = 1$, $a = 3$, $c = 0$.

9. In Example 9.6.2, remove the limitation on prey and predator populations by setting $c = s = 0$, so that $q = r = b = 1$, $a = 3$. Show that periodic trajectories exist.

10. In Example 9.6.2, increase the effect of the predator on the prey by increasing b to 6, so that $q = r = s = 1$, $a = 3$, $c = 2$, $b = 6$.

11. In Exercises 7 through 10, express the change in the phase portrait in biological terms.

Consider a predator–prey model in which we assume that there is unlimited food and shelter available, so that, left to itself, the prey will grow exponentially. Assume that, without the prey, the predator will die out slowly. [Perhaps there is an alternative food supply, but it lacks the proper nutrients.] Instead of a constant multiple of xy, use the more realistic uptake function discussed in this section, so that the model is

$$\begin{aligned} x' &= ax - b\left(\frac{x}{1+x}\right)y, \\ y' &= -qy + r\left(\frac{x}{1+x}\right)y. \end{aligned} \tag{17}$$

Note that again the x, y axes contain trajectories and we need only consider $x \geq 0$, $y \geq 0$.

12. Show that if $q = r$, then $(0, 0)$ is the only equilibrium of (17). Show that if $q \neq r$, then

$$(0, 0) \quad \text{and} \quad \left(\frac{q}{r-q}, \frac{ar}{b(r-q)}\right)$$

are equilibria of (17).

13. Show that, if $q > r$ in (17), then the predator y dies out and the prey population x goes to infinity if $x(0) > 0$, $y(0) > 0$.

For Exercises 14 through 16, determine the physically meaningful equilibria, and the behavior of solutions near them, and sketch the phase portrait.

14. Let $a = b = 1$, $q = 2$, $r = 1$ in (17).

15. Lower the predator death rate in (17) by setting $a = b = 1$, $q = 0.5$, $r = 1$.

16. Increase the predator uptake rate in (17) by setting $a = b = 1$, $q = 1$, $r = 2$.

17. Suppose $f(x, y)$, $g(x, y)$ are quadratic functions. Verify that Eqs. (11) through (15) imply that (1) has the form of (16).

All of the systems examined in this section have had the x- and y-axes as invariant curves. This need not always be the case, as the remaining exercises in this section show. Consider a species that has an immature and an adult phase. Let x be the number of immature organisms and y the number of adults. Assume that each group suffers a death rate that is proportional to its size. Also assume that each group produces members of the other group at a rate proportional to its own size (immatures grow up and adults give birth to immatures). Suppose also that limitations on food and shelter only affect the adults.

18. Explain why

$$x' = -ax + by,$$
$$y' = cx - dy - ey^2 \tag{18}$$

with a, b, c, d, e positive constants might be a reasonable model for this species made up of immatures and adults.

19. a) Show that the coordinate axes are not invariant curves for (18).

b) Show that $R = \{(x, y): x \geq 0, y \geq 0\}$ and $\tilde{R} = \{(x, y): x > 0, y > 0\}$ are invariant sets for (18).

c) Explain why (a) and (b) are biologically reasonable.

In Exercises 20 through 22, find all physical ($x \geq 0$, $y \geq 0$) equilibria and sketch the phase portrait for $x \geq 0$, $y \geq 0$. Note that these exercises differ only in the adult death rate.

20. Let $b = 2$, $a = c = e = d = 1$ in (18).

21. Let $d = b = 2$, $a = c = e = 1$ in (18).

22. Let $b = 2$, $d = 3$, $a = c = e = 1$ in (18).

9.7 Nonlinear Circuits

As noted earlier, many electrical devices such as diodes, transistors, etc., are inherently nonlinear. Often they can be modeled by a combination of capacitors, inductors, and nonlinear current-controlled resistors. (See Section 2.11.) As pointed out in Section 9.1, the circuit in Figure 9.7.1 may be modeled by the system

$$q' = i,$$
$$i' = \frac{-q}{CL} - \frac{f(i)}{L} + \frac{e}{L}, \tag{1}$$

where we assume that $f(i)$ is a differentiable function of i and e is a **constant** voltage source. (Note Exercises 21 and 22 at the end of this section.) Since Figure 9.7.1 is equivalent to Figure 9.7.2, we shall assume also throughout this section that

$$f(0) = 0. \tag{2}$$

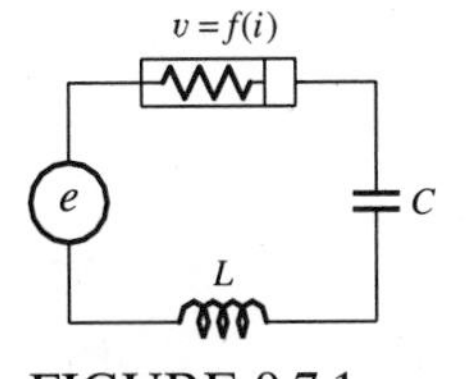

FIGURE 9.7.1

If we take $\hat{q} = q - Ce$, then system (1) becomes

$$\hat{q}' = i = h(\hat{q}, i),$$
$$i' = -\frac{\hat{q}}{CL} - \frac{f(i)}{L} = k(\hat{q}, i). \tag{3}$$

Hence the effect of the voltage source is merely to shift the entire phase portrait over Ce units in the q direction. Thus we may assume for the moment that $e = 0$. We are led then to consider (3) or, equivalently, (1) with $e = 0$. The only equilibrium of (3) is

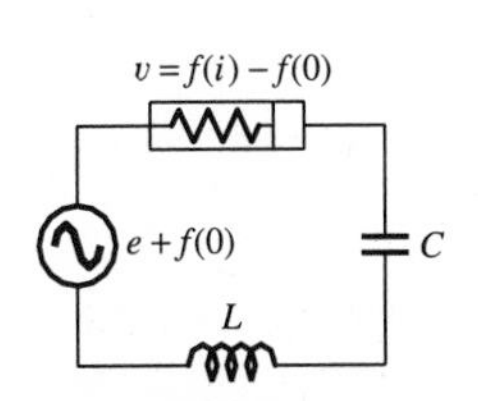

FIGURE 9.7.2

$$\hat{q} = 0, \qquad i = 0. \tag{4}$$

To determine the behavior near this equilibrium, by the method of Section 9.4, compute the eigenvalues of the coefficient matrix of the linearization of

(3) at $(0,0)$, that is

$$\begin{bmatrix} h_{\hat{q}}(0,0) & h_i(0,0) \\ k_{\hat{q}}(0,0) & k_i(0,0) \end{bmatrix} = \begin{bmatrix} 0 & 1 \\ -\dfrac{1}{CL} & -\dfrac{f'(0)}{L} \end{bmatrix}.$$

The characteristic polynomial,

$$p(\lambda) = \lambda\left(\lambda + \frac{f'(0)}{L}\right) + \frac{1}{CL} = \lambda^2 + \frac{f'(0)\lambda}{L} + \frac{1}{CL}, \tag{5}$$

has roots

$$\lambda = -\frac{f'(0)}{2L} \pm \sqrt{\frac{f'(0)^2}{4L^2} - \frac{1}{CL}}. \tag{6}$$

From (6) and Theorem 9.4.1, we have the following theorem.

■ **THEOREM 9.7.1 Equilibria of Nonlinear RLC Circuit**

Let $v = f(i)$ be the v–i characteristic for the nonlinear resistor in Figure 9.7.1. Then there is one equilibrium given by (4), which is $(0,0)$ and it is

i. An attractor if $f'(0) > 0$,
ii. A repellor if $f'(0) < 0$.

Furthermore, solutions near the equilibrium spiral about the equilibrium if $f'(0)^2 < 4L/C$. They do not spiral if $f'(0)^2 > 4L/C$. If $f'(0) = 0$, the solution near the origin will either spiral while near the origin, or be periodic around the origin. ■

It is of interest to see if periodic solutions can exist. To do so, we shall use a **Liapunov function** (Section 9.5) since it is too difficult, in general, to find invariant curves. Those who skipped Liapunov functions may wish to move directly to the exercises.

Suppose that (i, q) is a solution of (1) and assume $f(0) = 0$, $e = 0$. Let

$$F = i^2 + \frac{1}{CL}q^2. \tag{7}$$

Using (1) we find that

$$\frac{dF}{dt} = -2\frac{if(i)}{L}. \tag{8}$$

We shall now verify the following theorem.

■ **THEOREM 9.7.2 Stability of Equilibria**

Let $f(i)$ be as in Theorem 9.7.1 with $f(0) = 0$, $e = 0$.

a. If $if(i) \geq 0$ for i near zero, then the equilibrium is stable. That is, solutions starting near it stay close to it. If $if(i) > 0$ for i near zero and

$i \neq 0$, then there is a region about the equilibrium that contains no periodic trajectories.

b. If $if(i) < 0$ for i near zero and $i \neq 0$, then there is a region about the equilibrium that contains no periodic trajectories. ■

Verification Note that if K is a nonzero constant, then $F = K$ defines a family of ellipses in the i, q plane about the equilibrium. If $if(i) \geq 0$ for i near zero, we have that $F' \leq 0$ along any trajectory near the equilibrium. Thus the trajectory must either stay on that ellipse or move to an ellipse closer to the equilibrium.

If $if(i) < 0$, then the solutions must move out to a further ellipse since $F' > 0$ and thus the solutions cannot be periodic.

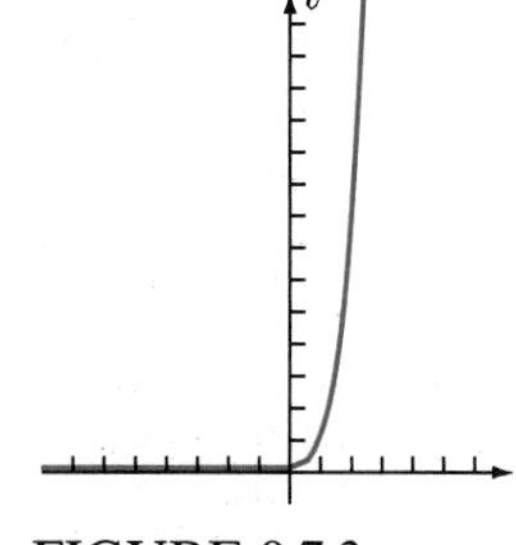

FIGURE 9.7.3

EXAMPLE 9.7.1 Phase Portrait for Nonlinear Resistor

Suppose that

$$f(i) = \begin{cases} 0 & i \leq 0, \\ i^3 & i \geq 0. \end{cases}$$

Determine the phase portrait for (1) if $C = 1$, $L = 1$, $e = 0$.

Solution The graph of $f(i)$, given in Figure 9.7.3, is similar to the characteristic curve of some diodes. The equilibrium is $q = 0$, $i = 0$. From (6), the roots are $\pm\sqrt{-1}$ since $f'(0) = 0$. Thus, by Theorem 9.7.1, near the equilibrium the trajectories will move around the equilibrium.

Now if F is as in (7), then, by (8),

$$F'(t) = \begin{cases} 0 & i \leq 0, \\ -2i^4 & i \geq 0. \end{cases}$$

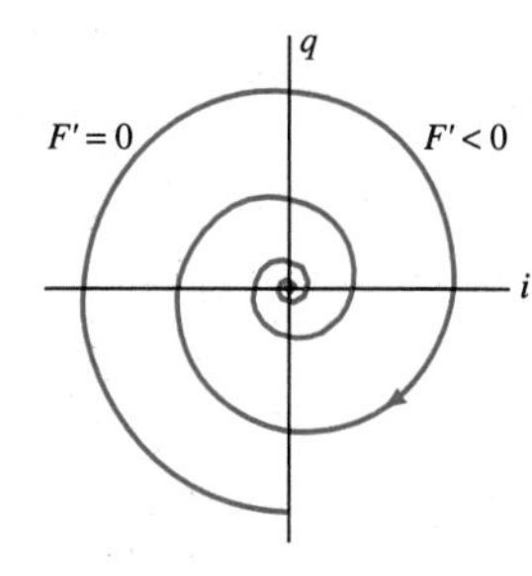

FIGURE 9.7.4

Thus, while the solution has $i \leq 0$, it keeps F constant and hence lies on an ellipse, while if $i > 0$, then F decreases. The result is that $(0, 0)$ is a type of spiraling attractor, as shown in Figure 9.7.4. In fact, Figure 9.7.4 represents the **global behavior**. That is, solutions spiral into the origin from any point in the i, q plane. ◄

Exercises

In Exercises 1 through 16, determine the behavior near the equilibrium $i = 0$, $q = 0$ for the circuit (1) with $C = 1$, $L = 1$, $e = 0$, and the indicated v–i characteristic.

1. $f(i) = 2i$

2. $f(i) = -3i$

3. $f(i) = \sin i$

4. $f(i) = \sin 3i$

5. $f(i) = 1 - e^i$

6. $f(i) = e^{-i} - 1$

7. $f(i) = \sinh 4i$

8. $f(i) = 6i + i^3$

9. $f(i) = 7i^3 - 4i$

10. $f(i) = i^3$

11. $f(i) = 3i^3 + i^5$

12. $f(i) = -2i + i^4$

13. $f(i) = i^5$

14. $f(i) = -i^5$

15. $f(i) = \begin{cases} 0 & i \geq 0, \\ i^2 & i < 0 \end{cases}$

16. $f(i) = \tan^{-1} i$

If $f(i) = \alpha i^2 + \beta$ for constants α, β, then it is possible to determine invariant curves by the integrating-factor method of Section 1.14. In Exercises 17 through 20, determine invariant curves for (1) if $e = 0$, $C = 1$, $L = 1$, and:

17. $f(i) = i^2$

18. $f(i) = -i^2$

19. $f(i) = 3i^2$

20. $f(i) = -3i^2$

Exercises 21 and 22 show how the analysis of equilibria can be used to study the dynamic case. The idea is as follows: Suppose that e is a constant in (1) and that there is an equilibrium (q_e, i_e) of (1), which is an attractor. (The subscript e is not denoting a partial derivative here.) If we change e by a little bit and hold it constant, then (q, i) should go to the new equilibrium. Intuitively, if $e(t)$ varies slowly, then a solution $(i(t), q(t))$ should stay close to the points $(q_{e(t)}, i_{e(t)})$. To simplify the exposition consider

$$\begin{aligned} x' &= -x + u(t), \\ y' &= -2y + u(t). \end{aligned} \tag{9}$$

Of course, (9) can be solved if $u(t)$ is known, but the ideas involved work for a general system. Suppose that $u(t)$ is constant; $u(t) \equiv u_0$. Then (9) has a single equilibrium $x_u = u_0$, $y_u = u_0/2$.

21. Show that the phase portrait for (9) at the equilibrium is given by Figure 9.7.5 if $u(t) \equiv u_0$.

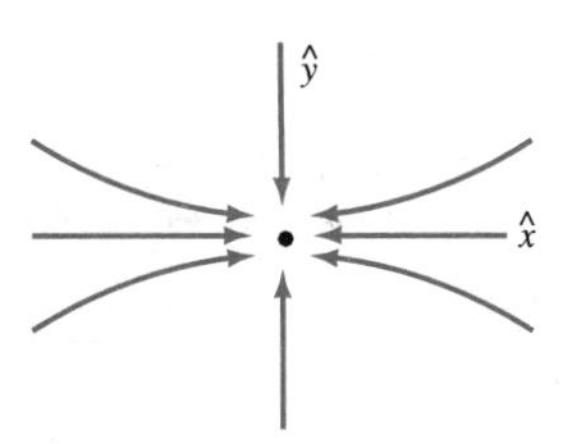

FIGURE 9.7.5

$\hat{x} = x - 1,$
$\hat{y} = y - 0.5.$

22. Suppose now that we consider (9) for $0 \leq t \leq 20$ with $x(0) = 1$, $y(0) = 0.5$, and

$$u(t) = \begin{cases} -1, & 0 \leq t < 10, \\ 1, & 10 \leq t \leq 20. \end{cases}$$

For $0 \leq t < 10$, (x, y) moves toward the "equilibrium" $(-1, -0.5)$. However, for $10 \leq t < 20$, the solution (x, y) moves toward the equilibrium $(1, 0.5)$. On the basis of Figure 9.7.5, we would expect the resulting motion to look like the trajectory in Figure 9.7.6. Verify this by solving (9) explicitly and plotting the resulting trajectory in the *xy* plane.

23. Theorem 9.7.2 is closely related to Theorem 9.7.1. Suppose that f is continuous and $f(0) = 0$.

a) Show that $if(i) \geq 0$ for i near zero and f differentiable implies $f'(0) \geq 0$.

b) Show that $if(i) \leq 0$ for i near zero and f differentiable implies $f'(0) \leq 0$.

c) Show that $f'(0) > 0$ implies $if(i) \geq 0$ for i near zero.

d) Show that $f'(0) < 0$ implies $if(i) \leq 0$ for i near zero.

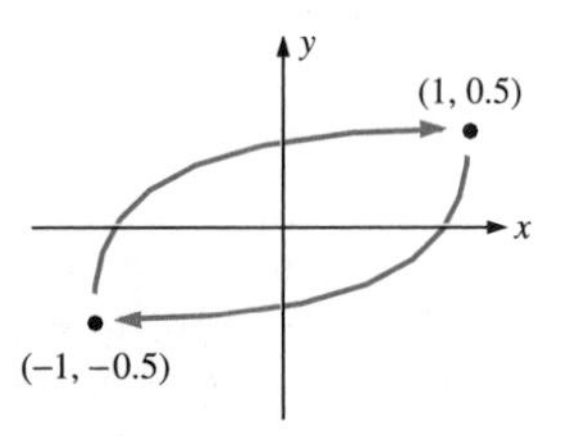

FIGURE 9.7.6

9.8 Mechanical Systems

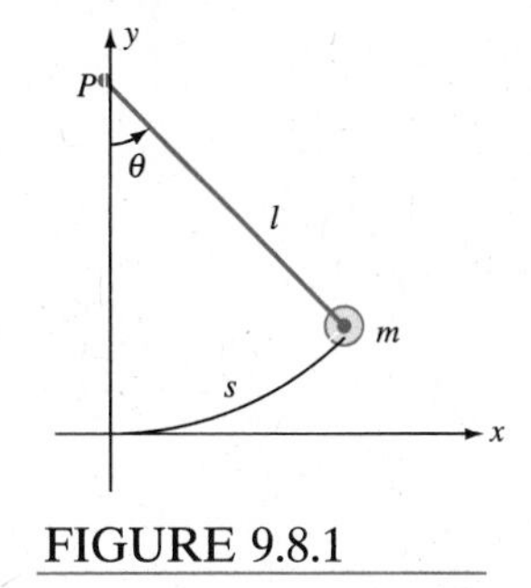

FIGURE 9.8.1

Most mechanics problems are linear only if one assumes that there are "small displacements" or "limited variations in velocity." In this section we shall analyze one nonlinear problem. An additional problem is considered in the exercises at the end of the section. The problem to be considered is a rigid pendulum of length l. (See Figure 9.8.1.) It is free to rotate around the point P: $(0, l)$. We suppose that the mass of the arm of the pendulum is negligible with respect to the mass m at the end of the pendulum. The mass moves along a circle, which we denote C.

Let (x, y) be the location of the mass at time t, s the distance along the circle C from the origin to the point (x, y), and θ the angle the pendulum makes with the y-axis. At a given s, the velocity along C is ds/dt, and it is in a direction tangential to the circle C, which the mass is following. The component of the force of gravity in this direction is $-mg \sin \theta$. As in Section 2.8 we assume that resistance is linear and in the opposite direction to the velocity, so that

$$\text{Force total} = \text{gravity} + \text{damping},$$

or

$$ms'' = -mg \sin \theta - \beta s'. \tag{1}$$

If θ is measured in radians, then $s = l\theta$ and equation (1) becomes $ml\theta'' = -mg \sin \theta - \beta l\theta'$ or

$$ml\theta'' + \beta l\theta' + mg \sin \theta = 0. \tag{2}$$

If θ is small, then $\sin \theta \approx \theta$, and (2) is often approximated by the linear constant coefficient equation

$$ml\theta'' + \beta l\theta' + mg\theta = 0, \tag{3}$$

which we studied in Chapter 2. For $\beta = 0$ we find that (3) predicts oscillations, with frequency $\frac{1}{2\pi}\sqrt{g/l}$ as shown in Section 2.8. We shall analyze the nonlinear differential equation (2). Let $\gamma = \beta/m$, $\delta = g/l$, $\phi = \theta'$; then (2) can be rewritten as the system

$$\phi' = -\gamma\phi - \delta \sin \theta, \tag{4a}$$

$$\theta' = \phi. \tag{4b}$$

The frictionless case $\beta = \gamma = 0$ will be analyzed here. The $\beta \neq 0$ case is covered in the exercises at the end of this section. If $\beta = 0$, the equations (4) become:

$$\phi' = -\delta \sin \theta = f(\phi, \theta), \tag{5a}$$

$$\theta' = \phi = g(\phi, \theta). \tag{5b}$$

The equilibria are $\phi = 0$, $\sin\theta = 0$, or

$$\phi = 0, \qquad \theta = n\pi, \qquad n = 0, \pm 1, \pm 2, \ldots . \tag{6}$$

Note that, in (4), the change of variables $\hat{\phi} = \phi$, $\hat{\theta} = \theta + 2\pi$, again produces the same system of differential equations. Thus, the phase portrait will repeat itself every 2π units in θ. In particular, the equilibria (ϕ_0, θ_0) and $(\phi_0, \theta_0 + 2n\pi)$ will look the same. To use Theorem 9.4.1 to determine the behavior near the equilibria, we need the eigenvalues of the linearized coefficient matrix

$$\begin{bmatrix} f_\phi & f_\theta \\ g_\phi & g_\theta \end{bmatrix} = \begin{bmatrix} 0 & -\delta\cos\theta \\ 1 & 0 \end{bmatrix};$$

that is, the roots of the characteristic polynomial

$$\lambda^2 + \delta\cos\theta. \tag{7}$$

There are two cases:

Case 1

$\phi = 0$, $\theta = (2m+1)\pi$, m an integer. Then the characteristic polynomial is $\lambda^2 - \delta$ with real roots $\pm\sqrt{\delta}$ of opposite sign. These equilibria, which correspond to the pendulum sticking straight up, are unstable (saddles).

Case 2

$\phi = 0$, $\theta = 2m\pi$, m an integer. Then the characteristic polynomial is $\lambda^2 + \delta$. The roots are $\pm i\sqrt{\delta}$. This suggests the possibility of, but does not prove that, there are periodic oscillations. As noted in Theorem 9.4.1, it does show that there are trajectories that circle the equilibrium for a while. To determine what these equilibria look like, we shall determine the invariant curves. From (4) with $\beta = 0$,

$$\frac{d\phi}{d\theta} = -\frac{\delta\sin\theta}{\phi}.$$

Solving by separation of variables yields the invariant curves

$$\frac{\phi^2}{2} = \delta\cos\theta + c. \tag{8}$$

If $(\phi(t), \theta(t))$ is a trajectory near the origin $(0, 0)$, we have at $t = 0$

$$c = \frac{\phi^2(0)}{2} - \delta\cos(\theta(0))$$

and c is negative if $\phi^2(0) < 2\delta\cos\theta(0)$. Suppose then that $c = -K^2$, where $K \neq 0$.

To show that the solutions passing through these $(\theta(0), \phi(0))$ are periodic, we shall use Theorem 9.5.2, and show that the invariant curve is a simple closed curve not containing an equilibrium. We already know that it does not contain an equilibrium since $K^2 > 0$. Suppose, then, that we have

$$\frac{\phi^2}{2} = \delta \cos \theta - K^2$$

and $K^2/\delta < 1$. Then

$$\phi = 0 \qquad \text{for } \theta = \cos^{-1}\left(\frac{K^2}{\delta}\right).$$

As θ increases, there are two values of ϕ,

$$\phi = \pm(2\delta \cos \theta - 2K^2)^{1/2},$$

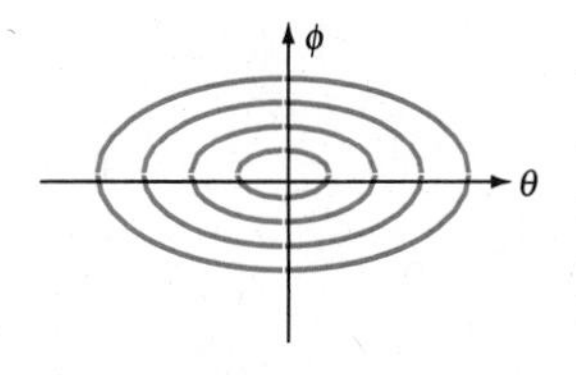

FIGURE 9.8.2

until θ is the next solution of $\cos \theta = K^2/\delta$, when $\phi = 0$ again. The resulting curves, shown in Figure 9.8.2, are simple closed curves. Thus every trajectory near enough to the origin is periodic, by Theorem 9.5.2. The actual phase portrait is shown in Figure 9.8.3.

Trajectories A in Figure 9.8.3 correspond to the situation when the pendulum is given enough initial velocity to keep swinging around the pivot point P. Note that it slows down as the mass approaches the top $[\theta = (2n + 1)\pi]$ and speeds up as the mass swings down. The trajectories B correspond to when the conditions are just right so that the pendulum approaches the unstable equilibrium. The trajectories B separate the trajectories C where the pendulum oscillates about the stable equilibrium from those like A, where it keeps circling the pivot point.

FIGURE 9.8.3

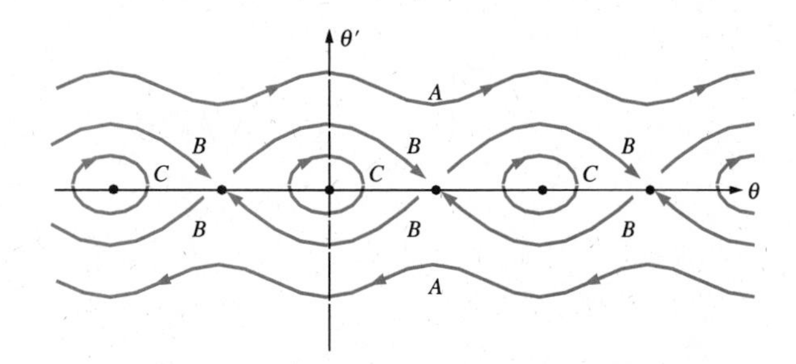

Exercises

For Exercises 1 through 4, suppose, in the pendulum problem, that there is a resistive force of $-\beta s'$. In Exercises 1 through 3, determine the nature of the equilibria under the given conditions.

1. $l = 32$ ft, $m = 1$ slug, $\beta = 1$ slug/s

2. $l = 32$ ft, $m = \frac{1}{2}$ slug, $\beta = 1$ slug/s

3. $l = 32$ ft, $m = \frac{1}{3}$ slug, $\beta = 1$ slug/s

4. Determine the values of β for which $(0, 0)$ is:

i) A spiral attractor;

ii) An attractor that does not spiral;

iii) Interpret the answers to (i) and (ii) in terms of underdamping and overdamping.

Rather than thinking of resistance as due to the movement of the mass through some medium, let us consider the pendulum as being attached at point P (perhaps by a coupling with ball bearings) to a shaft. Suppose that the resistance is proportional to the angular velocity of the pendulum relative to the shaft. If the shaft is not moving, then the resistance term is again a constant multiple of θ'. Suppose that the shaft is rotating at a fixed velocity of r rad/s. Then our model is

$$ml\theta'' + k(\theta' - r) + mg \sin \theta = 0. \tag{9}$$

5. Rewrite (9) as a system with variables θ, $\phi = \theta'$.

In Exercises 6 through 8, determine the equilibria and phase portrait for (9) under the given conditions. Try to physically interpret your results. Units are ft-slug-rad-s.

6. $k = 2$, $l = 2$, $m = 1$, $r = 8$

7. $k = 2$, $l = 2$, $m = 1$, $r = 16$

8. $k = 2$, $l = 2$, $m = 1$, $r = 32$
(*Hint*: Show that for any initial condition, eventually $15 \le \phi \le 49$ and $\theta \to \infty$.)

This section considered an equation of the form $x'' + \gamma x' + f(x) = 0$ while Section 9.7 considered $x'' + f(x') + \gamma x = 0$. More complicated circuits and mechanical systems can take the more general form

$$x'' + f(x, x') = 0. \tag{10}$$

9. Let $y = x'$ and rewrite (10) as a system in (x, y).

a) Show that the only equilibria of this system are of the form $(c, 0)$ where $x(t) \equiv c$ is an equilibrium of (10).

b) Show that the characteristic polynomial of the linearized coefficient matrix at the equilibrium $(c, 0)$ is $\lambda^2 + \lambda f_y(c, 0) + f_x(c, 0)$.

CHAPTER

10

Fourier Series

10.1 Orthogonal Functions

Until now we have considered only initial value problems for ordinary differential equations. Chapter 11 will discuss partial differential equations (PDEs) and present the fundamental technique of separation of variables (which is not the same separation of variables as in Section 1.4). The technique of separation of variables will utilize the properties of boundary value problems and Fourier series developed in this chapter.

There are many types of Fourier series. The simplest, and the ones we discuss first, are the trigonometric Fourier series. An example is the Fourier sine series

$$\sum_{n=1}^{\infty} \frac{1}{n} \sin nx = \sin x + \frac{1}{2} \sin 2x + \frac{1}{3} \sin 3x + \cdots. \tag{1}$$

We shall see that Fourier series have important physical interpretations in applications. However, Fourier series are based on a different type of theory from the more familiar power series of Chapter 4. This section will introduce some of the basic concepts needed to understand and work with Fourier series.

The reader who is familiar with inner (also known as scalar or dot) products of vectors will notice that that theory is similar in many ways to the one we now develop.

Throughout this chapter we assume that functions are piecewise differentiable unless stated otherwise. Recall from Section 3.1 that a **piecewise continuous** function is continuous except for jump discontinuities and that there can be only a finite number of jump discontinuities in a finite interval. A function f is **piecewise differentiable** if f and f' are piecewise continuous.

Equivalently, a piecewise differentiable function $f(x)$ is a function such that on any finite interval it can be broken into a finite number of pieces, each of which is continuous and has a continuous derivative. In addition, the only discontinuities of $f(x)$ and $f'(x)$ are jump discontinuities.

EXAMPLE 10.1.1 Not Piecewise Differentiable Function

Determine if $f(x) = x^{1/3}$ is piecewise differentiable on $[-1, 1]$.

SOLUTION $f(x) = x^{1/3}$ is continuous for all x. However, $f'(x) = \frac{1}{3}x^{-2/3}$ is not continuous at $x = 0$. The left- and right-hand limits of $f'(x) = \frac{1}{3}x^{-2/3}$ are not finite at $x = 0$. Thus $x = 0$ is not a jump discontinuity of $f'(x)$, and hence $f(x)$ is not piecewise differentiable. ◄

EXAMPLE 10.1.2 Piecewise Differentiable Function

Show that $f(x) = |x|$ is piecewise differentiable.

SOLUTION $f(x)$ is continuous for all x. Also

$$f'(x) = \begin{cases} -1 & \text{for } x < 0, \\ 1 & \text{for } x > 0 \end{cases}$$

is continuous everywhere except at $x = 0$. At $x = 0$ we have

$$\lim_{x \to 0^-} f'(x) = -1, \qquad \lim_{x \to 0^+} f'(x) = 1,$$

and $x = 0$ is a jump discontinuity of $f'(x)$. Thus $f(x)$ is piecewise differentiable. ◄

Figure 10.1.1 shows a more general function that is piecewise differentiable.

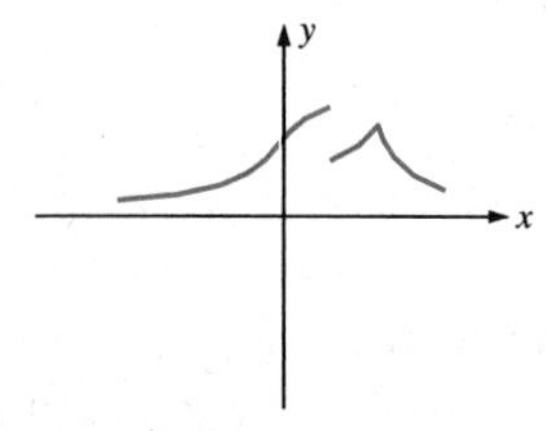

FIGURE 10.1.1

Piecewise differentiable function.

The **inner product** of two functions $f(x)$, $g(x)$ on the interval $[a, b]$ is a number, denoted (f, g), given by

$$(f, g) = \int_a^b f(x)g(x)\,dx. \tag{2}$$

Two functions f, g are **orthogonal** with respect to the inner product if their inner product is zero. That is,

$$(f, g) = \int_a^b f(x)g(x)\,dx = 0. \tag{3}$$

EXAMPLE 10.1.3 **Orthogonal Functions**

The two functions $f(x) = x$ and $g(x) = x^2$ are orthogonal on the interval $[-1, 1]$ since

$$(x, x^2) = \int_{-1}^{1} x x^2 \, dx = \int_{-1}^{1} x^3 \, dx = \frac{x^4}{4}\bigg|_{-1}^{1} = \frac{1}{4} - \frac{1}{4} = 0.$$

Orthogonal Set of Functions

A set of functions $\{\phi_1(x), \phi_2(x), \ldots\}$ is an **orthogonal set of functions** on the interval $[a, b]$ if any two functions in the set are orthogonal to each other:

$$(\phi_n, \phi_m) = \int_a^b \phi_n(x)\phi_m(x) \, dx = 0 \qquad (n \neq m). \tag{4}$$

We will consider only orthogonal sets where none of the functions are identically zero on $[a, b]$.

EXAMPLE 10.1.4 **Orthogonal Set**

Show that $\{\sin x, \sin 2x, \sin 3x, \ldots\}$ is an orthogonal set of functions on $[0, \pi]$.

Solution We take two representative functions from the set $\{\sin nx, \sin mx\}$, with $n \neq m$ and n, m integers, and verify that the functions are orthogonal.

$$\begin{aligned}(\sin nx, \sin mx) &= \int_0^{\pi} \sin nx \sin mx \, dx \\ &= \frac{1}{2}\int_0^{\pi} [\cos(n-m)x - \cos(n+m)x] \, dx \\ &= \frac{1}{2}\left[\frac{\sin(n-m)x}{n-m} - \frac{\sin(n+m)x}{n+m}\right]\bigg|_{x=0}^{\pi} = 0. \end{aligned} \tag{5}$$

The coefficients of a series with respect to an orthogonal set have a useful form, which we now derive. Suppose that $\{\phi_1(x), \phi_2(x), \ldots\}$ is an orthogonal set of functions on the interval $[a, b]$ and that

$$f(x) = \sum_{n=1}^{\infty} c_n \phi_n(x). \tag{6}$$

We wish to get a formula for the coefficients c_n in terms of $f(x)$ and the orthogonal functions $\phi_n(x)$. Select a member of the orthogonal set, say

$\phi_m(x)$, and take its inner product with $f(x)$, using (6). That is, multiply both sides of (6) by $\phi_m(x)$ and integrate over the interval to get

$$(f, \phi_m) = \int_a^b f(x)\phi_m(x)\,dx$$
$$= \int_a^b \left[\sum_{n=1}^{\infty} c_n \phi_n(x)\right] \phi_m(x)\,dx. \tag{7}$$

Assume that the integration and summation can be interchanged in (7) to give

$$(f, \phi_m) = \sum_{n=1}^{\infty} c_n \int_a^b \phi_n(x)\phi_m(x)\,dx = \sum_{n=1}^{\infty} c_n(\phi_n, \phi_m). \tag{8}$$

But the ϕ_i form an orthogonal set, so that $(\phi_n, \phi_m) = 0$ if $n \neq m$. Thus (7), (8) becomes

$$(f, \phi_m) = c_m(\phi_m, \phi_m). \tag{9}$$

Solving (9) for c_m leads to the following fundamental theorem.

■ **THEOREM 10.1.1 Representation of a Function by a Series of Orthogonal Functions**

Suppose that $f(x)$ is piecewise differentiable on the interval $[a, b]$ and

$$f(x) = \sum_{n=1}^{\infty} c_n \phi_n(x),$$

where $\{\phi_n(x)\}$ is an orthogonal set of functions on $[a, b]$. Then

$$c_n = \frac{(f, \phi_n)}{(\phi_n, \phi_n)} = \frac{\int_a^b f(x)\phi_n(x)\,dx}{\int_a^b \phi_n^2(x)\,dx}, \tag{10}$$

or equivalently,

$$f(x) = \sum_{n=1}^{\infty} \frac{(f, \phi_n)}{(\phi_n, \phi_n)} \phi_n(x).$$ ■

A rigorous proof of Theorem 10.1.1 involves technical considerations that are beyond the level of this course. Among these considerations are in what sense (6) converges and showing that the summation and integration in (7) can be interchanged. Also, when we write $f(x) = \sum_{n=1}^{\infty} c_n \phi_n(x)$, we do not require that the series actually converges to $f(x)$ for every x. Sufficient conditions to guarantee the interchange of the integration and summation in (7) will be given in the next two sections for Fourier sine and cosine series; these sections also discuss in what sense $\sum_{n=1}^{\infty} c_n \phi_n(x)$ equals $f(x)$. We need only piecewise continuity of f and the ϕ_m for this theorem. The differentiability is used later.

EXAMPLE 10.1.5 **Coefficients of a Sine Series**

Suppose that

$$f(x) = \sum_{n=1}^{\infty} c_n \sin nx \tag{11}$$

on the interval $[0, \pi]$. Find the coefficients c_n.

SOLUTION From Example 10.1.4, we have that $\{\sin nx: n \geq 1\}$ is an orthogonal set of functions on $[0, \pi]$. Thus from (10) we have

$$c_n = \frac{(f(x), \sin nx)}{(\sin nx, \sin nx)} = \frac{\int_0^{\pi} f(x) \sin nx\, dx}{\int_0^{\pi} \sin^2 nx\, dx} = \frac{2}{\pi}\int_0^{\pi} f(x) \sin nx\, dx,$$

since

$$(\sin nx, \sin nx) = \int_0^{\pi} \sin^2 nx\, dx$$
$$= \frac{1}{n}\left(-\frac{1}{2}\sin nx \cos nx + \frac{nx}{2}\right)\Bigg|_{x=0}^{\pi} = \frac{\pi}{2}.$$

EXAMPLE 10.1.6 **Representation of a Constant by a Sine Series**

Express $f(x) = 1$ as a series in terms of the orthogonal set of functions $\{\sin nx: n \geq 1\}$ on $[0, \pi]$, assuming that formula (10) is valid.

SOLUTION From (10) and Example 10.1.5 with $f(x) = 1$ we have

$$c_n = \frac{(1, \sin nx)}{(\sin nx, \sin nx)} = \frac{2}{\pi}\int_0^{\pi} 1 \sin nx\, dx = \frac{2}{\pi}\left(-\frac{\cos nx}{n}\right)\Bigg|_{x=0}^{\pi}$$

$$= \frac{2}{\pi}\left(-\frac{\cos n\pi}{n} + \frac{1}{n}\right) = \begin{cases} 0 & \text{if } n \text{ is even,} \\ \dfrac{4}{n\pi} & \text{if } n \text{ is odd.} \end{cases}$$

Thus (11) is

$$1 = \frac{4}{\pi} \sin x + \frac{4}{3\pi} \sin 3x + \frac{4}{5\pi} \sin 5x + \cdots. \tag{12}$$

This series may be expressed as

$$1 = \sum_{n=0}^{\infty} \frac{4}{(2n+1)\pi} \sin (2n+1)x.$$

◄

Clearly the series (12) is not valid at $x = 0$, since the right-hand side is zero when $x = 0$ but the left-hand side is 1. Thus some discussion on the convergence of these infinite series will be necessary.

A set of orthogonal functions $\{\phi_1(x), \phi_2(x), \ldots\}$ on the interval $[a, b]$ is **complete** if any function $f(x)$ that is piecewise continuous on $[a, b]$ may be expressed as an infinite series of the orthogonal functions:

$$f(x) = \sum_{n=1}^{\infty} c_n \phi_n(x).$$

One may show that an equivalent definition of completeness is

$$(f, \phi_n) = 0 \text{ for all } \phi_n \text{ implies } f(x) = 0 \text{ on } [a, b]. \tag{13}$$

Knowing that a set of orthogonal functions is complete is important because it will mean that there are enough functions in the set to express all of the functions we are interested in. Showing completeness is difficult. Results guaranteeing completeness are given in the following sections (such as Theorem 10.2.1).

EXAMPLE 10.1.7 A Complete Set

The orthogonal set $\{\sin nx : n \geq 1\}$ considered in Example 10.1.4 is complete on the interval $[0, \pi]$, so that (11) is valid. ◄

EXAMPLE 10.1.8 A Set Which Is Not Complete

We can show that $\{\sin nx : n \geq 1\}$ is also an orthogonal set of functions on $[-\pi, \pi]$. However, the function $f(x) = 1$ cannot be expressed as an infinite series. Suppose that

$$1 = \sum_{n=1}^{\infty} c_n \sin nx. \tag{14}$$

Then, from (10), a calculation gives

$$c_n = \frac{\int_{-\pi}^{\pi} 1 \sin nx\, dx}{\int_{-\pi}^{\pi} \sin^2 nx\, dx} = 0.$$

Thus (14) says that $1 = 0$ for all x, which is not true. We conclude that $\{\sin nx: n \geq 1\}$ is not a complete set of orthogonal functions on $[-\pi, \pi]$, since (14) does not hold for $f(x) = 1$. ◄

10.1.1 Weighted Inner Products

In several important applications to be discussed later, we must use a slightly more general inner product.

Let $w(x)$ be a nonnegative function on the interval $[a, b]$ that is zero at most a finite number of times. Then the **inner product** of two functions $f(x), g(x)$ **with respect to the weight function** $w(x)$ on the interval $[a, b]$ is

$$(f, g) = \int_a^b f(x)g(x)w(x)\, dx. \tag{15}$$

If $w(x) = 1$, then the weighted inner product (15) is the same as the inner product (2) discussed earlier. We again say that $f(x)$, $g(x)$ are **orthogonal** if $(f, g) = 0$.

A set of not identically zero functions $\{\phi_n(x)\}$ is an **orthogonal set** on the interval $[a, b]$ **with weight** $w(x)$ if any two functions in the set are orthogonal with weight $w(x)$. That is,

$$(\phi_n, \phi_m) = \int_a^b \phi_n(x)\phi_m(x)w(x)\, dx = 0 \qquad (n \neq m). \tag{16}$$

■ **THEOREM 10.1.2 Representation of a Function Using Orthogonal Functions with a Weight**

If $\{\phi_1(x), \phi_2(x), \ldots\}$ are orthogonal functions with respect to the weight function $w(x)$ on the interval $[a, b]$ and

$$f(x) = \sum_{n=1}^{\infty} c_n \phi_n(x), \tag{17}$$

then

$$c_n = \frac{(f, \phi_n)}{(\phi_n, \phi_n)} = \frac{\int_a^b f(x)\phi_n(x)w(x)\,dx}{\int_a^b \phi_n^2(x)w(x)\,dx}. \tag{18}$$

■

Verification This theorem is verified as before when we had $w(x) = 1$. We take the inner product of both sides with ϕ_m. That is, multiplying both sides of (17) by $\phi_m(x)$ and the weight $w(x)$ and integrating from a to b gives

$$\int_a^b f(x)\phi_m(x)w(x)\,dx = \sum_{n=1}^{\infty} c_n \int_a^b \phi_n(x)\phi_m(x)w(x)\,dx.$$

Since $\{\phi_1(x), \phi_2(x), \ldots\}$ is an orthogonal set with respect to the weight $w(x)$, all the integrals are zero except for when $n = m$. Thus we have

$$\int_a^b f(x)\phi_m(x)w(x)\,dx = c_m \int_a^b \phi_m(x)\phi_m(x)w(x)\,dx,$$

and (18) follows.

EXAMPLE 10.1.9 **Orthogonal Set of Functions with a Weight**

Show that $\{e^{-x}\sin nx\colon n \geq 1\}$ is an orthogonal set of functions with respect to the weight function $w(x) = e^{2x}$ on $[0, \pi]$.

SOLUTION Take $\phi_n(x) = e^{-x}\sin nx$, $\phi_m(x) = e^{-x}\sin mx$, with $n \neq m$. Then

$$\begin{aligned}(\phi_n, \phi_m) &= \int_0^{\pi} \phi_n(x)\phi_m(x)w(x)\,dx \\ &= \int_0^{\pi} e^{-x}\sin nx\, e^{-x}\sin mx\, e^{2x}\,dx \\ &= \int_0^{\pi} \sin nx \sin mx\,dx = 0,\end{aligned}$$

using (5). ◄

Exercises

In Exercises 1 through 8, verify that the given functions are orthogonal on the given interval.

1. $\{1, \sin x\}$, $[-\pi, \pi]$

2. $\{x^2 + 1, x^3\}$, $[-1, 1]$

3. $\{1, \cos x\}$, $[0, \pi]$

4. $\{\sin x, \cos x\}$, $[-\pi, \pi]$

5. $\{\cos 3x, \cos x\}$, $[0, \pi]$

6. $\{e^x, xe^{-x}\}$, $[-1, 1]$

7. $\{1 + x, x - x^2\}$, $[-2, 2]$

8. $\{x \cos x, 1\}$, $\left[-\frac{\pi}{2}, \frac{\pi}{2}\right]$

In Exercises 9 through 14, show that the given set of functions is orthogonal on the indicated interval.

9. $\{\sin x, \sin 2x, \sin 3x, \dots\}$ on $[-\pi, \pi]$

10. $\{1, \cos x, \cos 2x, \dots\}$ on $[-\pi, \pi]$

11. $\{1, \cos x, \cos 2x, \dots\}$ on $[0, \pi]$

12. $\left\{\sin \frac{\pi x}{L}, \sin \frac{2\pi x}{L}, \sin \frac{3\pi x}{L}, \dots\right\}$ on $[0, L]$

13. $\left\{\sin \frac{\pi(x-a)}{b-a}, \sin \frac{2\pi(x-a)}{b-a}, \sin \frac{3\pi(x-a)}{b-a}, \dots\right\}$ on $[a, b]$

14. $\left\{1, \cos \frac{\pi(x-a)}{b-a}, \cos \frac{2\pi(x-a)}{b-a}, \cos \frac{3\pi(x-a)}{b-a}, \dots\right\}$ on $[a, b]$

In Exercises 15 through 18, verify that the given pair of functions is orthogonal with respect to the given weight function $w(x)$ on the indicated interval.

15. $\{x, x^3\}$, $w(x) = x^5$, $[-1, 1]$

16. $\{e^{-2x} \sin x, e^{-2x} \sin 2x\}$, $w(x) = e^{4x}$, $[0, \pi]$

17. $\{x, 1 + x\}$, $w(x) = 1 - x$, $[-1, 1]$

18. $\{1, \cos 3x\}$, $w(x) = \sin x$, $[0, \pi]$

If α is a real number and $f(x)$, $g(x)$, $h(x)$ are piecewise continuous functions on the interval $[a, b]$, then using definition (2) it can be shown that an inner product has the following **inner product properties**:

1. $(f, g) = (g, f)$

2. $(\alpha f, g) = \alpha(f, g) = (f, \alpha g)$

3. $(f + g, h) = (f, h) + (g, h)$

4. $(f, f) = 0$ implies $f = 0$ on $[a, b]$ except possibly for a finite number of points.

The **length**, or **norm**, $\|f\|$ of a function f with respect to the inner product is

$$\|f\| = (f, f)^{1/2} = \left[\int_a^b f(x)^2\, dx\right]^{1/2}.$$

19. Show that $(f + g, h) = (f, h) + (g, h)$.

20. Show that $\|f\| = 0$ implies that $f = 0$ on $[a, b]$ except possibly for a finite number of points.

21. Show that if $(f, g) = 0$, then $\|f + g\|^2 = \|f\|^2 + \|g\|^2$.

22. Show that $(f, g) = (g, f)$.

23. If α is a number and $f(x)$ is a function, verify that $\|\alpha f\| = |\alpha|\, \|f\|$. Thus $g(x) = (1/\|f\|)f(x)$ has the property that $\|g\| = 1$ and g is said to be **normalized**. An orthogonal set of normalized functions is called an **orthonormal set of functions**.

24. Verify Exercises 19 through 23 for the weighted inner product (15).

25. From the formulas

$$\sin(\theta + \phi) = \sin\theta \cos\phi + \cos\theta \sin\phi,$$

$$\cos(\theta + \phi) = \cos\theta \cos\phi - \sin\theta \sin\phi,$$

derive the following formulas:

a) $\sin\theta \cos\phi = \frac{1}{2}[\sin(\theta + \phi) + \sin(\theta - \phi)]$

b) $\cos\theta \cos\phi = \frac{1}{2}[\cos(\theta + \phi) + \cos(\theta - \phi)]$

c) $\sin\theta \sin\phi = \frac{1}{2}[\cos(\theta - \phi) - \cos(\theta + \phi)]$

10.2 Fourier Series

In Exercise 44 it is shown that the set of functions

$$\{1, \cos x, \sin x, \cos 2x, \sin 2x, \ldots\} \tag{1}$$

is an orthogonal set of functions on the interval $[-\pi, \pi]$.

The infinite series

$$a_0 + \sum_{n=1}^{\infty} (a_n \cos nx + b_n \sin nx) \tag{2}$$

with a_n, b_n constants is a **Fourier series** with coefficients a_n, b_n. If the Fourier series converges to a function $f(x)$, then

$$f(x) = a_0 + \sum_{n=1}^{\infty} (a_n \cos nx + b_n \sin nx), \tag{3}$$

and by (10) of Section 10.1 we have that

$$a_0 = \frac{1}{2\pi} \int_{-\pi}^{\pi} f(x)\, dx \tag{4}$$

$$a_n = \frac{1}{\pi} \int_{-\pi}^{\pi} f(x) \cos nx\, dx \; (n \geq 1) \tag{5}$$

$$b_n = \frac{1}{\pi} \int_{-\pi}^{\pi} f(x) \sin nx\, dx, \tag{6}$$

since

$$\int_{-\pi}^{\pi} \sin^2 nx\, dx = \int_{-\pi}^{\pi} \cos^2 nx\, dx = \pi, \qquad \int_{-\pi}^{\pi} 1^2\, dx = 2\pi,$$

provided the technical assumptions of Theorem 10.2.1 hold. If the series in (3) converges, and (4), (5), (6) hold, the Fourier series (3) is called a **representation** of $f(x)$. We shall see that the Fourier series representation of $f(x)$ does not have to converge to $f(x)$ for every x.

Since all of the functions $\{\cos nx, \sin nx : n \geq 1\}$ have period 2π, a function given by a convergent Fourier series (3) must also have period 2π. That is, $f(x + 2\pi) = f(x)$. If the function $f(x)$ is defined only on $[-\pi, \pi]$, then the function must be extended periodically for (3) to be valid for all x.

Using the change of variables $x = (\pi z)/L$, it is easy to show that

$$\left\{1, \cos \frac{\pi z}{L}, \sin \frac{\pi z}{L}, \cos \frac{2\pi z}{L}, \sin \frac{2\pi z}{L}, \ldots\right\} \tag{7}$$

is an orthogonal set on the interval $[-L, L]$. Note that (7) is the same as (1) if $L = \pi$.

The fundamental theorem on the convergence of Fourier series is the following:

THEOREM 10.2.1 Convergence of Fourier Series

Suppose that $f(x)$ is a periodic function with period $2L$ and $f(x)$ is piecewise differentiable on $[-L, L]$. Then $f(x)$ has a Fourier series representation

$$f(x) = a_0 + \sum_{n=1}^{\infty} \left(a_n \cos \frac{n\pi x}{L} + b_n \sin \frac{n\pi x}{L} \right), \tag{8}$$

where

$$\begin{aligned} a_0 &= \frac{1}{2L} \int_{-L}^{L} f(x)\, dx, \\ a_n &= \frac{1}{L} \int_{-L}^{L} f(x) \cos \frac{n\pi x}{L}\, dx, \qquad n = 1, 2, \ldots, \\ b_n &= \frac{1}{L} \int_{-L}^{L} f(x) \sin \frac{n\pi x}{L}\, dx, \qquad n = 1, 2, \ldots. \end{aligned} \tag{9}$$

If x is a point of continuity of $f(x)$, then the series in (8) converges to the value of $f(x)$.

If $f(x)$ has a jump discontinuity at x, then the series converges to $\frac{1}{2}[f(x^+) + f(x^-)]$.

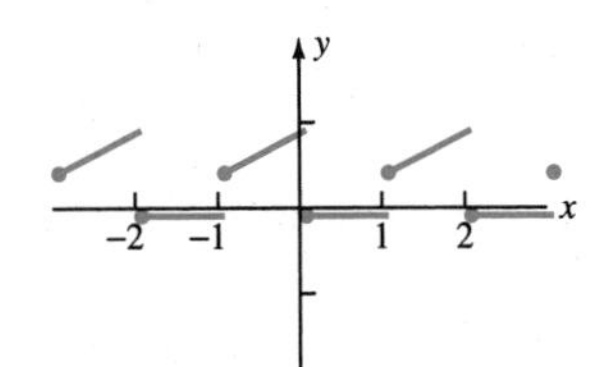

FIGURE 10.2.1

Here $f(x^+)$ denotes the limit from the right and $f(x^-)$ denotes the limit from the left. Note that $\frac{1}{2}[f(x^+) + f(x^-)]$ is just the midpoint of $f(x^+)$, $f(x^-)$, which is the middle of the jump. ■

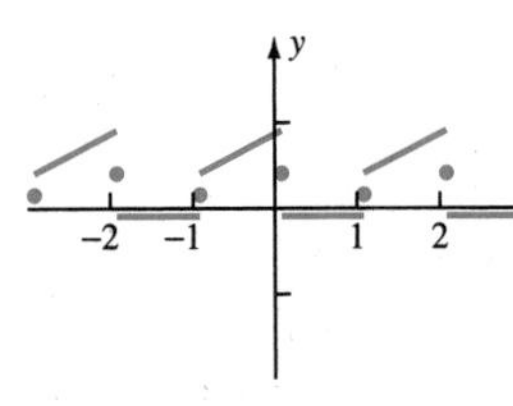

FIGURE 10.2.2

EXAMPLE 10.2.1 Fourier Series

The function of period 2 in Figure 10.2.1 has a Fourier series representation that converges to the function in Figure 10.2.2. ◄

EXAMPLE 10.2.2 Fourier Series

Suppose that $f(x)$ has period 2 and that

$$f(x) = \begin{cases} 0 & -1 \le x \le 0, \\ 1 & 0 \le x \le 1. \end{cases}$$

Compute and graph the Fourier series representation of $f(x)$.

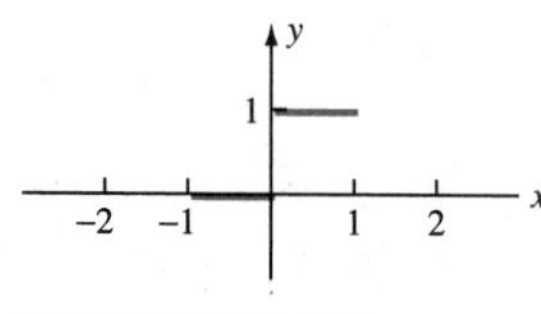

FIGURE 10.2.3

Graph of $f(x)$.

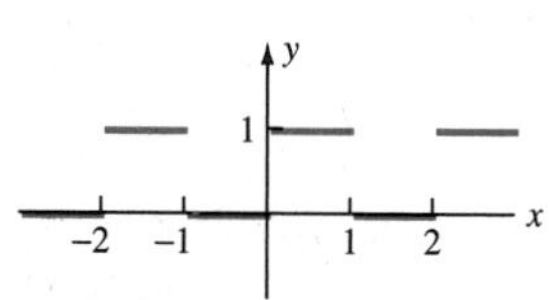
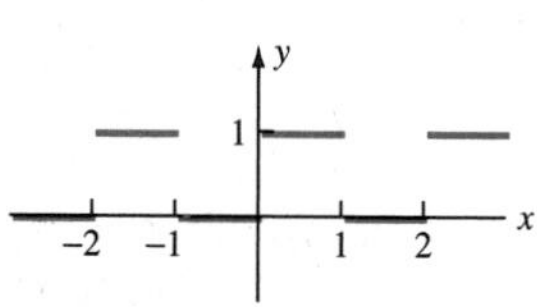

FIGURE 10.2.4

Graph of periodic extension.

FIGURE 10.2.5

Graph of Fourier series.

Solution Note that $f(x)$ is piecewise differentiable, so that the representation exists by Theorem 10.2.1. We have $2L = 2$ or $L = 1$.

The graph of the Fourier series representation can be obtained without computing the coefficients. Figure 10.2.3 shows $f(x)$, Figure 10.2.4 shows the periodic extension of $f(x)$, and Figure 10.2.5 shows the Fourier series representation.

We shall now compute the Fourier series. By (9), the coefficients a_n for $n \geq 1$ are

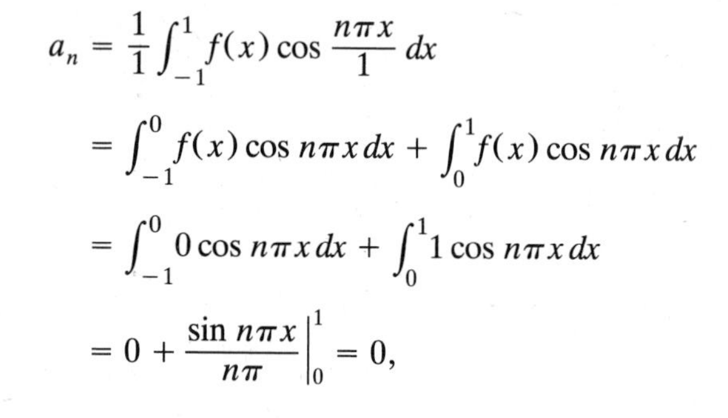

$$\begin{aligned} a_n &= \frac{1}{1}\int_{-1}^{1} f(x)\cos\frac{n\pi x}{1}\,dx \\ &= \int_{-1}^{0} f(x)\cos n\pi x\,dx + \int_{0}^{1} f(x)\cos n\pi x\,dx \\ &= \int_{-1}^{0} 0\cos n\pi x\,dx + \int_{0}^{1} 1\cos n\pi x\,dx \\ &= 0 + \left.\frac{\sin n\pi x}{n\pi}\right|_0^1 = 0, \end{aligned}$$

while

$$a_0 = \frac{1}{2}\int_{-1}^{1} f(x)\,dx = \frac{1}{2}\int_{0}^{1} 1\,dx = \frac{1}{2}.$$

Similarly, if $n \geq 1$, using (9) with $L = 1$, we have

$$\begin{aligned} b_n &= \frac{1}{1}\int_{-1}^{1} f(x)\sin n\pi x\,dx = \int_{0}^{1} 1\sin n\pi x\,dx \\ &= -\left.\frac{\cos n\pi x}{n\pi}\right|_0^1 = -\frac{\cos n\pi}{n\pi} + \frac{1}{n\pi} \\ &= [-(-1)^n + 1]\frac{1}{n\pi} = \begin{cases} \dfrac{2}{n\pi} & \text{if } n \text{ odd,} \\ 0 & \text{if } n \text{ even.} \end{cases} \end{aligned}$$

Thus

$$\begin{aligned} f(x) &= \frac{1}{2} + \frac{2}{\pi}\sin \pi x + \frac{2}{3\pi}\sin 3\pi x + \frac{2}{5\pi}\sin 5\pi x + \cdots \\ &= \frac{1}{2} + \sum_{n=0}^{\infty} \frac{2}{(2n+1)\pi}\sin(2n+1)\pi x. \end{aligned} \tag{10}$$

◀

In applications, truncated series consisting of a finite number of terms of the Fourier series are usually employed instead of the full infinite series.

Let

$$P_m(x) = a_0 + \sum_{n=1}^{m} \left(a_n \cos \frac{n\pi x}{L} + b_n \sin \frac{n\pi x}{L} \right). \tag{11}$$

The finite **trigonometric sum** $P_m(x)$ (it is also sometimes called a **trigonometric polynomial**) provides an approximation for the function $f(x)$ given by the Fourier series. Figures 10.2.6, 10.2.7, and 10.2.8 graph the function $f(x)$ from Example 10.2.2 and several of the approximating $P_m(x)$. We see that $P_m(x) \to f(x)$ as $m \to \infty$ at points where $f(x)$ is continuous. However, $P_m(x)$ approaches the average $(0 + 1)/2 = \frac{1}{2}$ at points where $f(x)$ has a jump discontinuity.

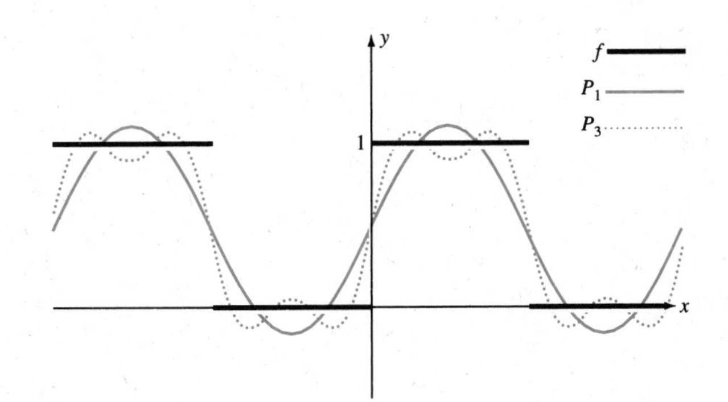

FIGURE 10.2.6

Function $f(x)$ from Example 10.2.2 and $P_1(x)$, $P_3(x)$.

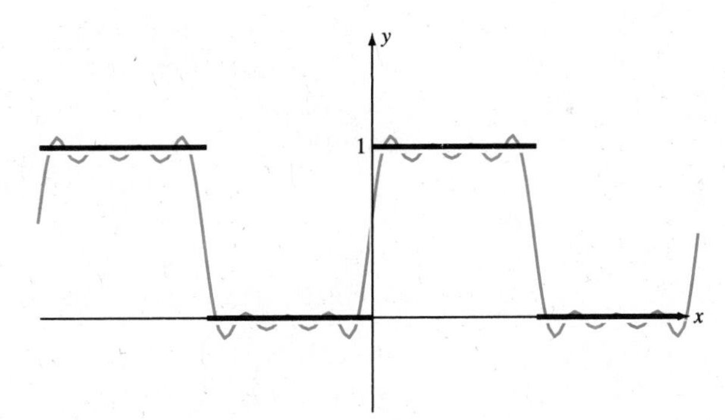

FIGURE 10.2.7

The approximation $P_7(x)$ for $f(x)$ of Example 10.2.2.

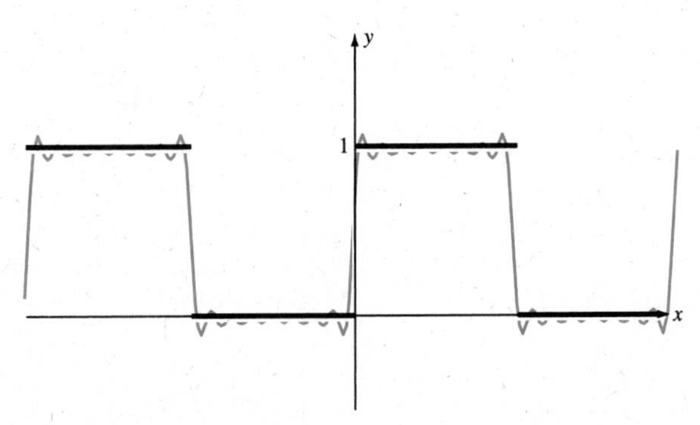

FIGURE 10.2.8

The approximation $P_{15}(x)$ for $f(x)$ of Example 10.2.2.

Gibbs' Phenomenon

Notice that while the approximation $P_m(x)$ of $f(x)$ gets better as m increases, the approximation is not **uniform** on $[-L, L]$. That is, given any sufficiently small tolerance $\epsilon > 0$, we cannot find a large enough m so that $|P_m(x) - f(x)| < \epsilon$ for *all* x in the interval. For example, if $\epsilon = 0.01$, there will always be points near the jump discontinuities at 0, 1 where the error is larger then $\epsilon = 0.01$. In fact, we can always find points where the error is close to 0.09.

The precise behavior of the difference between a function $f(x)$ and its approximation $P_m(x)$ is as follows: If $f(x)$ is continuous at x, then $P_m(x) \to f(x)$ as $m \to \infty$. However, if $f(x)$ has a jump discontinuity at a point c, then the behavior is different. $P_m(x)$ is continuous everywhere and has trouble approximating $f(x)$ near c. There is a theorem that says that for every large m, there are two places near the discontinuity where

Gibbs' Phenomenon

The undershoot or overshoot [difference between $f(x)$ and $P_m(x)$] near a jump is approximately 9% of the jump.

This is called **Gibbs' phenomenon**. For Example 10.2.2, where the jump is 1 at $c = 0$, this says that no matter how large we make m, there will always be values of x just past the discontinuity at 0 where the value of $P_m(x)$ is about 1.09 and values just before the discontinuity where it is about -0.09.

Stronger conditions than those of Theorem 10.2.1 are needed to guarantee uniform convergence.

■ **THEOREM 10.2.2 Uniform Convergence**

If $f(x)$ is periodic with period $2L$ and $f'(x)$ is continuous, then the series (3) converges uniformly to $f(x)$. ■

If $f'(x)$ is continuous, not only is $f(x)$ continuous, but also there are no "corners" on the graph. A continuous function whose derivative is piecewise continuous but not continuous appears in Example 10.2.3.

10.2.1 Even and Odd Functions

A function $g(x)$ on $[-L, L]$ is **even** if $g(-x) = g(x)$ and **odd** if $g(-x) = -g(x)$. Some examples of even functions are

$$x^2, \quad x^2 + 1, \quad x^4, \quad \cos\frac{n\pi x}{L}, \quad |x|, \quad \cosh x,$$

while some odd functions are

$$x, \quad x^3, \quad x^5 - 9x^7, \quad \sin\frac{n\pi x}{L}, \quad x|x|, \quad \sinh x, \quad x^{1/3}.$$

■ **PROPOSITION 10.2.1 Even-Odd Parity**

Products and sums of functions with the same domain obey the following rules:

even × even is even,
odd × odd is even,
odd × even is odd,
odd + odd is odd,
even + even is even. ■

Verification We shall verify the second rule. Verification of the rest is left to the exercises. Suppose that f, g are both odd functions. Then

$$(fg)(-x) = f(-x)g(-x) = [-f(x)][-g(x)] = f(x)g(x) = (fg)(x),$$

so that fg is even.

Knowing that a function is even or odd will be useful because of the following two facts.:

1. If $g(x)$ is an odd function, then

$$\int_{-L}^{L} g(x)\,dx = 0. \tag{12}$$

2. If $g(x)$ is an even function, then

$$\int_{-L}^{L} g(x)\,dx = 2\int_{0}^{L} g(x)\,dx. \tag{13}$$

We leave the verification of these facts to the exercises and turn to applying them to Fourier series.

■ **PROPOSITION 10.2.2 Fourier Series of Even and Odd Functions**

Suppose that $f(x)$ is a piecewise continuous function of period $2L$.

1. If $f(x)$ is an even function, then all of the $b_n = 0$ in its Fourier series expansion. Thus the sine terms are absent. Only a constant and cosine occur if $f(x)$ is even.
2. If $f(x)$ is an odd function, then all of the $a_n = 0$ in its Fourier series expansion. Thus the constant and cosine terms are absent. Only sines occur if $f(x)$ is odd. ■

Verification If $f(x)$ is even, the $f(x)\sin(n\pi x/L)$ is odd, so that

$$b_n = \frac{1}{L}\int_{-L}^{L} f(x)\sin\frac{n\pi x}{L}\,dx = 0$$

by (12). Similarly, if $f(x)$ is odd, then $f(x)\cos(n\pi x/L)$ is odd and $a_n = 0$.

This useful property of even and odd functions will be used several times in this chapter.

EXAMPLE 10.2.3 Fourier Series of an Even Function

Let $f(x)$ be a periodic function of period 2π such that $f(x) = |x|$ for $-\pi \le x \le \pi$. Graph $f(x)$ and compute the Fourier series expansion of $f(x)$.

SOLUTION Figure 10.2.9 gives the graph of $f(x)$. Since $f(-x) = f(x)$, we have that $f(x)$ is an even function and

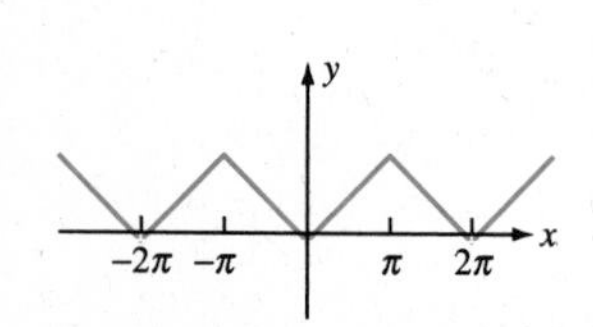

FIGURE 10.2.9

Function $f(x)$ in Example 10.2.3.

$$b_n = 0$$

for all n. Using (9) and (13), we compute for $n \ge 1$ that

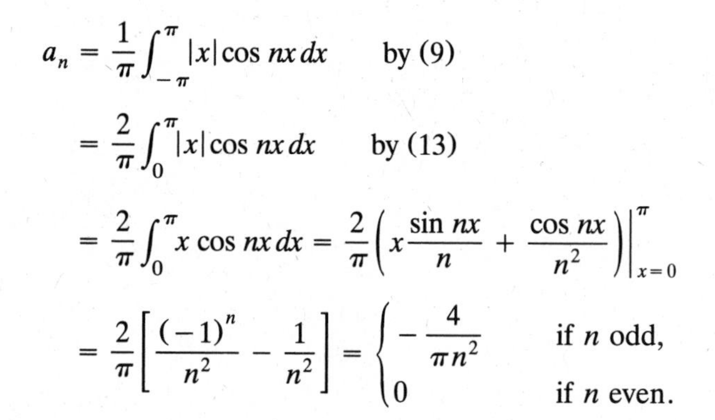

$$\begin{aligned} a_n &= \frac{1}{\pi}\int_{-\pi}^{\pi} |x|\cos nx\,dx \qquad \text{by (9)} \\ &= \frac{2}{\pi}\int_0^{\pi} |x|\cos nx\,dx \qquad \text{by (13)} \\ &= \frac{2}{\pi}\int_0^{\pi} x\cos nx\,dx = \frac{2}{\pi}\left(x\frac{\sin nx}{n} + \frac{\cos nx}{n^2}\right)\Bigg|_{x=0}^{\pi} \\ &= \frac{2}{\pi}\left[\frac{(-1)^n}{n^2} - \frac{1}{n^2}\right] = \begin{cases} -\dfrac{4}{\pi n^2} & \text{if } n \text{ odd}, \\ 0 & \text{if } n \text{ even}. \end{cases} \end{aligned}$$

Also,

$$\begin{aligned} a_0 &= \frac{1}{2\pi}\int_{-\pi}^{\pi} |x|\,dx \qquad \text{by (9)} \\ &= \frac{2}{2\pi}\int_0^{\pi} |x|\,dx \qquad \text{by (13)} \\ &= \frac{1}{\pi}\int_0^{\pi} x\,dx = \frac{\pi}{2}. \end{aligned}$$

Thus

$$f(x) = \frac{\pi}{2} - \frac{4}{\pi}\cos x - \frac{4}{9\pi}\cos 3x + \cdots,$$

or equivalently,

$$f(x) = \frac{\pi}{2} - \sum_{n=0}^{\infty} \frac{4}{\pi(2n+1)^2} \cos(2n+1)x.$$

It is sometimes more convenient to compute the Fourier coefficients over a different interval.

PROPOSITION 10.2.3 Integral Is the Same over Different Intervals with Same Length

Suppose that $f(x)$ is piecewise continuous and periodic with period $2L$. Then for any number a,

$$\int_{-L}^{L} f(x) \cos \frac{n\pi x}{L}\, dx = \int_{a-L}^{a+L} f(x) \cos \frac{n\pi x}{L}\, dx.$$

A similar statement holds for $\sin x$.

Technical Point Since the Fourier coefficients of a function $f(x)$ are given by integrals involving $f(x)$, the coefficients are unaltered if the value of $f(x)$ is changed at a finite number of points in a finite interval. Accordingly, we follow our practice begun with the Laplace transform in Chapter 3 of considering two functions equal if on any finite interval they agree except at a finite number of points.

Exercises

In Exercises 1 through 10, determine whether the given function is even, odd, or neither.

1. $x^3 - x^7$
2. $x^2 + 6x^4$
3. $1 + 3x^2$
4. $x + x^2$
5. $x^2 \sin x$
6. $x^3 \cos x$
7. $\cosh x$
8. $\sinh x$
9. e^x
10. $x \sinh x$

In Exercises 11 through 18, graph for all x the Fourier series representation of the given function. Do not determine the Fourier coefficients.

11. $f(x) = \begin{cases} 3 & -5 < x < 0, \\ 4 & 0 < x < 5 \end{cases}$
12. $f(x) = \begin{cases} -1 & -3 < x < 1, \\ 2 & 1 < x < 3 \end{cases}$
13. $f(x) = \begin{cases} x & -1 < x < 0, \\ 0 & 0 < x < 1 \end{cases}$
14. $f(x) = \begin{cases} 0 & -2 < x < 0, \\ x^2 & 0 < x < 2 \end{cases}$
15. $f(x) = e^{-x}, \quad -1 < x < 1$
16. $f(x) = \dfrac{1}{1+x^2}, \quad -1 < x < 1$

17. $f(x) = \begin{cases} 0 & -\pi < x < 0, \\ \sin x & 0 < x < \pi \end{cases}$

18. $f(x) = \begin{cases} 0 & -\pi < x < 0, \\ \cos x & 0 < x < \pi \end{cases}$

In Exercises 19 through 34, determine the Fourier series representation of the given periodic function on the indicated interval.

19. $f(x) = \begin{cases} 1 & 0 < x \le \pi, \\ 0 & x = 0, \\ -1 & -\pi \le x < 0, \end{cases}$
period 2π on $[-\pi, \pi]$

20. $f(x) = 1 - x$ for $-1 \le x < 1$, period 2 on $[-1, 1]$

21. $f(x) = |\cos x|$ on $\left[-\frac{\pi}{2}, \frac{\pi}{2}\right]$

22. $f(x) = \cos^2 3x$ on $\left[-\frac{\pi}{6}, \frac{\pi}{6}\right]$

23. $f(x) = e^x$ for $-1 \le x < 1$, period 2 on $[-1, 1]$

24. $f(x) = x^2$ for $-1 \le x \le 1$, period 2 on $[-1, 1]$

25. $f(x) = |\sin x|$, period π on $\left[-\frac{\pi}{2}, \frac{\pi}{2}\right]$

26. $f(x) = \begin{cases} 0 & -3 < x < 1, \\ 1 & 1 < x < 2, \\ 0 & 2 < x < 3, \end{cases}$
period 6 on $[-3, 3]$

27. $f(x) = \sin^2 x$ on $\left[-\frac{\pi}{2}, \frac{\pi}{2}\right]$

28. $f(x) = e^x \sin x$ for $-\pi < x < \pi$, period 2π on $[-\pi, \pi]$

29. $f(x) = \begin{cases} 1 & -2 < x < 0, \\ 2 & 0 < x < 2, \end{cases}$
period 4 on $[-2, 2]$

30. $f(x) = \begin{cases} 0 & -1 < x < 0, \\ x & 0 < x < 1, \end{cases}$
period 2 on $[-1, 1]$

31. $f(x) = \begin{cases} 0 & -\pi < x < 0, \\ x^2 & 0 < x < \pi, \end{cases}$
period 2π on $[-\pi, \pi]$

32. $f(x) = 3 + \cos 2x$ on $[-\pi, \pi]$.

33. Same function $f(x)$ as in Exercise 26, but use the interval $[-6, 6]$.

34. Same function $f(x)$ as in Exercise 29, but use the interval $[-4, 4]$.

In Exercises 35 through 37, assume that $f(x)$, $g(x)$ are functions defined on an interval $[-M, M]$.

35. Show that if $f(x)$ and $g(x)$ are both even functions, then fg and $f + g$ are even functions.

36. Show that if $f(x)$ is even and $g(x)$ is odd, then fg is odd.

37 a) Show that $f(x) = f_1(x) + f_2(x)$ where $f_1(x) = \frac{1}{2}[f(x) + f(-x)]$ and $f_2(x) = \frac{1}{2}[f(x) - f(-x)]$.

b) Show that $f_1(x)$ is an even function.

c) Show that $f_2(x)$ is an odd function.

38. Let $f(x) = x^{1/3}$ for $-1 < x \le 1$ and have period 2. Show that $f(x)$ is a piecewise continuous function but $f'(x)$ is not piecewise continuous.

39. Let $f(x) = (\sin x)^{1/3}$. Show that $f(x)$ is a piecewise continuous function with period 2π but $f'(x)$ is not piecewise continuous.

40. Show that the set of functions in (7) is an orthogonal set of functions on $[-L, L]$ if the set in (1) is an orthogonal set of functions on $[-\pi, \pi]$.

41. Verify that (13) holds for even functions $g(x)$.

42. Verify that (12) holds for odd functions $g(x)$.

43. Verify Proposition 10.2.3.

44. Show that $\{1, \cos x, \sin x, \cos 2x, \sin 2x, \ldots\}$ form an orthogonal set on $[-\pi, \pi]$.

In Exercises 45 through 49, if you have access to a computer system with graphic capabilities, graph the function $f(x)$ and the approximating trigonometric polynomials $P_0(x)$, $P_1(x)$, $P_2(x)$, $P_5(x)$, $P_{10}(x)$ for the Fourier representations computed in the indicated exercises.

45. Exercise 19

46. Exercise 20

47. Exercise 26

48. Exercise 29

49. Exercise 30

10.3 Half-Range Expansions

In Section 10.2 we considered periodic functions $f(x)$ of period $2L$ and computed their Fourier series. In later sections in Chapter 10 and in Chapter 11 we shall have a function $f(x)$ defined only on an interval $[0, L]$. Depending on the application, we will want to get a series for $f(x)$ involving only sine or cosine terms.

Fourier Cosine Series

Suppose that we have a function $f(x)$ defined on the interval $[0, L]$. We shall show how to construct the cosine series first. Since we are interested in the values of the series only on the interval $[0, L]$, we can define $f(x)$ any way we wish outside this interval. In order to get a series with just cosine terms, we will define an **even periodic extension** of $f(x)$.

First define $\tilde{f}(x)$ on $[-L, L]$ by the **even extension** of $f(x)$:

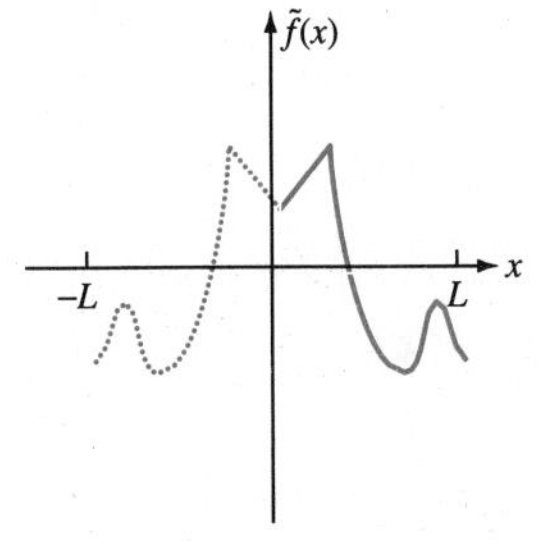

FIGURE 10.3.1

Even extension of $f(x)$.

$$\tilde{f}(x) = \begin{cases} f(x) & 0 \le x \le L, \\ f(-x) & -L \le x < 0, \end{cases} \tag{1}$$

as graphed in Figure 10.3.1.

Note that $\tilde{f}(x)$ is an even function on $[-L, L]$. Now extend $\tilde{f}(x)$ to all of $(-\infty, \infty)$ by making $\tilde{f}(x)$ an even periodic function with period $2L$. The Fourier series for $\tilde{f}(x)$ is

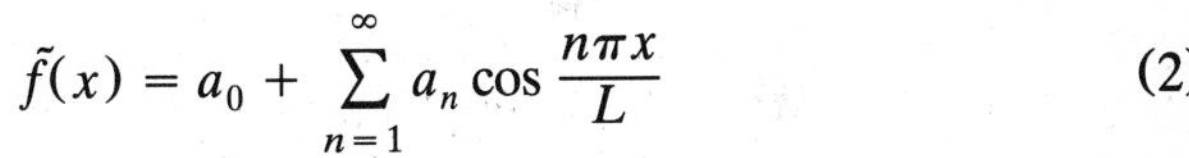

$$\tilde{f}(x) = a_0 + \sum_{n=1}^{\infty} a_n \cos \frac{n\pi x}{L} \tag{2}$$

with

$$a_0 = \frac{1}{2L} \int_{-L}^{L} \tilde{f}(x)\, dx, \tag{3}$$

$$a_n = \frac{1}{L} \int_{-L}^{L} \tilde{f}(x) \cos \frac{n\pi x}{L}\, dx \qquad (n \ge 1). \tag{4}$$

The sine terms are absent since $\tilde{f}(x)$ is an even function. Also, $\tilde{f}(x)$ being even implies by (13) of Section 10.2 that

$$a_0 = \frac{1}{L} \int_0^L \tilde{f}(x)\, dx = \frac{1}{L} \int_0^L f(x)\, dx, \tag{5}$$

$$a_n = \frac{2}{L} \int_0^L \tilde{f}(x) \cos \frac{n\pi x}{L}\, dx = \frac{2}{L} \int_0^L f(x) \cos \frac{n\pi x}{L}\, dx. \tag{6}$$

In summary:

■ **THEOREM 10.3.1 Cosine Series**

If $f(x)$ is a piecewise differentiable function on $[0, L]$, then the Fourier cosine series for $f(x)$ is

$$f(x) = a_0 + \sum_{n=1}^{\infty} a_n \cos \frac{n\pi x}{L}, \tag{7}$$

where

$$\begin{aligned} a_0 &= \frac{1}{L}\int_0^L f(x)\,dx, \\ a_n &= \frac{2}{L}\int_0^L f(x)\cos\frac{n\pi x}{L}\,dx \qquad (n \geq 1). \end{aligned} \tag{8}$$

■

EXAMPLE 10.3.1 Fourier Cosine Series

Let $f(x) = x$ on $[0, 2]$. Find the Fourier cosine series for $f(x)$.

SOLUTION From (7) and (8), we have, with $L = 2$ and $f(x) = x$,

$$a_0 = \frac{1}{2}\int_0^2 x\,dx = 1$$

and

$$a_n = \frac{2}{2}\int_0^2 x\cos\frac{n\pi x}{2}\,dx = \frac{4[\cos(n\pi) - 1]}{n^2\pi^2} = \begin{cases} 0 & n \text{ even}, \\ -\dfrac{8}{n^2\pi^2} & n \text{ odd}. \end{cases}$$

Thus

$$f(x) = x = 1 - \sum_{n=0}^{\infty} \frac{8}{(2n+1)^2\pi^2} \cos\frac{(2n+1)\pi x}{2}. \tag{9}$$

◄

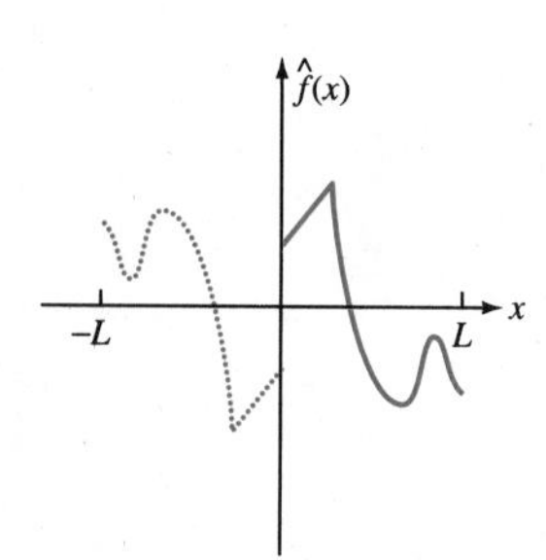

FIGURE 10.3.2

Odd extension of $f(x)$.

Fourier Sine Series

Given a function $f(x)$ on $[0, L]$, we can get a Fourier sine series by defining an **odd extension** $\hat{f}(x)$ of $f(x)$ on $[-L, L]$ (see Figure 10.3.2),

$$\hat{f}(x) = \begin{cases} f(x) & 0 < x < L, \\ -f(-x) & -L < x < 0, \end{cases}$$

and then extending $\hat{f}(x)$ to be periodic with period $2L$ on $(-\infty, \infty)$. Using the properties of odd functions, one can show that the Fourier series for $\hat{f}(x)$ has only sine terms and the coefficients depend just on $f(x)$.

THEOREM 10.3.2 Sine Series

If $f(x)$ is piecewise differentiable on $[0, L]$, then the Fourier sine series for $f(x)$ is

$$f(x) = \sum_{n=1}^{\infty} b_n \sin \frac{n\pi x}{L}, \tag{10}$$

where

$$b_n = \frac{2}{L} \int_0^L f(x) \sin \frac{n\pi x}{L}\, dx. \tag{11}$$

■

EXAMPLE 10.3.2 Fourier Sine Series

Find the Fourier sine series for $f(x) = x$ on $[0, 2]$.

SOLUTION From (10) and (11) with $L = 2$ and $f(x) = x$, we have

$$b_n = \frac{2}{2} \int_0^2 x \sin \frac{n\pi x}{2}\, dx = \frac{(-1)^{n+1} 4}{n\pi},$$

so that

$$f(x) = x = \sum_{n=1}^{\infty} \frac{(-1)^{n+1} 4}{n\pi} \sin \frac{n\pi x}{2}. \tag{12}$$

■

Convergence

Convergence of the Fourier cosine and sine series given by Theorems 10.3.1 and 10.3.2 is covered by Theorems 10.2.1 and 10.2.2, since the series are actually Fourier series of the extensions.

EXAMPLE 10.3.3 Graph of Fourier Cosine and Sine Series

Determine the functions that the Fourier cosine and sine series of Examples 10.3.1 and 10.3.2 converge to on $[0, 2]$.

Solution The graph of $f(x) = x$ on $[0, 2]$ is given in Figure 10.3.3.

The even extension $\tilde{f}(x)$ of this function is shown in Figure 10.3.4. The even periodic extension is then the function in Figure 10.3.5.

Since the even periodic extension has no jump discontinuities in $[0, 2]$, the Fourier cosine series (9) converges to the same function as in Figure 10.3.3 for all x in $[0, 2]$.

However, the odd extension of $f(x) = x$ on $[0, 2]$ is given in Figure 10.3.6. Figure 10.3.7 gives the odd periodic extension.

The odd periodic extension has a jump discontinuity at $x = 2$. Thus the Fourier sine series (12) converges on $[0, 2]$ to the function in Figure 10.3.8.

Comment In the terminology of Section 10.1, both $\{1, \cos \pi x/L, \cos 2\pi x/L, \ldots\}$ and $\{\sin \pi x/L, \sin 2\pi x/L, \ldots\}$ are complete sets of orthogonal functions on $[0, L]$.

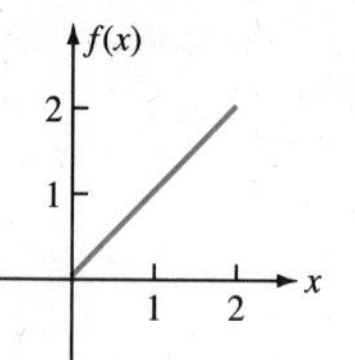

FIGURE 10.3.3

Graph of $f(x) = x$, $0 \le x \le 2$.

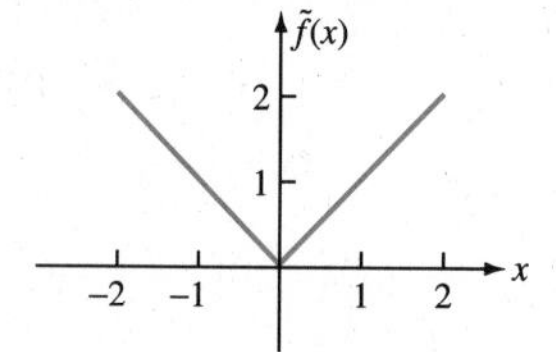

FIGURE 10.3.4

Even extension of Example 10.3.1 on $[-2, 2]$.

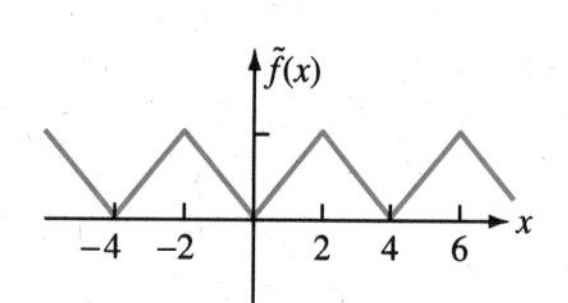

FIGURE 10.3.5

Even periodic extension of Example 10.3.1.

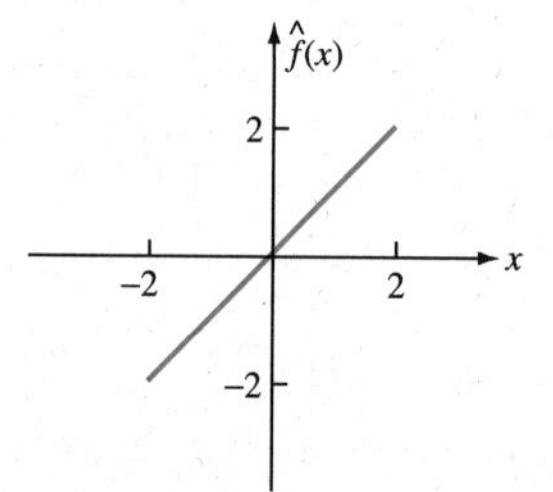

FIGURE 10.3.6

Odd extension of Example 10.3.2.

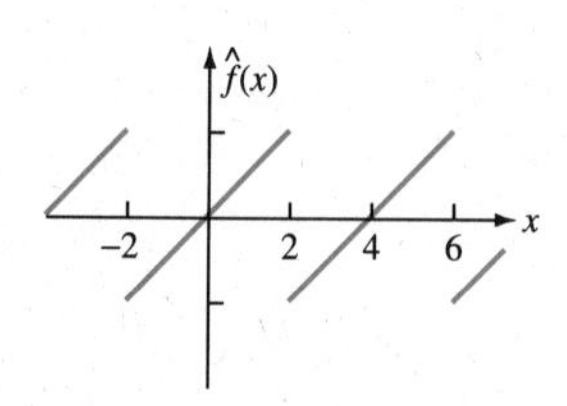

FIGURE 10.3.7

Odd periodic extension of Example 10.3.2.

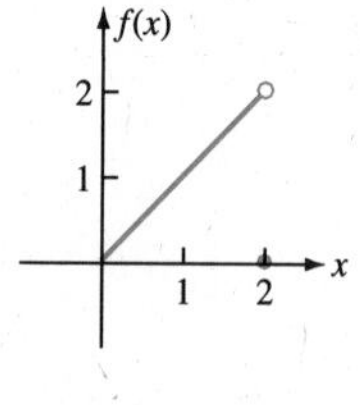

FIGURE 10.3.8

Limit of Fourier series (12) on $[0, 2]$.

Exercises

In Exercises 1 through 8, graph for all x the Fourier cosine series representation of the given function on the indicated interval. Do *not* determine any Fourier coefficients.

1. $f(x) = 1 + x, \quad [0, 2]$

2. $f(x) = x^3, \quad [0, 1]$

3. $f(x) = e^x, \quad [0, 1]$

4. $f(x) = e^{-x}, \quad [0, \pi]$

5. $f(x) = \sin x, \quad [0, \pi]$

6. $f(x) = \sin 2x, \quad [0, \pi]$

7. $f(x) = \begin{cases} 1 + x, & 0 \le x \le 1, \\ 0, & 1 < x \le 2, \end{cases} \quad [0, 2]$

8. $f(x) = \begin{cases} 0, & 0 \le x \le 3, \\ x, & 3 < x \le 4, \end{cases} \quad [0, 4]$

In Exercises 9 through 16, graph for all x the Fourier sine series representation of the given function on the indicated interval. Do *not* determine any Fourier coefficients.

9. $f(x) = 1 + x, \quad [0, 2]$

10. $f(x) = x^3, \quad [0, 1]$

11. $f(x) = e^x, \quad [0, 1]$

12. $f(x) = e^{-x}, \quad [0, \pi]$

13. $f(x) = \cos x, \quad [0, \pi/2]$

14. $f(x) = \cos x, \quad [0, \pi]$

15. $f(x) = 1 + x^2, \quad [0, 1]$

16. $f(x) = \sin x, \quad [0, \pi/2]$

In Exercises 17 through 29, find the Fourier cosine series on the indicated interval. Also give the value of the series at any points on the interval where the series does not converge to the original function.

17. $f(x) = 1, \quad [0, 1]$

18. $f(x) = \sin x, \quad [0, \pi]$

19. $f(x) = \cos x, \quad [0, \pi]$

20. $f(x) = \sin 3x, \quad [0, \pi]$

21. $f(x) = \cos 3x, \quad [0, \pi]$

22. $f(x) = \begin{cases} 1 & 0 \le x \le 1, \\ 0 & 1 < x \le 2, \end{cases} \quad [0, 2]$

23. $f(x) = \begin{cases} 0, & 0 \le x \le 1, \\ 1 & 1 < x \le 2, \end{cases} \quad [0, 2]$

24. $f(x) = \begin{cases} x & 0 \le x \le 1, \\ 0 & 1 < x \le 2, \end{cases} \quad [0, 2]$

25. $f(x) = \begin{cases} 0 & 0 \le x \le 1, \\ x & 1 < x \le 2, \end{cases} \quad [0, 2]$

26. $f(x) = 1 - x, \quad [0, 2]$

27. $f(x) = \begin{cases} 1 & 0 \le x \le 1, \\ 0 & 1 < x \le 4, \end{cases} \quad [0, 4]$

28. $f(x) = \begin{cases} 0 & 0 \le x \le 1, \\ 1 & 1 < x \le 4, \end{cases} \quad [0, 4]$

29. $f(x) = \begin{cases} 0 & 0 \le x \le 1, \\ 1 & 1 < x \le 2, \\ 0 & 2 < x \le 3, \end{cases} \quad [0, 3]$

In Exercises 30 through 42, find the Fourier sine series on the indicated interval. Also give the value of the series at any points on the interval where the series does not converge to the original function.

30. $f(x) = 1, \quad [0, 1]$

31. $f(x) = \sin x, \quad [0, \pi]$

32. $f(x) = \cos x, \quad [0, \pi]$

33. $f(x) = \sin 3x, \quad [0, \pi]$

34. $f(x) = \cos 3x, \quad [0, \pi]$

35. $f(x) = \begin{cases} 1 & 0 \le x \le 1, \\ 0 & 1 < x \le 2, \end{cases} \quad [0, 2]$

36. $f(x) = \begin{cases} 0 & 0 \le x \le 1, \\ 1 & 1 < x \le 2, \end{cases} \quad [0, 2]$

37. $f(x) = \begin{cases} x & 0 \le x \le 1, \\ 0 & 1 < x \le 2, \end{cases} \quad [0, 2]$

38. $f(x) = \begin{cases} 0 & 0 \le x \le 1, \\ x & 1 < x \le 2, \end{cases} \quad [0, 2]$

39. $f(x) = 1 - x, \quad [0, 2]$

40. $f(x) = \begin{cases} 1 & 0 \le x \le 1, \\ 0 & 1 < x \le 4, \end{cases} \quad [0, 4]$

41. $f(x) = \begin{cases} 0 & 0 \le x \le 3, \\ 1 & 3 < x \le 4, \end{cases} \quad [0, 4]$

42. $f(x) = \begin{cases} 0 & 0 \le x \le 1, \\ 1 & 1 < x \le 2, \\ 0 & 2 < x \le 3, \end{cases} \quad [0, 3]$

43. Suppose that $f(x)$ is piecewise differentiable on $[0, L]$ and is also continuous on $[0, L]$. Show that the Fourier sine series for $f(x)$ converges to $f(x)$ for all x in $[0, L]$ if and only if $f(0) = 0$ and $f(L) = 0$.

44. Suppose that $f(x)$ is piecewise differentiable on $[0, L]$ and is also continuous on $[0, L]$. Show that the Fourier cosine series for $f(x)$ converges to $f(x)$ for all x in $[0, L]$.

45. Verify that $\tilde{f}(x)$ in (1) is an even function.

46. Show that if $f(x)$ is a piecewise differentiable function on $[0, L]$ and all of the coefficients of the Fourier cosine series for $f(x)$ are zero, then $f(x) = 0$ for all except possibly a finite number of points in $[0, L]$.

The results of this section can be extended to intervals $[a, b]$ instead of $[0, L]$. In Exercises 47 and 48, suppose that $f(x)$ is a piecewise differentiable function on the interval $[a, b]$. (Note Exercises 10.1.13 and 10.1.14.)

47. Show that

$$f(x) = a_0 + \sum_{n=1}^{\infty} a_n \cos \frac{n\pi(x-a)}{b-a},$$

where

$$a_0 = \frac{1}{b-a}\int_a^b f(x)\,dx,$$

$$a_n = \frac{2}{b-a}\int_a^b f(x)\cos\frac{n\pi(x-a)}{b-a}\,dx \quad (n > 0).$$

48. Show that

$$f(x) = \sum_{n=1}^{\infty} b_n \sin \frac{n\pi(x-a)}{b-a},$$

where

$$b_n = \frac{2}{b-a}\int_a^b f(x)\sin\frac{n\pi(x-a)}{b-a}\,dx.$$

10.4 Application to Initial Value Problems

Before continuing with the development of the theory of Fourier series and boundary value problems in this chapter and the application of Fourier series to the solution of partial differential equations in Chapter 11, we wish to give one application of Fourier series to ordinary differential equations.

In this section we show how to find a particular solution of

$$y'' + py' + qy = f(t) \tag{1}$$

when p, q are *constants* and $f(t)$ is a periodic piecewise differentiable function with period $2L$. From Sections 1.10 and 2.12 we know that (1) can be interpreted as an *RLC* circuit with voltage source $f(t)$ or as a spring–mass system exposed to an external force $f(t)$. Often we can use the Laplace transform of Chapter 3 to solve (1). However, we shall see that using Fourier series has certain advantages. The procedure we shall employ to find a particular solution of equation (1) is the following.

Procedure for Solving $y'' + py' + qy = f$ with Fourier Series

Suppose that $f(t)$ is piecewise differentiable with period $2L$. The solution $y_h(t)$ of the associated homogeneous equation $y'' + py' + qy = 0$ is found as in Section 2.5. The particular solution $y_p(t)$ is found as follows.

1. Let $f(t)$ have the Fourier series representation

$$f(t) = a_0 + \sum_{n=1}^{\infty} \left(a_n \cos \frac{n\pi t}{L} + b_n \sin \frac{n\pi t}{L}\right). \tag{2}$$

2. Substituting the series in (2) for $f(t)$ in (1) gives

$$y'' + py' + qy = a_0 + \sum_{n=1}^{\infty} \left(a_n \cos \frac{n\pi t}{L} + b_n \sin \frac{n\pi t}{L}\right). \tag{3}$$

Use the superposition principle to reduce solving (3) to Steps 3 and 4.

3. Find a particular solution $y_0(t)$ of

$$y'' + py' + qy = a_0. \tag{4}$$

4. For each $n \geq 1$, find a particular solution $y_n(t)$ of

$$y'' + py' + qy = a_n \cos \frac{n\pi t}{L} + b_n \sin \frac{n\pi t}{L}. \tag{5}$$

5. A particular solution of $y'' + py' + qy = f$ is then

$$y_p(t) = \sum_{n=0}^{\infty} y_n(t). \tag{6}$$

In general, given $f(t)$, we would need to find the Fourier series representation of $f(t)$ using the techniques of Section 10.2. In this section we assume that the series for $f(t)$ has already been determined, so that we may concentrate on Steps 2 through 5 of the procedure.

First consider Step 3 of the procedure. Assume that $q \neq 0$. Then $r = 0$ is not a root of the characteristic equation

$$r^2 + pr + q = 0 \tag{7}$$

for $y'' + py' + qy = f$. The method of undetermined coefficients (Section 2.9) then says that there is a particular solution of (4) of the form $y_0(t) = A_0$, where A_0 is a constant. Substituting this formula for the solution y into (4) gives

$$qA_0 = a_0,$$

so that

$$y_0(t) = \frac{a_0}{q}. \tag{8}$$

If $q = 0$ and $p \neq 0$, then $y_0(t)$ has the form tA_0, and we find that $y_0(t) = ta_0/p$. If $q = 0$ and $p = 0$, then $y_0(t)$ has the form $t^2 A_0$, and we find that $y_0(t) = \frac{1}{2}a_0 t^2$.

Now consider Step 4. Undetermined coefficients applied to (5) give a particular solution of the form

$$y_n(t) = A_n t^k \cos \frac{n\pi t}{L} + B_n t^k \sin \frac{n\pi t}{L}, \tag{9}$$

where

$$k = \begin{cases} 0 & \text{if } \frac{n\pi}{L}i \text{ is not a root of } r^2 + pr + q = 0, \\ 1 & \text{if } \frac{n\pi}{L}i \text{ is a root of } r^2 + pr + q = 0. \end{cases}$$

For convenience, let $\beta_n = n\pi/L$.

Case 1 (k = 0)

Substituting (9) with $k = 0$ into (5) for y, performing the differentiation, and equating the coefficients of $\sin \beta_n t$ and $\cos \beta_n t$ will give two linear equations to solve for A_n, B_n:

$$-\beta_n^2 A_n + \beta_n p B_n + q A_n = a_n,$$
$$-\beta_n^2 B_n - \beta_n p A_n + q B_n = b_n.$$

Solving for A_n, B_n determines a solution $y_n(t)$ of equation (5).

Case 2 (k = 1)

Suppose that $p = 0$ and for some n we have that $\beta_n i = (n\pi/L)i$ is a root of $r^2 + q = 0$. Then the differential equation for $y_n(t)$ is

$$y'' + \beta_n^2 y = a_n \cos \beta_n t + b_n \sin \beta_n t. \tag{10}$$

The method of undetermined coefficients gives a particular solution of the form

$$y_n(t) = A_n t \cos \beta_n t + B_n t \sin \beta_n t. \tag{11}$$

Substituting (11) into (10) for y and then equating the coefficients of $\sin \beta_n t$ and $\cos \beta_n t$ gives

$$-2\beta_n A_n = b_n,$$
$$2\beta_n B_n = a_n,$$

so that

$$A_n = -\frac{b_n}{2\beta_n}, \qquad B_n = \frac{a_n}{2\beta_n}.$$

EXAMPLE 10.4.1 Fourier Series Method for Periodic Forcing Functions

Using the Fourier series method of this section, find a particular solution of

$$y'' + 4y' + y = f(t), \tag{12}$$

where $f(t)$ is a periodic function with period 2π and Fourier series representation

$$f(t) = 3 + \sum_{n=1}^{\infty} \frac{2}{n^2} \sin nt. \tag{13}$$

Solution In Step 3 of the procedure, we first find a particular solution of

$$y'' + 4y' + y = 3. \tag{14}$$

There is a particular solution of the form $y_0 = A_0$. Substituting $y = A_0$ into (14) gives $A_0 = 3$, so that

$$y_0(t) = 3. \tag{15}$$

Next, in Step 4, we find a particular solution of

$$y'' + 4y' + y = \frac{2}{n^2} \sin nt. \tag{16}$$

There is a particular solution of (16) of the form

$$y_n(t) = A_n \cos nt + B_n \sin nt, \tag{17}$$

since ni is not a root of $r^2 + 4r + 1 = 0$ for any integer value of n. Substituting (17) for y in (16) and equating the coefficients of $\sin nt$ and $\cos nt$ gives

$$\begin{aligned} \cos nt:&\ -n^2A_n + 4nB_n + A_n = 0, \\ \sin nt:&\ -n^2B_n - 4nA_n + B_n = \frac{2}{n^2}. \end{aligned}$$

Solving for A_n, B_n, we get

$$A_n = \frac{-8}{n\left[(1-n^2)^2 + 16n^2\right]}, \qquad B_n = \frac{2(1-n^2)}{n^2\left[(1-n^2)^2 + 16n^2\right]}. \tag{18}$$

Thus a particular solution of (12) is

$$\begin{aligned} y(t) &= \sum_{n=0}^{\infty} y_n(t) \\ &= 3 + \sum_{n=1}^{\infty} \frac{-8}{n\left[(1-n^2)^2 + 16n^2\right]} \cos nt + \frac{2(1-n^2)}{n^2\left[(1-n^2)^2 + 16n^2\right]} \sin nt. \end{aligned}$$

◄

The next theorem shows that the above procedure is, in fact, a valid one.

■ **THEOREM 10.4.1 Convergence of Solution by Fourier Series Method**

Suppose that p, q are constants, $f(t)$ is piecewise differentiable and periodic with period $2L$, and the $y_n(t)$ are computed as described in the procedure for solving $y'' + py' + qy = f(t)$ by Fourier series. Then (6) gives a differentiable solution of (1) and the series (6) converges for every t. ■

10.4.1 Frequency Analysis

One of the major uses of Fourier series is in determining the frequencies in the response of a physical system and deciding whether or not there will be resonance at any of these frequencies. There are several forms that this analysis can take.

Suppose that $g(t)$ is a piecewise differentiable function with period $2L$. Then $g(t)$ has a Fourier series representation

$$g(t) = a_0 + \sum_{n=1}^{\infty} a_n \cos \frac{n\pi t}{L} + b_n \sin \frac{n\pi t}{L}. \qquad (19)$$

If n is such that a_n or b_n is nonzero, then we say that $n/(2L)$ is one of the **frequencies** of $g(t)$. (Recall that a term $\cos \beta t$ or $\sin \beta t$ has frequency $\beta/(2\pi)$.)

The **frequencies** of the Fourier series (19) are those $n/(2L)$ for which a_n or b_n are not zero.

In practice, a function $g(t)$ could be the sum of several terms, each giving a Fourier series (19) with different values of L. In this section, except for Exercises 15 and 16, we shall assume that our functions can be given by a single Fourier series (19).

EXAMPLE 10.4.2 Frequencies

Determine the frequencies of

$$g(t) = \sum_{n=1}^{\infty} \frac{1}{2n+1} \cos (2n+1)t.$$

SOLUTION We have $L = \pi$. Thus the frequencies are

$$\frac{(2n+1)}{2\pi}, \qquad n = 1, 2, \ldots.$$ ◄

EXAMPLE 10.4.3 **Frequencies of Forced Response**

Determine the frequencies of the forced response for

$$y'' + 6y = \sum_{n=1}^{\infty} \frac{1}{n^2 + 1} \cos 3nt.$$

SOLUTION The particular solution y_p will contain terms of the form $A_n \cos 3nt + B_n \sin 3nt$, where A_n and B_n are not both zero for any value of n, $n > 1$. Hence the frequencies are $\{(3n)/2\pi \colon n = 1, 2, \ldots\}$. ◀

Note that the free response had the additional frequency of $\sqrt{6}/(2\pi)$.

EXAMPLE 10.4.4 **Resonance**

Suppose that $f(t)$ has period $2L$. For what values of L would

$$y'' + 16y = f(t)$$

possibly exhibit **resonance**?

SOLUTION Suppose that $f(t)$ is piecewise differentiable with period $2L$. Then

$$f(t) = a_0 + \sum_{n=1}^{\infty} a_n \cos \frac{n\pi t}{L} + b_n \sin \frac{n\pi t}{L},$$

where any of the a_n, b_n might be nonzero. The roots of the characteristic polynomial $r^2 + 16$ are $\pm 4i$. Resonance will occur if

$$\frac{n\pi}{L} = 4.$$

Thus the values of L for which resonance might occur are

$$\left\{ L = \frac{n\pi}{4}, n = 1, 2, \ldots \right\}.$$ ◀

Exercises

In Exercises 1 through 10, find a particular solution using the Fourier series technique of this section.

1. $y'' - y = \sum_{n=1}^{\infty} \frac{1}{n} \sin nt$

2. $y'' - y = \sum_{n=2}^{\infty} \frac{1}{n} \cos nt$

3. $y'' + y = \sum_{n=1}^{\infty} \frac{1}{n+1} \cos 2nt$

4. $y'' + y = \sum_{n=1}^{\infty} \frac{1}{n^2} \sin 2nt$

5. $y'' + 4y = \sum_{n=1}^{\infty} \frac{1}{n} \sin nt$

6. $y'' + 4y = \sum_{n=1}^{\infty} \frac{1}{n^2} \cos nt$

7. $y'' + y = \sum_{n=1}^{\infty} \frac{2}{n^2} \sin nt$

8. $y'' + y = 3 + \sum_{n=1}^{\infty} \frac{2}{n^3} \cos nt$

9. $y'' + 2y' + y = \sum_{n=1}^{\infty} \frac{1}{n} \sin 4nt$

10. $y'' - 2y' + y = \sum_{n=1}^{\infty} \frac{1}{n} \cos 4nt$

For Exercises 11 through 20, assume that $f(t)$ is piecewise differentiable with period $2L$ and that the differential equation is

$$y'' + b^2 y = f(t). \tag{20}$$

11. If $b^2 = 12$, what values of L could lead to resonance?

12. If $b^2 = 8$, what values of L could lead to resonance?

13. If $f(t) = \sum_{n=1}^{\infty} \frac{1}{n^2} \sin 3n\pi t$, what values of b will lead to resonance? What are the frequencies of $f(t)$?

14. If $f(t) = \sum_{n=2}^{\infty} \frac{1}{n^3} \cos(2n+1)\pi t$, what values of b will lead to resonance? What are the frequencies of $f(t)$?

15. If $f(t) = \sum_{n=1}^{\infty} \frac{1}{n^2} \sin n\pi t + \sum_{n=1}^{\infty} \frac{1}{n^3} \sin \frac{n\pi t}{\sqrt{3}}$, what values of b will lead to resonance? What are the frequencies of $f(t)$?

16. If $f(t) = \sum_{n=1}^{\infty} \frac{1}{n+1} \cos \frac{2nt}{5} + \sum_{n=2}^{\infty} \frac{(-1)^n}{n^2} \sin \frac{nt}{\sqrt{3}}$, what values of b will lead to resonance? What are the frequencies of $f(t)$?

17. Suppose that you know that the period of $f(t)$ is between 5 and 6. What values of b might lead to resonance?

18. Suppose that you know that the period of $f(t)$ is between 8π and 9π. What values of b might lead to resonance?

19. Suppose that you know that $1 \le b \le 2$. For what periods might $f(t)$ cause resonance?

20. Suppose that you know that $4 \le b \le 5$. For what periods might $f(t)$ cause resonance?

10.5 Regular Sturm-Liouville Problems

In Chapter 11 we shall use Fourier series to solve partial differential equations by the method of separation of variables. However, different partial differential equations will require using different sets of orthogonal functions. Fortunately, the method will tell us which orthogonal functions and which inner product to use. The key turns out to be solving a boundary value problem with a parameter in it. The general theory for such problems is called **Sturm-Liouville theory**. In order to understand the general statements of this theory, it is helpful to first work a specific example.

EXAMPLE 10.5.1 Boundary Value Problem

Find the values of λ for which the boundary value problem

$$y''(x) + \lambda y(x) = 0, \qquad y(0) = y(L) = 0, \tag{1}$$

has a nonzero solution and determine the nonzero solutions.

SOLUTION From Section 2.6 we know that the general solution of the homogeneous linear constant coefficient differential equation $y'' + \lambda y = 0$ can be found from the roots of the characteristic polynomial $r^2 + \lambda$. There are three cases to consider.

Case I ($\lambda = 0$)
If $\lambda = 0$, then the roots are $r = 0, 0$ and

$$y = c_1 + c_2 x. \tag{2}$$

Applying the boundary conditions of (1) to the solution (2) gives

$$0 = y(0) = c_1,$$
$$0 = y(L) = c_1 + c_2 L,$$

so that $c_1 = c_2 = 0$ and $y = 0$. Thus, there are no nonzero solutions of (1) if $\lambda = 0$.

Case II ($\lambda < 0$)
Suppose that $\lambda < 0$, so that $\lambda = -\alpha^2$ for some $\alpha \neq 0$. Then the roots of $r^2 - \alpha^2$ are $r = \pm\alpha$ and the general solution of $y'' + \lambda y = 0$ is

$$y = c_1 e^{\alpha x} + c_2 e^{-\alpha x}. \tag{3}$$

Applying the boundary conditions of (1) to (3) gives

$$0 = y(0) = c_1 + c_2, \tag{4}$$
$$0 = y(L) = c_1 e^{\alpha L} + c_2 e^{-\alpha L}. \tag{5}$$

From (4), $c_2 = -c_1$, so that (5) becomes

$$0 = c_1(e^{\alpha L} - e^{-\alpha L}). \tag{6}$$

If $c_1 \neq 0$ in (6), then $e^{\alpha L} = e^{-\alpha L}$ or $e^{2\alpha L} = 1$, which violates the assumption that $\alpha \neq 0$. Thus the only solution of (6) is $c_1 = 0$. But (4) implies that $c_2 = 0$ also, so that again $y = 0$.

Case III ($\lambda > 0$)
Suppose that $\lambda > 0$, so that $\lambda = \alpha^2$ for some $\alpha > 0$. The roots of $r^2 + \alpha^2$ are $r = \pm\alpha i$, and the general solution of $y'' + \alpha^2 y = 0$ is

$$y = c_1 \cos \alpha x + c_2 \sin \alpha x. \tag{7}$$

Applying the boundary conditions of (1) to (7) gives

$$0 = y(0) = c_1, \tag{8}$$
$$0 = y(L) = c_1 \cos \alpha L + c_2 \sin \alpha L. \tag{9}$$

Equation (8) gives $c_1 = 0$. Thus (9) becomes

$$c_2 \sin \alpha L = 0. \tag{10}$$

To get a nonzero solution, we must have $c_2 \neq 0$. Then (10) implies that

$$\sin \alpha L = 0.$$

Thus, from the properties of the sine function,

$$\alpha L = n\pi, \qquad n \text{ an integer},$$

or

$$\alpha = \frac{n\pi}{L}.$$

Then the values of λ for which (1) has a nonzero solution are

$$\lambda_n = \alpha^2 = \frac{n^2\pi^2}{L^2}, \qquad n = 1, 2, 3, \ldots. \tag{11}$$

The nonzero solutions of (1) for λ_n are

$$y_n(x) = c_2 \sin \frac{n\pi x}{L}, \tag{12}$$

where c_2 is an arbitrary constant. ◄

This example has several features that are typical of a Sturm-Liouville problem.

First, the values of λ given by (11) for which the boundary value problem has a nonzero solution form a sequence of real numbers that goes to ∞ as n goes to ∞. These λ_n are called the **eigenvalues** of the Sturm-Liouville problem. The nonzero solution $\sin(n\pi x/L)$ of the boundary value problem is called the **eigenfunction** corresponding to the eigenvalue $\lambda_n = n^2\pi^2/L^2$.

Second, the nonzero solutions $y_n(x)$ given by (12) form an orthogonal set of functions on the interval $[0, L]$. These orthogonal functions are called the **eigenfunctions** of the Sturm-Liouville problem. Note that the eigenfunctions are unique only up to a constant multiple (since the differential equations and boundary conditions are homogeneous).

This example of a boundary value problem can be generalized.

Regular Sturm-Liouville Problems

Suppose that $r(x)$, $p(x)$, $q(x)$ are given continuous functions on the interval $I = [a, b]$ and that $r(x) > 0$ and $p(x) > 0$ on I. Then a **regular Sturm-Liouville problem** is a differential equation of the form

$$\frac{d}{dx}\left[r(x)\frac{dy}{dx}\right] + [q(x) + \lambda p(x)]y = 0 \tag{13}$$

with boundary conditions

$$\alpha_1 y(a) + \beta_1 y'(a) = 0, \tag{14}$$

$$\alpha_2 y(b) + \beta_2 y'(b) = 0, \tag{15}$$

where neither (14) nor (15) is trivial. That is, α_1, β_1 are not both zero and α_2, β_2 are not both zero.

EXAMPLE 10.5.2 Sturm-Liouville Problem

Example 10.5.1 is a Sturm-Liouville problem where $[a, b] = [0, L]$, $r(x) = p(x) = 1$, $q(x) = 0$, and $\alpha_1 = \alpha_2 = 1$, $\beta_1 = \beta_2 = 0$. ◄

EXAMPLE 10.5.3 Sturm-Liouville Problem

The boundary value problem

$$\begin{gathered}(e^x y')' + [x + \lambda(1 + x^2)]y = 0,\\ y(0) = 0, \\ y(1) - y'(1) = 0,\end{gathered} \tag{16}$$

is a Sturm-Liouville problem on $[0, 1]$ with $r(x) = e^x$, $q(x) = x$, $p(x) = 1 + x^2$, $\alpha_1 = 1$, $\beta_1 = 0$, $\alpha_2 = 1$, $\beta_2 = -1$. ◄

The key theorem that relates Sturm-Liouville problems to our earlier study of orthogonal functions is the following.

THEOREM 10.5.1 Generalized Fourier Series Using Eigenfunctions of Sturm-Liouville Problem

Suppose that $r(x)$, $p(x)$, $q(x)$ are continuous functions on $[a, b]$ and that $r(x) > 0$, $p(x) > 0$ on $[a, b]$. Then the following statements are true for the Sturm-Liouville problem (13), (14), (15):

1. The values of λ for which the Sturm-Liouville problem has a nonzero solution are called the **eigenvalues** of the Sturm-Liouville problem. The eigenvalues are real numbers and can be arranged in a sequence $\{\lambda_n: n \geq 1\}$ such that $\lambda_1 < \lambda_2 < \cdots$ and $\lim_{n\to\infty} \lambda_n = \infty$.
2. For each eigenvalue λ_n, the nonzero solution of (13), (14), (15), denoted $y_n(x)$, is called an **eigenfunction** of the Sturm-Liouville problem. The eigenfunction is unique up to a constant multiple.

3. The eigenfunctions $\{y_n: n \geq 1\}$ form a set of orthogonal functions on the interval $a \leq x \leq b$ with weight function $p(x)$:

$$\int_a^b y_n(x) y_m(x) p(x)\, dx = 0 \qquad (n \neq m). \tag{17}$$

The weight $p(x)$ is the function that multiplies the eigenvalue λ in the defining equation (13).

4. The eigenfunctions $\{y_n: n \geq 1\}$ are a complete set of orthogonal functions with respect to the inner product on $[a, b]$ with weight function $p(x)$. That is, any piecewise differentiable function $f(x)$ can be represented by an infinite series of the orthogonal eigenfunctions:

$$f(x) = \sum_{n=1}^{\infty} c_n y_n(x). \tag{18}$$

The series in (18) is called a **generalized Fourier series.**

5. Properties 3 and 4 can be combined to yield a formula for the **generalized Fourier coefficients** c_n:

$$c_n = \frac{\displaystyle\int_a^b f(x) y_n(x) p(x)\, dx}{\displaystyle\int_a^b y_n^2(x) p(x)\, dx}. \tag{19}$$

(See Theorem 10.1.2.) ■

EXAMPLE 10.5.4 **Fourier Sine Series**

In Example 10.5.1 we had $p(x) = 1$ and found that $y_n(x) = \sin(n\pi x/L)$. From Section 10.3 we know that this is an orthogonal set of functions on $[0, L]$ with respect to the weight $p(x) = 1$. This gives rise to the Fourier sine series:

$$f(x) = \sum_{n=1}^{\infty} c_n \sin \frac{n\pi x}{L},$$

where

$$c_n = \frac{2}{L} \int_0^L f(x) \sin \frac{n\pi x}{L}\, dx.$$ ◄

EXAMPLE 10.5.5 **Eigenfunctions for Sturm-Liouville Problem**

For Example 10.5.3 we had $p(x) = 1 + x^2$. Thus the eigenfunctions will be orthogonal with respect to the weight function $p(x) = 1 + x^2$. However, we have no methods to solve the differential equation (16) and to determine the eigenvalues and corresponding eigenfunctions. There are analytic and numerical methods for solving such boundary value problems, but they are beyond the scope of this text.

Exercises

In Exercises 1 through 9, find the eigenvalues and eigenfunctions of the given Sturm-Liouville problem.

1. $y'' + \lambda y = 0; \quad y(0) = 0, \quad y'(\pi) = 0$

2. $y'' + \lambda y = 0; \quad y'(0) = 0, \quad y(\pi) = 0$

3. $y'' + \lambda y = 0; \quad y'(0) = 0, \quad y'(\pi) = 0$

4. $y'' + 4\lambda y = 0; \quad y(0) = 0, \quad y(\pi) = 0$

5. $y'' + (\lambda + 3)y = 0; \quad y(0) = 0, \quad y(\pi) = 0$

6. $y'' + (\lambda + 3)y = 0; \quad y'(0) = 0, \quad y'(\pi) = 0$

7. $4y'' + \lambda y = 0; \quad y(0) = 0, \quad y'(\pi) = 0$

8. $4y'' + \lambda y = 0; \quad y'(0) = 0, \quad y(\pi) = 0$

9. $4y'' + \lambda y = 0; \quad y(0) = 0, \quad y(\pi) = 0$

10. Suppose $y = c_1e^{rx} + c_2e^{sx}$ with $r \neq s$ and both r, s nonzero. Show that if y satisfies the boundary conditions $y'(0) = 0$, $y'(L) = 0$, then $c_1 = c_2 = 0$.

11. Assuming $\lambda > 0$, find the eigenvalues and eigenvectors of $y'' + \lambda y = 0$, $y(0) = 0$, $y(\pi) + y'(\pi) = 0$. Show that the eigenvalues satisfy $\tan \sqrt{\lambda}\,\pi = -\sqrt{\lambda}$. By sketching the graphs of $y = \tan \pi x$ and $y = -x$, show that the equation $\tan \sqrt{\lambda}\,\pi = -\sqrt{\lambda}$ has a sequence of solutions $\sqrt{\lambda_n}$, $n \geq 1$, such that $\lim_{n\to\infty} \sqrt{\lambda_n} = \infty$.

12. Assuming $\lambda > 0$, find the eigenvalues and eigenvectors of $y'' + \lambda y = 0$, $y'(0) = 0$, $y(\pi) + y'(\pi) = 0$. Show that the eigenvalues satisfy $\tan \sqrt{\lambda}\,\pi = 1/\sqrt{\lambda}$. By sketching the graphs of $y = \tan \pi x$ and $y = 1/x$, show that the equation $\tan \sqrt{\lambda}\,\pi = 1/\sqrt{\lambda}$ has a sequence of solutions $\sqrt{\lambda_n}$, $n \geq 1$, such that $\lim_{n\to\infty} \sqrt{\lambda_n} = \infty$.

In Exercises 13 through 18, let $y_n(x)$ be a complete set of eigenfunctions from the given exercise in this section. Let $f(x)$ be a given piecewise differentiable function. Determine the formula for c_n if

$$f(x) = \sum_{n=1}^{\infty} c_n y_n(x).$$

13. Exercise 1

14. Exercise 2

15. Exercise 3

16. Exercise 4

17. Exercise 11

18. Exercise 12

CHAPTER

11

Partial Differential Equations

11.1 Introduction

In the ordinary differential equations studied in Chapters 1 through 10, the solution depended on one variable—for example, $y = y(x)$ or $x = x(t)$. A **partial differential equation (PDE)** is an equation relating the partial derivatives of a function of more than one variable. Examples are

$$\frac{\partial u}{\partial t} = u^2 \frac{\partial u}{\partial x}, \tag{1}$$

$$\frac{\partial^2 u}{\partial x^2} = 5\frac{\partial u}{\partial y} + \frac{\partial^2 u}{\partial x\, \partial y} + 3\frac{\partial^2 u}{\partial y^2}. \tag{2}$$

Looking at the partial derivatives, we see in (1) that u depends on x and t, $u(x, t)$. In (2) u depends on x and y, $u(x, y)$.

Most techniques for solving partial differential equations reduce the partial differential equation in some manner to an ordinary differential equation. Then, techniques for ordinary differential equations are used. In this chapter, we will discuss only one technique for solving partial differential equations. The technique is known as the method of **separation of variables.** Although this method does not work for every linear partial differential equation, it is valid for many PDEs, including some that model fundamental problems of physical interest, such as the wave and heat equations.

In the next three sections we describe three fundamental situations in which partial differential equations arise and then solve the resulting PDEs by the method of separation of variables. The three examples are chosen both because they are very important PDEs and because they illustrate some of the different types of PDEs that occur.

Exercises

In Exercises 1 through 14, verify that the given functions are solutions of the given partial differential equations.

1. $\dfrac{\partial u}{\partial t} = 5\dfrac{\partial u}{\partial x}$, $u(x, t) = x + 5t$
2. $\dfrac{\partial u}{\partial t} = 5\dfrac{\partial u}{\partial x}$, $u(x, t) = (x + 5t)^2$
3. $\dfrac{\partial u}{\partial t} = 5\dfrac{\partial u}{\partial x}$, $u(x, t) = (x + 5t)^3$
4. $\dfrac{\partial u}{\partial t} = 5\dfrac{\partial u}{\partial x}$, $u(x, t) = \sin(x + 5t)$
5. $\dfrac{\partial u}{\partial t} = 5\dfrac{\partial u}{\partial x}$, $u(x, t) = e^x e^{5t}$
6. $\dfrac{\partial u}{\partial t} = 5\dfrac{\partial u}{\partial x}$, $u(x, t) = \cos(x + 5t)$
7. $\dfrac{\partial^2 u}{\partial x^2} + \dfrac{\partial^2 u}{\partial y^2} = 0$, $u(x, y) = 8e^y \sin x$
8. $\dfrac{\partial^2 u}{\partial x^2} + \dfrac{\partial^2 u}{\partial y^2} = 0$, $u(x, y) = 4e^{-y} \sin x$
9. $\dfrac{\partial^2 u}{\partial x^2} + \dfrac{\partial^2 u}{\partial y^2} = 0$, $u(x, y) = e^{5y} \sin 5x$
10. $\dfrac{\partial^2 u}{\partial x^2} + \dfrac{\partial^2 u}{\partial y^2} = 0$, $u(x, y) = 2e^x \sin y$
11. $\dfrac{\partial^2 u}{\partial x^2} + \dfrac{\partial^2 u}{\partial y^2} = 0$, $u(x, y) = 7e^{-x} \sin y$
12. $\dfrac{\partial^2 u}{\partial x^2} + \dfrac{\partial^2 u}{\partial y^2} = 0$, $u(x, y) = 3e^x \cos y$
13. $\dfrac{\partial u}{\partial t} = u\dfrac{\partial u}{\partial x} + \sin x - t^2 \sin x \cos x$, $u(x, t) = t \sin x$
14. $\dfrac{\partial^2 u}{\partial x^2} + \dfrac{\partial^2 u}{\partial y^2} = -2 \sin x \cos y$, $u(x, y) = \sin x \cos y$

11.2 One-Dimensional Diffusion of a Pollutant

One important use of PDEs is in modeling the diffusion (spreading out) of some quantity, such as heat, a chemical, or a population.

11.2.1 Physical Derivation of the PDE

Derivation Using Thin Regions and Differentials

FIGURE 11.2.1

We consider the concentration of a pollutant in a region chosen for simplicity to be one-dimensional, as illustrated in Figure 11.2.1. For example, the one-dimensional region could be a long, narrow river, where we assume that the concentration of the pollutant is uniform across the river but might vary significantly upstream and downstream, perhaps due to strong sources of pollution at specific locations. We let x measure the distance along the river. We assume that the cross-sectional area A is constant. We introduce the variable $u(x, t)$, the concentration of a pollutant (measured, for example, in grams per liter) at position x at time t. Although it represents a physical problem in real three-dimensional space, we call our problem **one-dimensional** because the variable u depends on only one spatial variable x. Initially we shall assume that there is no current in the river. Technically, we are assuming that the pollutant spreads by diffusion but not by transport. That is, we actually have a lake or reservoir. Later in this section, the effect of current (transport) will be included under the heading of "Fluid Media."

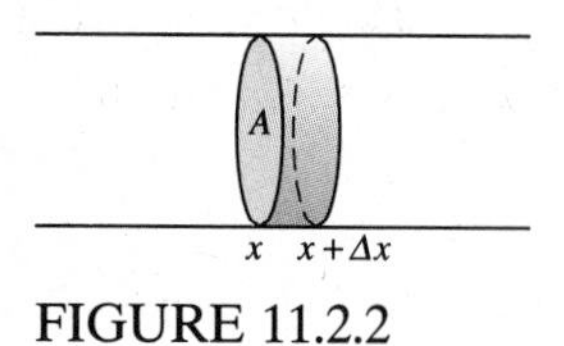

FIGURE 11.2.2

One way to derive a partial differential equation modeling the change in pollutant level is to consider how the amount of pollution changes over time in a simple thin section of the river, the region from x to $x + \Delta x$ (see Figure 11.2.2). To be thin, Δx must be small. The volume of this thin region is $A\,\Delta x$. Since the thin region is thin, the concentration is approximately constant throughout the region. The amount of pollutant in this thin region can be accurately approximated by the concentration u at a point (x, t) in the thin region times the volume $A\,\Delta x$:

$$\begin{bmatrix} \text{Amount of pollutant} \\ \text{in thin region} \end{bmatrix} = u(x,t)A\,\Delta x.$$

The amount of a pollutant changes as a result of the pollutant's entering and leaving the boundaries of the thin region (at x and $x + \Delta x$) and perhaps as a result of sources of the pollutant in the region. Our word equation is

$$\begin{bmatrix} \text{Rate of change} \\ \text{of the amount} \\ \text{of pollutant} \end{bmatrix} = \begin{bmatrix} \text{net amount of} \\ \text{pollutant entering} \\ \text{per unit time} \end{bmatrix} + \begin{bmatrix} \text{amount of pollutant} \\ \text{generated inside region} \\ \text{per unit time} \end{bmatrix}. \quad (1)$$

We introduce the **flux** $q(x, t)$ of the pollutant:

$$q(x,t) = \begin{bmatrix} \text{amount of pollutant flowing to the right at } x \\ \text{per unit time per unit surface area} \end{bmatrix},$$

so that

$$q(x+\Delta x,t) = \begin{bmatrix} \text{amount of pollutant flowing to the right at } x+\Delta x \\ \text{per unit time per unit surface area} \end{bmatrix}.$$

Thus the net amount of pollutant entering the thin region is $q(x,t)A - q(x + \Delta x, t)A$. The minus sign indicates that if the pollutant is flowing to the right at $x + \Delta x$, then it contributes to a decrease in the pollutant inside the region. The **source** $g(x, t)$ of the pollutant per unit volume per unit time is

$$g(x,t) = \begin{bmatrix} \text{amount of pollutant generated inside region} \\ \text{per unit time per unit volume} \end{bmatrix}.$$

As an approximation, the total source equals the source per unit volume $g(x, t)$ multiplied by the volume $A\,\Delta x$. Thus, (1) becomes

$$\frac{\partial}{\partial t}[u(x,t)A\,\Delta x] = q(x,t)A - q(x+\Delta x,t)A + g(x,t)A\,\Delta x. \quad (2)$$

The partial derivative in time is used because it stands for the ordinary derivative with respect to time with x fixed. The flows per unit surface area must be multiplied by the surface area A.

Dividing (2) by $A\,\Delta x$ yields the approximation

$$\frac{\partial u}{\partial t} = \frac{q(x,t) - q(x+\Delta x,t)}{\Delta x} + g(x,t). \tag{3}$$

We now take the limit of both sides of (3) as $\Delta x \to 0$. We claim (as shown in the next subsection) that the errors in this approximation vanish as $\Delta x \to 0$. Since we recognize the limit in the right-hand side of (3) to be the partial derivative with respect to x (keeping t fixed),

$$\frac{\partial q}{\partial x} = \lim_{\Delta x \to 0} \frac{q(x+\Delta x,t) - q(x,t)}{\Delta x},$$

we derive

$$\frac{\partial u}{\partial t} = -\frac{\partial q}{\partial x} + g(x,t). \tag{4}$$

We usually assume that the sources of pollution $g(x,t)$ are known. The mathematical and physical problem is still interesting if there are no sources, in which case $g(x,t) = 0$.

Equation (4) represents a fundamental law for the rate of change of the concentration of the pollutant (and many other physical quantities). Unfortunately, (4) is one partial differential equation in two unknowns $u(x,t)$ and $q(x,t)$. Before describing a relation for $q(x,t)$, we briefly rederive (4) in a manner that removes some the approximate nature of various expressions.

Derivation Using Large Regions and Integrals

We again consider a one-dimensional region (with constant cross-sectional area A) as illustrated in Figure 11.2.1. Instead of considering a thin region, we can express the total amount of pollutant in any finite region from $x = a$ to $x = b$

$$\text{Amount of pollutant} = \int_a^b u(x,t)A\,dx.$$

Thus, an exact expression for the rate of change of the amount of pollution in this finite region is

$$\begin{bmatrix}\text{Rate of change}\\ \text{of the amount}\\ \text{of pollutant}\end{bmatrix} = \begin{bmatrix}\text{net amount of}\\ \text{pollutant entering}\\ \text{per unit time}\end{bmatrix} + \begin{bmatrix}\text{amount of pollutant}\\ \text{generated inside region}\\ \text{per unit time}\end{bmatrix}$$

or

$$\frac{d}{dt}\int_a^b u(x,t)A\,dx = [q(a,t)A - q(b,t)A] + \int_a^b g(x,t)A\,dx. \tag{5}$$

We can cancel the cross-sectional area A at any time, since A is a constant.

One way to derive a partial differential equation for the chemical concentration $u(x,t)$ is to reduce all expressions in (5) to integrals. According to the rules for differentiating an integral with respect to a different variable,

$$\frac{d}{dt}\int_a^b u(x,t)\,dx = \int_a^b \frac{\partial u(x,t)}{\partial t}\,dx. \tag{6}$$

With a small trick, the boundary contribution $q(a,t) - q(b,t)$ can be expressed as an integral:

$$\int_a^b \frac{\partial q(x,t)}{\partial x}\,dx = q(x,t)\Big|_{x=a}^{x=b} = q(b,t) - q(a,t). \tag{7}$$

Using (6) and (7), (5) becomes (after canceling the cross-sectional area)

$$\int_a^b \left(\frac{\partial u}{\partial t} + \frac{\partial q}{\partial x} - g\right) dx = 0. \tag{8}$$

Since the limits of the integral in (8) are arbitrary, it follows that the integral can be zero only if the integrand is zero:

$$\frac{\partial u}{\partial t} + \frac{\partial q}{\partial x} - g = 0. \tag{9}$$

Equation (9) is the same as (4).

We now turn to finding an expression for q. Then we will eliminate q from (9) in order to get a PDE just in the unknown $u(x,t)$.

Fick's Law

In solids, "chemicals diffusing" means that their concentrations spread out. Chemicals do not diffuse if the concentration of the chemical is the same everywhere. Instead, experimentally, it is observed that concentrations of chemicals (atoms) flow from regions of high concentration to regions of low concentration. In many situations, **Fick's law** holds:

$$\left[\begin{array}{c}\text{Amount of pollutant}\\ \text{flowing to the right at } x\\ \text{per unit time}\\ \text{per unit surface area}\end{array}\right] \text{ is proportional to } \left[\begin{array}{c}\text{rate of change}\\ \text{of the concentration}\\ \text{per unit distance}\end{array}\right],$$

or

$$q = -k\frac{\partial u}{\partial x}. \tag{10}$$

The minus sign in Fick's law accounts for the flow from high concentration to low. If the chemical has higher concentration to the right ($\partial u/\partial x > 0$), then atoms of the chemical migrate to the left, and vice versa. The proportionality constant k (measured experimentally) is called the **diffusivity**. In fluids (liquids or gases), in situations in which the flow of the fluid can be ignored, the spreading of chemicals also satisfies (10).

Using Fick's law (10) to eliminate q, the fundamental equation (4) or (9) for the rate of change of the chemical concentration becomes

$$\frac{\partial u}{\partial t} = \frac{\partial}{\partial x}\left(k\frac{\partial u}{\partial x}\right) + g,$$

or

$$\frac{\partial u}{\partial t} = k\frac{\partial^2 u}{\partial x^2} + g \tag{11}$$

when k is constant (which is frequently the case). Equation (11) is a partial differential equation for the concentration of the pollutant. It is known as the **diffusion equation** with sources g.

Linearity

If $g = g(x,t)$, (11) is a linear partial differential equation. It is linear because the partial differential operator

$$L(u) \equiv \frac{\partial u}{\partial t} - k\frac{\partial^2 u}{\partial x^2}$$

can be shown to satisfy the **linearity property**

$$L(c_1u_1 + c_2u_2) = c_1L(u_1) + c_2L(u_2).$$

The linear partial differential equation is **homogeneous** if $g = 0$ and **nonhomogeneous** if $g \neq 0$.

Fluid Media

Suppose we are interested in the change of the concentration of a pollutant, but that there is a one-dimensional fluid medium with fluid velocity $c(x,t)$ in which the chemical particles are **convected** (carried along). In terms of the river we discussed earlier, we are now allowing for a current (velocity) at position x at time t. In a fluid medium, Fick's law for the flow of the pollutant per unit time per unit surface area is supplemented by the amount of the chemical carried by the fluid:

$$q = -k\frac{\partial u}{\partial x} + c(x,t)u(x,t). \tag{12}$$

To understand this extra term, $c(x,t)u(x,t)$, consider the flow of pollutant at a fixed point x over a very short time interval Δt. During this short time, the fluid will have moved a distance $c(x,t)\,\Delta t$ to the right. As an approximation, the volume of fluid crossing x will be the distance multiplied by the surface area: $c(x, t)A\Delta t$. The amount of chemical crossing the boundary will be the volume multiplied by the concentration: $c(x, t)A\Delta t u(x, t)$. Thus, the flow of pollutant per unit surface area per unit time due to the fluid flow is

$$\frac{c(x,t)A\,\Delta t u(x,t)}{A\,\Delta t} = c(x,t)u(x,t).$$

If (12) is substituted into the fundamental equation (4) or (9) for the rate of change of the concentration, the following partial differential equation is derived:

$$\frac{\partial u}{\partial t} + \frac{\partial}{\partial x}(cu) = k\frac{\partial^2 u}{\partial x^2} + g. \tag{13}$$

Equation (13) accounts for both diffusion and convection of a chemical pollutant.

Application to Thermal Conduction (Heat Equation)

A similar partial differential equation describes the diffusion of thermal energy in a one-dimensional solid:

$$\frac{\partial u}{\partial t} = k\frac{\partial^2 u}{\partial x^2} + g.$$

Here $u(x,t)$ is the temperature at position x at time t, k the thermal diffusivity, and $g(x,t)$ results from possible sources of thermal energy distributed throughout the solid. Sometimes this partial differential equation is called the **heat equation** instead of the diffusion equation. The derivation of this equation is outlined in the exercises and is similar to the derivation of the diffusion of a chemical pollutant.

11.2.2 Method of Separation of Variables

Boundary and Initial Conditions

The method of separation of variables is applied to linear partial differential equations that are homogeneous ($u = 0$ is a solution). Here, we examine the diffusion of a pollutant without sources, where the concentration $u(x, t)$ satisfies the diffusion equation

$$\frac{\partial u}{\partial t} = k\frac{\partial^2 u}{\partial x^2}, \tag{14}$$

in a finite region ($0 \le x \le L$). There is a unique solution if we specify the concentration of the pollutant at the boundaries $x = 0$ and $x = L$ (called the **boundary conditions**),

$$\begin{aligned} u(0,t) &= A, \\ u(L,t) &= B, \end{aligned} \tag{15}$$

and we give the initial ($t = 0$) distribution of the pollutant (called the **initial conditions**),

$$u(x,0) = f(x). \tag{16}$$

Note that (14) involves the second partial derivative with respect to the space variable x and (15) has two boundary conditions, and that (14) involves the first partial derivative with respect to time and (16) has only one initial condition. A different physical problem would be to suppose that instead of the boundary level of pollution being given, there was an insulated boundary in which there was no flow of pollutant across the boundary. Because of Fick's law, the boundary condition for an **insulated boundary** would be

$$\frac{\partial u}{\partial x} = 0 \quad \text{at an insulated boundary.}$$

Product Solutions

The method of separation of variables works only if the boundary condition is homogeneous, meaning that $u(x,t) = 0$ satisfies the boundary condition. For that reason, we first study the initial value problem with the level of pollution zero at the boundaries:

$$\frac{\partial u}{\partial t} = k\frac{\partial^2 u}{\partial x^2} \quad \text{(PDE)}, \tag{17}$$

$$u(0,t) = 0 \quad \text{(BC1)}, \tag{18}$$

$$u(L,t) = 0 \quad \text{(BC2)}, \tag{19}$$

$$u(x,0) = f(x) \quad \text{(IC)}. \tag{20}$$

The problem consists of a PDE, (17), two boundary conditions, (18) and (19), and an initial condition, (20). If the boundary conditions are not homogeneous, the method we will describe must be modified.

For the moment, we ignore the initial condition (20). A key idea in separation of variables is to look for simple nonzero **product solutions** of the partial differential equation (17). Thus, we look for solutions in the product form

$$u(x,t) = \phi(x)h(t). \tag{21}$$

Here $\phi(x)$ is a function only of x and $h(t)$ is a function only of t. Substituting (21) into (17) for u and performing the differentiations yields

$$\phi(x)\frac{dh}{dt} = kh(t)\frac{d^2\phi}{dx^2}. \tag{22}$$

If we divide by $k\phi(x)h(t)$, we can **separate** the variables (hence the name the method of separation of variables):

$$\frac{1}{kh}\frac{dh}{dt} = \frac{1}{\phi}\frac{d^2\phi}{dx^2}. \tag{23}$$

Dividing by k was convenient but not necessary. The left-hand side of (23) is a function only of t and the right-hand side is a function only of x. Since x and t are independent variables, the only way a function of x can equal a function of t is if both functions are the same constant, the **separation constant**. For convenience, we call this constant $-\lambda$:

$$\frac{1}{kh}\frac{dh}{dt} = \frac{1}{\phi}\frac{d^2\phi}{dx^2} = -\lambda. \tag{24}$$

From (24), we derive two ordinary differential equations, one for $h(t)$ and the other for $\phi(x)$:

$$\frac{d^2\phi}{dx^2} = -\lambda\phi, \tag{25}$$

$$\frac{dh}{dt} = -\lambda kh. \tag{26}$$

Equation (26) is a first-order linear homogeneous differential equation with constant coefficients. The solution of the time-dependent equation (26) is

$$h(t) = ce^{-\lambda kt}. \tag{27}$$

The time-dependent part exponentially decays if $\lambda > 0$, exponentially grows if $\lambda < 0$, and is constant in time if $\lambda = 0$. In the next subsection, we will determine the separation constant λ. For the diffusion of a pollutant, physical intuition suggests that exponential growth in time is impossible. We will show this mathematically.

Boundary Value Problem for the Separation Constants

The spatial functions $\phi(x)$ of the product solutions satisfy (25). In addition, we insist that the product solutions satisfy the homogeneous boundary conditions (18) and (19). For example, if the product form (21) is substituted into (18), we obtain

$$\phi(0)h(t) = 0. \tag{28}$$

We are looking for nonzero product solutions, and so we need $h(t) \not\equiv 0$. Thus, from (28), we conclude that $\phi(x)$ itself must satisfy the boundary condition:

$$\phi(0) = 0.$$

The same argument at $x = L$ shows that ϕ also satisfies

$$\phi(L) = 0.$$

The ordinary differential equation for ϕ does not satisfy an initial value problem, but instead satisfies a **boundary value problem** consisting of the differential equation

$$\frac{d^2\phi}{dx^2} = -\lambda\phi \tag{29}$$

with two boundary conditions,

$$\phi(0) = 0 \tag{30}$$

and

$$\phi(L) = 0. \tag{31}$$

In some sense, it is good that this is a boundary value problem. One solution, called the **trivial solution,** of this boundary value problem is $\phi(x) = 0$ for all x. This solution is not very interesting physically, since it corresponds to the chemical concentration $u(x,t) = 0$ for all x and t. Note that this satisfies the differential equation and the boundary conditions. If the boundary value problem had a unique solution, the unique solution would be $u(x,t) = 0$. Fortunately, we will show that boundary value problems may have nonunique solutions. This is different from the usual initial value problems for ordinary differential equations. We will show there are only certain special values of λ, known as **eigenvalues**, for which there are nontrivial solutions of the boundary value problem (29) through (31). Corresponding to an eigenvalue, there are nontrivial solutions $\phi(x)$ called **eigenfunctions**:

Eigenvalues:	λ for which (29) through (31) has a nonzero solution.
Eigenfunctions:	Nonzero solutions of (29) through (31).

Boundary value problems like (29), (30), (31) were discussed in Section 10.5.

To solve the boundary value problem and find the eigenvalues and corresponding eigenfunctions, we first find the general solution of the ordinary differential equation (29), and then attempt to satisfy the two boundary conditions (30) and (31). Equation (29) is a linear constant coefficient differential equation (Chapter 2). The characteristic equation [based on $\phi(x) = e^{rx}$] for (29) is

$$r^2 = -\lambda. \tag{32}$$

We first show that in this example, $\lambda = 0$ is not an allowable separation constant. If $\lambda = 0$, then (29) is $d^2\phi/dx^2 = 0$. The general solution is

$$\phi(x) = c_1 + c_2 x.$$

The boundary condition at $x = 0$, (30), implies that

$$\phi(0) = 0 = c_1.$$

The boundary condition at $x = L$, (31), implies that

$$\phi(L) = 0 = c_2 L,$$

so that $c_2 = 0$. Thus, $\phi(x) = 0$ for all x if $\lambda = 0$. We conclude that $\lambda = 0$ is not an eigenvalue.

Next we assume $\lambda > 0$. (In the exercises and in Section 10.5 it is shown that there are no nontrivial solutions *in this example* for $\lambda < 0$.) If $\lambda > 0$, the roots of the characteristic equation (32) are $\pm i\sqrt{\lambda}$. The general solution of (29) is then

$$\phi(x) = c_1 \cos\sqrt{\lambda}\,x + c_2 \sin\sqrt{\lambda}\,x. \tag{33}$$

The boundary condition, (30), at $x = 0$ implies that

$$\phi(0) = 0 = c_1.$$

Since $c_1 = 0$, we have

$$\phi(x) = c_2 \sin\sqrt{\lambda}\,x. \tag{34}$$

The boundary condition at $x = L$, (31), implies that

$$\phi(L) = 0 = c_2 \sin\sqrt{\lambda}\,L. \tag{35}$$

If $c_2 = 0$, then $\phi(x) = 0$ for all x. But we are looking for nonzero solutions. Assuming that $c_2 \neq 0$, (35) implies that

$$\sin\sqrt{\lambda}\,L = 0. \tag{36}$$

The sine function is zero at integral multiples of π. Thus, (36) implies that the eigenvalues (separation constants) satisfy

$$\sqrt{\lambda}\,L = n\pi \qquad \text{for } n = 1, 2, 3, \ldots,$$

or equivalently,

$$\lambda = \left(\frac{n\pi}{L}\right)^2 \qquad \text{for } n = 1,2,3,\ldots. \tag{37}$$

We exclude $n = 0$, since then $\lambda = 0$, and we have shown that $\lambda > 0$. Thus, from (34), the eigenfunctions (spatial part of the product solutions) satisfy

$$\phi(x) = c_2 \sin\frac{n\pi x}{L} \qquad \text{for } n = 1,2,3,\ldots. \tag{38}$$

The constant c_2 is arbitrary. If there is a nontrivial solution of the homogeneous differential equation satisfying the homogeneous boundary conditions, then any multiple will also satisfy the boundary value problem as a result of linearity.

Summary of the Boundary Value Problem

$$\frac{d^2\phi}{dx^2} = -\lambda\phi \qquad \text{(DE)}$$

$$\phi(0) = 0, \quad \phi(L) = 0 \qquad \text{(BC)}$$

Eigenvalues: $\lambda = \left(\frac{n\pi}{L}\right)^2$ for $n = 1,2,3,\ldots,$

Eigenfunctions: $\sin\frac{n\pi x}{L}$ for $n = 1,2,3,\ldots.$

From (27) and (38), we have obtained product solutions

$$u(x,t) = B\sin\frac{n\pi x}{L}e^{-\lambda kt} = B\sin\frac{n\pi x}{L}e^{-(n\pi/L)^2kt}, \tag{39}$$

where B is an arbitrary constant, of the diffusion equation

$$\frac{\partial u}{\partial t} = k\frac{\partial^2 u}{\partial x^2}. \tag{40}$$

It is easy to verify (after the fact) by direct substitution that the product solution (39) (for any one fixed value of n) satisfies the linear partial differential equation (40). It is even easier to check that these product solutions satisfy the boundary conditions $u(0) = 0$ and $u(L) = 0$. For different values of n, the constant B may be different, and we denote it B_n instead of B.

Initial Value Problem

These product solutions are analogous to solutions of homogeneous linear differential equations. For an nth-order homogeneous linear differential equation, there are n linearly independent solutions, and the general solution is a linear combination of these n solutions. For the linear homogeneous partial differential equation (40), there are an infinite number of solutions (39) for $n = 1, 2, 3, \ldots$. A linear combination of all of these product solutions will also be a solution:

$$u(x,t) = \sum_{n=1}^{\infty} B_n \sin \frac{n\pi x}{L} e^{-(n\pi/L)^2 kt}, \tag{41}$$

where the B_n are arbitrary constants. For nth-order linear ordinary differential equations, the n constants are determined from the n initial conditions. For the diffusion equation (40), the arbitrary constants B_n are determined from the initial condition (20), the spatial dependence of the chemical concentration at $t = 0$,

$$u(x,0) = f(x), \qquad \text{for } 0 < x < L, \tag{42}$$

where $f(x)$ is given.

There are difficult questions, such as whether the infinite series (41) converges. Even if the series converges, it is not a trivial question to prove that the infinite series actually solves the partial differential equation. We will avoid these questions, and simply say that in scientific and engineering practice, these series are very useful.

We now ensure that the solution (41) satisfies the initial condition (20). Letting $t = 0$ gives

$$u(x,0) = f(x) = \sum_{n=1}^{\infty} B_n \sin \frac{n\pi x}{L}, \qquad \text{for } 0 < x < L. \tag{43}$$

From Section 10.3, we recognize (43) as the Fourier sine series representation of the function $f(x)$ over the interval $0 < x < L$. Given the initial condition $f(x)$, the coefficients must equal the coefficients of the Fourier sine series, so that

$$B_n = \frac{2}{L} \int_0^L f(x) \sin \frac{n\pi x}{L} \, dx. \tag{44}$$

We note that the integral is over the physical region $0 < x < L$. Equation (44) can be derived by multiplying (43) by $\sin m\pi x/L$, integrating from $x = 0$ to $x = L$, and using the fact that the eigenfunctions $\sin n\pi x/L$, $n = 1, 2, 3, \ldots$, form an orthogonal set.

Summary

The solution of the initial value problem for the diffusion equation (40) is (41), where the coefficients are given by (44), the Fourier sine coefficients of the initial condition $f(x)$.

The solution (41) has the property that $u(x,t) \to 0$ as $t \to \infty$ for all initial conditions. Thus, no matter how polluted the region is initially, the pollution-free boundary condition eventually dominates the interior. The chemicals initially present eventually diffuse out of the region.

Each product solution exponentially decays in time at its own rate. If t is large but finite, a good approximation is to just keep the first term with a nonzero coefficient, since all the others decay faster. If $B_1 \neq 0$, this approximation (for large time) would be

$$u(x,t) \approx B_1 \sin \frac{\pi x}{L} e^{-(\pi/L)^2 kt}.$$

EXAMPLE 11.2.1 **Initial Value Problem for the Diffusion Equation**

Solve the diffusion equation (40) with $u(0,t) = 0$ and $u(L,t) = 0$, subject to the initial condition $u(x,0) = 1$.

Solution From (41), the solution of the partial differential equation (based on the method of separation of variables) is

$$u(x,t) = \sum_{n=1}^{\infty} B_n \sin \frac{n\pi x}{L} e^{-(n\pi/L)^2 kt}.$$

Here, $u(x, 0) = f(x) = 1$. The initial condition is satisfied if

$$1 = \sum_{n=1}^{\infty} B_n \sin \frac{n\pi x}{L}, \qquad \text{for } 0 < x < L.$$

According to the theory of Fourier sine series, the coefficients satisfy

$$B_n = \frac{2}{L}\int_0^L 1 \cdot \sin \frac{n\pi x}{L}\, dx = -\frac{2}{n\pi} \cos \frac{n\pi x}{L}\Big|_{x=0}^{L}$$

$$= \frac{2}{n\pi}(1 - \cos n\pi) = \begin{cases} 0 & \text{if } n \text{ even}, \\ \dfrac{4}{n\pi} & \text{if } n \text{ odd}. \end{cases}$$

Summing over just the odd terms ($n = 2i + 1$), we have

$$u(x,t) = \sum_{i=0}^{\infty} \frac{4}{(2i+1)\pi} \sin \frac{(2i+1)\pi x}{L} e^{-[(2i+1)\pi/L]^2 kt}. \tag{45}$$

◄

Discussion

It is instructive to examine this solution more carefully. Note that we can consider this problem either as the diffusion of a uniformly distributed pollutant or as the cooling of a uniformly heated metal rod that is insulated along the sides so that heat can flow only out the ends.

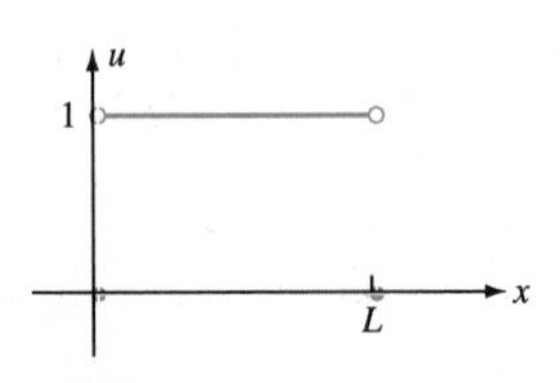

FIGURE 11.2.3

Initial distribution of pollutant.

The initial distribution of pollutant is graphed in Figure 11.2.3. There are two standard ways to show the variation with respect to time. One is to graph the pollutant profile (u as a function of x) at several selected times, as in Figure 11.2.4. Alternatively, we can give a 3-D plot of $u(x, t)$ as a function of x and t, as in Figure 11.2.5.

In practice, it is impossible to sum the infinite number of terms required by (45). Instead, we often sum a large number of terms of an eigenfunction expansion (and in some situations, like this one, only the first few terms may be needed). Let $P_m(x, t)$ be the sum of the first m nonzero terms in (45):

$$P_m(x, t) = \sum_{i=0}^{m-1} \frac{4}{(2i + 1)\pi} \sin \frac{(2i + 1)\pi x}{L} e^{-[(2i+1)\pi/L]^2 kt}. \tag{46}$$

This P_m is the same P_{2m-1} of Section 10.2. In Figures 11.2.6 through 11.2.8, we give 3-D graphs of P_m for $m = 1, 3, 8$. Notice that it takes several terms to try to approximate $u(x, t)$ for t close to zero but fewer terms for larger t. This is typical in diffusion problems. Finally, in Figure 11.2.9, we give a 3-D graph

FIGURE 11.2.4

$u(x, t)$ for several t values.

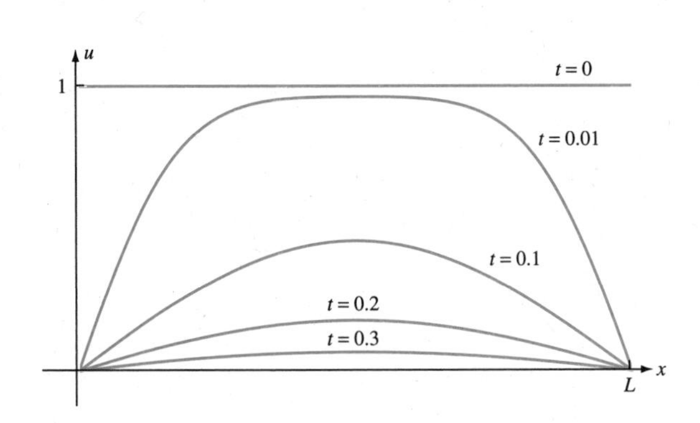

FIGURE 11.2.5

Graph of $u(x, t)$ in (45) with $0 \leq x \leq L$, $0 \leq t \leq 0.5$, and $k = L = 1$.

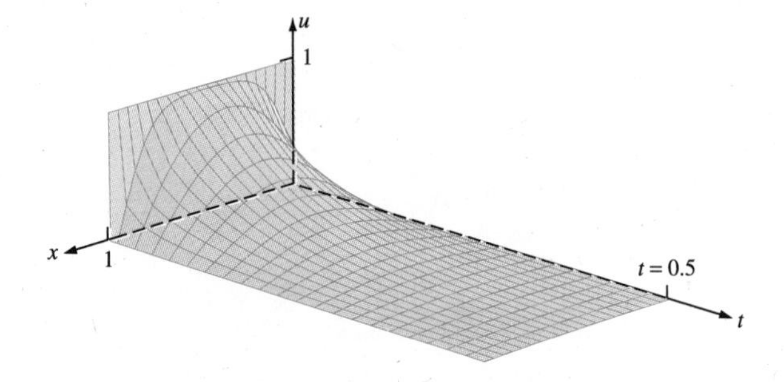

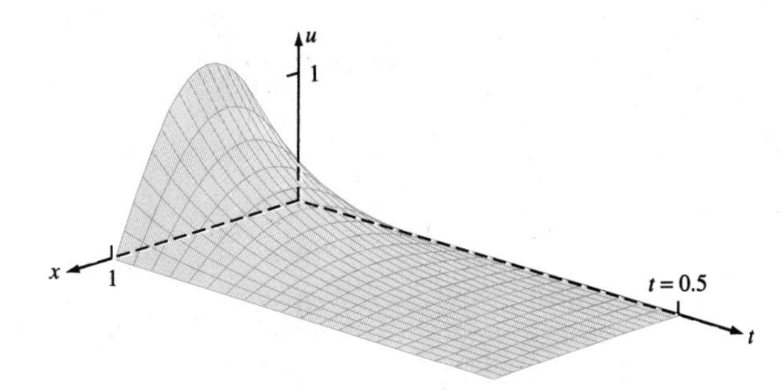

FIGURE 11.2.6

Graph of $P_1(x, t)$.

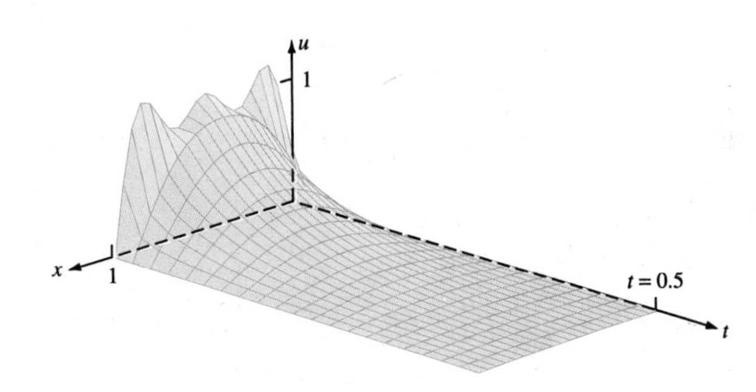

FIGURE 11.2.7

Graph of $P_3(x, t)$.

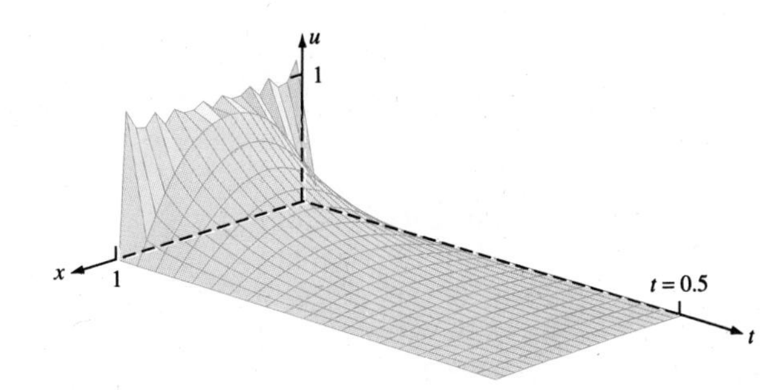

FIGURE 11.2.8

Graph of $P_8(x, t)$.

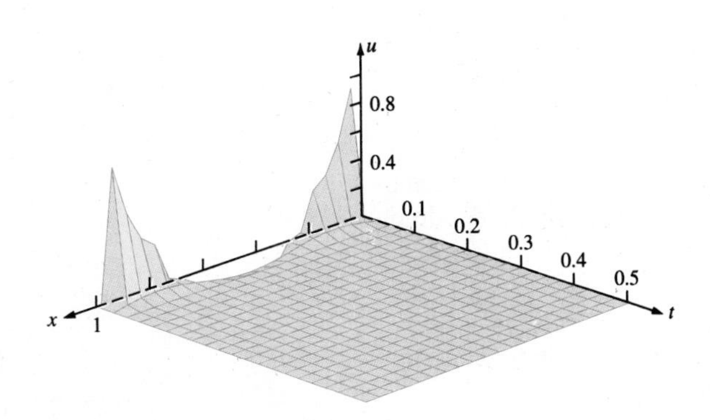

FIGURE 11.2.9

Error graph, $u(x, t) - P_1(x, t)$.

of the difference between $u(x,t)$ and $P_1(x,t)$. This is a graph of the error in using just the first term as an approximation. Notice how the error goes to zero as t increases. (Actually u also goes to zero, but more slowly. For larger t, the key point is that the percentage error goes to zero also.)

Exercises

In Exercises 1 through 10, the eigenvalues λ of the boundary value problem $d^2\phi/dx^2 = -\lambda\phi$ depend on the boundary condition. Determine the eigenvalues and eigenfunctions. You may assume $\lambda \geq 0$, but check $\lambda = 0$ separately.

1. $\phi(0) = 0$ and $\phi(\pi) = 0$

2. $\phi(0) = 0$ and $\phi(1) = 0$

3. $\phi(0) = 0$ and $\phi(H) = 0$

4. $\phi'(0) = 0$ and $\phi'(\pi) = 0$

5. $\phi'(0) = 0$ and $\phi'(1) = 0$

6. $\phi(0) = 0$ and $\phi'(L) = 0$

7. $\phi'(0) = 0$ and $\phi(L) = 0$

8. $\phi(-L) = \phi(L)$ and $\phi'(-L) = \phi'(L)$

9. $\phi(-\pi) = \phi(\pi)$ and $\phi'(-\pi) = \phi'(\pi)$

10. $\phi(0) = 0$ and $\phi(L) + \phi'(L) = 0$ [*Hint*: Determine eigenvalues graphically from the equation λ must satisfy.]

11. Show that there are no negative eigenvalues λ for $d^2\phi/dx^2 = -\lambda\phi$ with $\phi(0) = 0$ and $\phi(L) = 0$.

12. Show that there are no negative eigenvalues λ for $d^2\phi/dx^2 = -\lambda\phi$ with $\phi'(0) = 0$ and $\phi'(L) = 0$.

13. Solve the diffusion equation $\partial u/\partial t = k(\partial^2 u/\partial x^2)$ with $u(0,t) = 0$ and $u(L,t) = 0$ subject to the initial condition $u(x,0) = 5\sin(\pi x/L) + 8\sin(3\pi x/L)$.

14. Solve the diffusion equation $\partial u/\partial t = k(\partial^2 u/\partial x^2)$ with $u(0,t) = 0$ and $u(L,t) = 0$ subject to the initial condition $u(x,0) = 2\sin(\pi x/L) - 5\sin(3\pi x/L) + 9\sin(4\pi x/L)$.

15. Solve the diffusion equation $\partial u/\partial t = k(\partial^2 u/\partial x^2)$ with $u(0,t) = 0$ and $u(L,t) = 0$ subject to the initial condition $u(x,0) = 5\cos(\pi x/L)$.

16. Solve the diffusion equation $\partial u/\partial t = k(\partial^2 u/\partial x^2)$ with $u_x(0,t) = 0$ and $u_x(L,t) = 0$ subject to the initial condition $u(x,0) = f(x)$.

17. Solve the diffusion equation $\partial u/\partial t = k(\partial^2 u/\partial x^2)$ with $u_x(0,t) = 0$ and $u_x(L,t) = 0$ subject to the initial condition $u(x,0) = 1$.

18. Solve the diffusion equation $\partial u/\partial t = k(\partial^2 u/\partial x^2)$ with $u_x(0,t) = 0$ and $u_x(L,t) = 0$ subject to the initial condition $u(x,0) = 4 + 2\cos(\pi x/L)$.

19. Solve the diffusion equation $\partial u/\partial t = k(\partial^2 u/\partial x^2)$ with $u_x(0,t) = 0$ and $u_x(L,t) = 0$ subject to the initial condition $u(x,0) = \sin(\pi x/L)$.

20. Suppose a one-dimensional region (with a small cross-sectional area) from $x = -L$ to $x = L$ is twisted into the shape of a ring, so that the position $x = -L$ corresponds also to the position $x = L$.

a) The concentration of the pollutant will be the same at $x = -L$ and $x = L$. Also, the flow of the pollutant will be the same at $x = -L$ and $x = L$. Show that

$$u(-L,t) = u(L,t),$$
$$\frac{\partial u}{\partial x}(-L,t) = \frac{\partial u}{\partial x}(L,t).$$

b) Solve the diffusion equation $\partial u/\partial t = k(\partial^2 u/\partial x^2)$ with $u(-L,t) = u(L,t)$ and $u_x(-L,t) = u_x(L,t)$ subject to the initial condition $u(x,0) = f(x)$.

21. If there are sources of pollution that are proportional to the concentration with proportionality constant 2, then $u(x,t)$ satisfies $\partial u/\partial t = k(\partial^2 u/\partial x^2) + 2u$. Solve this equation with $u(0,t) = 0$ and $u(L,t) = 0$ subject to the initial condition $u(x,0) = f(x)$.

22. Negative sources of pollution correspond to the chemical being removed from the region. If there are negative sources of pollution that are proportional to the concentration (with proportionality constant -2), then $u(x,t)$ satisfies $\partial u/\partial t = k(\partial^2 u/\partial x^2) - 2u$. Solve this equation with $u(0,t) = 0$ and $u(L,t) = 0$ subject to the initial condition $u(x,0) = f(x)$.

23. Solve the diffusion equation $\partial u/\partial t = k(\partial^2 u/\partial x^2)$ with $u(0,t) = 5$ and $u(L,t) = 8$ subject to the initial condition $u(x,0) = f(x)$. This problem cannot be solved by the method of separation of variables without some modification.

a) Find an equilibrium distribution of pollution $u_0(x)$, satisfying $d^2u_0/dx^2 = 0$ with $u_0(0) = 5$ and $u_0(L) = 8$.

b) What problem (partial differential equation, boundary condition, and initial condition) does the difference between $u(x,t)$ and $u_0(x)$ satisfy? [*Hint*: Let $v(x,t) = u(x,t) - u_0(x)$.]

c) Solve the problem in part (b) for $v(x,t)$. Find $u(x,t)$.

24. The heat equation for the temperature $u(x,t)$ is derived by considering the conservation of thermal energy. The flow of thermal energy per unit surface area per unit time is $q(x,t)$. Let ρ be the constant mass density and c be the constant specific heat (the amount of thermal energy necessary to raise a unit mass of a substance one degree of temperature). In this problem, consider a thin slice (from x to $x + \Delta x$) of a one-dimensional rod of constant cross-sectional area A.

a) Show that the thermal energy in the thin slice is approximately $cu(x,t)A\,\Delta x$.

b) Derive an equation for $u(x,t)$ in terms of $q(x,t)$, assuming that there are no sources of thermal energy.

c) **Fourier's law of heat conduction** is that $q = -\beta(\partial u/\partial x)$, where β is called the thermal conductivity. Briefly explain the minus sign.

d) Using parts (b) and (c), show that the temperature satisfies the diffusion equation. Determine a formula for the thermal diffusivity.

11.3 Laplace's Equation

In the last section we considered the one-dimensional diffusion equation. In this section we will consider steady-state diffusion in two- and three-dimensional regions.

11.3.1 Multidimensional Diffusion Equation and Steady-State

Let t be time, (x,y,z) a point in the region, and u the concentration of a chemical at (x,y,z) at time t. If the chemical concentration $u(x,y,z,t)$ depends on more than one spatial dimension, then the second spatial derivative in the one-dimensional diffusion equation, $\partial u/\partial t = k(\partial^2 u/\partial x^2) + f$, is replaced by the sum of the second spatial derivatives

$$\frac{\partial u}{\partial t} = k\nabla^2 u + f, \tag{1}$$

where the **Laplacian** ∇^2 is defined as

$$\nabla^2 u \quad \frac{\partial^2 u}{\partial x^2} + \frac{\partial^2 u}{\partial y^2} + \frac{\partial^2 u}{\partial z^2}. \tag{2}$$

Here $f(x, y, z, t)$ represents possibly spatially distributed sources of pollution. The chemical concentration satisfies (1), which is known as the **multidimensional diffusion equation.**

The time-dependent nature of the chemical concentration can be complicated even if the sources are time-independent. In some situations, there is major interest in the **equilibrium** or **steady-state** chemical concentration, in which $\partial u/\partial t = 0$. Since $\partial u/\partial t = 0$ in (1), the steady-state chemical concentration satisfies **Poisson's equation**:

$$\nabla^2 u = -\frac{1}{k} f, \tag{3}$$

where f does not depend on t. Equation (3) is a linear partial differential equation that is nonhomogeneous if $f \neq 0$.

If there are no sources $f = 0$, the steady-state chemical concentration satisfies **Laplace's equation**:

$$\nabla^2 u = 0. \tag{4}$$

There will be a unique solution if the chemical concentration is given along the entire boundary, called a boundary condition. For example, it can be shown that if $u = 0$ along the entire boundary, then the only steady-state solution is $u = 0$ in the entire region. This is a rather simple (but still important) case, and it should be understood physically. Other boundary conditions usually (but not always) result in a unique solution. Laplace's equation can be solved by the method of separation of variables only for certain simple geometries: rectangles, boxes, circles, spheres, and a few others. In other geometric regions, numerical methods are usually used. Fortunately, these simple geometries are of great physical interest, and the solutions in these geometries provide guidelines to the numerical solution in other geometries. In this section we consider (4) on a rectangle.

11.3.2 Method of Separation of Variables

Laplace's Equation on a Rectangle and Boundary Conditions

Steady-state diffusion of a chemical without any sources satisfies Laplace's equation

$$\nabla^2 u = 0.$$

We consider a two-dimensional rectangular region, $0 \leq x \leq L$ and $0 \leq y \leq H$, as shown in Figure 11.3.1. Then Laplace's equation becomes

$$\frac{\partial^2 u}{\partial x^2} + \frac{\partial^2 u}{\partial y^2} = 0. \tag{5}$$

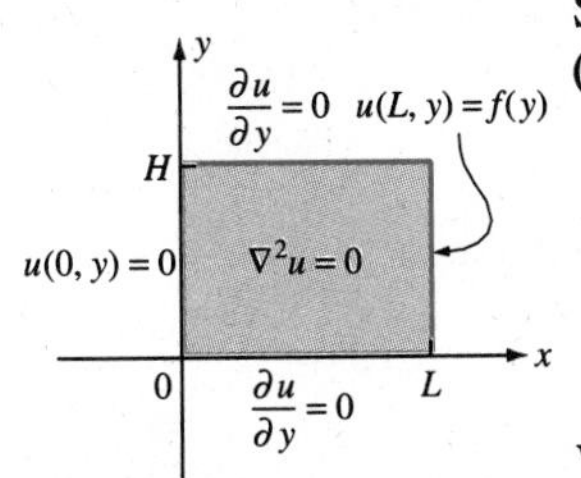

FIGURE 11.3.1

Region including boundary conditions.

Some conditions must be imposed on the four sides. We assume that the top $(y = H)$ and bottom $(y = 0)$ are insulated, so that

$$\begin{aligned} \frac{\partial u}{\partial y}(x,0) &= 0, \\ \frac{\partial u}{\partial y}(x,H) &= 0. \end{aligned} \tag{6}$$

We also assume that the left side $(x = 0)$ is at a zero level of concentration and the right side $(x = L)$ is at a prescribed (but variable) level of concentration:

$$\begin{aligned} u(0,y) &= 0, \\ u(L,y) &= f(y). \end{aligned} \tag{7}$$

Three of the boundary conditions are homogeneous, and one $(x = L)$ is nonhomogeneous.

Product Solutions

To use the method of separation of variables, we first ignore the nonhomogeneous boundary condition. We seek product solutions of Laplace's equation

$$u(x,y) = \phi(y)h(x). \tag{8}$$

and insist that the product solutions satisfy the homogeneous boundary conditions. The assumed product solutions (8) satisfy Laplace's equation (5) if

$$\phi(y)\frac{d^2h}{dx^2} + h(x)\frac{d^2\phi}{dy^2} = 0.$$

Dividing by $\phi(y)h(x)$ enables the variables to be separated:

$$-\frac{1}{h(x)}\frac{d^2h}{dx^2} = \frac{1}{\phi(y)}\frac{d^2\phi}{dy^2}. \tag{9}$$

The left side depends only on x and the right side only on y. Thus, they equal a constant:

$$-\frac{1}{h}\frac{d^2h}{dx^2} = \frac{1}{\phi}\frac{d^2\phi}{dy^2} = -\lambda. \tag{10}$$

We have again introduced the separation constant and called it $-\lambda$. Two ordinary differential equations (one in x and one in y) follow from (9):

$$\frac{d^2h}{dx^2} = \lambda h, \tag{11}$$

$$\frac{d^2\phi}{dy^2} = -\lambda\phi. \tag{12}$$

Three homogeneous boundary conditions follow from (6) and (7):

$$h(0) = 0, \tag{13}$$

$$\frac{d\phi}{dy}(0) = 0, \quad \text{and} \quad \frac{d\phi}{dy}(H) = 0. \tag{14}$$

Boundary Value Problem for the Separation Constant

The separation constant is determined by a boundary value problem, (12) with two homogeneous boundary conditions (14). In this case the independent variable is y.

Unlike in the example in Section 11.2, we will see that in this case $\lambda = 0$ is an allowable separation constant (eigenvalue). If $\lambda = 0$, then the general solution of (12) is

$$\phi(y) = c_1 y + c_2.$$

Both boundary conditions, $\phi_y(0) = 0$ and $\phi_y(H) = 0$, are satisfied by $c_1 = 0$, so that any constant is an eigenfunction corresponding to $\lambda = 0$:

$$\phi(y) = c_2. \tag{15}$$

This is *not* the trivial solution $\phi(y) = 0$ for all y. We conclude that $\lambda = 0$ is an eigenvalue for this example. If $\lambda = 0$, the x-dependent part of the product solution satisfies (11), so that

$$h(x) = c_3 x + c_4.$$

The homogeneous condition $h(0) = 0$ implies that $c_4 = 0$. Thus,

$$h(x) = c_3 x. \tag{16}$$

Product solutions of Laplace's equation (solving the three homogeneous boundary conditions) corresponding to $\lambda = 0$ follow from (15) and (16):

$$u(x, y) = Ax. \tag{17}$$

Clearly, this is a nontrivial solution of Laplace's equation, that also satisfies the three homogeneous boundary conditions.

It can be shown (see Exercise 27) that the only solution for $\lambda < 0$ is the trivial solution $u(x, y) = 0$. Most of the solutions of interest will correspond to $\lambda > 0$. Equation (12) is a linear differential equation with constant coefficients. Its characteristic equation $r^2 + \lambda = 0$ has roots $r = \pm i\sqrt{\lambda}$ if $\lambda > 0$. The general solution of (12) is then

$$\phi(y) = c_1 \cos\sqrt{\lambda}\, y + c_2 \sin\sqrt{\lambda}\, y. \tag{18}$$

The boundary conditions, $\phi_y(0) = 0$ and $\phi_y(H) = 0$, require that we calculate $d\phi/dy$:

$$\frac{d\phi}{dy} = -c_1\sqrt{\lambda}\sin\sqrt{\lambda}\, y + c_2\sqrt{\lambda}\cos\sqrt{\lambda}\, y.$$

The boundary condition at $y = 0$ implies $\phi_y(0) = 0 = c_2\sqrt{\lambda}$. Thus,

$$c_2 = 0,$$

and hence

$$\phi(y) = c_1 \cos \sqrt{\lambda}\, y. \tag{19}$$

The boundary condition at $y = H$ implies

$$\frac{d\phi}{dy}(H) = 0 = -c_1\sqrt{\lambda}\, \sin \sqrt{\lambda}\, H. \tag{20}$$

If $c_1 = 0$, then $c_1 = c_2 = 0$ and $\phi(y) = 0$ for all y, which is the trivial solution. Thus,

$$\sin \sqrt{\lambda}\, H = 0. \tag{21}$$

The sine function is zero at integral multiples of π. Thus, (21) implies that the eigenvalues (separation constants) satisfy

$$\sqrt{\lambda}\, H = n\pi \qquad \text{for } n = 1, 2, 3, \ldots,$$

or equivalently,

$$\lambda = \left(\frac{n\pi}{H}\right)^2 \qquad \text{for } n = 1, 2, 3, \ldots. \tag{22}$$

We exclude $n = 0$, since we have already shown that $\lambda = 0$ is an eigenvalue. The corresponding eigenfunctions (spatial part of the product solutions) follow from (19):

$$\phi(y) = c_1 \cos \frac{n\pi y}{H} \qquad \text{for } n = 1, 2, 3, \ldots. \tag{23}$$

The constant c_1 is arbitrary. Here, the eigenfunctions are cosines (whereas in the previous section the eigenfunctions were sines). This significant difference is caused by the boundary conditions being different. In summary, note we include $\lambda = 0$:

Summary of the Boundary Value Problem

$$\frac{d^2\phi}{dy^2} = -\lambda\phi \qquad \text{(DE)}$$

$$\frac{d\phi}{dy}(0) = 0, \quad \frac{d\phi}{dy}(H) = 0 \qquad \text{(BC)}$$

Eigenvalues: $\lambda = \left(\frac{n\pi}{H}\right)^2$ for $n = 0, 1, 2, 3, \ldots,$

Eigenfunctions: $\cos \frac{n\pi y}{H}$ for $n = 0, 1, 2, 3, \ldots.$

The x part of the product solutions, h, satisfies the linear homogeneous differential equation

$$\frac{d^2h}{dx^2} = \lambda h = \left(\frac{n\pi}{H}\right)^2 h \tag{24}$$

with $h(0) = 0$. The characteristic polynomial for (24) is $r^2 - (n\pi/H)^2$, which has roots $r = \pm n\pi/H$. The general solution ($n \neq 0$) of (24) is thus

$$h(x) = \alpha e^{n\pi x/H} + \beta e^{-n\pi x/H},$$

where α, β are arbitrary constants. The boundary condition $h(0) = 0$ implies $0 = \alpha + \beta$, so that

$$h(x) = \alpha(e^{n\pi x/H} - e^{-n\pi x/H}) = 2\alpha \sinh\frac{n\pi x}{H} \qquad \text{for } n = 1,2,3,\dots. \tag{25}$$

It is not necessary to use the hyperbolic functions, but they make the solution simpler to write.

From (23) and (25), we have obtained product solutions (corresponding to $\lambda > 0$) of Laplace's equation

$$u(x,y) = A\cos\frac{n\pi y}{H}\sinh\frac{n\pi x}{H}, \qquad \text{for } n = 1,2,3,\dots, \tag{26}$$

where A is an arbitrary constant. It is easy to verify (after the fact) by direct substitution that the product solution (26) (for any one fixed value of n) satisfies Laplace's equation ($\partial^2u/\partial x^2 + \partial^2u/\partial y^2 = 0$). The three homogeneous boundary conditions can also be verified. For different values of n, the constant A may be different, and we can denote it A_n instead of A.

We do not wish to forget that we also obtained a product solution that corresponded to $\lambda = 0$,

$$u(x,y) = Ax. \tag{27}$$

Nonhomogeneous Boundary Condition

These solutions can be added together to get a more general solution:

$$u(x,y) = A_0x + \sum_{n=1}^{\infty} A_n\cos\frac{n\pi y}{H}\sinh\frac{n\pi x}{H}, \tag{28}$$

where A_0 and the A_n are arbitrary constants. These constants are determined from the nonhomogeneous boundary condition (7) for the chemical concentration at $x = L$:

$$u(L,y) = f(y), \qquad \text{for } 0 < y < H. \tag{29}$$

Here $f(y)$ is given.

Substituting $x = L$ into the solution (28) and applying the nonhomogeneous boundary condition (29) yields

$$u(L, y) = f(y) = A_0 L + \sum_{n=1}^{\infty} A_n \sinh \frac{n\pi L}{H} \cos \frac{n\pi y}{H}, \qquad \text{for } 0 < y < N. \tag{30}$$

We will be able to determine the coefficients A_0 and A_n, since (30) is the Fourier cosine series representation of the function $f(y)$ over the interval $0 < y < H$ as discussed in Section 10.3. Given the nonhomogeneous boundary condition $f(y)$, the coefficients must equal the coefficients of the Fourier cosine series:

$$\begin{aligned} A_0 L &= \frac{1}{H} \int_0^H f(y)\, dy, \\ A_n \sinh \frac{n\pi L}{H} &= \frac{2}{H} \int_0^H f(y) \cos \frac{n\pi y}{H}\, dy, \qquad n = 1, 2, 3, \ldots. \end{aligned} \tag{31}$$

The coefficient A_n may be easily obtained, since $\sinh n\pi L/H \neq 0$. Thus

$$\begin{aligned} A_0 &= \frac{1}{LH} \int_0^H f(y)\, dy, \\ A_n &= \frac{2}{H \sinh \frac{n\pi L}{H}} \int_0^H f(y) \cos \frac{n\pi y}{H}\, dy, \qquad n = 1, 2, 3, \ldots. \end{aligned} \tag{32}$$

From our study of a Fourier cosine series, we know that the constant term in (30) is necessary in order to represent an arbitrary function using cosines. The eigenfunctions $\cos n\pi x/L$, $n = 0, 1, 2, 3, \ldots$, form an orthogonal set, but they cannot represent an arbitrary function unless the $n = 0$ term is included.

EXAMPLE 11.3.1 Laplace's Equation Inside a Rectangle

Solve Laplace's equation (5) with the boundary conditions

$$\frac{\partial u}{\partial y}(x, 0) = 0,$$

$$\frac{\partial u}{\partial y}(x, H) = 0$$

and

$$\begin{aligned} u(0, y) &= 0, \\ u(L, y) &= f(y) = \cos \frac{\pi y}{H}. \end{aligned} \tag{33}$$

Solution From (28), the solution of Laplace's equation that satisfies the three homogeneous boundary conditions is

$$u(x, y) = A_0 x + \sum_{n=1}^{\infty} A_n \cos \frac{n\pi y}{H} \sinh \frac{n\pi x}{H}.$$

Here $u(L, y) = f(y) = \cos \pi y/H$. The nonhomogeneous boundary condition (33) is satisfied if

$$\cos \frac{\pi y}{H} = u(L, y) = A_0 L + \sum_{n=1}^{\infty} A_n \cos \frac{n\pi y}{H} \sinh \frac{n\pi L}{H}.$$

Thus all $A_n = 0$ except

$$A_1 = \frac{1}{\sinh \frac{\pi L}{H}}.$$

Thus the solution of Laplace's equation satisfying the given boundary conditions is

$$u(x, y) = \frac{1}{\sinh \frac{\pi L}{H}} \cos \frac{\pi y}{H} \sinh \frac{\pi x}{H}. \tag{34}$$

◄

Discussion

Although this solution involves a single term, it is still interesting to graph it. There are two ways to graph u. In Figure 11.3.2, contours of constant values of the pollution are sketched. Concentration levels of the pollutant range from -1 to $+1$, since that is the range of the given nonhomogeneous

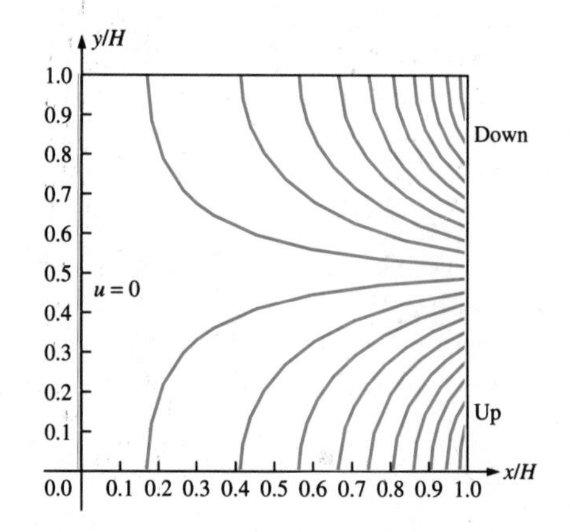

FIGURE 11.3.2

Contour plot of $u(x, y)$ given by (34).

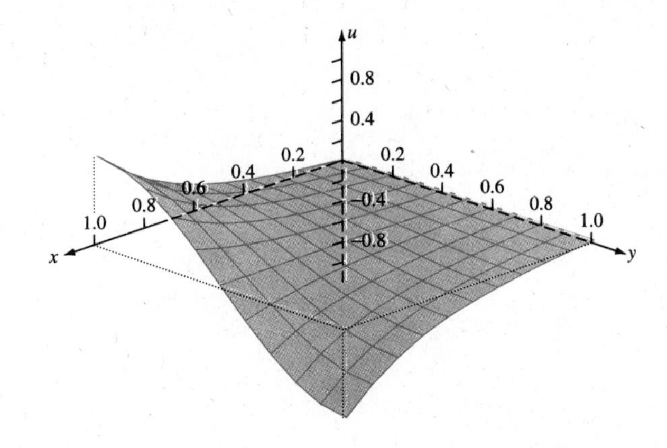

FIGURE 11.3.3

Plot of $u(x, y)$ in (34).

boundary condition. From the boundary condition, we have $u = 0$ at the side $x = 0$. In Figure 11.3.3, a 3-D sketch is shown of u as a function of x and y.

Exercises

In Exercises 1 through 23, solve Laplace's equation $\partial^2u/\partial x^2 + \partial^2u/\partial y^2 = 0$ in the rectangle $0 < x < L$ and $0 < y < H$.

1. $\frac{\partial u}{\partial y}(x,0) = 0, \quad \frac{\partial u}{\partial y}(x,H) = 0,$
$u(0,y) = f(y), \quad u(L,y) = 0$

2. $\frac{\partial u}{\partial y}(x,0) = 0, \quad \frac{\partial u}{\partial y}(x,H) = 0,$
$u(0,y) = f(y), \quad \frac{\partial u}{\partial x}(L,y) = 0$

3. $\frac{\partial u}{\partial y}(x,0) = 0, \quad \frac{\partial u}{\partial y}(x,H) = 0,$
$\frac{\partial u}{\partial x}(0,y) = 0, \quad u(L,y) = f(y)$

4. $\frac{\partial u}{\partial y}(x,0) = 0, \quad \frac{\partial u}{\partial y}(x,H) = 0,$
$\frac{\partial u}{\partial x}(0,y) = f(y), \quad u(L,y) = 0$

5. $\frac{\partial u}{\partial y}(x,0) = 0, \quad \frac{\partial u}{\partial y}(x,H) = 0,$
$u(0,y) = 0, \quad \frac{\partial u}{\partial x}(L,y) = f(y)$

6. $\frac{\partial u}{\partial y}(x,0) = 0, \quad \frac{\partial u}{\partial y}(x,H) = 0,$
$\frac{\partial u}{\partial x}(0,y) = 0, \quad \frac{\partial u}{\partial x}(L,y) = f(y)$. Under what condition concerning $f(y)$ will there be a solution of this problem?

7. $\frac{\partial u}{\partial y}(x,0) = f(x), \quad \frac{\partial u}{\partial y}(x,H) = 0,$
$\frac{\partial u}{\partial x}(0,y) = 0, \quad \frac{\partial u}{\partial x}(L,y) = 0$. Under what condition concerning $f(x)$ will there be a solution of this problem?

8. $u(x,0) = 0, \quad u(x,H) = 0,$
$u(0,y) = f(y), \quad u(L,y) = 0$

9. $u(x,0) = 0, \quad u(x,H) = 0,$
$u(0,y) = 0, \quad u(L,y) = f(y)$

10. $u(x,0) = 0, \quad u(u,H) = f(x),$
$u(0,y) = 0, \quad u(L,y) = 0$

11. $u(x,0) = f(x), \quad u(x,H) = 0,$
$u(0,y) = 0, \quad u(L,y) = 0$

12. $\frac{\partial u}{\partial y}(x,0) = f(x), \quad u(x,H) = 0,$
$u(0,y) = 0, \quad u(L,y) = 0$

13. $\frac{\partial u}{\partial y}(x,0) = 0, \quad u(x,H) = f(x),$
$u(0,y) = 0, \quad u(L,y) = 0$

14. $u(x,0) = f(x), \quad \frac{\partial u}{\partial y}(x,H) = 0,$
$u(0,y) = 0, \quad u(L,y) = 0$

15. $u(x,0) = 0, \quad \frac{\partial u}{\partial y}(x,H) = f(x),$
$u(0,y) = 0, \quad u(L,y) = 0$

16. $u(x,0) = f(x), \quad u(x,H) = 0,$
$u(0,y) = 0, \quad \frac{\partial u}{\partial x}(L,y) = 0$

17. $u(x,0) = 0, \quad u(x,H) = f(x),$
$u(0,y) = 0, \quad \frac{\partial u}{\partial x}(L,y) = 0$

18. $\frac{\partial u}{\partial y}(x,0) = g(x), \quad u(x,H) = f(x),$
$u(0,y) = 0, \quad u(L,y) = 0$

19. $u(x,0) = f(x), \quad u(x,H) = g(x),$
$\frac{\partial u}{\partial x}(0,y) = 0, \quad \frac{\partial u}{\partial x}(L,y) = 0$

20. $u(x,0) = 0, \quad u(x,H) = f(x),$
$u(0,y) = 0, \quad u(L,y) = g(y).$
(*Hint*: Divide into two problems.)

21. $u(x,0) = f(x), \quad u(x,H) = 0,$
$u(0,y) = 0, \quad u(L,y) = g(y).$
(*Hint*: Divide into two problems.)

22. $\frac{\partial u}{\partial y}(x,0) = g(x), \quad \frac{\partial u}{\partial y}(x,H) = 0,$
$u(0,y) = f(y), \quad u(L,y) = 0.$
(*Hint*: Divide into two problems.)

23. $\frac{\partial u}{\partial y}(x,0) = g(x), \quad \frac{\partial u}{\partial y}(x,H) = 0,$
$u(0,y) = 0, \quad u(L,y) = f(y).$
(*Hint*: Divide into two problems.)

In Exercises 24 through 26, solve Laplace's equation $\partial^2 u/\partial x^2 + \partial^2 u/\partial y^2 = 0$ in the semi-infinite strip $0 < x < L$ and $0 < y < \infty$.

24. $u(0,y) = 0, \quad u(L,y) = 0, \quad u(x,0) = f(x)$

25. $u(0,y) = 0, \quad u(L,y) = 0, \quad \frac{\partial u}{\partial y}(x,0) = f(x)$

26. $\frac{\partial u}{\partial x}(0,y) = 0, \frac{\partial u}{\partial x}(L,y) = 0, u(x,0) = f(x)$

27. Suppose that $\lambda < 0$. Show that the only solution of $\phi''(x) = -\lambda\phi(x)$, $\phi'(0) = 0$, $\phi'(H) = 0$, is $\phi = 0$.

11.4 Wave Equation

Another important use of PDEs is in studying the vibration of flexible shapes. In this section we consider a fundamental PDE called the wave equation.

11.4.1 Vibrating String

The vertical displacement $u(x,t)$ of a highly stretched string approximately satisfies the **wave equation**

$$\frac{\partial^2 u}{\partial t^2} = c^2 \frac{\partial^2 u}{\partial x^2},$$

where $c^2 = T/\rho$. Here T is the tension in the string and ρ is the constant mass density (mass per unit length) of the string. We will show that $c = \sqrt{T/\rho}$ is the velocity at which waves moves.

The physics of a vibrating string is actually quite complicated. The derivation we give neglects some subtleties. We assume that we have an elastic string that is tightly stretched with a constant horizontal tensile force T and is at rest. We introduce a small vertical displacement $u(x, t)$ depending on x, which because of vibrations will also depend on t. Thus for a given t, $u(x, t)$ gives the shape of the string. We assume that the string vibrates only vertically (actually there is a small horizontal motion as well), and in addition we assume that the horizontal tensile force remains constant (this is only approximately valid). These two assumptions are good approximations only if the string is nearly horizontal, that is, if the slope of the string is small everywhere along the string.

We consider the vertical force equation for a small segment between x and $x + \Delta x$ as shown in Figure 11.4.1. The mass of the small segment is $\rho\,\Delta x$, and the position of the segment will be treated as a point mass located at $u(x, t)$. We will assume that the only forces are the tensile forces (see Figure 11.4.1) acting on the two ends of the string in the directions indicated. The angle $\theta(x, t)$ that the string makes with the horizontal is important. The vertical components of the two tensile forces ($T \sin \theta$) may be different if the string is not exactly straight. According to Newton's law ($ma = F$),

$$(\rho\,\Delta x)\frac{\partial^2 u}{\partial t^2} = -T \sin \theta(x, t) + T \sin \theta(x + \Delta x, t). \tag{1}$$

As shown in Figure 11.4.1, the right end of the segment is pulled upward (when $\sin \theta > 0$) by the rest of the string and the left end is pulled downward.

We divide (1) by Δx and take the limit as $\Delta x \to 0$:

$$\rho\frac{\partial^2 u}{\partial t^2} = T \lim_{\Delta x \to 0} \frac{\sin \theta(x + \Delta x, t) - \sin\ \theta(x, t)}{\Delta x} = T\frac{\partial}{\partial x}[\sin \theta(x, t)]. \tag{2}$$

We have recognized the definition of the derivative of $\sin \theta(x, t)$ with respect to x, keeping t fixed. A simple relationship exists between the angle $\theta(x, t)$ and the slope $\partial u/\partial x$ of the string, namely

$$\tan \theta(x, t) = \frac{\partial u}{\partial x}. \tag{3}$$

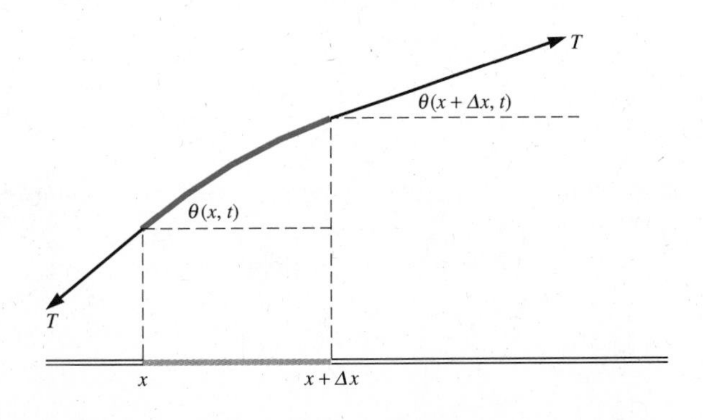

FIGURE 11.4.1

Forces for vibrating string.

As mentioned earlier, the validity of (1) depends on many assumptions, such as constant tension, that are a good approximation only for small displacements from a horizontally stretched string. Thus, (1) and (2) should be used only if the angle $\theta(x,t)$ is small. However, if the angle is small (equivalent to small slopes of the string), $\tan\theta = \sin\theta/\cos\theta \approx \sin\theta$. Thus, we may approximate $\sin\theta$ in (2) by (3). Thus, (2) becomes

$$\rho\frac{\partial^2 u}{\partial t^2} = T\frac{\partial}{\partial x}\left(\frac{\partial u}{\partial x}\right).$$

In this manner, under the assumption that the slope is small, we have derived that the vertical displacement $u(x, t)$ satisfies the wave equation:

$$\frac{\partial^2 u}{\partial t^2} = c^2\frac{\partial^2 u}{\partial x^2}, \tag{4}$$

where we have introduced the constant c, which will be explained later,

$$c = \sqrt{\frac{T}{\rho}}.$$

Electromagnetic Waves

Under various assumptions, derived from Maxwell's equations, the components of the electric and magnetic fields can be shown to satisfy the wave equation in three-dimensional space:

$$\frac{\partial^2 u}{\partial t^2} = c^2\nabla^2 u,$$

where c is related to the speed of light and $\nabla^2 u$ is the Laplacian,

$$\nabla^2 u = \frac{\partial^2 u}{\partial x^2} + \frac{\partial^2 u}{\partial y^2} + \frac{\partial^2 u}{\partial z^2}.$$

11.4.2 Method of Separation of Variables

The wave equation,

$$\frac{\partial^2 u}{\partial t^2} = c^2\frac{\partial^2 u}{\partial x^2}, \tag{5}$$

describes the vertical motion of a stretched string. One of the simplest physical problems is a string of length L $(0 \le x \le L)$ fixed at both ends. Thus, the vertical displacement will be zero at both ends:

$$\begin{aligned} u(0,t) &= 0, \\ u(L,t) &= 0. \end{aligned} \tag{6}$$

Since the wave equation has two partial derivatives with respect to t, as with Newton's laws for a vertically moving particle, the appropriate initial conditions specify the initial displacement $f(x)$ and the initial vertical velocity $g(x)$:

$$\begin{aligned} &\text{Initial displacement (shape):} && u(x,0) = f(x), \\ &\text{Initial velocity:} && \frac{\partial u}{\partial t}(x,0) = g(x). \end{aligned} \tag{7}$$

Product Solutions

The wave equation (5) is a linear homogeneous PDE with homogeneous boundary conditions (6). We seek product solutions of the wave equation (5) in the form

$$u(x,t) = \phi(x)h(t) \tag{8}$$

that also satisfy the two homogeneous boundary conditions (6). We temporarily ignore the two initial conditions.

When (8) is substituted into (5) for u, we obtain

$$\phi(x)\frac{d^2h}{dt^2} = c^2h(t)\frac{d^2\phi}{dx^2}.$$

Dividing by $c^2\phi(x)h(t)$ puts the equations in the form

$$\frac{1}{c^2h(t)}\frac{d^2h}{dt^2} = \frac{1}{\phi(x)}\frac{d^2\phi}{dx^2}.$$

The left side depends only on t, and the right side depends only on x. Thus, they equal a constant, which we call the separation constant and denote $-\lambda$:

$$\frac{1}{c^2h(t)}\frac{d^2h}{dt^2} = \frac{1}{\phi(x)}\frac{d^2\phi}{dx^2} = -\lambda. \tag{9}$$

Two ordinary differential equations follow from (9),

$$\frac{d^2h}{dt^2} = -\lambda c^2 h \tag{10}$$

and

$$\frac{d^2\phi}{dx^2} = -\lambda\phi. \tag{11}$$

The boundary conditions (6) imply that $\phi(0)h(t) = 0$ and $\phi(L)h(t) = 0$ for all t. Since $h(t)$ is not identically zero by assumption, we have

$$\phi(0) = 0, \qquad \phi(L) = 0. \tag{12}$$

The boundary value problem consisting of (11) and (12) is solved in Section 11.2.2. The eigenvalues are

$$\lambda = \left(\frac{n\pi}{L}\right)^2, \qquad n = 1, 2, 3, \ldots, \tag{13}$$

and the corresponding eigenfunctions are

$$\phi(x) = \sin\frac{n\pi x}{L}, \qquad n = 1, 2, 3, \ldots. \tag{14}$$

The time-dependent part of the product solution $h(t)$ solves (10). The characteristic equation $r^2 + \lambda c^2 = 0$ for (10) has roots $r = \pm ic\sqrt{\lambda} = \pm i(n\pi c/L)$. Thus,

$$h(t) = c_1 \cos\frac{n\pi ct}{L} + c_2 \sin\frac{n\pi ct}{L}. \tag{15}$$

Combining (14) and (15) in (8), we get two families of product solutions,

$$\sin\frac{n\pi x}{L}\cos\frac{n\pi ct}{L},$$
$$\sin\frac{n\pi x}{L}\sin\frac{n\pi ct}{L}.$$

We should include both families and sum, so that

$$u(x,t) = \sum_{n=1}^{\infty} A_n \sin\frac{n\pi x}{L}\cos\frac{n\pi ct}{L} + \sum_{n=1}^{\infty} B_n \sin\frac{n\pi x}{L}\sin\frac{n\pi ct}{L}. \tag{16}$$

The two sets of coefficients, A_n, B_n, are determined from the two initial conditions. Letting $t = 0$ in (16), the initial position $u(x,0) = f(x)$ yields

$$f(x) = \sum_{n=1}^{\infty} A_n \sin\frac{n\pi x}{L}.$$

Thus, the coefficients A_n are the Fourier sine coefficients of the initial position,

$$A_n = \frac{2}{L}\int_0^L f(x)\sin\frac{n\pi x}{L}\,dx, \qquad n = 1, 2, 3, \ldots. \tag{17}$$

We now calculate the velocity $\partial u/\partial t$ from (16):

$$\frac{\partial u}{\partial t} = -\sum_{n=1}^{\infty}\frac{n\pi c}{L}A_n \sin\frac{n\pi x}{L}\sin\frac{n\pi ct}{L} + \sum_{n=1}^{\infty}\frac{n\pi c}{L}B_n \sin\frac{n\pi x}{L}\cos\frac{n\pi ct}{L}.$$

Letting $t = 0$ and using the initial velocity $u_t(x, 0) = g(x)$ yields

$$g(x) = \sum_{n=1}^{\infty}\frac{n\pi c}{L}B_n \sin\frac{n\pi x}{L}.$$

Thus $(n\pi c/L)B_n$ are the Fourier sine coefficients of the initial velocity,

$$\frac{n\pi c}{L}B_n = \frac{2}{L}\int_0^L g(x)\sin\frac{n\pi x}{L}\,dx, \qquad n = 1,2,3,\ldots,$$

or

$$B_n = \frac{2}{n\pi c}\int_0^L g(x)\sin\frac{n\pi x}{L}\,dx, \qquad n = 1,2,3,\ldots. \tag{18}$$

EXAMPLE 11.4.1 **Plucked String**

Solve the wave equation (5) with fixed ends $u(0,t) = 0$ and $u(L,t) = 0$, such that the string is initially at rest $[u_t(x,0) = g(x) = 0]$ with the initial position tent-shaped

$$u(x,0) = f(x) \doteq \begin{cases} x, & x < \dfrac{L}{2}, \\ L - x, & x > \dfrac{L}{2}, \end{cases} \tag{19}$$

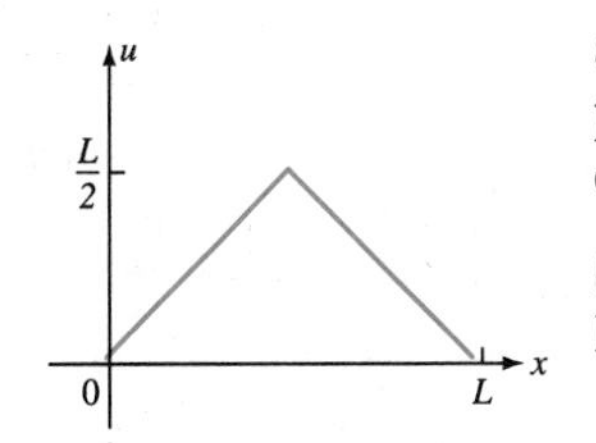

FIGURE 11.4.2

Graph of $u(x,0) = f(x)$ in (19).

as graphed in Figure 11.4.2. One can think of (19) as arising by some form of plucking of the stretched string in the middle. We note that the initial condition $f(x)$ is symmetric around the middle $x = L/2$.

Solution From (16), the solution of the wave equation that satisfies the homogeneous boundary conditions $u(0,t) = 0$, $u(L,t) = 0$ is

$$u(x,t) = \sum_{n=1}^{\infty} A_n \sin\frac{n\pi x}{L}\cos\frac{n\pi ct}{L} + \sum_{n=1}^{\infty} B_n \sin\frac{n\pi x}{L}\sin\frac{n\pi ct}{L}.$$

Since the string is initially at rest $[u_t(x,0) = g(x) = 0]$, it follows from (18) that $B_n = 0$. From (17), the coefficients A_n are determined from the initial position:

$$A_n = \frac{2}{L}\int_0^L f(x)\sin\frac{n\pi x}{L}\,dx, \qquad n = 1,2,3,\ldots, \tag{20}$$

where $f(x)$ is given by (19) and is symmetric around the middle $x = L/2$.

The integral in (20) may be simplified a little using symmetry properties. The eigenfunctions $\sin n\pi x/L$ have some symmetry properties with respect to the middle $x = L/2$. In particular, $\sin n\pi x/L$ is symmetric around $x = L/2$ for n odd, and $\sin n\pi x/L$ is antisymmetric around $x = L/2$ for n even. Since $f(x)$ is symmetric around $x = L/2$, it follows that $f(x)\sin n\pi x/L$ is symmetric around $x = L/2$ for n odd, and $f(x)\sin n\pi x/L$ is antisymmetric around $x = L/2$ for n even. Thus,

$$A_n = \begin{cases} 0, & n \text{ even}, \\ \dfrac{4}{L}\displaystyle\int_0^{L/2} x \sin \frac{n\pi x}{L}\, dx, & n \text{ odd}. \end{cases} \tag{21}$$

The integral in (21) or (20) can be evaluated using integration by parts, tables of integrals, or symbolic integration available on commercial packages such as Maple, Macsyma, or Mathematica to give

$$A_n = \begin{cases} 0, & n \text{ even}, \\ \dfrac{4L}{n^2\pi^2} \sin \dfrac{n\pi}{2}, & n \text{ odd}. \end{cases} \tag{22}$$

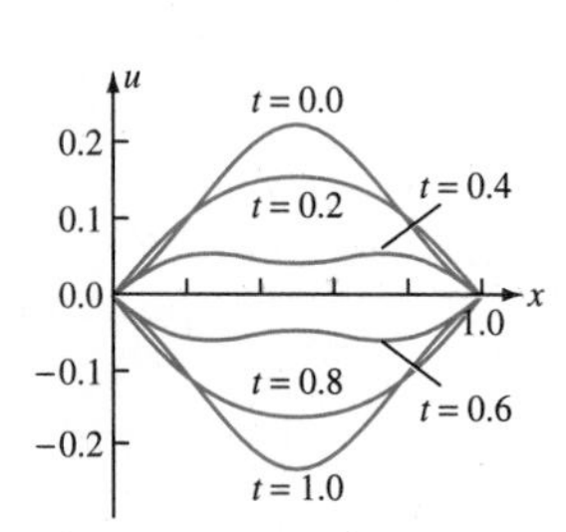

FIGURE 11.4.3

Graph of $P_3(x, t)$ with $L = c = 1$.

The solution of the wave equation is the infinite series

$$u(x,t) = \sum_{n=1}^{\infty} A_n \sin \frac{n\pi x}{L} \cos \frac{n\pi ct}{L},$$

where A_n is given by (22). The solution $u(x, t)$ may be approximated by $P_m(x,t)$, the first m terms of the infinite series:

$$P_m(x,t) = \sum_{n=1}^{m} A_n \sin \frac{n\pi x}{L} \cos \frac{n\pi ct}{L}.$$

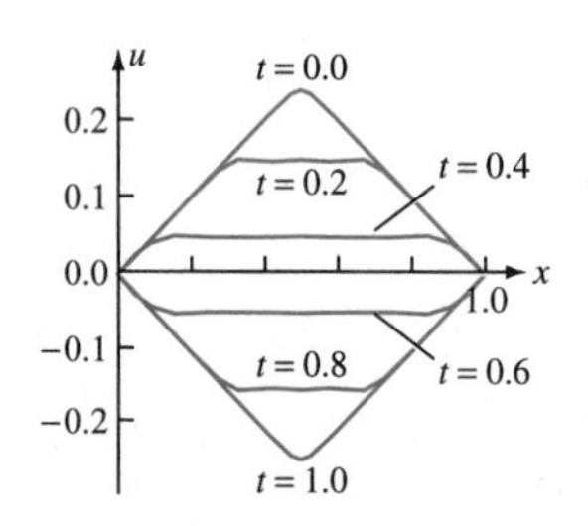

FIGURE 11.4.4

Graph of $P_{11}(x, t)$ with $L = c = 1$.

This approximation is graphed in Figures 11.4.3 and 11.4.4 for various values of m. The solution (the graph as $m \to \infty$) appears to be made up of line segments (at least for the better approximations, in which m is large) similar to the initial condition. This will be shown later in this section when we discuss traveling waves. ◄

Vibrating Strings and Music

The solution (16) consists of the superposition of the product solutions known as modes of vibration. The **fundamental mode of vibration**

$$u(x,t) = \sin \frac{\pi x}{L} \cos \frac{\pi ct}{L}$$

corresponding to $n = 1$ is graphed in Figure 11.4.5. The **natural frequency of vibration** corresponding to the fundamental mode $c/2L$ (where $c = \sqrt{T/\rho}$) is called the **fundamental** or **first harmonic**.

From this we can investigate how the sound produced depends on the three important parameters L, T, and ρ. For example, as any violinist can tell you, tightening the string increases the tension and hence increases the fundamental frequency. This is heard as a `higher" note, since a higher

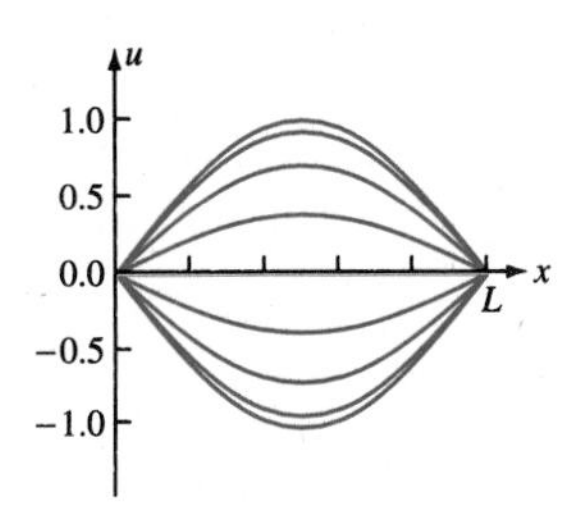

FIGURE 11.4.5

Fundamental mode of vibration ($n = 1$).

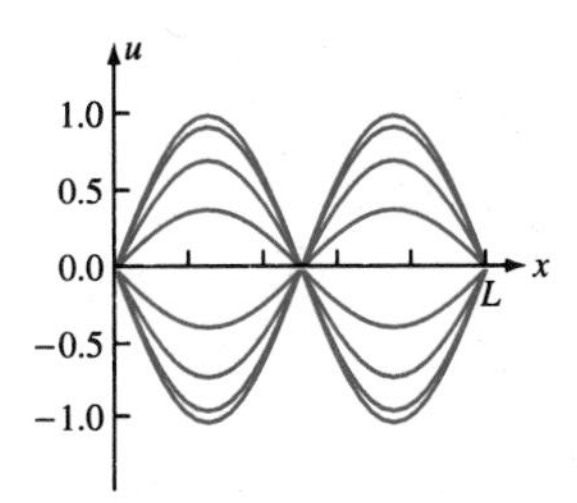

FIGURE 11.4.6

Second mode of vibration ($n = 2$).

frequency is what is meant by a higher note. The bass player easily changes the length. The fundamental frequency is lower with increased length. The strings in a piano have different lengths and different mass densities. And the piano is tuned to desired frequencies by tightening the strings (varying the tension).

The second mode of vibration for a vibrating string ($n = 2$)

$$u(x,t) = \sin \frac{2\pi x}{L} \cos \frac{2\pi ct}{L}$$

is graphed in Figure 11.4.6. It has a natural frequency that is twice the fundamental frequency and hence is called the second harmonic. Similarly, there are third, fourth, and higher harmonics. Vibrating strings have the property that the higher modes of vibration oscillate at the higher harmonics. In percussion instruments (drums), a membrane vibrates (which solves a two-spatial-dimensional wave equation), and it can be shown that the higher modes of vibration are not higher harmonics (provided the drum is round).

The sound heard from a vibrating string involves the fundamental and all its higher harmonics. The particular sound heard depends on the amplitude of each mode, which is determined from the initial conditions. For real musical instruments, the effect of the specific manner of plucking, bowing, or striking the vibrating string must be considered. They contribute to the sound, as does the shape of the resonating chamber. The difference between an expensive violin and a cheap one is more complicated. Modern synthesizers are able to reproduce frequency combinations with desirable amplitude distributions.

11.4.3 Traveling Waves

Using a fundamental trigonometric identity [$\sin a \cos b = \frac{1}{2}\sin(a+b) + \frac{1}{2}\sin(a-b)$], the first mode of vibration can be shown to be the superposition of a right traveling wave (velocity $c > 0$) and a left traveling wave (velocity $-c$):

$$\sin \frac{\pi x}{L} \cos \frac{\pi ct}{L} = \frac{1}{2} \sin \frac{\pi}{L}(x+ct) + \frac{1}{2} \sin \frac{\pi}{L}(x-ct). \tag{23}$$

This is also true for any mode of vibration, and thus it is true for the solution obtained by the method of separation of variables. Remarkably, the wave equation

$$\frac{\partial^2 u}{\partial t^2} = c^2 \frac{\partial^2 u}{\partial x^2} \tag{24}$$

has a general solution that involves two arbitrary functions F, G:

$$u(x,t) = F(x-ct) + G(x+ct). \tag{25}$$

The first corresponds to an arbitrary left traveling wave, and the second to an arbitrary right traveling wave. The left- and right-going waves are called **permanent** waves, since their shape does not change in time. It can be verified by direct substitution that (25) directly solves (24). Equation (25) can be used to solve the initial value problem on the infinite domain (no boundary conditions):

$$u(x,0) = f(x) \qquad \text{for } -\infty < x < \infty,$$

$$\frac{\partial u}{\partial t}(x,0) = g(x) \qquad \text{for } -\infty < x < \infty.$$

The initial condition for the position is satisfied if

$$u(x,0) = f(x) = F(x) + G(x). \tag{26}$$

The initial condition for the velocity is not as easy to analyze:

$$\frac{\partial u}{\partial t}(x,0) = g(x) = \frac{\partial}{\partial t}F(x-ct)\bigg|_{t=0} + \frac{\partial}{\partial t}G(x+ct)\bigg|_{t=0}.$$

Some simplification occurs if we note that, based on the chain rule,

$$\frac{\partial}{\partial t}F(x-ct)\bigg|_{t=0} = -c\frac{dF(x-ct)}{d(x-ct)}\bigg|_{t=0} = -c\frac{dF(x)}{dx},$$

$$\frac{\partial}{\partial t}G(x+ct)\bigg|_{t=0} = c\frac{dG(x+ct)}{d(x+ct)}\bigg|_{t=0} = c\frac{dG(x)}{dx}.$$

Thus

$$\frac{g(x)}{c} = -\frac{dF(x)}{dx} + \frac{dG(x)}{dx}.$$

By differentiating (26), we get

$$f'(x) = \frac{dF(x)}{dx} + \frac{dG(x)}{dx}.$$

By adding these two equations and dividing by 2, we obtain

$$\frac{dG(x)}{dx} = \frac{1}{2}\left[f'(x) + \frac{1}{c}g(x)\right].$$

Then integration gives

$$G(x) = \frac{1}{2}f(x) + \frac{1}{2c}\int_0^x g(s)\,ds + k,$$

where k is an arbitrary constant that we will show does not affect the solution of the wave equation. From (26), we find $F(x)$:

$$F(x) = \frac{1}{2}f(x) - \frac{1}{2c}\int_0^x g(s)\,ds - k.$$

The general solution of the wave equation from (25) is

$$u(x,t) = \frac{1}{2}f(x-ct) - \frac{1}{2c}\int_0^{x-ct} g(s)\,ds + \frac{1}{2}f(x+ct) + \frac{1}{2c}\int_0^{x+ct} g(s)\,ds.$$

The k's have canceled. The integrals can be combined to give what is called **D'Alembert's solution** of the wave equation:

$$u(x,t) = \frac{1}{2}[f(x-ct) + f(x+ct)] + \frac{1}{2c}\int_{x-ct}^{x+ct} g(s)\,ds. \tag{27}$$

The first term is easier to interpret, since the initial position splits in half, half going to the left without changing shape and half going to the right without changing shape. The second term is more complicated but still represents the sum of a wave moving to the left and a wave moving to the right.

EXAMPLE 11.4.2 Plucked String Revisited

Let us reconsider the plucked string problem in which the wave equation (24) is to be solved with fixed ends, $u(0,t) = 0$ and $u(L,t) = 0$, such that the string is initially at rest $[u_t(x,0) = g(x) = 0]$ with the initial position tent-shaped:

$$u(x,0) = f(x) = \begin{cases} x, & x < \dfrac{L}{2}, \\ L - x, & x > \dfrac{L}{2}. \end{cases} \tag{28}$$

SOLUTION This problem is formulated for a finite string ($0 < x < L$), while the traveling wave method and in particular D'Alembert's solution (27) is best suited for the infinite spatial domain ($-\infty < x < \infty$). For the wave equation with these boundary conditions, a problem can be formulated on the infinite domain that is equivalent to the given problem for the finite-length string. The boundary condition $u(0,t) = 0$ can be satisfied by extending the given initial conditions as odd periodic functions around $x = 0$, since then the solutions will also be odd periodic functions. Then the antisymmetry around $x = 0$ will guarantee that $u(0,t) = 0$. The right boundary condition $u(L,t) = 0$ will also be satisfied. For the infinite domain, $u_t(x,0) = g(x) = 0$ for all x. The initial position that occurs by extending $f(x)$ given by (28) as an odd periodic function is graphed in Figure 11.4.7.

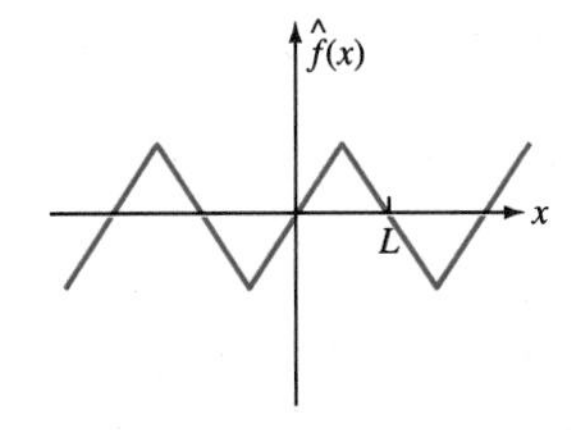

FIGURE 11.4.7

Odd periodic extension of $f(x)$ in (28).

According to D'Alembert's solution (27), since $g(x) = 0$ for all x,

$$u(x,t) = \frac{1}{2}[f(x-ct) + f(x+ct)],$$

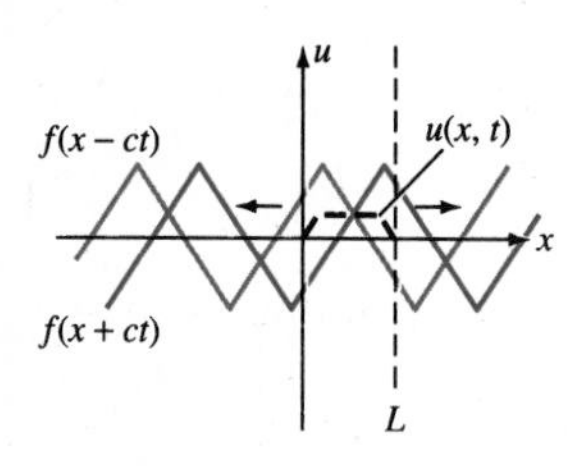

FIGURE 11.4.8

D'Alembert's solution

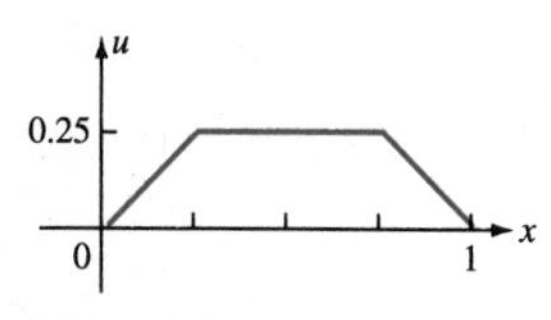

FIGURE 11.4.9

$u(x, t)$, t a little past 0.

FIGURE 11.4.10

where $f(x)$ is the initial conditions on the extended infinite domain graphed in Figure 11.4.7. The solution to the wave equation may be obtained by taking one-half this initial condition, shifting a distance ct to the right, and adding it to one-half this initial condition, shifted to the left the same distance. The solution will be continuous and composed of segments of straight lines, since it is the sum of two functions with this property. Figure 11.4.8 shows $f(x - ct)$, $f(x + ct)$, and $u(x, t)$ for a small positive value of t. Figure 11.4.9 gives $u(x, t)$ for the same value of t. As t increases, the curve gets flatter and then reappears below the axis. The pattern repeats. A three-dimensional graph of $u(x, t)$ for three-fourths of a period is given in Figure 11.4.10. Note that in the graphs the boundary conditions, $u(0, t) = 0$ and $u(L, t) = 0$, are satisfied. This exact solution should be compared to the solution graphed in Figures 11.4.3 and 11.4.4, which was an approximation using terms of the Fourier series. ◀

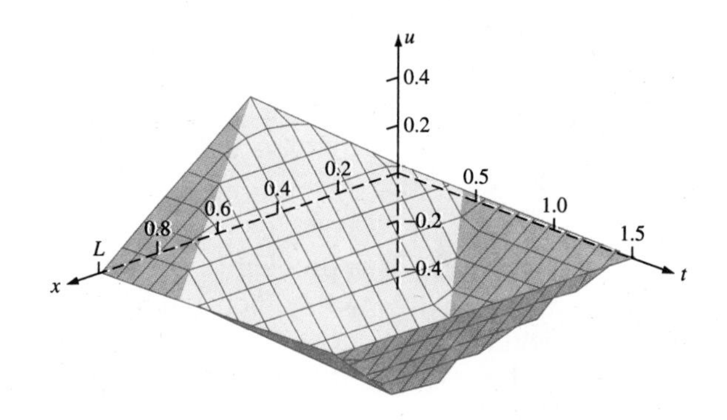

Exercises

In Exercises 1 through 4, use separation of variables to solve the wave equation $\partial^2 u/\partial t^2 = c^2(\partial^2 u/\partial x^2)$ on the interval $0 < x < L$.

1. $u(0, t) = 0, \quad u(L, t) = 0,$
   $u(x, 0) = f(x), \quad \dfrac{\partial u}{\partial t}(x, 0) = 0$

2. $u(0, t) = 0, \quad u(L,t) = 0,$
   $u(x, 0) = 0, \quad \dfrac{\partial u}{\partial t}(x, 0) = g(x)$

3. $\dfrac{\partial u}{\partial x}(0, t) = 0, \quad \dfrac{\partial u}{\partial x}(L, t) = 0,$
   $u(x, 0) = f(x), \quad \dfrac{\partial u}{\partial t}(x, 0) = 0$

4. $\frac{\partial u}{\partial x}(0,t) = 0, \quad \frac{\partial u}{\partial x}(L,t) = 0$
$u(x,0) = 0, \quad \frac{\partial u}{\partial t}(x,0) = g(x)$

In Exercises 5 and 6, use separation of variables to solve the wave equation $\partial^2 u/\partial t^2 = c^2(\partial^2 u/\partial x^2)$ on the interval $0 < x < L$.

5. $u(0,t) = 0, \quad \frac{\partial u}{\partial x}(L,t) = 0,$
$u(x,0) = f(x), \quad \frac{\partial u}{\partial t}(x,0) = 0.$
What are the frequencies of vibration?

6. $\frac{\partial u}{\partial x}(0,t) = 0, \quad u(L,t) = 0,$
$u(x,0) = 0, \quad \frac{\partial u}{\partial t}(x,0) = g(x).$
What are the frequencies of vibration?

In Exercises 7 and 8, use separation of variables to solve the wave equation with an extra restoring force $\partial^2 u/\partial t^2 = c^2(\partial^2 u/\partial x^2) - ku$ on the interval $0 < x < L$.

7. $u(0,t) = 0, \quad u(L,t) = 0,$
$u(x,0) = f(x), \quad \frac{\partial u}{\partial t}(x,0) = 0$

8. $u(0,t) = 0, \quad u(L,t) = 0,$
$u(x,0) = 0, \quad \frac{\partial u}{\partial t}(x,0) = g(x)$

In Exercises 9 and 10, use (27) to solve the wave equation $\partial^2 u/\partial t^2 = c^2(\partial^2 u/\partial x^2)$ on the infinite domain $(-\infty < x < \infty)$.

9. $u(x,0) = f(x), \quad \frac{\partial u}{\partial t}(x,0) = 0$

10. $u(x,0) = 0, \quad \frac{\partial u}{\partial t}(x,0) = g(x)$

In Exercises 11 through 16 you are given a specific initial shape $f(x)$ for the wave equation $\partial^2 u/\partial t^2 = c^2 \partial^2 u/\partial x^2$, with boundary conditions $u(0,t) = 0$, $u(L,t) = 0$ and initial conditions $u(x,0) = f(x)$, $u_t(x,0) = 0$. For convenience take $L = \pi$.

a) Find the solution of the wave equation for the given initial shape $f(x)$.

b) Determine the frequencies of the terms that are actually present in the solution.

c) If you have access to a computer, let $c = 1$ and graph the sum of the first four terms with nonzero coefficients at times $t = 0$, P/4, P/2, 3P/4, P where P is the period of the lowest frequency term that appears in the series for u.

11. $f(x) = |\sin 2x|$

12. $f(x) = \sin 2x$

13. $f(x) = \begin{cases} \sin 2x & 0 < x < \frac{\pi}{2} \\ 0 & \frac{\pi}{2} < x < \pi \end{cases}$

14. $f(x) = \begin{cases} 0 & 0 < x < \frac{\pi}{2} \\ \sin 2x & \frac{\pi}{2} < x < \pi \end{cases}$

15. $f(x) = \begin{cases} \sin 3x & 0 < x < \frac{\pi}{3} \\ 0 & \frac{\pi}{3} < x < \frac{2\pi}{3} \\ 0 & \frac{2\pi}{3} < x < \pi \end{cases}$

16. $f(x) = \begin{cases} 0 & 0 < x < \frac{\pi}{3} \\ 0 & \frac{\pi}{3} < x < \frac{2\pi}{3} \\ \sin 3x & \frac{2\pi}{3} < x < \pi \end{cases}$

17. $f(x) = \begin{cases} 0 & 0 < x < \frac{\pi}{3} \\ \sin 3x & \frac{\pi}{3} < x < \frac{2\pi}{3} \\ 0 & \frac{2\pi}{3} < x < \pi \end{cases}$

APPENDIX

Existence and Uniqueness

In this appendix we discuss how one could prove the Basic Existence and Uniqueness Theorem for first-order initial value problems given in Section 1.5 (Theorem 1.5.1). The same approach may be modified to prove the other existence and uniqueness theorems in this book, such as those for systems and higher-order equations. The discussion will also introduce several ideas, such as iterative methods and integral equations, that are of great importance in applied mathematics.

Given the first-order initial value problem,

$$\frac{dy}{dx} = f(x, y), \qquad y(x_0) = y_0, \tag{1}$$

how are we to prove the existence and uniqueness of a solution? There are two key steps. The first is to rewrite (1) as an **integral equation** by integrating both sides of (1),

$$\int_{x_0}^{x} \frac{dy}{ds}\, ds = \int_{x_0}^{x} f(s, y(s))\, ds,$$

evaluating the integral on the left side,

$$\int_{x_0}^{x} \frac{dy}{ds}\, ds = y(x) - y(x_0) = y(x) - y_0,$$

and solving for y to get an integral equation for y:

$$y(x) = \int_{x_0}^{x} f(s, y(s))\, ds + y_0. \tag{2}$$

Let $I = [\alpha, \beta]$ be an interval containing x_0. Note that if $y(x)$ is a differentiable function defined on I that satisfies the differential equation (1), then it satisfies the integral equation (2). Conversely, if $y(x)$ is a continuous function defined on I that satisfies the integral equation (2) for $x \in I$, and if $f(s, y(s))$ is continuous for $s \in I$, then by the fundamental theorem of calculus, the right-hand side of (2) is continuously differentiable. Thus $y(x)$ is continuously differentiable, and $y(x)$ satisfies (1). We have established that:

Finding a differentiable solution of the initial value problem (1) is equivalent to finding a continuous solution of the integral equation (2).

The second key in establishing existence and uniqueness is to try and solve the integral equation (2) by the **method of successive approximation**, or **Picard iteration.** [Emile Picard was a French mathematician (1856–1941) who extensively developed the idea of successive approximations. The method had been used earlier by Liouville.] In the method of successive approximations, one solves an equation of the form

$$y = G(y) \tag{3}$$

for y by making an initial guess ϕ_0 for y and then defining a sequence of approximations by

$$\begin{aligned} \phi_1 &= G(\phi_0) \\ \phi_2 &= G(\phi_1) \\ &\vdots \\ \phi_{n+1} &= G(\phi_n) \\ &\vdots \end{aligned} \tag{4}$$

If the approximations ϕ_n converge, say $\phi_n \to \phi$, and if $G(\phi_n) \to G(\phi)$, then (4) implies that $\lim_{n\to\infty} \phi_{n+1} = \lim_{n\to\infty} G(\phi_n)$, so that $\phi = G(\phi)$ and ϕ is a solution of (3).

In our problem, (3) is the integral equation (2) and y is an unknown function, so that the iterates ϕ_n will also be functions. We take ϕ_0 to be the constant function y_0:

$$\phi_0(x) = y_0. \tag{5}$$

The successive approximations are then

$$\phi_1(x) = \int_{x_0}^{x} f(s, \phi_0(s))\, ds + y_0 \tag{6}$$

$$\phi_2(x) = \int_{x_0}^{x} f(s, \phi_1(s))\, ds + y_0 \tag{7}$$

$$\vdots$$

$$\phi_{n+1}(x) = \int_{x_0}^{x} f(s, \phi_n(s))\, ds + y_0 \tag{8}$$

$$\vdots$$

In order for this approach to work, there are several requirements that must be met:

1. For each $\phi_n(x)$ it must be possible, at least theoretically, to evaluate the integral in (8) to get the next iterate $\phi_{n+1}(x)$.
2. The iterates $\phi_n(x)$ must converge to a function $\phi(x)$.
3. $\int_{x_0}^{x} f(s, \phi_n(s))\, ds + y_0$ must converge to $\int_{x_0}^{x} f(s, \phi(s))\, ds + y_0$ as $n \to \infty$.
4. $\phi(x)$ must be continuous on an interval containing x_0.

Before showing how to satisfy all these requirements, we will give a specific example of the iteration process.

EXAMPLE A.1 Method of Successive Approximation (Picard Iteration)

Find the first four iterates on $[0, 2]$ for the initial value problem

$$\frac{dy}{dx} = y^2, \qquad y(0) = 1, \tag{9}$$

using the successive approximations (5) through (8).

SOLUTION In this example we have $x_0 = 0$, $y_0 = 1$, and $f(s, y(s)) = y(s)^2$. Thus the equations (5) through (8) defining the iterates become

$$\phi_0(x) = 1$$

$$\phi_1(x) = \int_0^x \phi_0(s)^2\, ds + 1 = \int_0^x ds + 1 = x + 1$$

$$\phi_2(x) = \int_0^x \phi_1(s)^2\, ds + 1 = \int_0^x (1+s)^2\, ds + 1 = 1 + x + x^2 + \frac{1}{3}x^3$$

$$\phi_3(x) = \int_0^x \phi_2(s)^2\, ds + 1 = \int_0^x \left(1 + s + s^2 + \frac{1}{3}s^3\right)^2 ds + 1$$

$$= 1 + x + x^2 + x^3 + \frac{2}{3}x^4 + \frac{1}{3}x^5 + \frac{1}{9}x^6 + \frac{1}{63}x^7.$$

Additional iterates could be computed in a similar fashion. ◄

FIGURE A.1

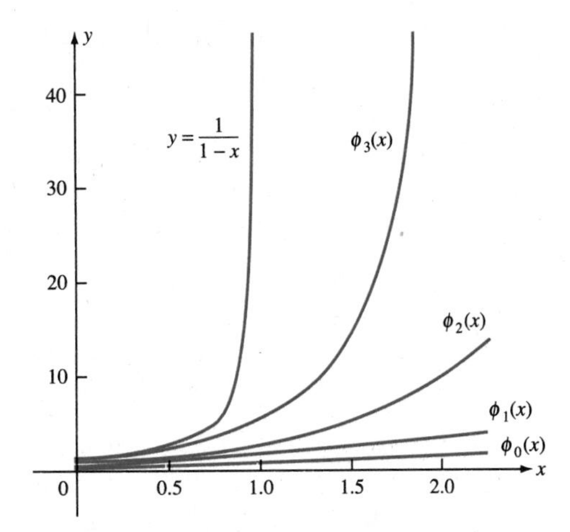

It is instructive to examine this example more carefully. The actual solution of (9) can be found by separation of variables to be

$$y(x) = \frac{1}{1-x}. \tag{10}$$

Notice that while the iterates $\phi_n(x)$ are defined on $[0, 2]$ [in fact, they are defined on $(-\infty, \infty)$ for all n], the solution (10) is defined only on $[0, 1)$ [or more generally $(-\infty, 1)$]. This shows that in our proof we will have to make restrictions on the interval of definition in order to obtain convergence of the $\phi_n(x)$ on all of that interval. Next, notice that the solution (10) has the series expansion on $[0, 1)$

$$y(x) = \frac{1}{1-x} = \sum_{i=0}^{\infty} x^i = 1 + x + x^2 + x^3 + x^4 + \cdots. \tag{11}$$

Looking at the $\phi_n(x)$ we computed in Example A.1, we see that $\phi_0(x)$ gives the first term of the expansion (11), $\phi_1(x)$ agrees with the first two terms, $\phi_2(x)$ has the same first three terms, and $\phi_3(x)$ has the same first four terms. The solution (10) and $\phi_0(x)$, $\phi_1(x)$, $\phi_2(x)$, $\phi_3(x)$ are graphed in Figure A.1.

The existence and uniqueness theorem is:

■ **THEOREM A.1 Existence and Uniqueness**

Suppose that $f(x, y)$ and $f_y(x, y)$ are continuous at and near (x_0, y_0). Then there is an interval I containing x_0 on which there is a unique solution to the initial value problem $dy/dx = f(x, y)$, $y(x_0) = y_0$. ■

In an actual proof, we have to be more precise about the sets involved. To begin, let R be a rectangular region centered at (x_0, y_0),

$$R = \{(x, y): |x - x_0| \le a, \quad |y - y_0| \le b\}, \qquad a > 0, b > 0. \tag{12}$$

We shall assume that:

Assumption 1 $f(x, y)$ is continuous on the rectangular region R.

Assumption 2 $f_y(x, y)$ is continuous on the rectangular region R.

From advanced calculus, we know that a continuous function on a closed and bounded set, such as R, is bounded. Thus there exist constants M, L such that

$$|f(x, y)| \leq M, \qquad \text{all } (x, y) \in R, \tag{13}$$

$$|f_y(x, y)| \leq L, \qquad \text{all } (x, y) \in R. \tag{14}$$

Before giving the technical version of Theorem A.1, we need one more estimate, which will be discussed more fully in the exercises. If (x, y_1), (x, y_2) are both in R and Assumptions 1 and 2 hold, then we may use the mean value theorem and (14) on $f(x, y_1) - f(x, y_2)$ to get

$$|f(x, y_1) - f(x, y_2)| = |f_y(x, \hat{y})|\,|y_1 - y_2| \leq L|y_1 - y_2|, \tag{15}$$

where $\hat{y}$ is a number between y_1 and y_2. The technical version of Theorem A.1 is

■ THEOREM A.2 Technical Version of Theorem A.1

Let $f(x, y)$ satisfy Assumptions 1 and 2. Let $\bar{a}$ be the smaller of a and b/M, where M is from (13). Then there is a unique solution to $dy/dx = f(x, y)$, $y(x_0) = y_0$, defined for $|x - x_0| \leq \bar{a}$. ■

The proof consists of using Picard iteration and showing that Requirements 1 through 4 are met.

Exercises

Exercises 1 through 6 illustrate the use of iteration on algebraic equations. These exercises are best done with a programmable calculator or computer.

As mentioned in the introduction to this appendix, iteration can sometimes be used to solve algebraic equations, $h(y) = 0$, by rewriting them as $y = G(y)$, making an initial guess ϕ_0, and then using the iteration $\phi_{n+1} = G(\phi_n)$ to generate a sequence of approximations $\{\phi_n\}$ to the solution of $y = G(y)$. There are many ways to get $y = G(y)$ from $h(y) = 0$. For example,

$$y^3 + 4y - 1 = 0 \tag{16}$$

could be rewritten as

$$y = \frac{1}{y^2 + 4} \tag{17}$$

or

$$y = \frac{1}{4}(1 - y^3). \tag{18}$$

Since the polynomial $y^3 + 4y - 1$ is continuous on [0, 1] and changes sign in going from 0 to 1, it is clear that the polynomial has a root in the interval [0, 1].

1. Take ϕ_0 in $[0, 1]$ and find a solution of (16) using iteration on (17).

2. Take ϕ_0 in $[0, 1]$ and find a solution of (16) using iteration on (18).

3. Using iteration and ϕ_0 in $[0, 1]$, find a solution of $y^5 + y^3 + 9y - 1 = 0$.

4. Using iteration and ϕ_0 in $[0, 1]$, find a solution of $y^5 + 2y^3 + 14y - 1 = 0$.

5. Observe that $y^5 - 1 = 0$ has 1 as a solution and may be rewritten as $y = G(y) = y^{-4}$. Take any $\phi_0 \neq 1$ and observe that the iteration $\phi_{n+1} = G(\phi_n)$ does not converge to 1.

6. Observe that $y^5 - 1 = 0$ has 1 as a solution and may be rewritten as $y = G(y) = y^5 + y - 1$. Take any $\phi_0 \neq 1$ and observe that the iteration $\phi_{n+1} = G(\phi_n)$ does not converge to 1.

In Exercises 7 through 16,

a) Rewrite the given initial value problem as an integral equation.

b) Using Picard iteration, find the first m approximations to the solution of the given initial value problem.

c) Find the exact solution of the given initial value problem.

d) Graph the m approximations from part (b) and the exact solution from part (c) on the same set of coordinates. (Best done with a computer.)

7. $\dfrac{dy}{dx} = 3y, \quad y(0) = 1, \quad m = 4$

8. $\dfrac{dy}{dx} = -y, \quad y(0) = 2, \quad m = 4$

9. $\dfrac{dy}{dx} = y + x, \quad y(0) = 0, \quad m = 4$

10. $\dfrac{dy}{dx} = -y + x, \quad y(0) = 0, \quad m = 4$

11. $\dfrac{dy}{dx} = 2xy, \quad y(0) = 1, \quad m = 4$

12. $\dfrac{dy}{dx} = -xy, \quad y(0) = 2, \quad m = 4$

13. $\dfrac{dy}{dx} = \dfrac{1}{y}, \quad y(0) = 1, \quad m = 3$

14. $\dfrac{dy}{dx} = y^2 - 1, \quad y(0) = 0, \quad m = 3$

15. $\dfrac{dy}{dx} = y^2 - 2y + 1, \quad y(1) = 0, \quad m = 3$

16. $\dfrac{dy}{dx} = \dfrac{x}{y^2}, \quad y(1) = 1, \quad m = 2$

Solutions to Selected Exercises

Chapter 1

Section 1.2

1, 3, 5, 7. Substitute the expression for y in the differential equation.

9. $y = 3e^x + c$

11. $y = -\frac{5}{6}\sin 6x + c$

13. $y = 8\int_0^x \cos t^{-\frac{1}{2}}\,dt + c$

15. $y = \int_0^x \ln(4 + \cos^2 t)\,dt + c$

17. $y = \frac{1}{5}x^5 - \frac{17}{5}$

19. $y = \int_2^x \frac{\ln t}{4 + \cos^2 t}\,dt + 5$

21. $y = \int_1^x \frac{e^t}{1+t}\,dt + 3$

23. $v_0 = \sqrt{2250}$ m/s $= 15\sqrt{10}$ m/s

25. $x = 0.72$ km

27. $x = \frac{v_0^2}{5000}$ km

29. $x = 2(\frac{62}{3})^{\frac{3}{2}} - 116$

31. **(b)** $x = \frac{1}{k}\ln\frac{4}{3}$ m

33. $v_0 = \sqrt{1960}$ m/s

35. $t = \frac{20}{\sqrt{9.8}}$ s

37. $c = y_0 - \int_0^{x_0} f(\bar{x})\,d\bar{x}$

Section 1.3

1. $y = cx - 1$

3. $y = e^x + c$

5. $y = ce^{(x^2/2)+4x} - 3$

7. $y = 3x + c$

9. $y = (9 - 4x)^{-1/4}$

11. $y = \dfrac{1}{1 - \int_0^x \cos t^2\, dt}$

13. $\displaystyle\int_2^y \frac{dt}{\cos t^{-\frac{1}{2}}} = \frac{1}{2}(x^2 - 1)$

15. $u^3 + 12u = t^3 + 3t + 13$

17. $y = \tan\left(\dfrac{x^3}{3} + x + \tan^{-1} 2\right)$

19. $-\ln|y| + \ln|y - 1| = x + c;$

$y = (1 - ke^x)^{-1}$ and $y = 0$

21. $\ln|y - 1| - \ln|y - 2| - (y - 2)^{-1} = x + c;$

$y = 1,\ y = 2$

23. $xe^x ye^y = \tilde{c}$

25. $\left(\dfrac{z-3}{z+3}\right) = \tilde{c}\left(\dfrac{t+2}{t-2}\right)^{3/2}$

27. $-e^{-x} - \dfrac{e^{-4y}}{4} = c;$

$y = -\frac{1}{4}\ln(-4e^{-x} + k)$

29. $z' = bf(z) + a$

31. $\frac{1}{2}\tan^{-1}(2x + 8y - 2) = x + c;$

$y = \frac{1}{4}[1 - x + \frac{1}{2}\tan(2x + k)]$

33. $(x + y + 1)e^{-x-y} = -x + c$

Section 1.4

1. **3.**

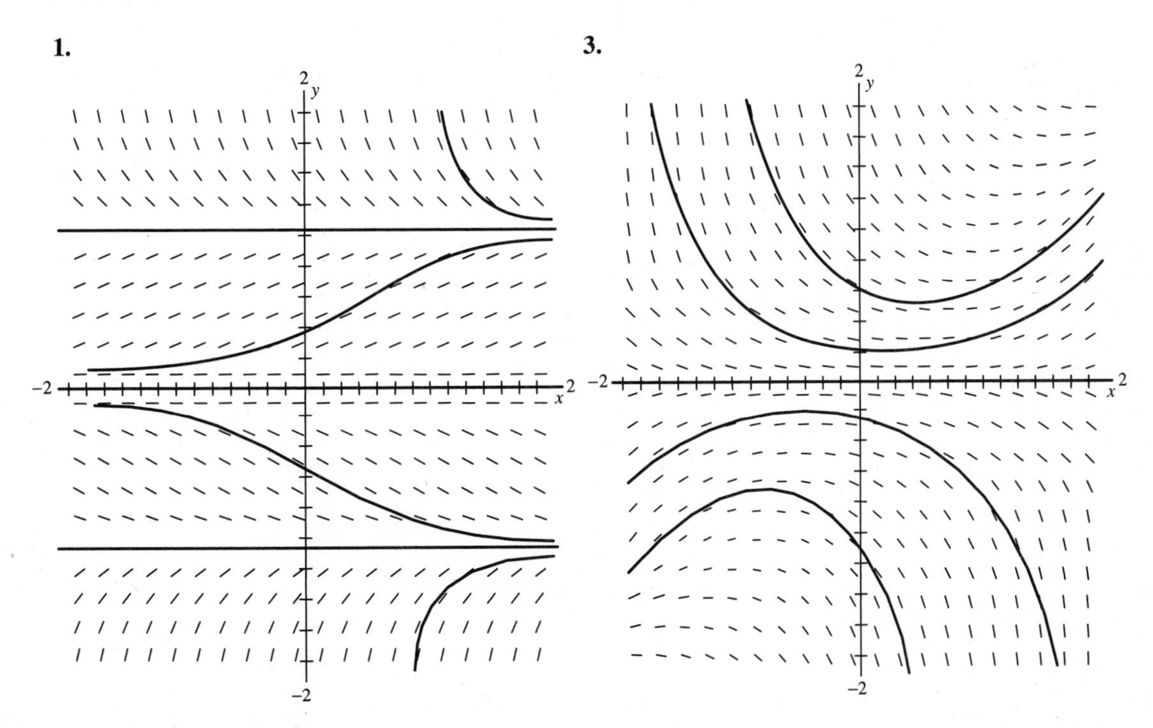

5.

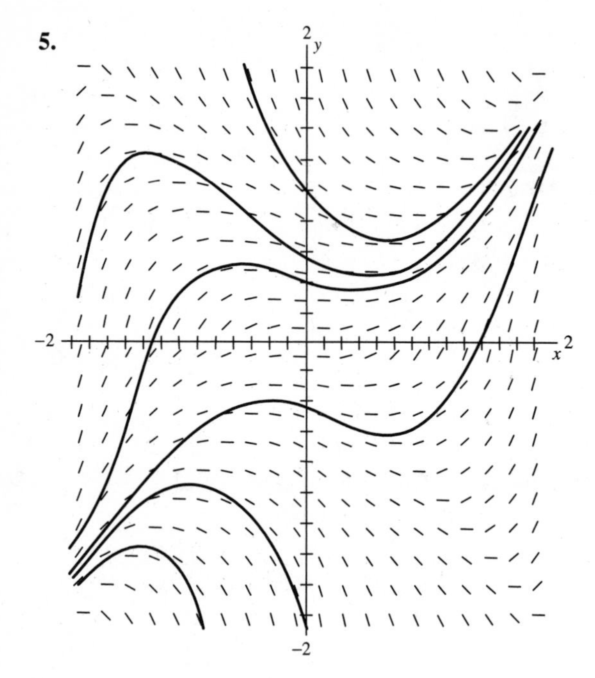

7.

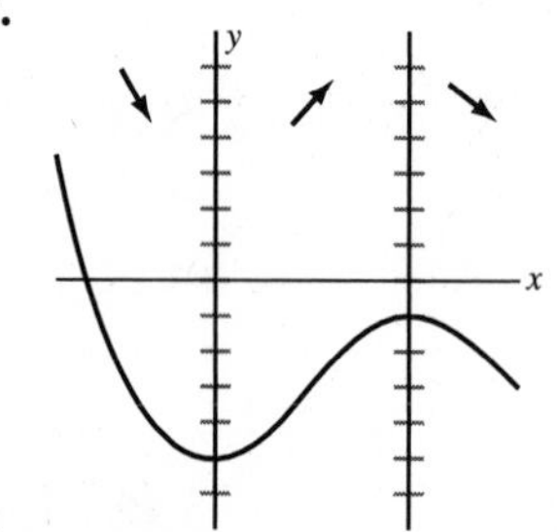

9.

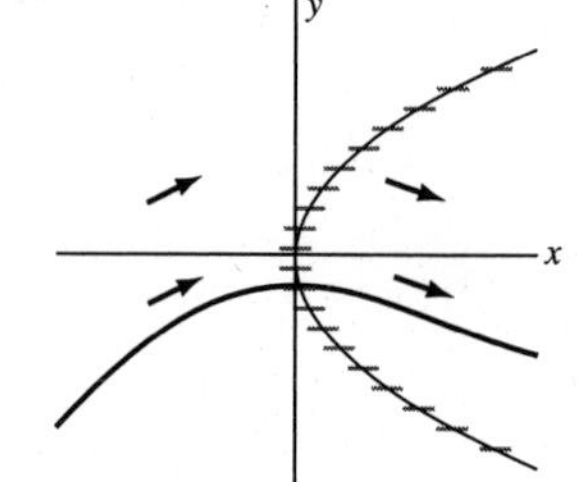

11.

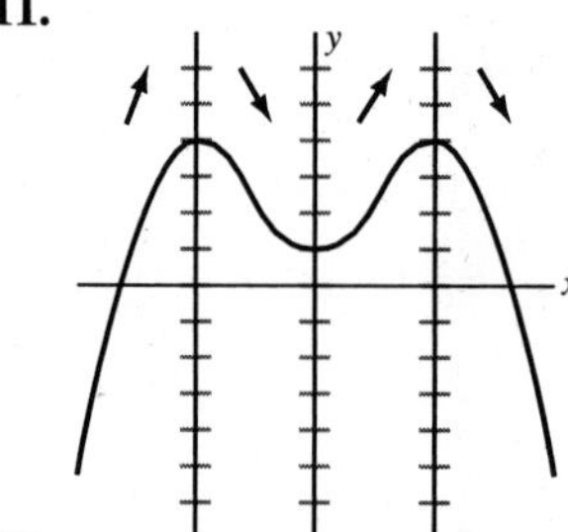

13.

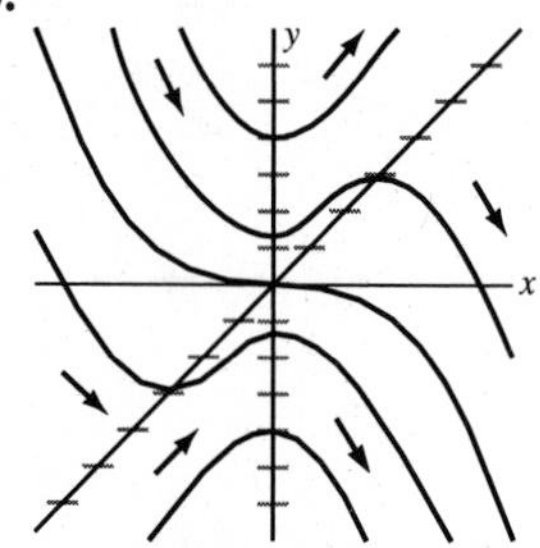

15.

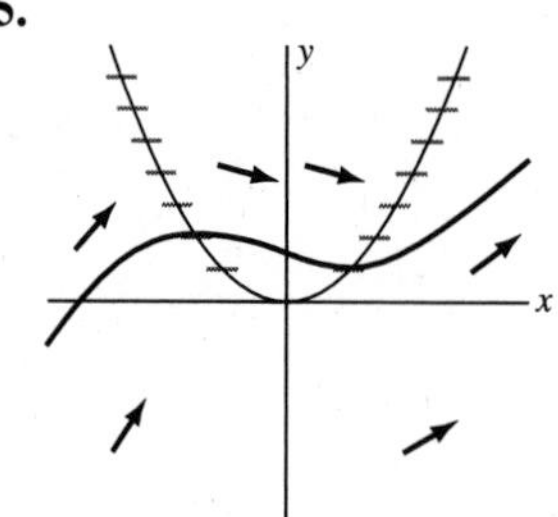

17.

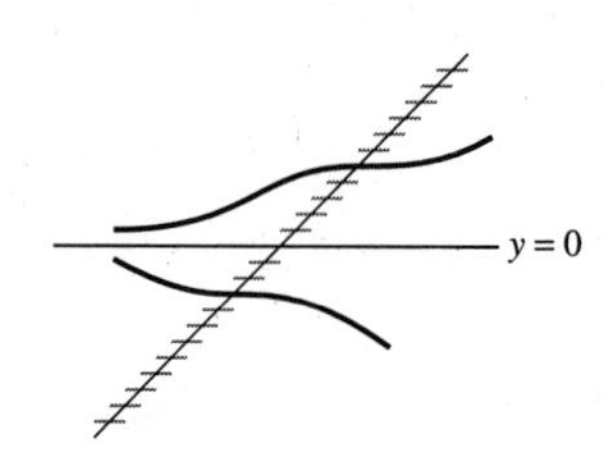

19.

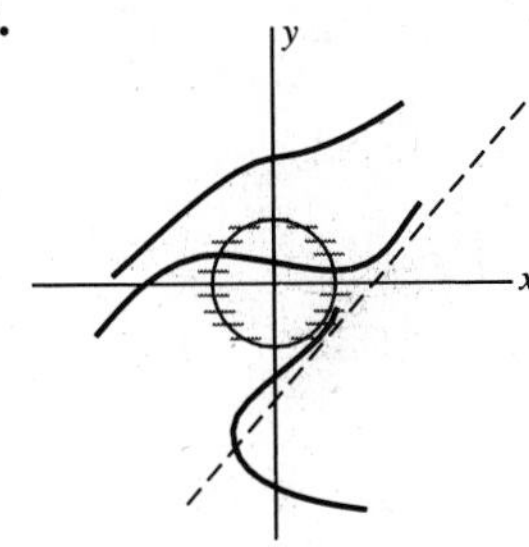

21.

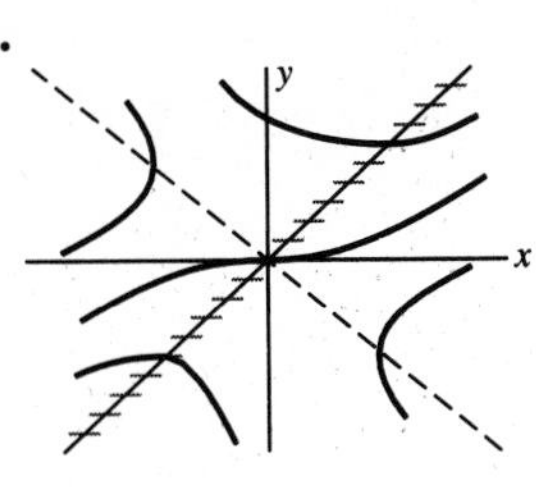

Section 1.5

1. All (x_0, y_0)

3. All (x_0, y_0)

5. $x_0 + y_0 \neq 0$

7. $y_0 \neq 1$

9. $x_0^2 + 2y_0^2 < 1$

11. $f(x, y) = x^{-1/3}$, $f_y = 0$ (if $x \neq 0$). Both continuous if $x \neq 0$.

13. $y_0 = 0$

15. $y_0 = 0$

17. $x_0 = n\pi$, $n = 0, \pm 1, \pm 2, \ldots$

19. None

21. $y = 1$ and $y = \left(\frac{4}{5}x - \frac{4}{5}x_0\right)^{5/4} + 1$, f_y not continuous at $y_0 = 1$.

23. Explodes for $y_0 > 0$

25. No solutions if $y_0 \neq 0$; $y = 0$ is unique solution if $y_0 = 0$.

Section 1.6.2

1. $y = ce^{3x}$

3. $y = c\exp(x^2)$

5. $y = cx^{-\frac{1}{2}}$

7. $y = ce^{-\sin x}$

9. $y = c\exp(-\int_0^x \cos t^{-\frac{1}{2}}\, dt)$

11. $y = 9e^{-5x}$

13. $y = 7e^{9(x-3)}$

15. $y = 10\exp\left(-\int_5^x \frac{\sin t}{4 + e^t}\, dt\right)$

17. $y = 3\exp\left(\frac{1}{x} - 1\right)$

Section 1.6.3

19. $y = x^{-1}e^x + (1 - e)x^{-1}$

21. $y = 3e^x + c$ (*Note*: This can be done by just antidifferentiating both sides.)

23. $y = x(x^2 + 1)^{-1} + c(x^2 + 1)^{-1}$

25. $y = \frac{x}{4} - \frac{1}{16} + \frac{1}{16}e^{-4x}$

27. $y = 4x^2$

29. $y = x^{-1}\ln x + cx^{-1}$

31. $y = \frac{1}{4}x + cx^{-3}$. One solution continuous at $(0, 0)$.
For the rest, $|y| \to \infty$ as $x \to 0$.

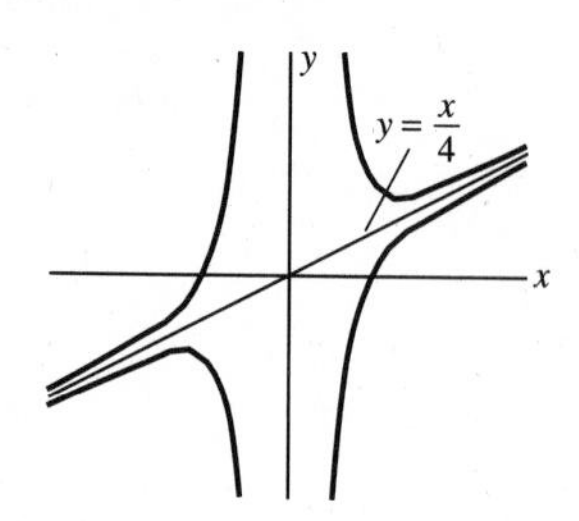

33. $y = x^2 + cx$. All solutions continuous, and pass through $(0, 0)$.

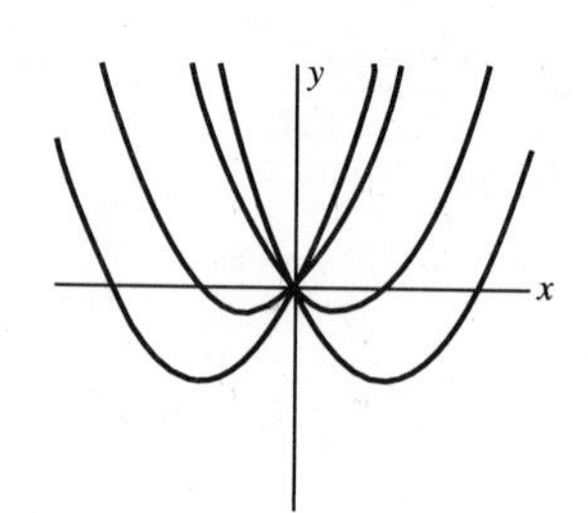

35. $y = e^{-x^2} \int_0^x e^{t^2}\, dt + ce^{-x^2}$

37. $y = e^{-x^3/3} \int_0^x e^{t^3/3} t\, dt + ce^{-x^3/3}$

39. $y = e^{-e^x} \int_0^x 3e^{e^t}\, dt + ce^{-e^x}$

41. $y = e^{-x} \int_2^x \frac{e^t}{t+1}\, dt + e^{-(x-2)}$

43. $y = \frac{1}{3}x^{\frac{1}{3}} \int_5^x t^{-\frac{1}{3}} \sin t\, dt$

45. $y = \frac{1}{7}x^{-\frac{1}{7}} \int_0^x t^{-\frac{6}{7}} e^t\, dt + cx^{-\frac{1}{7}}$

47. $x = -y - 1 + ce^y$

49. *Hint*: $e^{\int p(x)\,dx + c_1} = e^{\int p(x)\,dx} e^{c_1} = ue^{c_1}$, so the new integrating factor is just a constant times the old one.

51. $(uy)' = u(y' + py) \Leftrightarrow (kuy)' = ku(y' + py)$

53. $G(x, \bar{x}) = \exp\left(\int_x^{\bar{x}} \frac{\sin t}{t}\, dt\right) = \exp\left(-\int_{\bar{x}}^x \frac{\sin t}{t}\, dt\right)$

55. $\frac{du}{dx} = qe^{\int p\,dx}, u = \int qe^{\int p\,dx}\, dx,\ y = e^{-\int p\,dx} \int qe^{\int p\,dx}\, dx$

Section 1.7

1. $y = ce^{8x}$

3. $y = ce^{-2x}$

5. $y = ce^{-7x}$

7. $y = ce^{-x}$

9. $y = ce^{5t}$

11. $y = \frac{8}{3} + ce^{-3x}$

13. $y = \frac{9}{4} + ce^{4x}$

15. $y = -\frac{15}{4} + ce^{\frac{4}{5}x}$

17. $y = -2 + ce^{-\frac{2}{3}x}$

19. $y = -9 + ce^{2x}$

21. $y = -\frac{17}{3} + ce^{-x}$

23. $y = \frac{8}{3}e^{-4x} + ce^{-7x}$

25. $y = \frac{3}{7}e^{-5x} + ce^{2x}$

27. $y = \frac{3}{8}e^{4x} + ce^{-4x}$

29. $y = e^{-2x} + ce^{-3x}$

31. $y = \frac{5}{7}x + \frac{16}{49}$

33. $y = 7x - \frac{7}{2}$

35. $y = 2x^2 + x - 9$

37. $y = \frac{1}{3}x^3 - \frac{1}{3}x^2 + \frac{2}{9}x - \frac{2}{27}$

39. $y = \frac{1}{5}x - \frac{1}{25}$

41. $y = -\frac{9}{20}\cos 6x + \frac{3}{20}\sin 6x$

43. $y = -\frac{5}{13}\cos x + \frac{1}{13}\sin x$

45. $y = \frac{1}{10}\cos 2x + \frac{13}{10}\sin 2x$

47. $y = \frac{6}{37}\cos x + \frac{1}{37}\sin x - \frac{5}{61}\cos 5x + \frac{6}{61}\sin 5x$

49. $y = -\frac{5}{9} - \frac{1}{15}\cos 3x - \frac{1}{5}\sin 3x$

51. $y = \frac{1}{2}e^{3x} - \frac{1}{2}\cos x + \frac{1}{2}\sin x$

53. $y = 8xe^{-3x} + ce^{-3x}$

55. $\frac{dv}{dx} = 7,\ v = 7x + c,\ y = 7xe^{2x} + ce^{2x}$

57. $y = 4xe^x + ce^x$

59. $y = 5xe^{-x} + ce^{-x}$

61. $y = 8xe^{7x} + ce^{7x}$

63. $y = \begin{cases} 2e^{-2x}, \\ 2e^{-1}e^{-x} = 2e^{-(x+1)} \end{cases}$
$0 \le x \le 1,$
$1 \le x \le 2$

65. $y = \begin{cases} x, & 0 \le x \le 1, \\ 2 - x, & 1 \le x \le 2 \end{cases}$

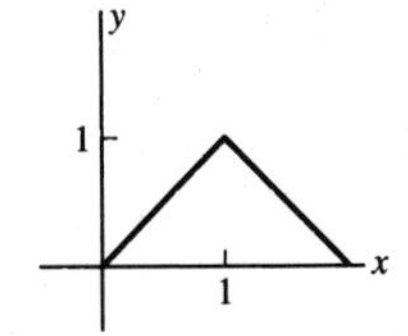

67. $y = \begin{cases} 2e^{-x}, & 0 \le x \le 1, \\ x + 2e^{-1} - 1, & 1 \le x \le 2 \end{cases}$

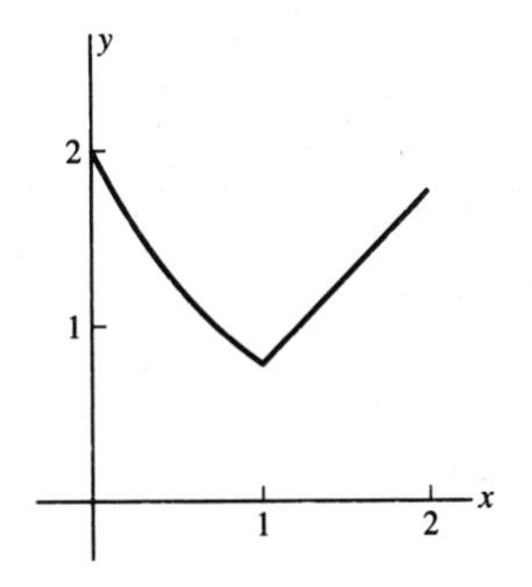

Section 1.8

1. $t_{\text{doubling}} = \dfrac{\ln 2}{0.015} \approx 46.2$ yr

3. $k = 3 \ln 2 \approx 2.08 = 208\%$ per day

5. $x = 1500(\frac{4}{3})^4$

7. 9.97 yr

9. $Q' = k_1 Q - k_2 Q + k$

11. 3% per year implies $t_{\text{doubling}} = \dfrac{\ln 2}{0.03} \approx 23.10$ yr,
3% yield implies $t_{\text{doubling}} = \dfrac{\ln 2}{\ln(1.03)} \approx 23.45$ yr.

13. $t_{\text{doubling}} = \dfrac{\ln 2}{0.064} \approx 10.83$ yr

15. $k = \ln(1.1) \approx 0.0953 = 9.53\%$ per year

17. 110.04 g

19. **a)** 31.063 yr
b) 75.018 yr

21. **a)** $\dfrac{10 \ln 17}{\ln 17 - \ln 15} \approx 226.4$ min

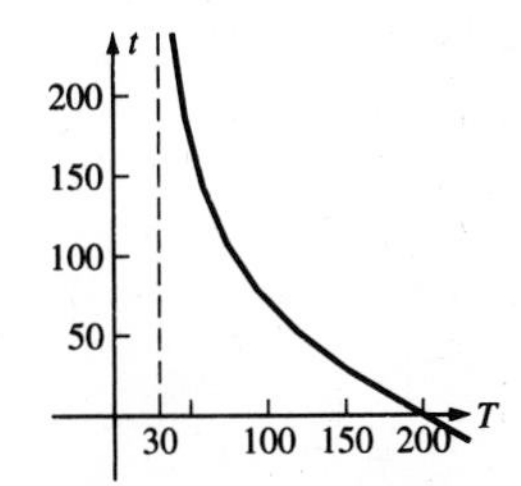

b) $t(T) = \dfrac{10}{\ln 15 - \ln 17} \ln(\dfrac{T - 30}{170})$

23. $k = -\dfrac{1}{10} \ln 3 = -0.1099$

25. **a)** $\dfrac{dT}{dt} = -k(T - Q_0) = -k[T - (20 + 10t)]$

b) $T = 10 + 10t + 30e^{-t}$

c)

27. **a)** $\dfrac{dA}{dt} = 0.06(A - 500),\ A(0) = 2000$

b) $A = 500 + 1500e^{0.06t}$, \$3233.18

c) \$411.06

29. $Q' = 0.08Q + 365B,\ Q(0) = 1000$

a) $B = \$3.79$

b) $B(x) = (800 - 80e^{0.08x})(365e^{0.08x} - 365)^{-1}$

31. $Q' = 0.2Q + 400\cos 2\pi t,\ Q(0) = 100$

a) $Q = \dfrac{1}{\pi^2 + 0.01}(-20\cos 2\pi t + 200\pi \sin 2\pi t) + \left(100 + \dfrac{20}{\pi^2 + 0.01}\right)e^{0.2t}$

b)

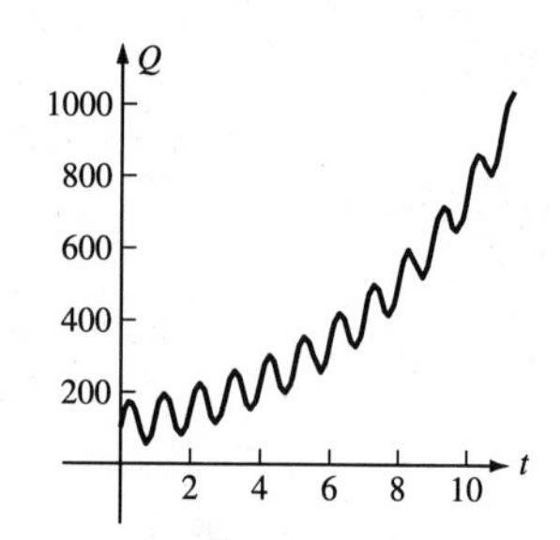

Section 1.9

1. **a)** $Q(t) = 120 - 60e^{-t/150}$ lb,

$c(t) = \dfrac{Q}{300} = \dfrac{2}{5} - \dfrac{1}{5}e^{-t/150}$ lb/gal

b) $t = -150\ln(0.5) \approx 104$ min

3. **a)** $Q(t) = 25 - e^{-t/25}15$,

$c(t) = 0.25 - e^{-t/25}0.15$

b) $\lim_{t\to\infty} c(t) = 0.25 > 0.2$

5. **a)** and **b)**

$$Q(t) = \begin{cases} 10\left(\dfrac{100}{100 + t}\right)^{1/5}, & 0 \le t \le 100, \\ 10e^{3(100-t)/500}(2^{-1/5}), & 100 \le t \end{cases}$$

$$c(t) = \begin{cases} \dfrac{Q(t)}{500 + 5t}, & 0 \le t \le 100 \\ \dfrac{Q(t)}{1000}, & 100 \le t \end{cases}$$

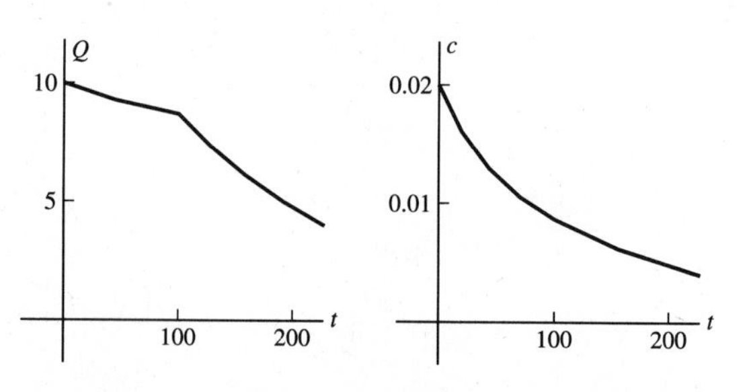

7. $Q(t) = 7(1000 + 3t) - 500{,}000(1000 + 3t)^{-\frac{2}{3}}$,
$c(t) = 7 - 500{,}000(1000 + 3t)^{-\frac{5}{3}}$

9. $\dfrac{dS_1}{dt} = 13 \cdot 3 - 7\dfrac{S_1}{150 + 6t}, \; \dfrac{dS_2}{dt} = 7\dfrac{S_1}{150 + 6t} - 28\dfrac{S_2}{250 - 21t}$

11. $\dfrac{dS_1}{dt} = 21 \cdot 5 - 18\dfrac{S_1}{230 + 3t}, \; \dfrac{dS_2}{dt} = 9\dfrac{S_1}{230 + 3t} - 4\dfrac{S_2}{275 + 5t}$

13. $\dfrac{dS_1}{dt} = 11 \cdot 5 - 18\dfrac{S_1}{100 - 7t}, \; \dfrac{dS_2}{dt} = 18\dfrac{S_1}{100 - 7t} - 18\dfrac{S_2}{200}, \; \dfrac{dS_3}{dt} = 18\dfrac{S_2}{200}$

Section 1.10

1. $\dfrac{di}{dt} + 2i - 1 = 0$. Equilibrium is $i = \frac{1}{2}$
$i = \frac{1}{2} + ce^{-2t} = \frac{1}{2} + [i(0) - \frac{1}{2}]e^{-2t}$

3. $\dfrac{di}{dt} + i - \sin t = 0$,
$i = \frac{1}{2}\sin t - \frac{1}{2}\cos t + [i(0) + \frac{1}{2}]e^{-t}$

5. $i(t) = \begin{cases} 9(1 - e^{-t}) & 0 \le t \le 10 \\ 9(e^{10} - 1)e^{-t} & t \ge 10 \end{cases}$

7. $\dfrac{dq}{dt} + q = 3\sin t,\; q(0) = 1$
$q = \frac{3}{2}(\sin t - \cos t) + \frac{5}{2}e^{-t}$

Section 1.11

1. **a)** $v = -1960(1 - e^{-t/2})$
b) $v = -1947$
c) -1960

3. $-14{,}416$ cm/s

5. **a)** $v = -\sqrt{32}\,\dfrac{(1 + ke^{-2\sqrt{32}t})}{(1 - ke^{-2\sqrt{32}t})}, \; k = (1000 - \sqrt{32})(1000 + \sqrt{32})^{-1}$
b) $v = -\sqrt{32}$

Section 1.12

1. $y = -x + c_2$

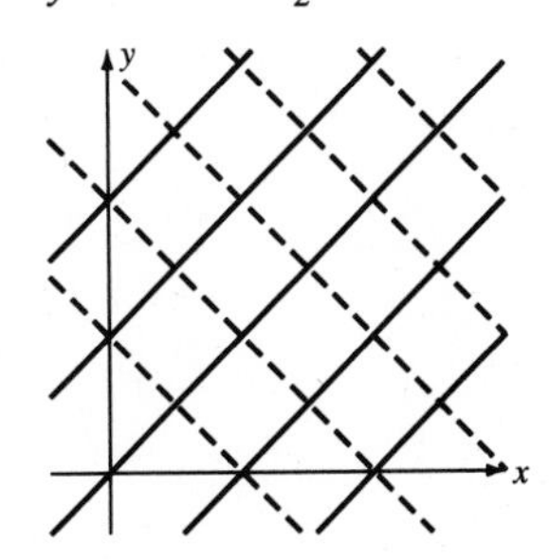

3. $y^2 - x^2 = c_2$

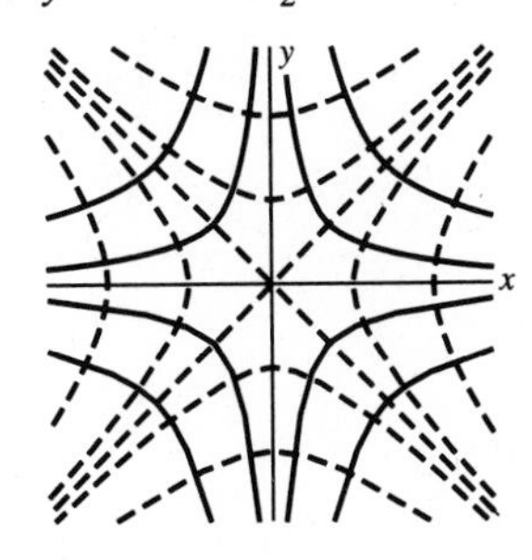

5. $2y^2 \ln|y| - y^2 = x^2 - 2x^2\ln|x| + c_2$

7. $y = -\dfrac{\ln|x|}{2} + c_2$

9. $y = \pm\,\frac{4}{3}x^{3/2} + c_2$

11. $y + \dfrac{y^3}{3} = -x + c_2$

13. $y^2 = 2\ln|\sin x| + c_2$

15. $2y^2 = -x^2 + c_2$

17. $y = (c_2 - \frac{3}{2}\ln|x|)^{-1}$

Section 1.13

1. $x^2 + xy + y^2 = c$

3. $y \ln x + y^3 = c$

5. Not exact

7. $x^{-2} + e^{x^2}y + y^2 = c$

9. $u^2 + u\theta + u^2\theta^2 + \theta^3 = c$

11. Not exact

13. $x^2 + x^3y^2 - y^5 = c$

15. $x^2e^{xy} + y^3 = c$

17. $\ln(x + y) + y^3 = c$

19. $\sin(x + y) + \cos(y^2) = c$

23. Need $F(x, y) = F(x_0, y_0)$ for all (x, y) on the path followed.

25. Exact $\Leftrightarrow (-F_y)_y = (F_x)_x$

Section 1.14

1. $x^2 + yx + y^{-2} = c,\ u(y) = y^{-3}$

3. $x^{-2} + xy + 3y^2 = c,\ u(x) = x^{-3}$

5. $ye^x + x = c,\ u(x) = e^x$

7. $x^2e^{-y} + 3y^2 = c,\ u(y) = e^{-y}$

9. $x^3y^3 + y^5 = c,\ u(y) = y^2$

11. $x^4y^5 + y^6 = c,\ u(y) = y^4$

13. $x^3y^3 + y^2 + x^2 = c,\ u(y) = y^2$

15. The method of this section gives the solution as $e^{\int p(x)\,dx}y - \int e^{\int p(x)\,dx}q(x)\,dx = c.$

19. $y^2e^x + e^x = c,\ u(x) = e^x,\ x + \ln(y^2 + 1) = c,\ \tilde{u}(y) = (y^2 + 1)^{-1}$

21. $u = xy^2,\ x^2y^3 + x^3y^4 = c$

23. $u = x^{-3}y^{-3},\ x^{-3}y^{-3} + x^{-2}y = c$

Chapter 2

Section 2.1

1. All (x_0, y_0, z_0)

3. All (x_0, y_0, z_0)

5. $z_0^2 + y_0^2 \neq 0$

7. All (x_0, y_0, z_0)

9. $x_0 \neq 0$ and $z_0 \neq -1$

11. $y_0 \neq 0$ and $z_0 \neq -1$

13. $y = x^2 + cx$

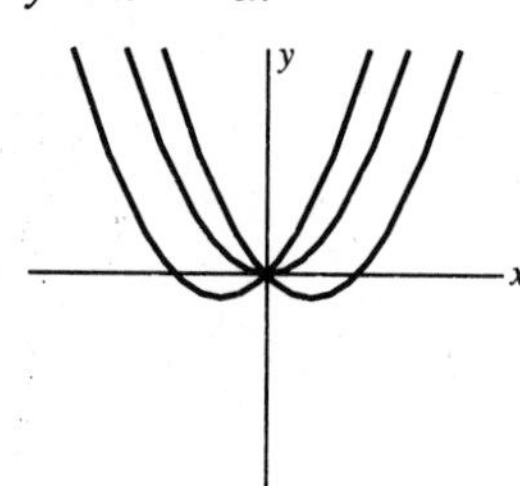

15. $y = x^2 + c$

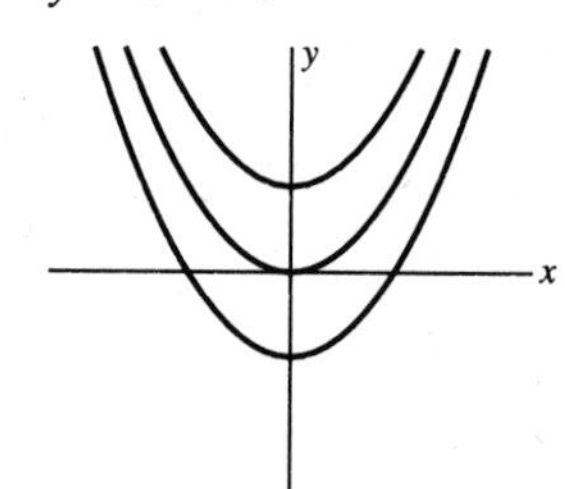

17. $y = \dfrac{x^3}{6} + \dfrac{x^2}{2} + cx + 1$

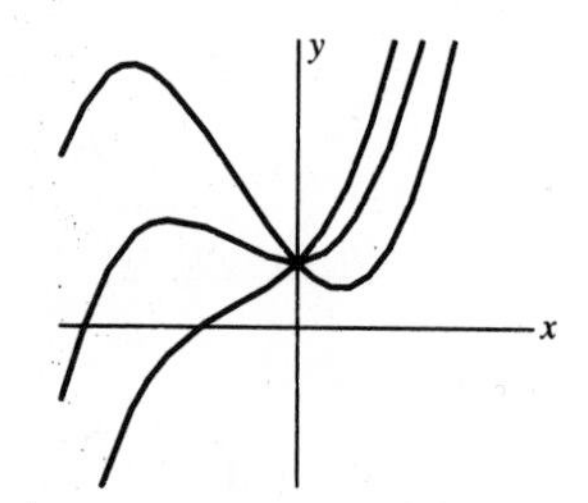

19. $y = \dfrac{x^3}{6} + \dfrac{x^2}{2} - x + c$

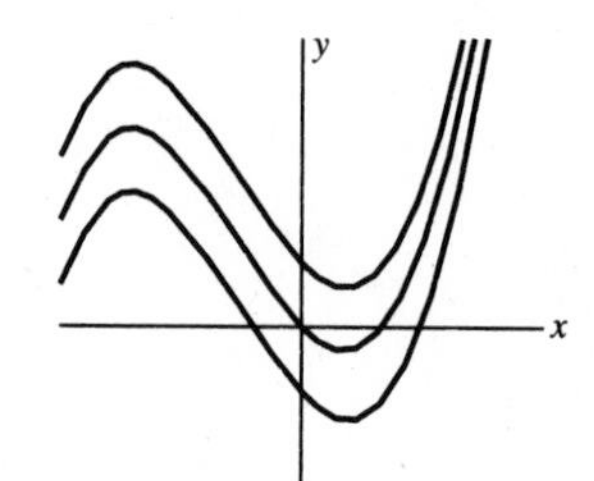

21. $f(x, y, z) = \dfrac{xz}{y}$, not continuous at $(0, 0, 0)$

23. Theorem 2.1.1 implies that for any (x_0, y_0, z_0), there is a solution y such that $(x_0, y_0, z_0) = (x, y(x), y'(x))$ when $x = x_0$.

Section 2.2

1. $y = c_1 \sin x + c_2 \cos x + 1$, $y = 1 - \cos x$

3. $y = c_1 e^x + c_2 e^{2x} + x + \frac{3}{2}$, $y = -\frac{1}{2} e^{2x} + x + \frac{3}{2}$

5. $y = c_1 e^{-x} \cos x + c_2 e^{-x} \sin x + 3$,
$y = -2e^{-x} \cos x - e^{-x} \sin x + 3$

7. $y = c_1 \cos x + c_2 \sin x + x \sin x$,
$y = \cos x - \sin x + x \sin x$

9. b) $\tilde{y} = 2e^x + e^{2x} + e^{-x} = e^x + e^{2x} + 2\cosh x = y$

11. $y = c_1 \cos x + c_2 \sin x - \frac{1}{3} \sin 2x - \frac{1}{3} \cos 2x$

17. $y_p = \frac{1}{4} e^x$

19. $y_p = \frac{1}{5} \sin x - 4x^2$

21. $y = c(\cos x - 2e^x)$ for any c

Section 2.3

1. Yes, $W[\sin x, \cos x] = -1 \neq 0$.

3. x, no; only one solution.

5. a) $W[y_1, y_2](1) = -1 \neq 0$ **b)** $y_3 = 2y_1 + 2y_2$

7. $W[y_1, y_2](x_0) = 1$

9. Manipulate definition (26)

15. $W = y_1 y_2' - y_2 y_1'$, $\dfrac{dW}{dx} = 0$

17. $W = 1$, $p(x) = 0$, $W(x) = W(x_0)$

19. $W = e^{3x}$, $p(x) = -3$, $W(x) = W(x_0)e^{3x}$

21. $W = -x^{-4}$, $p(x) = \dfrac{4}{x}$, $W(x) = W(x_0)x^{-4}$

Section 2.4

1. Yes

3. b) $\tilde{y} = y_1 + \frac{1}{3} y_3$

5. b) $y = x + c_1 + c_2 e^x + c_3 x e^x$

7. b) $y = -e^{-x} + c_1 e^x + c_2 x e^x + c_3 x^2 e^x$

Section 2.5

1. $y = c_1 x^{-1} \ln x + c_2 x^{-1}$

3. $y = c_1 e^{-5x} + c_2 x e^{-5x}$

5. $y = c_1(x - 1) + c_2 e^{-x}$

7. $y = c_2 x \int_1^x \frac{1}{t^2} e^{-\frac{1}{2}t^2} \, dt + c_1 x$

9. $r = 2$, $y = c_1 x^2 + c_2 x^2 \ln x$

11. $r = -3$, $y = c_1 x^{-3} + c_2 x^{-3} \ln x$

13. $r = 2$, $y = c_1 \sin 2x + c_2 \cos 2x$

15. $r = 2$, $y = c_1 e^{2x} + c_2 x e^{2x}$

17. $r = -1$, $y = c_1 e^{-x} + c_2 x e^{-x}$

19. $r = \pm 2$, $y = c_1 e^{2x} + c_2 e^{-2x}$

21. Use Wronskian

Section 2.6

1. $y = c_1 e^{2x} + c_2 e^{-3x}$

3. $y = c_1 \cos x + c_2 \sin x$

5. $y = c_1 e^{-2x} \cos x + c_2 e^{-2x} \sin x$

7. $y = c_1 e^x + c_2 e^{2x}$

9. $y = c_1 + c_2 e^x$

11. $y = c_1 + c_2 x$

13. $y = c_1 e^{-x} + c_2 e^{x/3}$

15. $y = \frac{1}{3} e^x - \frac{1}{3} e^{-2x}$

17. $y = c_1 \cos \sqrt{2}x + c_2 \sin \sqrt{2}x$

19. $y = -e^{-3x} + 3e^{-x}$

21. $y = c_1 e^{-5x} + c_2 x e^{-5x}$

23. $y = c_1 e^{7x} + c_2 x e^{7x}$

25. $y = c_1e^{3x}\cos 4x + c_2e^{3x}\sin 4x$

27. $y = c_1e^{\sqrt{12}x} + c_2e^{-\sqrt{12}x}$

29. $y = c_1e^{-2x}\cos 2x + c_2e^{-2x}\sin 2x$

31. $y = c_1\cos\sqrt{8}x + c_2\sin\sqrt{8}x$

33. $y = e^{-3x}(c_1e^{\sqrt{2}x} + c_2e^{-\sqrt{2}x})$

35. $W[e^{\alpha x}\cos\beta x, e^{\alpha x}\sin\beta x] = \beta e^{2\alpha x} \neq 0$ if $\beta \neq 0$

37. $b = -2ar_1,\ c = ar_1^2$

41. $y'' + 3y' + 2y = 0$

43. $y'' - 6y' + 9y = 0$

45. $y'' + 16y = 0$

47. $y'' + 9y = 0$

49. $y'' = 0$

51. $y'' - 2y' + 2y = 0$

Section 2.7

1. $y = c_1e^{2x} + c_2xe^{2x} + c_3x^2e^{2x}$

3. $y = c_1e^{x} + c_2e^{-x} + c_3e^{2x} + c_4e^{-2x}$

5. $y = c_1 + c_2e^{x} + c_3e^{-2x}$

7. $y = c_1e^{-x} + c_2xe^{-x} + c_3x^2e^{-x} + c_4x^3e^{-x}$

9. $y = c_1e^{3x}$

11. $y = c_1e^{-x}\cos x + c_2e^{-x}\sin x + c_3e^{x}$

13. $y = c_1e^{-x}\cos x + c_2e^{-x}\sin x + c_3xe^{-x}\cos x + c_4xe^{-x}\sin x$

15. $y = c_1e^{x} + c_2xe^{x} + c_3x^2e^{x} + c_4x^3e^{x}$

17. $y = c_1e^{x} + c_2xe^{x} + c_3x^2e^{x} + c_4e^{-x} + c_5xe^{-x} + c_6x^2e^{-x}$

19. $y = c_1e^{x} + c_2e^{-x} + c_3\cos x + c_4\sin x$

21. $y = c_1\cos 5x + c_2\sin 5x + c_3x\cos 5x + c_4x\sin 5x$

23. $y = c_1e^{x} + c_2e^{-2x}\cos x + c_3e^{-2x}\sin x$

25. $y''' - 4y'' + 4y' = 0;\ r(r-2)^2$

27. $y'''' + 50y'' + 625y = 0;\ (r^2+25)^2 = (r+5i)^2(r-5i)^2$

29. $y''' + 6y'' + 12y' + 8y = 0;\ (r+2)^3$

31. $y'''' - 4y''' + 8y'' - 8y' + 4y = 0;\ (r^2 - 2r + 2)^2$

33. $y'''' + 2y'' + y = 0;\ (r^2+1)^2$

35. $y = e^{8x}(c_1 + c_2x + c_3x^2 + c_4x^3) + e^{8x}(c_5\cos 7x + c_6\sin 7x)$

37. $y = c_1 + c_2x + c_3x^2 + e^{4x}(c_4\cos 5x + c_5\sin 5x) + (c_6 + c_7x)\cos 5x + (c_8 + c_9x)\sin 5x$

39. $y = e^{2x}[(c_1 + c_2x + c_3x^2)\cos x + (c_4 + c_5x + c_6x^2)\sin x]$

41. $y = e^{2x}(c_1\cos 6x + c_2\sin 6x) + e^{-2x}(c_3\cos 6x + c_4\sin 6x) + (c_5 + c_6x)\cos 6x + (c_7 + c_8x)\sin 6x$

Section 2.8.2

1. $x = R\cos(5t - \phi)$ with $R = \sqrt{58}$; $\tan\phi = -\frac{7}{3}$, 4th quadrant; $\phi = \arctan(-\frac{7}{3}) = -1.1659$ radians

3. $x = R\cos(14t - \phi)$ with $R = 2$, $\tan\phi = \dfrac{1}{\sqrt{3}}$, 1st quadrant, $\phi = \dfrac{\pi}{6}$ radian

5. $x = R\cos(5t - \phi)$ with $R = 6\sqrt{2}$, $\tan\phi = -1$, 2nd quadrant, $\phi = \dfrac{3\pi}{4}$ radians

7. $x = R\cos(6t - \phi)$ with $R = 2$, $\tan\phi = -\dfrac{1}{\sqrt{3}}$, 4th quadrant, $\phi = -\dfrac{\pi}{6}$ radian

9. $x = R\cos(2t - \phi)$ with $R = 8$, $\tan\phi = -\sqrt{3}$, 2nd quadrant, $\phi = \dfrac{2\pi}{3}$ radians

11. $30x'' + 1470x = 0$, $x(0) = 10$, $x'(0) = 0$, $x = 10\cos 7t$

13. $2x'' + 128x = 0$, $x(0) = 2$, $x'(0) = 1$, $x = 2\cos 8t + \frac{1}{8}\sin 8t$

15. $k = 1000\ \pi^2\ \text{ft/s}^2$

17. $x = c_1 \cos 2t + c_2 \sin 2t$, $c_1 = R\cos\phi = 1$, $c_2 = R\sin\phi = \sqrt{3}$, $x(0) = c_1 = 1$, $x'(0) = 2c_2 = 2\sqrt{3}$

19. **a)** $x = 10\sqrt{\frac{m}{k}} \sin \sqrt{\frac{k}{m}}\, t$ **b)** Amplitude $= 10\sqrt{\frac{m}{k}}$ **c)** Decreases **d)** Increases

21. **a)** $x = \cos\sqrt{\frac{k}{m}}\, t + \sqrt{\frac{m}{k}} \sin\sqrt{\frac{k}{m}}\, t$ **b)** Amplitude $= \sqrt{1 + \frac{m}{k}}$
c) Increasing m increases amplitude, increasing k decreases amplitude toward 1.

23. $\int 0\, dt = \int (mx''x' + kxx')\, dt \Rightarrow c = \frac{1}{2}m(x')^2 + \frac{1}{2}kx^2$

25. $x(t) = c_1 \cos\sqrt{7}\, t + c_2 \sin\sqrt{7}\, t$

27. $x(t) = c_1 e^{\sqrt{7}t} + c_2 e^{-\sqrt{7}t}$

29. $x(t) = 2\cos\sqrt{5}\, t + \frac{3}{\sqrt{5}} \sin\sqrt{5}\, t = R\cos(\sqrt{5}\, t - \phi)$, where $R = \sqrt{\frac{29}{5}}$, $\tan\phi = \frac{3}{2\sqrt{5}}$, $\phi = 0.59091$ (1st quadrant).

31. If the constant $= c$, then $\frac{c}{2\pi} =$ cycles per second. Thus $c =$ cycles per 2π seconds or radians per second (the circular frequency). $x(t) = r\cos\theta$, where $\theta = ct + \theta_0$ and r is a constant.

Section 2.8.3

33. $10x'' + 40x' + 30x = 0$, $x(0) = 3$, $x'(0) = -5$, $x = e^{-3t} + 2e^{-t}$, overdamped

35. $x'' + 49x = 0$, $x(0) = 1$, $x'(0) = 7$, $x = \sin 7t + \cos 7t$, $x = \sqrt{2}\cos\left(7t - \frac{\pi}{4}\right)$, harmonic

37. $x'' + 7x' + 12x = 0$, $x(0) = -1$, $x'(0) = 1$, $x = 2e^{-4t} - 3e^{-3t}$, overdamped.

39. $x'' + 4x' + 5x = 0$, $x(0) = 2$, $x'(0) = 0$, $x = 4e^{-2t}\sin t + 2e^{-2t}\cos t$, damped oscillation,
$x = e^{-2t}\sqrt{20}\cos(t - \phi)$, $\phi = \tan^{-1}2 = 1.107$ radian.

41. Differentiate x, set equal to zero, and show that there is at most one solution.

43. $0 < m < \delta^2/4k$, overdamped; $m = \delta^2/4k$, critically damped; $\frac{\delta^2}{4k} < m$ is underdamped oscillation. As $m \to +\infty$, solution decays slower, period $\to \infty$, frequency $\to 0$.

45. $\delta = 12\pi$.

47. $x(t) = e^{-t}\left(c_1 \cos\frac{\sqrt{2}}{2}t + c_2 \sin\frac{\sqrt{2}}{2}t\right)$, underdamped

49. $x(t) = e^{-\frac{5}{6}t}\left(c_1 \cos\frac{\sqrt{23}}{6}t + c_2 \sin\frac{\sqrt{23}}{6}t\right)$, underdamped

51. $x(t) = c_1 e^{-2t} + c_2 e^{-\frac{2}{3}t}$, overdamped

53. $x(t) = c_1 e^{-\frac{2}{3}t} + c_2 t e^{-\frac{2}{3}t}$, critically damped

55. $x(t) = c_1 e^{-t} + c_2 e^{-t/3}$, overdamped

57. $t = \frac{1}{r_1 - r_2} \ln \frac{1 - \frac{v_0}{r_1}}{1 - \frac{v_0}{r_2}}$

Section 2.9

1. Yes

3. No, $\ln |x|$

5 No; $\dfrac{\sin x}{\cos x}$ not allowed

7. No, not constant coefficients

9. No, negative power of x

11. Yes, $\sinh 3x = \frac{1}{2}e^{3x} - \frac{1}{2}e^{-3x}$

13. $y(x) = \frac{2}{3} - \frac{2}{27}x + \frac{1}{9}x^3 + c_1 \cos 3x + c_2 \sin 3x$

15. $y(x) = \frac{81}{64}x + \frac{7}{16}x^2 + c_1 + c_2 e^{-8x}$

17. $y(x) = \frac{5}{2}e^{2x} + c_1 e^{3x} + c_2 e^{4x}, c_1 = -1, c_2 = -\frac{1}{2}$

19. $y = -2xe^x + c_1 e^{2x} + c_2 e^x$

21. $y(x) = e^x + c_1 e^x \cos 2x + c_2 e^x \sin 2x$

23. $y(x) = -\frac{5}{6}xe^{-3x} + c_1 e^{3x} + c_2 e^{-3x}$

25. $y(x) = \frac{3}{5}\sin x - \frac{3}{10}\cos x + c_1 e^{-x}\cos 2x + c_2 e^{-x}\sin 2x$, $c_1 = \frac{13}{10}$, $c_2 = \frac{17}{20}$

27. $y(x) = -\frac{2}{3}x \cos 3x + c_1 \cos 3x + c_2 \sin 3x$

29. $y(x) = -\frac{1}{3}\cos 2x + c_1 \cos x + c_2 \sin x$, $c_1 = \frac{1}{3}$, $c_2 = 2$

31. $y(x) = \frac{1}{5}xe^{3x} - \frac{6}{25}e^{3x} + c_1 e^{2x} + c_2 e^{-2x}$

33. $y_p = (A_0 x + A_1 x^2)e^x$, $y = -\frac{1}{4}xe^x - \frac{1}{4}x^2 e^x + c_1 e^{3x} + c_2 e^x$

35. $y(x) = \frac{3}{8}x \sin 4x + c_1 \cos 4x + c_2 \sin 4x$

37. $y_p = A_1 e^x \cos 3x + A_2 e^x \sin 3x$, $y = -\frac{1}{5}e^x \cos 3x + c_1 e^x \cos 2x + c_2 e^x \sin 2x$

39. $y_p = x(A_0 + A_1 x + A_2 x^2) + A_3 e^x$, $y = \frac{3}{8}x - \frac{3}{4}x^2 + x^3 + \frac{1}{5}e^x + \frac{29}{32} - \frac{17}{160}e^{-4x}$

41. $y_p = Ax^2 e^{-x}$, $y = \frac{3}{2}x^2 e^{-x} + c_1 e^{-x} + c_2 xe^{-x}$

43. $y_p = A_1 e^x + A_2 xe^x + A_3 e^{3x}$, $y = e^x + xe^x + 2e^{3x} + c_1 e^{2x} + c_2 xe^{2x}$

45. $y_p = A_0 + A_1 x + A_2 x^2 + A_3 \cos 2x + A_4 \sin 2x$, $y = 6 - 5x + 2x^2 + \frac{1}{5}\sin 2x + c_1 e^{-x} + c_2 e^{-4x}$

47. $y_p = A_0 + A_1 x + A_2 x^2$, $y = \frac{11}{27} - \frac{2}{9}x + \frac{1}{3}x^2 + ce^{-3x}$

49. $y_p = Ax \sin 2x + Bx \cos 2x$, $y = -\frac{1}{4}x \cos 2x + c_1 \cos 2x + c_2 \sin 2x$

51. $y_p = (A_1 + A_2 x)e^x$, $y = -3e^x + xe^x + ce^{2x/3}$

53. $y_p(x) = A_0 + A_1 x + A_2 x^2 + A_3 x^3 + A_4 x^4 + A_5 x^5$

55. $y_p(x) = x(A_0 + A_1 x + A_2 x^2 + A_3 x^3)$

57. $y_p(x) = Ae^{4x}$

59. $y_p(x) = Axe^{2x}$

61. $y_p(x) = Ax^2 e^{-3x}$

63. $y_p(x) = x^2(A_0 + A_1 x + A_2 x^2 + A_3 x^3)e^{-x}$

65. $y_p(x) = x(A_0 + A_1 x + A_2 x^2 + A_3 x^3 + A_4 x^4 + A_5 x^5)e^{4x}$

67. $y_p = x^2(A_0 + A_1 x + A_2 x^2 + A_3 x^3 + A_4 x^4)e^{3x}$

69. $y_p = x(A_0 + A_1 x + A_2 x^2)e^{2x} + A_3 e^{5x}$

71. $y_p(x) = A \cos 5x + B \sin 5x$

73. $y_p(x) = (A_0 + A_1 x + A_2 x^2)\cos 4x + (B_0 + B_1 x + B_2 x^2)\sin 4x$

75. $y_p(x) = x(A \cos 5x + B \sin 5x)$

77. $y_p = Axe^x \sin 2x + Bxe^x \cos 2x$

79. $y_p = (A_0 + A_1 x)e^{-x} \sin 2x + (B_0 + B_1 x)e^{-x} \cos 2x$

81. $y_p(x) = (A_0 + A_1 x + A_2 x^2)e^x \cos x + (B_0 + B_1 x + B_2 x^2)e^x \sin x$

83. $y_p = A_1 xe^{-x} \cos x + B_1 xe^{-x} \sin x + A_2 e^x \cos x + B_2 e^x \sin x$

85. $y_p(x) = Axe^{-x} + Bxe^x + e^x(D \cos x + E \sin x)$

87. $y_p = x(A_0 + A_1x)\cos 4x + x(A_2 + A_3x)\sin 4x + A_4e^{-x}\sin 4x + A_5e^{-x}\cos 4x + A_6e^{-4x}$

89. $y_p = (A_0 + A_1x + A_2x^2 + A_3x^3)e^{-x}\sin x + (A_4 + A_5x + A_6x^2 + A_7x^3)e^{-x}\cos x + A_8xe^x\sin x + A_9xe^x\cos x$

91. $y_p(x) = (A_0 + A_1x)e^x\cos 3x + (B_0 + B_1x)e^x\sin 3x$

93. $y_p(x) = x(A_0 + A_1x + A_2x^2)e^{-2x}\cos 2x + x(B_0 + B_1x + B_2x^2)e^{-2x}\sin 2x$

95. $y(x) = x(A_0 + A_1x + A_2x^2 + A_3x^3)e^{-2x}\cos 3x + x(B_0 + B_1x + B_2x^2 + B_3x^3)e^{-2x}\sin 3x$

97. **a)** $y(x) = x^2(A_0 + A_1x + A_2x^2 + A_3x^3)$ **b)** $y(x) = \frac{1}{20}x^5 + \frac{7}{6}x^3 - x^2$

99. $y = e^{2x} + c_1e^x + c_2xe^x + c_3x^2e^x$

101. $y = \frac{1}{2}\cos x + c_1 + c_2e^x + c_3e^{-x}$

103. $y_p = Ax\cos x + Bx\sin x + Cx,\ y = -x\cos x + 3x - 2\sin x$

105. $y = -\frac{4}{45}e^x - \frac{1}{3}xe^x + c_1e^{2x} + c_2e^{-2x} + c_3\cos 2x + c_4\sin 2x$

107. $y = -\frac{1}{40}e^{3x} + \frac{1}{12}xe^{2x} + c_1e^x + c_2e^{-x} + c_3e^{2x} + c_4e^{-2x}$

109. $y_p = x^3(A_0 + A_1x + A_2x^2)e^x$

111. $y_p = x^4(A_0 + A_1x + A_2x^2 + A_3x^3)e^x + (A_4 + A_5x + A_6x^2)e^{-x}$

113. $y_p = x(Ae^{-x}\cos x + Be^{-x}\sin x)$

115. $r^4 + 4r^3 + 8r^2 + 8r + 4 = (r^2 + 2r + 2)^2,\ y_p = x^2[Ae^{-x}\cos x + Be^{-x}\sin x]$

Section 2.10

1. $6x^2 + x^4$

3. $3e^{-x} - 3xe^{-x}$

5. e^{-x}

7. $L_1L_2 = xD^2 + D,\ L_2L_1 = xD^2$

9. $D^3 + 3D^2 + D + 3$

11. $D^3 - D$

13. $L_1L_2 = 2\cos xD^2 + \sin xD^3, L_2L_1 = \sin x(D^3 + D)$

15. $L_2L_1(y) = L_2(f) = 0$

17. $L_1(y) = 3 \Rightarrow y_1 = c_1e^{-x} + 3,\ L_2(y_2) = y_1 \Rightarrow$

$y_2' - y_2 = c_1e^{-x} + 3 \Rightarrow y_2 = c_2e^x - \frac{c_1}{2}e^{-x} - 3$

21. $h = f - g,\ L(h) = L(f) - L(g) = 0.$

23. $p(D)f = m(D)q(D)f + r(D)f = m(D)0 + r(D)f = r(D)f$

25. **b)** $(D^5 + 3D^2 + 1) = (D^3 - 2D^2 + 3D - 1)(D^2 + 2D + 1) - D + 2 \Rightarrow$
$(D^5 + 3D^2 + 1)xe^{-x} = (-D + 2)xe^{-x} = -e^{-x} + 3xe^{-x}$

27. $L_2 = D - 1,\ y = c_1xe^x$

29. $L_2 = D^3,\ y = c_1 + c_2x + c_3x^2$

31. $L_2 = D + 2,\ y = c_1x^2e^{-2x}$

33. $L_2 = (D - 2)(D - 3),\ y = c_1xe^{2x} + c_2xe^{3x}$

Section 2.11.1

1. $m = \dfrac{1}{400\pi^2}$

3. $k = 144(22)^2\pi^2$

5. $\dfrac{10}{(140)^2\pi^2} < m < \dfrac{10}{(20)^2\pi^2}$

7. $\sqrt{\frac{8}{15}}\,\dfrac{1}{2\pi}$

9. $m = (240\pi^2)^{-1}g$

13. $mg = kd,\ mx'' = F_T = -k(x + d) + mg = -kx,\ mx'' + kx = 0$

15. $x = \dfrac{F_0}{k - m\omega^2}(\cos\omega t - \cos\omega_0 t) + x_0\cos\omega_0 t$

17. $x = \dfrac{F_0}{k - m\omega^2}\left(\sin\omega t - \dfrac{\omega}{\omega_0}\sin\omega_0 t\right) + x_0\cos\omega_0 t$

19. $x = \frac{F_0}{2m\omega} t \sin \omega t$

21. $x = -\frac{8}{21} \cos 5t + c_1 \cos 2t + c_2 \sin 2t$

23. $x = -\frac{8}{29} \cos 5t + c_1 e^{2t} + c_2 e^{-2t}$

25. $x = \frac{3}{4} t \sin 2t + c_1 \cos 2t + c_2 \sin 2t$

27. $x = \frac{F_0}{2m\omega} t \sin \omega t + c_1 \cos \omega t + c_2 \sin \omega t$

Section 2.11.2

29. $x(t) = \frac{3}{73} \cos 4t + \frac{8}{73} \sin 4t = \frac{\sqrt{73}}{73} \cos(4t - \phi)$, where $\tan \phi = \frac{8}{3}$, $\phi = 1.212$

31. $x(t) = \frac{9}{29} \cos 2t + \frac{8}{29} \sin 2t = \frac{\sqrt{145}}{29} \cos(2t - \phi)$, where $\tan \phi = \frac{8}{9}$, $\phi = 0.72661$

33. $x(t) = -\frac{3}{208} \cos t + \frac{15}{208} \sin t = \frac{3\sqrt{26}}{208} \cos(t - \phi)$, where $\tan \phi = -5$, $\phi = 1.7682$

35. $x(t) = -\frac{3}{10} \cos t + \frac{1}{10} \sin t = \frac{\sqrt{10}}{10} \cos(t - \phi)$, $\tan \phi = -\frac{1}{3}$, $\phi = 2.8198$

37. $x(t) = -\frac{2}{5} \cos t + \frac{1}{5} \sin t = \frac{\sqrt{5}}{5} \cos(t - \phi)$, $\tan \phi = -\frac{1}{2}$, $\phi = 2.6779$

39. $x(t) = -\frac{1}{2} \cos t = \frac{1}{2} \cos(t - \phi)$, $\tan \phi = 0$, $\phi = \pi$

41. If $\gamma < \frac{\sqrt{2}}{2}$, maximum occurs at $z = 1 - 2\gamma^2$ or $\omega = \omega_0 \sqrt{1 - 2\gamma^2}$. As $\gamma \to 0$, $\omega \to \omega_0$. If $\gamma \geq \frac{\sqrt{2}}{2}$, the maximum occurs at $\omega = 0$ since $\frac{dy}{d\omega} \neq 0$ for all $\omega > 0$ and $y \to 0$ as $\omega \to \infty$.

Section 2.11.3

47. $x = \frac{1}{13} + \frac{38}{13} e^{-3t} \cos 2t + \frac{57}{13} e^{-3t} \sin 2t$; steady state $= \frac{1}{13}$, transient $= \frac{38}{13} e^{-3t} \cos 2t + \frac{57}{13} e^{-3t} \sin 2t$

49. $x(t) = \frac{1}{2} t - \frac{3}{4} + c_1 e^{-t} + c_2 e^{-2t}$, $c_1 = 1$, $c_2 = -\frac{1}{4}$; transient $= e^{-t} - \frac{1}{4} e^{-2t}$

51. **a)** $x = \frac{3}{25} - \frac{1}{17} e^{-2t} + \frac{393}{425} e^{-3t} \sin 4t + \frac{399}{425} e^{-3t} \cos 4t$ **b)** all but $\frac{3}{25}$

Section 2.12

1. $q = \frac{6}{13} \cos t + \frac{9}{13} \sin t - \frac{1}{2} e^{-t} + \frac{1}{26} e^{-5t}$,
$i = -\frac{6}{13} \sin t + \frac{9}{13} \cos t + \frac{1}{2} e^{-t} - \frac{5}{26} e^{-5t}$

3. $q = 3 + 6e^{-0.5t} - 7e^{-t}$
$i = -3e^{-0.5t} + 7e^{-t}$

5. $L < 10(60\pi)^{-2}$ or $L > 10(40\pi)^{-2}$

7. $q(t) =$
$$\begin{cases} \frac{1}{2} - \frac{1}{2} e^{-t} \cos t - \frac{1}{2} e^{-t} \sin t, & 0 \leq t \leq \pi \\ \frac{1}{2}(e^{-\pi} + 1) e^{-t}(\cos t + \sin t), & \pi \leq t \leq 2\pi \end{cases}$$

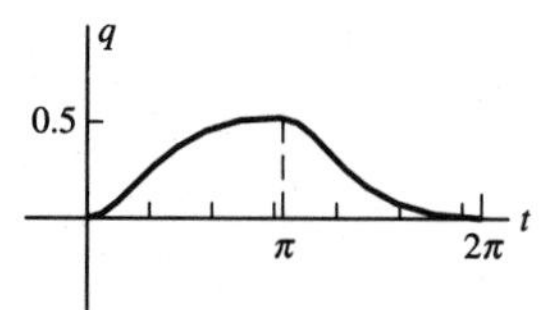

9. $\frac{dE}{dt} = Lii' + \frac{1}{c} qq' = \left(Li' + \frac{1}{c} q\right) i = 0i = 0$

11. $q(0) = -\frac{4}{35}$, $i(0) = 0$

17. $I = \frac{3}{\sqrt{340}}$

Section 2.13

1. $y = c_1x + c_2x^{-1}$

3. $y = c_1x^{-1} + c_2x^{-1}\ln x$

5. $y = c_1x^{-1/2} + c_2x^{-1/2}\ln x$

7. $y = 2x^{-1} - x^{-2}$

9. $y = c_1\cos(2\ln x) + c_2\sin(2\ln x)$

11. $y = x^{-1}[c_1\cos(\sqrt{7}\ln x) + c_2\sin(\sqrt{7}\ln x)]$

13. $y = c_1x^r + c_2x^r\ln x$

15. i) $x\dfrac{dy}{dx} = x\dfrac{dy}{ds}\dfrac{ds}{dx} = ks\dfrac{dy}{ds}\dfrac{1}{k} = s\dfrac{dy}{dx}$ (etc.)

17. $y = c_1x + c_2x^{-1} + c_3\cos(\ln x) + c_4\sin(\ln x)$

19. $y = c_1x + c_2x\ln x + c_3x(\ln x)^2$

Section 2.14

1. $y = \frac{1}{3}e^{2x} + c_1e^x + c_2e^{-x}$, yes

3. $y = (\sin x)\ln|\sin x| - x\cos x + c_1\sin x + c_2\cos x$, no

5. $y = (-\ln|\sec x + \tan x| + \sin x)\cos x + (-\cos x)\sin x + c_1\cos x + c_2\sin x$
$= -(\cos x)\ln|\sec x + \tan x| + c_1\cos x + c_2\sin x$, no

7. $y = -\frac{1}{2}e^{-2x}\ln(1 + e^{2x}) + e^{-x}\tan^{-1}e^x + c_1e^{-2x} + c_2e^{-x}$, no

9. $y = -\frac{2}{7}x^{7/2}e^{3x} + \frac{2}{5}x^{5/2}xe^{3x} + c_1e^{3x} + c_2xe^{3x} = \frac{4}{35}x^{7/2}e^{3x} + c_1e^{3x} + c_2xe^{3x}$, no

11. $y = -\dfrac{\ln x}{4}e^{-x/2} - \dfrac{x^{-1}}{4}xe^{-x/2} + c_1e^{-x/2} + c_2xe^{-x/2} = -\dfrac{\ln x}{4}e^{-x/2} + \tilde{c}_1e^{-x/2} + c_2xe^{-x/2}$, no

13. $y = (-\frac{1}{25}e^{-5x})e^{3x} - \frac{1}{5}xe^{-2x} + c_1e^{3x} + c_2e^{-2x} = -\frac{1}{5}xe^{-2x} + c_1e^{3x} + \tilde{c}_2e^{-2x}$, yes

15. $y = \dfrac{x^3}{2} + c_1x + c_2x^2$

17. $y = -x^{-1} - x^{-1}\ln x + c_1 + c_2x^{-1} = -x^{-1}\ln x + c_1 + \tilde{c}_2x^{-1}$

19. Use (9) with $v_i = \int_0^x v_i'(s)\,ds$ and $y_p = v_1y_1 + v_2y_2$

21. $\int_0^x \sinh(x - s)e^{-s^2}\,ds + c_1e^x + c_2e^{-x}$

23. $-\int_0^x [e^{3(s-x)} - e^{2(s-x)}]\dfrac{1}{s+1}\,ds + c_1e^{-2x}$
$+c_2e^{-3x}$

25. $y = v_1x + v_2x^{-1}$, where $v_1 = \frac{1}{2}\int_1^x \dfrac{e^t}{t^2}\,dt + c_1$, $v_2 = -\frac{1}{2}e^x + c_2$

Section 2.15

1. $y = \left(-\dfrac{e^{2x}}{2}\right)1 + \left(\dfrac{e^x}{2}\right)e^x + \left(\dfrac{e^{3x}}{6}\right)e^{-x} + c_1 + c_2e^x + c_3e^{-x} = \dfrac{e^{2x}}{6} + c_1 + c_2e^x + c_3e^{-x}$

3. $y = \left(-\dfrac{x^4}{8} - \dfrac{x^2}{2} + \dfrac{x^3}{3}\right)1 + \left(\dfrac{x^3}{3} - \dfrac{x^2}{2}\right)x - \dfrac{x^2}{4}\cdot x^2 - (x+1)e^{-x}e^x + c_1 + c_2x + c_3x^2 + c_4e^x$
$= \dfrac{-x^4}{24} - \dfrac{x^3}{6} + \tilde{c}_1 + \tilde{c}_2x + \tilde{c}_3x^2 + c_4e^x$

7. $y = -\dfrac{x}{2}e^x - \dfrac{1}{3}e^{-x}e^{2x} + \dfrac{e^{2x}}{12}e^{-x} + c_1e^x + c_2e^{2x} + c_3e^{-x} = -\dfrac{x}{2}e^x + \tilde{c}_1e^x + c_2e^{2x} + c_3e^{-x}$

9. $y = -\frac{1}{6}e^{-x}e^x - \frac{1}{2}e^xe^{-x} + \frac{1}{6}e^{2x}e^{-2x} + c_1e^x + c_2e^{-x} + c_3e^{-2x} = -\frac{1}{2} + c_1e^x + c_2e^{-x} + c_3e^{-2x}$

11. $y = \dfrac{x}{8}e^x - \frac{1}{8}e^{2x}e^{-x} + \frac{1}{32}e^{4x}e^{-3x} + c_1e^x + c_2e^{-x} + c_3e^{-3x} = \frac{1}{8}xe^x + \tilde{c}_1e^x + c_2e^{-x} + c_3e^{-3x}$

15. $y = \frac{1}{2}\ln xe^{-x} + x^{-1}(xe^{-x}) - \frac{1}{4}x^{-2}x^2e^{-x} + c_1e^{-x} + c_2xe^{-x} + c_3x^2e^{-x} = \frac{1}{2}\ln xe^{-x} + \tilde{c}_1e^{-x} + c_2xe^{-x}$
$+ c_3x^2e^{-x}$

17. $y = \frac{1}{13}x^{13/2}e^{2x} - \frac{2}{11}x^{11/2}xe^{2x} + \frac{1}{9}x^{9/2}x^2e^{2x} + c_1e^{2x} + c_2xe^{2x} + c_3x^2e^{2x} = \frac{8}{1287}x^{13/2}e^{2x} + c_1e^{2x} + c_2xe^{2x} + c_3x^2e^{2x}$

19. $y = \dfrac{x^3}{12}e^x - \dfrac{x^2}{4}xe^x + \dfrac{x}{4}x^2e^x + c_1e^x + c_2xe^x + c_3x^2e^x = \dfrac{x^3}{12}e^x + c_1e^x + c_2xe^x + c_3x^2e^x$

21. $y = \dfrac{x^4}{2}1 + \frac{2}{3}x^3x - \frac{8}{7}x^{7/2}x^{1/2} + c_1 + c_2x + c_3x^{1/2} = \frac{1}{42}x^4 + c_1 + c_2x + c_3x^{1/2}$

Chapter 3

Section 3.1.1

1. $\lim_{t\to 2^+} f(t)$ does not exist.

3. Continuous except at $t = 1$; $\lim_{t\to 1^+} f(t)$, $\lim_{t\to 1^-} f(t)$ exist.

5. $(1 - e^{-s})/s$

7. $(1 - 2e^{-s} + e^{-2s})/s^2$

9. $\int_0^\infty e^{-st}\sin at\,dt = \lim_{B\to\infty} (s^2 + a^2)^{-1} \times (-ae^{-st}\cos at - se^{-st}\sin at)\big|_{t=0}^{B}$

15. $\dfrac{3s}{s^2 - 4}$

17. $\dfrac{30}{s^2 + 36}$

19. $-\dfrac{1}{s^2} + \dfrac{3}{s}$

21. $\dfrac{2}{s} + \dfrac{s}{s^2 + 25}$

23. $\dfrac{3}{s^2 - 9}$

25. $\dfrac{2}{s + 1} + \dfrac{6}{s - 3}$

27. $\dfrac{3}{s^2} - \dfrac{1}{s} + \dfrac{s}{s^2 - 4}$

29. $F(s) = \dfrac{42}{s^4} + \dfrac{11}{s^2} + \dfrac{8}{s}$

31. $F(s) = \dfrac{3(4!)}{s^5} + \dfrac{24}{s^4}$

33. $F(s) = \dfrac{5}{(s - 3)^2 + 25}$

35. $F(s) = \dfrac{s - 4}{(s - 4)^2 + 49}$

37. $F(s) = \dfrac{5}{(s + 3)^2 + 25}$

39. $F(s) = \dfrac{s + 4}{(s + 4)^2 + 49}$

41. $F(s) = \dfrac{6!}{(s - 2)^7}$

43. $F(s) = \dfrac{6!}{(s + 2)^7}$

45. $F(s) = \dfrac{5!}{(s - 3)^6} + \dfrac{6}{(s - 3)^4} + \dfrac{1}{s - 3}$

47. $t - 1$

49. $\frac{1}{3}\sin 3t$

51. $1 + t$

53. $3 - 7t + 19\sin t$

55. $3\cos 4t + \frac{7}{4}\sin 4t$

57. $f(t) = 5e^{-3t} + 7e^{5t}$

59. $f(t) = \dfrac{1}{5!}t^5$

61. $f(t) = \frac{2}{3}\sin 3t$

63. $f(t) = \dfrac{3}{9!}t^9$

65. $f(t) = \frac{3}{2}\sqrt{\frac{2}{7}}\sin\sqrt{\frac{7}{2}}\,t$

67. $f(t) = \frac{1}{2}t^2e^{4t}$

69. $f(t) = e^{5t}\cos 3t$

71. $f(t) = \frac{7}{4}e^{7t}\sin 4t$

73. $f(t) = e^{-3t}\cos\sqrt{5}\,t$

77. Yes.

79. $|f(t)e^{-st}| \le Me^{\alpha t}e^{-st} = Me^{(\alpha - s)t}$. $\int_0^\infty e^{(\alpha - s)t}\,dt$ converges if $s > \alpha$

Section 3.1.2

81. $F(s) = \dfrac{1}{(s-5)^2}$

83. $F(s) = \dfrac{s^2 - 25}{(s^2+25)^2}$

85. $F(s) = \dfrac{2s}{(s^2+9)^2} - \dfrac{72s}{(s^2+9)^3}$

87. $F(s) = \dfrac{6(s-5)}{[(s-5)^2+9]^2}$

89. $F(s) = \dfrac{(s+4)^2 - 25}{[(s+4)^2+25]^2}$

91. $f(t) = \dfrac{1}{5!}t^5 e^{-3t}$

93. $f(t) = \frac{5}{6} t \sin 3t$

95. $f(t) = t \cos 3t$

97. $f(t) = \frac{5}{18}(\frac{1}{3}\sin 3t - t\cos 3t)$

99. $f(t) = \frac{1}{2}(\sin t - t\cos t) + t \sin t$

101. Use change of variables $\tau = ct$.

103. If $F(s) = 1/(s-a)$,
$$F^{(n)}(s) = (-1)^n \frac{n!}{(s-a)^{n+1}}$$

Section 3.2

1. $\frac{1}{6}(e^{3t} - e^{-3t})$

3. $5\cos\sqrt{7}t - \dfrac{1}{\sqrt{7}}\sin\sqrt{7}t$

5. $\cosh t + 2\sinh t$ or $\frac{3}{2}e^t - \frac{1}{2}e^{-t}$

7. $1 + e^t$

9. $-\frac{1}{3}e^{-2t} - \frac{2}{3}e^t$

11. $-\frac{4}{5}e^{-3t} + \frac{4}{5}e^{2t}$

13. $e^t \cos 5t + \frac{1}{5}e^t \sin 5t$

15. $2e^{3t}\cos 3t + \frac{11}{3}e^{3t}\sin 3t$

17. $3e^{-5t}\cos t - 17e^{-5t}\sin t$

19. $3e^{3t} - 2e^{2t}$

21. $\frac{1}{2} + \frac{1}{2}e^{2t} - e^t$

23. $-\frac{2}{7} + \frac{1}{6}e^{-t} + \frac{47}{42}e^{-7t}$

25. $-3 + 2e^t + e^{-t}$

27. $\dfrac{2s+1}{s^3+4s^2+13s} = \dfrac{1}{13s} - \dfrac{1}{13}\dfrac{-22+s}{s^2+4s+13}$, $f(t) = \dfrac{1}{13} - \dfrac{1}{13}e^{-2t}(-8\sin 3t + \cos 3t)$

29. $\dfrac{s^2-3}{s^3+2s^2+26s} = -\dfrac{3}{26s} + \dfrac{1}{26}\dfrac{6+29s}{s^2+2s+26}$, $f(t) = -\dfrac{3}{26} + \dfrac{1}{26}e^{-t}(-\dfrac{23}{5}\sin 5t + 29\cos 5t)$

31. $\dfrac{s-8}{(s-5)(s^2+4)} = -\dfrac{3}{29(s-5)} + \dfrac{1}{29}\dfrac{44+3s}{s^2+4}$, $f(t) = -\dfrac{3}{29}e^{5t} + \dfrac{1}{29}(22\sin 2t + 3\cos 2t)$

33. $\dfrac{s^2}{(s-3)(s^2-2s+26)} = \dfrac{9}{29(s-3)} + \dfrac{2}{29}\dfrac{39+10s}{s^2-2s+26}$,
$f(t) = \dfrac{9}{29}e^{3t} + \dfrac{2}{29}e^t\left(\dfrac{49}{5}\sin 5t + 10\cos 5t\right)$

35. $\frac{1}{24}\sin 3t + \frac{9}{8}\cos 3t - \frac{1}{8}\sin t - \frac{1}{8}\cos t$

37. $\frac{4}{3}\cosh 2t - \frac{1}{3}\cosh t = \frac{2}{3}e^{2t} + \frac{2}{3}e^{-2t} - \frac{1}{6}e^t - \frac{1}{6}e^{-t}$

39. $\frac{1}{8}\cosh 2t - \frac{1}{8}\cos 2t = \frac{1}{16}e^{2t} + \frac{1}{16}e^{-2t} - \frac{1}{8}\cos 2t$

41. $\frac{4}{5!}t^5 e^{-3t} = \frac{1}{30}t^5 e^{-3t}$

43. $\dfrac{1}{3!3^5}t^4 e^{-t/3}$

45. $(\frac{1}{4}t - \frac{1}{12})e^{-3t} + (\frac{1}{4}t + \frac{1}{12})e^{3t}$

47. $-\frac{1}{2}te^{-t} + \frac{1}{2}\sin t$

49. $-\frac{1}{2}e^{-t} + e^t(\frac{1}{2} - t + \frac{3}{2}t^2)$

51. $\dfrac{s^2-s}{(s^2+4)^2} = \dfrac{1}{s^2+4} - \dfrac{s+4}{(s^2+4)^2} = \dfrac{1}{2}\dfrac{2}{s^2+4} - \dfrac{1}{4}\dfrac{2\cdot 2s}{(s^2+4)^2} - \dfrac{1}{2}\dfrac{2\cdot 4}{(s^2+4)^2}$,
$f(t) = \frac{1}{2}\sin 2t - \frac{1}{4}t\sin 2t - \frac{1}{2}(\frac{1}{2}\sin 2t - t\cos 2t) = \frac{1}{4}\sin 2t - \frac{1}{4}t\sin 2t + \frac{1}{2}t\cos 2t$

53. $\dfrac{s^3-1}{(s^2+9)^2} = \dfrac{s^3+9s-9s-1}{(s^2+9)^2} = \dfrac{s}{s^2+9} - \dfrac{9s+1}{(s^2+9)^2} = \dfrac{s}{s^2+9} - \dfrac{3}{2}\dfrac{2\cdot 3s}{(s^2+9)^2} - \dfrac{1}{18}\dfrac{2\cdot 9}{(s^2+9)^2}$,

$f(t) = \cos 3t - \frac{3}{2}t\sin 3t - \frac{1}{18}\left(\dfrac{1}{3}\sin 3t - t\cos 3t\right)$

55. $\dfrac{-36s+36}{(s^2+36)^2} + \dfrac{s-1}{s^2+36}$, $f(t) = \frac{1}{2}(\frac{1}{6}\sin 6t - t\cos 6t) - 3t\sin 6t - \frac{1}{6}\sin 6t + \cos 6t$

57. $f(t) = -\frac{2}{25}e^{-t}(\frac{1}{5}\sin 5t - t\cos 5t) + \frac{3}{10}te^{-t}\sin 5t$

59. $\dfrac{-6s-25}{(s^2+6s+25)^2} + \dfrac{1}{s^2+6s+25}$,

$f(t) = -\dfrac{7}{32}e^{-3t}\left(\dfrac{1}{4}\sin 4t - t\cos 4t\right) - \dfrac{3}{4}te^{-3t}\sin 4t + \dfrac{1}{4}e^{-3t}\sin 4t$

61. $f(t) = -\frac{1}{2}e^{-t} + e^{t}(\frac{1}{2} - t + \frac{3}{2}t^2)$

63. $\dfrac{12}{8^6} + e^{8t}\left(-\dfrac{12}{8^6} + \dfrac{12}{8^5}t - \dfrac{3}{2048}t^2 + \dfrac{1}{256}t^3 - \dfrac{1}{128}t^4 + \dfrac{1}{80}t^5\right)$

65. $f(t) = e^{-t}(\frac{5}{4} + t) + e^{t}(-\frac{5}{4} + \frac{3}{2}t)$

67. $\dfrac{1}{2^2 5^8}\left[527\cos t - 336\sin t + e^{7t}\left(-527 + 161t\cdot 5^2 - 15{,}000t^2 + \dfrac{5^6 7}{3}t^3\right)\right]$

Section 3.3

1. $-\frac{1}{4} + \frac{3}{8}e^{2t} - \frac{1}{8}e^{-2t}$

3. e^t

5. $\frac{1}{6} + \frac{5}{2}e^{-2t} - \frac{5}{3}e^{-3t}$

7. $\frac{1}{2} + \frac{3}{2}e^{2t} - 2e^t$

9. $\frac{1}{10}e^{-t} + \frac{7}{10}\sin 3t + \frac{9}{10}\cos 3t$

11. $\frac{2}{13} + \frac{22}{39}e^{-2t}\sin 3t + \frac{11}{13}e^{-2t}\cos 3t$

13. $-1 + \frac{9}{4}e^t - \frac{1}{4}e^{-t} + 2\cos t + \frac{5}{2}\sin t$

15. $t - e^t$

17. $\frac{1}{2}t^2e^{-t} + te^{-t}$

19. $\frac{1}{2}e^{-t}\cos t + \frac{1}{2}e^{-t}\sin t - \frac{1}{2}e^{-2t}$

21. $y(t) = -\frac{1}{2}t\cos t + \frac{1}{2}\sin t$

23. $y(t) = \frac{1}{6}t\sin 3t + \frac{1}{3}\sin 3t$

25. $y(t) = \frac{1}{4}t\sin 2t$

27. $y(t) = -\frac{1}{2}t\cos t + \frac{1}{2}t\sin t + \frac{1}{2}\sin t$

Section 3.4

1. $f(t) = 3[H(t-2) - H(t-5)] + tH(t-5)$
$= 3H(t-2) + (t-3)H(t-5)$

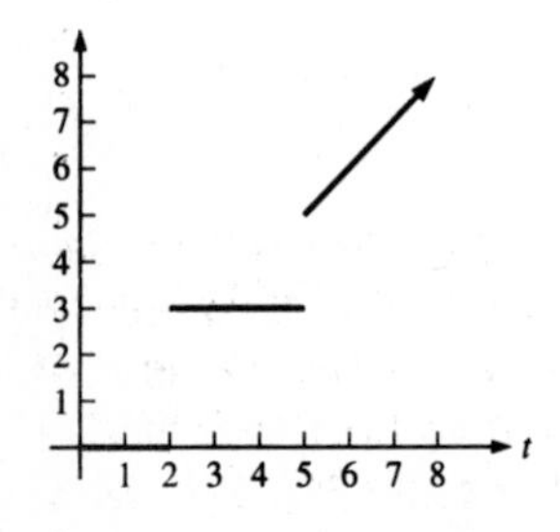

3. $f(t) = \sin t\,[1 - H(t-\pi)] + \sin t[H(t-2\pi) - H(t-3\pi)]$

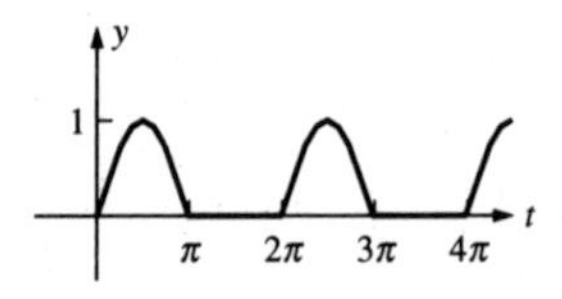

5. $f(t) = 1[1 - H(t-1)] + 1[H(t-2) - H(t-3)] = 1 - H(t-1) + H(t-2) - H(t-3)$

7. $f(t) = (t-1)[H(t-1) - H(t-2)] + [H(t-2) - H(t-3)]$
$+(4-t)[H(t-3) - H(t-4)]$
$= (t-1)H(t-1) + (2-t)H(t-2) + (3-t)H(t-3) - (4-t)H(t-4)$

9. $f(t) = (2-t)[H(t-1) - H(t-3)] + (t-4)[H(t-3) - H(t-5)]$

11. $Y(s) = e^{-2s}\left[\dfrac{1}{s^2} + \dfrac{2}{s}\right]$

13. $Y(s) = e^{-s}\left[\dfrac{6}{s^4} + \dfrac{6}{s^3} + \dfrac{3}{s^2} + \dfrac{2}{s}\right]$

15. $Y(s) = e^{-\pi s}\left[-\dfrac{1}{s^2+1}\right]$

17. $Y(s) = e^{-2\pi s}\left[\dfrac{s}{s^2+1}\right]$

19. $Y(s) = e^{-3s}\left[\dfrac{e^6}{s-2}\right]$

21. $Y(s) = e^{-2s}e^{10}\left[\dfrac{1}{(s-5)^2} + \dfrac{2}{s-5}\right]$

23. $y(t) = [-e^{-(t-2)} + 1]H(t-2) + [-e^{-(t-3)} + 1]H(t-3)$
$= [1 - e^{2-t}]H(t-2) + [1 - e^{3-t}]H(t-3)$

25. $y(t) = e^{-(t-3)} \sin(t-3)H(t-3)$

27. $y(t) = \sin(t-1)H(t-1) - \frac{1}{2}\sin(2(t-2))H(t-2)$

29. $y(t) = [4 + 6\cos(3(t-1))]H(t-1)$

31. $y(t) = [1 - \cos(t-2)]H(t-2) - [1 - \cos(t-5)]H(t-5)$

33. $y(t) = e^{3t} + \left[-\dfrac{1}{9} - \dfrac{t}{3} + \dfrac{1}{9}e^{3t}\right] - 2\left[-\dfrac{1}{9} - \dfrac{(t-1)}{3} + \dfrac{1}{9}e^{3(t-1)}\right]H(t-1)$
$+\left[-\dfrac{1}{9} - \dfrac{(t-2)}{3} + \dfrac{1}{9}e^{3(t-2)}\right]H(t-2)$

35. $y(t) = [\frac{1}{10} + \frac{9}{10}e^{-t}\cos 3t + \frac{3}{10}e^{-t}\sin 3t]$
$-[-\frac{1}{10}e^{-(t-3)}\cos(3t-9) - \frac{1}{30}e^{-(t-3)}\sin(3t-9) + \frac{1}{10}]H(t-3)$

37. $y(t) = \frac{1}{4}[1 - \cos 2t] - \frac{1}{4}[1 - \cos(2(t-1))]H(t-1)$
$+ \frac{1}{4}[1 - \cos(2(t-2))]H(t-2) - \frac{1}{4}[1 - \cos(2(t-3))]H(t-3)$

39. $y(t) = \frac{1}{9} - \frac{1}{9}\cos 3t - [\frac{1}{9} - \frac{1}{9}\cos 3(t-2)]H(t-2)$
$+ \frac{1}{10}e^{-2}[e^{-(t-2)} + \frac{1}{3}\sin 3(t-2) - \cos 3(t-2)]H(t-2)$

41. $y' - 5y = e^{3t}(1 - H(t-4))$
$y(t) = -\frac{1}{2}e^{3t} + \frac{1}{2}e^{5t} + e^{12}(\frac{1}{2}e^{3(t-4)} - \frac{1}{2}e^{5(t-4)})H(t-4)$

43. $y' + 3y = \cos tH\left(t - \dfrac{\pi}{2}\right)$
$y(t) = \left(\dfrac{1}{10}e^{-3(t-\frac{\pi}{2})} - \dfrac{1}{10}\cos\left(t - \dfrac{\pi}{2}\right)\right.$
$\left.+ \dfrac{3}{10}\sin\left(t - \dfrac{\pi}{2}\right)\right) - H\left(t - \dfrac{\pi}{2}\right) = \left(\dfrac{1}{10}e^{-3(t-\frac{\pi}{2})} - \dfrac{1}{10}\sin t - \dfrac{3}{10}\cos t\right) - H\left(t - \dfrac{\pi}{2}\right)$

45. $y' + 2y = \sin tH(t-3)$
$y(t) = \frac{1}{5}[(2\cos 3 + \sin 3)\sin(t-3) + (-\cos 3 + 2\sin 3)\cos(t-3)$
$- \frac{1}{5}e^{-2(t-3)}(-\cos 3 + 2\sin 3)]H(t-3)$

47. $y'' + 9y = \sin t[1 - H(t-4)]$
$y(t) = \frac{1}{8}\sin t + \frac{7}{24}\sin 3t + \frac{1}{8}[\frac{1}{3}\cos 4\sin 3(t-4)$
$+\sin 4\cos 3(t-4) - \cos 4\sin(t-4) - \sin 4\cos(t-4)]H(t-4)$

49. $y(t) = \frac{1}{4}[1 - \cos 2t] + \frac{1}{2}\sum\limits_{n=1}^{\infty}(-1)^n[1 - \cos(2(t-n))]H(t-n)$

Section 3.5

1. $G(s) = \dfrac{1 + e^{-\pi s}}{(s^2 + 1)(1 - e^{-\pi s})}$

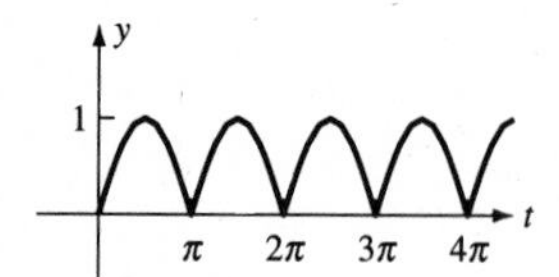

3. $G(s) = \left[\dfrac{2}{s^3} - \left(\dfrac{4}{s^3} + \dfrac{4}{s^2}\right)e^{-s} + \left(\dfrac{2}{s^3} + \dfrac{4}{s^2} + \dfrac{2}{s}\right)e^{-2s}\right]\dfrac{1}{1 - e^{-2s}}$

5. $G(s) = \dfrac{e^{1-s} - 1}{(1 - s)(1 - e^{-s})}$

7. $G(s) = \left[\dfrac{1}{s} - \dfrac{2}{s}e^{-s} + \dfrac{1}{s}e^{-2s}\right]\dfrac{1}{1 - e^{-2s}}$

9. $\displaystyle\sum_{n=0}^{\infty}\left[(t - n) + \frac{(t - n)^2}{2}\right]H(t - n)$

11. $\displaystyle\sum_{n=0}^{\infty}\cos(2(t - n))H(t - n)$

13. $\displaystyle\sum_{n=0}^{\infty}\left[(t - 2n) + \frac{(t - 1 - 2n)^2}{2}H(t - 1 - 2n)\right]H(t - 2n)$

$\displaystyle = \sum_{n=0}^{\infty}(t - 2n)H(t - 2n) + \frac{(t - 1 - 2n)^2}{2}H(t - 1 - 2n)$

15. $\displaystyle\sum_{n=0}^{\infty}\left[1 + \sin\left(t - \frac{\pi}{2} - n\pi\right)H\left(t - \frac{\pi}{2} - n\pi\right)\right]H(t - n\pi)$

$\displaystyle = \sum_{n=0}^{\infty}H(t - n\pi) + \sin\left(t - \frac{\pi}{2} - n\pi\right)H\left(t - \frac{\pi}{2} - n\pi\right)$

17. $\displaystyle\sum_{n=0}^{\infty}(-1)^n\left[\tfrac{1}{2}(t - 5n)^2H(t - 5n) + \tfrac{1}{6}(t - 5n - 2)^3H(t - 5n - 2)\right]$

19. $G(s) = \dfrac{1}{s^2}\operatorname{csch} s = \dfrac{2}{s^2(e^s - e^{-s})} = \dfrac{2e^{-s}}{s^2(1 - e^{-2s})}$

$\displaystyle g(t) = \sum_{n=0}^{\infty}2(t - 2n - 1)H(t - 2n - 1)$

23. $\displaystyle\frac{1}{3}\sum_{n=0}^{\infty}\left\{\left[2\sin\left(t - \frac{n\pi}{2}\right) - \sin(2t - n\pi)\right]H\left(t - \frac{n\pi}{2}\right) + \left[2\sin\left(t - \frac{\pi}{2} - \frac{n\pi}{2}\right) - \sin(2t - \pi - n\pi)\right]H\left(t - \frac{\pi}{2} - \frac{n\pi}{2}\right)\right\}$

25. $\displaystyle e^{-t} + \sum_{n=0}^{\infty}\sinh(t - n)H(t - n) - e\sinh(t - n - 1)H(t - n - 1)$

27. $\frac{1}{4}\sum_{n=0}^{\infty}[\cosh(2(t-2n))-1]H(t-2n)-2[\cosh(2(t-2n-1))-1]H(t-2n-1)$
$+[\cosh(2(t-2n-2))-1]H(t-2n-2)$

29. $\sum_{n=0}^{\infty}\frac{1}{s^{n+1}}=\frac{1}{s}\sum_{n=0}^{\infty}\frac{1}{s^n}=\frac{1}{s\left(1-\frac{1}{s}\right)}=\frac{1}{s-1}$

Section 3.6

1. $t * e^t = -t - 1 + e^t$

3. t

5. $t - \sin t$

7. $y(t)=\int_0^t f(\tau)\left(\frac{1}{3}e^{4(t-\tau)}-\frac{1}{3}e^{t-\tau}\right)d\tau$

9. $y(t)=\int_0^t f(\tau)\left(\frac{1}{6}e^{3(t-\tau)}-\frac{1}{6}e^{-3(t-\tau)}\right)d\tau$

11. $y(t)=\int_0^t f(\tau)\frac{1}{2}\sin 2(t-\tau)\,d\tau+\frac{7}{2}\sin 2t$

13. $y(t)=\int_0^t f(\tau)e^{-7(t-\tau)}\,d\tau+2e^{-7t}$

15. $y(t)=\int_0^t f(\tau)\frac{1}{3}e^{t-\tau}\sin 3(t-\tau)\,d\tau$

17. $y(t)=\int_0^t f(\tau)\frac{1}{3}e^{-2(t-\tau)}\sin 3(t-\tau)\,d\tau$

19. $X(s)=\frac{1}{s^2-s+1},\ x(t)=\frac{2}{\sqrt{3}}e^{t/2}\sin\left(\frac{\sqrt{3}}{2}t\right)$

21. $X(s)=\frac{2(s+1)}{s^2},\ x(t)=2+2t$

23. $X(s)=\frac{s^2+4}{s^2(s^2+1)},\ x(t)=4t-3\sin t$

25. $x(t)=3-\frac{3}{2}t^2$

27. $|F(s)|\le\int_0^{\infty}e^{-st}|f(t)|\,dt\le\int_0^{\infty}e^{-st}Me^{\alpha t}\,dt=\frac{M}{s-\alpha}$ for $s>\alpha$.

Section 3.7

1. $y(t)=e^{-8(t-1)}H(t-1)+e^{-8(t-2)}H(t-2)=e^{8-8t}H(t-1)+e^{16-8t}H(t-2)$

3. $y(t)=\frac{1}{10}e^{-3(t-1)}\sin(10(t-1))H(t-1)-\frac{1}{10}e^{-3(t-7)}\sin(10(t-7))H(t-7)$

5. $y(t)=\frac{1}{3}(1-e^{-3t})+\frac{1}{2}[e^{-(t-3)}-e^{-3(t-3)}]H(t-3)$

7. $y(t)=1+\sin(t-2\pi)H(t-2\pi)=1+\sin t\,H(t-2\pi)$

9. **(a)** The following are graphs of the solutions of Exercises 1 through 4.

(1)

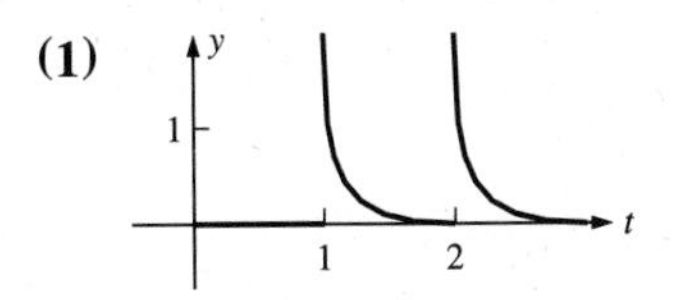

(2)

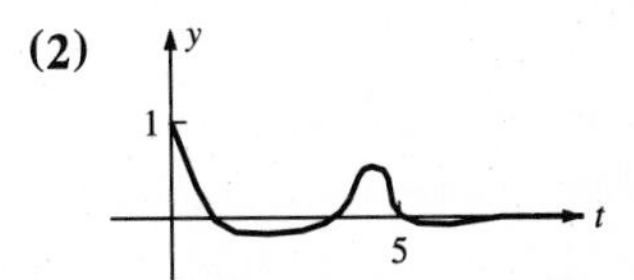

(3)

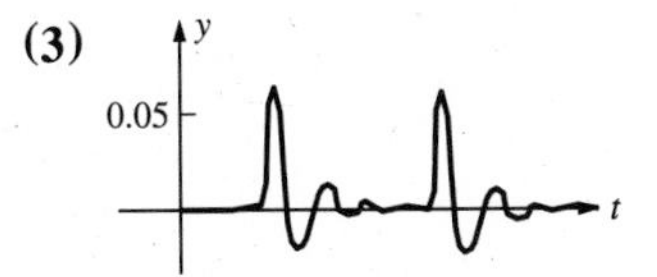

(4)

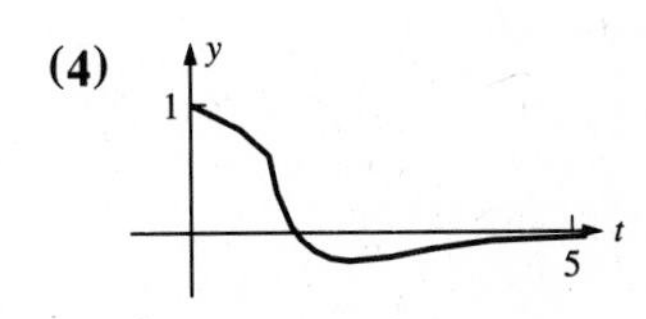

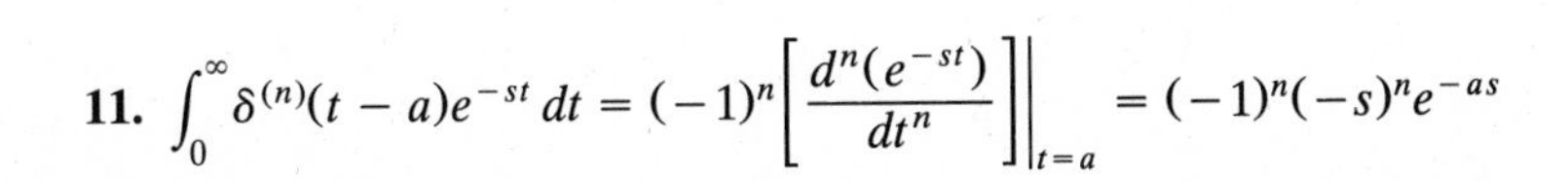

11. $\int_0^{\infty}\delta^{(n)}(t-a)e^{-st}\,dt=(-1)^n\left[\frac{d^n(e^{-st})}{dt^n}\right]\Bigg|_{t=a}=(-1)^n(-s)^ne^{-as}$

Chapter 4

Section 4.2

1. Yes

3. Yes

5. Yes (all but the first coefficient are zero)

7. No (negative power)

9. Yes

11. No (negative power)

13. Yes

15. Fails

17. $r = \infty$, $(-\infty, \infty)$

19. $r = 0$, $\{0\}$

21. $e^{2x} = \sum_{n=0}^{\infty} \frac{2^n e^2}{n!}(x-1)^n$

23. $\cos x = 1 - \frac{x^2}{2} + \frac{x^4}{4!} - \cdots = \sum_{n=0}^{\infty} (-1)^n \frac{x^{2n}}{(2n)!}$

25. $x^3 = x^3$

27. $x = x$

29. $\frac{1}{1-x} = -\frac{1}{4}\sum_{n=0}^{\infty}\left(\frac{x-5}{-4}\right)^n$, $|x-5| < 4$

31. $\frac{1}{4-x} = \frac{1}{4}\sum_{n=0}^{\infty}\left(\frac{x}{4}\right)^n$, $|x| < 4$

33. $\frac{1}{4-x} = -\frac{1}{2}\sum_{n=0}^{\infty}\left(\frac{x-6}{-2}\right)^n$, $|x-6| < 2$

35. $\frac{1}{1+x^2} = \sum_{n=0}^{\infty} (-1)^n x^{2n}$

37. $e^{x^3} = \sum_{n=0}^{\infty} \frac{(x^3)^n}{n!} = \sum_{n=0}^{\infty} \frac{x^{3n}}{n!}$

39. No. No center.

41. $p_2(x) = e^2 + 2e^2(x-1) + 2e^2(x-1)^2$

43. $p_5(x) = 1 - \frac{1}{2}\left(x - \frac{\pi}{2}\right)^2 + \frac{1}{24}\left(x - \frac{\pi}{2}\right)^4$

45. $p_2(x) = x$

Section 4.3

1. All x; $(-\infty, \infty)$

3. $x \neq 0$ and $x \neq 1$; $(-\infty, 0) \cup (0, 1) \cup (1, \infty)$

5. $x \neq 2$, $x \neq 3$, and $x \neq n\pi$ for $n = 0, \pm 1, \pm 2, \ldots$

7. $|x| < 1$

9. $|x - 2| < 3$

11. $|x| < 1$ (i a root of $x^2 + 1$)

13. $|x - 3.3| < 1.3$

15. All x are ordinary points.

17. All x are ordinary points except $x = \pm 1$, which are singular points.

19. All real x are ordinary points; the only singular points are $x = \pm 3i$.

21. Continuous for all x; guaranteed region of convergence is $|x| < 3$.

23. Continuous for all x except $x = \pm 3$; guaranteed region of convergence is $|x| < 3$.

Section 4.4

1. $y = 2x - \frac{x^3}{6} + \cdots$

3. $y = 2 + x - \frac{2x^3}{3} + \cdots$

5. $y = (x-1) + \frac{(x-1)^3}{3} + \cdots$

7. $y = c_0 + c_1 x - c_0\frac{x^2}{2} - c_1\frac{x^3}{6} + c_0\frac{x^4}{24} + \cdots = c_0\left[1 - \frac{x^2}{2} + \frac{x^4}{24} + \cdots\right] + c_1\left[x - \frac{x^3}{6} + \cdots\right]$, $(y_p = 0)$

9. $y = c_0 + c_1 x + \frac{x^2}{2} + \frac{x^3}{6} + (1 - 2c_0)\frac{x^4}{24} + \cdots$
$= \left[\frac{x^2}{2} + \frac{x^3}{6} + \frac{x^4}{24} + \cdots\right] + c_0\left[1 - \frac{x^4}{12} + \cdots\right] + c_1[x + \cdots]$

11. $y = c_0 + c_1\left(x - \frac{\pi}{2}\right) + c_0\frac{[x - (\pi/2)]^2}{2} + (c_1 - 1)\frac{[x - (\pi/2)]^3}{6} + \cdots$
$= \left\{-\frac{[x - (\pi/2)]^3}{6} + \cdots\right\} + c_0\left\{1 + \frac{[x - (\pi/2)]^2}{2} + \cdots\right\}$
$+ c_1\left\{[x - (\pi/2)] + \frac{[x - (\pi/2)]^3}{6} + \cdots\right\}$

13. $y = c_0 + (1 - 2c_0)(x - 1) + (-2 + 6c_0)\frac{(x - 1)^2}{2} + (8 - 24c_0)\frac{(x - 1)^3}{6} + (-40 + 120c_0)\frac{(x - 1)^4}{24} + \cdots$
$= \left[(x - 1) - (x - 1)^2 + \frac{4}{3}(x - 1)^3 - \frac{5}{3}(x - 1)^4 + \cdots\right]$
$+ c_0[1 - 2(x - 1) + 3(x - 1)^2 - 4(x - 1)^3 + 5(x - 1)^4 + \cdots]$

15. $y = c_0 + c_1 x + c_0\frac{x^2}{2} + (c_0 + c_1)\frac{x^3}{6} + (3c_0 + 2c_1 + 2)\frac{x^4}{24} + \cdots$
$= \left[\frac{x^4}{12} + \cdots\right] + c_0\left[1 + \frac{1}{2}x^2 + \frac{1}{6}x^3 + \frac{1}{8}x^4 + \cdots\right] + c_1\left[x + \frac{1}{6}x^3 + \frac{1}{12}x^4 + \cdots\right]$

17. $y = 2 - z - \frac{3}{2}z^2 + \frac{1}{2}z^3 - \frac{7}{8}z^4 + \cdots$

19. $y = (\frac{1}{2}z^2 - \frac{2}{3}z^3 + \frac{3}{4}z^4 + \cdots) + c_0(1 - z + z^2 - z^3 + z^4 + \cdots)$

21. $y = \left(\frac{z^2}{2} - \frac{z^3}{3} + \frac{z^4}{8} + \cdots\right) + c_0\left(1 - \frac{z^2}{2} + \frac{z^3}{6} + \cdots\right) + c_1\left(z - \frac{z^2}{2} + \frac{z^4}{24} + \cdots\right)$

Section 4.5

1. $(n + 1)c_{n+1} + 3c_n = 0$ for $n \geq 0$, $y = 2 - 6x + 9x^2 - 9x^3 + \frac{27}{4}x^4 + \cdots$ (*Note:* $y = 2e^{-3x}$)

3. $c_{n+1} + c_n = 0$ for $n \geq 0$, $y = 2 - 2(x - 1) + 2(x - 1)^2 - 2(x - 1)^3 + 2(x - 1)^4 + \cdots$, $y = 2x^{-1}$

5. $c_2 = \frac{1}{2}$, $(n + 2)(n + 1)c_{n+2} + c_{n-1} = 1$ for $n \geq 1$, $y = \frac{1}{2}x^2 + \frac{1}{6}x^3 + \frac{1}{12}x^4 + \cdots$

7. $2c_2 + 6c_0 = 0$, $(n + 2)(n + 1)c_{n+2} + (6 - 2n)c_n = 0$ for $n \geq 1$, $y = x - \frac{2}{3}x^3$

9. $c_0 = 0$, $c_1 = 1$, $c_2 = \frac{1}{2}$, $(n + 1)c_{n+1} + 3c_{n-2} = \frac{1}{n!}$ for $n \geq 2$, $y = x + \frac{1}{2}x^2 + \frac{1}{6}x^3 + \cdots$

11. $c_0 = 0$, $c_1 = 0$, $c_2 = 0$, $c_3 = 1/6$, $(n + 2)(n + 1)c_{n+2} + c_{n-2} = \begin{cases} 0 & \text{if } n \text{ even} \\ \dfrac{(-1)^{(n-1)/2}}{n!} & \text{if } n \text{ odd} \end{cases}$

13. $c_0 = 2$, $c_1 + c_0 = 0$, $2c_2 + c_1 = 0$, $3c_3 + c_2 + c_0 = 1$, $(n + 1)c_{n+1} + c_n + c_{n-2} = 0$ if $n > 2$; $c_0 = 2$, $c_1 = -2$, $c_2 = 1$, $c_3 = -2/3$; $y = 2 - 2x + x^2 - 2/3x^3 + 2/3x^4 + \cdots$

15. $y = c_0\left[1 + \frac{x^2}{2} + \frac{x^4}{4!} + \cdots\right] + c_1\left[x + \frac{x^3}{3!} + \cdots\right] = c_0 \cosh x + c_1 \sinh x$;
recursion is $(n + 2)(n + 1)c_{n+2} - c_n = 0$, $n \geq 0$

17. $y = c_0\left[1 - \frac{x^2}{2} + \frac{x^4}{8} + \cdots\right] + c_1\left[x - \frac{x^3}{6} + \cdots\right]$; recursion is $(n + 2)(n + 1)c_{n+2} + (n^2 - n + 1)c_n = 0$.

19. $y = c_0\left[1 - \frac{x^3}{3} + \cdots\right] + c_1\left[x - \frac{x^4}{6} + \cdots\right]$; recursion is $c_2 = 0$, $(n+2)(n+1)c_{n+2} + 2c_{n-1} = 0$ for $n \geq 1$.

21. $y = c_0\left[1 - \frac{x^6}{2} + \frac{x^{12}}{8} + \cdots\right]$; recursion is $c_1 = \cdots = c_5 = 0$, $(n+1)c_{n+1} + 3c_{n-5} = 0$, $n \geq 5$.

23. Let $h(x)$ be the solution of $ay'' + by' + cy = 0$, $y(0) = 1$, $y'(0) = 0$, and $g(x)$ be the solution of $ay'' + by' + cy = 0$, $y(0) = 0$, $y'(0) = 1$. Solutions are unique and series expansions are unique.

25. $y = \sum_{n=0}^{\infty} c_n(x-4)^n$, where $c_2 = 8c_0$, $c_3 = \frac{8}{3}c_1 + \frac{4}{3}c_0$, $c_{n+2} = \dfrac{16c_n + 8c_{n-1} + c_{n-2}}{(n+1)(n+2)}$ for $n \geq 2$;
$y(x) = c_0[1 + 8(x-4)^2 + \frac{4}{3}(x-4)^3 + \cdots] + c_1[(x-4) + \frac{8}{3}(x-4)^3 + \cdots]$

27. $y = \sum_{n=0}^{\infty} c_n(x-2)^n$, where $c_2 = -c_0$, $c_{n+2} = \dfrac{-2c_n - c_{n-1}}{(n+1)(n+2)}$ for $n \geq 1$, $c_3 = -\frac{1}{3}c_1 - \frac{1}{6}c_0$;
$y(x) = c_0[1 - (x-2)^2 - \frac{1}{6}(x-2)^3 + \cdots] + c_1[(x-2) - \frac{1}{3}(x-2)^3 + \cdots]$

29. $y = \sum_{n=0}^{\infty} c_n(x-1)^n$, where $c_{n+2} = \dfrac{2(n+1)c_{n+1} + (2n-3)c_n}{(n+1)(n+2)}$ for $n \geq 0$; $c_2 = c_1 - \frac{3}{2}c_0$, $c_3 = \frac{1}{2}c_1 - c_0$,
$y(x) = c_0[1 - \frac{3}{2}(x-1)^2 - (x-1)^3 + \cdots] + c_1[(x-1) + (x-1)^2 + \frac{1}{2}(x-1)^3 + \cdots]$

31. $y = \sum_{n=0}^{\infty} c_n(x-2)^n$, where $c_2 = -3c_0 + 3c_1$,
$c_{n+2} = \dfrac{6(n+1)c_{n+1} + (n-6)c_n - 3c_{n-1}}{(n+1)(n+2)}$ for $n \geq 1$; $c_3 = \frac{31}{6}c_1 - \frac{39}{6}c_0$,
$y(x) = c_0[1 - 3(x-2)^2 - \frac{39}{6}(x-2)^3 + \cdots] + c_1[(x-2) + 3(x-2)^2 + \frac{31}{6}(x-2)^3 + \cdots]$

Section 4.6

1. a) $x = -1$ and $x = -2$ are regular singular points.
b), c) $x = -1$: $(x+1)^2y'' + 4y = 0$, $y = (x+1)^r$, $r(r-1) + 4 = 0$
$x = -2$: $(x+2)^2y'' + 16y = 0$, $y = (x+2)^r$, $r(r-1) + 16 = 0$

3. a) $x = 0$ and $x = -3$ are regular singular points.
b), c) $x = 0$: $x^2y'' + \frac{1}{9}xy' + \frac{7}{9}y = 0$, $y = x^r$, $r(r-1) + \frac{1}{9}r + \frac{7}{9} = 0$
$x = -3$: $(x+3)^2y'' + \frac{7}{9}y = 0$, $y = (x+3)^r$, $r(r-1) + \frac{7}{9} = 0$

5. a) $x = 0$ is a regular singular point, but $x = -2$ is an irregular singular point.
b), c) $x = 0$: $x^2y'' + \frac{1}{4}xy' + y = 0$, $y = x^r$, $r(r-1) + \frac{1}{4}r + 1 = 0$

7. a) $x = -3$ and $x = -5$ are regular singular points.
b), c) $x = -5$: $(x+5)^2y'' + \frac{5}{2}(x+5)y' + \frac{25}{2}y = 0$, $y = (x+5)^r$, $r(r-1) + \frac{5}{2}r + \frac{25}{2} = 0$
$x = -3$: $(x+3)^2y'' - \frac{3}{2}(x+3)y' + \frac{9}{2}y = 0$, $y = (x+3)^r$, $r(r-1) - \frac{3}{2}r + \frac{9}{2} = 0$

9. a) $x = 1$, $x = -1$, and $x = -4$ are regular singular points.
b), c) $x = -1$: $(x+1)^2y'' - \frac{1}{6}(x+1)y' = 0$, $y = (x+1)^r$, $r(r-1) - \frac{1}{6}r = 0$
$x = 1$: $(x-1)^2y'' + \frac{1}{10}(x-1)y' + \frac{4}{25}y = 0$, $y = (x-1)^r$, $r(r-1) + \frac{1}{10}r + \frac{4}{25} = 0$
$x = -4$: $(x+4)^2y'' + \frac{1}{15}(x+4)y' - \frac{1}{25}y = 0$, $y = (x+4)^r$, $r(r-1) + \frac{1}{15}r - \frac{1}{25} = 0$

11. a) $x = 0$ and $x = 2$ are regular singular points.
b), c) $x = 0$: $x^2y'' + \frac{1}{4}xy' + \frac{1}{4}y = 0$, $y = x^r$, $r(r-1) + \frac{1}{4}r + \frac{1}{4} = 0$
$x = 2$: $(x-2)^2y'' + \frac{3}{4}y = 0$, $y = (x-2)^r$, $r(r-1) + \frac{3}{4} = 0$

13. a) $x = 0$, $x = -4$, and $x = 5$ are regular singular points.

b), c) $x = 0$: $x^2y'' - \frac{1}{20}xy' + \frac{1}{20}y = 0$, $y = x^r$, $r(r-1) - \frac{1}{20}r + \frac{1}{20} = 0$

$x = 5$: $(x-5)^2y'' + \frac{1}{45}(x-5)y' + \frac{1}{45}y = 0$, $y = (x-5)^r$, $r(r-1) + \frac{1}{45}r + \frac{1}{45} = 0$

$x = -4$: $(x+4)^2y'' + \frac{1}{36}(x+4)y' = 0$, $y = (x+4)^r$, $r(r-1) + \frac{1}{36}r = 0$

Section 4.7

1. $36r(r-1) + 5 = 0$, $r = \frac{1}{6}, \frac{5}{6}$

$y = \sum_{n=0}^{\infty} c_n x^{n+r}$; for $n \geq 1$, $c_n = -\dfrac{n+r-1}{36(n+r)(n+r-1)+5}c_{n-1}$

$r = \frac{1}{6}$: for $n \geq 1$, $c_n = -\dfrac{n-\frac{5}{6}}{6n(6n-4)}c_{n-1}$, $c_1 = -\frac{1}{72}c_0$

$r = \frac{5}{6}$: for $n \geq 1$, $c_n = -\dfrac{n-\frac{1}{6}}{6n(6n+4)}c_{n-1}$, $c_1 = -\frac{1}{72}c_0$

$y = c_0x^{\frac{1}{6}}(1 - \frac{1}{72}x + \cdots) + \bar{c}_0x^{\frac{5}{6}}(1 - \frac{1}{72}x + \cdots)$

3. $6r(r-1) + 2r = 0$, $r = 0, \frac{2}{3}$

$y = \sum_{n=0}^{\infty} c_n x^{n+r}$; for $n \geq 1$, $c_n = \dfrac{6(n+r-1)(n+r-2) + 15(n+r-1) + 3}{(n+r)[6(n+r)-4]}c_{n-1}$

$r = 0$: for $n \geq 1$, $c_n = \dfrac{n(6n-3)}{n(6n-4)}c_{n-1}$, $c_1 = \frac{3}{2}c_0$

$r = \frac{2}{3}$: for $n \geq 1$, $c_n = \dfrac{(n+\frac{2}{3})(6n+1)}{(n+\frac{2}{3})6n}c_{n-1}$, $c_1 = \frac{7}{6}c_0$

$y = c_0x^0(1 + \frac{3}{2}x + \cdots) + \bar{c}_0x^{\frac{2}{3}}(1 + \frac{7}{6}x + \cdots)$

5. $2r(r-1) - 5r + 5 = 0$, $r = 1, \frac{5}{2}$

$y = \sum_{n=0}^{\infty} c_n x^{n+r}$; for $n \geq 1$, $c_n = -\dfrac{4}{2(n+r)(n+r-1) - 5(n+r) + 5}c_{n-1}$

$r = 1$: for $n \geq 1$, $c_n = -\dfrac{4}{n(2n-3)}c_{n-1}$, $c_1 = 4c_0$

$r = \frac{5}{2}$: for $n \geq 1$, $c_n = -\dfrac{4}{n(2n+3)}c_{n-1}$, $c_1 = -\frac{4}{5}c_0$

$y = c_0x^1(1 + 4x + \cdots) + \bar{c}_0x^{\frac{5}{2}}(1 - \frac{4}{5}x + \cdots)$

7. $2r(r-1) - 3r + 3 = 0$, $r = 1, \frac{3}{2}$

$y = \sum_{n=0}^{\infty} c_n x^{n+r}$, $c_1 = 0$; for $n \geq 2$, $c_n = -\dfrac{2}{2(n+r)(n+r-1) - 3(n+r) + 3}c_{n-2}$

$r = 1$: for $n \geq 2$, $c_n = -\dfrac{2}{n(2n-1)}c_{n-2}$, $c_2 = -\frac{1}{3}c_0$

$r = \frac{3}{2}$: for $n \geq 2$, $c_n = -\dfrac{2}{n(2n+1)}c_{n-2}$, $c_2 = -\frac{1}{5}c_0$

$y = c_0x^1(1 - \frac{1}{3}x^2 + \cdots) + \bar{c}_0x^{\frac{3}{2}}(1 - \frac{1}{5}x^2 + \cdots)$

9. $3r(r-1) + 2r = 0$, $r = 0, \frac{1}{3}$

$y = \sum_{n=0}^{\infty} c_n x^{n+r}$, $c_1 = -\dfrac{1}{(r+1)(3r+2)}c_0$; for $n \geq 2$, $c_n = \dfrac{-c_{n-1} - 4c_{n-2}}{(n+r)[3(n+r)-1]}$

$r = 0$: $c_1 = -\frac{1}{2}c_0$; for $n \geq 2$, $c_n = \dfrac{-c_{n-1} - 4c_{n-2}}{n(3n-1)}$

$r = \frac{1}{3}$: $c_1 = -\frac{1}{4}c_0$; for $n \geq 2$, $c_n = \dfrac{-c_{n-1} - 4c_{n-2}}{n(3n+1)}$

$y = c_0x^0(1 - \frac{1}{2}x + \cdots) + \bar{c}_0x^{\frac{1}{3}}(1 - \frac{1}{4}x + \cdots)$

11. $r_1 = 1$: for $n \geq 1$, $c_n = \dfrac{1}{n(n+1)}c_{n-1}$, $c_0 = 1$, $y_1 = x^1 \sum_{n=0}^{\infty} c_n x^n$
$r_2 = 0$: $y_2 = y_1 \ln x + \sum_{n=0}^{\infty} b_n x^{n+0}$; for $n \geq 0$, $n(n+1)b_{n+1} = b_n - (2n+1)c_n$, $n = 0 \Rightarrow b_0 = c_0 = 1$ (let $b_1 = 0$)

13. Double root $r_1 = 3$: for $n \geq 1$, $c_n = \dfrac{4}{n^2}c_{n-1}$, $(c_0 = 1)$, $y_1 = x^3 \sum_{n=0}^{\infty} c_n x^n$
Second solution: $y_2 = y_1 \ln x + x^3 \sum_{n=0}^{\infty} b_n x^n$, $0b_0 = 0$ (let $b_0 = 0$); for $n \geq 1$, $n^2 b_n = 4b_{n-1} - 2nc_n$

15. $r_1 = 4$: for $n \geq 1$, $c_n = \dfrac{-1}{n(n+1)}c_{n-1}$, $c_0 = 1$, $y_1 = x^4 \sum_{n=0}^{\infty} c_n x^n$
$r_2 = 3$: $y_2 = y_1 \ln x + \sum_{n=0}^{\infty} b_n x^{n+3}$
for $n \geq 1$, $n(n-1)b_n = -b_{n-1} + (1-2n)c_{n-1}$, $n = 1 \Rightarrow b_0 = -c_0 = -1$ (let $b_1 = 0$)

17. $r_1 = 3$: $c_1 = 0$; for $n \geq 2$, $c_n = \dfrac{-1}{n(n+1)}c_{n-2}$, $(c_0 = 1)$, $y_1 = x^3 \sum_{n=0}^{\infty} c_n x^n$
$r_2 = 2$: $0\bar{c}_1 = 0$ (let $\bar{c}_1 = 0$); for $n \geq 2$, $\bar{c}_n = \dfrac{-1}{n(n-1)}\bar{c}_{n-2}$, $\bar{c}_0 = 1$, $y_2 = x^2 \sum_{n=0}^{\infty} \bar{c}_n x^n$

19. $r_1 = \frac{1}{2}$, $r_2 = 0$, $y_1 = x^{1/2} \sum_{n=0}^{\infty} x^n = \dfrac{x^{1/2}}{1-x}$, $y_2 = \sum_{n=0}^{\infty} x^n = \dfrac{1}{1-x}$

21. $r_1 = r_2 = 0$, $y_1 = \sum_{n=0}^{\infty} x^n = \dfrac{1}{1-x}$, $y_2 = y_1 \ln x = \dfrac{\ln x}{1-x}$ ($\bar{c}_0$ arbitrary, so choose $\bar{c}_0 = 0$)

23. $r_1 = 0$, $r_2 = -2$, $y_1 = 1$, $y_2 = x^{-2}(1-2x)$ $(a = 0, \tilde{c}_3 = 0 \Rightarrow \tilde{c}_n = 0, n \geq 3)$

Section 4.8

1. May be verified like (26). **4.** Subtract (30) from (29). **5.** $J_{3/2}(x) = \sqrt{\dfrac{2}{\pi x}}\left[\dfrac{\sin x}{x} - \cos x\right]$

6. $J_{5/2}(x) = \sqrt{\dfrac{2}{\pi}}[3x^{-5/2}\sin x - 3x^{-3/2}\cos x - x^{-1/2}\sin x]$

8. $J_{3/2}(x) = 0 \Leftrightarrow \tan x = x$

10.

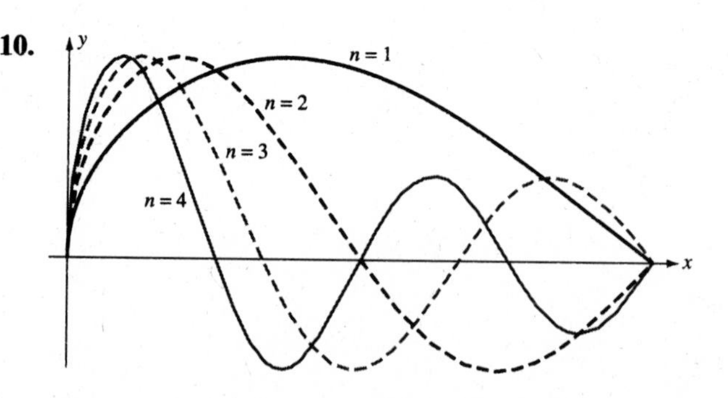

11. $q_5 = 0.673$, $q_{10} = 0.626$, $q_{20} = 0.602$, $q_{30} = 0.593$, $q_{50} = 0.587$, $q_{100} = 0.582$, $q_{200} = 0.579$, $q_{300} = 0.578$

Chapter 5

Section 5.1

5. Two

7. Three

9. Degenerate

11. Three

13. Two

15. $z' - x + y = t,$
$y' + z = 0,$
$x' - z = 0$

17. $x_3' + x_5' = \sin t,$
$x_4' - x_3 + x_2 = t,$
$x_1' - x_3 = 0,$
$x_3' - x_4 = 0,$
$x_2' - x_5 = 0$

19. $x_3' = 3x_3 + 4x_4 - x_1,$
$x_4' = x_3 - x_4 + x_2,$
$x_1' = x_3,$
$x_2' = x_4$

21. Yes;
$x' = 11x - 5y - 2t + 5\cos t,$
$y' = -5x + 2y + t - 2\cos t$

23. Yes; $x' = \frac{1}{10}x + \frac{39}{10}y + \frac{7}{10}t$
$y' = -\frac{1}{10}x + \frac{21}{10}y + \frac{3}{10}t$

Section 5.2

1. $x = \frac{3}{2}c_1e^t + c_2e^{2t}$, $y = c_1e^t + c_2e^{2t}$

3. $x = 8 + 2t - c_1e^t - 3c_2e^{-t},$
$y = -6 - t + c_1e^t + c_2e^{-t}$

5. (General solution):
$x = c_2e^t\cos 2t - c_1e^t\sin 2t,$
$y = c_1e^t\cos 2t + c_2e^t\sin 2t,$
$x = -e^t\sin 2t$, $y = e^t\cos 2t$

7. $x = c_1e^{3t}\cos t + c_2e^{3t}\sin t - \frac{4}{13}\sin t - \frac{8}{39}\cos t,$
$y = -(c_1 + c_2)e^{3t}\cos t +$
$(c_1 - c_2)e^{3t}\sin t - \frac{4}{39}\cos t + \frac{7}{39}\sin t$

9. $x = c_1e^t$

11. Let $z = L_2x + L_3y$ and note that $L_1z = 0$.

13. $x = c_1 - \frac{3}{2}c_2e^{-5t}$, $y = c_1 + c_2e^{-5t}$

15. $x = -\frac{4}{3} + c_1e^{3t}$, $y = \frac{5}{3} - \frac{7}{8}c_1e^{3t} + c_2e^{-t}$

17. $x = c_1e^{3t} + c_2e^{2t} + c_3e^{-t}$, $y = -c_1e^{3t} + c_2e^{2t} + c_3e^{-t}$,
$z = c_2e^{2t} - 2c_3e^{-t}$, $c_1 = -\frac{1}{2}$, $c_2 = \frac{1}{3}$, $c_3 = \frac{1}{6}$

19. $x = -\dfrac{c_1}{2} + c_2e^{3t} - c_3e^{4t}$, $y = -\dfrac{c_1}{2} + c_2e^{3t} + c_3e^{4t}$, $z = c_1 + c_2e^{3t}$

21. $x = c_1e^t + c_2e^{3t} + c_3e^{-3t}$, $y = \frac{7}{2}c_1e^t - \frac{1}{4}c_2e^{3t} + \frac{1}{2}c_3e^{-3t} + c_4e^{-t}$

23. $x = -1 + 2t - 3c_2 + 2c_1 - c_3e^t - 3c_4e^{-t} = 2t + \tilde{c}_2 - c_3e^t - 3c_4e^{-t}$
$y = -t + 2c_2 - c_1 + c_3e^t + c_4e^{-t} = -t + \tilde{c}_1 + c_3e^t + c_4e^{-t}$

25. $x = c_1 + \frac{5}{14}e^{2t} - \frac{3}{2}c_2e^{-5t} + c_3e^t,$
$y = c_1 + c_2e^{-5t} - \frac{1}{14}e^{2t}$

Section 5.3

1. $x = 3e^t - e^{2t},\ y = 2e^t - e^{2t}$

3. $x = 8 + 2t - \frac{1}{2}e^t - \frac{3}{2}e^{-t},\ y = -6 - t + \frac{1}{2}e^t + \frac{1}{2}e^{-t}$

5. $x = -e^t \sin 2t,\ y = e^t \cos 2t$

7. $x = \frac{1}{2}e^{3t}\cos t + \frac{87}{78}e^{3t}\sin t - \frac{4}{13}\sin t - \frac{8}{39}\cos t,$
$y = -\frac{63}{39}e^{3t}\cos t - \frac{8}{13}e^{3t}\sin t + \frac{7}{39}\sin t - \frac{4}{39}\cos t$

9. $x = e^{5t} + e^{-t},\ y = 2e^{5t} - 2e^{-t},\ z = 3e^{5t} + e^{-t}$

11. $x = 2 + te^t + e^t,\ y = 6 + 2te^t,\ z = 1 + te^t$

13. $x = \frac{1}{2}e^t + \frac{1}{2}\cos t - \frac{3}{2}\sin t - 1 + t,$
$y = \frac{3}{4}e^t + \cos t - \frac{1}{2}\sin t + \frac{5}{4}e^{-t} - 3 + t$

15. $x = 3 + 2t + e^t - 3e^{-t},\ y = -t + 6 - e^t + e^{-t}$

17. $x = 1 - \frac{5}{14}e^{2t} - 3e^{-5t} - e^t,\ y = 1 + 2e^{-5t} + \frac{1}{14}e^{2t}$

Section 5.4

1. $x' = -x + 50, \quad x(0) = 0,$
$y' = x - 2y, \quad y(0) = 0;$
$x = 50 - 50e^{-t},$
$y = 25 - 50e^{-t} + 25e^{-2t}$

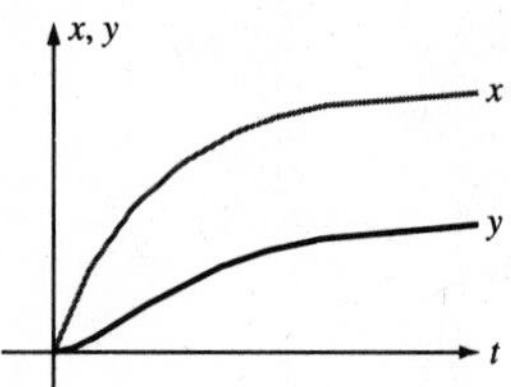

3. 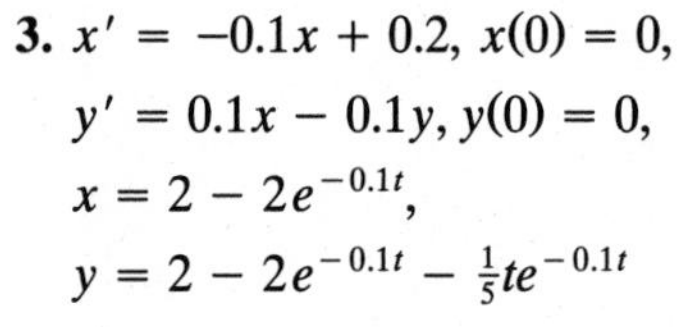
$x' = -0.1x + 0.2,\ x(0) = 0,$
$y' = 0.1x - 0.1y,\ y(0) = 0,$
$x = 2 - 2e^{-0.1t},$
$y = 2 - 2e^{-0.1t} - \frac{1}{5}te^{-0.1t}$

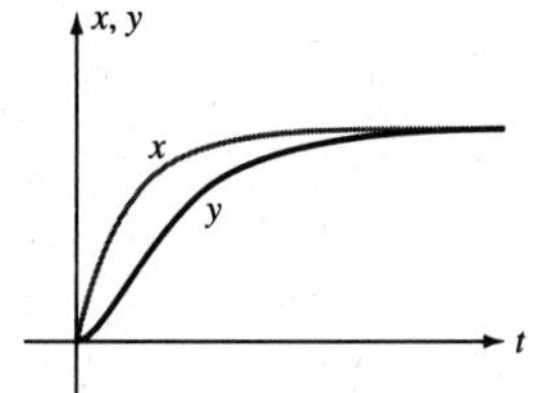

5. 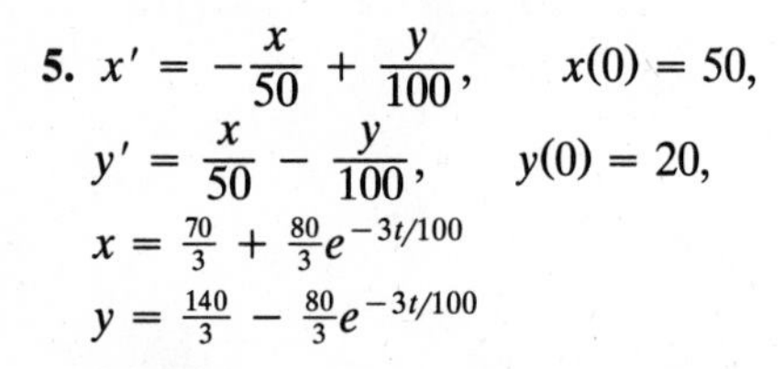
$x' = -\frac{x}{50} + \frac{y}{100}, \quad x(0) = 50,$
$y' = \frac{x}{50} - \frac{y}{100}, \quad y(0) = 20,$
$x = \frac{70}{3} + \frac{80}{3}e^{-3t/100}$
$y = \frac{140}{3} - \frac{80}{3}e^{-3t/100}$

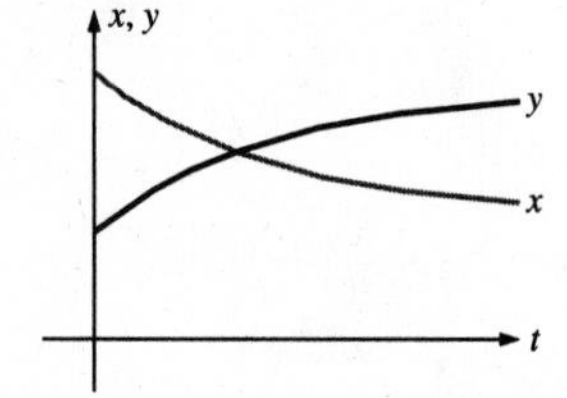

7. 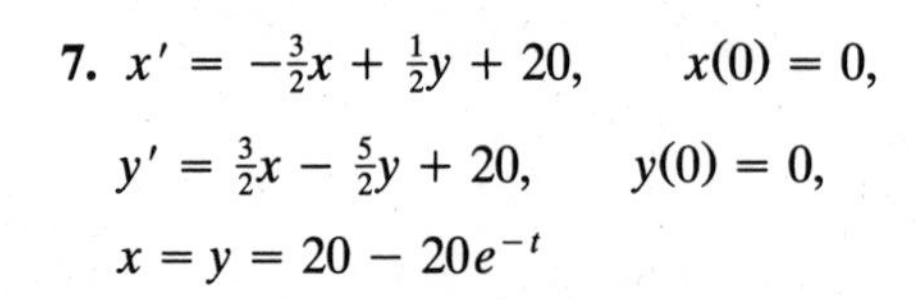
$x' = -\frac{3}{2}x + \frac{1}{2}y + 20, \quad x(0) = 0,$
$y' = \frac{3}{2}x - \frac{5}{2}y + 20, \quad y(0) = 0,$
$x = y = 20 - 20e^{-t}$

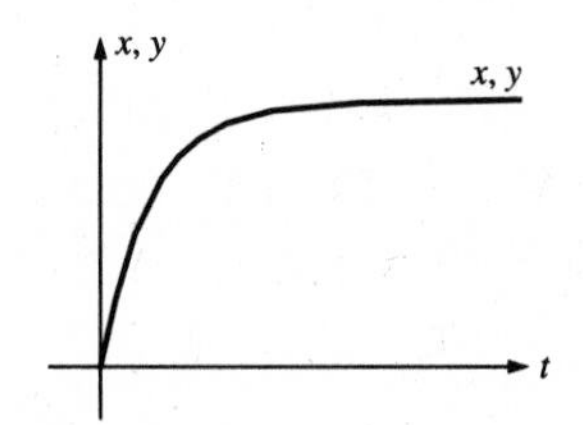

9. $x' = -9\dfrac{x}{100+5t} + 7, \qquad x(0) = 0,$

$y' = 9\dfrac{x}{100+5t} - 6\dfrac{y}{100+3t}, \qquad y(0) = 0$

11. $x' = -5\dfrac{x}{100} + 10,$ $\quad x, y, z$ amounts in tanks A, B, C, $x(0) = y(0) = z(0) = 0.$

$y' = 2\dfrac{x}{100} + 3\dfrac{z}{100} - 5\dfrac{y}{100},$

$z' = 3\dfrac{x}{100} - 3\dfrac{z}{100}$

13. **i)** $x' = \alpha\delta - \dfrac{\beta}{V_1}x,$

$y' = \beta\dfrac{x}{V_1} - \gamma\dfrac{y}{V_2}$

Let $\tau = \dfrac{\beta}{V_1}, k = \dfrac{\gamma}{V_2},$

$x' = -\tau x + \alpha\delta,$

$y' = \tau x - ky$

ii) $y'' + (\tau + k)y' + \tau k y = \alpha\delta\tau$

iii) $-\tau, -k$

iv) $\gamma \neq 0$ and $\beta \neq 0$ produces equilibrium (some outflow from both tanks). Equilibrium $x = \dfrac{\alpha\delta}{\tau}, y = \dfrac{\alpha\delta}{k}$

15. $x = c_1e^{-t} + c_2e^{-3t}, \qquad y = c_1e^{-t} - c_2e^{-3t}$

17. $x = 3c_1e^{t} - 3c_2e^{-5t}, \quad y = c_1e^{t} + c_2e^{-5t}, \quad x(0) > 0, \quad y(0) > 0 \Rightarrow c_1 > 0, \ \lim_{t\to\infty} \dfrac{x}{y} = 3.$

20. $a' = 0.1a + 0.2(b - 2000), \qquad a(0) = 1000$

$b' = 0.05b + 3650 - 0.2(b - 2000), \qquad b(0) = 2000$

Section 5.5

1. $x = \frac{1}{2}\sin t - \frac{1}{6}\sin 3t,$

$y = \frac{1}{2}\sin t + \frac{1}{6}\sin 3t;$

Resonance at frequencies of $1/2\pi$, $3/2\pi$ cycles/s

3. $x = te^{-2t}, \ y = -te^{-2t}$

5. $x = \cos 2t + \sin 2t - \cos 4t - \sin 4t,$

$y = \cos 2t + \sin 2t + \cos 4t + \sin 4t;$

Resonance at $1/\pi$, $2/\pi$ cycles/s

7. $x = e^{-t} - e^{-2t} - 2e^{-4t} + e^{-5t},$

$y = e^{-t} + e^{-2t} + 2e^{-4t} + e^{-5t}$

9. **a)** $0 \le \delta^2 < 4k;$

b) $4k \le \delta^2 < 4k + 8l;$

c) $4k + 8l < \delta^2$

11. $m_1x_1'' + \delta_1x_1' + (k_1 + k_2)x_1 - k_2x_2 = 0,$

$m_ix_i'' + \delta_ix_i' + (k_i + k_{i+1})x_i - k_ix_{i-1} - k_{i+1}x_{i+1} = 0,$

$i = 2, \ldots, n-1, \ m_nx_n'' + \delta_nx_n' + (k_n + k_{n+1})x_n - k_nx_{n-1} = 0$

15. $\mathbf{M} = \begin{bmatrix} m_1 & 0 \\ 0 & m_2 \end{bmatrix}$, $\boldsymbol{\delta} = \begin{bmatrix} \delta & 0 \\ 0 & \boldsymbol{\delta} \end{bmatrix}$, $\mathbf{K} = \begin{bmatrix} k_1 + k_2 & -k_2 \\ -k_2 & k_2 + k_3 \end{bmatrix}$

17. $\mathbf{M} = \begin{bmatrix} m_1 & 0 & 0 \\ 0 & m_2 & 0 \\ 0 & 0 & m_3 \end{bmatrix}$, $\mathbf{K} = \begin{bmatrix} k_1 + k_2 & -k_2 & 0 \\ -k_2 & k_2 + k_3 & -k_3 \\ 0 & -k_3 & k_3 + k_4 \end{bmatrix}$

Section 5.6

1. $(4D + 4)i_1 - 3i_3 = e_1,$
$-3i_1 + (5D + 5)i_3 = e_2$

3. $(4D^2 + \frac{1}{4})q_1 - 3i_3 = e_1,$
$-3D^2 q_1 + (5D + 5)i_3 = e_2$

5. $i_3 = C_1 e^{-t/2} + C_2 t e^{-t/2},$
$q_1 = (2C_1 - 4C_2)e^{-t/2} + 2C_2 t e^{-t/2}$

7. $\alpha + \beta i$ is root of $2RD^2 + D + R/2$.

9. $q_1 = c_1 e^{-t/2}$, $q_3 = -c_1 e^{-t/2}$

Section 5.7

1. Not possible

3. Not possible

5. -5

7. $\begin{bmatrix} 2 & 4 & 0 \\ 2 & -2 & 2 \end{bmatrix}$

9. Not possible

11. $\begin{bmatrix} -1 & 3 \\ -1 & 2 \end{bmatrix}$

13. Not possible

15. $\begin{bmatrix} 1 & 2 & 0 \\ 1 & -1 & 1 \end{bmatrix}$

17. $\begin{bmatrix} 1 & 0 \\ 0 & 1 \end{bmatrix} = \mathbf{I}$

19. y arbitrary, $x = 1 - \frac{3}{2}y$;
$\mathbf{x} = \begin{bmatrix} 1 - \frac{3}{2}y \\ y \end{bmatrix} = \begin{bmatrix} 1 \\ 0 \end{bmatrix} + y\begin{bmatrix} -\frac{3}{2} \\ 1 \end{bmatrix}$

21. $x = \frac{1}{2}y$, y arbitrary; $\mathbf{x} = \begin{bmatrix} y/2 \\ y \end{bmatrix} = y\begin{bmatrix} \frac{1}{2} \\ 1 \end{bmatrix}$

23. Inconsistent

25. $x = 0$, $y = -1/2$, $z = -3/2$

27. $x = 1$, $y = 0$, $z = -2$

29. $\mathbf{x} = \begin{bmatrix} -1/2 \\ 0 \\ -3/2 \\ -3 \end{bmatrix} + y\begin{bmatrix} 1 \\ 1 \\ 0 \\ 0 \end{bmatrix}$, y arbitrary

31. Not generally; $(\mathbf{A} + \mathbf{B})^2 =$
$(\mathbf{A} + \mathbf{B})(\mathbf{A} + \mathbf{B}) = \mathbf{A}^2 + \mathbf{BA} + \mathbf{AB} + \mathbf{B}^2$

33. $\mathbf{A}(c_1\mathbf{x}_1 + c_2\mathbf{x}_2) =$
$c_1\mathbf{Ax}_1 + c_2\mathbf{Ax}_2 = c_1\mathbf{0} + c_2\mathbf{0} = \mathbf{0}$

35. $\mathbf{A}(\mathbf{x}_1 + \mathbf{x}_2) = \mathbf{Ax}_1 + \mathbf{Ax}_2 = \mathbf{b}_1 + \mathbf{b}_2$

37. $\mathbf{B}^{-1}\mathbf{A}^{-1}(\mathbf{AB}) = \mathbf{B}^{-1}\mathbf{IB} = \mathbf{B}^{-1}\mathbf{B} = \mathbf{I}$

39. $\mathbf{A}^{-1} = \begin{bmatrix} -1 & 2 \\ 1 & -1 \end{bmatrix}$

41. $\mathbf{A}^{-1} = \begin{bmatrix} -2 & 1 \\ 1 & 0 \end{bmatrix}$

43. $\mathbf{x} = \begin{bmatrix} 1 \\ 2 \\ 3 \\ 0 \\ 0 \end{bmatrix} + x_4\begin{bmatrix} -1 \\ 0 \\ -1 \\ 1 \\ 0 \end{bmatrix} + x_5\begin{bmatrix} 0 \\ -1 \\ -1 \\ 0 \\ 1 \end{bmatrix}$

45. $\mathbf{x} = \begin{bmatrix} 0 \\ 1 \\ 0 \\ 0 \end{bmatrix} + x_1\begin{bmatrix} 1 \\ 0 \\ 0 \\ 0 \end{bmatrix} + x_4\begin{bmatrix} 0 \\ -2 \\ -3 \\ 1 \end{bmatrix}$

47. $\mathbf{x} = \begin{bmatrix} 0 \\ 2 \\ 0 \\ 1 \end{bmatrix} + x_1 \begin{bmatrix} 1 \\ 0 \\ 0 \\ 0 \end{bmatrix} + x_3 \begin{bmatrix} 0 \\ -2 \\ 1 \\ 0 \end{bmatrix}$

49. Not consistent

51. $\mathbf{x} = \begin{bmatrix} 0 \\ 0 \\ 0 \\ 0 \\ 3 \end{bmatrix} + x_1 \begin{bmatrix} 1 \\ 0 \\ 0 \\ 0 \\ 0 \end{bmatrix} + x_2 \begin{bmatrix} 0 \\ 1 \\ 0 \\ 0 \\ 0 \end{bmatrix} + x_3 \begin{bmatrix} 0 \\ 0 \\ 1 \\ 0 \\ 0 \end{bmatrix} + x_4 \begin{bmatrix} 0 \\ 0 \\ 0 \\ 1 \\ 0 \end{bmatrix}$

53. $\mathbf{x} = \begin{bmatrix} 2 \\ 0 \\ 0 \\ 5 \\ 0 \end{bmatrix} + x_2 \begin{bmatrix} -1 \\ 1 \\ 0 \\ 0 \\ 0 \end{bmatrix} + x_3 \begin{bmatrix} -2 \\ 0 \\ 1 \\ 0 \\ 0 \end{bmatrix} + x_5 \begin{bmatrix} -1 \\ 0 \\ 0 \\ -1 \\ 1 \end{bmatrix}$

Section 5.8

1. -2 **3.** -5 **5.** 2

7. 9 **9.** No **11.** Yes, det $= -4$

15. $\lambda^2 - 4\lambda + 3$ **17.** $\lambda^2 - 18$ **19.** $\lambda^3 - 3\lambda^2 + 2\lambda$

21. $\lambda^3 - \lambda^2 - 5\lambda + 5$

Section 5.9

1. $\mathbf{A}'(t) = \begin{bmatrix} 0 & 1 \\ 2t & 0 \end{bmatrix}$

3. b) $\mathbf{x} = c_1 \begin{bmatrix} e^{-t} \\ -e^{-t} \end{bmatrix} + c_2 \begin{bmatrix} e^{3t} \\ e^{3t} \end{bmatrix}$

c) $\mathbf{x} = \begin{bmatrix} \frac{3}{2}e^{-t} + \frac{1}{2}e^{3t} \\ -\frac{3}{2}e^{-t} + \frac{1}{2}e^{3t} \end{bmatrix}$

5. b) $\mathbf{x} = c_1 \begin{bmatrix} e^{3t} \\ e^{3t} \end{bmatrix} + c_2 \begin{bmatrix} 1 \\ -2 \end{bmatrix}$

c) $\mathbf{x} = \begin{bmatrix} \frac{7}{3}e^{3t} + \frac{14}{3} \\ \frac{7}{3}e^{3t} - \frac{28}{3} \end{bmatrix}$

7. b) $\mathbf{x} = c_1 \begin{bmatrix} \cos 2t \\ \sin 2t \end{bmatrix} + c_2 \begin{bmatrix} -\sin 2t \\ \cos 2t \end{bmatrix}$

c) $\mathbf{x} = \begin{bmatrix} 19\cos 2t + 37\sin 2t \\ 19\sin 2t - 37\cos 2t \end{bmatrix}$

11. $\mathbf{A}(t) = \begin{bmatrix} 0 & 1 \\ -c(t) & -b(t) \end{bmatrix}$,

$\mathbf{x}(t) = \begin{bmatrix} y(t) \\ z(t) \end{bmatrix}$, $\mathbf{a} = \begin{bmatrix} a_0 \\ a_1 \end{bmatrix}$

13. Note that $z_i = y_i'$

15. a) $\mathbf{E} = \begin{bmatrix} 1 & 1 \\ 1 & -1 \end{bmatrix}$, $\mathbf{F} = \begin{bmatrix} 2 & -3 \\ 2 & -1 \end{bmatrix}$, $\mathbf{g} = \begin{bmatrix} t \\ 1 \end{bmatrix}$

b) $\mathbf{E}^{-1} = \frac{1}{2}\begin{bmatrix} 1 & 1 \\ 1 & -1 \end{bmatrix}$,

$\mathbf{A} = \begin{bmatrix} -2 & 2 \\ 0 & 1 \end{bmatrix}$, $\mathbf{f} = \frac{1}{2}\begin{bmatrix} t+1 \\ t-1 \end{bmatrix}$

17.

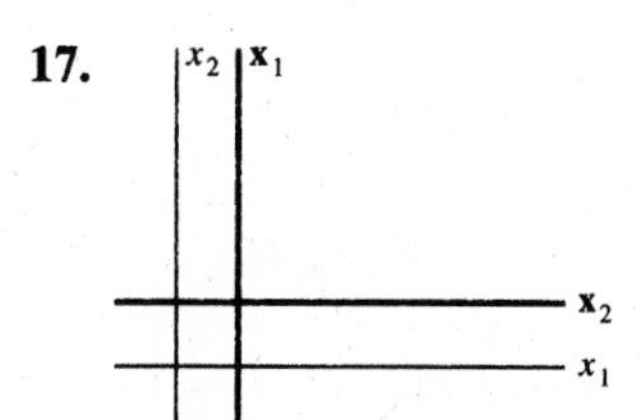

19. $(c_1\mathbf{x}_1 + c_2\mathbf{x}_2)' = c_1\mathbf{x}_1' + c_2\mathbf{x}_2' =$
$c_1\mathbf{A}\mathbf{x}_1 + c_2\mathbf{A}\mathbf{x}_2 = \mathbf{A}(c_1\mathbf{x}_1 + c_2\mathbf{x}_2)$

21. Differentiate $\mathbf{A}\mathbf{A}^{-1} = \mathbf{I}$

Section 5.10

1. $\mathbf{x} = c_1 e^{t} \begin{bmatrix} -\frac{1}{2} \\ 1 \end{bmatrix} + c_2 e^{2t} \begin{bmatrix} -1 \\ 1 \end{bmatrix}$

3. $\mathbf{x} = \frac{4}{5}\begin{bmatrix} 2 \\ 1 \end{bmatrix} + \frac{6}{5}e^{5t}\begin{bmatrix} -\frac{1}{2} \\ 1 \end{bmatrix}$

5. $\mathbf{x} = c_1 e^{2t} \begin{bmatrix} -\frac{1}{2} \\ 1 \end{bmatrix} + c_2 e^{7t} \begin{bmatrix} 2 \\ 1 \end{bmatrix}$

7. $\mathbf{x} = c_1 e^{t} \begin{bmatrix} 2 \\ 1 \end{bmatrix} + c_2 e^{6t} \begin{bmatrix} -\frac{1}{2} \\ 1 \end{bmatrix}$

9. $\mathbf{x} = c_1 \begin{bmatrix} -\frac{1}{2} \\ -\frac{1}{2} \\ 1 \end{bmatrix} + c_2 e^{2t} \begin{bmatrix} -1 \\ 1 \\ 0 \end{bmatrix} + c_3 e^{3t} \begin{bmatrix} 1 \\ 1 \\ 1 \end{bmatrix}$

11. $\mathbf{x} = c_1 \begin{bmatrix} -1 \\ 0 \\ 1 \end{bmatrix} + c_2 e^{2t} \begin{bmatrix} 1 \\ 0 \\ 1 \end{bmatrix} + c_3 e^{3t} \begin{bmatrix} 0 \\ 1 \\ 0 \end{bmatrix}$

13. $\mathbf{x} = c_1 \begin{bmatrix} 1 \\ 1 \\ 0 \end{bmatrix} + c_2 e^{3t} \begin{bmatrix} 1 \\ -1 \\ 1 \end{bmatrix} + c_3 e^{6t} \begin{bmatrix} -\frac{1}{2} \\ \frac{1}{2} \\ 1 \end{bmatrix}$

15. $\mathbf{x} = c_1 e^{t} \begin{bmatrix} -1 \\ 0 \\ 1 \end{bmatrix} + c_2 e^{t} \begin{bmatrix} 0 \\ 1 \\ 0 \end{bmatrix} + c_3 e^{3t} \begin{bmatrix} 1 \\ 0 \\ 1 \end{bmatrix}$

17. $\mathbf{x} = c_1 e^{t} \begin{bmatrix} 1 \\ 0 \\ 1 \\ 1 \end{bmatrix} + c_2 e^{t} \begin{bmatrix} 1 \\ 1 \\ 0 \\ 0 \end{bmatrix} + c_3 e^{-t} \begin{bmatrix} -1 \\ 0 \\ -1 \\ 1 \end{bmatrix} + c_4 e^{-t} \begin{bmatrix} -1 \\ 1 \\ 0 \\ 0 \end{bmatrix}$

19. $p(\lambda) = \lambda^3 - \lambda^2 - 5\lambda + 5$

21. $e^{t} \begin{bmatrix} 1 \\ 0 \\ 0 \\ 0 \end{bmatrix}, e^{t} \begin{bmatrix} 0 \\ 1 \\ 0 \\ 0 \end{bmatrix}$

23. $e^{2t} \begin{bmatrix} -1 \\ 1 \\ 0 \\ 0 \end{bmatrix}, e^{2t} \begin{bmatrix} 0 \\ 0 \\ -1 \\ 1 \end{bmatrix}$

25. $e^{2t} \begin{bmatrix} 1 \\ 0 \\ 0 \\ 0 \\ 0 \end{bmatrix}, e^{2t} \begin{bmatrix} 0 \\ -1 \\ 1 \\ 0 \\ 0 \end{bmatrix}, e^{2t} \begin{bmatrix} 0 \\ -1 \\ 0 \\ 1 \\ 0 \end{bmatrix}, e^{2t} \begin{bmatrix} 0 \\ -1 \\ 0 \\ 0 \\ 1 \end{bmatrix}$

27. $e^{4t} \begin{bmatrix} -1 \\ 1 \\ 0 \\ 0 \end{bmatrix}, e^{4t} \begin{bmatrix} -1 \\ 0 \\ 1 \\ 0 \end{bmatrix}, e^{4t} \begin{bmatrix} 0 \\ 0 \\ 0 \\ 1 \end{bmatrix}$

29. $\begin{bmatrix} 0 \\ 0 \\ -1 \\ 1 \end{bmatrix}$

31. $\lambda = \pm 2i,\ \mathbf{x} = c_1 \begin{bmatrix} -2\sin 2t \\ \cos 2t \end{bmatrix} + c_2 \begin{bmatrix} 2\cos 2t \\ \sin 2t \end{bmatrix}$

33. $\lambda = -1 \pm i,\ \mathbf{x} = e^{-t}\left(c_1 \begin{bmatrix} -\sin t \\ \cos t \end{bmatrix} + c_2 \begin{bmatrix} \cos t \\ \sin t \end{bmatrix} \right),\ c_1 = -1,\ c_2 = 1$

35. $\lambda = 2 \pm 2i,\ \mathbf{x} = c_1 e^{2t} \begin{bmatrix} -\cos 2t + 2\sin 2t \\ \cos 2t \end{bmatrix} + c_2 e^{2t} \begin{bmatrix} -2\cos 2t - \sin 2t \\ \sin 2t \end{bmatrix}$

37. $\lambda = 3 \pm 2i,\ \mathbf{x} = c_1 e^{3t} \begin{bmatrix} -\dfrac{\cos 2t}{2} + \dfrac{\sin 2t}{2} \\ \cos 2t \end{bmatrix} + c_2 e^{3t} \begin{bmatrix} -\dfrac{\cos 2t}{2} - \dfrac{\sin 2t}{2} \\ \sin 2t \end{bmatrix}$

39. $\lambda = \alpha \pm \beta i$ eigenvalues, trace $(\mathbf{A}) = (\alpha + \beta i) + (\alpha - \beta i) = 2\alpha$. Solutions involve functions of the form $e^{\alpha t} \cos \beta t$.

41. $\mathbf{A}(\mathbf{a} + i\mathbf{b}) = (\alpha + i\beta)(\mathbf{a} + i\mathbf{b}) \Rightarrow \mathbf{Aa} + i\mathbf{Ab} = (\alpha\mathbf{a} - \beta\mathbf{b}) + i(\beta\mathbf{a} + \alpha\mathbf{b})$

$$\Rightarrow \mathbf{Aa} = \alpha\mathbf{a} - \beta\mathbf{b} \text{ and } \mathbf{Ab} = \beta\mathbf{a} + \alpha\mathbf{b}$$

43. $\left[\begin{array}{cccc|c} 2 & 2 & 2 & 0 & 0 \\ -4 & -2 & 0 & 2 & 0 \\ 2 & 0 & -2 & -2 & 0 \\ 0 & 2 & 4 & 2 & 0 \end{array}\right] \to \left[\begin{array}{cccc|c} 1 & 0 & -1 & -1 & 0 \\ 0 & 1 & 2 & 1 & 0 \\ 0 & 0 & 0 & 0 & 0 \\ 0 & 0 & 0 & 0 & 0 \end{array}\right],\ \mathbf{a} = \begin{bmatrix} a_1 \\ a_2 \end{bmatrix},\ \mathbf{b} = \begin{bmatrix} b_1 \\ b_2 \end{bmatrix}$

Choose b_1, b_2 arbitrary but not both zero. Let $a_1 = b_1 + b_2$, $a_2 = -2b_1 - b_2$.

45. If $\mathbf{A}$ has more than one conjugate pair of eigenvalues, $\mathbf{A}$ has *at least four* eigenvalues, and so $\mathbf{A}$ is at least 4×4.

47. $\lambda_1 = 2$, $\mathbf{u} = \begin{bmatrix} 0 \\ 0 \\ 1 \end{bmatrix}$, $\lambda_2 = 1 + i$, $\mathbf{u} = \begin{bmatrix} \frac{1}{2} - \frac{1}{2}i \\ -1 \\ 1 \end{bmatrix}$

$$\mathbf{x} = c_1 e^{2t}\begin{bmatrix} 0 \\ 0 \\ 1 \end{bmatrix} + c_2 e^{t}\begin{bmatrix} \frac{1}{2}\cos t + \frac{1}{2}\sin t \\ -\cos t \\ \cos t \end{bmatrix} + c_3 e^{t}\begin{bmatrix} -\frac{1}{2}\cos t + \frac{1}{2}\sin t \\ -\sin t \\ \sin t \end{bmatrix}$$

49. $e^{2t}\begin{bmatrix} 0 \\ 1 \\ 2 \end{bmatrix}, e^{t}\begin{bmatrix} 3\cos t - \sin t \\ \cos t \\ 0 \end{bmatrix}, e^{t}\begin{bmatrix} \cos t + 3\sin t \\ \sin t \\ 0 \end{bmatrix}$

51. $e^{-t}\begin{bmatrix} 1 \\ 3 \\ -1 \end{bmatrix}, \begin{bmatrix} -2\sin 4t \\ \cos 4t \\ 2\cos 4t - 3\sin 4t \end{bmatrix}, \begin{bmatrix} 2\cos 4t \\ \sin 4t \\ 3\cos 4t + 2\sin 4t \end{bmatrix}$

53. $\begin{bmatrix} 0 \\ \cos 3t \\ -\sin 3t \\ 0 \end{bmatrix}, \begin{bmatrix} 0 \\ \sin 3t \\ \cos 3t \\ 0 \end{bmatrix}, e^{t}\begin{bmatrix} \cos 2t \\ \sin 2t \\ \cos 2t - \sin 2t \\ \cos 2t \end{bmatrix}, e^{t}\begin{bmatrix} \sin 2t \\ -\cos 2t \\ \cos 2t + \sin 2t \\ \sin 2t \end{bmatrix}$

55. $\begin{bmatrix} \cos 3t \\ \cos 3t - \sin 3t \\ 0 \\ \cos 3t \end{bmatrix}, \begin{bmatrix} \sin 3t \\ \cos 3t + \sin 3t \\ 0 \\ \sin 3t \end{bmatrix}, \begin{bmatrix} \cos 2t - 2\sin 2t \\ -3\sin 2t \\ \cos 2t \\ 0 \end{bmatrix}, \begin{bmatrix} 2\cos 2t + \sin 2t \\ 3\cos 2t \\ \sin 2t \\ 0 \end{bmatrix}$

57. $e^{-t}\begin{bmatrix} 0 \\ 1 \\ 0 \\ 0 \end{bmatrix}, \begin{bmatrix} 2 \\ 1 \\ 3 \\ 0 \end{bmatrix}, e^{-t}\begin{bmatrix} \cos 3t \\ \cos 3t - \sin 3t \\ \cos 3t + \sin 3t \\ \cos 3t \end{bmatrix}, e^{-t}\begin{bmatrix} \sin 3t \\ \sin 3t + \cos 3t \\ \sin 3t - \cos 3t \\ \sin 3t \end{bmatrix}$

59. $\lambda = 1$, $\mathbf{x} = e^{t}\left(\begin{bmatrix} 2u_2 + v_2 \\ u_2 \end{bmatrix} + t\begin{bmatrix} 2v_2 \\ v_2 \end{bmatrix}\right) = u_2 e^{t}\begin{bmatrix} 2 \\ 1 \end{bmatrix} + v_2 e^{t}\begin{bmatrix} 1 + 2t \\ t \end{bmatrix}$

61. $\lambda = 0$, $\mathbf{x} = e^{0t}\left(\begin{bmatrix} u_2 - v_2 \\ u_2 \end{bmatrix} + t\begin{bmatrix} v_2 \\ v_2 \end{bmatrix}\right) = u_2\begin{bmatrix} 1 \\ 1 \end{bmatrix} + v_2\begin{bmatrix} -1 + t \\ t \end{bmatrix}$

63. $\lambda = 2$, $\mathbf{x} = e^{2t}\left(\begin{bmatrix} \frac{1}{4}u_2 - \frac{1}{16}v_2 \\ u_2 \end{bmatrix} + t\begin{bmatrix} \frac{1}{4}v_2 \\ v_2 \end{bmatrix}\right) = u_2 e^{2t}\begin{bmatrix} \frac{1}{4} \\ 1 \end{bmatrix} + v_2 e^{2t}\begin{bmatrix} -\frac{1}{16} + \frac{1}{4}t \\ t \end{bmatrix}$

65. $\lambda = 1$, $\mathbf{x} = e^{t}\left(\begin{bmatrix} u_3 + v_3 \\ u_2 \\ u_3 \end{bmatrix} + t\begin{bmatrix} v_3 \\ v_3 \\ v_3 \end{bmatrix}\right) = u_2 e^{t}\begin{bmatrix} 0 \\ 1 \\ 0 \end{bmatrix} + u_3 e^{t}\begin{bmatrix} 1 \\ 0 \\ 1 \end{bmatrix} + v_3 e^{t}\begin{bmatrix} 1 + t \\ t \\ t \end{bmatrix}$

67. $\lambda = 1$, $\mathbf{x} = e^{t}\left(\begin{bmatrix} 2w_3 \\ v_3 - 2w_3 \\ u_3 \end{bmatrix} + t\begin{bmatrix} 0 \\ 2w_3 \\ v_3 \end{bmatrix} + t^2\begin{bmatrix} 0 \\ 0 \\ w_3 \end{bmatrix}\right) = u_3 e^{t}\begin{bmatrix} 0 \\ 0 \\ 1 \end{bmatrix} + v_3 e^{t}\begin{bmatrix} 0 \\ 1 \\ t \end{bmatrix} + w_3 e^{t}\begin{bmatrix} 2 \\ 2t - 2 \\ t^2 \end{bmatrix}$

69. $\lambda = 1$,

$$\mathbf{x} = e^{t}\left(\begin{bmatrix} u_2 - \frac{1}{2}v_2 + \frac{1}{2}w_2 \\ u_2 \\ \frac{1}{2}v_2 \end{bmatrix} + \begin{bmatrix} v_2 - w_2 \\ v_2 \\ w_2 \end{bmatrix}t + \begin{bmatrix} w_2 \\ w_2 \\ 0 \end{bmatrix}t^2\right) = u_2 e^{t}\begin{bmatrix} 1 \\ 1 \\ 0 \end{bmatrix} + v_2 e^{t}\begin{bmatrix} -\frac{1}{2} + t \\ t \\ \frac{1}{2} \end{bmatrix} + w_2 e^{t}\begin{bmatrix} \frac{1}{2} - t + t^2 \\ t^2 \\ t \end{bmatrix}$$

71. $\mathbf{x} = u_2 e^{-t}\begin{bmatrix} \frac{1}{2} \\ 1 \end{bmatrix} + v_2 e^{-t}\begin{bmatrix} \frac{1}{4} + \frac{1}{2}t \\ 1 \end{bmatrix}$

73. $\mathbf{x} = u_2 e^{5t}\begin{bmatrix} 0 \\ 1 \end{bmatrix} + v_2 e^{5t}\begin{bmatrix} 1 \\ t \end{bmatrix}$

75. $\mathbf{x} = u_2 e^{2t}\begin{bmatrix} 1 \\ 1 \\ 0 \end{bmatrix} + u_3 e^{2t}\begin{bmatrix} 0 \\ 0 \\ 1 \end{bmatrix} + v_3 e^{2t}\begin{bmatrix} -1+t \\ t \\ t \end{bmatrix}$

77. $\mathbf{x} = u_2 e^{t}\begin{bmatrix} \frac{1}{2} \\ 1 \\ 0 \end{bmatrix} + u_3 e^{t}\begin{bmatrix} \frac{1}{2} \\ 0 \\ 1 \end{bmatrix} + v_3 e^{t}\begin{bmatrix} -\frac{1}{2}+t \\ t \\ t \end{bmatrix}$

Section 5.11

1. $x_{1p} = Ae^t + Be^{-t}$,
$x_{2p} = Ce^t + De^{-t}$;
$\mathbf{x} = \mathbf{x}_p + \mathbf{x}_h$
$= \begin{bmatrix} -\frac{1}{3}e^{-t} \\ -e^t + \frac{2}{3}e^{-t} \end{bmatrix} + c_1\begin{bmatrix} -1 \\ 1 \end{bmatrix} + c_2 e^{2t}\begin{bmatrix} 1 \\ 1 \end{bmatrix}$

3. $x_{1p} = Ae^{-t}$,
$x_{2p} = Be^{-t}$;
$\mathbf{x} = \mathbf{x}_p + \mathbf{x}_h$
$= \begin{bmatrix} 0 \\ 2e^{-t} \end{bmatrix} + c_1\begin{bmatrix} \frac{1}{2} \\ 1 \end{bmatrix} + c_2 e^{t}\begin{bmatrix} 1 \\ 1 \end{bmatrix}$

5. $x_{1p} = A\sin t + B\cos t$,
$x_{2p} = C\sin t + D\cos t$;
$\mathbf{x} = \mathbf{x}_p + \mathbf{x}_h$
$= \begin{bmatrix} -\sin t \\ \sin t - \cos t \end{bmatrix} + c_1 e^{t}\begin{bmatrix} 1 \\ -1 \end{bmatrix} + c_2 e^{3t}\begin{bmatrix} 1 \\ 1 \end{bmatrix}$

7. $x_{1p} = Ae^{2t} + Bte^{2t} \qquad B = D = \frac{1}{2}$
$\Rightarrow$
$x_{2p} = Ce^{2t} + Dte^{2t} \qquad A - C = \frac{1}{4}$ (let $A = 0$);
$\mathbf{x} = \begin{bmatrix} \frac{1}{2}te^{2t} \\ -\frac{1}{4}e^{2t} + \frac{1}{2}te^{2t} \end{bmatrix} + c_1 e^{2t}\begin{bmatrix} 1 \\ 1 \end{bmatrix} + c_2 e^{-2t}\begin{bmatrix} -1 \\ 1 \end{bmatrix}$

9. $x_{1p} = A\sin t + B\cos t + Ct\sin t + Dt\cos t + E\sin 2t + F\cos 2t$,
$x_{2p} = G\sin t + H\cos t + It\sin t + Jt\cos t + K\sin 2t + L\cos 2t$

11. $x_{1p} = Ae^t + Bte^t$,
$x_{2p} = Ce^t + Dte^t$

13. $x_{1p} = Ae^{2t} + Bte^{2t} + Ct^2e^{2t} + Dt^3e^{2t} + Fe^t$,
$x_{2p} = Ge^{2t} + Hte^{2t} + It^2e^{2t} + Jt^3e^{2t} + Ke^t$

15. $x_{1p} = A + Bt + Ct^2 + Dt^3 + Et^4$
$x_{2p} = F + Gt + Ht^2 + It^3 + Jt^4$

17. $x_{1p} = A\sin t + B\cos t + Ct\sin t + Dt\cos t$
$+ Et^2\sin t + Ft^2\cos t$
$x_{2p} = G\sin t + H\cos t + It\sin t$
$+ Jt\cos t + Kt^2\sin t + Lt^2\cos t$

19. $x_{1p} = Ae^{-t} + Be^{-t}\cos 3t + Ce^{-t}\sin 3t$
$x_{2p} = De^{-t} + Ee^{-t}\cos 3t + Fe^{-t}\sin 3t$

Section 5.12

1. $e^{\mathbf{A}t} = \begin{bmatrix} -e^t + 2e^{2t} & -e^t + e^{2t} \\ 2e^t - 2e^{2t} & 2e^t - e^{2t} \end{bmatrix}$,

$\mathbf{x} = \begin{bmatrix} -1 - 2e^t + 4e^{2t} \\ 2 + 4e^t - 4e^{2t} \end{bmatrix}$

3. $e^{\mathbf{A}t} = \frac{1}{5}\begin{bmatrix} 4 + e^{5t} & 2 - 2e^{5t} \\ 2 - 2e^{5t} & 1 + 4e^{5t} \end{bmatrix}$,

$\mathbf{x} = \begin{bmatrix} -5e^{-t} + \frac{1}{2}e^t + \frac{22}{5} + \frac{1}{10}e^{5t} \\ -2e^{-t} + \frac{11}{5} - \frac{1}{5}e^{5t} \end{bmatrix} + e^{\mathbf{A}t}\mathbf{c}$

5. $e^{\mathbf{A}t} = -\frac{1}{5}\begin{bmatrix} -e^{2t} - 4e^{7t} & 2e^{2t} - 2e^{7t} \\ 2e^{2t} - 2e^{7t} & -4e^{2t} - e^{7t} \end{bmatrix}$,

$\mathbf{x} = \frac{1}{25}\begin{bmatrix} 5te^{2t} - 4e^{2t} + 4e^{7t} \\ -10te^{2t} - 2e^{2t} + 2e^{7t} \end{bmatrix} + e^{\mathbf{A}t}\mathbf{c}$

7. $e^{\mathbf{A}t} = \begin{bmatrix} 1 & t \\ 0 & 1 \end{bmatrix}$.

(Note: $\mathbf{A}^m = \mathbf{0}$ if $m \geqslant 2$.) $e^{-\mathbf{A}t} = \begin{bmatrix} 1 & -t \\ 0 & 1 \end{bmatrix}$.

9. $e^{\mathbf{A}t} = \begin{bmatrix} \cos t & \sin t \\ -\sin t & \cos t \end{bmatrix}$

11. $\mathbf{A}^2 + \mathbf{AB} + \mathbf{BA} + \mathbf{B}^2 = (\mathbf{A} + \mathbf{B})^2 \Rightarrow \mathbf{AB} = \mathbf{BA}$

13. $\mathbf{A}[\mathbf{u}_1, \mathbf{u}_2] = [\mathbf{Au}_1, \mathbf{Au}_2] =$

$[\lambda_1\mathbf{u}_1, \lambda_2\mathbf{u}_2] = [\mathbf{u}_1, \mathbf{u}_2]\begin{bmatrix} \lambda_1 & 0 \\ 0 & \lambda_2 \end{bmatrix}$

15. Use (1).

17. $e^{\mathbf{A}t} = \frac{1}{5}\begin{bmatrix} 4 + e^{5t} & 2 - 2e^{5t} \\ 2 - 2e^{5t} & 1 + 4e^{5t} \end{bmatrix}$

19. $e^{\mathbf{A}t} = -\frac{1}{5}\begin{bmatrix} -e^{2t} - 4e^{7t} & 2e^{2t} - 2e^{7t} \\ 2e^{2t} - 2e^{7t} & -4e^{2t} - e^{7t} \end{bmatrix}$

21. $e^{\mathbf{A}t} = \frac{1}{10}\begin{bmatrix} e^{6t} + 9e^{-4t} & 3e^{6t} - 3e^{-4t} \\ 3e^{6t} - 3e^{-4t} & 9e^{6t} + e^{-4t} \end{bmatrix}$

23. a) $e^{\mathbf{A}t} = \begin{bmatrix} -e^t + 2e^{2t} & -e^t + e^{2t} \\ 2e^t - 2e^{2t} & 2e^t - e^{2t} \end{bmatrix}$

b) $e^{\mathbf{A}t} = \frac{1}{5}\begin{bmatrix} 4 + e^{5t} & 2 - 2e^{5t} \\ 2 - 2e^{5t} & 1 + 4e^{5t} \end{bmatrix}$

c) $e^{\mathbf{A}t} = \begin{bmatrix} -2 + 3e^t & -1 + e^t \\ 6 - 6e^t & 3 - 2e^t \end{bmatrix}$

Chapter 6

Section 6.2

1. 3

3. 0

5. 5

7. 0

9. a) 59,049

11. a) 4.15×10^{-20}

13. 4.109, 6.129, 30.390, 193.137

15. $y(2) = 551{,}627$. Appears to be going from -5 to infinity.

16.

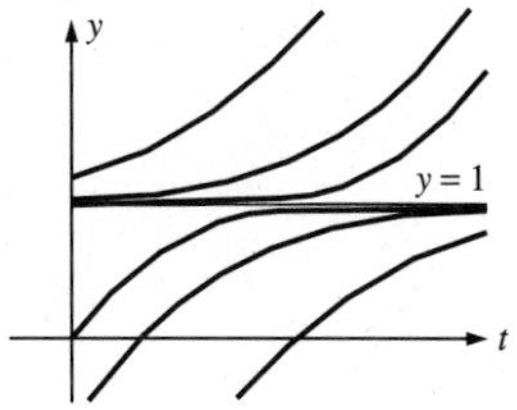

17. $y_N = y_{20} = 0.601$ with $h = 0.1$

19. $\dfrac{y_{N_1} - y_{N_2}}{y_{N_2} - y_{N_3}} = \dfrac{2^r - 1}{1 - (\frac{1}{2})^r} \approx 2^r$

20. 1.7106, 1.8411, 1.9165 (The next ratio is 1.957.)

Section 6.3

9. b) $y = x + 1 + 3e^{-30x}$

11. Numerical solution oscillates between approximately 0.07 and 3.06. Actual nonequilibrium solutions are monotonic.

Section 6.4

	Euler	Modified Euler	Second-Order Taylor
1. y_N	6.1917	7.3046	7.3046
$\lvert e_N \rvert$	1.1973	0.0844	0.0844
3. y_N	0.6529	0.60698	0.60141
$\lvert e_N \rvert$	0.0463	0.00044	0.00513
5. y_N	0.7634	0.761582	0.761598
$\lvert e_N \rvert$	0.0018	0.000012	0.000004
7. y_N	0.8672	0.865763	0.865775
$\lvert e_N \rvert$	0.0014	0.000006	0.000006

Note: In Ex. 7, $y = \sin^{-1}\left(\dfrac{e^{2x} - 1}{e^{2x} + 1}\right)$.

11. $y_5 = 7.3604$, $e_5 = 0.0286$

13. $y_{10} = 2.0588$

Section 6.5

1. $y_1 = 1.010025042$, $y_{10} = 2.297442014$

3. $y_1 = 0.571202424$, $y_{10} = 0.913669564$

5. $y_1 = 1.051795725$, $y_{10} = 1.355751179$

7. $y_1 = 20.1$, $y_{10} = 39{,}062{,}499$

8. $y_1 = 2.6916$, $y_{100} = 1.000000$

Section 6.6

1. $y_{10} = 24.30623$

3. $y_{10} = 0.86706$

5. $y(1) = 1$ (to 18 digits)

	$h = 0.1$	$h = 0.05$	$h = 0.04$	$h = 0.01$
y_N for Adams–Bashforth	-6×10^7	-4×10^7	1.7×10^6	1
y_N for Adams–Moulton	5×10^9	5	0.98880	1

7. 0.66547 [$y(1) = 0.66666\ldots$]

Section 6.7

1. Euler, $x_5 = 5.813$, $y_5 = 12.983$;
Modified Euler, $x_5 = 10.8926$, $y_5 = 25.2104$.

3. Euler, $x_{10} = 1.20608$, $y_{10} = 1.05994$;
Modified Euler, $x_{10} = 1.21620$, $y_{10} = 1.07288$

5. Euler, $x_{10} = -0.68137$, $y_{10} = 1.33303$;
Modified Euler, $x_{10} = -0.61452$, $y_{10} = 1.24624$

	Euler y_{10}	Error in Euler	Modified Euler y_{10}	Error in Modified Euler
7.	0.8825	−0.041	0.84247	−0.001
9.	6.1917	1.197	7.30463	0.0844

11. $y_{10} = 7.38889$, $e_{10} = 0.00017$

Chapter 7

Section 7.2

1. $x = 0,\ x = 1$

3. $x = 1,\ x = 2$

5. $x = n\pi$

7. $x = 0,\ x = \dfrac{a}{b}$

Section 7.3

1. $x = 0$ is stable, $x = 1$ is unstable.

3. $x = 1$ is unstable; for $x = 2$ the linear analysis is inconclusive.

5. $x = n\pi$ is unstable if n is even and $x = n\pi$ is stable if n is odd.

7. $x = \frac{1}{3} \ln 7$ is unstable.

Section 7.4

1. $x = 0$ is stable, $x = 1$ is unstable.

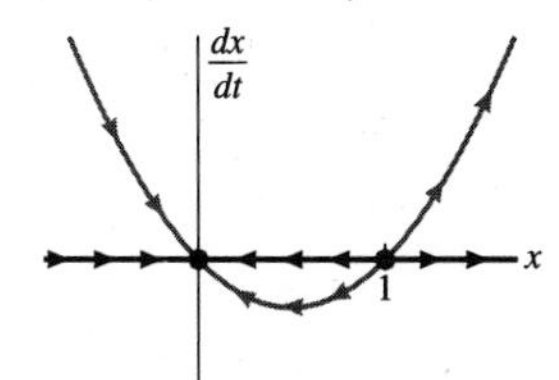

3. $x = 1$ is unstable, for $x = 2$ the linear analysis is inconclusive (but $x = 2$ is unstable).

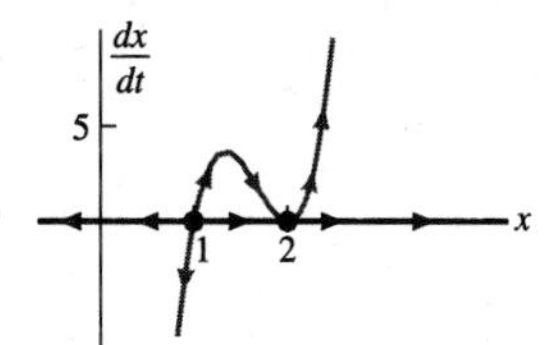

5. $x = n\pi$ is unstable if n is even and $x = n\pi$ is stable if n is odd.

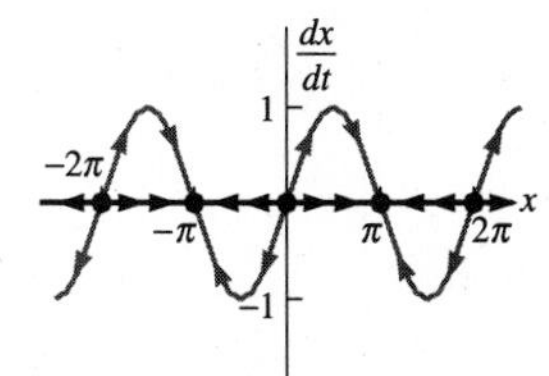

7. No equilibria

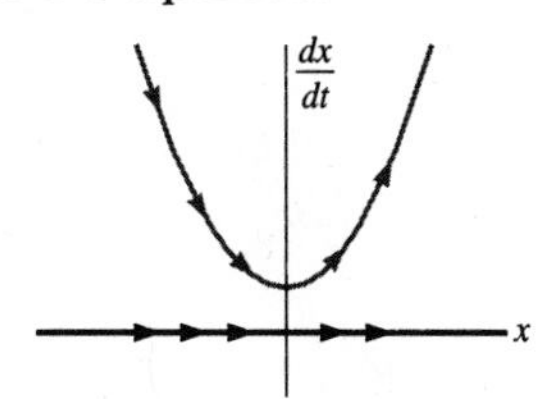

9. For $x = 0$ the linear analysis is inconclusive (but $x = 0$ is unstable).

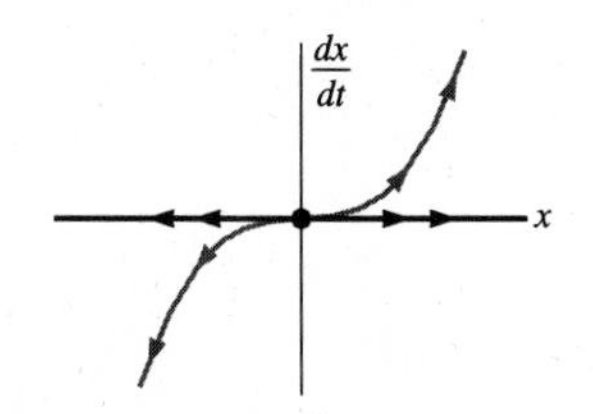

11. For $x = 3$ the linear analysis is inconclusive (but $x = 3$ is unstable).

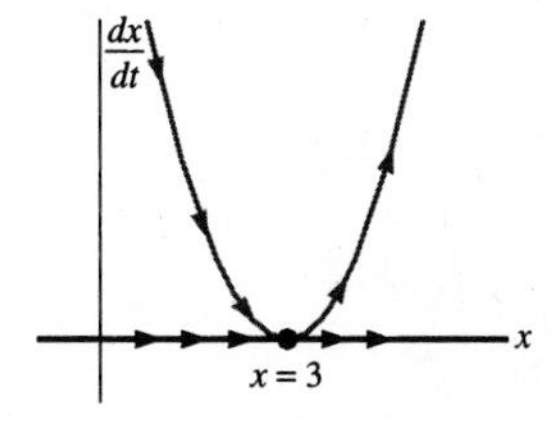

13.

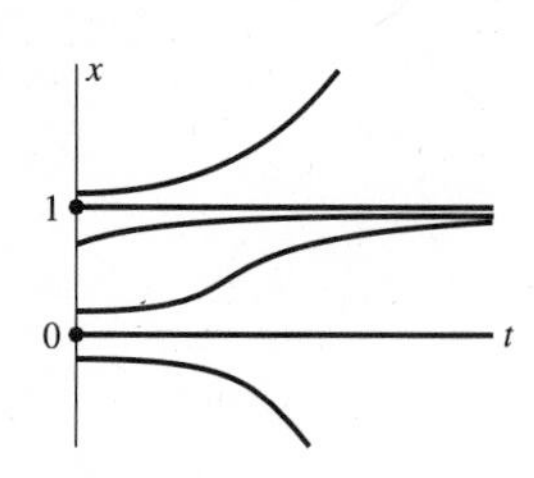

15.

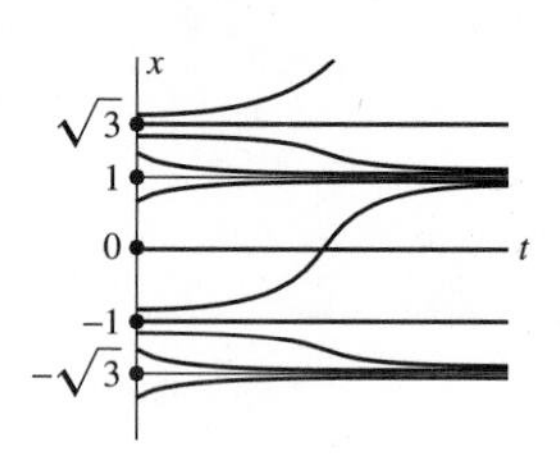

Section 7.5

1. **a)** Growth rate $= a - bx^2$. The growth rate is positive for $x < \sqrt{a/b}$ and the growth rate is negative for $x > \sqrt{a/b}$.

b)

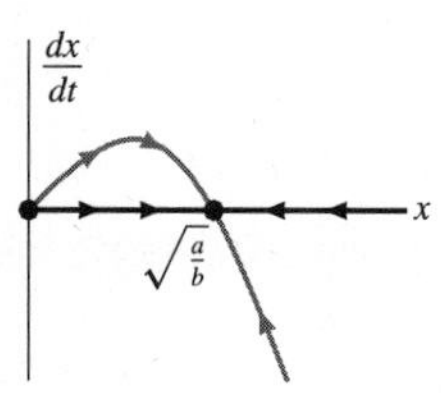

c)

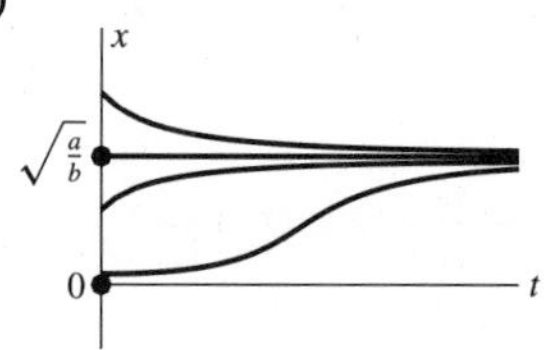

7. **a)** Growth rate $= a + bx$. The growth rate increases as x increases.

b)

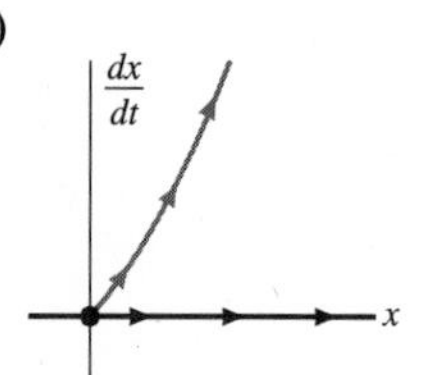

c)

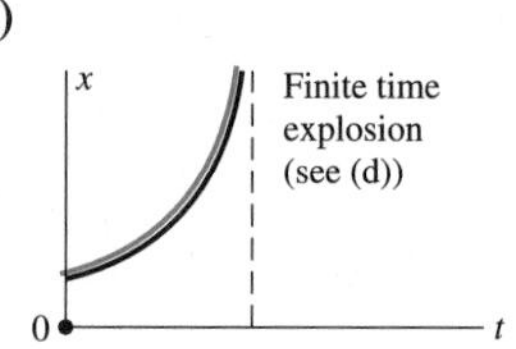

d) $x(t) = \dfrac{-\dfrac{a}{b}x_0}{x_0 - e^{-at}\left(x_0 + \dfrac{a}{b}\right)}$

9. **a)** $\alpha = \beta = \dfrac{a}{2b}$, $t_0 = \dfrac{1}{a}\ln\dfrac{a - bx_0}{bx_0}$

c) $\lim_{t\to\infty} \tanh t = 1$, $\lim_{t\to-\infty} \tanh t = -1$.

d) and **e)**

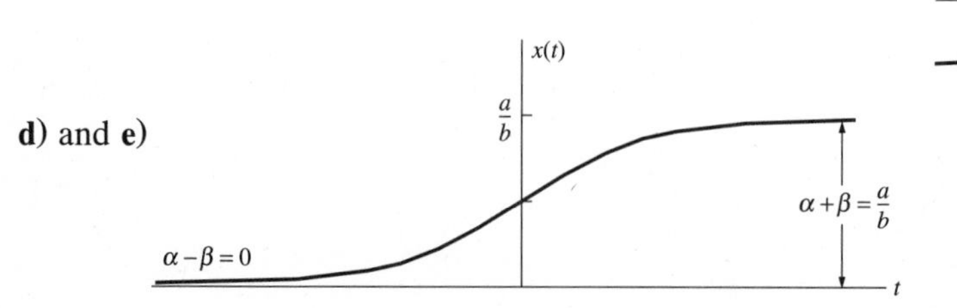

11. $f(0) = a \Rightarrow a = 0$. $f'(Q) = b + 2cQ$. Condition (c) $\Rightarrow c < 0$. Condition (b) $\Rightarrow b > 0$. Thus $dQ/dt = rQ - kQ^2$ where $r = b$, $k = -c$.

Section 7.6

1. a) $\dfrac{di}{dt} = i^2$

b)

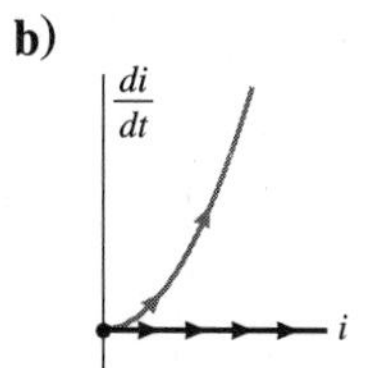

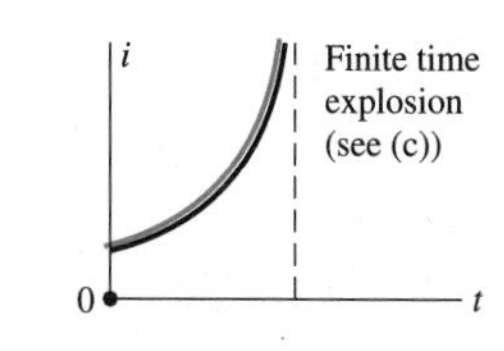

c) $i(t) = \dfrac{i(0)}{1 - i(0)t}$

3. a) $\dfrac{di}{dt} = -i^2$

b)

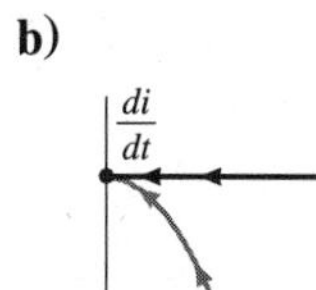

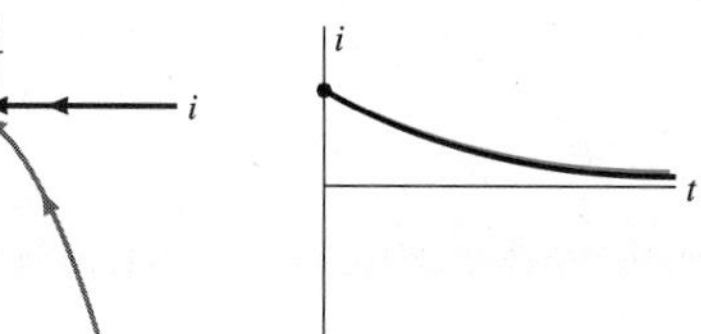

c) $i(t) = \dfrac{i(0)}{1 + i(0)t} = \dfrac{4}{1 + 4t}$

7. $i + \frac{1}{2}i^2 + \ln|i - 1| = 5 - t$

Section 7.7

1. a) $x = 3$ is stable, $x = 4$ is unstable.

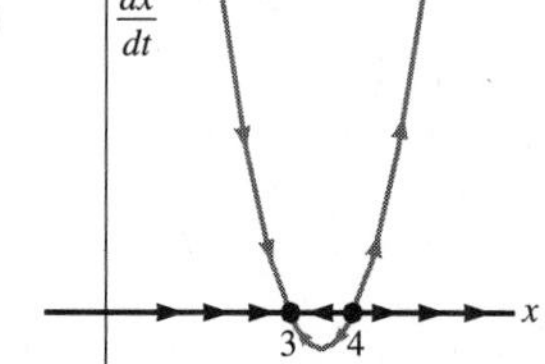

c) The solution corresponding to the initial condition $x(0) = 0$ goes through the origin.

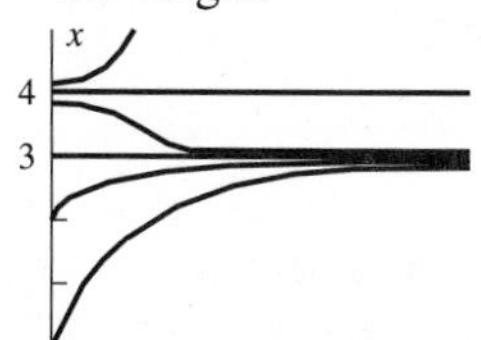

d) $\lim_{t \to \infty} x(t) = 3$, $b = 1$.

3. a) $x = \alpha$ is stable, $x = \beta$ is unstable.

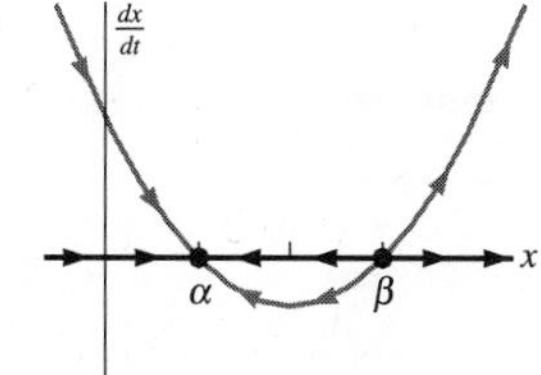

c) The solution corresponding to the initial condition $x(0) = 0$ goes through the origin.

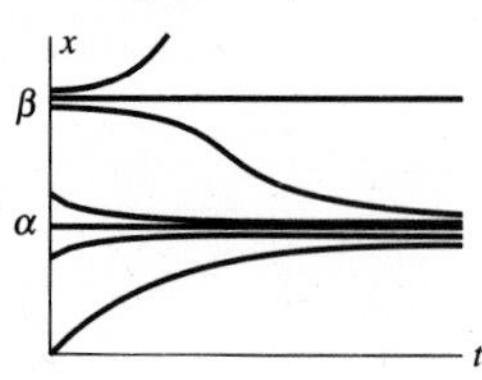

d) The limiting concentration of E product is α. $b = \beta - \alpha$, $a = 0$.

5. a) $x = 1, 2, 3$ are stable, unstable, stable, respectively.

b)

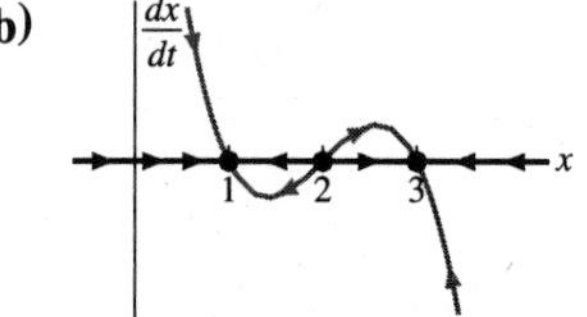

c) The solution corresponding to the initial condition $x(0) = 0$ goes through the origin.

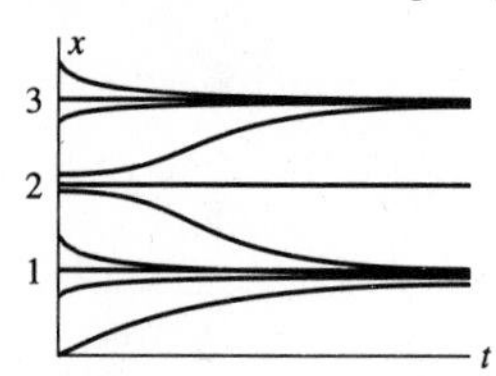

d) The limiting concentrations are $E = 1$, $a = 0$, $b = 1$, $c = 2$.

7. a) $x = \alpha, \beta, \gamma$ are stable, unstable, stable, respectively.

b)

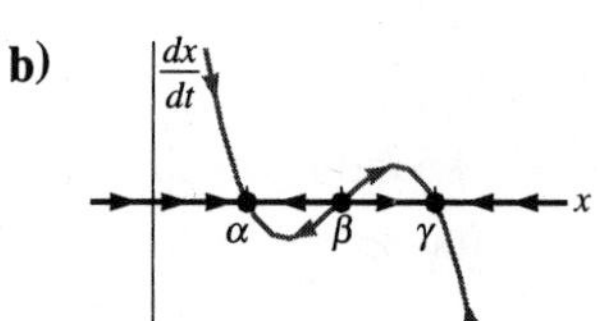

c) The solution corresponding to the initial condition $x(0) = 0$ goes through the origin.

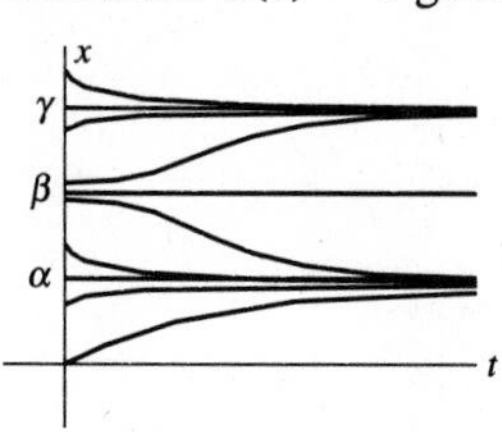

d) The limiting concentrations are $E = \alpha$, $a = 0$, $b = \beta - \alpha$, $c = \gamma - \alpha$.

9. **a)** α, β are the equilibrium concentrations. $x = \alpha$ is stable. Stability test fails for $x = \beta$.

b)

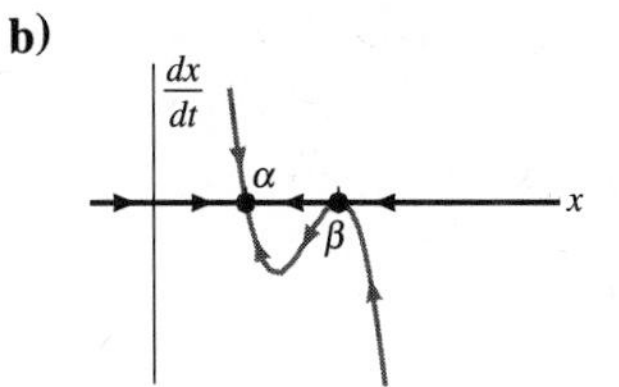

c) The solution corresponding to the initial condition $x(0) = 0$ goes through the origin.

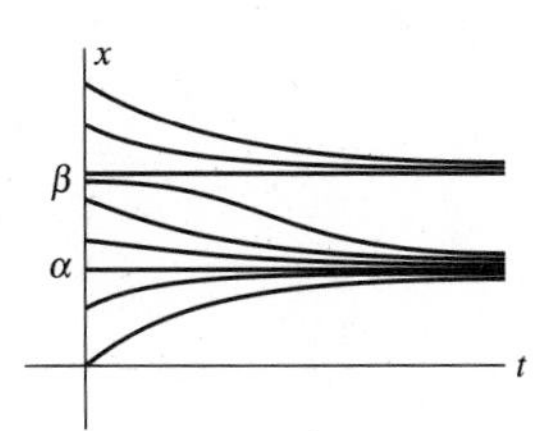

d) The limiting concentrations are $E = \alpha$, $a = 0$, $b = c = \beta - \alpha$.

Chapter 8

Section 8.2

1. $x_k = x_0(\frac{1}{2})^k$

3. $x_k = x_0(-3)^k$

5. $x_k = -1 + 2^{k+1}$

7. $x_k = x_0(\frac{3}{2})^k$

9. $x_k = -3 + c2^k$

11. $x_k = \frac{3}{2} + c(-1)^k$

13. $x_k = -\frac{2}{3} + \frac{20}{3}4^k$

15. $x_k = \frac{14}{9} + c(-\frac{4}{5})^k$

17. $T_{k+1} - T_k = -0.4(T_k - 100)$ with $T_0 = 60$. $T_k = 100 - 40(0.6)^k$

Section 8.3

1. Fixed point $x = 3$ is unstable.

3. Fixed point $x = 8$ is stable.

5. Fixed points are $x = 0$ and $x = \frac{1}{4}$. $f(x) = \frac{4}{3}x(1 - x)$ and $f'(x) = \frac{4}{3}(1 - 2x)$. $f'(0) = \frac{4}{3}$, so that $x = 0$ is unstable. $f'(\frac{1}{4}) = \frac{2}{3}$, so that $x = \frac{1}{4}$ is stable.

7. Fixed points are $x = 0$ and $x = \frac{7}{10}$. $f(x) = \frac{10}{3}x(1 - x)$ and $f'(x) = \frac{10}{3}(1 - 2x)$. $f'(0) = \frac{10}{3}$, so that $x = 0$ is unstable. $f'(\frac{7}{10}) = -\frac{4}{3}$, so that $x = \frac{7}{10}$ is unstable.

9. There are no fixed points.

11. $y_k = y_0(\frac{1}{7})^k$, $r = -\ln 7$

13. $y_k = y_0(-\frac{1}{2})^k$, $r = -\ln 2$

15.

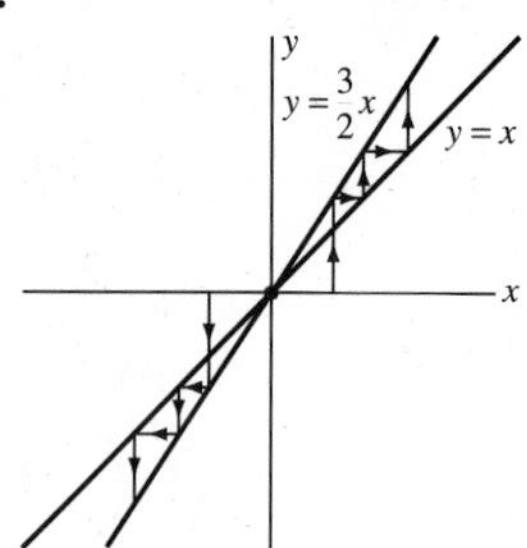

17.

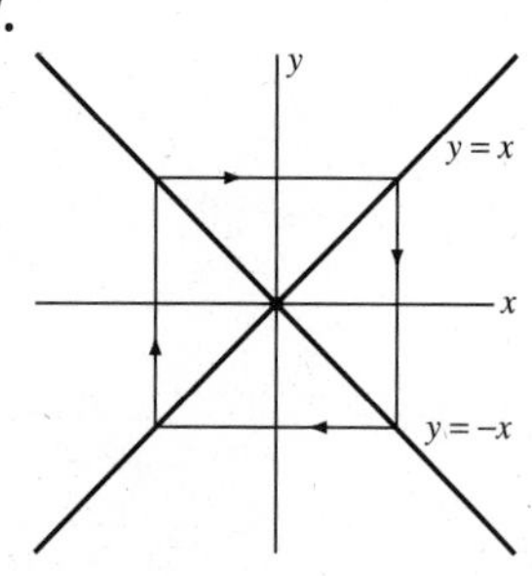

19.

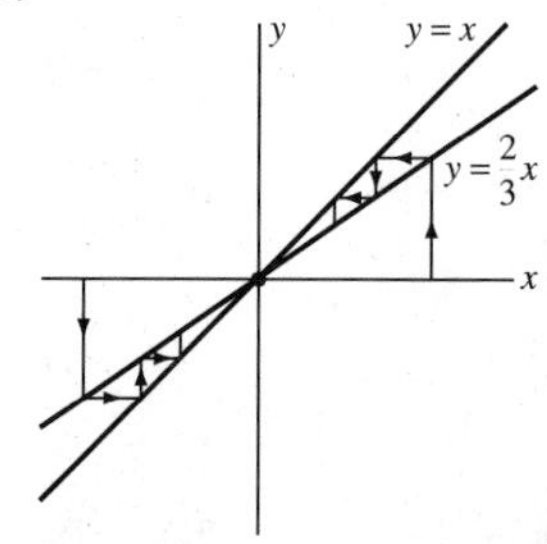

21.

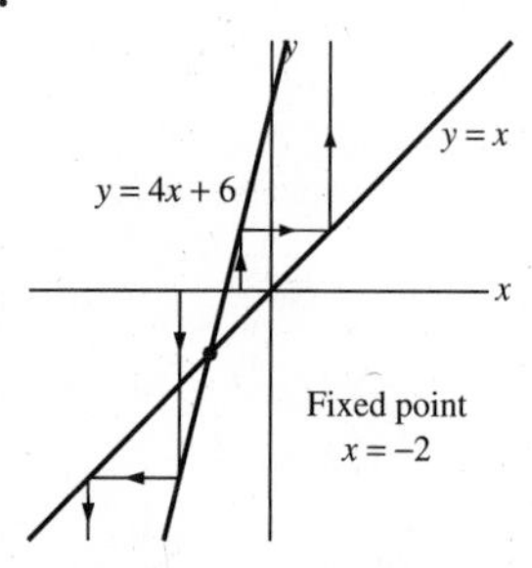

23.

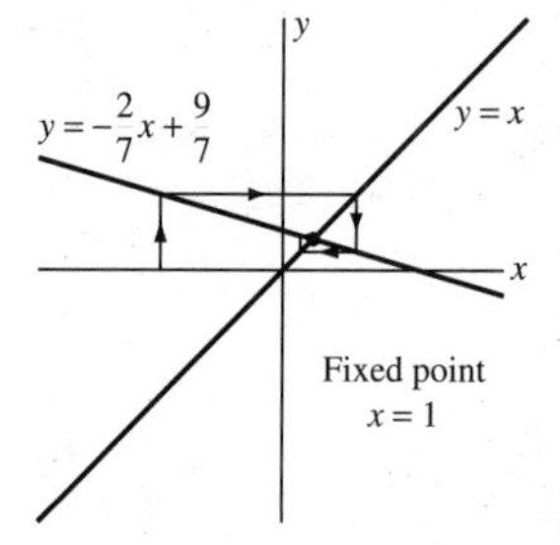

25. $\bar{x} = 0$ is unstable

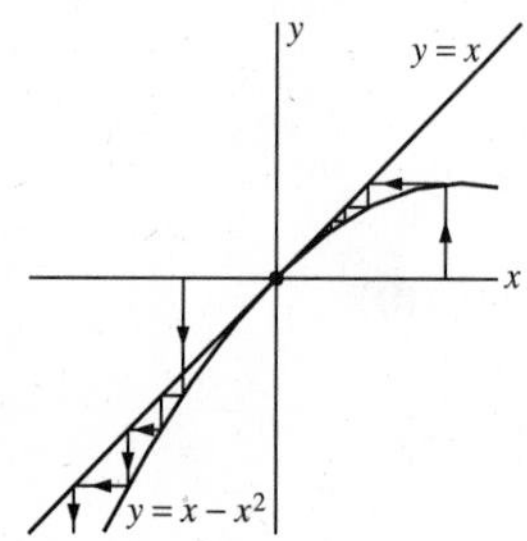

27. $\bar{x} = 0$ is stable

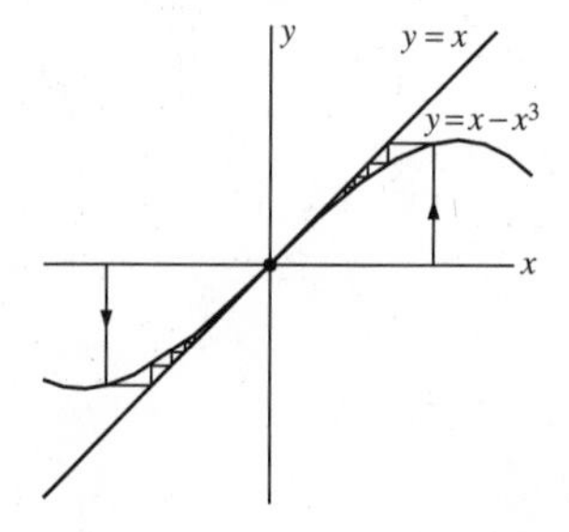

Section 8.4

1.

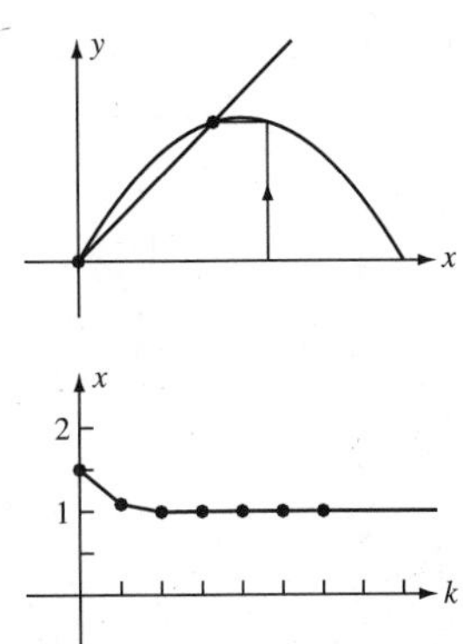

3. First bifurcation at $\gamma = 2$ (period 2 created), $\gamma = 2.44949\ldots$ (period 4 created), $\gamma = 2.54409\ldots$ (period 8 created). $(\gamma_2 - \gamma_1)/(\gamma_3 - \gamma_2) = 0.44949/0.0946 \approx 4.771\ldots$, while the universal factor is $4.6692\ldots$, which occurs in the limit as $n \to \infty$.

7. Two of the fixed points are $x = 0$ and $x = 1$, which are fixed points of the original map and not part of a period 3 solution. However, the other six fixed points represent two different period 3 orbits. The stability can be determined by the slope of the function (third iterate) at the fixed point of the third iterate. For each point on a period 3 orbit, the slope must be the same, since $(d/dx)f(f(f(x))) = \prod f'(x_i)$. From the figure it can be seen that the positive slope must be greater than 1 and hence represents an unstable period 3 solution. It is not clear from the figure if the orbit with negative slope has slopes more negative than -1, in which case that period 3 solution is unstable. When the period 3 solution is first created, one solution will have a positive slope less than 1, and hence it will be stable. The period 3 solutions are created as a stable and unstable pair. However, the stable period 3 orbit itself becomes unstable at some value of the parameter that can be determined.

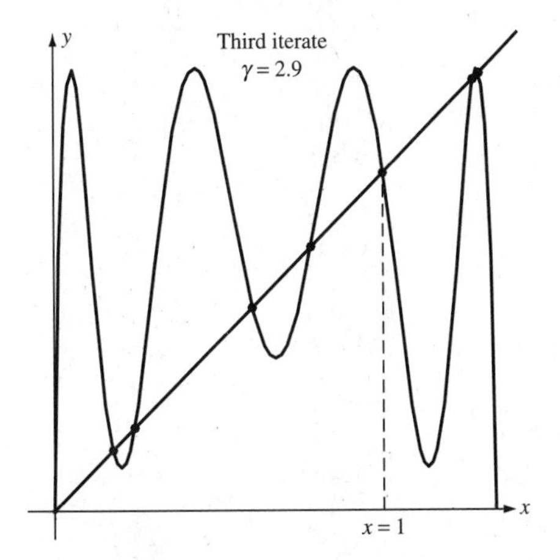

9. When $\gamma = 2.83$, the third iterate has a fixed point that corresponds to part of a period 3 orbit. When $\gamma = 2.845$, the third iterate has a period 2 solution that corresponds to part of a period 6 orbit of the original difference equation. When $\gamma = 2.849$, the third iterate has a period 4 solution that corresponds to part of a period 12 orbit of the original difference equation.

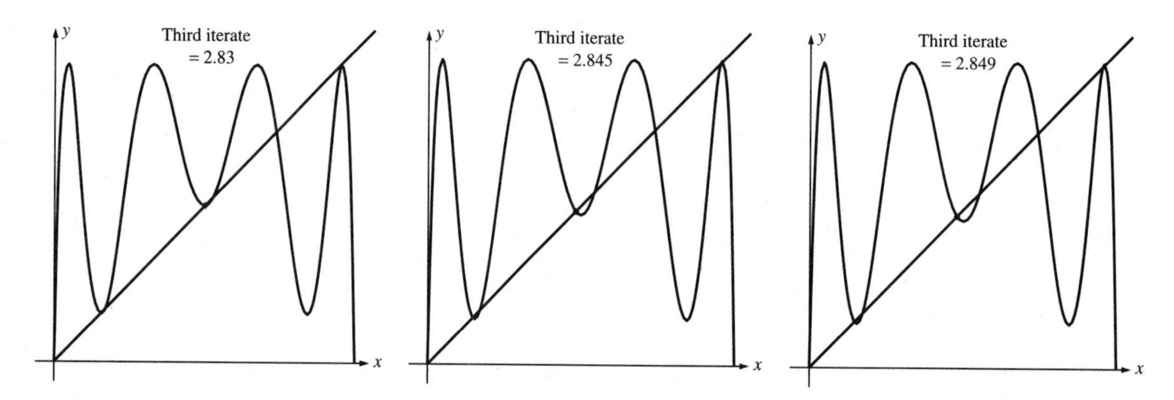

11. Two of the fixed points are $x = 0$ and $x = 1$, which are fixed points of the original map and not part of period 5 solutions. However, the other 10 fixed points represent two different period 5 orbits. Their stability can be determined by the slope of the function (fifth iterate) at the fixed point of the fifth iterate. For each point on a period 5 orbit, the slope must be the same, since $(d/dx)f(f(f(f(f(x))))) = \Pi f'(x_i)$. From the figure it can be seen that the positive slope must be greater than 1 and hence represents an unstable period 5 solution. The negative slope is more negative than -1, so that this period 5 orbit is also unstable. When the period 5 solution is first created, one solution will have a positive slope less than 1, and hence it will be stable. The period 5 solutions are created as a stable and unstable pair. However, this stable period 5 orbit becomes unstable by the time $\gamma = 2.9$.

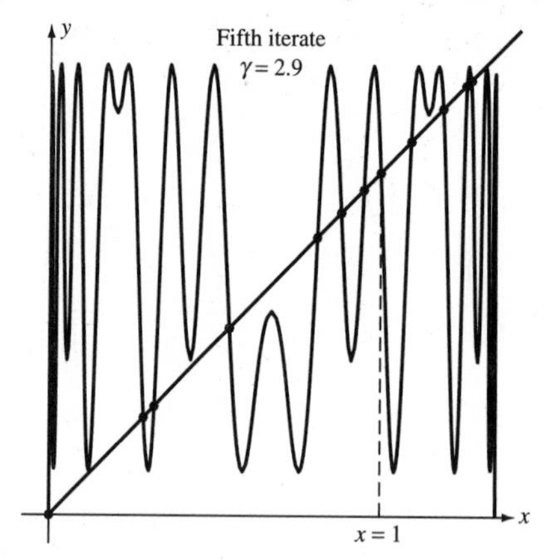

13. A period 3 orbit is stable if $|(d/dx)f(f(f(x)))| < 1$ and unstable if $|(d/dx)f(f(f(x)))| > 1$. It can be shown that $(d/dx)f(f(f(x))) = \Pi_{i=1}^{3} f'(x_i)$, so that stability depends on $\Pi_{i=1}^{3} |f'(x_i)|$.

15. The fixed point is stable if $r < 0$ [corresponding to $|f'(\bar{x})| < 1$] and unstable if $r > 0$ [corresponding to $|f'(\bar{x})| > 1$].

17. The fixed point is stable if $r < 0$ [corresponding to $\Pi_{i=1}^{n} |f'(\bar{x}_i)| < 1$] and unstable if $r > 0$ [corresponding to $\Pi_{i=1}^{n} |f'(\bar{x}_i)| > 1$].

Chapter 9

Section 9.2

1. a) $xy = c$ **b)** $x = y = 0$

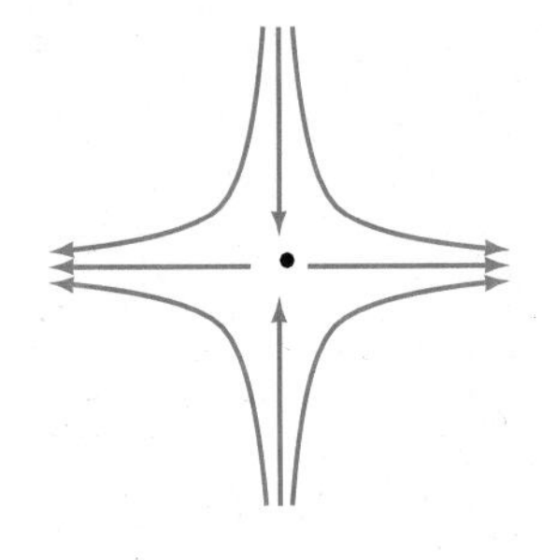

3. a) $\frac{y^2}{2} = x + c_1$ **b)** $x = 0$, $y = c_2$
(whole y axis is equilibrium points)

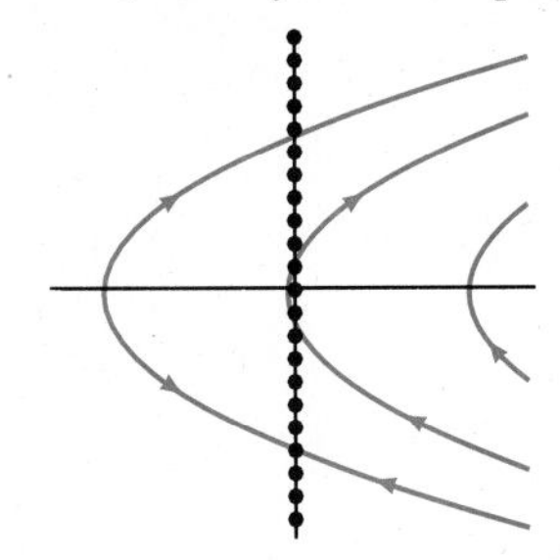

5. a) $y = \frac{1}{3}x + c$ **b)** no equilibria **c)** upward

7. $x^2 - 2x + y^2 - 2y = c$
(counterclockwise ellipses)

9. a) $y = cx^3$ **b)** y axis is equilibria

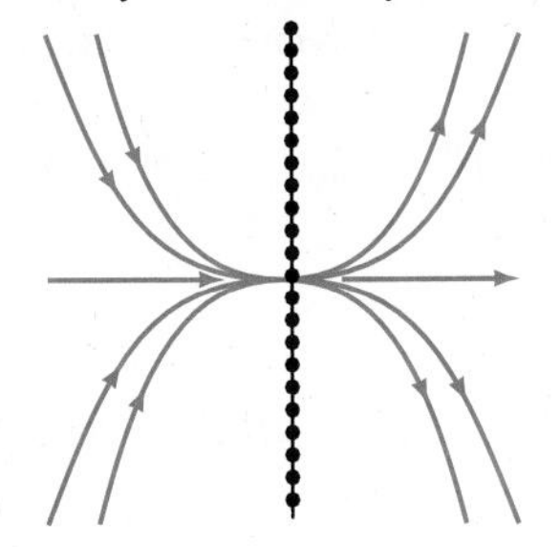

11. a) $yx = c$ **b)** y axis is equilibria

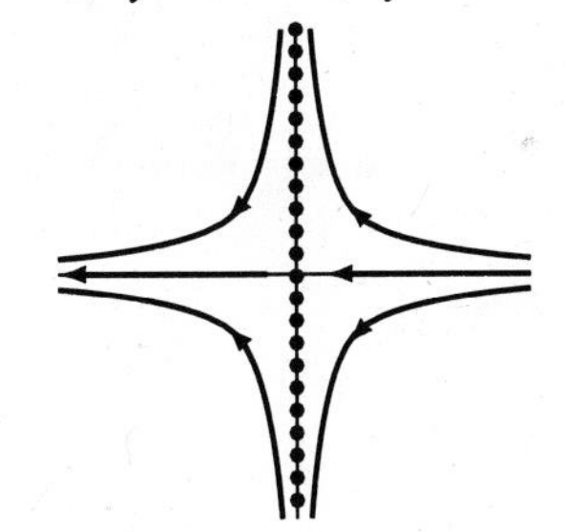

13. $y^2 = 2\cos x + c$

15. $y^2 = -2x^3/3 + c$

17. $y^2 = -\ln(1 + x^2) + c$

19. $y^2 = 2x^3/3 + c$

21. b) $y = 0$, $x = \pm 1$ **c)**

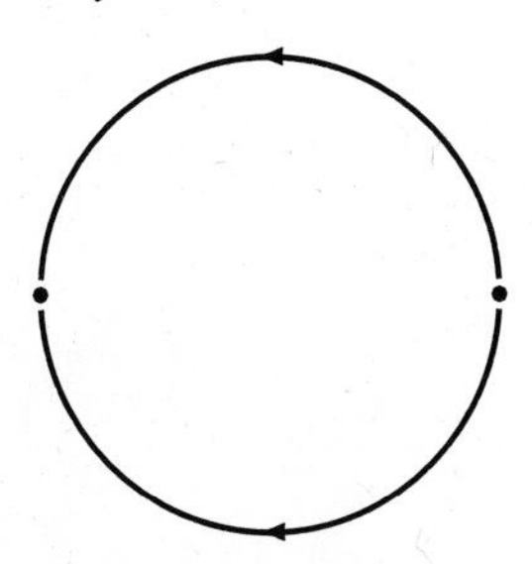

23. b) $y = 1$, $x = 0$ **c)**

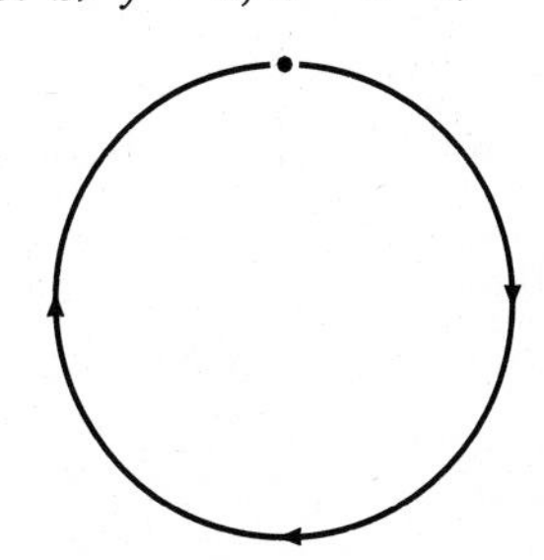

25. $(x(t_0), y(t_0))$ is a constant solution so that $x(t) \equiv x(t_0)$, $y(t) \equiv y(t_0)$ by Theorem 9.2.1.

27.

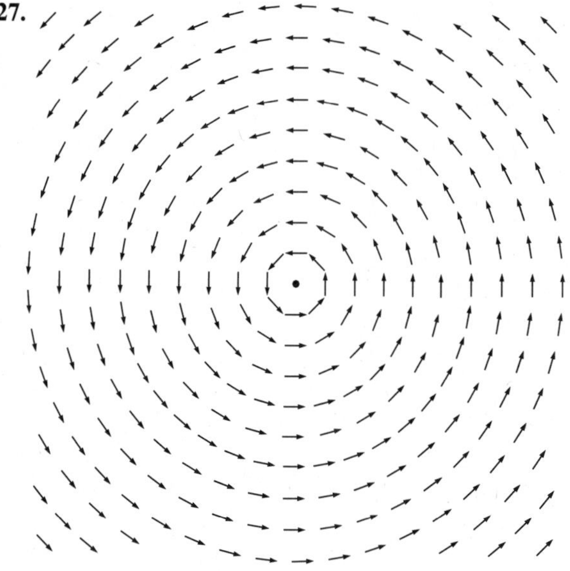

29.

Section 9.3

1. $\lambda = 1, 2$; unstable; repeller

3. $\lambda = \pm 2i$; stable; ellipses traversed counterclockwise

5. $\lambda = 1 \pm i$; unstable; outward spiral counterclockwise

7. $\lambda = -1, -3$; attractor (stable)

9. $\lambda = 2, -1$; saddle (unstable)

11. $\lambda = -2, -4$; attractor (stable)

13. $\lambda = -1 \pm 4i$; stable, inward spiral clockwise

15. $\lambda = 1, 7$; unstable; repeller

23. $\tan^{-1} \frac{y}{x} + \ln\left(\frac{y^2}{x^2} + 1\right) = -2 \ln x + c$

Section 9.4

1. $(0, 0)$, repeller; $(\frac{1}{2}, -1)$, saddle

3. $(0, 0)$, repeller; $(1, 1)$, saddle

5. $(0, 0)$, no info; $(-1, 1)$, saddle; $(1, 1)$, saddle

7. $(1, 1)$ and $(-1, -1)$ are saddles; $(1, -1)$ and $(-1, 1)$ pure imaginary roots $\Rightarrow$ circles equilibrium, long-run behavior unclear

9. $(0, 0)$, spiral repeller; $(-2, 2)$, saddle

11. $(0, 0)$, spiral attractor

13. $(0, 0)$, saddle; $(\frac{1}{8}, 0)$, attractor; $(1, -7)$, saddle

15. $(0, 0)$, saddle; $(-3, 0)$, attractor; $(1, \frac{4}{3})$, spiral repeller

17. $(0, 0)$, saddle; $(-\frac{1}{8}, 0)$, attractor; $(1, 9)$, repeller

In 25 through 27, $z = x - r$, $w = y - s$.

25. $z' = 3z + 4w + 2,$

$w' = z - 3w$

27. $z' = z + w,$

$w' = -z - w$

Section 9.5

The equilibrium, and simple closed invariant curves not containing the equilibrium are:

1. $(0,0)$; $x^2 + 4y^2 = c,\ c > 0$

3. $(1,2)$; $x^2 - 2x + 1 + 2y^2 - 8y = c,\ c > -8$

5. $(1,1)$; $(x-1)^4 + (y-1)^4 = c,\ c > 0$

7. $(0,0)$; $8x^2 - 8xy + 5y^2 = c,\ c > 0$

9. $(0,0)$; $2y^2 + x^4 = c,\ c > 0$

11. $(0,0)$; $4y^2 + 8y^4 + \frac{5}{4}x^4 = c,\ c > 0$

13. $(0,0)$; $3y^2 + \frac{3}{2}x^2 + \frac{5}{4}x^4 = c,\ c > 0$

15.

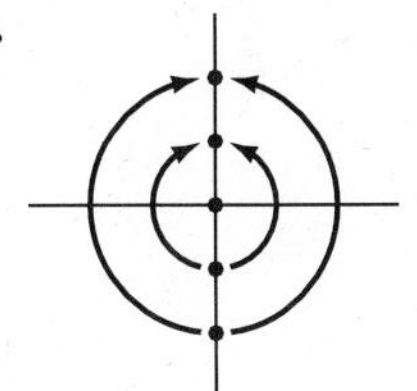

17.

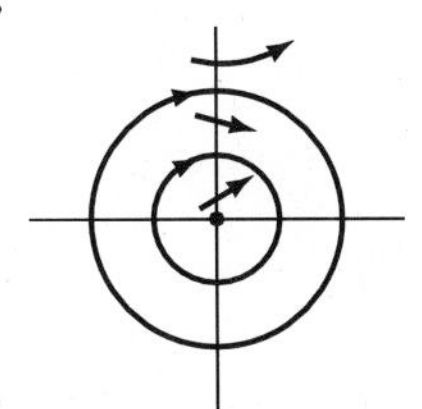

18. $r = \cos\theta + c$

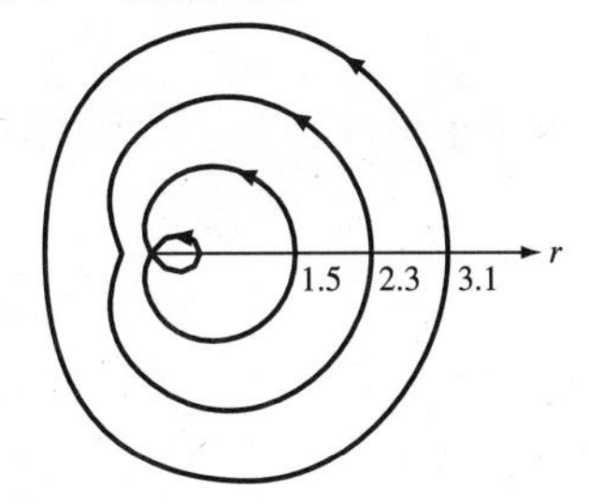

19. $r(t) = \cos\theta(t) + [r(0) - 1]$, $\theta = t$ (Note figure for Exercise 18.)

Section 9.6

1. $(0,0)$, $(0,1)$, $(1,0)$; are repeller, saddle, and test fails.

3. $(0,0)$, $(0,1)$, $(\frac{1}{4},0)$; are repeller, attractor, saddle.

4. Extinction of x. Equilibrium value of y is great enough so that the $-2xy$ term eventually dominates the growth term for x.

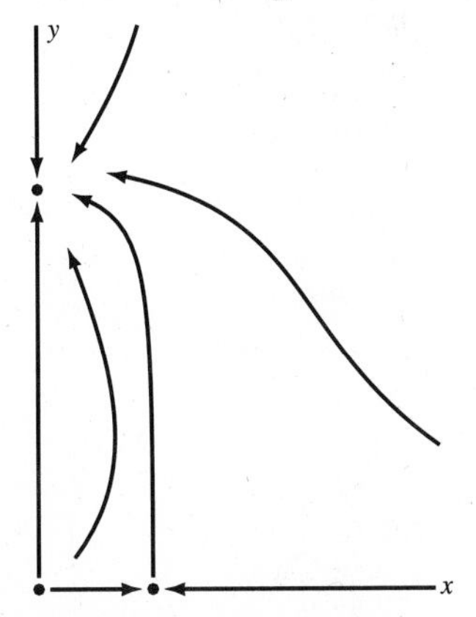

6.

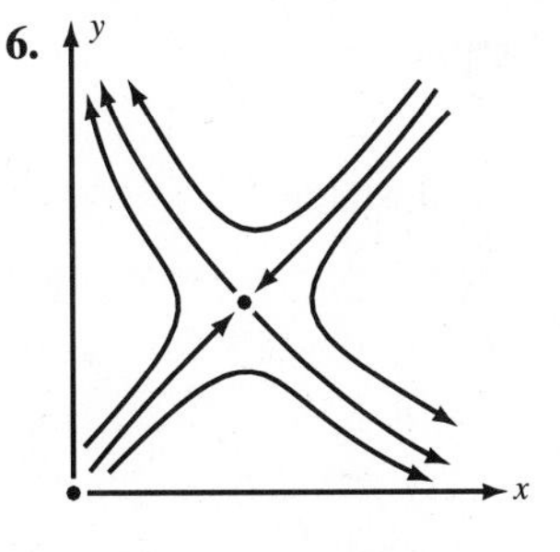

7. $(0,0)$, $(\frac{1}{2},0)$; are saddle, and attractor (predators die out).

9. $(0,0)$, is saddle. Invariant curves are $3\ln|y| + \ln|x| = y + x + c$. Apply Theorem 9.5.2. Periodic around $(1,3)$.

13. $(0,0)$ is a saddle. Use $y' = -qy + r\frac{x}{1+x}y \le (r-q)y$, $x' = ax - b\frac{x}{1+x}y \ge ax - by$.

15. $(0,0)$, $(1,2)$; are saddle, spiral repeller.

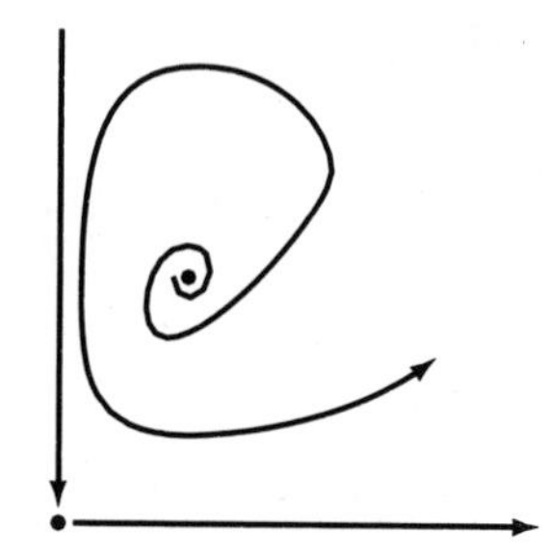

19. a, c. Velocity vectors (x', y') point into R from positive x and y axes.

21. $(0,0)$; test fails

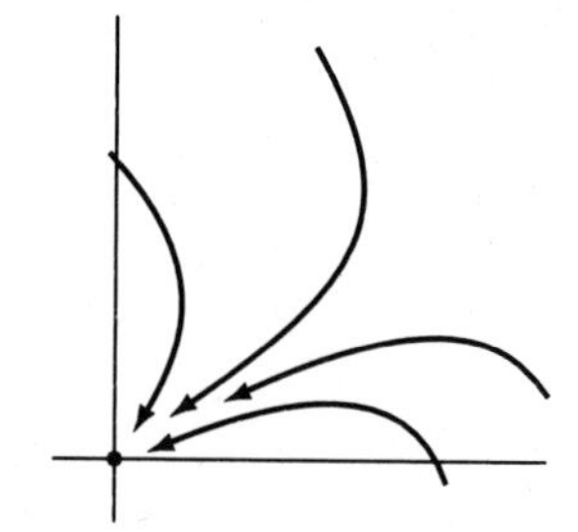

Section 9.7

1. $f'(0) = 2$, stable, attractor

3. $f'(0) = 1$, spiral attractor

5. $f'(0) = -1$, spiral repeller

7. $f'(0) = 4$, stable, attractor

9. $f'(0) = -4$, repeller

11. $f'(0) = 0$, $if(i) = 3i^4 + i^6 \ge 0$, stable

13. $f'(0) = 0$, $if(i) = i^6 \ge 0$, stable

15. $f'(0) = 0$, $if(i) \ge 0$, stable

17. $2i^2 + 2q - 1 = ce^{-2q}$

19. $18i^2 + 6q - 1 = ce^{-6q}$

23. a) $f(i) \le 0$ for $i \le 0$ and $f(i) \ge 0$ for $i \ge 0 \Rightarrow f'(0) \ge 0$.

b) $f(i) \ge 0$ for $i < 0$ and $f(i) \le 0$ for $i > 0 \Rightarrow f'(0) \le 0$.

c) $f'(0) > 0$, $f(0) = 0 \Rightarrow f(i) < 0$ for $i < 0$ (i near 0), $f(i) > 0$ for $i > 0$ (i near 0) $\Rightarrow if(i) \ge 0$.

d) $f'(0) < 0$, $f(0) = 0 \Rightarrow f(i) > 0$ for $i < 0$, $f(i) < 0$ for $i > 0$ (i near 0).

Section 9.8

1. $(0, 2n\pi)$ are spiral attractors, $(0, (2n+1)\pi)$ are saddles

3. $(0, 2n\pi)$ are attractors, $(0, (2n+1)\pi)$ are saddles

5. $\theta' = \phi,\ \phi' = [-mg \sin\theta - k\phi + rk]/ml$

6. $(\phi, \theta) = \left(0, \frac{\pi}{6} + 2n\pi\right)$ spiral attractor (n integer), $(\phi, \theta) = \left(0, \frac{5\pi}{6} + 2n\pi\right)$ unstable, saddle

7. $\phi = 0,\ \theta = \pi/2 + 2n\pi$, one zero eigenvalue, test fails

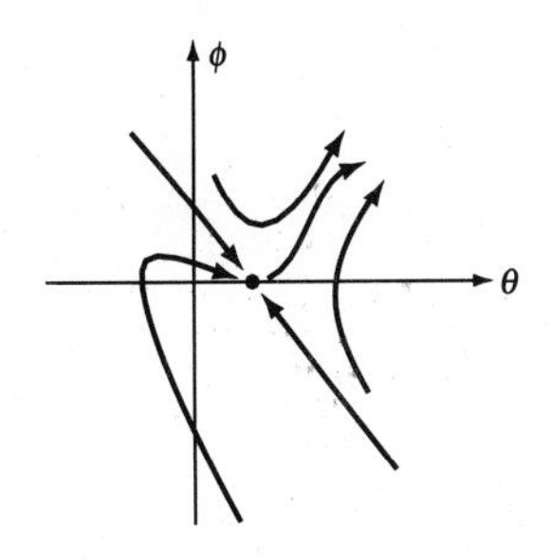

Chapter 10

Section 10.1

1. $\int_{-\pi}^{\pi} 1 \sin x\, dx = 0$

3. $\int_0^{\pi} 1 \cos x\, dx = 0$

5. $\int_0^{\pi} \cos 3x \cos x\, dx = \frac{1}{2}\int_0^{\pi}(\cos 4x + \cos 2x)\, dx = 0$

7. $\int_{-2}^{2}(1 + x)(x - x^2)\, dx = 0$

9. $\int_{-\pi}^{\pi} \sin nx \sin mx\, dx = \int_{-\pi}^{\pi} - \frac{1}{2}[\cos(n + m)x - \cos(n - m)x]\, dx = 0, n \neq m$

11. $\int_0^{\pi} \cos nx \cos mx\, dx = \int_0^{\pi} \frac{1}{2}[\cos(n + m)x + \cos(n - m)x]\, dx = 0, n \geq 0, n \neq m$

13. $\left(\text{Let } z = \frac{x - a}{b - a}\right)$. $\int_a^b \sin \frac{n\pi(x - a)}{b - a} \sin \frac{m\pi(x - a)}{b - a}\, dx = (b - a)\int_0^1 \sin n\pi z \sin m\pi z\, dz = 0$ if $n \neq m$

15. $\int_{-1}^{1} xx^3x^5\, dx = 0$

17. $\int_{-1}^{1} x(1 + x)(1 - x)\, dx = 0$

19. $(f + g, h) = \int_a^b[f(x) + g(x)]h(x)\, dx = \int_a^b f(x)h(x)\, dx + \int_a^b g(x)h(x)\, dx = (f, h) + (g, h)$

21. $\|f + g\|^2 = (f + g, f + g) = (f, f) + (f, g) + (g, f) + (g, g) = (f, f) + (g, g) = \|f\|^2 + \|g\|^2$

Section 10.2

1. Odd

3. Even

5. $f(-x) = (-x)^2 \sin(-x) = -f(x)$, odd

7. Even

9. Neither

11.

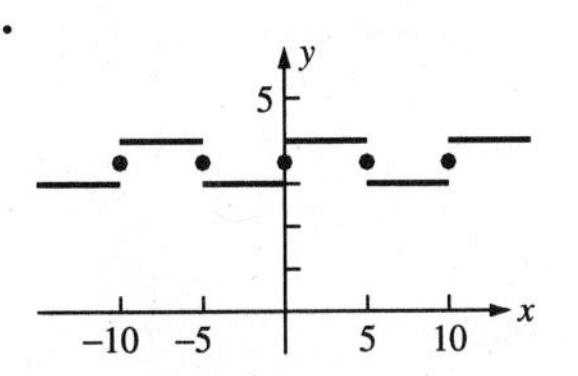

13.

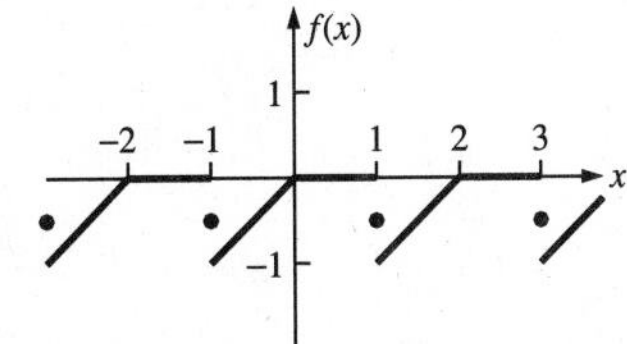

15.

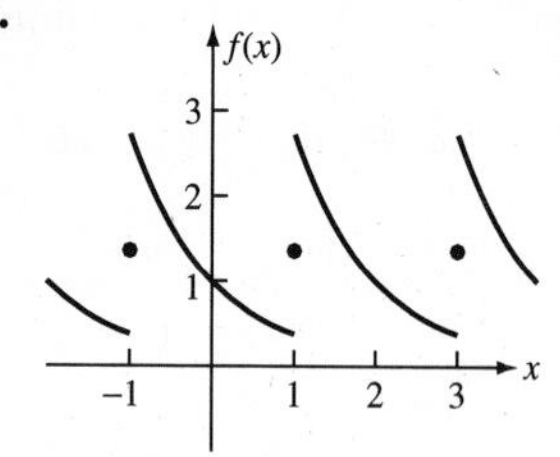

17.

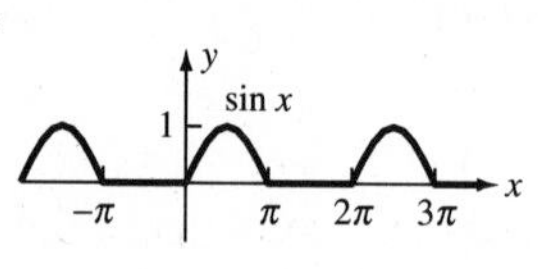

19. Theorem 10.2.1. $f(x)$ odd $\Rightarrow a_n = 0$. $b_n = \dfrac{1}{\pi}\displaystyle\int_{-\pi}^{\pi} f(x)\sin nx\,dx$
$= \dfrac{2}{\pi}\displaystyle\int_0^{\pi} f(x)\sin nx\,dx = \dfrac{2}{\pi}\left(\dfrac{-\cos n\pi}{n} + \dfrac{1}{n}\right)$
$= \dfrac{2}{\pi n}[-(-1)^n + 1]$ is 0 if n is even and $\dfrac{4}{n\pi}$ if n is odd. $f(x) = \displaystyle\sum_{i=0}^{\infty} \dfrac{4}{(2i+1)\pi}\sin(2i+1)x$.

21. Theorem 10.2.1. Even function $\Rightarrow b_n = 0$. $a_0 = \dfrac{2}{\pi}\displaystyle\int_0^{\pi/2}\cos x\,dx = \dfrac{2}{\pi}$. For $n \geq 1$,
$$a_n = \frac{4}{\pi}\int_0^{\pi/2}\cos x\cos 2nx\,dx = (-1)^{n+1}\frac{4}{\pi(4n^2-1)}.$$

23. $a_0 = \dfrac{1}{2}\displaystyle\int_{-1}^{1} e^x\,dx = \dfrac{e-e^{-1}}{2} = \sinh 1$. For $n \geq 1$, $a_n = \displaystyle\int_{-1}^{1} e^x\cos n\pi x\,dx = 2(-1)^n\dfrac{\sinh 1}{1+n^2\pi^2}$,
$b_n = \displaystyle\int_{-1}^{1} e^x\sin n\pi x\,dx = 2(-1)^{n+1}n\pi\dfrac{\sinh 1}{1+n^2\pi^2}$.

25. $a_0 = \dfrac{1}{\pi}\displaystyle\int_{-\pi/2}^{\pi/2}|\sin x|\,dx = \dfrac{2}{\pi}\int_0^{\pi/2}|\sin x|\,dx = \dfrac{2}{\pi}$. For $n \geq 1$, $a_n = \dfrac{4}{\pi}\displaystyle\int_0^{\pi/2}\sin x\cos 2nx\,dx$
$= \dfrac{4}{(1-4n^2)\pi}$, $b_n = 0$ (even).

27. $f(x) = \frac{1}{2} - \frac{1}{2}\cos 2x$

29. $a_0 = \dfrac{1}{4}\left[\displaystyle\int_{-2}^{0} 1\,dx + \int_0^2 2\,dx\right] = \dfrac{3}{2}$. For $n \geq 1$, $a_n = \dfrac{1}{2}\left[\displaystyle\int_{-2}^{0} 1\cos\dfrac{n\pi x}{2}\,dx + \int_0^2 2\cos\dfrac{n\pi x}{2}\,dx\right] = 0$,
$$b_n = \frac{1}{2}\left[\int_{-2}^{0} 1\sin\frac{n\pi x}{2}\,dx + \int_0^2 2\sin\frac{n\pi x}{2}\,dx\right] = \frac{1}{n\pi}(1-\cos n\pi) = \begin{cases} 0 & \text{if } n \text{ even,} \\ \dfrac{2}{n\pi} & \text{if } n \text{ odd.}\end{cases}$$

31. $a_0 = \dfrac{1}{2\pi}\displaystyle\int_0^{\pi} x^2\,dx = \dfrac{\pi^2}{6}$. For $n \geq 1$, $a_n = \dfrac{1}{\pi}\displaystyle\int_0^{\pi} x^2\cos nx\,dx = (-1)^n 2n^{-2}$, $b_n = \dfrac{1}{\pi}\displaystyle\int_0^{\pi} x^2\sin nx\,dx$
$= \dfrac{1}{\pi n^3}[2(-1)^n - n^2\pi^2(-1)^n - 2]$.

33. On $[-6,6]$, $f(x) = \begin{cases} 1 & \text{if } 1 < x < 2, \\ 1 & \text{if } -5 < x < -4, \\ 0 & \text{otherwise.}\end{cases}$ $a_0 = \dfrac{1}{12}\left[\displaystyle\int_{-5}^{-4} 1\,dx + \int_1^2 1\,dx\right] = \dfrac{1}{6}$. For $n \geq 1$,
$$a_n = \frac{1}{6}\left[\int_{-5}^{-4} 1\cos\frac{n\pi x}{6}\,dx + \int_1^2 1\cos\frac{n\pi x}{6}\,dx\right] = \frac{1}{n\pi}\left[\sin\frac{5}{6}n\pi + \sin\frac{1}{3}n\pi - \sin\frac{2}{3}n\pi - \sin\frac{1}{6}n\pi\right],$$
$$b_n = \frac{1}{6}\left[\int_{-5}^{-4} 1\sin\frac{n\pi x}{6}\,dx + \int_1^2 1\sin\frac{n\pi x}{6}\,dx\right]$$
$$= -\frac{1}{n\pi}\left[\cos\frac{1}{3}n\pi + \cos\frac{2}{3}n\pi - \cos\frac{5}{6}n\pi - \cos\frac{1}{6}n\pi\right]$$

35. $(fg)(-x) = f(-x)g(-x) = f(x)g(x) = (fg)(x)$.
$(f+g)(-x) = f(-x) + g(-x) = f(x) + g(x) = (f+g)(x)$.

37. b) $f_1(-x) = \frac{1}{2}[f(-x) + f(--x)] = \frac{1}{2}[f(-x) + f(x)] = f_1(x)$.

c) $f_2(-x) = \frac{1}{2}[f(-x) - f(--x)] = \frac{1}{2}[f(-x) - f(x)] = -f_2(x)$.

39. $f(x)$ is continuous everywhere. Discontinuities of $f'(x) = \frac{1}{3}(\sin x)^{-2/3}\cos x$ at $x = n\pi$ are not jump discontinuities.

41. $\int_{-L}^{L} g(x)\,dx = \int_0^L g(x)\,dx + \int_{-L}^{0} g(x)\,dx = \int_L^0 g(-z)\,d(-z) + \int_0^L g(x)\,dx = --\int_0^L g(z)\,dz + \int_0^L g(x)\,dx = 2\int_0^L g(x)\,dx$.

43. $g(x) = f(x)\cos\dfrac{n\pi x}{L}$ is periodic with period $2L$. Let $a = 2mL + b$ with m an integer and $0 \le b \le L$.

Then $\displaystyle\int_{a-L}^{a+L} g(x)\,dx = (z = x - 2mL);\ \int_{b-L}^{b+L} g(z + 2mL)\,dz = \int_{b-L}^{b+L} g(z)\,dz =$

$\displaystyle\int_{b-L}^{L} g(z)\,dz + \int_{L}^{b+L} g(z)\,dz = (w = z - 2L);\ \int_{b-L}^{L} g(z)\,dz + \int_{-L}^{b-L} g(w + 2L)\,dw$

$\displaystyle= \int_{b-L}^{L} g(z)\,dz + \int_{-L}^{b-L} g(w)\,dw = \int_{-L}^{L} g(z)\,dz.$

45. a) $P_0(x) = 0$

b)

c) $P_2(x) = P_1(x)$; see part (b)

d)

e)

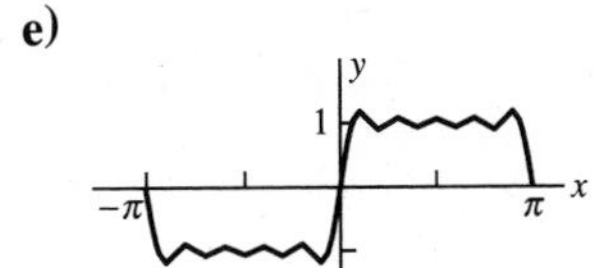

47. a) $P_0(x) = \frac{1}{6}$ constant

b)

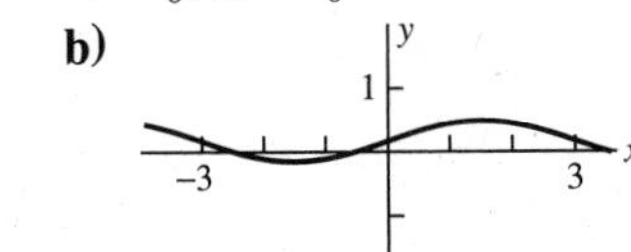

c)

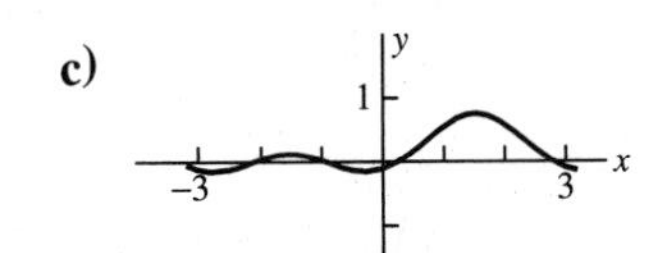

d)

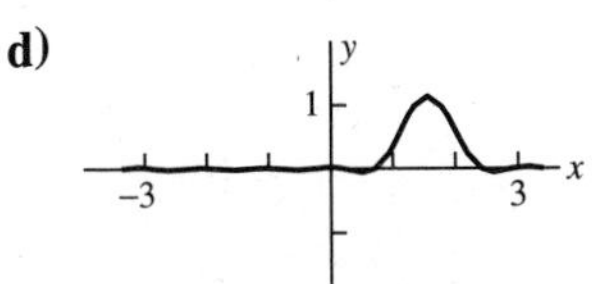

e)

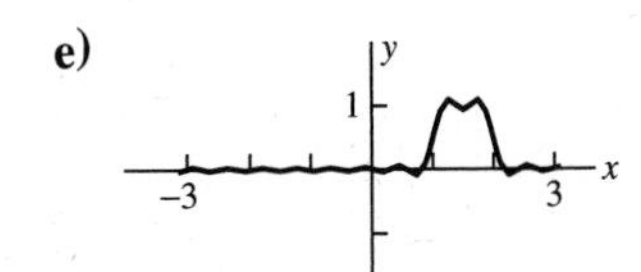

49. a) $P_0(x) = \frac{1}{4}$

b)

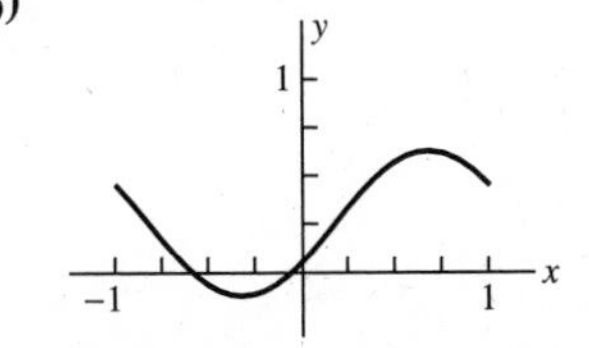

c)

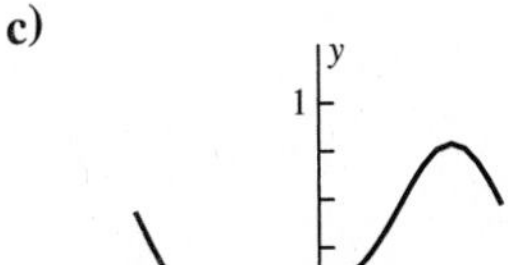
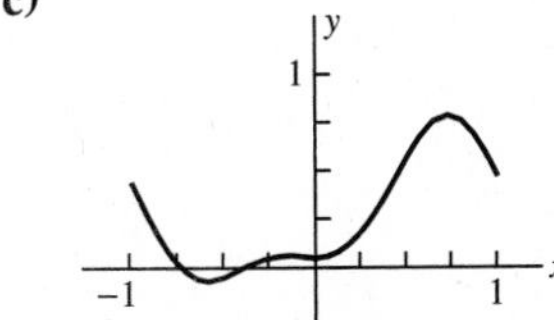

d)

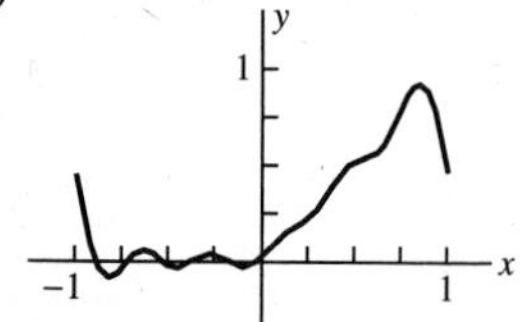

e)

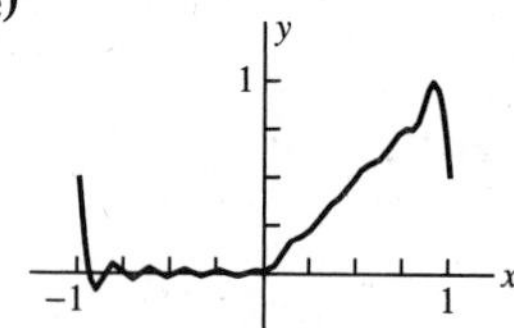

Section 10.3

1.

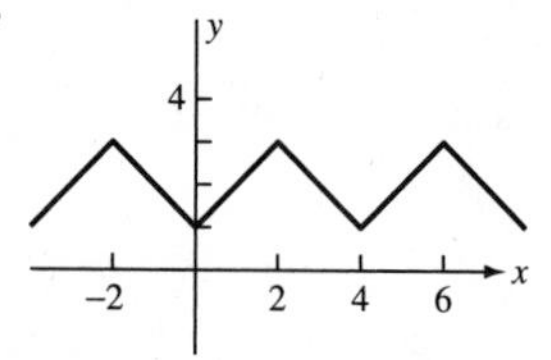

3.

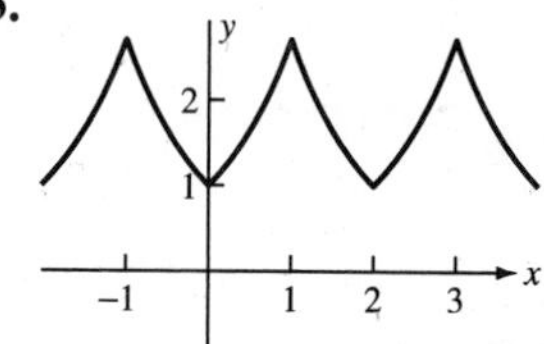

5.

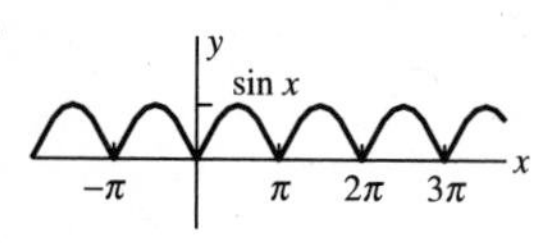

7.

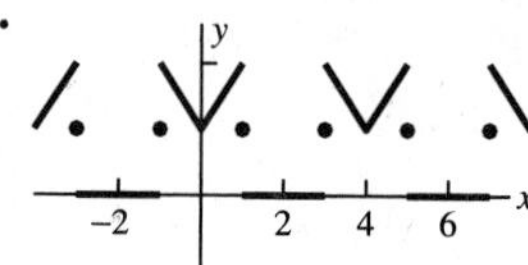

9.

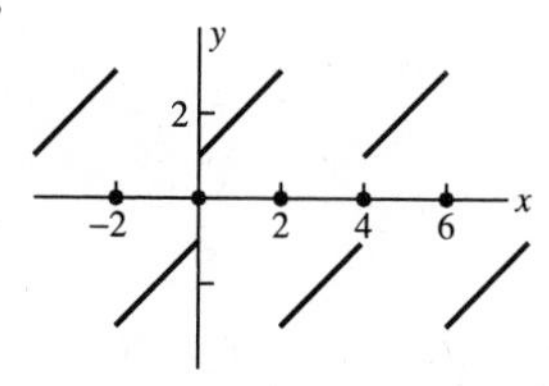

11.

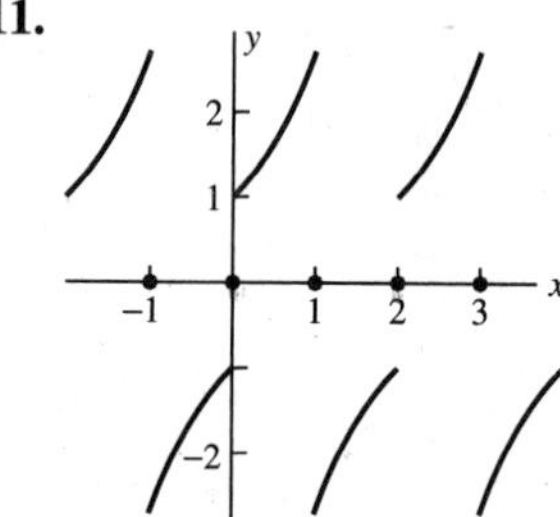

13.

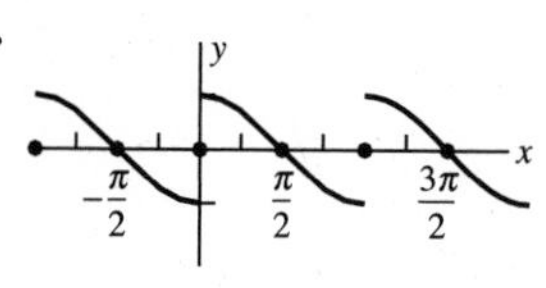

15.

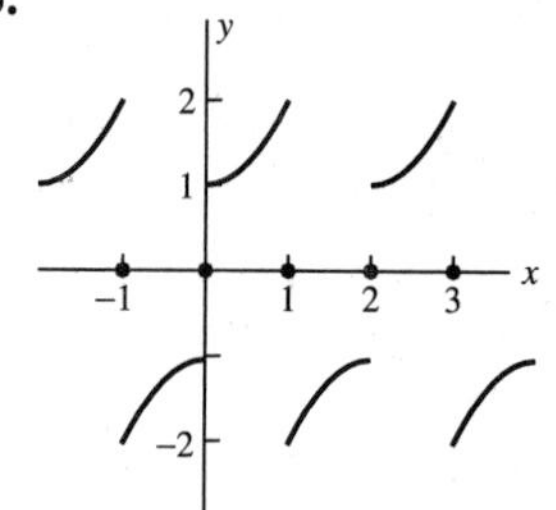

17. $a_0 = 1,\ a_n = 2\int_0^1 \cos n\pi x\,dx = 0;\ f(x) = 1$

19. $a_1 = 1,\ a_n = 0\ n \neq 1\ f(x) = \cos x$

21. $a_3 = 1,\ a_n = 0\ n \neq 3\ f(x) = \cos 3x$

23. $a_0 = \frac{1}{2},\ a_n = \int_1^2 \cos\frac{n\pi x}{2}\,dx = -\frac{2}{n\pi}\sin\frac{n\pi}{2};\ \text{f}(1) = \frac{1}{2}$

25. $a_0 = \frac{1}{2}\int_1^2 x\,dx = \frac{3}{4},\ a_n = \int_1^2 x\cos\frac{n\pi x}{2}\,dx = \frac{2}{n^2\pi^2}\left[2(-1)^n - 2\cos\frac{n\pi}{2} - n\pi\sin\frac{n\pi}{2}\right];\ f(1) = \frac{1}{2}$

27. $a_0 = \frac{1}{4}$, $a_n = \frac{1}{2}\int_0^1 \cos\frac{n\pi x}{4}\,dx = \frac{2}{n\pi}\sin\frac{n\pi}{4}$; $f(1) = \frac{1}{2}$

29. $a_0 = \frac{1}{3}$, $a_n = \frac{2}{3}\int_1^2 \cos\frac{n\pi x}{3}\,dx = \frac{2}{n\pi}\left(\sin\frac{2n\pi}{3} - \sin\frac{n\pi}{3}\right)$; $f(1) = f(2) = \frac{1}{2}$

31. $b_1 = 1$, $b_n = 0$ $n \neq 1$; $f(x) = \sin x$

33. $b_3 = 1$, $b_n = 0$ $n \neq 3$; $f(x) = \sin 3x$

35. $b_n = \int_0^1 \sin\frac{n\pi x}{2}\,dx = \frac{2}{n\pi}\left(1 - \cos\frac{n\pi}{2}\right)$; $f(1) = \frac{1}{2}$, $f(0) = 0$

37. $b_n = \int_0^1 x\sin\frac{n\pi x}{2}\,dx = -\frac{2}{n^2\pi^2}\left(-2\sin\frac{n\pi}{2} + n\pi\cos\frac{n\pi}{2}\right)$; $f(1) = \frac{1}{2}$

39. $b_n = \int_0^2 (1 - x)\sin\frac{n\pi x}{2}\,dx = \frac{2}{n\pi}[(-1)^n + 1]$; $f(0) = f(2) = 0$

41. $b_n = \frac{1}{2}\int_3^4 \sin\frac{n\pi x}{4}\,dx = \frac{2}{n\pi}\left[\cos\frac{3n\pi}{4} - (-1)^n\right]$; $f(4) = 0$, $f(3) = \frac{1}{2}$

43. For an odd extension, $f(L^+) = f(-L^-)$ is needed for continuity on $(-\infty, \infty)$, but then $f(-L) = -f(L)$ implies $f(L) = -f(L)$ and $f(L) = 0$. Similarly for $f(0) = 0$.

45. If $x > 0$, then $\tilde{f}(-x) = f(x) = \tilde{f}(x)$. If $x < 0$, then $\tilde{f}(-x) = f(-x) = \tilde{f}(x)$.

47. Use change of variables $z = \frac{x - a}{b - a}$ to give an integral on $[0, 1]$.

Section 10.4

1. $y_p(t) = -\sum_{n=1}^{\infty} \frac{1}{n^3 + n}\sin nt$

3. $y_p(t) = -\sum_{n=1}^{\infty} \frac{1}{(n + 1)(4n^2 - 1)}\cos 2nt$

5. $y_p(t) = \frac{1}{3}\sin t - \frac{1}{8}t\cos 2t - \sum_{n=3}^{\infty} \frac{2}{n(n^2 - 4)}\sin nt$

7. $y_p(t) = -t\cos t - \sum_{n=2}^{\infty} \frac{2}{n^2(n^2 - 1)}\sin nt$

9. $y_p(t) = \sum_{n=1}^{\infty} \frac{1 - 16n^2}{nR_n}\sin 4nt - \frac{8}{R_n}\cos 4nt$,

$R_n = (1 - 16n^2)^2 + 64n^2$

11. $L = \frac{n\pi}{\sqrt{12}}$

13. $b = 3n\pi$, $n \geq 1$; $\left\{\frac{3n}{2} : n \geq 1\right\}$

15. $b = n\pi$, $b = \frac{n\pi}{\sqrt{3}}$, $n \geq 1$;

$\left\{\frac{n}{2} : n \geq 1\right\} \cup \left\{\frac{n}{2\sqrt{3}} : n \geq 1\right\}$

17. $\frac{n\pi}{6} < b < \frac{n\pi}{5}$

19. $\frac{n\pi}{2} <$ period $< \frac{n\pi}{1}$

Section 10.5

1. $\lambda = (n + \frac{1}{2})^2$, $n \geq 0$; $y_n(x) = \sin(n + \frac{1}{2})x$

3. $\lambda = n^2$, $n \geq 0$; $y_n(x) = \cos nx$

5. $\lambda + 3 = n^2(\lambda = n^2 - 3)$, $n \geq 1$; $y_n(x) = \sin nx$

7. $\frac{\lambda}{4} = \left(n + \frac{1}{2}\right)^2$ or $\lambda = (2n + 1)^2$,

$n \geq 0$; $y_n(x) = \sin\left(n + \frac{1}{2}\right)x$

9. $\frac{\lambda}{4} = n^2$, $y_n(x) = \sin\frac{n}{2}x$, $n \geq 1$

11. $y_n = \sin\sqrt{\lambda_n}x$. $\sqrt{\lambda_n}$ from graph:

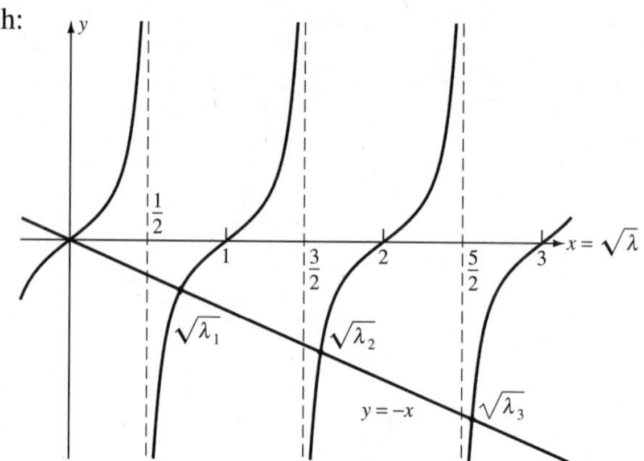

13. $c_n = \dfrac{\int_0^\pi f(x)\sin\left(n+\frac{1}{2}\right)x\,dx}{\int_0^\pi \sin^2\left(n+\dfrac{1}{2}\right)x\,dx} = \dfrac{2}{\pi}\displaystyle\int_0^\pi f(x)\sin\left(n+\frac{1}{2}\right)x\,dx,\ n \geq 0$

15. $c_n = \dfrac{\int_0^\pi f(x)\cos nx\,dx}{\int_0^\pi \cos^2 nx\,dx}$, $c_0 = \dfrac{1}{\pi}\displaystyle\int_0^\pi f(x)\,dx$, $c_n = \dfrac{2}{\pi}\displaystyle\int_0^\pi f(x)\cos nx\,dx,\ n \geq 1$

17. $c_n = \dfrac{\int_0^\pi f(x)\sin\sqrt{\lambda_n}x\,dx}{\int_0^\pi \sin^2\sqrt{\lambda_n}x\,dx}$

Chapter 11

Section 11.2

1. $\lambda = n^2,\ n \geq 1;\ \phi_n(x) = \sin nx$

3. $\lambda = \left(\dfrac{n\pi}{H}\right)^2,\ n \geq 1;\ \phi_n(x) = \sin\dfrac{n\pi x}{H}$

5. $\lambda = n^2\pi^2,\ n \geq 0;\ \phi_n(x) = \cos n\pi x$

7. $\lambda = \left(n+\dfrac{1}{2}\right)^2\left(\dfrac{\pi}{L}\right)^2,\ n \geq 0;$

$$\phi_n(x) = \cos\left(n+\frac{1}{2}\right)\frac{\pi x}{L} = \cos\frac{(2n+1)\pi x}{2L}$$

9. $\lambda = n^2,\ n \geq 0;\ \phi_0(x) = 1,\ \phi_n(x) = \sin nx$ and $\cos nx$ $(n \geq 1)$

13. $u(x,t) = 5\sin\dfrac{\pi x}{L}e^{-k(\pi/L)^2 t} + 8\sin\dfrac{3\pi x}{L}e^{-k(3\pi/L)^2 t}$

15. $u(x,t) = \displaystyle\sum_{n=1}^{\infty} B_n \sin\frac{n\pi x}{L}e^{-k(n\pi/L)^2 t}$, where $B_n = \dfrac{2}{L}\displaystyle\int_0^L 5\cos\frac{\pi x}{L}\sin\frac{n\pi x}{L}\,dx$. By symmetry around $x = \dfrac{L}{2}$ or evaluating these integrals, $B_n = 0$ (for n odd) and $B_n = \dfrac{1}{\pi}\left(\dfrac{20n}{n^2-1}\right)$ (for n even).

$$\int_0^L \cos\frac{\pi x}{L}\sin\frac{n\pi x}{L}\,dx = -\frac{L}{2}\,\frac{\cos\pi\frac{1+n}{L}x}{\pi(1+n)}\Bigg|_0^L - \frac{L}{2}\,\frac{\cos\pi\frac{-1+n}{L}x}{\pi(-1+n)}\Bigg|_0^L$$

17. $u(x,t) = 1$

19. $u(x,t) = \sum_{n=0}^{\infty} A_n \cos \frac{n\pi x}{L} e^{-k(n\pi/L)^2 t}$, where $A_0 = \frac{2}{\pi}$, and $(n \geq 1)$ $A_n = \frac{2}{L}\int_0^L \sin \frac{\pi x}{L} \cos \frac{n\pi x}{L}\, dx$. By symmetry around $x = \frac{L}{2}$ or evaluating these integrals, $A_n = 0$ (for n odd) and $A_n = -\frac{1}{\pi}\left(\frac{4}{n^2 - 1}\right)$ (for n even).

21. $u(x,t) = e^{2t} \sum_{n=1}^{\infty} B_n \sin \frac{n\pi x}{L} e^{-k(n\pi/L)^2 t}$, where $B_n = \frac{2}{L}\int_0^L f(x) \sin \frac{n\pi x}{L}\, dx$.

23. **a)** $u_0(x) = 5 + \frac{3}{L}x$

b) $\frac{\partial v}{\partial t} = k\frac{\partial^2 v}{\partial x^2}$ with $v(0, t) = 0$ and $v(L, t) = 0$ and the initial condition $v(x, 0) = f(x) - \left(5 + \frac{3}{L}x\right) \equiv g(x)$

c) $v(x, t) = \sum_{n=1}^{\infty} B_n \sin \frac{n\pi x}{L} e^{-k(n\pi/L)^2 t}$, where $B_n = \frac{2}{L}\int_0^L g(x) \sin \frac{n\pi x}{L}\, dx$. $u(x, t) = 5 + \frac{3}{L}x + v(x,t) =$
$5 + \frac{3}{L}x + \sum_{n=1}^{\infty} B_n \sin \frac{n\pi x}{L} e^{-k(n\pi/L)^2 t}$, where $B_n = \frac{2}{L}\int_0^L \left[f(x) - \left(5 + \frac{3}{L}x\right)\right] \sin \frac{n\pi x}{L}\, dx$

Section 11.3

1. $u(x, y) = A_0(x - L) + \sum_{n=1}^{\infty} A_n \cos \frac{n\pi y}{H} \sinh \frac{n\pi(x-L)}{H}$, where $A_0 = -\frac{1}{LH}\int_0^H f(y)\, dy$ and

$(n \geq 1)$ $A_n = -\frac{2}{H \sinh \frac{n\pi L}{H}} \int_0^H f(y) \cos \frac{n\pi y}{H}\, dy$

3. $u(x, y) = A_0 + \sum_{n=1}^{\infty} A_n \cos \frac{n\pi y}{H} \cosh \frac{n\pi x}{H}$, where $A_0 = \frac{1}{H}\int_0^H f(y)\, dy$ and

$(n \geq 1) A_n = \frac{2}{H \cosh \frac{n\pi L}{H}} \int_0^H f(y) \cos \frac{n\pi y}{H}\, dy$

5. $u(x, y) = A_0 x + \sum_{n=1}^{\infty} A_n \cos \frac{n\pi y}{H} \sinh \frac{n\pi x}{H}$, where $A_0 = \frac{1}{H}\int_0^H f(y)\, dy$ and

$(n \geq 1)$ $A_n = \frac{2}{n\pi \cosh \frac{n\pi L}{H}} \int_0^H f(y) \cos \frac{n\pi y}{H}\, dy$

7. $u(x, y) = A_0 + \sum_{n=1}^{\infty} A_n \cos \frac{n\pi x}{L} \cosh \frac{n\pi(y - H)}{L}$, where $0 = \frac{1}{L}\int_0^L f(x)\, dx$ and

$(n \geq 1)$ $A_n = -\frac{2}{n\pi \sinh \frac{n\pi L}{H}} \int_0^L f(x) \cos \frac{n\pi x}{L}\, dx$. [No solutions if $\int_0^L f(x)\, dx \neq 0$, whereas if $\int_0^L f(x)\, dx = 0$, then A_0 is arbitrary and there is an arbitrary additive constant to the solution]

9. $u(x,y) = \sum_{n=1}^{\infty} B_n \sin \frac{n\pi y}{H} \sinh \frac{n\pi x}{H}$, where $B_n = \dfrac{2}{H \sinh \frac{n\pi L}{H}} \int_0^H f(y) \sin \frac{n\pi y}{H}\, dy$

11. $u(x,y) = \sum_{n=1}^{\infty} B_n \sin \frac{n\pi x}{L} \sinh \frac{n\pi (y-H)}{L}$, where $B_n = -\dfrac{2}{L \sinh \frac{n\pi H}{L}} \int_0^L f(x) \sin \frac{n\pi x}{L}\, dx$

13. $u(x,y) = \sum_{n=1}^{\infty} B_n \sin \frac{n\pi x}{L} \cosh \frac{n\pi y}{L}$, where $B_n = \dfrac{2}{L \cosh \frac{n\pi H}{L}} \int_0^L f(x) \sin \frac{n\pi x}{L}\, dx$

15. $u(x,y) = \sum_{n=1}^{\infty} B_n \sin \frac{n\pi x}{L} \sinh \frac{n\pi y}{L}$, where $B_n = \dfrac{2}{n\pi \cosh \frac{n\pi H}{L}} \int_0^L f(x) \sin \frac{n\pi x}{L}\, dx$

17. $u(x,y) = \sum_{n=1}^{\infty} B_n \sin \frac{(n-\frac{1}{2})\pi x}{L} \sinh \frac{(n-\frac{1}{2})\pi y}{L}$,
where $B_n = \dfrac{2}{L \sinh \frac{(n-\frac{1}{2})\pi H}{L}} \int_0^L f(x) \sin \frac{(n-\frac{1}{2})\pi x}{L}\, dx$

19. $u(x,y) = A_0(y-H) + B_0 y + \sum_{n=1}^{\infty} \cos \frac{n\pi x}{L}\left(A_n \sinh \frac{n\pi(y-H)}{L} + B_n \sinh \frac{n\pi y}{L}\right)$, where $B_0 = \frac{1}{LH}\int_0^L g(x)\, dx$, $A_0 = -\frac{1}{LH}\int_0^L f(x)\, dx$, $(n \ge 1)$ $A_n = -\dfrac{2}{L \sinh \frac{n\pi H}{L}} \int_0^L f(x) \cos \frac{n\pi x}{L}\, dx$, and $(n \ge 1) B_n = \dfrac{2}{L \sinh \frac{n\pi H}{L}} \int_0^L g(x) \cos \frac{n\pi x}{L}\, dx$

21. $u(x,y) = \sum_{n=1}^{\infty} A_n \sin \frac{n\pi y}{H} \sinh \frac{n\pi x}{H} + \sum_{n=1}^{\infty} B_n \sin \frac{n\pi x}{L} \sinh \frac{n\pi(y-H)}{L}$,
where $A_n = \dfrac{2}{H \sinh \frac{n\pi L}{H}} \int_0^H g(y) \sin \frac{n\pi y}{H}\, dy$ and $B_n = -\dfrac{2}{L \sinh \frac{n\pi H}{L}} \int_0^L f(x) \sin \frac{n\pi x}{L}\, dx$

23. $u(x,y) = A_0 x + \sum_{n=1}^{\infty} A_n \cos \frac{n\pi y}{H} \sinh \frac{n\pi x}{H} + \sum_{n=1}^{\infty} B_n \sin \frac{n\pi x}{L} \cosh \frac{n\pi(y-H)}{L}$,
where $A_0 = \frac{1}{LH}\int_0^H f(y)\, dy$, $(n \ge 1) A_n = \dfrac{2}{H \sinh \frac{n\pi L}{H}} \int_0^H f(y) \cos \frac{n\pi y}{H}\, dy$,
and $B_n = -\dfrac{2}{n\pi \sinh \frac{n\pi H}{L}} \int_0^L g(x) \sin \frac{n\pi x}{L}\, dx$

25. $u(x,y) = \sum_{n=1}^{\infty} B_n \sin \frac{n\pi x}{L} e^{-n\pi y/L}$, where $B_n = -\frac{2}{n\pi}\int_0^L f(x) \sin \frac{n\pi x}{L}\, dx$

Section 11.4

1. $u(x,t) = \sum_{n=1}^{\infty} A_n \sin \frac{n\pi x}{L} \cos \frac{n\pi ct}{L}$, where $A_n = \frac{2}{L}\int_0^L f(x) \sin \frac{n\pi x}{L}\, dx$

3. $u(x,t) = A_0 + \sum_{n=1}^{\infty} A_n \cos\frac{n\pi x}{L}\cos\frac{n\pi ct}{L}$, where $A_0 = \frac{1}{L}\int_0^L f(x)\,dx$ and $(n \geq 1)$ $A_n = \frac{2}{L}\int_0^L f(x)\cos\frac{n\pi x}{L}\,dx$

5. Frequencies $(n \geq 1)$ are $\frac{(n-\frac{1}{2})c}{2L} = \frac{(2n-1)c}{4L}$

7. $u(x,t) = \sum_{n=1}^{\infty} A_n \sin\frac{n\pi x}{L}\cos\sqrt{k + \left(\frac{n\pi c}{L}\right)^2}\,t$, where $A_n = \frac{2}{L}\int_0^L f(x)\sin\frac{n\pi x}{L}\,dx$

9. $u(x,t) = \frac{1}{2}f(x-ct) + \frac{1}{2}f(x+ct)$

11. a) $f(x)$ symmetric about $\pi/2$. $A_n = 0$ for n even since $\sin nx$ symmetric about $\pi/2$ if n even. For n odd, $A_n = \frac{2}{\pi}\int_0^{\pi} |\sin 2x|\sin nx dx = \frac{-2}{\pi}\left[\frac{4}{n^2-4}\right]\sin\frac{n\pi}{2}$.

b) Frequencies $= \frac{nc}{2\pi}$: n odd.

13. a) For $n = 2$, $A_2 = \frac{2}{\pi}\int_0^{\pi/2}\sin 2x\sin 2x dx = \frac{1}{2}$. For $n \neq 2$, $A_n = \frac{2}{\pi}\int_0^{\pi/2}\sin 2x \sin nx dx$ $= \frac{1}{\pi}\left[\frac{\sin(n-2)x}{n-2} - \frac{\sin(n+2)x}{n+2}\right]\Big|_{x=0}^{\pi/2} = \frac{-4}{\pi(n^2-4)}\sin\frac{n\pi}{2}$ if n is odd. $A_n = 0$ if $n > 2$ and even.

b) Frequencies $= \frac{nc}{2\pi}$ for n odd and $\frac{c}{\pi}$.

15. a) $A_3 = \frac{2}{\pi}\int_0^{\pi/3}\sin^2(3x)dx = \frac{1}{3}$. For $n \neq 3$, $A_n = \frac{2}{\pi}\int_0^{\pi/3}\sin 3x\sin nx dx$ $= \frac{1}{\pi}\left[\frac{\sin(n-3)x}{n-3} - \frac{\sin(n+3)x}{n+3}\right]\Big|_{x=0}^{\pi/3} = \frac{-6}{\pi(n^2-9)}\sin\frac{n\pi}{3}$. $A_n = 0$ if $n \neq 3$ and n a multiple of 3.

b) Frequencies $= \frac{nc}{2\pi}$, $n \neq 6, 9, 12, \ldots$.

17. a) $A_3 = \int_{\pi/3}^{2\pi/3}\sin^2(3x)dx = \frac{1}{3}$. For $n \neq 3$, $A_n = \frac{2}{\pi}\int_{\pi/3}^{2\pi/3}\sin 3x\sin nx dx = \frac{6}{\pi(n^2-9)}\left(\sin\frac{2n\pi}{3} + \sin\frac{n\pi}{3}\right)$. $A_n = 0$ if $n \neq 3$ and n is a multiple of 3.

b) Frequencies $= \frac{nc}{2\pi}$, $n \neq 6, 9, 12, \ldots$.

Appendix A

1. 0.2462661722

3. 0.1109574581

7. a) $y(x) = \int_0^x 3y(s)\,ds + 1$

b) $\phi_0(x) = 1$, $\phi_1(x) = 1 + 3x$, $\phi_2(x) = 1 + 3x + \frac{9}{2}x^2$, $\phi_3(x) = 1 + 3x + \frac{9}{2}x^2 + \frac{9}{2}x^3$

c) $y(x) = e^{3x}$

9. **a)** $y(x) = \int_0^x (y(s) + s)\, ds$

b) $\phi_0(x) = 0,\ \phi_1(x) = \frac{1}{2}x^2,\ \phi_2(x) = \frac{1}{2}x^2 + \frac{1}{6}x^3,\ \phi_3(x) = \frac{1}{2}x^2 + \frac{1}{6}x^3 + \frac{1}{24}x^4$

c) $y(x) = -1 - x + e^x$

11. **a)** $y(x) = \int_0^x 2sy(s)\, ds + 1$

b) $\phi_0(x) = 1,\ \phi_1(x) = 1 + x^2,\ \phi_2(x) = 1 + x^2 + \frac{1}{2}x^4,\ \phi_3(x) = 1 + x^2 + \frac{1}{2}x^4 + \frac{1}{6}x^6$

c) $y(x) = e^{x^2}$

13. **a)** $y(x) = \int_0^x y(s)^{-1}\, ds + 1$

b) $\phi_0(x) = 1,\ \phi_1(x) = x + 1,\ \phi_2(x) = \ln(x + 1) + 1$

c) $y(x) = \sqrt{2x + 1}$

15. **a)** $y(x) = \int_1^x y(s)^2 - 2y(s) + 1\, ds$

b) $\phi_0(x) = 0,\ \phi_1(x) = x - 1,\ \phi_2(x) = \frac{1}{3}x^3 - 2x^2 + 4x - \frac{7}{3}$

c) $y = 1 - \frac{1}{x}$

Index